Introduction to
BIOTECHNOLOGY

ASHIM K. CHAKRAVARTY

Visiting Professor
Department of Biotechnology
Centre for Life Sciences
University of North Bengal, Siliguri

OXFORD
UNIVERSITY PRESS

OXFORD
UNIVERSITY PRESS

Oxford University Press is a department of the University of Oxford.
It furthers the University's objective of excellence in research, scholarship,
and education by publishing worldwide. Oxford is a registered trade mark of
Oxford University Press in the UK and in certain other countries.

Published in India by
Oxford University Press
22 Workspace, 2nd Floor, 1/22 Asaf Ali Road, New Delhi 110 002

First published in 2013
Third impression 2022

ISBN-13: 978-0-19-808181-4
ISBN-10: 0-19-808181-2

Typeset in BaskervilleMT
by Anvi Composers, New Delhi 110063
Printed in India by Manipal Technologies Limited, Manipal

For product information and current price, please visit www.india.oup.com

To

Robert Auerbach
University of Wisconsin, Madison, USA

William E. Clark
University of California, Los Angeles

Oliver Smithies
(Nobel Laureate, 2007) University of North Carolina,
Chapel Hill (UW, Madison, 1960–1988)

Sivatosh Mookerjee
JNU, New Delhi (formerly Presidency College, Calcutta)

Jitendra N. Rudra
Presidency College, Calcutta

who inspired me

Features of

Learning Objectives
Provides an outline of the topics discussed within the chapter.

Illustrations
Numerous neatly drawn and well-labelled illustrations provide visual explanation for better understanding.

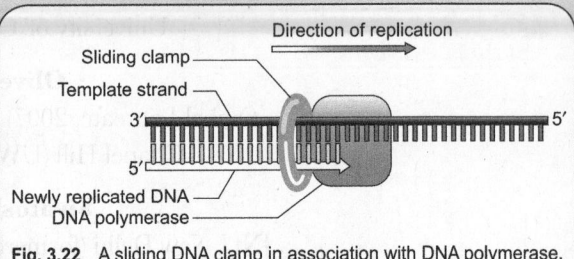

Fig. 3.22 A sliding DNA clamp in association with DNA polymerase.

Exhibit 5.B   Range of Variation of Fundamental Physical Quantities and
Some Practical Units

There are tremendous variations in mass and length (or distance between two bodies) in the objects of the physical world. The magnitude of variation in the physical quantities is often referred to as *atomic to astronomical* or *microscopic to macroscopic*.

Very small objects, such as ions, atoms, and molecules are referred to as microscopic objects. Very small distances are known as microscopic or atomic distances, for example, radius of an atom or an electron, or size of a dust particle.

Heavy bodies of millions and billions tonnes of mass, such as the sun and earth are known as macroscopic entities. Similarly very large distances like the distance between planets and distance of the sun from the earth are known as macroscopic or astronomical distances.

The usual SI units are used for measurements of microscopic and macroscopic objects and distances, with multiples of 1/10 or 10 of the metric system as shown in Tables 5.B1 and 5.B2. The notation with multiples of 10 is often called *exponential notation*.

Exhibits and Textboxes
Numerous exhibits provide additional information related to the topics discussed.
Various interesting facts related to the field are highlighted in textboxes.

Summary
Summary highlights the important points discussed in each chaper.

SUMMARY

♦ Chemical elements present in organisms are of low atomic number and hence low atomic weight. Carbon, hydrogen, oxygen, and nitrogen together make up over 99 per cent of the total atoms (mass) of most cells. Covalently linked carbon atoms form the backbone of biomolecules; functional groups, formed by other atoms, are added to the backbone. Tetrahedral covalent bonds of carbon atoms, other atoms, and functional groups make the three-dimensional structure of a biomolecule.

♦ Biochemical reaction mechanisms are basically similar to chemical reactions. In reactions, electron-rich *nucleophiles* contribute electrons to electron-deficient *electrophiles*. Functional groups containing oxygen, nitrogen, and sulphur are important biological nucleophiles; positively charged hydrogen atoms (protons) and positively charged metal ions (cations) act as electrophiles in biochemical reactions.

♦ Several atoms of H, O, C, and N make up simpler monomeric biological molecules like monosaccharides, amino acids, fatty acids, and nucleosides with molecular weights in the range of hundreds. These different types of monomers are polymerized into macromolecules of polysaccharides, proteins, lipids, and nucleic acids, respectively. The synthesis of macromolecules is a major energy-consuming activity of living cells. Macromolecules may be organized further into supramolecular structures, such as ribosomes, membranes, and other cellular organelles.

♦ Carbohydrates mean carbon with water. The empirical formula of carbohydrate is $(CH_2O)n$, where n indicates the number of monosaccharide units in a carbohydrate

the Book

Biographies

Biographies of prominent scientists give an insight into their lives and works.

J.B.S. Haldane

John Burdon Sanderson Haldane
(5 November 1892–1 December 1964)

During the early 1930s, when the molecular nature of enzymes was not yet fully understood, J.B.S. Haldane wrote a treatise entitled *Enzymes*. This book put forward the remarkable suggestion that weak-bonding interactions between an enzyme and its substrate might be used to distort the substrate and catalyse the reaction. Interestingly this insight is the basis of our current understanding of enzymatic catalysis. He was an outstanding scientist who contributed significantly to physiology, genetics, biochemistry, statistics, and biometry. He was talented and had interest in many subjects including subjects outside the arena of science. Haldane wrote on varied subjects including science fiction, stories for children, and popular science.

For his outstanding contributions, Haldane received much recognition. Haldane was elected a Fellow of the Royal Society in 1932. The French Government awarded

J.B.S Haldane was born in Oxford to a renowned physiologist father John Scott Haldane from an

Exercises

Objective questions and review questions enable facilitate students to test their understanding of concepts.

Objective Questions

1. Multiple Choice Questions
   (a) Bioprocess engineering involves only
       (i) bacteria
       (ii) eukaryotic cells
       (iii) enzymes
       (iv) all of the above
   (b) Catabolism in microbial growth systems uses
       (i) electron donor
       (ii) C and N sources
       (iii) electron
       (iv) electron donor and electron acceptor
   (c) A buffer system normally uses a
       (i) strong acid
       (ii) weak acid
       (iii) both of these
       (iv) none of these

Review Questions

1. Define pH.
2. How does a hydronium ion (H_3O^+) form in water?
3. Pure water at 25°C shows neutral pH 7. Why is this not so at other temperatures?
4. Briefly explain why addition of NaOH to water increases the pH.
5. Mention one property of a strong acid.
6. How is a specific cytoplasmic pH maintained in a cell?
7. Show how Hendersson–Hasselbach's equation is derived from the original equation for the dissociation constant of a weak acid.
8. What are the basic properties which a unit for measurement of a physical quantity must have?

Appendixes I and II

Appendix I—'Lab work' familiarizes the students with laboratory methods.
Appendix II—'Fun Experiments' provides some interesting and easy-to-do experiments for students.

EXPERIMENT 6   CALLUS CULTURE

Callus is a mass of undifferentiated cells originating from a plant explant in culture condition.

Callus cultures have multiple applications in biotechnology and agriculture. (i) Plants can be regenerated from a callus to get somaclonal variety of plants (genetically similar); (ii) Callus provides an excellent source of cell material for biochemical studies; (iii) Callus is very useful in secondary metabolite production, for biotechnological industries; (iv) it provides the opportunity to screen cells with useful traits; (v) callus cultures are necessary to initiate production of somatic embryos, cell suspension cultures, protoplast cultures.

Morphologically callus varies extensively—**compact** with firm cell to cell contact at one extremity to **friable** aggregate of poorly associated cells at other extremity.

Equipment and Materials for Callus Culture

Media preparation, glassware sterilization, and maintenance of sterile conditions, use of culture hood for callus culture are similar to that used for plant tissue (explants) culture as outlined in the earlier experiment (Experiment 5).

Glossary

Provides an extensive list of terms to help students prepare for various competitive exams.

Abzymes Antibodies capable of catalysing specific chemical reactions like enzyme. A catalytic abzyme is produced against the expected transition state off a specific substrate.

Accession number Unique identifying code number of each cloned DNA sequence is catalogued in databases such as GenBank. This number is used to retrieve database information on a particular sequence.

Acridine A chemical, with dye like properties, that intercalates in the double-stranded (ds) DNA and causes mutation involving usually a single base pair.

Active site The part of an enzyme that binds to its substrate for catalytic action.

Adenosine triphosphate (ATP) Triphosphates attached to the nitrogenous base adenine. It stores and transfers energy in cell system.

Adenovirus Virus that causes the common cold.

Adult stem cells (ASCs) Stem cells with totipotency that derive from adult tissues, including bone marrow.

Aerobes Organisms that grow in the presence of molecular

Alkaline phosphatase An enzyme that removes phosphate groups from the 5′ends of DNA molecules.

Allele One of the alternative forms of a gene present at a particular locus on a chromosome (DNA).

Allele-specific oligonucleotide (ASO) Oligonucleotide specific to a disease gene for analysing a person's DNA using PCR.

Alternative splicing Splicing joins certain exons with *cut out* other exons being treated as introns. This way multiple mRNAs of different sizes from the same gene originate and produce different proteins with different functions. The net result is that alternative splicing produces several different proteins from the same gene sequence.

Amino acids Organic acids bearing both amino and carboxyl groups, and act as the monomeric units (building block) of proteins. Twenty different amino acids join together by covalent bonds to make a polypeptide chain or protein.

Aminoacyl transfer RNA (tRNA) Transfer RNA (tRNA) molecule with an activated amino acid at one end and anti-

Preface

Applications of biotechnology was initially restricted to food production, agriculture, and medicine. Nowadays, biotechnological techniques are being used in advanced technologies, providing practical applications of organisms, their cellular components, and metabolic products. It has become essential for medical diagnosis, manufacturing pharmaceuticals, remediation of polluted environment, and better productivity in agriculture, aquaculture, and modern food industries. Biotechnology is now considered as the technology of the 21st century as it has an unlimited commercial potential to produce an extensive range of valuable and useful products. Thus, the demand for trained manpower to handle manufacturing, research, trades, and economics of biotechnological products in new industries is on the rise. Since 1970s, a few of the economically developed countries have started using the discoveries in genetics, cell biology, and molecular biology to launch new research activities in the field of modern biotechnology.

In an attempt to introduce genetic engineering as a thrust area, the Government of India constituted the National Biotechnology Board (NBTB) in March, 1982 under the Department of Science and Technology (DST). In 1986, a separate Department of Biotechnology (DBT) was formed under the Ministry of Science and Technology. The NBTB in collaboration with the University Grants Commission (UGC) formulated the first biotechnology education course for Indian universities. Since 1985–1986, biotechnology is being offered in universities and institutes in MSc and MTech courses. However, it has now been included as a separate subject in the curricula of undergraduate and postgraduate courses. Some M Sc courses specializing in other biological sciences offer biotechnology as an introductory paper as well.

Biotechnology is a multi-disciplinary course, having its foundation in many basic science subjects, such as biology, microbiology, cell biology, biochemistry, genetics, immunology, biophysics, chemistry and chemical engineering, statistics, and computer science.

About the Book

Introduction to Biotechnology is designed to serve as a textbook for students pursuing biotechnology and microbiology courses. It presents the basic concepts, methodologies, and applications of biotechnology.

Starting with the fundamentals of biotechnology, the book provides a detailed discussion on the structure and functioning of cells, organelles, chromosomes, structure and function of biomolecules, and genetics and molecular biology. It also presents an in-depth understanding of recombinant DNA technology, genomics and proteomics, bioinformatics, enzyme biotechnology, microbiology, plant and animal biotechnology, immunology, and environmental biotechnology. The book also covers bioethics and IPR. Owing to its wide coverage, the book would be useful as a reference text for postgraduate students as well.

All chapters present the subject in depth following a student-friendly approach. A systematic reading of the text from the beginning will allow the students to gain a clear understanding of all the aspects of biotechnology.

Key Features

The text includes certain important features to facilitate better understanding of the topics discussed in the chapters.

- **Learning objectives** at the beginning of each chapter list the main topics included in the chapter.
- **Boxed items** in the text present special topics and material in support of the content in the chapters.
- **Tables and figures** interspersed throughout the chapters enable easy understanding of the concepts discussed.
- **Summary** at the end of each chapter highlights the important concepts discussed and enables recapitulation.
- **Objective and review questions** in each chapter provide ample practice to the students.
- **Colour plates** present a set of important figures and photographs in colour.
- **Biographies** of great scientists who made outstanding contributions in the field of biotechnology have been provided in the chapters to stimulate young minds.
- **Appendix I** on Lab Experiments familiarizes students with the experimental methods used in research and industries. The step-wise procedure will help students to conduct experiments effectively in a fairly equipped laboratory.
- **Appendix II** presents *Fun Experiments*. Students will enjoy doing these easy-to-do experiments on their own.
- **Glossary** defines important terms in biotechnology and related fields. The list of terms is quite extensive and students will find it useful while preparing for exams.
- **Bibliography** at the end of the book familiarizes the readers with important texts and articles on different topics of biotechnology. The list will be a ready reference for teachers and students to pursue their studies further on certain aspects of biotechnology that may be of interest to them.

Organization of the Book

The book is organized into 16 chapters and 2 appendices.

Chapter 1   traces the brief history, different aspects, scope, and applications of biotechnology. It further elaborates on the biotechnology programmes in India.

Chapter 2   describes the structure of a cell—the basic unit of the living world and its specializations in bacterial, plant, and animal cells. The text imparts a detailed knowledge of different cell types and their organelles necessary in modern biotechnology for obtaining and improving biotechnological products.

Chapter 3   is devoted to chromosome and DNA in the perspective of DNA replication—the most basic function of life. Chromosomes and their constituent molecule DNA sit at the centre of the cell, as well biotechnology.

Chapter 4   starts with the organization of basic atoms—carbon, hydrogen, oxygen, and nitrogen in the macromolecular structures of the four basic biochemicals, namely carbohydrates, proteins, lipids, and nucleic acids. The discussion extends to details of variations in the structure and functions of each type.

Chapter 5   presents the essential technologies involved in mass-scale industrial production of biochemicals mainly from cellular sources. This is one of the most necessary topics in modern biotechnology. The chapter presents a detailed discussion on biological optimum conditions for cell growth and catalytic reactions, quantitative physical measures and variables, stoichiometry—material and energy balance in biochemical reactions, methods for statistical data analysis for probabilistic risk assessment, and types of bioreactors.

Chapter 6   is devoted entirely to genetic engineering which is the central point of the growing sphere of biotechnology. Genetic engineering has made it possible to fish out a desirable gene virtually from any living organism and to insert it into any other organism to come up with a useful mass product of the gene that may be a hormone or a pharmaceutical product for remedial measures. The technology is often termed as recombinant DNA or in short rDNA technology. The chapter begins with Mendel's formulation of the laws of inheritance. This is followed by a discussion on the different methodologies involved in genetic engineering, transgenic plants and animals, and applications of genetic engineering.

Chapter 7   deals with genomics, proteomics, and bioinformatics. These are comparatively recent sub-disciplines of genetics as well as biotechnology. Genomics is devoted to the mapping, sequencing, and comparative analyses of genomes of different organisms. Functional genomics initiated proteomics under the structure and function of all the proteins encoded by a genome are discussed. Bioinformatics has been developed as an interdisciplinary field involving biology, computer science, mathematics, and statistics to analyse enormous sequence data of genomics and proteomics and arrangement of genome contents. The information helps in the identification of genes and new proteins to be used for pharmaceuticals, industrial products, and in agriculture.

Chapter 8   on enzyme biotechnology discusses the classification and working mechanisms of enzymes and their multiple uses. Enzyme immobilization technology is presented in detail. Today, enzymes find multiple applications such as in detergents and food processing. Thus, the enzyme industry in recent years has been boosted for high yield of desired enzymes by involving genetically engineered microorganisms.

Chapter 9   presents the structure of proteins, and different methods of purification and characterization of proteins. Protein engineering and designing for new and better functions are also discussed.

Chapters 10–12   present microbial, plant, and animal biotechnologies, respectively. These chapters are mostly devoted to cell and tissue culture techniques for respective cellular materials, specific culture media, and scale-up of cell culture for industrial purpose and commercial products.

Chapter 13 covers environmental biotechnology primarily involving the use of organisms for remedial measures against pollution. The ever-increasing world population and industries have deteriorated the world environment to such an alarming extent that has already started spelling doom for many species including man.

Chapter 14 highlights the latest research happening in the field of biotechnology. The scope and applications of biotechnology in agriculture, biomimetics, biofuels, biosensors, monoclonal antibodies, stem cell technology, etc. are discussed.

Chapter 15 on bioethics and biosaftey presents risk assessment and containment of biological inventions and applications to ensure safety for now and in future. Human interference, in the name of rDNA technology, to create new genetic combinations in micro and other organisms, hither to unknown to the society, brings in the bioethical considerations.

Chapter 16 discusses intellectual property rights (IPR) and biopatent which are essential aspects of modern biotechnology. IPR is a social mechanism devised to ensure the rights of the inventor, private companies, and government institutes involved in bringing out new technologies and products. The ways and means of IPR and patenting for biological inventions form the contents of the chapter.

Appendix I presents laboratory experiments related to biotechnology such as bacterial culture, staining, tissue culture, DNA extraction, SDS-PAGE, and so on.

Appendix II presents a few simple and easy-to-do experiments which students will be able to perform on their own, while having some fun during the process.

Acknowledgements

I would like to thank Dr Ranadhir Chakrabarty, Head, Department of Biotechnology, North Bengal University for allowing me to enjoy certain facilities including the seminar library of the Department. I am grateful to Dr Tamal Mazumdar, Assistant Professor, North Bengal St. Xavier's College for helping me in handling the intricacies of working with computers and the initial typing of two chapters.

I appreciate the patience and constant effort of the editorial staff at Oxford University Press, India. The credit for suggesting certain pedagogical features for the book goes to the editors, which has definitely added much value to the text. My heartfelt thanks to all at OUP India for their cooperation and interest since the inception of the idea of writing this book.

I had many illuminating discussions with my son Anupam Kumar Chakravarty, a graduate student, Cornell University, New York. My daughter Anindita (Chakravarty) Roy Chowdhury eased me out in several occasions over phone with answers to my queries in certain areas of Physics. Mr Anirban Roy Chowdhury and Ms Ushati Chakravarty had deep interest and pride in this project that were much encouraging. I like to thank all of them. I also like to express my indebtedness to my wife Dr Alo Chakravarty for bearing with my off-routine hours and negligence of family chores in the course of writing this book.

Ashim K. Chakravarty

Brief Contents

Detailed Contents

List of Colour Plates

Plate 1

- Bt. Cotton
- Bioreactor with Control Panel
- Desktop Bioreactor

Plate 2

- Helios Gene Gun
- Blue Rose
- Giant and Normal Mice

Plate 3

- Plant Tissue Culture Rack
- Callus Culture
- Plant Differentiation in In-vitro Culture

Plate 4

- Laminar Air Flow Hood for Tissue Culture
- ELISA Reader
- Golden Rice

1

Biotechnology—An Overview

INTRODUCTION

Biotechnology is the application of the knowledge of living systems for practical or industrial purposes. The term 'biotechnology' was first coined by Karl Ereky, the Hungarian agricultural economist, in 1919.

An all-inclusive definition of biotechnology is *any technique that uses living organisms, to make or modify a product, to improve plants or animals, or to develop microorganisms for scientific uses*. From this definition, it is clear that the concept of biotechnology is not new and has been in practice since the dawn of civilization, in one way or the other.

1.1 HISTORY OF BIOTECHNOLOGY

In earlier times (and even today), farmers and gardeners crossed plants to produce hybrids for better yields or utilities. They utilized the grafting technique, using two types of plants, to produce a new variety. Veterinarians and animal breeders kept careful pedigree records of farm animals and race-horses and their offsprings that had desirable traits or genetic qualities, so that they could select the appropriate individuals for mating purposes.

Since time immemorial, biotechnological methods have been used in the kitchen intuitively and empirically. Marination, caramelization, food preservation, pickling, fermentation, tenderization of meat using papaya extracts, making curd and cheese, flavour enhancement using natural ingredients and spices, etc. are some familiar examples of processes that are still used in the food industry.

Louis Pasteur (1822–1895), the renowned French scientist, showed how invisible germs spoil milk, wine, and cheese, and how the simple act of boiling inactivates them. The technique of 'pasteurization', named after him, was developed, particularly for preventing bacterial contamination of milk. Pasteur also saved the wine industry from disastrous losses by introducing proper strains of yeast for fermentation, proper conditions of aeration, temperature, and storage techniques.

Technological research and induction of technologies for mass production have often been undertaken for defence purposes. The first large-scale microbial processes that were brought into practice were for brewing alcohol and for acetone–butanol production. During World War I, the classical method of making acetone by wood distillation could not meet the bulk requirement for it in the manufacture of an explosive, namely cordite. In 1912, Chaim Weizmann discovered the microorganism *Clostridium acetobutylicum* capable of producing acetone. Davis Lloyd George, the then British minister of ammunitions contacted Weizmann for the process of making acetone. Factories were set up in Canada, USA, and India (Nasik) and a supply of five million pounds of acetone for war purposes was ensured. In gratitude to Weizmann's contribution, the British Government made the 1917 Balfour Declaration in favour of a national home for Jews in Palestine. Israel came into being in 1948 and Weizmann was invited to be its first President. Thus, in a way the microbiological technology established by Weizmann earned the foundation of the national home for the Jews. The Jewish nation honoured him by naming the premier research institute at Rehovt after Weizmann.

Crude, tentative, and often bizarre human practices have given way to a sophisticated form of modern medicine. *Ayurveda* in India and similar practices of herbal medicine in Asian countries are age-old practices that have recently come under the purview of modern medicine.

1.2 NEW AGE OF BIOTECHNOLOGY

The explosion of knowledge in different branches of biology during the past few decades of the twentieth century and the simultaneous refinement of techniques to probe cells, cellular products, and gene contents have ushered in a new multidimensional age of biotechnology.

Initially, biotechnological research was carried out at the organism level; it has now reached the molecular level. The materials for modern biotechnological interests may range from a tissue to a cell to subcellular systems. The biological and biochemical techniques that deal at these micro levels have been perfected to such an extent that they can contribute positively in scaled-up versions for industrial purposes.

Modern biotechnology has its foundation in the principles and findings of basic sciences, such as biology, chemistry, biochemistry, biophysics, mathematics, statistics, computer science, and engineering, in addition to other disciplines, such as economics, management, law, philosophy, and ethics. Initially, biotechnology was considered to be just an extension of brewing or fermentation and baking. Later, with the advent of recombinant DNA (deoxyribonucleic acid) technology, *genetic engineering* gained prominence, and many other such technologies have since been developed. Biotechnology has already become a major technology of the 21st century. It will soon start off the third technological revolution following the industrial and information technology revolutions. Its multidisciplinary activities include recombinant DNA techniques, cloning, as well as the application of microbiology and cell culture technologies to the production of hybridoma and stem cell lines. All of these are used to produce a wide range of materials for the betterment of the human race. Besides revolutionizing the treatment of many diseases, it also provides clean technologies to mitigate environmental problems. Biotechnology will continue to influence the philosophy and future projections of mankind.

1.3 DIFFERENT ASPECTS OF BIOTECHNOLOGY

Beginning with cells (both prokaryotes and eukaryotes), modern biotechnology delves into the organization of genes in the chromosomes of cells and the replication of DNA. Functional expressions of genes take place via macromolecules, constituted by the following four categories of biochemicals—carbohydrates, proteins, lipids, and nucleic acids. Certain specific chemicals produced by cells are raised in enormous quantities for human needs by biochemical engineering in in-vitro set-ups, especially in bioreactors.

1.3.1 Bioreactor

A bioreactor is essentially a vessel in which a desired biological reaction can be conducted under regulated conditions. Bioreactors allow fermentation or biotransformation which is very important for biotechnological procedures.

1.3.2 Genetic Engineering

As has already been mentioned, genetic engineering is at the core of biotechnology. Thus, it needs an appropriate introduction.

Acquisition (or transfer) of a segment of DNA by a microorganism is very common in nature. The exchange of DNA segments normally occurs among the members of the same species or closely related ones, and rarely among unrelated organisms. Success in introducing and establishing a piece of foreign DNA into the genetic material (genome) of an unrelated organism was first achieved in 1973 and this opened a new chapter in biology—Genetic Engineering or *Recombinant DNA Technology*. Thus, a specific characteristic of one organism (determined by a transferred piece of DNA) can now be seen in a different organism as well. A living chimera has now become a reality.

In 1973, Stanley Cohen of Stanford University School of Medicine, Stanford, California, and Herbert Boyer of the University of California, School of Medicine, San Francisco, California, and their coworkers cleaved segments of DNA containing antibiotic-resistant genes from *Salmonella* and *Staphylococcus* by using enzymes called restriction endonucleases. Then they inserted these genes into a *plasmid* (extra-chromosomal circular DNA normally present in bacteria) by using another enzyme—DNA ligase—and introduced it into the bacterium *Escherichia coli*. The resulting *E. coli* became antibiotic resistant. This epoch-making discovery ushered in the era of genetic engineering. [*E. coli* lives in the human gut. A plasmid used to transfer foreign gene(s) is often called a *cloning vector*.]

Human insulin, human growth hormone (somatostatin), and interferon (to treat viral disease) are now produced in *E. coli* by genetic engineering. Antigenic proteins or glycoproteins expressed on the surface of an infecting agent can be obtained in large quantities or expressed on other harmless microbes with the help of genetic engineering and can be used as safe and effective vaccines.

1.3.3 Engineered DNA Molecules as Probe

A genetic disease is caused by a mutated (altered) gene, and one may inherit the disease from one's parents. Engineered DNA probes are being used to diagnose carrier status in parents who carry a mutated or defective gene but do not have the disease. Such parents are given genetic counselling to prevent the birth of defective children. Probes can also be used to detect viral genes present in human chromosomes that may lead to pathological conditions.

Sphere of genetic engineering

Although recombinant DNA (rDNA) technology is at the heart of genetic engineering, its sphere includes a wide range of technologies for directed manipulation of genes or genetic modification for desired results.

Transgenic technique

Introduction of foreign genes into organisms is not only limited to microbes, but is also possible in plants and animals. When genes from other organisms that cannot be introduced by conventional breeding are introduced into an organism, it is called a *transgenic organism*. These organisms, thus express the new gene and a new trait which is not present in their parental stock, but contributes to genetic modification in the offspring.

Transgenic organisms acquire new traits to get better yield and nutritive values, to become resistant to pests, herbicides, and certain stress factors, or to produce particular substances in large quantities. Examples of such substances produced by genetically modified organisms range from pharmaceuticals, enzymes, and spider silk (biosteel) to xenografts used to replace human organs or tissues.

Transgenic plants   These are developed by using plant cloning technologies that allow growth of plants from isolated plant cells in which the desired gene has already been inserted.

The first transgenic plant was the 'Flavr-Savr' tomato, designed to remain firm and maintain the right texture and improved flavour longer than the common tomato. Transgenic soybeans are resistant to herbicide applications and grow better. As they can tolerate herbicide application the cost of labour remains low. Transgenic potato, papaya, and squash plants with genes resistant to virus attacks have been produced. Transgenic *Bt* cotton plants contain the insecticidal gene of the bacterium *Bacillus thuringiensis*, which protects them from the attack of bollworms (larva of an insect pest). Transgenic rice plants containing a daffodil gene, required for the synthesis of vitamin A in the seed, are useful for preventing various diseases caused by vitamin A deficiency. Banana or other plants that produce edible fruits can be made transgenic with genes that code for antigens of disease-causing agents. These fruits can then serve as *edible vaccines*. (Also see colour plate 1).

Thus, biotechnology is regarded as a field that has improved the efficiency of agriculture, the nutritional quality of food, and the production of pharmaceuticals or therapeutic proteins.

Biotechnology can be used to enhance existing foods or to generate new ones. The high-quality enzymes used by the food processing industry are produced using biotechnology. Plant genetic engineering to alter texture or colour of flowers and ornamental plants is one of the two success stories of plant biotechnology, the other being pest resistance.

Transgenic animals   In order to ensure that a new or corrected gene is passed on to the progeny of a sexually reproducing animal, the gene needs to be inserted in the 'germ cells'—the cells that make the sperm and ovum. Transgenic animals, thus, are produced by a combination of in-vitro (outside the body) fertilization (IVF) techniques and recombinant DNA technology. Following IVF, the early embryo is transplanted into the oviduct of a pseudopregnant female and is allowed to develop further. The newborn is screened for the presence of the transgene by Southern blot analysis (Chapter 6) of the DNA extracted from its cells.

Transgenic animals have great potential to be used as model organisms for conducting research on gene function, drug testing, and synthesizing recombinant proteins for pharmaceutical use. For example, human protein C can be obtained from the milk of transgenic swine. This protein has anti-coagulant activity required for regulation of haemostasis (the process that stops bleeding). Similarly, for treating hemophiliacs (people who lack the coagulation factor in blood and can have uncontrolled bleeding), the blood clotting protein Factor IX is produced in transgenic sheep. The yield of recombinant human proteins is high in such systems. (Bacteria, yeast, and mammalian cell culture systems are also used for production of human growth hormone, cytokines, and vaccines, as mentioned earlier.) Numerous recombinant proteins used for therapeutic purposes are now commercially available. Transgenic pigs and salmon that grow better and heavier have been developed to meet the food requirements.

Gene therapy for humans   For transgenic animals, the genetic makeup of the descendants is altered by genetically engineering the germ cells. But such genetic alteration is not possible in human beings due to ethical reasons. As per international guidelines, recombinant DNA technology or genetic engineering can only be done in somatic cells of human beings to correct certain genetic diseases, such as anaemia, cystic fibrosis, and certain types of leukaemia. Somatic cells make the different tissues and organs of an individual, that is, the muscles, bone, nerves, blood, etc. Changing them affects the engineered person for his life span; however the germ cells remain unaltered. This is generally called *gene therapy* or somatic cell therapy, rather than 'genetic engineering'.

Bone marrow is a relatively easy target for gene therapy as it is easy to take out by aspiration and replace, and it generates itself inside the body. An engineered stem cell of bone marrow can multiply rapidly and produce a large number of progeny cells which can replace the defective cells. Using this technology, attempts have been made to cure severe combined immunodeficiency disease (SCID), caused by a deficiency in the enzyme adenosine deaminase.

1.3.4 Genomics and Bioinformatics

Since the mid-1980s, geneticists have been using recombinant DNA technology for finer genetic analyses. Collections of DNA sequences in clones formed a genomic library of an organism. These sequences were pieced together into overlapping sets to construct final genetic and physical maps of the entire *genome*. This was the beginning of *genomics*. Knowledge of the nucleotide sequences of entire genomes provided an enormous wealth of information. Functional genomics initiated *proteomics* which deals with amino acid sequence, structure, and function of all the proteins encoded by a genome.

The challenge of storing, classifying, and retrieving the data of billions of nucleotide base pairs for each of the thousand types of genomes and amino acid sequences of the thousands of different types of proteins gave birth to the discipline of *bioinformatics* (biology + informatics). This has been developed by geneticists, mathematicians, statisticians, computer scientists, and engineers, and has become a strong arm of biotechnology. Bioinformatics enabled the completion of the human genome project that involved sequencing of 3.2 billion nucleotide pairs in 2003, a few years earlier than the stipulated time.

1.3.5 Protein Engineering

Protein products of a genome perform multiple functions in the living world. The three-dimensional structure of proteins encoded by the genes is responsible for their diverse functions. The knowledge of the structure of proteins related to functions has paved the way for engineering protein structure at the genetic level to come up with altered proteins with desirable functional qualities.

Enzyme technology with all the back up of genetics, genetic engineering, genomics, proteomics, and biochemical engineering now caters to the multiple requirements of industrial biotechnology.

1.3.6 Cell Culture

In-vitro cell culture in defined media has been in practice for more than a hundred years. Robert Koch (1843–1910) was the pioneer in establishing a pure culture of bacteria. Later on, culture techniques for plant cells, plant protoplasts, and animal cells were developed. The main purpose of cell culture was to be able to grow a pure line of a cell type to study its characteristics and functions in defined media and to test the effect of a particular additive or a drug. The culture of cells isolated from multicellular organisms provides opportunities to study the functions and products of a cell type and the essential requirements including factors from other cell types, for the synthesis and growth of the cells.

The cell culture techniques, thus developed, are an integral part of industrial, agricultural, pharmaceutical, and environmental biotechnology. In fact, these techniques have initiated new branches of biotechnology—microbial, plant, and animal biotechnology. It is these cell culture techniques that are scaled up for industrial production in bioreactors guided by bioprocess engineering.

1.3.7 Environmental Biotechnology

Environmental depletion and pollution have become major problems that threaten the existence of many species and are projected to have a severe impact on mankind. The microbial world including genetically engineered microorganisms has a lot of potential to fight this menace. In fact, many of them have been employed and genetically improved for the purpose. The technologies to prevent and tackle pollution are improving day by day. All these activities are a part of environmental biotechnology.

1.3.8 Horizon of Biotechnology

So far we have discussed the trends of development in biotechnology which have been presented almost in the same order in this book. Besides these, other aspects, such as agricultural biotechnology, aquatic biotechnology, biofuels, monoclonal technology and vaccines, and biosensors exist under the purview of biotechnology. All these areas have the potential to be developed as frontier areas of biotechnology and with their commercial applications they will soon expand the horizon of biotechnology.

1.3.9 Bioethics and Biosafety

There are many hazardous industrial products that should be stored or disposed of properly to ensure the safety of living organisms as well as the environment. For this, some basic protocols have been laid down and it is necessary to follow them. Similarly, biosafety measures should always be maintained for biotechnological materials, especially genetically engineered micro and macro organisms. They should be handled with utmost care so that they remain exclusively confined to the laboratory and industrial establishments and do not leave these areas. They should be considered as hazardous for other organisms and ecosystems. One should be doubly careful when working with genetically altered organisms, as they can be more hazardous than industrial products. Industrial products do not multiply but biologically engineered products do!

There are some biosafety guidelines for the containment of genetically altered organisms that may be harmful. Decisions regarding whether genetically altered crops, fruits, vegetables, and animal products can be consumed should be made from their biosafety point of view.

Ethics provides the guidelines for doing things which are morally right. Bioethics deals with some fundamental questions confronting human society. In many respects, the decisions made now can affect the future of science, humanity, and the world we all live in. For anything new, risk assessment and ethical judgement are absolutely essential before it is used further, irrespective of the magnitude of the returns.

1.3.10 Intellectual Property Rights and Patents

Intellectual property rights are the mechanisms adopted by the society and the government to protect and reward inventors in order to encourage them, and others to develop new and more useful inventions.

In this context, the 'property' does not refer to tangible, material, or physical property, such as estate, land, goods, or money. Rather intellectual property refers to an idea, a design, an invention, a manuscript, and so on.

A *patent* is the right granted by the government to an inventor for using or selling the particular invention for commercial purposes for a specified period and simultaneously debars others from using the invention for the same benefits.

Earlier, living things were not patentable in any country, but biological materials obtained by non-biological means, in which 'the hand of human' had a part were patentable. The first patent for an organism altered by genetic manipulation was granted to Ananda Chakraborty, working at General Electric Company, USA, by the US Supreme Court in 1980, for the hydrocarbon degrading bacterial strain *Pseudomonas* to be used for cleaning up oil spillage (see *Superbug*, Exhibit 6.D in Chapter 6).

1.4 SCOPE, IMPORTANCE, AND COMMERCIAL POTENTIAL

In the past three decades, biotechnology has rapidly emerged as an area of intense activity in several fields ranging from agriculture, forestry, dairy, food processing and beverages, and mining to renewable energy and fuels, detergents, biochemicals, human health, and medicine. Today, it plays a very important role in employment, productivity, commerce, and the economy.

Genentech Inc., the first biotech company was established near San Francisco, California, in 1976. As a small company, Genentech met success and ushered the birth of this exciting industry. Some companies have even been established with the help of Nobel Laureates and scientists. Several thousand companies have now come up, some of which are being established in developing countries including India.

In India, there are more than a hundred biotechnology companies. Most of them are located in three main clusters: (1) Western Cluster—Mumbai, Pune, Ahmedabad, Aurangabad; (2) Southern Cluster—Bangalore, Chennai, Hyderabad; and (3) Northern Cluster—Delhi, Gurgaon, Noida. In terms of revenue earning, the Western Cluster is the dominant player.

Today, the hundreds of biotechnological products in the market generate about $70 billion in worldwide revenues. Enzymes and biological drugs, such as antibodies, growth factors, vaccines, and growth hormone, have the larger share in the sale. Over 350 biotechnology products are currently under research and development to bring relief from cancers, diabetes, heart disease, Parkinson's and Alzheimer's diseases, arthritis, AIDS, and many other diseases.

The US, Europe, and Japan have the largest share in the biotechnology market in the world. They have the largest number of companies, the US being at the top with about 1,500 companies. Other countries are rapidly establishing institutes and companies for biotechnology research and products; India and China are at the forefront. Already more than 4,900 biotechnology firms have been established in 54 countries. Basic science ideas for biotechnological applications are mostly generated in universities and research institutes.

The biotechnology industry provides exciting job opportunities for many scientists and skilled personnel. It offers a great range of high-tech employment to biologists and chemists, engineers, information technologists, management and salespeople. Biotech companies need a workforce of skilled people to maintain industrial laboratories, cell cultures, improve strains, maintain bioreactors, and the industrial ancillaries. These companies rope in scientists who are good at analysing information from databases, gene expression data from microarrays and SNPs, computer modelling data from DNA and protein structure analysis as well as chemical and biophysical data used to study molecular structure.

1.5 BIOTECHNOLOGY IN INDIA

A strong desire for developing research activities in the area of genetic engineering in India was noticed by the early 1980s. In the 69th session of the Indian Science Congress at Mysore in 1982, the importance of biotechnology was highlighted. Soon the Government of India constituted a National Biotechnology Board (NBTB) to encourage and coordinate research activities in biotechnology. In 1982–1983, three genetic engineering units were established in India at Bose Institute, Kolkata; Indian Institute of Science, Bangalore; and Jawaharlal Nehru University, New Delhi.

In 1986, a separate Department of Biotechnology (DBT) was created within the Ministry of Science and Technology. This is now located at CGO Complex, Lodi Road, New Delhi.

1.5.1 Biotechnology Education

DBT initiated an extensive teaching programme in biotechnology in order to develop manpower in this field. The course curricula were developed for three levels: (1) Master's [MSc., MSc. (Ag)/MTech.], (2) post-doctoral training, and (3) some certificate and diploma courses.

Now, in addition to the DBT supported institutions, many universities including private universities and colleges offer PG and UG courses in biotechnology. Boards like CBSE (Central Board of Secondary Education) offer biotechnology as an optional subject at the higher secondary level.

1.5.2 Biotechnology Research

The National Research Centre on Plant Biotechnology was established in 1993 in the Indian Agricultural Research Institute (IARI), New Delhi. The National Dairy Research Institute (NDRI), Karnal, and the Indian Veterinary Research Institute (IVRI), Izat Nagar, UP have also been marked for research work in biotechnology.

DBT provided generous funding for creation and development of seven Centres for Plant Molecular Biology (CPMB) at the following institutions: (1) Madurai Kamaraj University, Madurai, Tamil Nadu; (2) Jawaharlal Nehru University, New Delhi; (3) Osmania University, Hyderabad; (4) Tamil Nadu Agricultural University, Coimbatore; (5) Bose Institute, Kolkata; (6) National Botanical Research Institute, Lucknow; and (7) Delhi University (South Campus), New Delhi. Some of these centres have made reasonable progress in research towards their mandate to improve crops using biotechnological knowhow.

The Council for Scientific and Industrial Research (CSIR), New Delhi and the Indian Council of Agricultural Research (ICAR), New Delhi have collaborated with DBT to promote research activities in biotechnology. ICAR developed biotechnology wings in several of its crop research institutes—the Indian Pulses Research Institute at Kanpur, the Indian Grassland and Forage Research Institute at Jhansi, and so on.

In the interest of DBT, a national facility for plant tissue culture repository has come up at the National Bureau of Plant Genetic Resources (NBPGR) in New Delhi with the objective of germplasm conservation of clonally propagated crops. Three national gene banks for medicinal and aromatic plants have been created at NBPGR, New Delhi, the Central Institute of Medicinal and Aromatic Plants (CIMAP), Lucknow, and the Tropical Botanical Garden and Research Institute (TBGRI), Thiruvananthapuram. In addition to the conventional seed and field gene preservation, these national gene banks have facilities for tissue culture repository and cryobanks, that is preservation in liquid nitrogen at $-196°C$.

Several Indian biotechnology companies export ornamental plants produced by clonal propagation (micropropagation). DBT has helped in establishing three pilot projects at: (1) The Energy Research Institute (TERI), New Delhi, (2) National Chemical Laboratory (NCL), Pune, and (3) J.N. Vyas University, Jodhpur with the mandate for large-scale micropropagation of forest and fruit trees for afforestation programmes.

The United Nations Organisation (UNO) has set up the International Centre for Genetic Engineering and Biotechnology (ICGEB) at two centres—one at Trieste, Italy and the other at New Delhi.

The private sector including some big business houses has shown interest in this new upcoming technology. New companies or biotechnology wings have been set up and business in biotechnological products such as enzymes, certain drugs, vaccines, and diagnostic kits for certain diseases has begun.

The students are encouraged to consult multiple websites to get more information on the different aspects of biotechnology.

1.6 CONCLUSION

The environment with its plants, microorganisms, and fungi is well equipped to take care of the organic waste produced mainly by different organisms. However, industrial activities and new methodologies in farming generate a huge quantity of inorganic waste which the environment is not able to degrade and recycle completely. These wastes create problems of pollution all over the world to the extent of threatening the existence of many species including humans.

Environmental biotechnology is currently a flourishing area of activities for both scientists and social workers. Bioremediation and waste water treatment are being pursued very dynamically in the developed countries. All these techniques cannot easily be implemented in developing countries, mainly due to limited resources. In these cases, local solutions must be sought using local materials and innovations. As the developing countries become more industrialized, they should learn from the mistakes made by those countries with a long industrial history. It is probably the right time to curb human greed and the modern concept of consumerism. Economists might favour modern market force and activities. But biologists, biotechnologists, and environmentalists should make an attempt to convince people to protect the earth from the unnecessary introduction of manipulated genes in the living world and, polluting agents into the biosphere. The knowledge of biotechnology should not be used to introduce or manipulate genes across the species or phyla in order to increase production or profit. Its fundamental tenet should be protection of existing genomes and biodiversity. Modern India should exercise extra caution in this respect, as the country traditionally believes in nurturing nature and harbours a great treasure of biodiversity.

SUMMARY

♦ Biotechnology uses and modifies organisms and their products for practical and industrial purposes.
♦ Marination, caramelization, food preservation, pickling, fermentation, tenderizing meat using papaya extracts, making curd and cheese, enhancing flavour by using natural ingredients and spices are some familiar examples of age-old practices currently used in food biotechnology.
♦ Biotechnology is one of the major technologies of the 21st century. Its multidisciplinary activities include recombinant DNA techniques, cloning and the application of microbiology and cell culture technologies.
♦ Bioreactors are essential for modern biotechnology; they are used for fermentation and biotransformation and help in producing different biochemicals as products of genetically manipulated microorganisms.
♦ Success in introducing and establishing a piece of foreign DNA into the genetic material (genome) of an unrelated organism was achieved in 1973 and this opened a new chapter in biology—Genetic Engineering or Recombinant DNA Technology. Human insulin,

human growth hormone (somatostatin), and interferon (to treat viral disease) are made by genetically engineered *E. coli*.
♦ Introduction of foreign genes into organisms is not limited to microbes, but is also possible in plants and animals.
♦ Nowadays biological materials engineered by man are patentable.
♦ To become a biotechnologist, one needs to have an understanding of cells, genetics, biomolecules, bioreactors, genetic engineering, enzyme biotechnology, genomics, proteomics, bioinformatics, protein structure and engineering, microbial biotechnology, prokaryotic and eukaryotic cell culture technology, and environmental biotechnology. Moreover, one also needs to be informed about biosafety, bioethics, intellectual property rights, and the patent process.
♦ We must always remember that the use of genetically engineered plants and animals for agricultural and other purposes may disturb the existing ecological balance in an unpredictable way.

Schleiden and Schwann

Matthias Jakob Schleiden
(5 April 1804–23 June 1881)

Schleiden was born in Hamburg, Germany. His father was a respected doctor. He attended the University of Heidelberg from 1824 to 1827, and obtained a doctorate in law. Unhappy and depressed with the profession in law, he began studying natural science at Gottingen in 1833 and then moved to Berlin, where he focussed on Botany. He worked in the laboratory of the respected German physiologist and comparative anatomist Johannes Peter Müller. Schleiden was engaged with classifying plants, but did not restrict himself to morphological studies. He used a microscope to study minute structures. In the course of his work, he realized that the different parts of the plant organism are composed of cells. This realization led him to initiate the Cell Theory. He wrote *Beiträge zur Phytogenesis*, or *Contributions to Phytogenesis* (plant growth), in 1838. He is considered as the initiator of plant cytology. Schleiden described Robert Brown's (1832) discovery of the cell nucleus and suspected that the nucleus was in someway connected to cell division.

After obtaining a doctorate at the University of Jena, Germany in 1839, Schleiden was appointed extraordinary professor of Botany. Meanwhile Schleiden met Theodor Schwann in the laboratory, who extended the Cell Theory to the animal kingdom.

Schleiden moved to Dorpat (now Tartu in Estonia) in 1862, where he stayed for one year. Then, he moved from city to city, settling finally in Frankfurt, where he taught privately.

Other contributions of Schleiden include his observations on protoplasmic (more commonly cytoplasmic) streaming of colloidal complexes of organic and inorganic compounds inside the cell. He studied Mycorrhizae-symbiotic association of the fungi with plant roots. He authored several books, including a popular Botany textbook *Grundziüge der Wissenschaftlichen Botanik* (1842–43). He also delivered many popular lectures.

Theodor Schwann
(7 December 1810–11 January 1882)

Theodor Schwann was a German physiologist. He was born in Neuss, Germany; his father was a goldsmith, who later became a printer. He attended Jesuits College in Cologne and studied medicine at the universities of Bonn, Würzburg, and Berlin. In 1829, he enrolled at the University of Bonn, moved to Würzburg in 1831, and two years later to Berlin. He assisted the famous professor Johannes Müller at Berlin in the experimental work required for his book on physiology during 1834–1838. In Muller's laboratory he met and befriended Matthias J. Schleiden, cofounder of the Cell Theory.

Schwann investigated notochord development in tadpoles with the help of the microscope. He expanded Schleiden's Cell Theory to include the animal world, establishing Cell Theory as the fundamental concept in biology. He put forward his view in a publication, 'Microscopic Investigations on the Accordance in the Structure and Growth of Plants and Animals' in 1839.

In 1837, Schwann isolated pepsin, an enzyme essential for digestion. As professor of physiology at the Catholic University of Leuven (Louvain), Belgium (1839–48) Schwann observed the formation of yeast spores and realized that the fermentation of sugar and starch was the result of life processes. He also coined the term **metabolism** for the chemical changes that take place in living tissue. At the University of Liège in Belgium (1849–79), he taught physiology and anatomy and discovered the striated muscle in the upper esophagus and initiated research into muscle contraction and nerve structure. The cells of the myelin sheath which cover the peripheral axons of nerve cells are called Schwann cells in his honour. He is considered the founder of modern histology.

Schwann questioned the then belief in 'spontaneous generation of life' which led to the Germ Theory of Pasteur. Later in life he became increasingly concerned with theological issues.

Schleiden and Schwann formulated the Cell Theory by the end of 1830s. This principle of biology is equal in importance to the atomic theory of the physical sciences. Originally, the Cell Theory proposed two basic tenets: *all organisms consist of cells, and the cell is the basic structural and functional unit of life*. Rudolf Virchow, in 1858, added to the Cell Theory that *all cells arise from pre-existing cells*.

EXERCISES

Review Questions

1. Is animal breeding a form of biotechnology?
2. List a few biotechnological contributions of Louis Pasteur.
3. *Clostridium acetobutylicum* can produce butanol. Is the statement true or false?
4. Who was C. Weizmann? Write about his contributions.
5. What is a bioreactor?
6. Define genetic engineering.
7. What is bioethics?
8. What are the contributions of DBT in promoting biotechnology in India?
9. As a student of biotechnology, if you have to afforest an area in a short time, what steps will you implement?
10. Sum up the salient features of biotechnology that are in the service of society.
11. Discuss the negative impacts of biotechnology.

Cells and Organelles

INTRODUCTION

A cell is endowed with all the properties of a living system and constitutes the basic structural unit of all organisms. It is made up of protoplasm (living substance), which is differentiated into different micro-structures and organelles that carryout the complex and organized functions of life.

A cell carries out many functions. Some of its basic functions are mentioned below:

- A cell utilizes energy-rich substances as the source of energy for various metabolic and mechanical functions.
- It carries out a variety of chemical and biosynthetic reactions.
- It responds to external stimuli.
- It is capable of self-regulation.
- It carries genetic material (DNA) which inherits the characteristics of the previous generation and adapts to changes in the environment.
- It procreates, that is, reproduces itself.

Although cells across the plant and animal world bear basic similarities, they are differentiated into different forms for functional specialization. Even within an organism there are many different types of cells. There are more than 100 different and distinct cell types in the human body. Several different types of single-celled organisms can be found in a teaspoon of pond water.

2.1 VIRUSES

Although viruses are made up of organic compounds such as nucleic acids and proteins, they are not typical living organisms. They lack the cellular organization for synthesis and replication and depend on the cellular machinery and the enzyme system of the host cells for these processes. Viruses are often described as being at the border of the non-living and the living world. Viruses that parasitize bacterial cells are known as *bacteriophages*. Different varieties of viruses, during the course of infection and multiplication, produce a number of diseases in plants and animals. Some of the dreaded diseases in humans caused by viruses are smallpox, measles, rubella (German measles), influenza, mumps, hepatitis, yellow fever, polio, rabies, AIDS, and some types of cancers. Throughout the world, there is tremendous research in viruses because of their clinical importance and utility as a research tool in molecular genetics and genetic engineering.

2.1.1 Characteristics

Viruses characteristically differ among themselves in their size, nucleic acid content (DNA/RNA), organization of the capsid protein coat, and differential infectivity to specific host cells.

Size

Viruses range in size from about 17 nm (6.4 nm is the diameter of a molecule of haemoglobin) to about 300 nm (which is more than the size of small bacteria).

Structure and specificity

Hereditary material, either DNA or RNA, is present enclosed in a case made up of protein subunits. Some of these proteins can bind to specific receptors on the host cell membrane and infect the cells. Thus, a given type of virus can infect only certain specific types of cells. For example, cold viruses infect cells in the mucous membrane of the respiratory tract; measles and chicken pox viruses infect skin cells; and the polio virus infects the upper respiratory tract, intestinal cells, and nerve cells. Similarly, specific viruses (bacteriophages) infect bacteria; cyanophage viruses attack blue-green algae (a prokaryote); and other specific types of viruses infect fungi.

Form

The form of viruses differs depending on their type. However, the capsid protein case of the viruses occurs in a few basic forms (Fig. 2.1).

Icosahedral This form has 20 sides, each of which is an equilateral triangle, like Buckminster Fuller's geodesic domes; for example, adenovirus which causes colds in human beings, papovavirus which causes warts in human beings.

Helical This resembles rods with a fine regular helical pattern on their surface, for example, tobacco mosaic virus.

Phage This form (infects bacteria) has an icosahedral head containing the nuclear material (DNA or RNA) and an extended sheath or tail necessary for passing DNA to the host cell. Tail fibres at the tip of the tail help in the specific binding of a phage to the host cell.

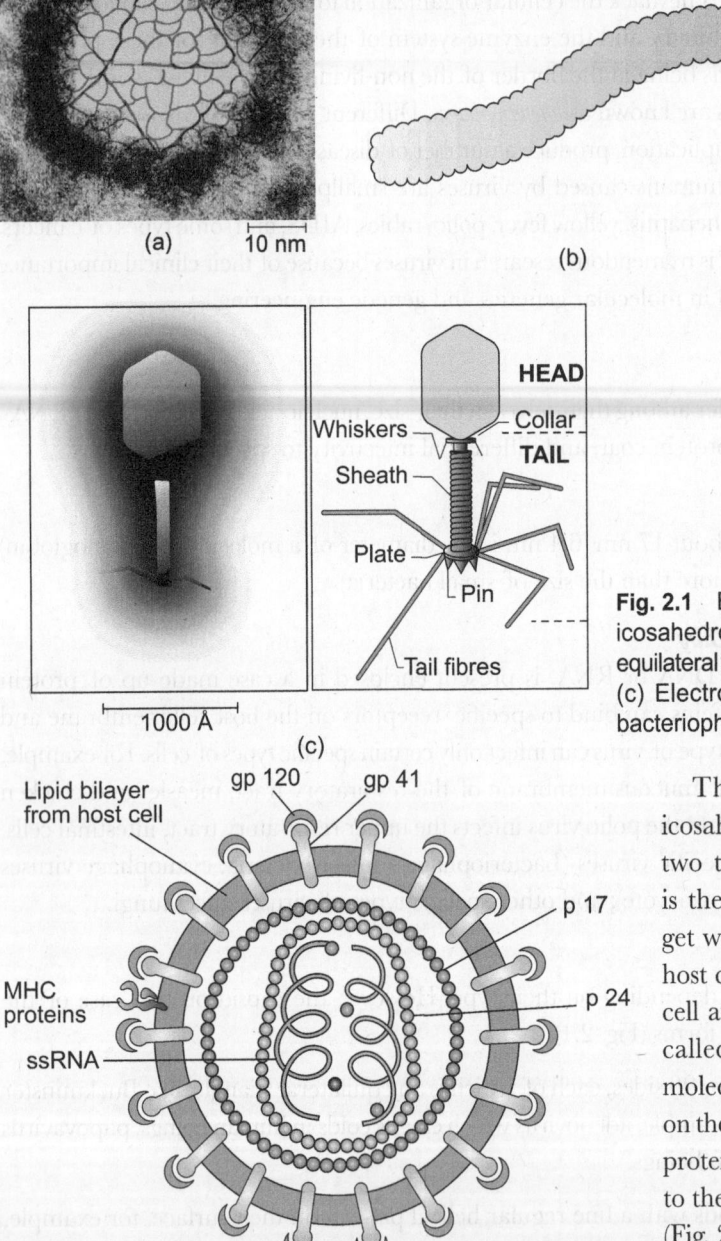

(a) |—— 10 nm ——|

(b)

|—— 1000 Å ——|

(c)

Fig. 2.1 Basic forms of viruses. (a) Adenovirus, icosahedron in shape and has 20 sides, each an equilateral triangle; (b) Tobacco mosaic virus, rod like; (c) Electron micrograph and line drawing of a bacteriophage (a virus which infects bacteria).

The first two categories of viruses, icosahedral and helical, again could be of two types—naked or enveloped. Naked is the more common form; some viruses get wrapped by the cell membrane of the host cell when they bud off from the host cell after replication is completed and are called enveloped viruses. Some protein molecules of viral origin may be expressed on the wrapping membrane along with host proteins and help in the binding of the virus to the specific receptors on other host cells (Fig. 2.2).

Fig. 2.2 Cross-sectional diagram of the human immunodeficiency virus (HIV) that causes AIDS. Note the combination of glycoproteins gp 120 and gp 41 (together gp 160) of viral origin protruding out of the lipid bilayer derived from the host cell. These surface proteins help in binding of the virus to the target cell (T cell) and infection. (Chakravarty, A. 2006. *Immunology and Immunotechnology*, Oxford University Press, New Delhi, p. 452)

2.1.2 Entry and Multiplication in Host Cells

After a virus and its host cell make contact through the coat protein and cell surface receptor respectively, the DNA/RNA in the virus enters the host cell. Fig. 2.3 In some cases, the protein coat of the virus is left outside the cell. In other cases, the intact virus enters the cell. Once inside, the protein coat is destroyed by enzymes. The DNA of a DNA-containing virus serves both as a template for replication of more viral DNA and for the transcription of messenger RNA. The mRNA codes for viral coat protein, viral enzymes, and in some cases, repressors and other regulatory proteins. For replication, transcription, and translation, the virus uses the cellular equipment of the host cell, including the ribosomes, tRNAs, amino acids, and nucleotides. The viruses use host enzymes as well as the enzymes encoded by their genes, and some break up host DNA and recycle the nucleotides to synthesize their own DNA.

In the case of RNA viruses, the viral RNA not only serves as a template for more RNA, but it also acts directly as messenger RNA. In the Retroviridae group of RNA viruses, viral RNA can also serve as a template for viral DNA production with the help of the enzyme reverse transcriptase polymerase. The phenomenon is called reverse transcription and this type of viral DNA when integrated in the host genome can cause cancer.

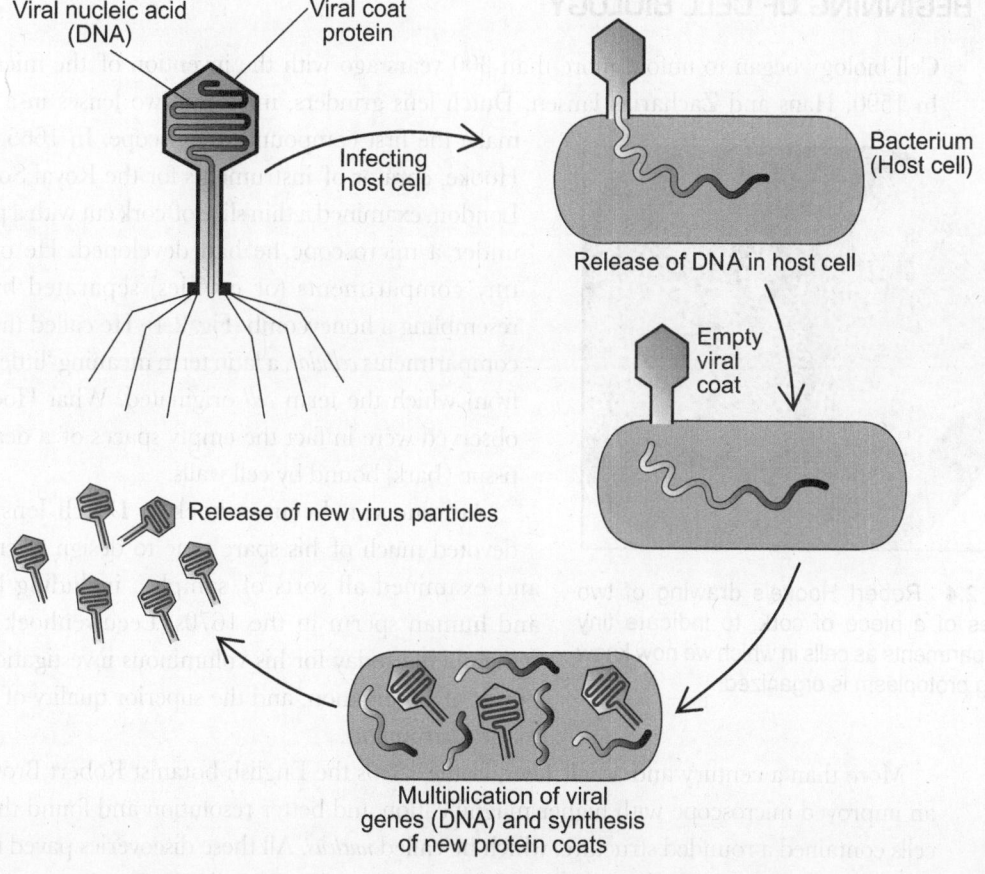

Viral nucleic acid (DNA)

Viral coat protein

Infecting host cell

Bacterium (Host cell)

Release of DNA in host cell

Empty viral coat

Release of new virus particles

Multiplication of viral genes (DNA) and synthesis of new protein coats

(a)

(Contd)

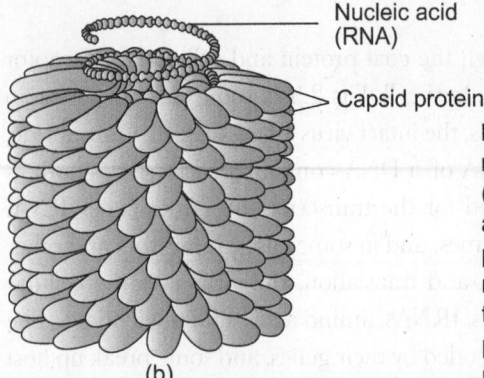

Nucleic acid
(RNA)

Capsid protein

(b)

Fig. 2.3 (a) The process of infection of the host cell by the nucleic acid molecule of a virus, multiplication of DNA/RNA (depending on the type of virus) within the host cell and assembly of viral nucleic acid and protein capsids as new viral particles to be released. (b) Diagram to show assembly of helical nucleic acid at the centre and surrounding protein capsids in the formation of a tobacco mosaic virus (RNA virus). Only a part of the virus after removal of the upper two rows of protein monomers (capsids) is depicted, to reveal the RNA.

Replicated viral nucleic acid and protein capsids assemble as viral particles within the host cell (Fig. 2.3). They then leave the cell, often lysing the cell membrane. Some viruses, such as influenza virus and HIV, bud off from the host cell membrane, and in the process, become wrapped in fragments of the membrane. Each new viral particle can initiate a new infective cycle in a cell.

2.2 BEGINNING OF CELL BIOLOGY

Cell biology began to unfold more than 300 years ago with the invention of the microscope. In 1590, Hans and Zacharias Jansen, Dutch lens grinders, mounted two lenses in a tube to make the first compound microscope. In 1665, Robert Hooke, curator of instruments for the Royal Society of London, examined a thin slice of cork cut with a penknife under a microscope he had developed. He observed tiny compartments (or cavities) separated by walls, resembling a honeycomb (Fig. 2.4). He called these little compartments *cellulae*, a latin term meaning 'little rooms', from which the term *cell* originated. What Hooke had observed were in fact the empty spaces of a dead plant tissue (bark) bound by cell walls.

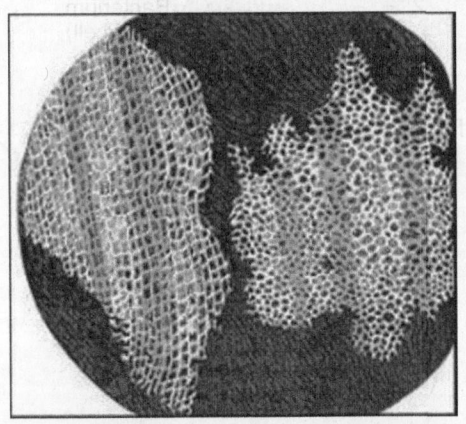

Fig. 2.4 Robert Hooke's drawing of two slices of a piece of cork, to indicate tiny compartments as cells in which we now know living protoplasm is organized.

Antoine van Leeuwenhoek, a Dutch lens maker, devoted much of his spare time to design microscopes and examined all sorts of samples, including bacteria and human sperm in the 1670s. Leeuwenhoek is most remembered today for his voluminous investigation, keen power of observation, and the superior quality of his self-made instruments.

More than a century and a half later, in the 1830s the English botanist Robert Brown used an improved microscope with higher magnification and better resolution and found that plant cells contained a rounded structure, which he called *nucleus*. All these discoveries paved the path for 'the cell theory'.

2.3 THE CELL THEORY

In 1838, Matthias Schleiden, a German botanist, concluded that all plant tissues are composed of cells and an embryonic plant always develops from a single cell. In the following year, zoologist Theodore Schwann extended Schleiden's observation to animal tissues and proposed the cell theory stating the cellular basis for all living organisms. *The cell theory* postulates that an organism consists of one or more cells and the cell is the basic structural unit of all organisms.

By 1858, Rudolf Virchow, a German physiologist, realized that new cells are formed by the division of pre-existing cells and added the third tenet of the modern cell theory—all cells arise only from pre-existing cells.

Thus, the cell theory highlights the underlying unity among the diverse organisms of the plant and animal kingdoms, and the unbroken continuity between the cells of modern organisms and the primitive cells that first appeared on the earth.

Postulates of the Cell Theory

(i)   An organism consists of one or more cells.

(ii) The cell is the basic structural unit of all organisms.

(iii) All cells arise only from pre-existing cells.

2.4 CELL SIZE AND SHAPE

Cells in general are tiny and invisible to the naked eye. Chicken egg (equivalent to an ovum, is a single cell), large amoeba, and certain other very large single-celled organisms are exceptions. Generally, the size of the cells is expressed in micrometres (μm). Most plant and animal cells are between 6 and 30 micrometres in diameter. Figure 2.5 depicts the comparative sizes of different biological materials including cellular organelles.

Why are cells so small?

The limitation in cell size is attributed mainly to two factors which are critical for the survival of cells:

(i) The physical relationship between volume and surface area: As Fig. 2.6 shows, when volume decreases, the surface area increases rapidly in proportion to volume. More surface area to volume ratio facilitates the to and fro movement of materials and diffusion of oxygen and carbon dioxide across the cell membrane.

(ii) Capacity of a nucleus to serve as the cell's control centre.

Cells tend to be spherical like drops of water and soap bubbles. Rectangular or other shapes of plant cells are due to cell walls. Many other shapes of plant and animal cells are due to attachments to and pressure from neighbouring cells in a tissue organization. As soon as they are dissociated from the tissue, most cells become spherical.

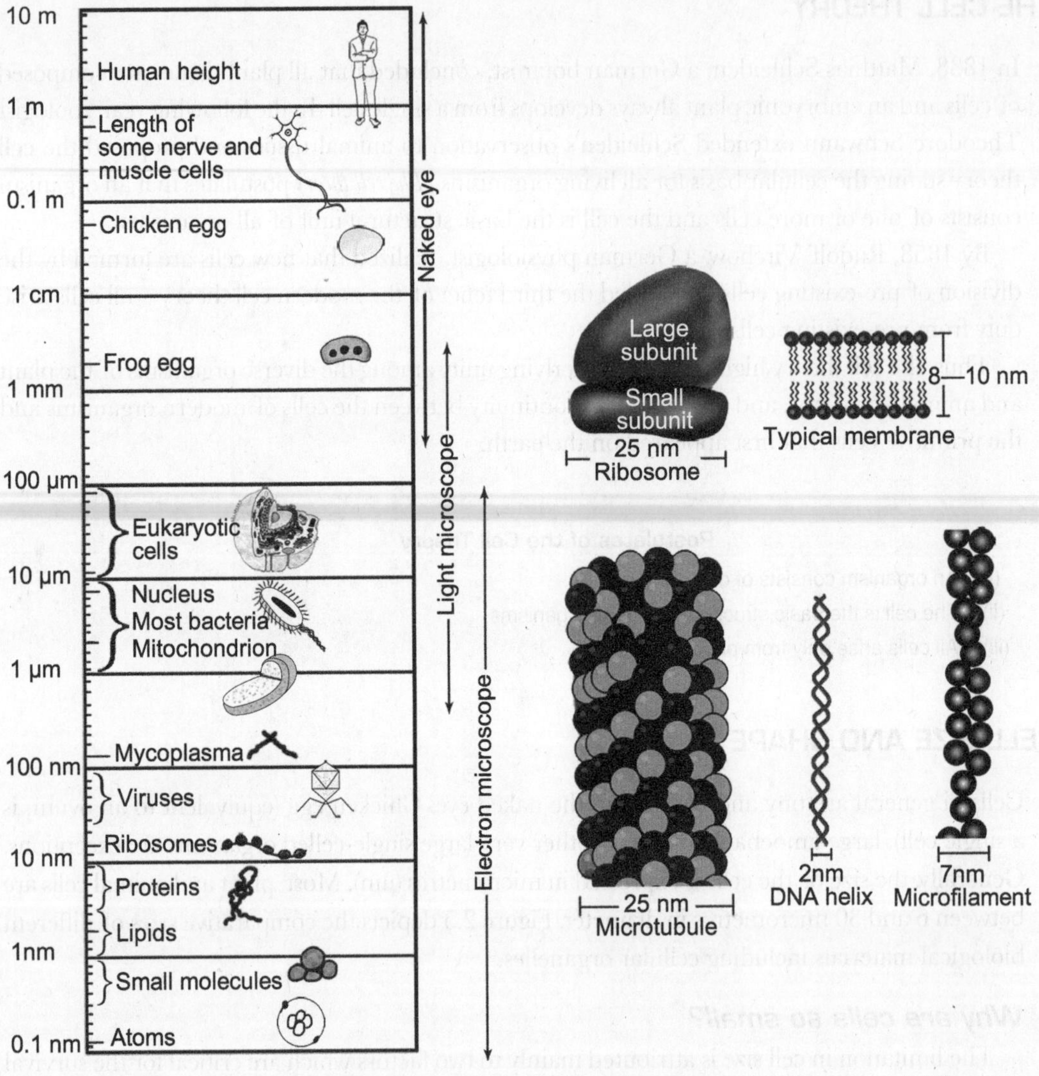

Fig. 2.5 Comparative size of biological objects including cell organelles in the range of different resolutions (nm = nanometre, μm = micrometre, mm = millimetre, cm = centimetre, and m = metre).

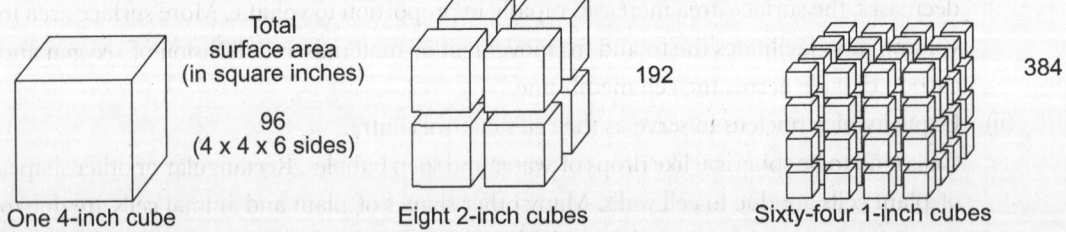

Fig. 2.6 Relationship between volume and surface area—as volume decreases surface area increases. A single 4-inch cube has a 96-square inch surface area (4″x 4″x 6 sides); when the same cube is divided into eight 2-inch cubes, the surface area increases to 192 square inches. (2″x 2″x 8 cubes x 6 sides); Similarly when the cube is divided into 64 1-inch cubes, the total surface area becomes 384 square inches.

2.5 CELL TYPES

There are two distinct types of cells–prokaryotic and eukaryotic cells. The difference lies in their structure, organelles, metabolism, and so on. Prokaryotes include all bacteria, and eukaryotes include plants and animals. These two divisions are based on the fundamental structural differences in cellular organization, particularly the nature of the nucleus (Gr. *karyon*)—*pro* (early) and *eu* (Gr. true).

2.5.1 Prokaryotic and Eukaryotic Cells

Bacteria are single-celled organisms. They are classified as *prokaryotes*, whereas plant and animal cells are *eukaryotes*. Besides certain specific differences in the organelles, the fundamental difference for the classification lies in the absence and presence of a nuclear membrane cover around the chromosome(s). Prokaryotic bacteria and blue-green algae lack the nuclear membrane, and they are primitive when compared to eukaryotes. Eukaryotic cells have a true nucleus in which the genetic material or chromosomes are separated by a nuclear membrane from the cytoplasm.

Prokaryotes originated about 3.5 billion years ago, much before eukaryotes. Prokaryotic cells are generally smaller and simpler than eukaryotic cells. In addition to the absence of a nuclear membrane, their genome is less complex and they lack many cytoplasmic organelles and a proper cytoskeleton (Table 2.1).

Table 2.1 Comparison of prokaryotic and eukaryotic cells

Feature	Prokaryote (bacteria and blue-green algae)	Eukaryote (plant and animal)
Cell size (Diameter)	In average 0.5 – 5 μm	5 – 40 μm 1000 or more times the volume of prokaryotic cells
Form	Unicellular or filamentous (held together in a common mucilaginous sheath)	Unicellular or filamentous or truly multicellular
Cell wall	Rigid, containing polysaccharide plus protein. Murein is the main strengthening compound.	Rigid in plants and fungi, containing the polysaccharide cellulose; chitin in fungi. No such wall in animal cells.
Cell membrane	Present	Present
Flagellum (or cilium)	Simple, lacking microtubule (not enclosed by cell surface membrane)	Complex, with 9+2 microtubules, 200 nm diameter. (Surrounded by cell surface membrane)
Nucleus	No nuclear envelope, no nucleolus	Nucleus bound by nuclear envelope, nucleolus present
Chromosome	Single, circular, containing only DNA	Multiple, containing DNA and protein
Endoplasmic reticulum	Absent	Usually present
Golgi body	Absent	Present

(Contd)

Table 2.1   (*Contd*)

Feature	Prokaryote (bacteria and blue-green algae)	Eukaryote (plant and animal)
Ribosome	Smaller 70S, 50S+30S subunits	Bigger 80S, 60S+40S subunits
Mitochondrion	Absent	Present
Plastid (chloroplast)	Absent	Present in plant cell
Lysosome	Absent	Present in animal cell
Vacuole	Absent	Present
Centriole	Absent	Present in animal cell
Nitrogen fixation	Few can	Absent
Cell division	Fission	Mitosis and meiosis
Exocytosis and endocytosis	No	Yes

A detailed description of prokaryotes and eukaryotes is provided in the following pages.

2.6 PROKARYOTES (BACTERIA)

Prokaryotes include various types of bacteria. They are the smallest cellular organisms, as small as 1 micrometre (10^{-3} mm) in diameter. A single gram of fertile soil can contain as many as 2.5 billion bacteria. A human being carries an estimated 10^{14} bacteria, more than the number of his or her own cells, that is, approximately 10^{13}. Bacteria are the smallest organisms that have all the machinery required for growth and self-replication at the expense of organic compounds synthesized by other organisms; thus, most bacteria are *heterotrophs*. Metabolically, bacteria are extremely versatile. Some are *autotrophs*, synthesizing their own organic compounds from inorganic elements through photosynthesis as in plants. Some others are *chemosynthetic autotrophs* that derive energy from the oxidation of inorganic molecules. Their ecological niches vary from hot springs at 90°C to the freezing temperatures of Antarctica. Bacteria are the most successful organisms on the planet due to their rapid rate of cell division, great metabolic versatility, and capacity to adapt to a large variety of niches. Species variation of bacteria is in the tune of several thousands. A bacterial population, growing under optimal conditions, can double in size every 20–30 minutes.

Being heterotrophs, bacteria along with fungi cause decay and recycling of organic material in the soil. Bacteria cause diseases in humans and animals by producing toxic metabolites.

Initially, prokaryotes were divided into *Archaebacteria* (evolutionarily an ancient group) and *Eubacteria* (true bacteria).

Bacterial cell (Eubacteria)

A bacterial cell is simple and lacks a nuclear membrane, endoplasmic reticulum, and mitochondria.

2.6.1 Cell Wall

A rigid cell wall surrounds the cytoplasmic membrane. The cell wall is complex and lacks cellulose which is found in plant cell walls. Instead, it is a network of another polysaccharide (*peptidoglycan*) connected by cross-linked peptides (a glycopeptide layer). Depending on the complexity, the thickness of the cell wall ranges from 10 to 80 nm. The cell wall protects against mechanical damage and osmotic rupture when salt concentration in the surrounding medium is too low (most bacterial cells are hypertonic). The cytoplasmic membrane acts as the osmotic barrier and participates in the active transport for maintaining appropriate intracellular concentration of specific ions and metabolites. The characteristic shapes of bacteria depend on the cell wall contour. There

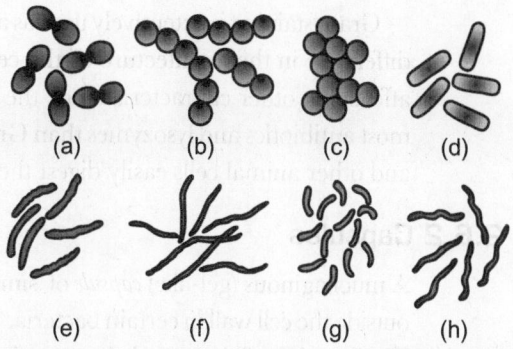

Fig. 2.7 Forms of eubacteria. (a) *Diplococci* (in pairs), (b) *Streptococci*, (c) *Staphylococci*, (d) *Bacilli*, (e) Fusiform bacilli, (f) Filamentous bacilli, (g) *Vibrios*, and (h) *Spirilla*.

are two principal forms of eubacteria: *coccus* (Gr. & L., berry, more or less spherical) and cylindrical *bacillus* (L., stick). The principal forms of eubacteria are depicted in Fig. 2.7.

The members of the *Mycoplasma* group are the smallest prokaryotes; without a cell wall they often parasitize eukaryotic cells.

Gram-positive and Gram-negative bacteria

In addition to the peptidoglycan constituting the bacterial cell wall, large molecules of lipopolysaccharide (LPS) can be found on the outer side of the cell wall in some bacteria (Fig. 2.8). Bacterial cell walls lacking this additional LPS layer combine firmly with dyes like gentian violet; those with the layer do not. Bacterial cells that bind with these dyes are called *Gram-positive*, whereas those that do not bind these dyes are called *Gram-negative*; they have been named after Hans Christian Gram, the Danish bacteriologist who developed the method empirically in 1884.

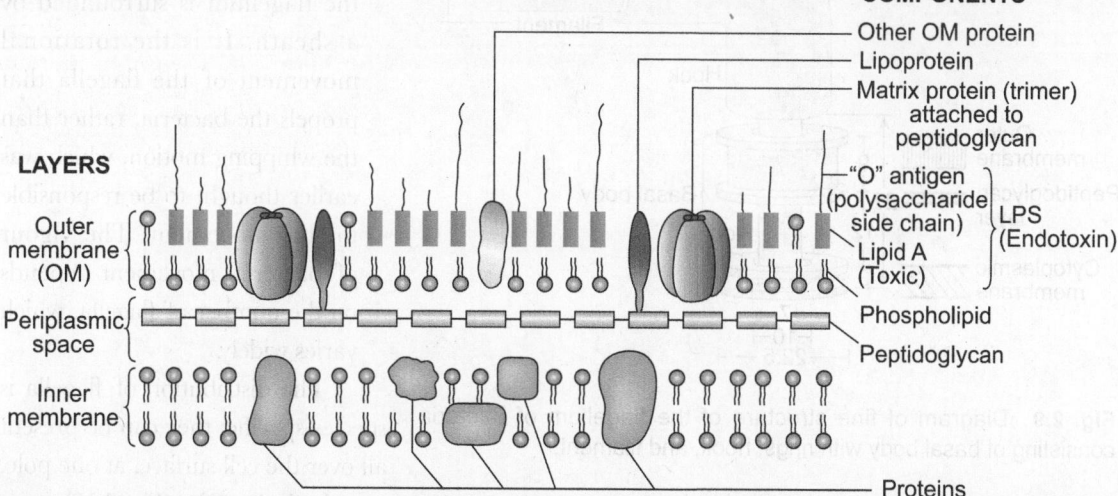

Fig. 2.8 Diagram of the cell wall of a Gram-negative bacterium.

Gram staining is extensively used as a basis for classifying bacteria, since it reflects a fundamental difference in the architecture of the cell wall (presence or absence of LPS layer), which in turn affects the other characteristics of the bacteria. Gram-positive bacteria are more susceptible to most antibiotics and lysozymes than Gram-negative bacteria. Lysozymes present in macrophages and other animal cells easily digest the cell wall of Gram-positive bacteria.

2.6.2 Capsules

A mucilaginous (gel-like) *capsule* of simple polysaccharides, secreted by the bacterium, is present outside the cell wall in certain bacteria. The thickness of the capsule varies with growth conditions. The function of the capsule is not entirely understood, but its presence usually makes the bacteria virulent. It appears that the capsule interferes with phagocytosis by white blood cells of the infected host.

2.6.3 Flagella

The term 'flagella' is derived from the Latin word *flagellum,* meaning whip. Flagella are long (3 μm–12 μm) helical and filamentous appendages present in some eubacteria and are responsible for motility. Since the flagella have a very small (12–25 nm) diameter, they cannot be seen by ordinary microscopes, but are easily visualized by dark field microscopy.

A flagellum consists of three parts—basal body, hook, and filament. The basal body is associated with the cell membrane and consists of two pairs of rings in Gram-negative bacteria (Fig. 2.9), and a single pair of rings (S and M rings) in Gram-positive bacteria. The hook is a small bent structure while the filament is extended. Both hook and filament are made up of the protein *flagellin.* In some Gram-negative bacteria, the flagellum is surrounded by a sheath. It is the rotational movement of the flagella that propels the bacteria, rather than the whipping motion, which was earlier thought to be responsible for the movement. The vigour of bacterial movement depends on the number of flagella, which varies widely.

The distribution of flagella is species-specific; they can be present all over the cell surface, at one pole, or at both the poles (Fig. 2.10).

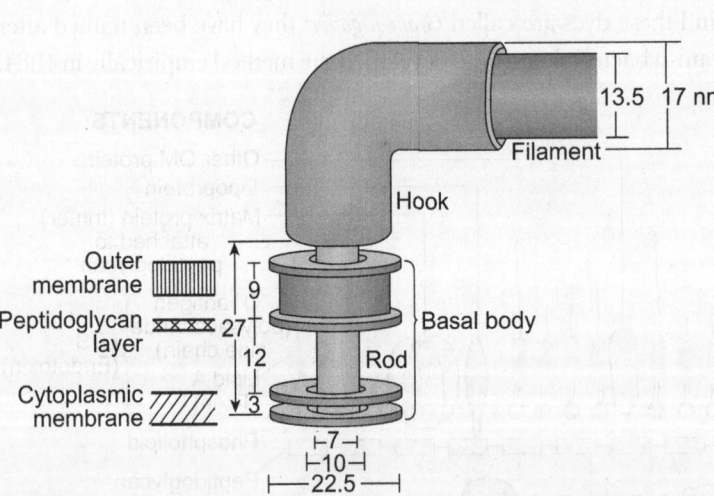

Fig. 2.9   Diagram of fine structure of the flagellum of bacteria consisting of basal body with rings, hook, and filament.

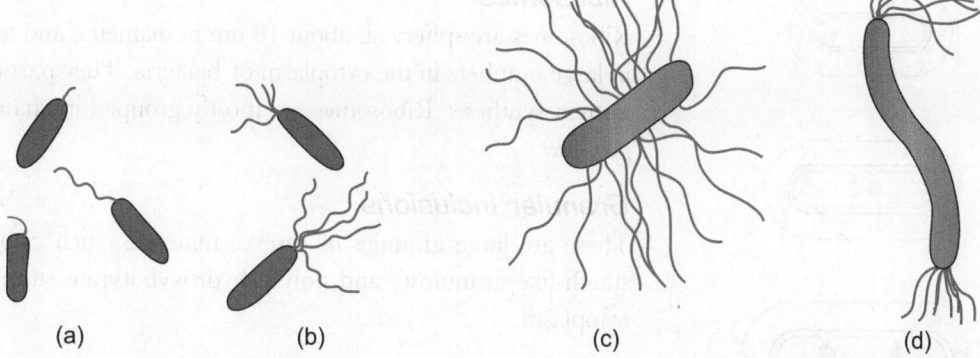

Fig. 2.10 Distribution of the flagella on bacteria. (a) Monotrichous—single polar flagellum, e.g., *Pseudomonas aeruginosa*; (b) Lophotrichous—cluster of polar flagella, e.g., *Pseudomonas fluorescens*; (c) Peritrichous—lateral flagella, e.g., *Escherichia coli*; (d) Amphitrichous—cluster of flagella at both ends, e.g., *Aquaspirillum serpens*.

2.6.4 Pili

The term 'pili' is derived from the Latin word *pilus*, meaning hair. Pili (singular – pilus), also known as fimbrial, are filamentous structures and resemble flagella, but are thinner, shorter, non-helical, and non-motile. They are made up of monomers of proteins known as *pilin*. They play a role in pathogenesis by attaching bacterial cells to the gastro-intestinal and genito-urinary tracts. A few special types of pili called *sex* or *F pili* are formed on the male bacterial cells during conjugation. One such sex pilus establishes contact with a female (F^-) bacterium and forms a conjugating tube through which genetic material from the male (F^+ or Hfr) strain enters the F^- strain.

2.6.5 Cytoplasmic Contents

The cytoplasm contains mesosomes, ribosomes, granular inclusions, nucleoid body, and circular DNA plasmids .

Mesosomes

Mesosomes are large irregular invaginations of the plasma membrane (Fig. 2.11). They are smaller in Gram-negative bacteria than in Gram-positive bacteria. Mesosomes evidently compartmentalize the cytoplasm somewhat like the endoplasmic reticulum in eukaryotic cells. They are known to participate in secretion and cell division.

Photosynthetic bacteria have a similar internal membrane structure, the *chromatophore*, which contains the photosynthetic pigments and components of the electron transport chain required for photosynthesis.

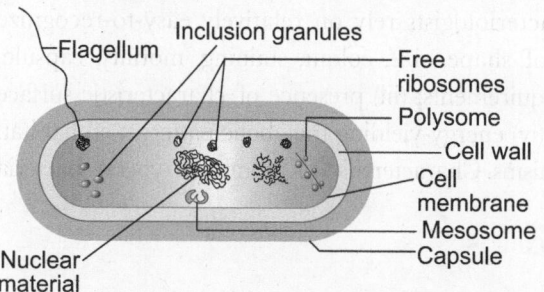

Flagellum
Inclusion granules
Free ribosomes
Polysome
Cell wall
Cell membrane
Mesosome
Capsule
Nuclear material

Fig. 2.11   General structure of a bacterial cell.

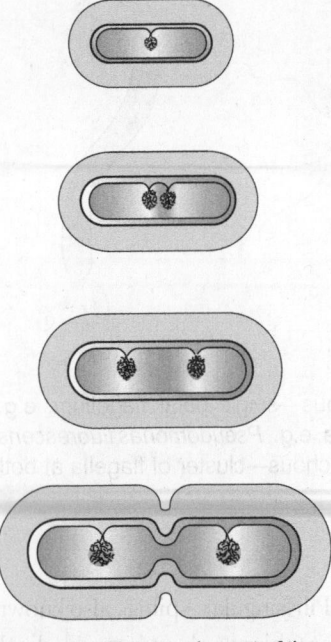

Fig. 2.12 The attachment of the bacterial chromosome to the cell membrane ensures separation of divided chromosomes in the new cells during cell division.

Ribosomes

Ribosomes are spherical, about 18 nm in diameter, and are found in large numbers in the cytoplasm of bacteria. They participate in protein synthesis. Ribosomes are mostly grouped in chains called *polysomes*.

Granular inclusions

These are large granules of reserve materials, such as glycogen, starch-like granulose, and poly-β-hydroxybutyrate stored in the cytoplasm.

Nuclear body (nucleoid)

A single highly folded circular DNA molecule constitutes the bacterial chromosome. The nuclear membrane, nucleolus, and mitotic apparatus that are found in eukaryotic cells are absent in bacteria. Hence, the nuclear body in bacteria is often referred to as *nucleoid*. During cell division, the attachment of the chromosome to the cell membrane helps in carrying the newly synthesized chromosome to the new cell (Fig. 2.12).

Unlike eukaryotic cells, the cytoplasm in bacteria has a high concentration of RNA (including ribosomes); hence basic dyes cannot reveal the nucleoid. The Feulgen staining method which is specific for DNA is used to selectively stain the bacterial nuclear material to study under the light microscope.

Plasmids

Plasmids are small circular pieces of DNA present in the cytoplasm; a bacterial cell can contain one or several plasmids. Plasmids carry different genes including those that code for resistance to different antibiotics. They are extensively used as vectors in molecular biology.

2.6.6 Types of Bacteria

To identify different types of bacteria, bacteriologists rely on relatively easy-to-recognize characteristics, such as (i) visible features of shape, size, colour, staining, motility, capsule, and colonial morphology; (ii) nutritional requirements; (iii) presence of characteristic surface macromolecules detected immunologically; (iv) energy-yielding metabolic patterns; (v) habitat, including the ability to parasitize higher organisms. Characteristics of prominent types of bacteria are discussed as follows.

Eubacteria

Eubacteria (true bacteria) constitute the largest group of bacteria and includes several forms differing principally in shape. Figure 2.7 shows some of the common types such as straight rod-shaped *Bacillus* like *E. coli* and spherical *coccus* like *Streptococci*.

Bacilli usually separate after cell division. In some groups, they stay together and spread out end to end in filaments giving a fungus-like (Gr. *Myco*) appearance, for example, *Mycobacterium tuberculosis* which causes tuberculosis.

Myxobacteria (Gr. myxa, slime)

Myxobacteria are distinguished by their secretion of a slime track on which the cells glide and by complex reproductive structures (cysts) known as fruiting bodies.

Spirochaetes

Spirochaetes are very long (5–500 μm) in relation to their width (about 0.5 μm) with a long flexible axial filament (similar to bacterial flagellum) wrapped around the cell body. Contractions of the axial filament cause the spirochaetes to move with a whirling motion.

Rickettesia and mycoplasma

Rickettesia and mycoplasma constitute the smallest prokaryotes. They do not have a cell wall and are mostly parasites of eukaryotic cells. These bacteria are difficult to culture in vitro.

Cyanobacteria (blue-green bacteria)

Cyanobacteria were earlier called blue-green algae because they produce oxygen during photosynthesis and structurally resemble the eukaryotic true algae (unicellular plants). These are filaments of cells which do not separate following cell division and are held together within a common mucilaginous (bacterial) capsule or sheath. Essentially the cells are separate entities. By the process of evolution and symbiosis, blue-green bacteria may have given rise to the chloroplasts in eukaryotic plants.

Photosynthetic bacteria that do not produce oxygen (O_2)

There are three photosynthetic forms of chlorophyll-containing eubacteria that consists of a magnesium-containing porphyrin ring and a lipid tail.

Green sulphur bacteria   An example of such bacteria is *Chlorobium*; the chlorophyll is chemically very similar to the chlorophyll a of higher plants.

Purple sulphur bacteria   These are bluish grey in colour.

Purple non-sulphur bacteria   The colour ranges from purple to red or brown. The bacteriochlorophyll chemically differs from chlorophyll a. The different colours of the purple bacteria are due to different yellow and red carotenoids which act as accessory pigments in photosynthesis.

Unlike plant photosynthesis where water serves as the hydrogen donor, in photosynthesis by sulphur bacteria, H_2S is the hydrogen donor, and it releases sulphur.

$$CO_2 + 2H_2S \xrightarrow{\text{Light}} (CH_2O) + H_2O + 2S$$

Photosynthesis is carried out anaerobically in sulphur bacteria and never yields molecular oxygen (O_2). This type of photosynthetic process evolved in the early stage of evolution when the atmosphere was more reducing in nature and did not contain much oxygen. Thus, these photosynthetic bacteria are considered evolutionarily ancient among eubacteria.

In non-sulphur photosynthetic bacteria, other organic substances, such as alcohols and fatty acids serve as electron donors for the photosynthetic process.

Chemosynthetic bacteria

Chemosynthetic bacteria obtain their energy from the oxidation of certain inorganic compounds, such as ammonia or ammonium compounds, sulphur, and hydrogen sulphide.

Nitrogen-fixing bacteria

Ammonia or ammonium in the soil is derived from the breakdown of organic materials, nitrogen-fixation by bacteria, and to a minor extent, from lightning or volcanic activities. Several species of soil bacteria oxidize ammonia or ammonium (nitrification) to nitrite (NO_2^-):

$$2NH_3 + 3O_2 \longrightarrow 2NO_2^- + 2H^+ + 2H_2O \quad \text{(energy released)}$$

Nitrite is toxic to higher plants, but it rarely accumulates. Members of another genus of bacteria oxidize the nitrite to nitrate, with the release of energy:

$$2NO_2^- + O_2 \longrightarrow 2NO_3^- \quad \text{(energy released)}$$

Although plants can utilize ammonium directly, nitrate is the form in which most nitrogen moves from the soil into the roots.

Once the nitrate is within the cell, it is reduced back to ammonium. The ammonium ions thus formed react with carbon-containing compounds to produce amino acids and other nitrogen-containing organic compounds. In contrast to nitrification, this assimilation process requires energy.

Sulphur bacteria

Chemosynthetic activities of sulphur bacteria oxidize elemental sulphur to sulphate to release energy:

$$2S + 2H_2O + 3O_2 \longrightarrow 2H_2SO_4 \quad \text{(energy released)}$$

The converted sulphur compound is then taken up by plant roots. Sulphur is required by plants for the synthesis of certain amino acids.

Some other sulphur bacteria, such as *Beggiatoa*, obtain energy by oxidizing hydrogen sulphide.

Archaebacteria

Archaebacteria represent an ancient group of bacteria, which are phylogenetically distant from eubacteria as based on 16S rRNA studies. They are found in extreme conditions that resemble the conditions that prevailed in ancient times (*archaic*). Table 2.2 presents the characteristic features of archaebacteria and eubacteria.

Table 2.2 Characteristic features of archaebacteria and eubacteria

Characteristics	Archaebacteria	Eubacteria
Muramic acid in cell wall	Absent	Mostly present
Membrane lipids	Ether-linked isopropenoid side chains	Ester-linked
Initiator tRNA for protein synthesis	met-tRNA	f-met-tRNA

Three major groups in archaebacteria have been described on the basis of their metabolic and ecological properties.

Methanogens   These are stringent anaerobes that obtain energy by oxidizing compounds like H_2, utilizing the electrons generated to reduce CO_2 to form CH_4 (methane).

$$CO_2 + 4H_2 \longrightarrow CH_4 + 2H_2O$$

These bacteria are also called methanogenic bacteria because they generate methane. Their habitats are marshes, swamps, pond and lake mud, marine sands, intestinal tract of humans and animals, rumen of cattle and sewage.

Halophiles   These are aerobic, Gram-negative bacteria and can be rods or cocci, and require 17–23 per cent sodium chloride (NaCl) for growth. The colonies appear red to orange due to the presence of carotenoids that protect them from the damaging effect of sunlight. The habitat is highly saline environment, for example, salt lakes (Dead Sea, Great Salt lake in the US), heavily salted fishes and meats.

High NaCl concentration prevents cell lysis, as the cell wall contains protein subunits that are rich in acidic charges that are shielded by Na^+. When Na^+ concentration falls, the COO^- subunits repel each other and the wall breaks apart leading to lysis of the cell.

At high NaCl concentration, the cells resist cell dehydration by maintaining high levels of the osmotic solute KCl.

Thermoacidophiles   These are Gram-negative and grow under highly acidic conditions at high temperatures. They constitute the following two groups of genera:

Thermoplasma   These are aerobic or anaerobic species that grow optimally at 55°C and pH 2. Morphologically they resemble *Mycoplasmas* which lack cell wall and are hence pleomorphic in shape.

Sulpholobus   These are spherical or lobe shaped bacteria that grow as chemolithotrophs with elemental sulphur as electron donor; alternatively, as chemo-organotrophs in the presence of organic substrates. Temperature optimum ranges from 70 to 87°C and pH optimum is 2. At pH greater than 4, the cells leak and disintegrate. Such bacteria are predominant in acidic hot springs.

Note: The earliest fossils of bacteria were found in rocks that are more than three billion (3×10^9) years old. However, the fossil record is not detailed enough to trace bacterial evolution.

2.7 EUKARYOTIC CELLS AND ORGANELLES

Eukaryotes include both plant and animal kingdoms. The organisms can be single celled, such as *Amoeba, Euglena, Paramaecium, Chlamydomonas*, or multicellular like plants and animals. The organisms are quite different from one another as all of them evolved separately through millions of years. But the cell theory proposes the underlying unity among diverse plants and animals through the similar structure of their cells. Thus, the cells of a plant and an animal bear structural and functional similarity. The differences are a few, for functional diversification (Fig. 2.13). Again, an animal or a human being has many different types of cells, which have all originated from a single

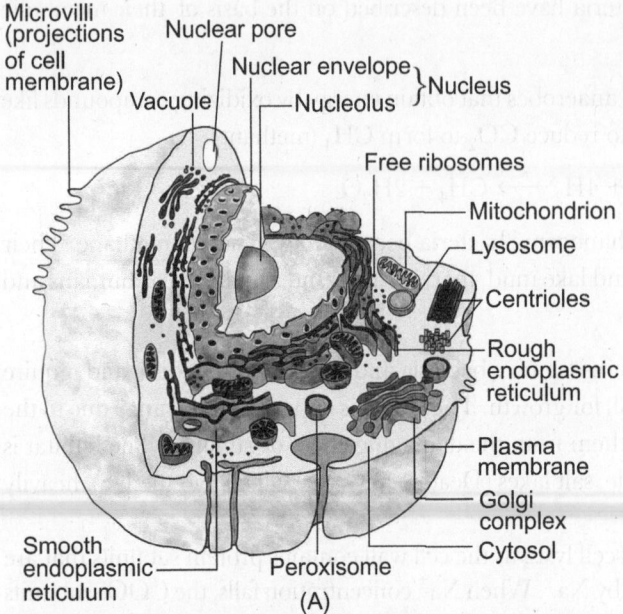

Microvilli (projections of cell membrane)
Nuclear pore
Nuclear envelope ⎫
Nucleolus ⎬ Nucleus
Vacuole
Free ribosomes
Mitochondrion
Lysosome
Centrioles
Rough endoplasmic reticulum
Plasma membrane
Golgi complex
Cytosol
Smooth endoplasmic reticulum
Peroxisome

(A)

Fig. 2.13   (A) Three-dimensional organization of a typical animal cell showing relative size and shape of organelles and other subcellular structures.

celled zygote and differentiated to carryout specialized functions.

Eukaryotic cells, in general, are much larger than prokaryotic cells, about thousand-fold larger. They are more complex and contain a variety of membrane-bound cytoplasmic organelles, such as nucleus, endoplasmic reticulum, Golgi body, mitochondria, chloroplasts, lysosomes, vacuoles, and peroxisomes that carryout the different functions of a cell efficiently. The nucleus contains several linear chromosomes and is bound by nuclear membrane. The nucleus is the site of DNA replication and RNA synthesis. The translation of RNA to protein takes place on ribosomes in the cytoplasm. The plasma membrane (or cell membrane) surrounds the living material of a cell which is sometimes called the protoplasm and consists of the cytoplasm and the nucleus.

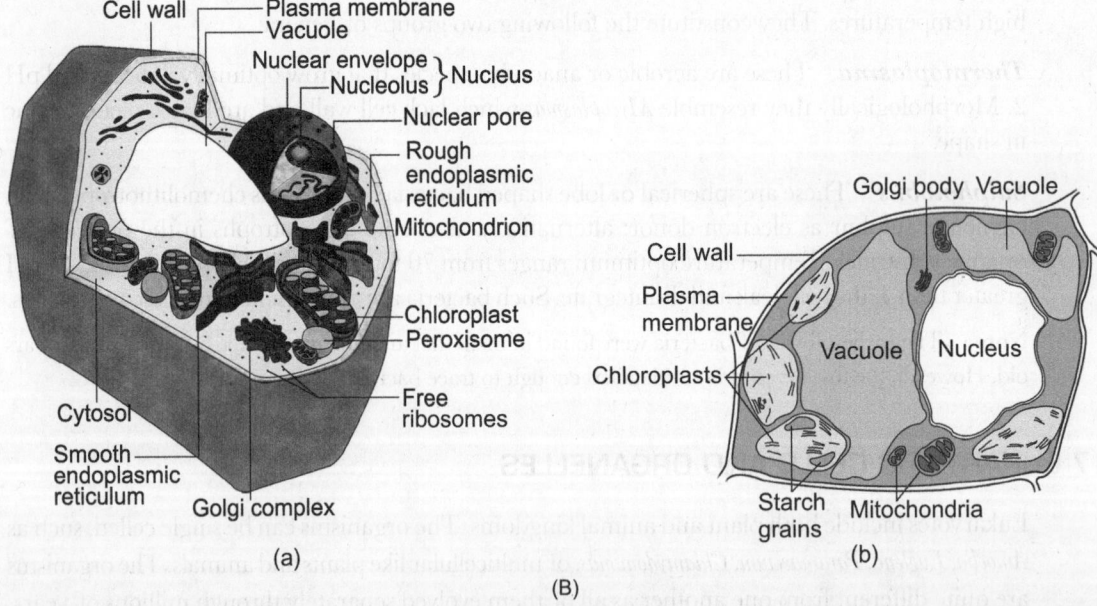

Cell wall
Plasma membrane
Vacuole
Nuclear envelope ⎫
Nucleolus ⎬ Nucleus
Nuclear pore
Rough endoplasmic reticulum
Mitochondrion
Chloroplast
Peroxisome
Free ribosomes
Cytosol
Smooth endoplasmic reticulum
Golgi complex

(a)

Golgi body   Vacuole
Cell wall
Plasma membrane
Vacuole   Nucleus
Chloroplasts
Starch grains   Mitochondria

(b)

(B)

Fig. 2.13   (B) A typical plant cell. (a) Three-dimensional organization of organelles, (b) Schematic diagram showing characteristic features. Note in comparison to an animal cell, the plant cell is characterized by the absence of lysosomes and centriole, and the presence of large vacuoles, chloroplasts, a cell wall, and starch granules.

Eukaryotic cells have a variety of *organelles* which are membrane-bound compartments that are specialized for different specific functions. The fundamental differences between the organelles of plant and animal cells are highlighted in Table 2.3.

Table 2.3 Characteristic differences between plant and animal cells

Characteristics	Plant	Animal
Cell wall	Present (cellulose)	Absent
Intercelluar connection through	Plasmodesmata in cell wall	Gap junctions (connexons) in plasma membrane
Plastids	Present (e.g. chloroplasts for photosynthesis)	Absent
Lysosomes	Absent	Usually present
Vacuoles	Usually large and single in mature cell	Small or absent
Centrioles	Absent	Present (helps in spindle formation during cell division)
Storage form of carbohydrate (sugar)	Starch	Glycogen

2.7.1 Plasma Membrane

The plasma membrane is a sturdy and slightly flexible cell boundary of about 90 Å thickness which encloses the cytoplasm of a cell. It acts as a selective barrier for the flow of materials into and out of a cell, to maintain appropriate conditions for cellular activities. The receptors and defined molecular structures present in the plasma membrane allow the cell to communicate with the surrounding environment and with other cells.

Earlier, quite a few hypotheses and models were proposed by different scientists to explain the structure of the plasma membrane. Electron microscopy contributed significantly to the development of the current model.

Today, the *Fluid Mosaic Model*, proposed by S. Jonathan Singer and Garth Nicolson in 1972, is widely accepted. Essentially this model describes the plasma membrane as a mosaic of protein molecules discontinuously embedded in a lipid bilayer with the lipid components in constant lateral motion (hence fluid). Many membrane protein molecules also move laterally within the membrane, except the ones that are anchored below the cytoskeletal elements.

Organization of the plasma membrane

Lipid bilayer   This is composed of two layers of amphipathic lipid molecules arranged back to back, as shown in Fig. 2.14. The lipid molecules orient themselves in the bilayer with their polar heads (hydrophilic) facing outwards for contact with the watery fluid on either side—the cytosol and the extracellular fluid, respectively. The hydrophobic non-polar fatty acid tails in each half of the bilayer point towards the centre or interior of the membrane.

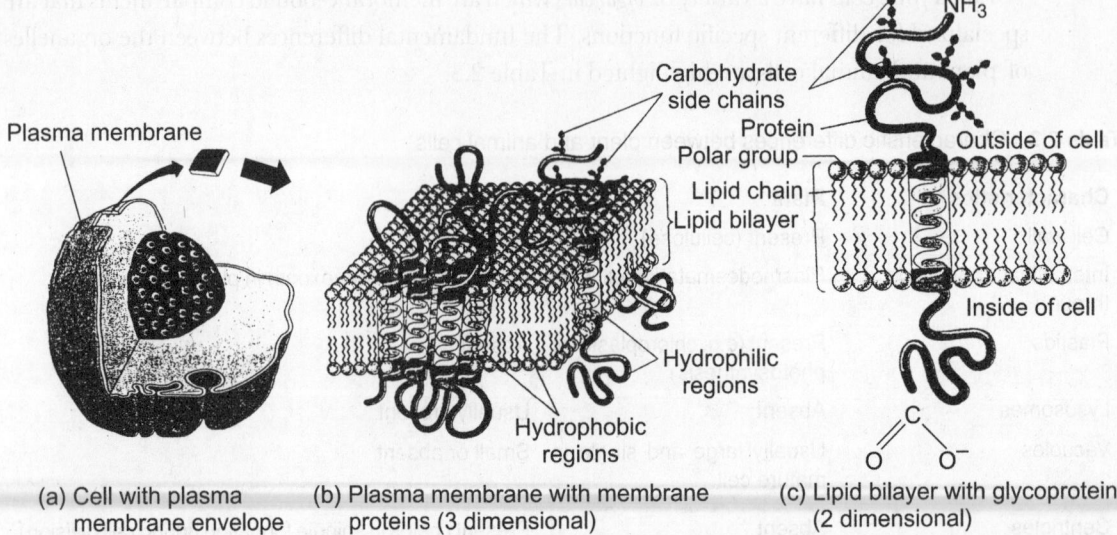

(a) Cell with plasma membrane envelope   (b) Plasma membrane with membrane proteins (3 dimensional)   (c) Lipid bilayer with glycoprotein (2 dimensional)

Fig. 2.14 Organization of plasma membrane. (a) Plasma membrane enveloping the cell. (b) A portion of the plasma membrane magnified to show the lipid bilayer with membrane proteins extending on both sides or on one side. The hydrophobic region of a protein remains associated with the hydrophobic interior of the lipid bilayer, and thus anchors the protein; the hydrophilic regions protrude out of the bilayer. (c) Two-dimensional view of the membrane. The short carbohydrate side chains are attached to the external side of the protein chains making them glycoproteins.

Lipid molecules are of three types depending on the chemical nature of the polar heads.

Phospholipids The phosphate groups in the polar head and constitute 75 per cent of the membrane lipids.

Glycolipids The carbohydrate groups form the polar head. They constitute about 5 per cent of membrane lipids and is present only in the half of the membrane that faces the exterior of the cell.

Cholesterol Tiny –OH group in the polar region, weakly amphipathic. It constitutes about 20 per cent of membrane lipids and is interspersed among the other lipids. The –OH group forms hydrogen bonds with the polar heads of phospholipids and glycolipids in the membranes of most animal cells.

Membrane proteins These are proteins embedded like mosaic in the lipid bilayer. Some protein molecules can extend from one side to the other side of the lipid bilayer as *integral proteins*. *Peripheral protein* molecules, on the other hand, associate with membrane lipids or integral proteins and extend half way into the inner or outer surface of the membrane. Like lipids, membrane proteins are also amphipathic (they have both hydrophobic and hydrophilic groups). Their hydrophilic regions protrude into the aqueous extracellular fluid or the cytosol, and the hydrophobic regions are associated with the fatty acid tails of lipids, providing anchorage to the molecules. Proteins in the plasma membrane are typically glycoproteins with short carbohydrate side chains attached to the hydrophilic region(s) on the external, but not the internal, side of the membrane (Fig. 2.14a and b). This differential distribution of carbohydrate side chains contributes to the membrane's asymmetry.

Membrane proteins carryout a variety of functions as: (a) *enzymes*, (b) *transport proteins* facilitating movement of nutrients such as sugars and amino acids, (c) *channel proteins* providing hydrophilic pathways through the hydrophobic lipid bilayer for ions and other small substances, (d) *receptors* for chemical signals, (e) *connexons* at gap junctions, and (f) *structural proteins* stabilizing the shape of cells.

Although the basic structure of the plasma membrane as a mosaic of proteins embedded in a lipid bilayer is similar in all living systems including bacteria, there are cell-specific variations in the types of lipids and in the number and types of proteins. The protein-to-lipid ratio is often distinctive for membranes of different cells or for different organelles. The organelles within the cell are enclosed by a membrane similar to the plasma membrane.

2.7.2 Gap Junctions

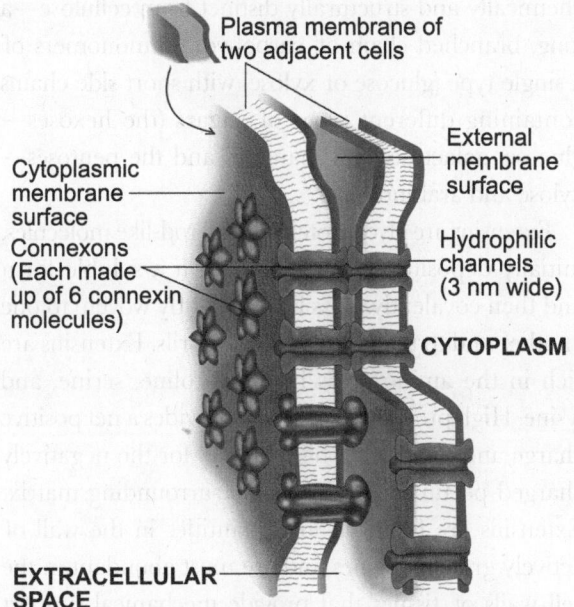

Plasma membrane of two adjacent cells

Cytoplasmic membrane surface

Connexons (Each made up of 6 connexin molecules)

External membrane surface

Hydrophilic channels (3 nm wide)

CYTOPLASM

EXTRACELLULAR SPACE

Fig. 2.15 Gap junction consists of a large number of hydrophilic channels formed by connexon molecules in the plasma membrane of adjoining cells. The channels are for direct electrical and chemical communication between cells.

Gap junctions in the plasma membrane of animal cells are regions of intercellular connections for the transfer of ions and small molecules between the cytoplasm of adjacent cells. Adjacent cells are thus in direct electrical and chemical communication with each other (Fig. 2.15). The connection channels from adjacent cells are aligned for continuity. Each channel is called *connexon* and is 7 nm in diameter. In vertebrates, each *connexon* is a circular assembly of six subunits of the protein *connexin* surrounding a hydrophilic channel about 3 nm in diameter. The channel is large enough to allow the passage of ions and small molecules. In invertebrate cells instead of connexin, the protein present is *innexin*.

Although gap junctions occur in most vertebrate and invertebrate cell types, they are abundant in tissues, such as muscle and nerve, where extremely rapid communication and flow of ions (electrical current) between cells are required (plant cell walls have similar structures called plasmodesmata).

2.7.3 Plant Cell Wall

The rigid, non-living, permeable wall, mainly made up of the polysaccharide *cellulose*, surrounds the plasma membrane of all plant cells except sperm and some eggs. The presence or absence of the cell wall is an important difference between plant and animal cells (bacterial cell wall mainly contains peptidoglycan, not cellulose). The following is a detailed description of the structural components of the plant cell wall.

Cellulose

Cellulose is the predominant polysaccharide of the plant cell wall. It is the single most abundant organic macromolecule on the earth. Cellulose is an unbranched polymer of β-D-glucose units linked together by β (1 → 4) bonds (Fig. 2.16).

The long ribbon-like polymer structure of cellulose is stabilized by intra-molecular hydrogen bonds. Many such polymers—typically 50 to 60— associate laterally to form the *microfibrils*. These microfibrils twist together in a rope-like fashion to form *macrofibrils*. Cellulose macrofibrils are as strong as steel of equivalent size.

Hemicellulose and extensins

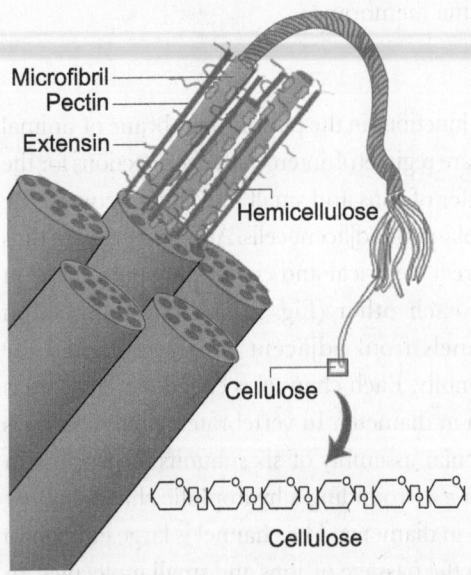

Microfibril
Pectin
Extensin

Hemicellulose

Cellulose

Cellulose

Fig. 2.16 Structural components of plant cell walls. Cellulose microfibrils complexed with hemicellulose and extensins, and embedded in a matrix of proteins. Several cellulose microfibrils are twisted together to form macrofibrils.

A complex network of branched polysaccharides called hemicellulose and glycoprotein called *extensins* enmesh the cellulose microfibrils (see Fig. 2.16). *Hemicellulose* is chemically and structurally distinct from cellulose—a long, branched chain of carbohydrate monomers of a single type (glucose or xylose) with short side chains containing different types of sugars (the hexoses— glucose, galactose, and mannose and the pentoses— xylose and arabinose).

Extensins are glycoproteins, rigid, rod-like molecules, initially deposited in the cell wall in a soluble form and then covalently cross linked (tightly woven) to one another and to the cellulose microfibrils. Extensins are rich in the amino acids hydroxyproline, serine, and lysine. High proportion of lysine provides a net positive charge and therefore a high affinity for the negatively charged pectin molecules in the surrounding matrix. Extensins are found in low quantities in the wall of actively growing tissues and are most abundant in the cell walls of tissues that provide mechanical support to the plant.

Pectins

Pectins are also branched polysaccharides with backbones of negatively charged galacturonic acid and rhamnose sugar. The side chains attached to the backbone contain the same monosaccharides as in hemicellulose. Pectin molecules form the *matrix* in which cellulose microfibrils, hemicelluloses, and extensin complexes are embedded. They also bind adjacent walls together. The highly branched structure and negative charge allow pectin to trap and bind water molecules and assume a gel-like consistency. (Due to its gel-forming capacity, pectin is added to fruit juice during the preparation of fruit jams and jellies.)

A typical cell wall contains 40 per cent cellulose, 20 per cent hemicellulose, 10 per cent extensins, and 25–30 per cent pectins.

Lignins

Lignins (L. *lignum*, wood) are very insoluble polymers of aromatic alcohols, occurring mainly in secondary cell walls in woody tissue. Lignin molecules are localized mainly between the cellulose fibrils and are also crosslinked with other cell wall constituents, especially the extensins. Functionally, lignins resist compression forces. Lignin constitutes about 25 per cent of the dry weight of woody plants, making it second only to cellulose as the most abundant organic compound on the earth.

Cell wall and turgor pressure

The cell wall helps plant cells withstand the considerable turgor pressure generated by the uptake of water mainly through the root system. This pressure cannot burst the plant cells due to the presence of a rigid cell wall. It is also responsible for the firmness (turgidity) of fully hydrated plant tissue and prevents a plant from wilting. This is vital for the plant. Turgor pressure drives the expansion of cells in the plants during development.

Plasmodesmata

Palsmodesmata (singular *plasmodesma*) are cytoplasmic channels through the openings in the cell wall (Fig. 2.17). Plasmodesmata thus permit direct inter-cellular communication between adjacent cells through the cell wall, just like gap junctions in animal cells. Each plasmodesma is lined with plasma membrane common to the two connected cells. It is cylindrical in shape with channel diameter varying from 20 to 200 nm. Most plasmodesmata have a narrow cylindrical *desmotubule* at the centre that is derived from the endoplasmic reticulum (ER) and seems to be continuous with the ER of both the cells.

The ring of cytoplasm between the *desmotubule* and the plasma membrane that lines the plasmodesma is called *annulus*. The cytoplasmic continuity of the annulus allows free passage of ions and molecules between neighbouring cells. Plasmodesmata reduce electrical resistance between adjacent cells.

Plasmodesmata are formed during cell division, when the new cell wall is being formed.

Primary and secondary cell wall

Primary cell wall is formed during cell division. Cellulose-synthesizing enzyme complexes called *rosettes*, located within the plasma membrane, generate cellulose microfibrils. Then other components of the cell wall are deposited. The primary cell wall is characterized by its flexibility and extensibility that enable it to expand during cell enlargement and elongation.

The *secondary cell wall* is formed by deposition of cell wall components in multiple layers on the inner surface of the primary wall after cell growth has ceased. Cellulose and lignins are major constituents of the secondary cell wall making it stronger, harder, and more rigid than the primary wall.

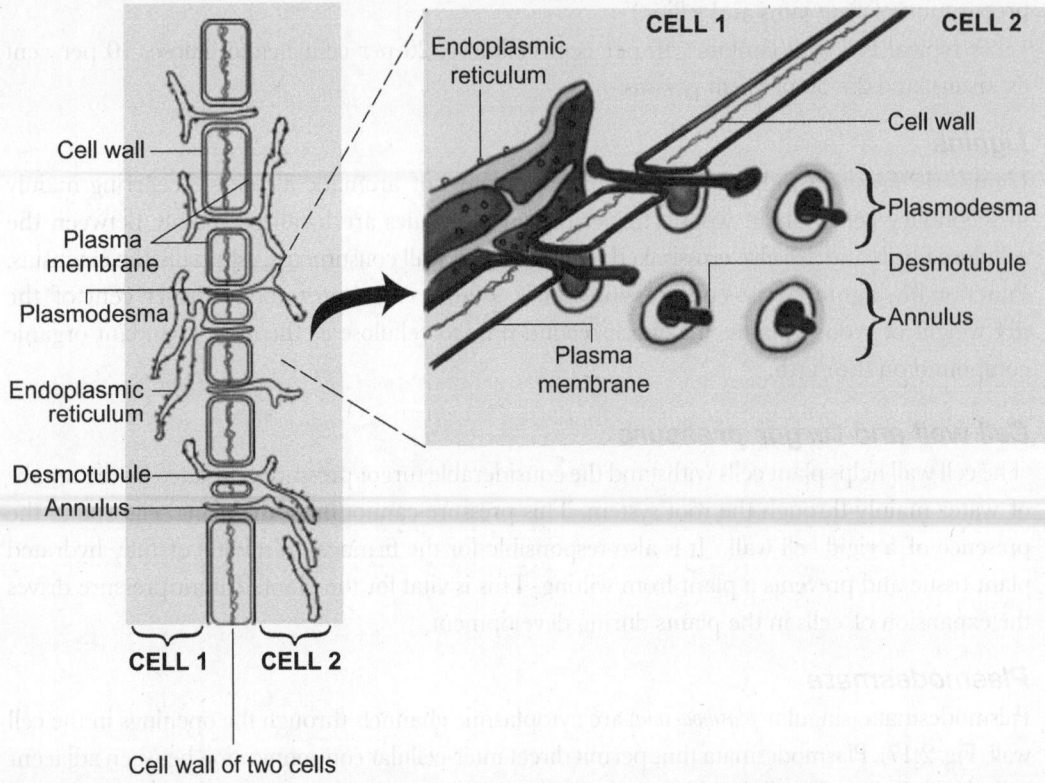

Fig. 2.17   Diagram showing plasmodesmata—cytoplasmic channels through the openings in the cell wall.

The cell wall acts as *a permeability barrier*; it does not allow large molecules to enter the cell. Water, gases, ions, and small water-soluble molecules, such as sugars and amino acids, easily diffuse through the wall. The cell wall normally allows the passage of globular molecules with molecular weight upto about 20,000 daltons. Plant hormones are usually small molecules with molecular weights less than 1000 daltons.

Plant Cell Wall in a Nutshell

The plant cell wall is a network of cellulose fibrils, polysaccharides, and glycoproteins. Apparently, it is non-living but is of utmost importance to plant cells for their safety, turgidity, selective permeability, and for maintaining inter-cellular communication through plasmodesmata.

2.7.4 Nucleus

The nucleus is a double membrane-bound spherical structure (sometimes elliptical or of other shape) which stores genetic information (DNA molecules are organized in chromosomes) and possesses the information for all the vital activities of a cell. The nucleus is the most prominent and characteristic feature of a eukaryotic cell (*eukaroyon* means 'true nucleus'). Usually, it is located more centrally in animal cells and towards one side in mature plant cells (Fig. 2.13). DNA replication, transcription, and RNA processing take place in the nucleus and remain separated from translation and other metabolic processes which take place in the cytoplasm. The content of the nucleus is fluid in nature, and is often called *nucleoplasm*.

The *nuclear envelope* is composed of two membranes—the inner and outer nuclear membranes, separated by a *perinuclear space*, that is 20 to 40 nm across (Fig. 2.18). Each membrane is structurally similar to the plasma membrane and is 7 to 8 nm thick. The inner nuclear membrane rests on a network of supporting fibres called *nuclear lamina*, which provide attachment sites for chromatin fibres. The outer nuclear membrane is continuous with the endoplasmic reticulum (ER), making the perinuclear space and content continuous with the lumen of the ER. Like the rough ER, the outer surface of the outer membrane is associated with ribosomes engaged in protein synthesis. Filaments of the cell's cytoskeleton remain in association with the outer membrane, anchoring the nucleus with other organelles or with the plasma membrane.

Nuclear pores

Multiple specialized openings known as *nuclear pores* are characteristic of the nuclear envelope. Freeze-fracture technique and electron microscopy revealed these structures in detail. Each pore is a small cylindrical channel that extends through both the membranes of the nuclear envelope, allowing direct continuity between the nucleoplasm (within the nucleus) and the cytosol (Fig. 2.19). The number of pores per unit surface area of the nuclear envelope (density of pores) varies with cell types and their activity. A typical mammalian nucleus may have 3000 to 4000 pores.

At the site of a pore, the inner and outer membranes of the nuclear envelope are fused together, forming a channel. Each channel (pore) is guarded by an intricate protein structure called *nuclear pore complex* (NPC). The diameter of the pore complex is 120 nm and it is an organization of as many as 100 distinct polypeptides.

NPC is made up of eight subunits with protruding parts on the cytoplasmic and nucleoplasmic sides (Fig. 2.19a). The protruding parts thus form two parallel rings in an octagonal pattern on the outer and inner sides of the nuclear pore. Eight *spokes*, each connected to one of the subunit structures (forming the two

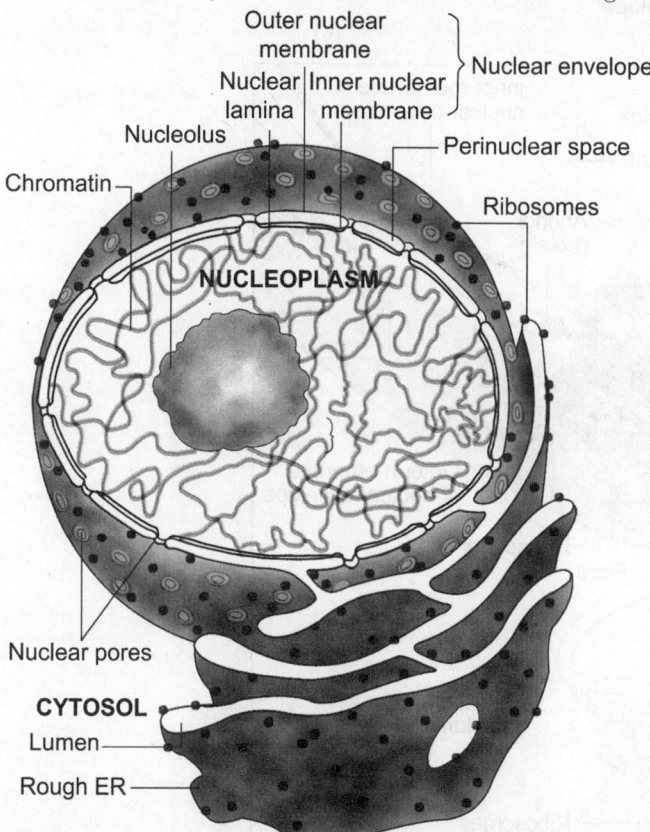

Fig. 2.18 Diagram showing the structural organization of a typical nucleus and nuclear envelope. Note the continuity between the outer nuclear membrane and the rough endoplasmic reticulum.

rings), extend to the central *transporter*. The spoke–ring assembly looks like a wheel and remains anchored to the nuclear envelope through 'anchor proteins' (Fig. 2.19c). The transporter seems to be a diaphragm-like structure that opens and closes to allow the passage of particles of different sizes. Protein filaments extend from both the cytoplasmic and the nucleoplasmic rings. The ones on the nucleoplasmic side form a basket-like structure (called a 'cage' or 'fishtrap') with specific functions yet to be assigned. The filaments on the cytoplasmic side possibly help in the import of large protein molecules from the cytoplasm.

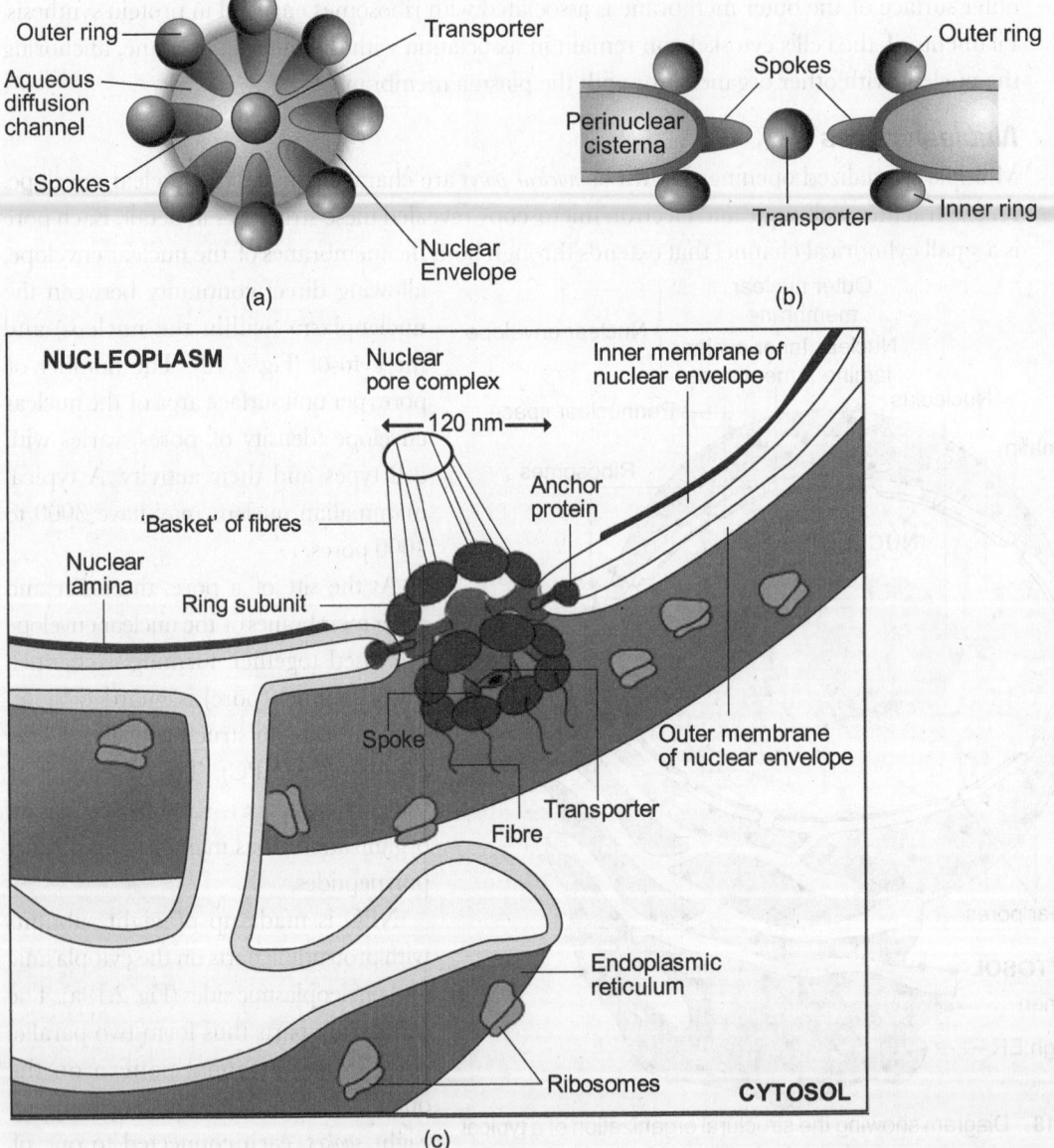

Fig. 2.19 Three different views of a nuclear pore guarded by nuclear pore complex (NPC); (a) Top view, (b) Cross-sectional view, and (c) 3-dimensional view.

Transport through nuclear pores

The evolution of the nuclear envelope in the eukaryotic cell is presumably advantageous for the safe custody of the chromosomes and their replication, and for transcription, and processing of RNA in the nuclear compartment without any interference from cytoplasmic organelles or enzymes. However, the nuclear membrane also poses several formidable transport problems for the import of a large number of molecules of different sizes from the cytosol and export of molecules from the nucleoplasm to the cytoplasm; such problems are unknown in prokaryotes in the absence of the nuclear membrane. For example, all the enzymes, histone proteins, and other proteins required for chromosome replication and transcription of DNA in the nucleus have to be imported from the cytoplasm, and all the RNA molecules and partially assembled ribosomes needed for protein synthesis have to be exported to the cytoplasm. In addition to all this macromolecular traffic, the nuclear pores mediate the transport of small molecules and ions.

Simple diffusion of small molecules   Experimental studies in which colloidal gold particles of various sizes were injected into the cytoplasm and subsequently traced by electron microscopy have revealed that these particles crossed the nuclear membrane. The larger the gold particle, the longer it took to enter the nucleus (that is, the rate of entry of particles was inversely related to their diameter). Particles larger than about 10 nm in diameter were excluded from entry. The overall diameter of an NPC is much larger than such gold particles. The conclusion drawn from this experiment was that the NPC contains tiny, aqueous diffusion channels through which small particles and molecules can freely move. Subsequently, by injecting radioactive proteins of various sizes, the size of the 'aqueous diffusion channel' was determined to be about 9 nm in diameter, a size that creates a permeability barrier for molecules significantly larger than 20,000 daltons. Histones, with molecular weights around 21,000 daltons are likely to passively diffuse through these channels. They are synthesized in the cytoplasm and are needed in the nucleus to complex with DNA (nucleohistone complex).

It now appears there may be eight separate 9 nm channels located between the spikes at the periphery of the NPC, and an additional 9 nm channel is likely to be present at the centre of the transporter. The nucleoside triphosphates required for DNA and RNA synthesis probably diffuse freely through these pores, along with other small molecules needed for metabolic pathways within the nucleus.

Active transport of large proteins and RNA molecules   Large molecules and particles such as enzymes involved in DNA and RNA synthesis, molecules imported from the cytoplasm or mRNA molecules bound to proteins (ribonucleoproteins), and ribosomal subunits exported to the cytoplasm, cannot pass through the 9 nm opening. These large molecules are actively transported through nuclear pores at the expense of energy. Like active transport across the plasma membrane (cell membrane), active transport through nuclear pores requires energy as well as carrier proteins. The details of the molecular mechanism of active transport through the nuclear membrane are summarized in Exhibit 2.A.

Exhibit 2.A   Transport through Nuclear Pore

Import Cycle (Cytosol to Nucleoplasm)

(i)   The proteins to be transported from the cytosol possess one or more *nuclear localization* signals (NLS), [which are usually 8 – 30 amino acids in length, and often contain proline and the basic (positively charged) amino acids lysine and arginine].

NLS marks the proteins as 'cargo' for transport through the nuclear pore and allows binding of a special type of receptor protein *importin*.

(ii)   The importin–cargo complex is then transported by the 'transporter' at the centre of the NPC.

(iii)   A GTP-binding protein Ran (Ran-GTP) in the nucleus binds to importin, triggering the release of the cargo protein (NLS containing protein).

(iv)   The Ran-GTP–importin complex is transported back through an NPC to the cytoplasm.

(v)   The hydrolysis of GTP to GDP and Pi is accompanied by the release of importin for reuse. (This GTP hydrolysis step provides the energy for nuclear import.)

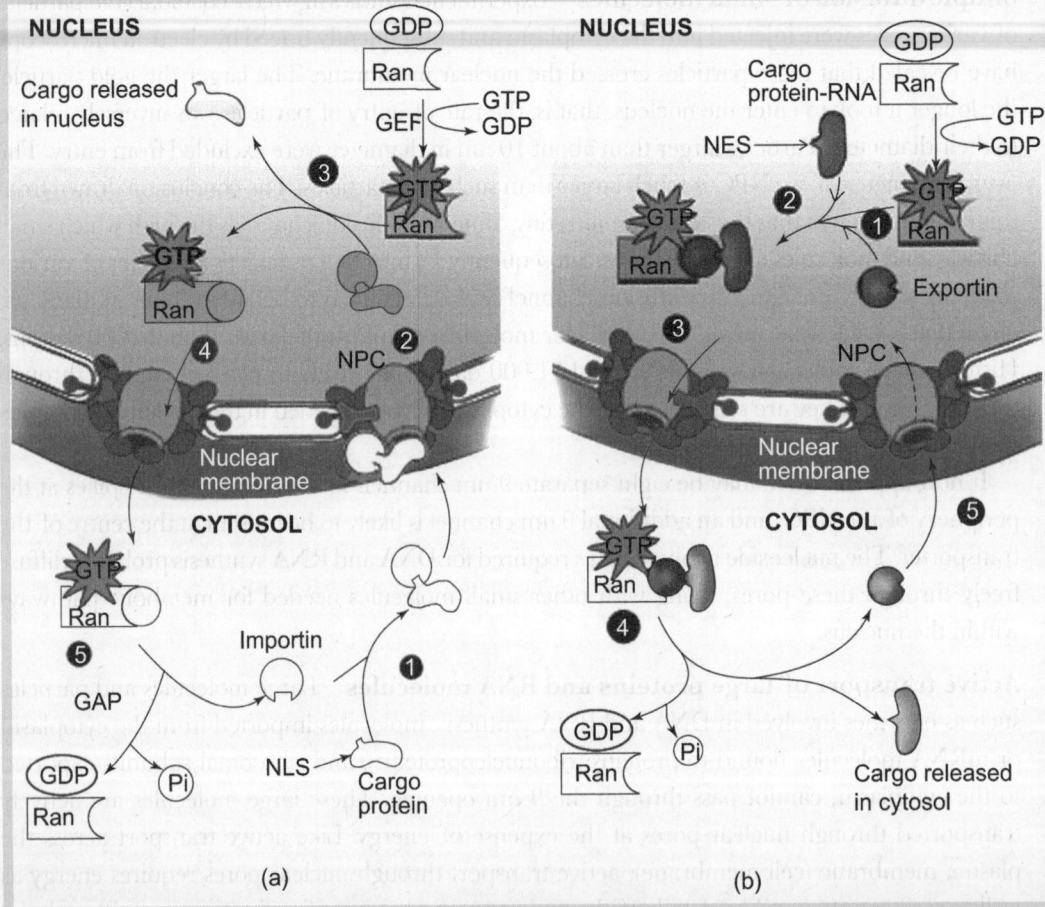

Fig. 2.A   Transport through the nuclear pore complex. (a) Import cycle and (b) Export cycle.

Export Cycle (Nucleoplasm to Cytosol)

The mechanisms involved in the export cycle are comparable to that in the import cycle. The differences are that the main cargo for export is RNA rather than protein and the transport receptor protein is called exportin. However, RNAs (cargo) are bound to a protein ('ribonucleoprotein') which contains amino acid sequences called nuclear export signals (NES) that are recognized by nuclear transport receptor proteins called *exportins*. The steps involved in the export cycle are as follows:

(i)   Ran-GTP binds to exportin in the nucleus, which activates the step.
(ii) Exportin–Ran-GTP binds to the cargo (protein–RNA complex).
(iii) Ran-GTP–exportin–cargo complex is exported through an NPC to the cytosol.
(iv) The exportin and its cargo are released from Ran, accompanied by hydrolysis of GTP to GDP and Pi.
(v)   The exportin then moves back into the nucleus to repeat the export cycle.

Governing direction of importin and exportin

The direction of importin- and exportin-mediated transport is governed by the differential interaction of Ran-GTP with these two classes of proteins, and the higher concentration of Ran-GTP in the nucleus. The Ran-GTP concentration is maintained at a higher level inside the nucleus by a *guanine-nucleotide exchange* factor (GEF) that promotes the binding of GTP to Ran in exchange for GDP; in the cytosol, a *GTPase activating protein* promotes the hydrolysis of GTP by Ran, thereby lowering the Ran-GTP concentration outside the nucleus.

Nuclear lamina

Nuclear lamina is a thin, dense meshwork of fibres that lines the inner surface of the inner nuclear membrane and provides support to the nuclear envelope. The thickness of the nuclear lamina varies from 10 to 40 nm. It is constructed by filaments of proteins called *lamins*. These filaments have attachment sites for proteins of the inner nuclear membrane, as well as for *chromatin* fibres (DNA–protein fibres of dispersed chromosomes in the nucleus).

Nuclear matrix

Nuclear matrix (or nucleoskeleton) is an insoluble network of fibres that retains the overall shape of the nucleus. The matrix provides attachment sites for chromatin fibres. Its existence was suggested by researchers in the early 1970s. The existence of the nuclear matrix with unequivocal functions has been questioned by many cell biologists.

2.7.5 Chromatin Fibres

The genome in a eukaryotic cell consists of multiple chromosomes. Each chromosome contains a linear molecule of DNA tightly bound to small basic proteins called *histones*. During the interphase between successive cell divisions, chromosomes are dispersed as DNA–histone complexes called *chromatin* that are not easy to visualize through the microscope. The histones are responsible for packing DNA molecules in an orderly fashion into chromosomes, which are visible within the nucleus during the course of cell division. The chromosomes are most condensed at the metaphase stage of cell division. The chromatin fibres account for up to 90 per cent of the nuclear mass in a cell.

Types of chromatin

Highly compact segments of chromatin that show up as dark spots in micrographs are called *heterochromatin*, whereas the loosely packed extended form of chromatin is called *euchromatin*. Heterochromatin contains DNA that is transcriptionally inactive, whereas euchromatin contains transcriptionally active DNA. Much of the chromatin in metabolically active cells is euchromatin.

Heterochromatin can be of two types—*constitutive* and *facultative heterochromatin*. The segments of chromatin involved in binding to the nuclear lamina are permanently in a highly condensed state in all cells of an organism and are called *constitutive heterochromatin*. This type of chromatin consists of repeated small sequences of DNA (satellite DNA) and is present in two major chromosomal regions—the centromere and the telomere.

Facultative heterochromatin represents chromosomal or chromatin regions that have become specifically inactivated in a cell type and can vary from time to time depending on the state of cellular activities. The amount of facultative heterochromatin is usually low in embryonic cells, indicating most regions of the chromatin are functionally (transcriptionally) active, but can be substantial in highly differentiated cells.

2.7.6 Nucleolus

Nucleolus is a prominent spherical membrane-free organelle in the eurakyotic nucleus. It is several micrometres in diameter and is responsible for producing ribosomes. There may be one or several nucleoli in a nucleus. It consists of fibrils and granules. *Fibrils* are stretch(es) of chromosomal DNA bearing multiple copies of rRNA genes arranged in tandem and known as the *nucleolus organizer region*. The rRNA genes are transcribed into *ribosomal RNAs* (rRNAs), which are packed with proteins imported from the cytoplasm to form ribosomal subunits as granules for subsequent transportation to the cytoplasm for protein synthesis.

The multiple copies of rRNA genes are an important example of repeated DNA sequences carrying genetic information. The number of copies of rRNA genes varies greatly from species to species. Animal cells often contain hundreds of copies and plant cells contain thousands of copies of rRNA genes. The multiple copies are grouped as one or more nucleolus organizer regions (NORs) of one or more chromosomes. A nucleolus may be derived from one or more NORs. For example, the human nucleus has a single large nucleolus containing loops of chromatin (NORs) derived from ten separate chromosomes.

The size of the nucleolus varies with its level of activity. The nucleolus or nucleoli tend to be large, upto 20–25 per cent of the total volume of the nucleus in cells with a high rate of protein synthesis and in need of many ribosomes.

The nucleolus usually disappears during mitosis when chromatin condenses into compact chromosomes. At the end of cell division, the NORs loop out to resume rRNA synthesis and subsequent formation of the nucleolus.

In addition to its primary function of producing ribosomes, the nucleolus contains some other molecules that are involved in other activities, such as nuclear export, chemical modification of small RNAs, and even regulation of cell division.

2.7.7 Mitochondrion

The mitochondrion (plural, mitochondria) is a sausage-shaped, prominent organelle, surrounded by two membranes. It is present in eukaryotic cells, both plant cells and animal cells. Mitochondria are often branched. Measuring several micrometres in length and 0.5 to 1 micrometre in width, mitochondria are large by cellular standards, rank next to the nucleus in size, and are comparable in size to a whole bacterial cell.

Most of the chemical reactions involved in the oxidation of sugars and fatty acids (cellular 'fuel' obtained from food materials) to extract energy and to conserve the same in the form of the high-energy compound adenosine triphosphate (ATP) occur within the mitochondria. The energy within ATP is utilized for all the cellular activities. Thus, the mitochondrion is often called the 'powerhouse' of the eukaryotic cell, and ATP is referred to as the energy 'currency' involved in most energy transactions in cells.

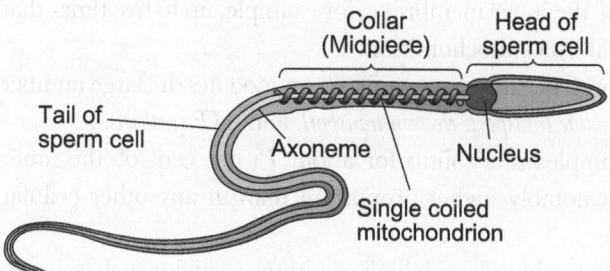

Fig. 2.20 A single mitochondrion coiled tightly around the axoneme of the tail of a sperm.

Calculations performed using the cross-sectional view of mitochondria in electron micrographs, have shown that the number of mitochondria per cell is highly variable, ranging from one or a few in many protists, fungi, algae, and even mammalian cells, to a few thousands per cell in some tissues of higher plants and animals. The number and location of mitochondria within a cell can often be related directly to its functional need or its requirement for energy. Usually, mitochondria are located within the cell just where the need for energy is greatest. A single spiral mitochondrion wraps around the central shaft or axoneme in a sperm (Fig. 2.20), which needs ATP for its flagellar movement, which propels it. Numerous mitochondria are located strategically to meet the special energy need of muscle cells.

Structure of mitochondria

Mitochondria are bound by two characteristic membranes (like the nucleus), each basically resembling a plasma membrane but with certain specialized features. The two membranes of a mitochondrion are called the outer and inner mitochondrial membranes (Fig. 2.21). They differ in certain characteristic features which are discussed as follows.

Outer membrane

- The outer membrane is not a significant permeability barrier for ions and small molecules due to the presence of *porins*.
- *Porins,* which are made up of transmembrane channel proteins permit the passage of solutes with molecular weights upto about 5000 daltons (similar proteins are present in the outer chloroplast membranes in plants and in the outer membrane of Gram-negative bacteria).

- Because of free movement of ions and small molecules through the porins, the *intermembrane space* between the outer and inner membrane is essentially continuous with the cytosol outside the mitochondrion (same in chloroplasts).
- Enzymes and other soluble proteins of large size incapable of passing through porins remain confined in the intermembrane space.

Inner membrane

- The inner membrane acts as a permeability barrier to most solutes (lacks porins).
- It effectively partitions the interior of the mitochondrion from the intermembrane space, thus forming two separate compartments—the intermembrane space and the interior.

Cristae

- Cristae are (singular *Crista*) formed by infolding of the inner membrane (Fig. 2.21). They greatly increase the surface area of the inner membrane, for example, up to five times that of the outer membrane in a typical liver mitochondrion.
- The large surface area due to the presence of cristae easily accommodates the large number of protein complexes needed for *solute transport, electron transport,* and *ATP synthesis.*
- The large number of protein complexes accounts for about 75 per cent of the inner membrane by weight, which is a notably higher proportion than in any other cellular membrane.
- The number of cristae is usually higher in cells with high respiratory activity, such as heart, kidney, and muscle cells, especially muscle cells in the flight muscles of birds.

Matrix

The interior of the mitochondrion, bound by the inner membrane, is filled with a semi-fluid *matrix*. The matrix contains many of the enzymes involved in mitochondrial function, a DNA molecule, and ribosomes. In most mammals, the mitochondrial genome consists of a circular DNA molecule (mt DNA) of about 15,000 to 20,000 base pairs which codes for ribosomal RNAs, transfer RNAs, and about a dozen polypeptide subunits of inner membrane proteins.

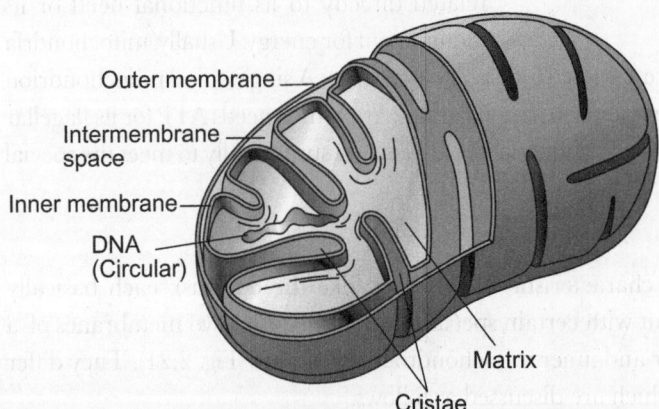

Outer membrane

Intermembrane space

Inner membrane

DNA (Circular)

Matrix

Cristae

Fig. 2.21   Mitochondrion with outer and inner cutaway view

In addition, numerous mitochondrial proteins are encoded by nuclear DNA and synthesized on cytoplasmic ribosomes and then imported into mitochondria.

Role of mitochondrion in aerobic respiration

Breakdown of glucose and other sugars (six-carbon compounds) begins in the cytosol by the process of *glycolysis*, and produces a three-carbon compound *pyruvate* and a limited amount of energy. Pyruvate is transported across the inner mitochondrial membrane and oxidized within the matrix to acetyl CoA, the primary substrate of the tricarboxylic acid (TCA) cycle. Acetyl CoA can also be formed by β-oxidation of fatty acids.

In the presence of oxygen, pyruvate is oxidized completely to carbon dioxide and the energy released in the process is used to drive ATP synthesis in the mitochondrion. The terminal electron acceptor is oxygen and the reduced form of the acceptor is water. The overall processes is called *aerobic respiration*. Energy release and ATP production in the aerobic process is much more than in the *anaerobic* one where oxygen does not participate in the glucose metabolism. In the anaerobic process, only glycolysis takes place and it is often called *fermentation*. Fermentation can be of two types depending on the end product: (i) *lactate fermentation* when the end product is lactate as in some animal cells and many bacteria and (ii) *alcoholic fermentation* when the end product is ethanol (an alcohol) as in most plant cells and in microorganisms like yeast.

Endosymbiont theory—Evolutionary origin of mitochondria and chloroplasts from ancient bacteria

Mitochondria and chloroplasts resemble bacterial cells in size, shape, and other molecular features. Both mitochondria and chloroplasts harbour their own circular DNA without histones, and ribosomes to carryout the synthesis of both RNA and proteins (although most of the proteins are encoded by nuclear DNA and imported from the cytosol). The processes of nucleic acid and protein synthesis in mitochondria and chloroplasts are strikingly similar to that in prokaryotic cells. Similarities are seen in rRNA sequences, ribosome size, sensitivity to inhibitors of RNA and protein synthesis, and protein factors required for the initiation of protein synthesis.

Based on these similarities, the *endosymbiont theory* proposes that both these organelles originated from prokaryotes that entered the cytoplasm of evolutionarily ancient single-celled organisms called *protoeukaryotes* and established a *symbiotic relationship;* prokaryotes entered the protoeukaryote cells by *phagocytosis*. Symbiosis is an intimate living relationship of two organisms or cells for mutual benefit.

The base sequences of contemporary mitochondrial ribosomal RNAs (rRNAs) have been compared with the base sequences of various bacterial rRNAs. The closest matches occur among *purple bacteria*, suggesting that the phagocytosed ancestor of mitochondria was an ancient member of this group of bacteria. Similarly, chloroplast rRNAs have closest matches with rRNA of *Cyanobacteria*, indicating chloroplast ancestry to these bacteria. (Cyanobacteria harbour chlorophyll and are called blue-green bacteria).

2.7.8 Endoplasmic Reticulum

The *endoplasmic reticulum* (ER) is a network of membrane-bound (plasma membrane like) tubular and flattened sacs or *cisternae* (singular *cisterna*) interconnected and extended throughout the cytoplasm of a eukaryotic cell (Fig. 2.22). The internal space enclosed by the ER membranes is called the lumen. The membrane of the ER is continuous with the outer membrane of the nuclear envelope. Thus, the space between the two nuclear membranes is continuous with the *lumen* of the ER. In general, the ER provides canal-type internal connections and a cytoskeleton type support for a cell.

Depending on the presence or absence of ribosomes on the membrane facing the cytosol, the ER is classified into *rough ER* and *smooth ER*, respectively. The numerous ribosomes on certain sections of the ER give a rough texture to it in the electron microphotograph; hence it is called rough ER.

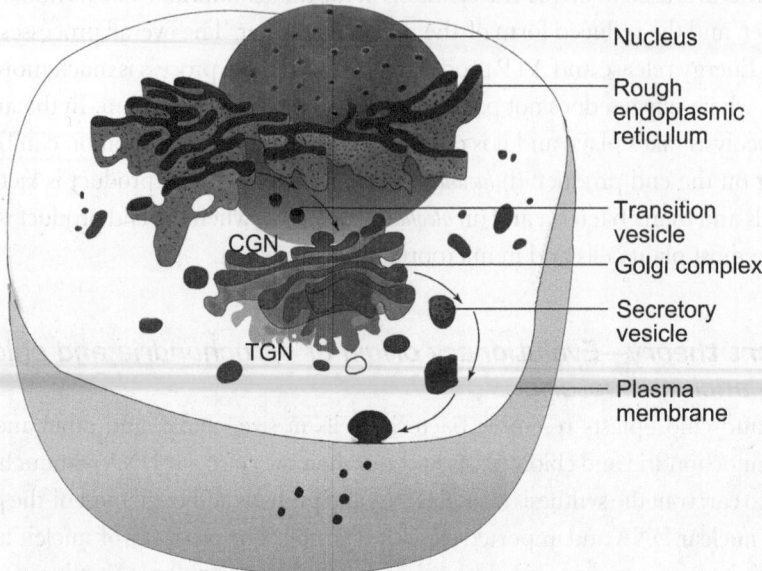

Fig. 2.22 Proteins synthesized on the rough ER pass to the Golgi complex for processing and packing into secretory vesicles. The vesicles from the ER fuse with the cis-Golgi network (CGN) membrane. On the opposite side, the trans-Golgi network (TGN) membrane buds off the vesicles containing processed proteins and secretory products.

Rough ER involved in the biosynthesis and processing of proteins

The ribosomes bound to the surface of the rough ER synthesize; (i) proteins that are present in all types of membranes (plasma membrane and membranes covering all types of organelles) and (ii) soluble proteins to be packed in the Golgi body and secreted by the cell.

Protein polypeptides in course of their synthesis on ribosomes are inserted into the rough ER lumen for further processing. However, the membrane-bound proteins remain anchored to the ER membrane by their hydrophobic regions or by covalent attachment to membrane lipids.

Further processing of proteins, such as (a) glycosylation (addition of carbohydrate groups for glycoproteins), (b) folding of polypeptides, (c) recognition and elimination of misfolded polypeptides, and (d) assembly of multimeric proteins (proteins constituted of multiple polypeptide chains), takes place in the lumen of the rough ER.

The ER is also a site for quality control. In an *ER-associated degradation* (ERAD) process, proteins that are not properly modified, folded, or assembled are exported from the ER for degradation by cytosolic *proteosomes*, before they get a chance to move on to the Golgi complex. Defects in the quality control process lead to many disorders in humans such as cystic fibrosis and familial hypercholesterolemia.

It must be noted that not all proteins are synthesized on the rough ER. *Synthesis of many proteins occurs on free ribosomes* in the cytosol that are not attached to the ER. Most of these proteins are intended for use within the cytosol.

Smooth ER functions

Smooth ER is involved in several necessary cellular processes, such as drug detoxification, carbohydrate metabolism, calcium storage, and biosynthesis of steroids (hormones).

Hydroxylation   The addition of hydroxyl groups to organic acceptor molecules is called hydroxylation. It is a reaction common to most pathways for *drug detoxification* and *steroid biosynthesis*. Hydroxylation depends on a member of the cytochrome P-450 family of proteins. Members of this protein family are especially prevalent in the smooth ER of hepatocytes (liver cells) which are an important site of detoxification. The proteins of this family are also present in lung and intestinal cells.

Carbohydrate metabolism   The smooth ER of liver cells is also involved in the enzymatic breakdown of stored glycogen in the liver. The enzyme glucose-6-phosphatase bound to the smooth ER membrane catalyses the removal of the phosphate group from glucose-6-phosphate; glucose is then moved out of the liver cell and into the blood by a permease (glucose transporter) for transport to other cells that need energy.

There is significant activity of glucose-6-phosphatase enzyme in liver, kidney, and intestinal cells, but not in muscle or brain cells. Muscle and brain cells retain glucose-6-phosphate to meet their own substantial energy need.

Calcium storage   Certain sections of smooth ER in some cells serve as a store of calcium ions. In such cells, high concentrations of calcium-binding proteins occur in the ER lumen. Calcium ions from the cytosol are pumped into the ER by ATP-dependent calcium ATPase. Binding of signalling molecules to the cell surface receptors triggers a cascade of biochemical reactions that causes the rapid release of calcium from the ER; the calcium levels in the ER are restored later by import of calcium ions from outside the cell through the plasma membrane. The quick release of calcium ions is a key event in many cell signalling processes.

Membrane synthesis   Endoplasmic reticulum is centrally involved in the biosynthesis of membranes. The ER is the primary source of membrane lipids, including phospholipids and cholesterol for all types of membranes in a cell, such as the plasma membrane and the membranes covering different organelles. Most of the enzymes required for the biosynthesis of various membrane phospholipids are present in the ER. However there are a few exceptions; for example, peroxisomal enzymes perform biosynthesis of cholesterol and dolichol, and the synthesis of chloroplast-specific lipids.

Biosynthesis of phospholipid molecules takes place only at the monolayer of the ER membrane that faces the cytosol, where the enzymes for synthesis are located. Translocation of newly synthesized phospholipids from the outer to the inner monolayer of the ER membrane is carried out by *phospholipid tanslocators* or *flippases*.

Vesicles bud off from the ER membrane and fuse with other organelles of the endo-membrane system, as source of the membrane. However, mitochondria and chloroplasts do not grow by fusion with ER-derived vesicles. Instead, *phospholipid exchange proteins* (or *phospholipid transfer proteins*) present in the cytosol transfer phospholipid molecules from the ER membrane to the outer membrane of mitochondria and chloroplasts. Each exchange protein recognizes a specific type of phospholipid, removes it from the ER membrane, and carries it through the cytosol to contribute it to the target membrane. Other membranes including the plasma membrane also grow by such a transfer of phospholipids, in addition to coalescing with membrane vesicles from the ER.

2.7.9 Golgi Complex

Golgi complex (or Golgi apparatus) is named after its Italian discoverer, Camillo Golgi who first described its structure in 1898 in the nerve cells of the owl. The Golgi complex remained highly controversial until the 1950s, when its existence was finally confirmed by electron microscopy. Since then, much has been understood about its structure and its role in the chemical modification, sorting, and packaging of the secretory proteins that it receives from the rough ER via *transition* or *transport vesicles*. The Golgi complex consists of a stack of flattened membrane-bound disc-shaped sacs or *cisternae*. It is closely linked, both physically and functionally, to the ER (Fig. 2.22). It is related more to the smooth ER.

Usually, one Golgi complex is present in an animal cell and several of them may be found in some plant cells. The size of the complex varies depending on the level of secretory activity in a cell.

The two faces of the Golgi complex

Each stack of sacs of the Golgi complex has two distinct sides or *faces* (Fig. 2.22). The *cis* (or forming) *face* is oriented to receive the transitional vesicles containing newly synthesized lipids and proteins which continuously bud off from the ER. These vesicles fuse with the *cis-Golgi network* (CGN) membrane. The opposite side of the Golgi complex is called the *trans* (or maturing) *face*. The network of membrane-bound sacs on this face is referred to as the *trans-Golgi network* (TGN). The transport vesicles continuously bud off from the tips of TGN cisternae, carrying processed proteins from the Golgi complex to secretory granules, endosomes, lysosomes, and the plasma membrane. The central sacs between the CGN and TGN are known as the *medial cisternae* of the Golgi stack, in which much of the processing of proteins occurs.

The CGN, TGN, and medial cisternae of the Golgi complex are biochemically and functionally distinct; each compartment contains specific receptor proteins and enzymes necessary for specific steps in protein and membrane processing.

Coated transport vesicles

The vesicles involved in lipid and protein transfer from the ER to the Golgi complex, between the Golgi stack cisternae, and from the Golgi complex to various destinations in the cell are referred to as *coated vesicles*. This is because of the characteristic coats or layers of protein that cover their cystosolic surface as they form.

A good number of functions have been assigned to coat proteins:

(i) The coat proteins in general promote membrane curvature which forces nearly flat membranes to form spherical vesicles.

(ii) They prevent premature fusion of a budding vesicle with nearby membranes.

(iii) The coat proteins regulate the interactions between budding vesicles and microtubules which are important for moving vesicles through the cytoplasm.

(iv) One very important function of the coat proteins is the sorting of molecules that are destined for different destinations into specific vesicles.

Thus, the specific set of proteins covering the exterior of a vesicle is an indicator of the origin and destination of the vesicle within the cell. The most studied coat proteins are *clathrin, COPI*, and *COPII* (*COP* = Coat protein). *Clathrin-coated vesicles* selectively transport proteins from the TGN to endosomes, and mediate endocytosis of receptor–ligand complexes from the plasma membrane. *COPI-coated vesicles* perform retrograde (return) transport of proteins from the Golgi back to the ER, as well as between cisternae of the Golgi complex. *COPII-coated vesicles* transport materials from the ER to the Golgi.

The coat proteins are removed from the vesicles before fusion with the target membrane.

Targeting and fusion of transport vesicles with the right organelle membrane

The molecular components on the target membrane facilitate sorting and targeting of vesicles, to avoid inadvertent fusing with wrong membranes. SNAP receptor proteins (SNAREs), v-SNAREs on transport vesicles, and t-SNAREs on target membranes are complementary molecules that along with tethering proteins on the target membrane allow a vesicle to recognize the correct target organelle to fuse with.

2.7.10 Lysosome and Cellular Digestion

The lysosome is a spherical single membrane-bound organelle, about 0.5 to1 minometre in diameter, containing *hydrolases*—enzymes capable of degrading (digesting) all the major classes of biological macromolecules such as lipids, carbohydrates, proteins, and nucleic acids.

The lysosome was discovered in the early 1950s by Christian de Duve and his colleagues. Its presence in the cell was suggested on the basis of its biochemical properties even before it had been observed by microscopy. As a result, the name of the organelle reflects its physiological function rather than the structural features as is the case with most other organelles. The name originated from the Greek root *lys*—meaning 'to loosen' or 'to digest'.

Lysosomal enzymes

Since de Duve's discovery, the list of lysosomal enzymes has expanded considerably, but they are all are *hydrolases*—hydrolytic enzymes with a pH optimum around 5. The list so far includes five phosphatases, 14 proteases and peptidases, two nucleases, six lipases, 13 glycosidases, and seven sulphatases. Together, these different types of lysosomal enzymes can digest almost all the biological molecules in a cell. For the safety of the cell, these enzymes are kept separately within lysosomes.

Membrane proteins on the luminal (interior) side of lysosomes are highly glycosylated, forming a nearly continuous carbohydrate coating that seems to provide protection from lysosomal proteases.

ATP-dependent proton pumps maintain the acidic optimum (pH 4.0–5.0) within the lysosome, which is needed for the enzymatic digestion of macromolecules.

Origin and development of lysosomes

The synthesis and packaging of lysosomal enzymes are similar to those of secretory proteins. They are synthesized by ribosomes on the rough ER from where they enter the ER lumen for chemical modification and transport to the Golgi complex. In the Golgi, a unique mannose-6-phosphate tag is added to mark the lysosomal enzymes for sorting and packing into vesicles. The vesicles are coated with the protein *clathrin* (see earlier 'coated transport vesicle') and bud from the TGN sacs of the Golgi complex. The lysosomal vesicles move to organelles called *early endosomes* formed from the plasma membrane at the time of endocytosis, lose the clathrin coat, and fuse with the endosomes. However, most of the lysosomal enzymes are delivered from the TGN to *late endosomes* (Fig. 2.23). Over time, early endosomes mature to form late endosomes. A late endosome either matures to form a new lysosome or delivers its enzymes to an active lysosome. A late endosome has a full set of hydrolases but is not engaged in digestive activity. *ATP-dependent proton pumps* lower the pH of the late endosomal lumen to 4–5, transforming the late endosome into a lysosome.

Different types of digestive processes by lysosomes

Depending on the origin of materials to be digested, lysosomes can be of two types—(i) *heterophagic lysosomes* containing substances of extracellular origin and (ii) *autophagic lysosomes* that contain materials of intracellular origin.

Heterophagy   Phagocytic vacuoles containing extracellular (foreign) material or small organisms transform into lysosomes. Digestion of the extracellular materials within lysosomes produces soluble products, such as sugars, amino acids, and nucleotides, which are transported across the lysosomal membrane into the cytosol and used for the synthesis of cellular macromolecules. Thus, the heterophagic lysosomes provide nutrition and defensive protection from pathogens (foreign materials). The indigestible materials from the extracellular substances remain in the lysosomes as *residual bodies*. In protozoa, lysosomes containing residual bodies regularly fuse with the plasma membrane to expel their contents to the outside by *exocytosis*. In vertebrates, such a mechanism is not evident, and residual bodies accumulate in the cytoplasm, contributing to cellular ageing.

Autophagy   Old and unwanted cell structures or damaged organelles are digested by lysosomes by a process called autophagy (Gr. self-eating). Instead of gradually accumulating lysosomal enzymes, like a phagocytic vacuole, autophagic vacuoles fuse with late endosomes or directly with active lysosomes. The soluble end products are reutilized by the cells. Autophagy is a normal event in most cells. But its rate may increase under certain situations. During the maturation of a red blood cell, autophagic digestion removes virtually all of the intracellular content including the nucleus and mitochondria, to make room for the large amount of haemoglobin that has to be accommodated in these cells. The rate of autophagy increases in a starving cell to provide materials to meet its energy requirements.

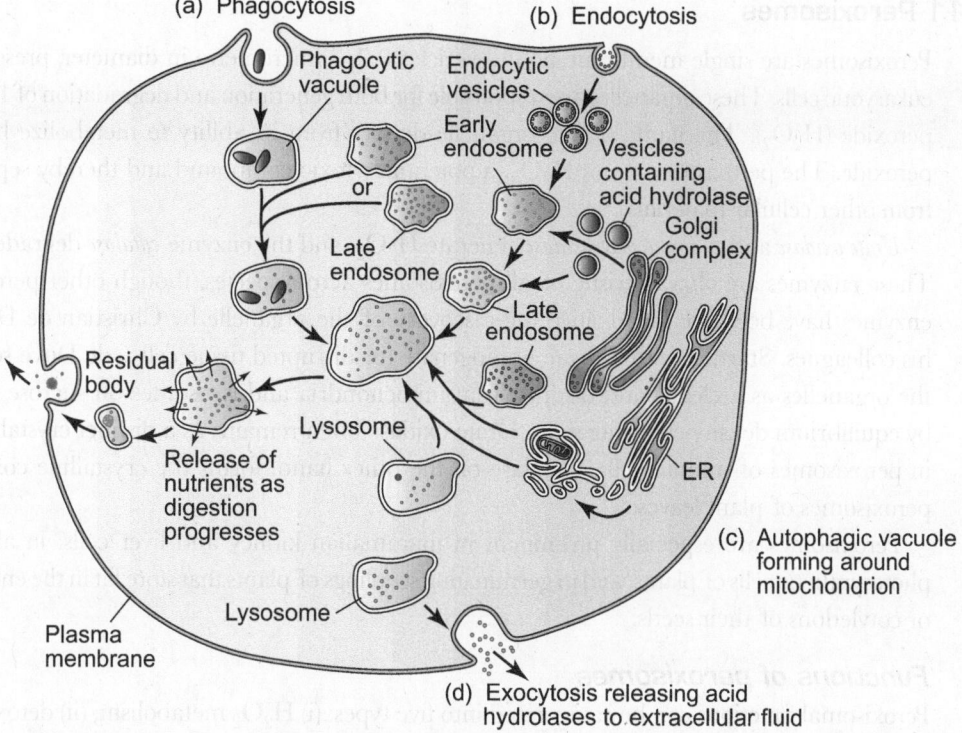

Fig. 2.23 Formation of lysosomes and their participation in (a) phagocytosis, (b) receptor-mediated endocytosis, (c) autophagy, and (d) extracellular digestion.

Extracellular digestion   In certain situations, the lysosomal enzymes are discharged to the outside of a cell by exocytosis to perform extracellular digestion. For example, lysosomes present in the head of the sperm cell release enzymes to digest the chemical and material barriers in its path so that it can penetrate the ovum during fertilization. Inadvertent release of lysosomal enzymes by inflammatory white blood cells in the joints results in rheumatoid arthritis. The steroid hormones, cortisone and hydrocortisone, prescribed as anti-inflammatory agents actually stabilize lysosomal membranes, thereby inhibiting the release of enzymes. Extracellular digestion by lysosomal enzymes secreted by macrophages cause damage to extracellular parasites; this is a ploy of the natural immune response.

Lysosomal deficiency and storage diseases

The role of lysosomes in regular recycling of cellular components for the well-being of the cell is well understood in the context of *lysosomal storage diseases*. More than 40 inheritable lysosomal storage diseases are known. Each of the diseases is characterized by the harmful accumulation of substances, usually polysaccharides or lipids, due to defective or missing digestive enzymes or defects in proteins involved in the transport of degradation products from the lysosomal lumen to the cytosol. The cells in which accumulation takes place are severely impaired or destroyed. Tissue wise, impairment leads to skeletal deformities, muscle weakness, and mental retardation. Most of the lysosomal storage diseases are not yet treatable.

2.7.11 Peroxisomes

Peroxisomes are single membrane-bound vesicles 0.2–2 micrometre in diameter, present in all eukaryotic cells. These organelles are responsible for both generation and degradation of hydrogen peroxide (H_2O_2). The name of the organelle derives from its ability to metabolize hydrogen peroxide. The peroxisome stores H_2O_2, a potentially toxic compound and thereby separates it from other cellular materials.

Urate oxidase and D-*amino acid oxidase* generate H_2O_2, and the enzyme *catalase* degrades H_2O_2. These enzymes are characteristic of all peroxisomes across species, though other peroxisomal enzymes have been identified since the discovery of the organelle by Christian de Duve and his colleagues. Starting with a tissue homogenate (or disrupted tissue cells), de Duve separated the organelles as a clear band distinct from mitochondria and lysosomes on sucrose gradient by equilibrium density centrifugation. Urate oxidase often remains as a distinct crystalline core in peroxisomes of animal cells. Catalase on the other hand, forms the crystalline core in the peroxisomes of plant leaves.

Peroxisomes are especially prominent in mammalian kidney and liver cells, in algae and photosynthetic cells of plants, and in germinating seedlings of plants that store fat in the endosperm or cotyledons of their seeds.

Functions of peroxisomes

Peroxisomal functions can be categorized into five types: (i) H_2O_2 metabolism, (ii) detoxification of harmful compounds, (iii) oxidation of fatty acids, (iv) metabolism of nitrogen containing compounds, and (v) catabolism of unusual substances. Hydrogen peroxide metabolism being the important function of this organelle, has been discussed below. The catalytic and oxidative enzymes present in peroxisomes perform the other types of function.

Hydrogen peroxide metabolism   The following two steps summarize the metabolism of hydrogen peroxide:

Generation of hydrogen peroxide   Oxidase enzymes, such as urate oxidase and D-amino acid oxidase, in peroxisomes transfer electrons from their respective substrates directly to molecular oxygen (O_2), forming H_2O_2.

Degradation of hydrogen peroxide   The synthesized harmful H_2O_2 is immediately broken down by *catalase* in peroxisomes to oxygen and water, before it can escape out of the organelle.

$$2H_2O_2 \xrightarrow{\text{Catalase}} O_2 + 2H_2O$$

Plant-specific peroxisomes   Certain peroxisomes in plant tissue play specific and special metabolic roles.

Leaf peroxisomes   In the electron micrograph of a mesophyll cell of leaf or a photosynthetic cell, often a characteristic large peroxisome containing a catalase crystal can be found in close proximity to a chloroplast and a mitochondrion. The spatial proximity of the three organelles allows metabolism of *glycolate* in a cyclical pathway involving the enzymes present in these organelles (see Exhibit 2.B).

Exhibit 2.B   The Glycolate Pathway

Glycolate produced in the chloroplast diffuses to a proximal peroxisome where oxidation of glycolate to glyoxylate is accompanied by the uptake of oxygen and the generation of H_2O_2. The H_2O_2 is immediately degraded to oxygen and water by catalase. Then an aminotransferase in the peroxisome transfers an amino group from glutamate to glyoxylate, forming glycine. Glycine diffuses from the peroxisome to a nearby mitochondrion. Two glycine molecules are used in the mitochondrion to produce a serine molecule which diffuses back in to the peroxisome to be converted to glycerate. Glycerate diffuses to the chloroplasts where it is phosphorylated by glycerate kinase to generate 3-phosphoglycerate, a key intermediate in the Calvin cycle that occurs in the chloroplast.

The glycolate pathway is also called the photorespiratory pathway, because it involves the light dependent uptake of O_2 and release of CO_2.

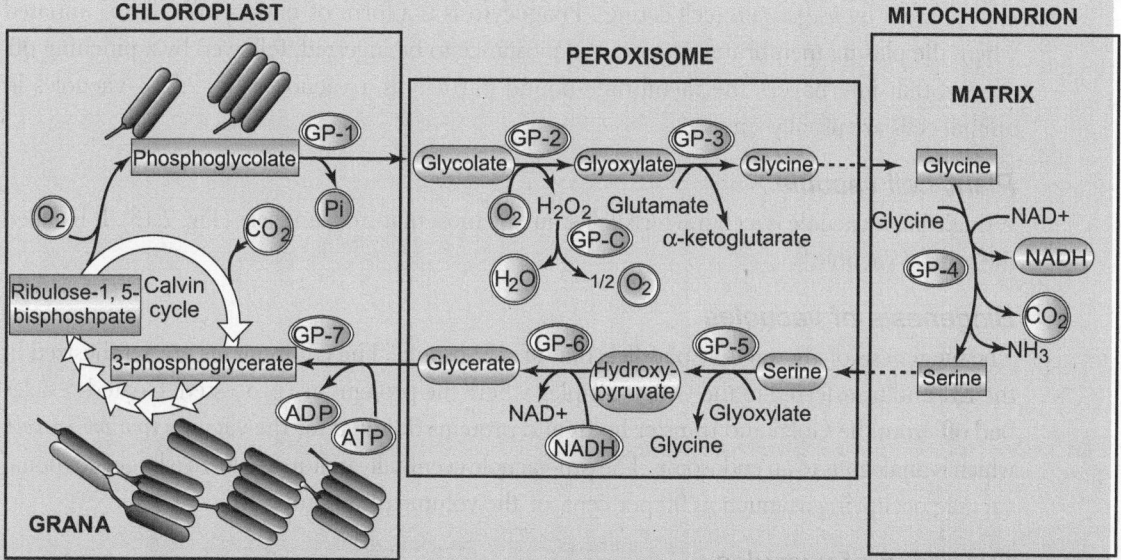

Fig. 2.B   A characteristic association of a large peroxisome containing catalase in close proximity of a chloroplast and a mitochondrion in the mesophyll or photosynthetic cell of a leaf, helps in the metabolism of glycolate.

2.7.12 Glyoxysomes

Triacylglycerol remains as storage fat in the endosperm and cotyledons of the seeds of many plant species. This substance is mobilized and converted to sucrose during the early postgermination stage. The process of mobilization involves β-oxidation of fatty acids and the *glyoxylate cycle*, and the necessary enzymes are localized in specialized peroxisomes called *glyoxysomes*. Glyoxysomes are present in the seedlings for a short period of time to deplete the stored fat, for growth. Once this role is fulfilled, the glyoxysomes are converted to regular peroxisomes.

Peroxisomes in root nodules for fixation of nitrogen   Another type of specialized peroxisome is present in the cells of *nodules* on plant roots—nodules harbour bacteria for the fixation of atmospheric nitrogen (conversion of N_2 into its organic form). Peroxisomes are involved in processing the fixed nitrogen.

Peroxisome biogenesis   The proliferation of organelles is called *biogenesis*. Earlier it was believed that peroxisomes originated from vesicles that had budded off from the Golgi complex in a manner similar to endosome and lysosome formation. The current view is that the biogenesis of peroxisomes occurs solely from the division of pre-existing peroxisomes, as in the case of mitochondria and chloroplasts.

2.7.13 Vacuole

Vacuoles are membrane-bound organelles in plant and animal cells which, like lysosomes, originate from pre-existing membranes. They help in temporary storage, transport, and a variety of other functions essential to the health of the cells.

Some protozoa and WBCs, for example, take up food or foreign particles from their environment by *phagocytosis* (cell eating). Phagocytosis is a form of endocytosis that is initiated when the plasma membrane encloses the substance to be ingested, followed by a pinching off process that internalizes the membrane-bound particle as a vacuole (Fig. 2.23). Vacuoles in animal cells are usually small.

Plant cell vacuole

A single large vacuole is a characteristic feature of most mature plant cells (Fig. 2.13). It is called the central vacuole.

Biogenesis of vacuoles

The biogenesis of the vacuole parallels that of a lysosome. The components are synthesized in the ER and transferred to the Golgi complex where the proteins are processed. Coated vesicles bud off from the Golgi and transfer lipids and proteins destined for the vacuole to a *pro-vacuole*, which is analogous to an endosome. The pro-vacuole eventually matures to form a large functional vacuole occupying as much as 90 per cent of the volume of a plant cell.

Functions of vacoules

The following are the main functions of the vacuole:

A plant cell vacuole confines hydrolytic enzymes to carry out a lysosome-like function during *intracellular digestion*. However, the central vacuole's real importance lies in the maintenance of *turgor pressure*, the osmotic pressure that prevents plant cells from collapsing. The vacuole has a high concentration of solutes and its membrane, called *tonoplast*, is differentially permeable. Water tends to move inward across the tonoplast, causing the vacuole to swell. As a consequence, the vacuole presses the cytoplasm and other cell constituents against the cell wall and *prevents the plant from wilting* (limp, flaccid appearance). When a piece of crisp celery or a plant is placed in salt water, the high concentration of salt on the outside of the cells will cause water to move out of the cells. As a result, turgor pressure will decrease and the plant tissue will soon become flaccid.

Turgor pressure also drives the expansion or *increase in the size of cells* during division and development. Higher turgor pressure is responsible for softening of the cell wall for expansion. The direction of expansion is controlled by selective softening of specific segments of the cell wall.

Turgor pressure *regulates the concentrations of various solutes* in the cytoplasm as well as the *cytosolic pH*. ATP-dependent proton pumps in the vacuolar membrane can compensate for a decline in cytosolic pH by pumping protons from the cytosol to the lumen of the vacuole.

The vacuole also serves as a means of storage for various substances. The vacuole stores proteins in seeds. During the germination of seeds, the storage proteins are hydrolyzed by vacuolar proteases to release amino acids for the biosynthesis of new proteins. The vacuoles of different plants or tissues also store other substances, such as malate in CAM plants (living in deserts and salt marshes having severely limited access to water), the anthocyanins that impart colour to flowers for attracting pollinators, toxic substances to deter predators, inorganic and organic nutrients, compounds that provide protection from UV light, and residual indigestible waste. Plant vacuoles serve as storage of soluble as well as insoluble waste. This is important, since unlike animals, most plants do not have a regular mechanism for excreting soluble cellular waste.

2.7.14 Ribosomes

Ribosomes are made up of rRNA and proteins, and provide the base for the translation of all types of proteins from mRNA. They are the smallest organelles, numerous, not surrounded by a membrane, and present in two different sizes in prokaryotic and eukaryotic cells.

Ribosomes are also found in the matrix of mitochondria and the stroma of chloroplasts, where they function in organelle-specific protein synthesis. The ribosomes of these organelles are strikingly similar to those found in bacteria and cyanobacteria; the similarity is even in the nucleotide sequence of ribosomal RNA (rRNA). These similarities provide strong support for the endosymbiotic prokaryote origins of mitochondria and chloroplasts (see Section 2.7.7).

Eukaryotic and prokaryotic ribosomes

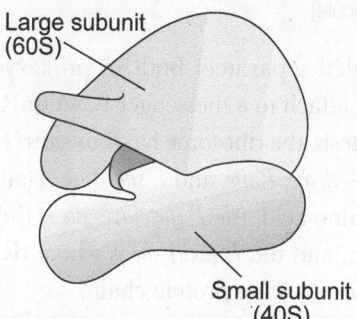

Large subunit (60S)

Small subunit (40S)

Fig. 2.24 A eukaryotic ribosome consisting of two dissociable subunits called large (60S) and small (40S) subunits differing in size, shape, and composition. (In case of prokaryotes the subunits are 50S and 30S.)

In the eukaryotic cytoplasm, ribosomes occur freely in the cytosol, or they may be bound to the outer membrane of the endoplasmic reticulum and the outer membrane of the nuclear envelope. Ribosomes are present mostly in the free state in the cytoplasm of prokaryotic cells. Prokaryotic and eukaryotic ribosomes resemble each other structurally, but differ in many ways. Prokaryotic ribosomes are smaller in size, contain fewer proteins, have smaller and fewer RNA molecules, and are sensitive to different inhibitors of protein synthesis.

Diameters of ribosomes of prokaryotic and eukaryotic cells are about 25 and 30 nm, respectively. A prokaryotic cell usually has thousands of ribosomes, whereas a eukaryotic cell contains hundreds of thousands or millions of them. Eubacteria and archaebacteria can be differentiated on the basis of characteristic differences in their ribosomal RNA sequences.

A ribosome consists of two dissociable subunits called the *large* and *small subunits* that differ in size, shape, and composition (Fig. 2.24). The large and small subunits of prokaryotic ribosomes have sedimentation coefficients of about 50S and 30S respectively and that of the two units together is 70S. For eukaryotic ribosomes, the corresponding values are about 60S and 40S for the large and small subunits respectively, and 80S for the intact ribosome. Table 2.4 lists the characteristics of prokaryotic and eukaryotic ribosomes and their subunits.

Table 2.4   Characteristics of prokaryotic and eukaryotic cytoplasmic ribosomes

Cell type	Size of ribosome		Subunit	Subunit size		No. of proteins in subunit	Subunit RNA	
	S value	Mol. wt		S value	Mol. wt		S value	No. of nucleotides
Prokaryote	70S	2.5×10^6	Large	50S	1.6×10^6	34	23S 5S	2900 120
			Small	30S	0.9×10^6	21	16S	1540
Eukaryote	80S	4.2×10^6	Large	60S	2.8×10^6	~46	25-28S 5.8S 5S	4700 160 120
			Small	40S	1.4×10^6	~32	18S	1900

Sedimentation Coefficient

Sedimentation coefficient (Svedberg unit, S) of a particle or macromolecule is a measure of the velocity at which a particle sediments on ultracentrifugation and is indirectly related to the mass of the particle. That is why the S values of the subunits do not add up to that of the whole ribosome. [In general, S value indicates the relative size of a macromolecule].

Ribosomal subunits are synthesized and assembled separately both in prokaryotes and eukaryotic cells. The subunits join together when they attach to a messenger RNA (mRNA) and begin to make a protein. In the context of protein synthesis, the ribosome has four sites (Fig. 2.25): an mRNA-binding site and three tRNA-binding sites—*A site*, *P site*, and *E site*. The *A* (aminoacyl) site binds a newly arriving tRNA with an attached amino acid, the *P (peptidyl) site* is the binding site for tRNA carrying the growing polypeptide chain, and the *E (exit) site* is where tRNAs exit the ribosome after contributing their amino acids to the growing protein chain.

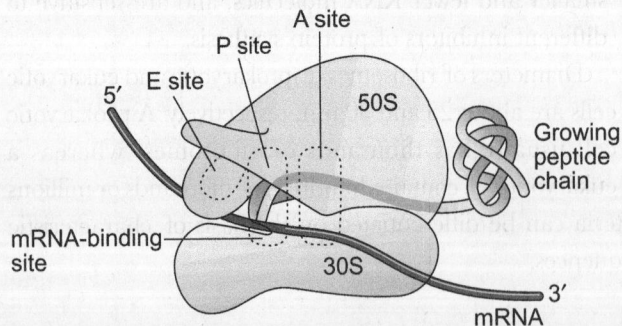

Fig. 2.25 The ribosomal subunits assemble together on an mRNA for protein (peptide chain) synthesis. There are three tRNA binding sites—A (aminoacyl) site binds a newly arriving tRNA with an activated amino acid, P (peptidyl) site binds tRNA carrying the growing polypeptide chain, and E (exit) site is the place where the tRNAs exit out of the ribosome after completion of the mission.

2.7.15 Chloroplasts

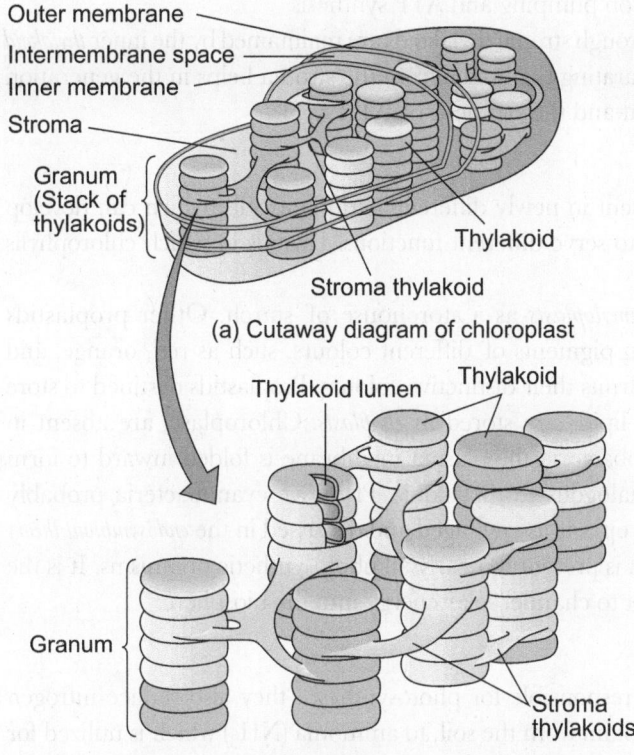

Outer membrane

Intermembrane space

Inner membrane

Stroma

Granum
(Stack of
thylakoids)

Thylakoid

Stroma thylakoid

(a) Cutaway diagram of chloroplast

Thylakoid lumen   Thylakoid

Granum

Stroma
thylakoids

(b) Grana and stroma thylakoids

Fig. 2.26 Characteristic feature of chloroplast. (a) Three-dimensional structure of a chloroplast with outer and inner membranes, stroma, and thylakoid system. (b) Thylakoids are organized into stacks of grana and interconnecting stroma thylakoids. Thylakoid membranes enclose the thylakoid lumen.

Chloroplasts are large organelles in eukaryotic plants that harbour a strong and elaborate system of *thylakoids* containing photosynthetic pigments, *chlorophylls* for solar energy transduction and carbon assimilation to manufacture sugar.

Chloroplasts are usually flattened ovoid structures, 5–10 micrometres long and a few micrometres in diameter, larger than any other structure in the cell except the nucleus and the vacuole. A mature leaf cell can have 20–100 chloroplasts, whereas an algal cell typically contains only one or a few ribbon-shaped elaborate chloroplasts. Not all plant cells contain chloroplasts.

Like mitochondria chloroplasts have two membranes, an *inner* and an *outer* *membrane* with a narrow *inter-membrane space* in between (Fig. 2.26). The inner membrane encloses the stroma, a gel-like matrix full of enzymes for the assimilation of carbon, nitrogen, and sulphur. The flattened interconnected sacs of *thylakoids* are suspended in the stroma. The outer membrane has *porins*, made by transmembrane proteins that permit easy movement of solutes up to 5,000 daltons molecular weight. Thus, the outer membrane is permeable to most small organic molecules and ions, whereas the inner membrane maintains a significant permeability barrier, as in the case of the mitochondrion. Transport proteins regulate the flow of most metabolites between the inter-membrane space and stroma. Three important metabolites—water, carbon dioxide, and oxygen—freely diffuse across both the outer and inner membranes.

Thylakoids

In addition to the outer and inner membrane, the membrane-bound thylakoids are considered as a third membrane system in chloroplasts. Although mitochondria and chloroplasts are quite similar, mitochondria lack an equivalent of the third membrane system. Thylakoids are flat, sac-like structures immersed in the stroma and arranged in stacks called *grana* (singular, *granum*). Grana look like stacks of coins and are interconnected by a network of longer thylakoids known as *stroma thylakoids*. The thylakoid system contains photosynthetic pigments called chlorophylls, all the enzymes necessary for the photoreactions, the carriers for electron transport, and the proteins that couple electron transport to proton pumping and ATP synthesis.

Interconnections between grana through stroma thylakoids are maintained by the inner *thylakoid lumen*. The thylakoid membrane separating the lumen from the stroma helps in the generation of an electrochemical proton gradient and the synthesis of ATP.

Origin of chloroplasts

Proplastids are smaller organelles present in newly differentiated plant cells, which can develop into any of several types of plastids to serve different functions. Plastids in which chlorophylls develop become *chloroplasts*.

Some proplastids develop into *amyloplasts* as a storehouse of starch. Other proplastids differentiate to *chromoplasts* containing pigments of different colours, such as red, orange, and yellow; chloroplasts give flowers and fruits their distinctive colours. Proplastids destined to store proteins are called *proteinoplasts* and lipids are stored in *elioplasts*. Chloroplasts are absent in photosynthetic prokaryotes. In cyanobacteria the plasma membrane is folded inward to form *photosynthetic membranes*, which are analogous to thylakoids. Primitive cyanobacteria probably were the evolutionary lineage to chloroplasts as envisaged and discussed in the *endosymbiont theory* of origin of chloroplasts. Chlorophyll is present in nearly all photosynthetic organisms. It is the primary energy-transduction pigment to channel solar energy into the biosphere.

Other functions

Although chloroplasts are primarily responsible for photosynthesis, they also reduce nitrogen from the nitrate (NO_3^-), that plants obtain from the soil, to ammonia (NH_3) which is utilized for protein synthesis. Sulphur metabolism also occurs in the chloroplasts.

2.7.16 Cytoskeleton

As we know, inside a cell all the organelles remain suspended in the cytoplasm. Earlier, structurally cytoplasm or cytosol was considered as a rather amorphous, gel-like substance containing soluble and freely diffusible proteins. More recently, advances in microscopy, particularly electron and digital video microscopy, and biochemical and immunological (particularly immunofluorescence) techniques have revealed a structured *cytoskeleton* in the cytoplasm of eukaryotic cells.

The cytoskeleton is a pervasive network of *microtubules*, *microfilaments*, and *intermediate filaments* in the cytoplasm that provides architectural framework and functional support to eukaryotic cells. It surrounds, but does not invade, the nucleus. The cytoskeleton performs multiple functions:

(i) It provides a eukaryotic cell its distinctive shape, strength, and high level of internal organization.

(ii) It plays an important role in cell movement and cell division—
    (a) locomotion of the cell itself as in case of amoeba or a phagocyte cell;
    (b) contraction of muscle cells;
    (c) beating of cilia and flagella (Fig. 2.27); and
    (d) the movement of chromosomes during cell division.

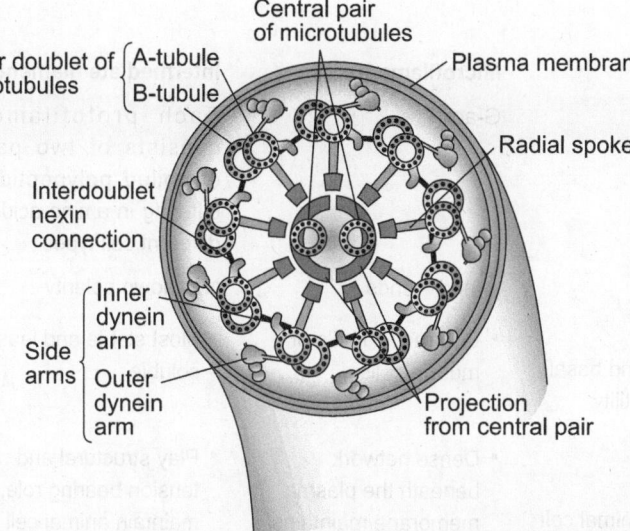

Outer doublet of microtubules
{ A-tubule
  B-tubule }

Central pair of microtubules

Plasma membrane

Radial spoke

Interdoublet nexin connection

Side arms
{ Inner dynein arm
  Outer dynein arm }

Projection from central pair

Fig.2.27 Diagram to show the enlarged cross-sectional view of the axoneme (shaft) of a flagellum or cilium (under electron microscope) as highly organized microtubules in nine pairs in an outer ring surrounding a pair of microtubules at the centre. The dots in each microtubule indicate the number of protofilaments that participate in their formation. The dynein side arms have ATPase activity and participate in bending and movement of the axoneme.

(iii) It provides a framework for positioning and actively moving organelles and mRNA within the cytosol.

(iv) It allows many enzymes to cluster and attach to it in close proximity to other enzymes involved in the same pathway, facilitating the channelling of intermediates.

(v) As much as 20–40 per cent of the water in the cytosol may be bound to the filaments and tubules of the cytoskeleton.

(vi) It is intimately related to cell signalling, cell–cell adhesion, and other events at the cell surface.

Three structural elements of the cytoskeleton—*microtubules*, *microfilaments*, and *intermediate filaments*—have a characteristic size, structure, and intracellular distribution. Each of these is formed by the polymerization of a different kind of protein subunit (Table 2.5). All three components are linked together structurally and functionally. Specific linker proteins of the *plakin* family, especially one member, *plectin*, is responsible for linking the three elements to each other. Thus, the coordinated activity of microtubules, microfilaments, and intermediate filaments

Table 2.5 Characteristics of different elements in the cytoskeleton

	Microtubules	Microfilaments	Intermediate filaments
Structure	Hollow cylinder, made up of 13 longitudinal protofilaments	Two interwined strands of F-actin	Eight protofilaments joined end to end
Diameter	Outer: 25 nm Inner: 15 nm	7 nm	8–12 nm

(Contd)

Table 2.5   *(Contd)*

	Microtubules	Microfilaments	Intermediate filaments
Monomers	α-tubulin β-tubulin	G-actin	Each protofilament consists of two pairs of coiled polypeptides, differing in amino acids in different cell types
Polarity	+ and – ends	+ and – ends	No known polarity
Functions	*Axonemal* • In cilia, flagella, and basal body—for cell motility	• Contractile fibrils of muscle cells	• Most stable and least soluble
	Cytoplasmic • Organization and maintenance of animal cell shape • Chromosome movement forming spindle fibres • Disposition and movement of organelles	• Dense network beneath the plasma membrane maintains shape change and cell movement (in animal cell) • Amoeboid movement • Cytoplasmic streaming • Produce cleavage furrows during animal cell division	• Play structural and tension bearing role, maintain animal cell shape • Formation of nuclear lamina by scaffolding • Functional diversity in different cell types—structural support for contractile machinery in muscle cells determine axon strength and size

adapt to stretching forces in such a way that the tension-bearing elements become aligned with the direction of stress.

Microtubules

Microtubules (MTs) are the largest structure present in the cytoskeleton. A microtubule is a straight hollow cylinder enclosing a lumen. The outer diameter is about 25 nm and the inner diameter is about 15 nm (Fig. 2.28a). MTs in eukaryotic cells can be broadly categorized into two types, depending on the degree of organization and the structural stability:

Axonemal microtubules   Highly organized, stable, present in cilia and flagella as in protozoa, in the basal bodies of sperm cells to which the tail is attached, and centrosome of animal cells.

Cytoplasmic microtubules   Loosely organized, dynamic network, pervading the cytosol.

Microtubules vary greatly in length. Some are less than 200 nm long; others like axonemal MTs can be many micrometres in length. MTs form and grow in length by the addition of α-and β-tubulin dimers at both the ends, more readily at the plus end. GTP is required for the process.

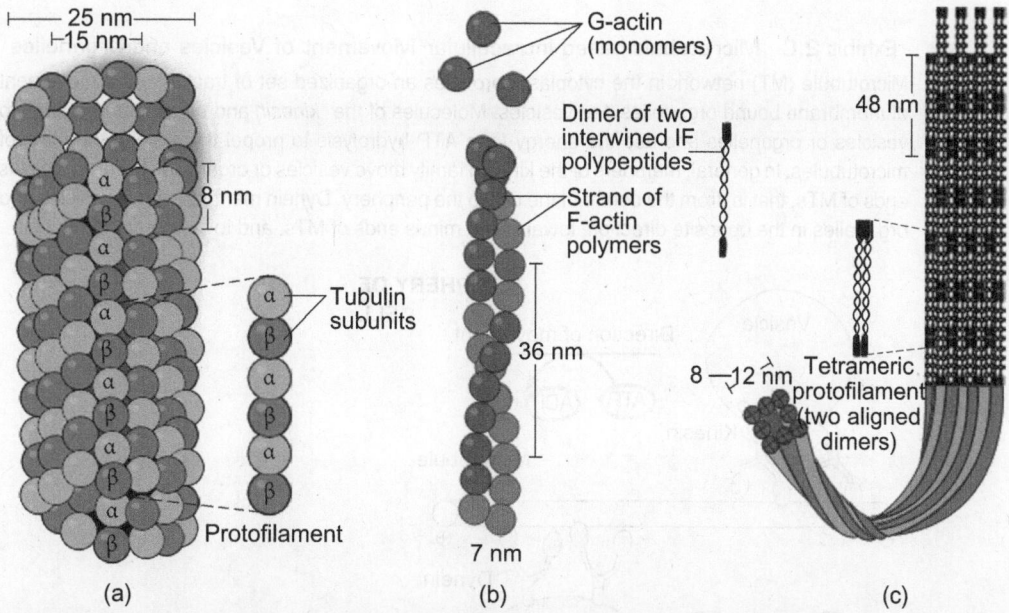

Fig. 2.28 Schematic diagram to illustrate the structures of microtubule, microfilament, and intermediate filament. (a) Microtubule— Longitudinally arranged 13 protofilaments form a hollow cylinder of a microtubule with an outer diameter of about 25 nm and an inner diameter of about 15 nm. Each protofilament is a polymer of tubulin dimers, each about 8 nm long, oriented in the same direction creating the polarity of the protofilament as well the whole microtubule. (b) Microfilament— A strand of F-actin polymer twisted into a helical structure about 7 nm in diameter. A half-turn of the helix takes place at every 36 nm. The F-actin polymer is made up of G-actin monomers, all oriented in the same direction to generate polarity. (c) Intermediate filament (IF)— Tetrameric protofilament, consisting of two pairs of coiled polypeptides, with a length of ~ 48 nm forms the structural unit. They assemble end-to-end and side-to-side to form eight protofilaments resulting in a total diameter of 8–12 nm for an IF.

Microfilaments

With a diameter of about 7 nm, microfilaments (MFs) are the smallest of the cytoskeletal elements. An MF is a strand of F-actin twisted into a helical structure. An *F-actin* (filamentous actin) polymer consists of monomers of *G-actin* (globular actin), all oriented in the same direction to give the microfilament its polarity (Fig. 2.28b). Actin is a single polypeptide of 375 amino acids, with a molecular weight of about 42,000 daltons. Once synthesized, it folds into a U-shaped (a kind of globular) molecule, with a central cavity that binds ATP or ADP. Actins are best known for their role in the contractile fibrils of muscle cells, where they interact with thicker filaments of myosin during the contraction of muscle cells. Actins can be broadly divided into two major groups—the muscle specific actins (α-actins) and non-muscle actins, (β- and γ-actins). A dense network of MFs just beneath the plasma membrane of animal cells, confers structural rigidity to the cells and facilitates shape change and cell movement. Parallel bundles of MFs in the core of microvilli, the finger-like tiny extensions on animal cells, provide the characteristic stiffness to the microvilli.

Exhibit 2.C Microtubule-based Intracellular Movement of Vesicles and Organelles

Microtubule (MT) network in the cytoplasm provides an organized set of tracks for the movement of membrane bound organelles and vesicles. Molecules of the *kinesin* and *dynein* family attach to vesicles or organelles and use the energy from ATP hydrolysis to propel them along the track of microtubules. In general, members of the kinesin family move vesicles or organelles towards the plus ends of MTs, that is, from the centre of the cell to the periphery. Dynein members move vesicles and organelles in the opposite direction, towards the minus ends of MTs, and to the centre of the cell.

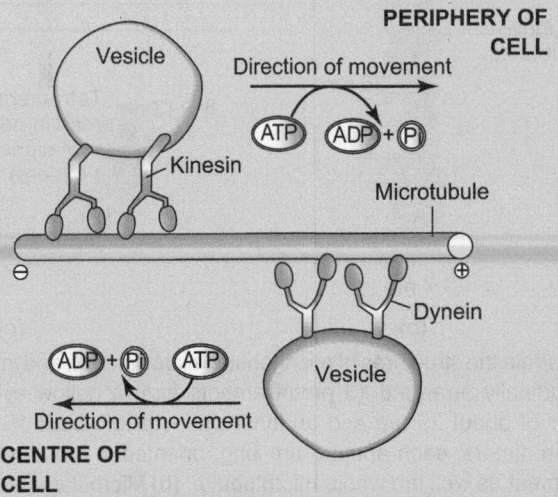

Fig. 2.C Kinesin and dynein molecules use the energy of ATP hydrolysis to move (walk) intracellular vesicles or organelles along microtubules. (Rate of movement of vesicle or an organelle is about 2 µm/sec.)

Intermediate filaments

Intermediate filaments (IFs) have a diameter of about 8–12 nm. The diameter of IFs ranges between that of the two other elements of the cytoskeleton—microtubules and microfilaments. IFs are the most stable and least soluble constituent of the cytoskeleton. Detergents or high or low ionic solutions remove most of the microtubules, microfilaments, and other proteins of the cytosol, but leaves the network of IFs in the original state. This suggests the importance of IFs as a scaffold for the entire cytoskeletal framework. The structural unit of an IF is the tetrameric protofilament, consisting of two pairs of coiled polypeptides, with a length of about 48 nm. The protofilaments assemble by end-to-end and side-to-side alignment into an IF, and an intermediate filament is considered to be eight protofilaments thick (Fig. 2.28c).

In contrast to microtubules and microfilaments, IFs differ markedly in amino acid composition from tissue to tissue. Broadly, six classes of proteins participate in the formation of IFs in different tissues.

State of cytoskeleton in the prokaryotes

Although cytoskeletal structure is considered to be unique to eukaryotic cells, recent studies have shown that bacillus type of bacteria have a cytoskeleton comprising microtubules (the *FtsZ* protein), microfilaments (*MreB* protein family), and intermediate filaments (*crescentin*). Although the amino acid sequences of bacterial and eukaryotic cytoskeletal proteins are fairly different, when they are assembled into polymers, their overall structure is strikingly similar.

2.7.17 Centrosome

Centrosome is a small zone of granular material located adjacent to the nucleus. It acts as a nucleation centre for the assembly of microtubules to form the *mitotic spindle* during cell division. During S phase, prior to mitosis the centrosome divides into two. The two centrosomes separate from each other, in course of their movement to the opposite poles of the nucleus at prophase. The growing microtubules from the two centrosomes fill the space between them to form the mitotic spindle. They then establish contact with the kinetochores around the centromeres of the replicated chromosomes at metaphase to pull the daughter chromosomes apart and move them to the opposite poles of the dividing cell in anaphase. A pair of small *centrioles*, which are a cylindrical assembly of microtubules, oriented at right angles to each other, is present within the centrosome of animal cells.

SUMMARY

♦ Cell is the structural and functional unit of all living organisms.

♦ Bacteria and blue-green algae are single celled organisms and are classified as *Prokaryotes;* they lack the nuclear membrane. (Nuclear material in bacteria is often referred to as nucleoid.) Plants and animals are *Eukaryotes* and their cells have a membrane-bound nucleus.

♦ Viruses are not typical living organisms, although they are made up of organic compounds like nucleic acids and protein. Viruses cannot multiply by themselves and always depend on host cells for this purpose. Hereditary material, either DNA or RNA, is enclosed in a case made up of protein subunits.

♦ The Cell Theory proposed by Schleiden and Schwann established the cellular basis of all living organisms.

♦ Eukaryotic cells have a variety of *organelles*, which are membrane-bound compartments that are specialized for different specific functions.

♦ Singer and Nicolson proposed the Fluid Mosaic Model of cell membrane in 1972. This model describes the plasma membrane as a mosaic of protein molecules discontinuously embedded in a lipid bilayer, with the lipid components in constant lateral motion.

♦ Cellulose is the predominant polysaccharide of the plant cell wall and the single most abundant organic macromolecule on the earth.

♦ The genome in a eukaryotic cell consists of multiple chromosomes; each containing a linear molecule of DNA tightly bound to small basic proteins called *histones*. During the inter-phase between divisions of cells, chromosomes are dispersed as DNA-histone complexes called *chromatin* and are not easily visible under the microscope.

♦ Highly compacted segments of chromatin that show up as dark spots in micrographs are called *heterochromatin*, whereas the loosely packed extended form of chromatin is called *euchromatin*. Heterochromatin contains DNA that is transcriptionally inactive, while euchromatin contains transcriptionally active DNA.

♦ Nucleolus is a prominent spherical membrane-free organelle in the eurakyotic nucleus, several micrometres in diameter, responsible for producing ribosomes.

♦ Most of the chemical reactions involved in the oxidation of sugars and fatty acids (cellular 'fuel' obtained from food materials) to extract energy and to conserve it in the form of the high-energy compound adenosine triphosphate (ATP) occur within the mitochondria. The energy within ATP is utilized for all the activities of the cell.

♦ The endosymbiont theory proposed the evolutionary origin of mitochondria and chloroplasts from ancient bacteria.

- Endoplasmic reticulum is a network of membrane-(plasma membrane like) bound tubular and flattened sacs or *cisternae* (singular *cisterna*) interconnected and extended throughout the cytoplasm of a eukaryotic cell.

- The Golgi complex consists of a stack of flattened membrane-bound disc shaped sacs or *cisternae* that is closely linked, both physically and functionally to the ER. It is responsible for the chemical modification, sorting, and packaging of the secretory proteins that it receives from the rough ER via *transition* or *transport vesicles*.

- Lysosome is a spherical single membrane-bound organelle about 0.5–1 micrometre in diameter, containing *hydrolases*. These are enzymes capable of degrading (digesting) all the major classes of biological macromolecules, such as lipids, carbohydrates, proteins, and nucleic acids.

- Peroxisomes are single membrane-bound vesicles 0.2–2 micrometres in diameter, present in all eukaryotic cells. These organelles are responsible for the generation and degradation of hydrogen peroxide

(H_2O_2). The name of the organelle is derived from its ability to metabolize hydrogen peroxide.

- Vacuoles are membrane-bound organelles in plant and animal cells, originating from pre-existing membranes, resembling lysosomes. They provide temporary storage, and help in transport and a variety of other functions essential to the health of cells.

- Ribosomes are made up of rRNA and proteins, and provide the base for translation of all kinds of proteins from mRNA. They are the smallest organelles, numerous, not surrounded by a membrane, and are present in two different sizes in prokaryotic and eukaryotic cells.

- Chloroplasts are large organelles in eukaryotic plants. They harbour a strong and elaborate system of *thylakoids* containing the photosynthetic pigments *chlorophyll* for solar energy transduction and carbon assimilation to manufacture sugar.

- Cytoskeleton is a pervasive network of *microtubules*, *microfilaments,* and *intermediate filaments* in the cytoplasm that provides the architectural framework and functional support to eukaryotic cells. It surrounds, but does not invade the nucleus.

Har Gobind Khorana

**Har Gobind Khorana
(9 January 1922–9 November 2011)**

Har Gobind Khorana was an Indian born American biochemist and molecular biologist. He shared the Nobel Prize in Physiology or Medicine, in 1968, with Robert W. Holley and Marshall W. Nirenberg.

Har Gobind Khorana was born in Raipur, a small village in Punjab, in British India, now located in Pakistan. His father was an agricultural taxation

clerk and encouraged education in his family. The youngest of five siblings, Har Gobind attended DAV High School in Multan. He was greatly influenced by one of his teachers, Ratan Lal. He went on to pursue bachelor's and master's degree in chemistry from Punjab University, Lahore. He found a great teacher and an accurate experimentalist in his supervisor Mahan Singh. He was awarded a fellowship from the Government of India which made it possible for him to go to England for a PhD degree at the University of Liverpool. Roger J.S. Beer was his supervisor for the doctoral work. After earning his PhD in organic chemistry in 1948, he moved to Eidgenössische Technische Hochschule in Zurich for his postdoctoral work in Vladimir Prelog's laboratory. Dr Prelog had a great influence on Khorana's philosophy towards science and work. After a brief break in India in the fall of 1949, Gobind (this is how he preferred to be called by his students) returned to England, where he obtained a fellowship to work with Dr G.W. Kenner and Professor (later Lord) A.R. Todd. He stayed in Cambridge from 1950 to 1952. He utilized the power of carbodiimides in synthesizing pyrophosphate linkages in Alexander Todd's laboratory. His interest in proteins and nucleic acids took root at that time.

A job offer from Dr Gordon M. Shrum of the University of British Columbia took Khorana to Vancouver. Dr Shrum's inspiration and counselling from Dr Jack Campbell initiated a young group of scientists including Khorana to begin work in the field of biologically interesting phosphate esters and nucleic acids. Starting his own laboratory, Khorana came up with the synthesis of deoxy- and ribosetriphosphates and coenzyme A. Subsequently, he moved to the Enzyme Research Institute at the University of Wisconsin, Madison in 1960, where he deciphered the genetic code for which he was awared the Nobel Prize. Khorana experimentally demonstrated the existence of three-nucleotide codes for specific amino acids. He assembled the first synthetic gene, using polymerase and ligase enzymes joining DNA pieces together in the test tube. In a way, Khorana was the pioneer to synthesize oligonucleotides, which now can be obtained commercially from different companies for sequencing, cloning, and genetic engineering. He anticipated the invention of PCR. In 1970, Khorana became the Alfred Sloan Professor of Biology and Chemistry at the Massachusetts Institute of Technology, where he worked until he retired in 2007. Since the mid-1970s he studied the biochemistry of the membrane protein bacteriorhodopsin responsible for converting light energy into photon gradient energy and the visual pigment rhodopsin.

He received many distinguished awards, such as Gairdner Foundation International Award, Louisa Gross Horwitz Prize from Columbia University, Albert Lasker Award for Basic Medical Research, and Padma Vibhushan from the Government of India.

Khorana was married in 1952 to Esther Elizabeth Sibler, who was of Swiss origin. She put a consistent sense of purpose in his life when he missed the country of his birth. Khorana died as a widower in Concord, Massachusetts on 9 November 2011, at the age of 89 years. He remained an active researcher until his death, testimony to which is his paper earlier in 2011.

EXERCISES

Objective Questions

1. Choose the most appropriate statement:
   (a) A cell utilizes energy for its mechanical functions.
   (b) A cell utilizes energy for respiration.
   (c) A cell utilizes energy for various metabolic and mechanical functions.

2. Choose the most appropriate statement:
   (a) Microtubules are present only in flagella of eukaryotic cells.
   (b) Microtubules are present in flagellum of both prokaryotic and eukaryotic cells.
   (c) Microtubules are present in the cilia, flagella, and centrosome of animal cells.
   (d) All of these

3. Nucleolus is the main site of synthesis of:
   (a) mRNA
   (b) DNA and mRNA
   (c) rRNA
   (d) all types of RNA

4. Mark the following statements as True of False. If false write the true statement.
   (a) Some cells are large enough to view with the naked eye.
   (b) In most cases, eukaryotic cells are larger than prokaryotic cells.

   (c) Prokaryotic cells do not possess mitochondria and plasma membrane.
   (d) The ribosomes found in the mitochondria of liver cells are more like those of bacteria.
   (e) Bacterial cell wall is often called capsule.
   (f) Gap junctions in the plasma membrane of animal cells are equivalent to plasmodesmata in plant cell wall.
   (g) Nucleolus is absent in plant cells.
   (h) Peroxisomes originate from vesicles budded off from the Golgi complex.

5. Indicate where the following cellular structures can be found in animal cells (A), bacterial cells (B), and/or plant cells (P).
   (a) Nuclear envelope
   (b) Cell wall
   (c) Microtubules
   (d) Chloroplasts
   (e) Central vacuole
   (f) Ribosomes
   (g) Golgi complex
   (h) Thylakoids
   (i) Centrosome
   (j) Plasmid
   (k) Nucleoid
   (l) ER

Review Questions

1. Distinguish between the following:
   (a) Bacterial cell and plant cell
   (b) Bacteria and bacteriophage
   (c) Gram-positive and Gram-negative bacteria
   (d) Rough ER and smooth ER
   (e) Lysosome and peroxisome
   (f) Eukaryotic ribosome and prokaryotic ribosome
   (g) Algae and blue-green algae
   (h) Intermediate filament and microfilament
   (i) Chlorophylls in bacteria and chlorophylls in plants
   (j) Flagella and pili in bacteria
   (k) Pectins and lignins
   (l) Autotrophy and heterotrophy

2. Elucidate the process of entry and multiplication of viruses in host cells.

3. How does the small size of cells in general contribute to their better survival?

4. What are chemosynthetic bacteria?

5. Provide examples of archaebacteria.

6. Write the functional aspects of the following:
   (a) Turgor pressure in plant cells
   (b) Nuclear lamina
   (c) Nuclear pore
   (d) Thylakoids
   (e) Golgi complex
   (f) Smooth ER
   (g) Autophagy
   (h) Mesosome

7. Indicate the functions of different membrane proteins in the plasma membrane.

8. Differentiate between the outer and inner membranes of a mitochondrion.

9. Describe the structural components of the plant cell wall.

10. Write about nuclear pores and transport through them.

11. Write about the biosynthesis and processing of proteins in the rough ER.

12. Vacuoles in eukaryotic cells are not simple vacuities and contribute in many ways to the metabolic and other functions of cells. Justify this statement with suitable examples.

13. Describe the structure of a chloroplast.

14. Characterize briefly the three types of structural elements in the cytoskeleton of eukaryotic cells.

15. Write a note on the following:
    (a) Cell theory
    (b) Fluid Mosaic Model of plasma membrane
    (c) Ribosomal RNA
    (d) Structure and functions of ribosome
    (e) Nitrogen-fixing bacteria
    (f) Chromatin
    (g) 'Endosymbiont theory' for origin of mitochondria and chloroplasts

3

Chromosome and DNA Replication

LEARNING OBJECTIVES

♦ Structure of eukaryotic chromosomes
♦ DNA compaction in prokaryotes
♦ DNA replication in eukaryotes and prokaryotes
♦ DNA polymerases

INTRODUCTION

DNA molecules are the custodian of genes, which are the units of inheritance. In eukaryotes, DNA in association with specific proteins is organized as a compact structure called chromosome. In prokaryotes (bacteria), the chromosome is a circular structure present in the cytoplasm without any nuclear envelope (see Chapter 2), and the amount of protein molecules associated with chromosomes is lower than in eukaryotes. Certain other self-replicating small circular DNA molecules, known as *plasmids*, also exist in prokaryotes. Plasmids are not essential for bacterial growth, but they carry certain traits such as antibiotic resistance. In eukaryotes, DNA remains associated with characteristic proteins in a chromosome. The proteins, particularly histones, help in compaction or packaging of DNA into chromosome. The structure of the DNA molecule is explained in detail in Chapter 4.

Packaging of DNA into chromosomes serves many purposes:

- It protects the DNA from damage.
- Compact form of DNA fits easily in a cell, e.g., the nucleus of a human cell that is approximately 5 micrometres in diameter has 46 chromosomes which together contain a total of 2 metres of DNA.
- Packaging facilitates efficient transmission of original and newly replicated DNA molecules in the daughter cells at the time of cell division.

• Chromosomal organization of each DNA molecule facilitates gene expression and regulation, as well as recombination between homologous chromosomes during meiosis.

DNA replication marks the beginning of cell division and ensures the continuation of life. Thus, replication of DNA needs to be discussed along with chromosome, the custodian of the DNA molecule.

Replication of DNA begins with the separation of the complementary strands of the DNA double helix followed by synthesis of new strands on each complementary strand. DNA replication results in the duplication of chromosomes. Thus, each replicative cycle of a cell produces two sets of chromosomes for two daughter cells. The whole process of cell division is finely regulated by specific enzymes and factors. When the fine-tuned regulation of cell division is disturbed due to mutation (change in original composition of DNA) or other external and cellular factors, cells may proceed towards apoptosis (programmed cell death) or in some cases towards malignancy or cancer.

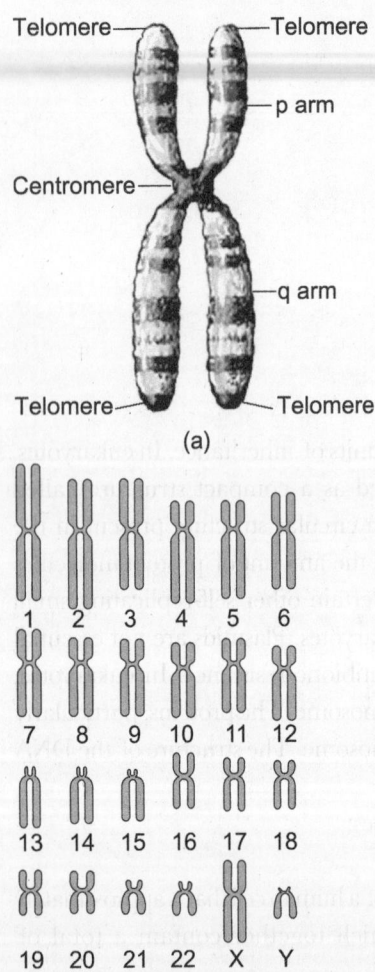

(a)

(b)

Fig. 3.1 (a) Structure of human chromosome (b) Human chromosome karyotype.

3.1 CHROMOSOME IN EUKARYOTIC CELLS

Usually eukaryotic cells are *diploid*, that is, they contain two copies of each chromosome ($2n$), one from each parent. The two copies of a given chromosome are called homologues. The reproductive cells (sperms and eggs) are *haploid*, i.e., they contain a single copy of each chromosome. The chromosome numbers are reduced to half (from $2n$ to n) in these cells through the process of meiotic cell division. When a cell has more than two copies of each chromosome it is called *polyploid*. A liver cell may have a $4n$ polyploid state. A megakaryotic cell in the bone marrow has approximately 128 copies of each chromosome.

The high number of chromosomes in a megakaryocyte allows the cell to have a higher metabolic rate so that it can synthesize the large amounts of RNA and proteins that are required for the production of the thousands of platelets that bud off from it. There are approximately 2,00,000 platelets per millilitre of human blood.

Centromere (Fig. 3.1a) is a specialized region on the chromosome that binds with aster spindle fibres to segregate the two copies of each replicated chromosome to the two poles of the dividing cell. It is made up of more than 40 kb of repetitive DNA sequences (less than 200 bp, in the yeast *S. cerevisiae*).

Telomeres (Fig. 3.1a) are located at the two ends of a linear chromosome. They protect the DNA molecule from degradation at the two ends and resist recombination at these sites. Furthermore, telomeres facilitate end *replication* through the recruitment of an unusual DNA polymerase enzyme called *telomerase* (see Exhibit 3.B).

Figure 3.1(b) shows the karyotype of human chromosomes. A karyotype is the number, size, and shape of the chromosomes of a somatic cell arranged in a standard manner.

3.1.1 DNA and Protein Association in Eukaryotic Chromosomes

Half of the molecular mass of a eukaryotic chromosome is protein. A given region of DNA with its associated protein is called chromatin. The majority of associated small, basic proteins are known as histones. Other DNA-associated proteins are called the non-histone proteins. All these DNA-binding proteins in the chromosomes regulate the compaction, replication, transcription (synthesis of mRNA), recombination, and repair mechanisms of DNA.

3.1.2 Nucleosomes—Building Units of Chromosomes

The major part of the DNA in eukaryotic cells is packaged into nucleosomes. A nucleosome consists of a DNA molecule wrapped around a core of eight (octamer) histone proteins (Fig. 3.2). The DNA in between two nucleosomes is known as linker DNA. Individually, a nucleosome under the electron microscope looks like a bead, and collectively with connecting linkers, the DNA appears as 'beads on a string'.

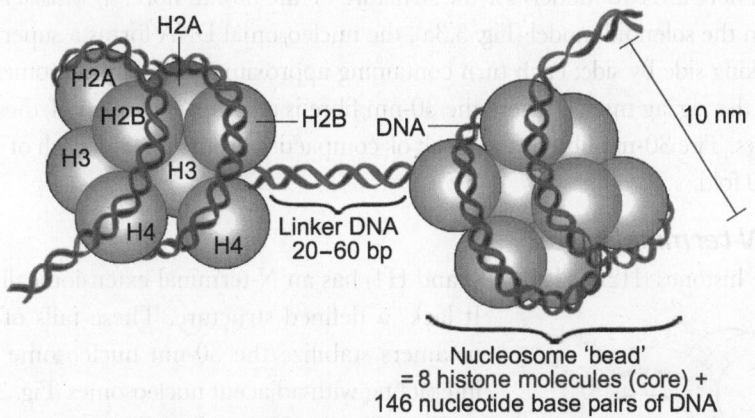

Nucleosome 'bead'
= 8 histone molecules (core) +
146 nucleotide base pairs of DNA

Fig. 3.2   Nucleosome structure. Each nucleosome consists of 146 base pairs of DNA wrapped around eight histone molecules.

The nucleosome *core* consists of a stretch of 146 nucleotide base pairs, called the core DNA which is tightly wrapped around the histone octamer forming 1.65 circular turns in a left-handed coil. The 146 base pair length of DNA is a constant feature of nucleosomes in all eukaryotic cells, but the length of linker DNA between nucleosomes is variable.

As has already been mentioned, some proteins called histones are present along with the DNA. There are four types of histone proteins: H2A, H2B, H3, and H4. Two copies (dimer) of each of these histone proteins constitute the octamer core of a nucleosome. These core histones are highly conserved, and thus, very similar in all eukaryotes. A single copy of histone H1 interacts with the linker DNA and ensures tightening of the association between the DNA and the nucleosome and better protection of the linker DNA. While the core histones are relatively small proteins ranging in size from 11 to 15 kiloDaltons (kDa), the size of histone H1 is approximately 20 kDa.

The histones proteins are positively charged due to a high content of positively charged amino acids. Thus, the histones can come in close association with the negatively charged DNA molecule. The formation of nucleosomes is the first step in the compaction process that allows the DNA molecule to be folded into highly compact structures that reduce the linear length by as much as 10,000 fold, into a chromosome.

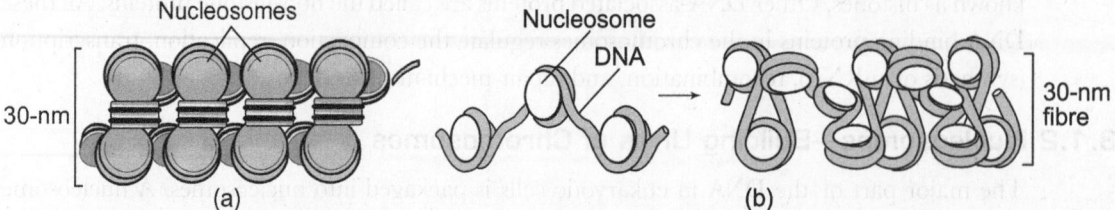

Fig. 3.3 Transition to a 30-nm chromation fibre with addition of histone H1 at linker region. This is known as compaction of DNA. (a) Solenoid model (b) Zig-zag model.

Binding of H1 leads to a higher level of DNA compaction, forming a 30-nm fibre of nucleosomal DNA (Fig. 3.3). There are two models for the structure of the 30-nm fibre: (i) *solenoid model* and (ii) *zigzag model*. In the solenoid model (Fig. 3.3a), the nucleosomal DNA forms a super helix by turning and stacking side by side; each turn containing approximately six nucleosomes. Figure 3.3(b) represents the zigzag model, where the 30-nm fibre is a compacted form of these zigzag nucleosome arrays. The 30-nm fibre is the result of compaction of the linear length of DNA by approximately 40 fold.

The histone N-terminal tails

Each of the core histones (H2A, H2B, H3, and H4) has an N-terminal extension called a *tail*. It lacks a defined structure. These tails of histone octamers stabilize the 30-nm nucleosome fibre by interacting with adjacent nucleosomes (Fig. 3.4).

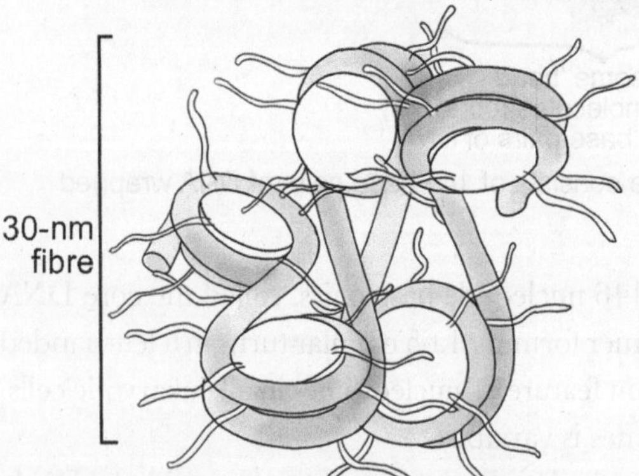

The tails are also the sites of extensive *biochemical modifications*, such as phosphorylation (addition of phosphate group) on serine, and acetylation and methylation on lysine residues. These modifications alter the function of individual nucleosomes and the associated DNA. For example, when acetyl or methyl groups are added to the lysine molecules in the tails, the associated DNA region becomes active in transcription, i.e., mRNA synthesis. With deacylation (removed of acetyl group), the section of DNA (or chromatin) associated with the nucleosome is transcriptionally repressed.

Fig. 3.4 Histone N-terminal tails stabilize the 30-nm chromatin fibres by interacting with adjacent nucleosomes.

Compaction of 30-nm fibre into chromosome

The higher order compaction of 30-nm fibre into chromosome is not yet fully understood. One popular model suggests that the 30-nm fibre forms loops that are held together at the base (i.e., at the centre of the chromosome) by a proteinaceous structure referred to as the *chromosome scaffold* (or nuclear scaffold). The proteins—topoisomerase II (Topo II) and structural maintenance of chromosome (SMC)—which contribute to the chromosome scaffold, have been identified.

Figure 3.5 depicts the stepwise formation of nucleosome from the DNA molecule, and finally its organization into the chromosome.

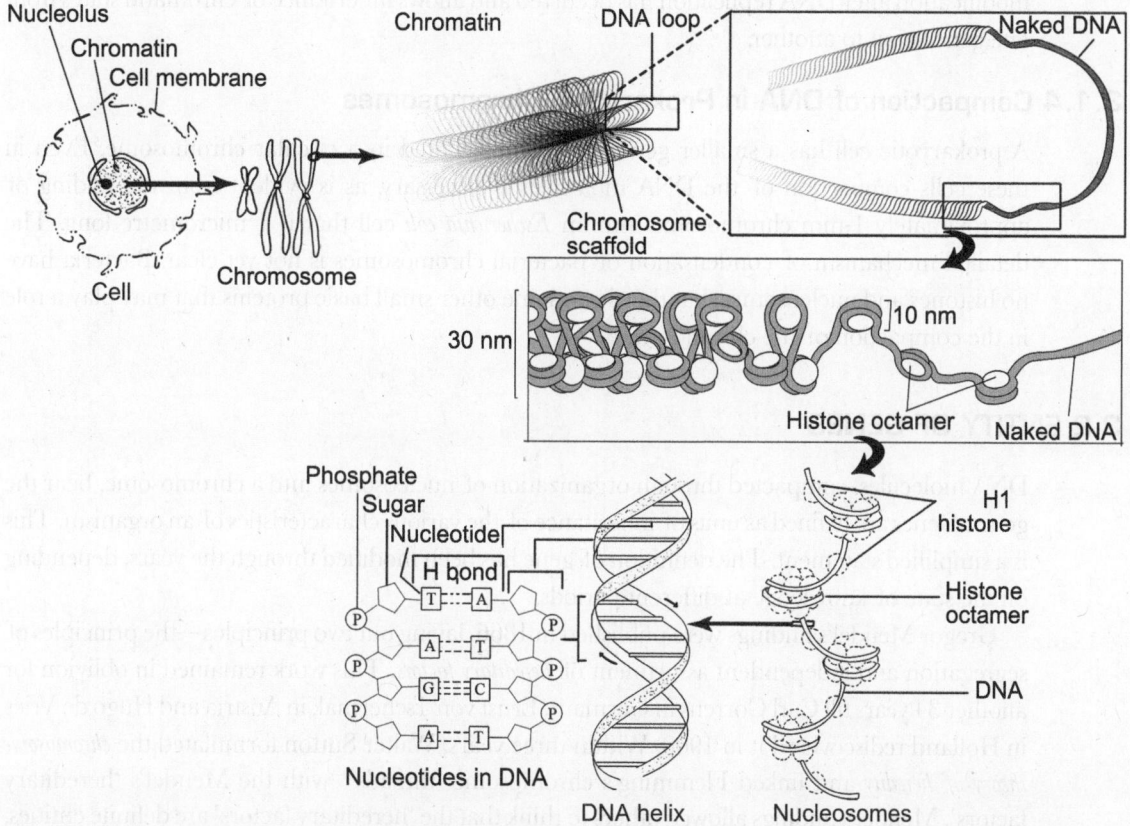

Fig. 3.5 A model showing that the majority of DNA in a eukaryotic chromosome is organized into large loops of 30-nm fibre, attached to the scaffold at the base.

3.1.3 Nucleosome Assembly after DNA Replication

The duplication of a chromosome requires replication of the DNA and the reassembly of the associated proteins, particularly histones, on each daughter DNA molecule.

The replication of DNA requires partial disassembly of the nucleosome ahead of the replication fork. Just behind the fork, the two newly replicated DNA molecules are immediately repackaged into nucleosomes in an ordered fashion. The first step in the assembly of nucleosomes on the DNA is the binding of an H3 and H4 histone tetramer. The two H2A.H2B dimers join to form the final nucleosome. H1 histone joins at linker DNA last, probably during the higher order compaction of chromatin.

As the replication fork passes, histone octamers in the original nucleosome get dissociated. At the time of reassembly of the histone octamers on replicated DNA molecules, old and newly synthesized histones take part in the formation of nucleosomes. The newly synthesized histones are imported from the cytoplasm to the nucleus through the nuclear pore (see Chapter 2). Retention of older histones in the newly formed nucleosomes allows propagation of the specific modifications of parental histones, at the N-terminal tail. This is to maintain similar states of modification after DNA replication has occurred and allows inheritance of chromatin states from one generation to another.

3.1.4 Compaction of DNA in Prokaryotic Chromosomes

A prokaryotic cell has a smaller genome, accommodated in a circular chromosome. Even in these cells compaction of the DNA molecules is necessary, as is evident from the folding of approximately 1-mm chromosome into an *Escherichia coli* cell that is 1 micrometre long. The detailed mechanism of condensation of bacterial chromosomes is not yet clear. Bacteria have no histones and nucleosomes, but they have some other small basic proteins that may play a role in the compaction of the chromosome.

3.2 ENTITY OF GENES

DNA molecules, compacted through organization of nucleosomes into a chromosome, bear the genes. Genes are defined as units of inheritance of the various characteristics of an organism. This is a simplified statement. The definition of gene has been modified through the years, depending on the state of knowledge at different periods.

Gregor Mendel's findings were published in 1866, laying out two principles—the principles of segregation and independent assortment of *hereditary factors*. This work remained in oblivion for another 34 years till Carl Correns in Germany, Ernst von Tschermak in Austria and Hugo de Vries in Holland rediscovered it in 1900. Within three years, Walter Sutton formulated the *chromosome theory of heredity* and linked Flemming's chromosome 'threads' with the Mendel's 'hereditary factors'. Mendel's findings allowed others to think that the 'hereditary factors' are definite entities.

Thomas Hunt Morgan with a band of his students at Columbia University during the first two decades of the 20th century was able to link specific traits or characters to specific chromosomes of the fruit fly *Drosophila melanogaster*.

Johann Friedrich Miescher, within three years of the first publication of Mendel's work, as early as in 1869 discovered 'nuclein' from salmon sperms and human pus cells obtained from surgical bandages. Just like Mendel, Meischer was ahead of his time. It took another 75 years to realize the role of nuclein in bearing genetic information as nucleic acid (DNA).

In 1944, Oswald Avery, Colin MacLeod, and Maclyn McCarty showed that DNA is the genetic material in the course of their experiments with transformation in bacteria. The scientific community remained largely unconvinced of this conclusion, because it was assumed that genes were made up of proteins. Just eight years later, Alfred Hershey and Martha Chase showed by radioactive labelling that it is the DNA of a bacterial phage (virus), and not the protein that enters a bacterial cell for infection and multiplication.

In the 1940s, George Beadle and Edward Tatum, working with the bread mold *Neurospora crassa*, formulated the *one gene–one enzyme* concept, mainly to demonstrate that a gene is responsible for the production of a single, specific protein. By 1953, James Watson and Francis Crick proposed the *double helix* structure of DNA, with features like replication, mutation, and other prerequisites for DNA to function as a gene.

The discoveries by Linus Pauling and Vernon Ingram further refined the one gene–one enzyme concept of Beadle and Tatum. A gene encodes the sequence of amino acids in a polypeptide chain, which may not necessarily be an enzyme or, a complete protein molecule like haemoglobin which consists of several polypeptide chains. Thus, the original hypothesis was refined into the *one gene–one polypeptide concept*. This concept has also become obsolete with the realization that genes also code for all types of RNA molecules, such as transfer RNAs, ribosomal RNAs, small nuclear RNAs, and micro RNAs that have unique cellular functions: these RNAs are not polypeptides. Today, a *gene* is defined as the functional unit of DNA that not only codes for the amino acid sequence of a polypeptide chain through mRNA, but also codes for several types of RNAs with different functions.

Besides encoding genes, the DNA molecule also has non-coding sequences that regulate gene transcription (may be called regulatory genes).

3.2.1 Gene in Viruses, Prokaryotes, and Eukaryotes

Dispositions of genes in DNA molecules of viruses, prokaryotes, and eukaryotes are different. It will be worthwhile to take a brief look at that.

Viruses

Viruses usually have a small DNA molecule (RNA molecule for RNA viruses) that contains the genes necessary for the synthesis of genetic material and coat proteins. They synthesize their genetic material and coat proteins with the help of enzymes and organelles of the host cell. Viral genes can be overlapping, i.e., the beginning of a gene may be located in a region corresponding to a part of the previous gene and it may end on a region that corresponds to a portion of the next gene (Fig. 3.6).

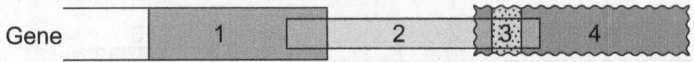

Fig. 3.6   Schematic representation of overlapping genes in virus genome.

Prokaryotes

Unlike viruses, bacteria do not have overlapping genes. The genes in the single circular chromosome of bacteria encode proteins and structural RNAs. The non-coding sequences participate in regulating gene transcription. A single site of regulation for the initiation of transcription often controls the expression of several genes downstream; thus, the regulatory regions in the chromosome are minimized. The gene density, which is the number of genes per length of genomic DNA, is higher in prokaryotes than in eukaryotes.

Eukaryotes

The decreased gene density observed in eukaryotes is mainly because of two factors: increase in gene size and increase in the length of DNA sequences between genes which are often referred to as *intergenic sequences*. This is in stark contrast to the situation of overlapping genes in viruses. The reasons for longer individual genes in eukaryotes are the following: (i) As the complexity of organisms increases, there is a significant increase in the requirement for regulatory regions of DNA to direct and regulate transcription. One or more regulatory sequences are needed for one gene. (ii) Protein encoding genes which are called *exons* often have discontinuous regions that are known as *introns* (Fig. 3.7). Initially, the interspersed non-protein encoding *introns* are transcribed as part of the primary mRNA transcript. Following transcription, these regions in the primary RNA loop out to bring consecutive exons or protein coding RNA parts together. The intervening loops containing introns are then removed by RNA *splicing* and exons are ligated, thereby forming the final mRNA for the translation of protein.

Approximately, 95 per cent human protein-encoding genes have introns. The remaining 5 per cent genes directly encode the protein.

Consistent with their higher gene density, simple eukaryotes have far fewer introns. For example, in the yeast *Saccharomyces cerevisiae*, only 3.5 per cent of genes have introns and they are shorter in length.

Intergenic DNA

Intergenic DNA of a genome (a chromosome set) is not associated with the expression of proteins or structural RNAs and has no known function. It constitutes more than 60 per cent of the human genome. There are two types of intergenic DNA—*unique and repetitive*.

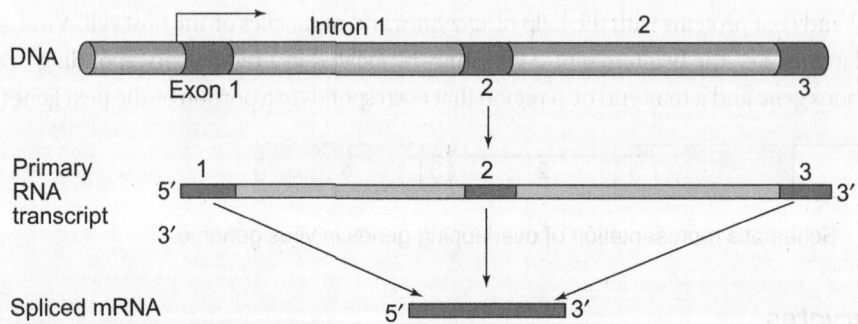

Fig. 3.7 DNA of eukaryotic gene with exons and introns, gets transcribed into primary mRNA transcript; introns are removed by splicing to form mRNA.

About one-fourth of the intergenic DNA is unique. It consists mostly of non-functional relics, such as non-functional mutant genes, gene fragments, and pseudogenes, which are presumed to be non-functional. The rest of the intergenic DNA, which corresponds to almost half of the human genome is composed of repetitive DNA. In general, the repetitive DNA is called *satellite* DNA. It is of the following two types.

Microsatellite DNA   It is composed of very short (less than 13 base pairs) sequences repeated many times and arranged tandemly (end wise one after the other in a line). The most common microsatellite sequences are dinucleotide repeats, for example, CACACACACACACACA.

Genome-wide repeats   These are larger satellite DNAs consisting of more than 100 base pairs and sometimes a repeat can be longer than 1 kb. These sequences may be in closely spaced clusters or dispersed throughout the genome. They can act as transposable elements or transposons.

Transposons

Transposable elements, in short, are known as *transposons*. They can 'move' from one place in the genome to another. Often in the process of moving out from a particular region, the original copy of a transposon may be left behind and the duplicated copy moves to a new position in the genome (Fig. 3.8). At the new location, a transposon needs to recombine at the site to integrate with the genomic DNA. The recombination and integration process of transposons is nearly identical to the entry of a virus into the genome of the host cell. They show little sequence selectivity in their choice of insertion sites.

As a result, transposons can insert within genes, disrupting gene function. If they insert in the regulatory sequences of a gene, the expression of that gene will be altered. Thus, transposons are the most common source of new mutations. They also cause certain genetic diseases in man.

Transposons are present in the genome of all life forms. Barbara McClintock's work with maize first suggested the existence of transposons. The transposon content in genomes of different organisms is highly variable. Although the human genome consists of 45–50 per cent of repetitive DNA with features of transposable elements, the movement of these elements is a relatively rare event.

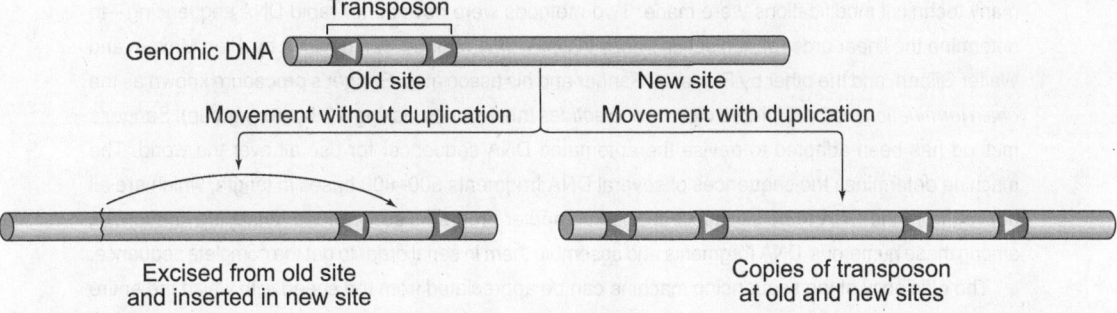

Fig. 3.8   Transposons move from one site to another in the genome.

3.2.2 Genome

The genome of an organism is the length of DNA (or for some viruses, RNA) associated with one haploid set of chromosomes. While the bacterial genome consists of a single chromosome the human genome is represented by 23 chromosomes (half of the diploid number, *2n* or 46 chromosomes). Thus, the genome contains one complete copy of all the genetic information of an organism.

Eukaryotic cells have both *nuclear genome* as well as *mitochondrial genome*. Plant cells contain a *chloroplast genome* in addition to these two types.

Genome size

Genome size in an organism is usually expressed as the total number of base pairs (bp) of nucleotides. The circular DNA molecule of *E. coli* consists of 4,639,221 bp. This is the size of the genome of *E. coli*. Such a large number of base pairs can be conveniently expressed in terms of kilobase (kb), megabase (Mb), and gigabase (Gb) to refer to a thousand, million, or billion base pairs, respectively. Thus, the size of the *E. coli* genome can be mentioned simply as 4.6 Mb.

There are exceptions where there is no direct correlation between genome size and complexity of organisms. In many cases, apparently similar organisms might have very different genome size. A fruit fly has a genome approximately 25 times smaller than a locust, and the rice genome is about 40 times smaller than that of wheat. In such cases, the number of genes rather than the size of the genome provides the correlation to complexity of the organisms. The number of genes per length of DNA is known as *gene density*. The reduction in gene density of higher or complex eukaryotes with 'intergenic' and 'intron' sequences has already been discussed. Gene density decreases since these sequences contribute to genome size without increasing gene number.

Exhibit 3.A   Genome Sequencing: Human Genome Project

The nucleotide sequence of the total genome in an organism is known as the *genome sequence*. Bacterial genomes were the first to be sequenced as they were relatively small.

Frederick Blattner and his team took six years to determine the complete base sequence of the *E. coli* genome in the early 1990s, when genome sequencing attempts began in all earnestness. Then many technical modifications were made. Two methods were devised for rapid DNA sequencing—to determine the linear order of nucleotide bases in DNA. One method was devised by Allan Maxam and Walter Gilbert, and the other by Frederick Sanger and his associates. Sanger's procedure known as the *chain termination method*, utilizes *dideoxynucleotides* (nucleotides lacking a 3´ hydroxy group). Sanger's method has been adapted to devise the automated DNA sequencer for use all over the world. The machine determines the sequences of several DNA fragments 500–800 bases in length, which are all derived from the DNA to be sequenced. Then computer programs search for overlapping sequences among these numerous DNA fragments and assemble them in serial order to get the complete sequence.

The efficiency of the sequencing machine can be appreciated from the speed with which the entire genomes of the yeast *Saccharomyces cerevisiae* (12.1 million bases), the roundworm *Caenorhabditis elegans* (97 million bases), the mustard plant *Arabidopsis thaliana* (125 million bases), the fruit fly *Drosophila melanogaster* (180 million bases), and many other organisms are being sequenced. One research institute reported sequencing of 15 different bacterial genomes in a month.

The grand success of sequencing the *human genome* of about 3.2 billion bases within a relatively short time speaks much about the performance of the DNA sequencer and Sanger's technique. It has already been pointed out that in the early 1990s it took six years to sequence the 4.6 Mb genome of *E. coli*. At that rate, a single laboratory would take almost 6000 years to sequence the entire human genome. However, scientists from different laboratories and countries came together in 1990 to establish the *Human Genome Project*. It was an international cooperation programme involving hundreds of scientists who shared their data in order to determine the entire sequence of the human genome. James D. Watson, who shared the Nobel Prize with Francis Crick and Maurice Wilkins for elucidation of the double helix structure of DNA in 1953, was the first Director of the National Centre for Genome Research of the National Institutes of Health, USA, from 1989 to 1992. A commercial company, Celera Genomics, also was involved in sequencing the human genome. All this effort helped sequence the entire human genome by April 2003, about two years ahead of schedule. Completion of the project also allowed the correct estimation of the number of genes in the human genome, around 3×10^4, which differed from the earlier estimate of 1×10^5.

The computer-based analysis of the vast torrent data for genome projects gave birth to a new discipline, bioinformatics.

3.3 CENTROMERES AND TELOMERES IN EUKARYOTIC CHROMOSOMES

Centromere and telomeres are specialized regions on chromosomes, made up of simple repeat sequences of DNA. The centromere is required for the proper segregation of replicated sister chromosomes, and the telomeres protect the ends of chromosomes.

The centromere is a slightly constricted specialized region on eukaryotic chromosomes that plays an important role in the separation of replicated chromosomes to the two poles of the dividing cell.

As shown in Fig. 3.9, usually a single centromere is present on a chromosome, and according to its location, a chromosome may be metacentric (centromere in the middle), acrocentric (at one side), or telocentric (at one end).

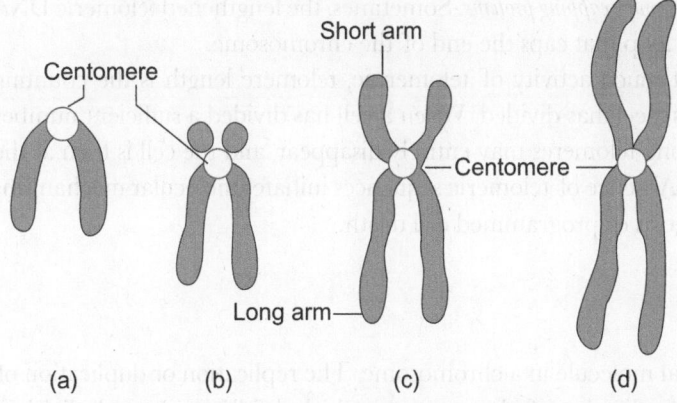

Fig. 3.9 Different types of chromosomes based on the position of centromere. (a) Telocentric, (b) Acrocentric, (c) Submetacentric, (d) Metacentric.

The centromere contains simple-sequence repetitive DNA (as in satellite DNA). It comprises a few hundred kb to several Mb of nucleotides. When DNA replication is completed, the two copies of each replicated chromosome are called sister chromatids, and they are held together at their centromere. With the completion of DNA replication during S phase, kinetochore proteins begin to associate with the centromere. Additional proteins are gradually added to form

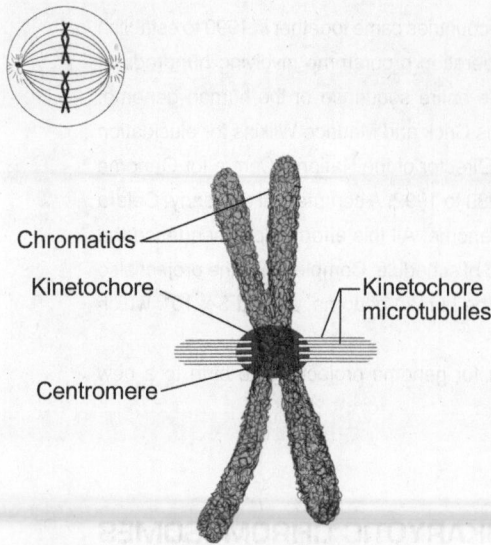

Chromatids

Kinetochore

Kinetochore microtubules

Centromere

Fig. 3.10 Schematic representation of the relationship of centromere and kinetochore.

a kinetochore surrounding the centromere. The spindle microtubules of the dividing cell attach to the kinetochore (Fig. 3.10) and then pull the sister chromatids (or daughter chromosomes) away from one another into the newly forming daughter cells.

Telomeres are simple-sequence repeats of DNA located at the two ends of a linear chromosome. Telomeres protect chromosomes, from recombination and, from degradation at the vulnerable ends of DNA during each round of replication.

Human telomeres contain 100–1500 repeat copies of the sequence 5′–TTAGGG–3′, which has been conserved over hundreds of millions of years of evolution. Other vertebrates, studied so far, have a repeat sequence identical to the human sequence in their telomeres. Even a unicellular organism *Tetrahymena* has telomeres with the repeat sequence TTGGGG, which differs from the human sequence by a single nucleotide. Such non-coding sequences at the ends of each chromosome ensure that no genetic information is lost, in case a DNA molecule is slightly shortened during the process of its replication.

Furthermore, a special DNA polymerase called *telomerase*, made up of protein and RNA, can extend the 3′ ends of chromosomes with the same repeat sequences. The complementary strand is extended by the regular DNA replication machinery in the 5′ to 3′ direction (Exhibit 3.B). Thus, the enzyme compensates for the gradual shortening that takes place at both ends of a chromosome during DNA replication. After telomerase-aided lengthening occurs, telomeres are protected by association of *telomere capping proteins*. Sometimes, the lengthened telomeric DNA folds back upon itself, making a loop that caps the end of the chromosome.

Interestingly, in cells without much activity of telomerase, telomere length is the counting device to know how many times a cell has divided. When a cell has divided a sufficient number of times, the gradually shortening telomeres may entirely disappear, and the cell is then at the risk of eroding its coding DNA. Absence of telomeric sequences initiates molecular mechanisms that steer the cell towards apoptosis or programmed cell death.

3.4 DNA REPLICATION

The DNA molecule is the central molecule in a chromosome. The replication or duplication of the DNA molecule leads to the replication of chromosomes which is followed by cell division. Thus, one of the most important endowments of life or living matter—the reproducibility—basically lies with the DNA molecule.

In their classic paper on 'Molecular Structure of Nucleic Acids: A Structure for Deoxyribose Nucleic Acid' published in *Nature*, Vol. 171 on 25 April 1953, Watson and Crick wrote a tiny paragraph:

It has not escaped our notice that the specific pairing we have postulated immediately suggests a possible copying mechanism for the genetic material.

They soon elaborated on it elsewhere. The essence of their suggestion was that prior to duplication, the hydrogen bonds connecting the bases are broken, and the two chains unwind and separate. Then each chain acts as a template for assembly of complementary nucleotides to form a new companion chain. Thus, a double helix DNA molecule allows formation of two pairs of chains or double helixes; in each pair, there is an old DNA strand and a newly synthesized one. This mode of DNA synthesis is called *semiconservative replication*, because half of the parent molecule is retained by each daughter molecule.

Within five years of the publication by Watson and Crick, Matthew Meselson and Franklin Stahl demonstrated by using the technique of equilibrium density centrifugation that DNA replication is indeed semiconservative. Hundreds of scientists in different laboratories across the globe contributed further towards understanding the molecular mechanism of DNA replication, which is a series of intricate events involving numerous enzymes and proteins, as well as RNA which serve as primers.

The basics of DNA replication in prokaryotes and eukaryotes are similar, indicating its origin fairly early in the evolution of life. In certain details and participatory molecules, DNA replication in eukaryotes shows elaborations and complications.

3.4.1 DNA Replication in Prokaryotes

John Cairns first attempted to visualize DNA replication in *E. coli*. He allowed the bacteria to grow for different lengths of time in a medium containing radioactive ^{3}H-thymidine as precursor for newly synthesized DNA and then employed autoradiography to get the picture of the duplicated chromosome after different intervals of time. The findings have been presented schematically in Fig. 3.11.

Replication is initiated at a point known as the origin of replication which has a special DNA sequence, that is mostly rich in AT base pairs. For initiation of the replication process, a specific group of initiator proteins bind to the origin and utilizing ATP-derived energy, unwind the double helix to provide access to DNA polymerase and other proteins that bind to single-stranded DNA.

DNA replication proceeds from the origin bidirectionally as indicated by thin arrows in the first figure. After half of the DNA molecule has been replicated, its shape resembles the Greek letter theta (θ) as shown in the second figure.

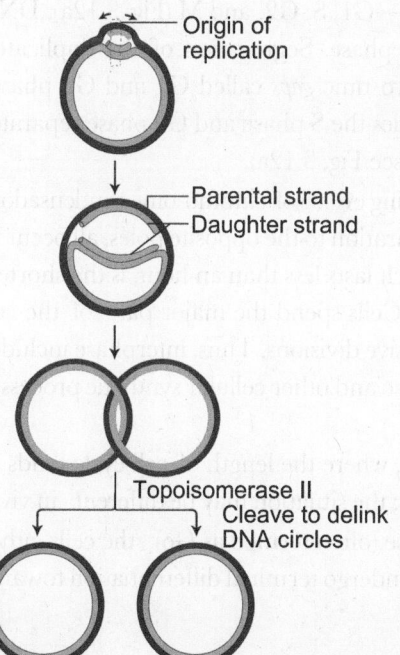

Origin of replication

Parental strand
Daughter strand

Topoisomerase II
Cleave to delink
DNA circles

Fig. 3.11 Replication of circular prokaryote DNA molecule. At the end, type II topoisomerases separate (or decatenate) the two interlinked circular replicated DNA molecules.

Often this mode of replication of a circular DNA is called theta replication. Similar replication also occurs in the circular DNAs of mitochondria, chloroplasts, plasmids, and some viruses. When replication of a circular DNA ends, the resulting two circular DNA molecules remain interlinked. Topoisomerase enzymes transiently cleave one strand of a double helix and by passing the unbroken strand through the break, delink the two DNA circles from each other. After cleaving the DNA, topoisomerase moves out and the cleaved point in the DNA chain closes up.

3.4.2 Chromosome Duplication, Segregation, and the Cell Cycle in Eukaryotes

It has already been pointed out that replication of the DNA molecule in a chromosome means duplication of a chromosome. Duplication of the chromosomes followed by their segregation into the daughter cells are the two major events in cell division. In bacterial (prokaryote) cells, these two events occur simultaneously. As the DNA is replicated, the resulting two copies are separated and moved to opposite sides of the dividing cell. In contrast, the replication of chromosomes and their segregation takes place in two distinct phases of the cell cycle, in eukaryotes.

Cell cycle

The events that occur when a cell undergoes a single round of cell division in a span of time constitute a *cell cycle*, see Fig. 3.12. After cell division of a eukaryotic cell, if the parental number of chromosomes is maintained in the daughter cells, the division is called *mitotic cell division*. (*Meiotic type of cell division* in reproductive germ cells reduces the chromosome number to half the parental number, i.e., from $2n$ to n.)

The mitotic cell cycle can be divided into four phases—G1, S, G2, and M (Fig. 3.12a). DNA synthesis (replication) occurs during the synthesis or S phase. Segregation of the duplicated chromosomes takes place in the *mitotic* or M phase. Two time *gaps* called G1 and G2 phases separate the M phase from the S phase; G1 phase precedes the S phase and G2 phase separates the end of S phase from the beginning of the M phase (see Fig. 3.12a).

Visually (when observed through a microscope) the striking events of chromosome condensation, orientation at the equator of the mitotic spindle, and separation to the opposite poles, all occur in the mitotic (M) phase (Fig. 3.12b). Usually, M phase which lasts less than an hour, is the shortest phase of the cell cycle whose duration is 18–24 hours. Cells spend the major part of the cell cycle in the growth phase called *interphase* between successive divisions. Thus, interphase includes G1, S, and G2 phase. DNA synthesis occurs in the S phase and other cellular synthetic processes and growth continue throughout the interphase.

Most studies of the cell cycle use cells grown in vitro, where the length of cell cycle tends to be similar for different cell types under similar conditions; the situation may be different in vivo.

Right after M phase, at the beginning of the G1 phase (often termed as Go), the cells either enter into another cell cycle or exit from the cell cycle to undergo terminal differentiation towards a particular cell type.

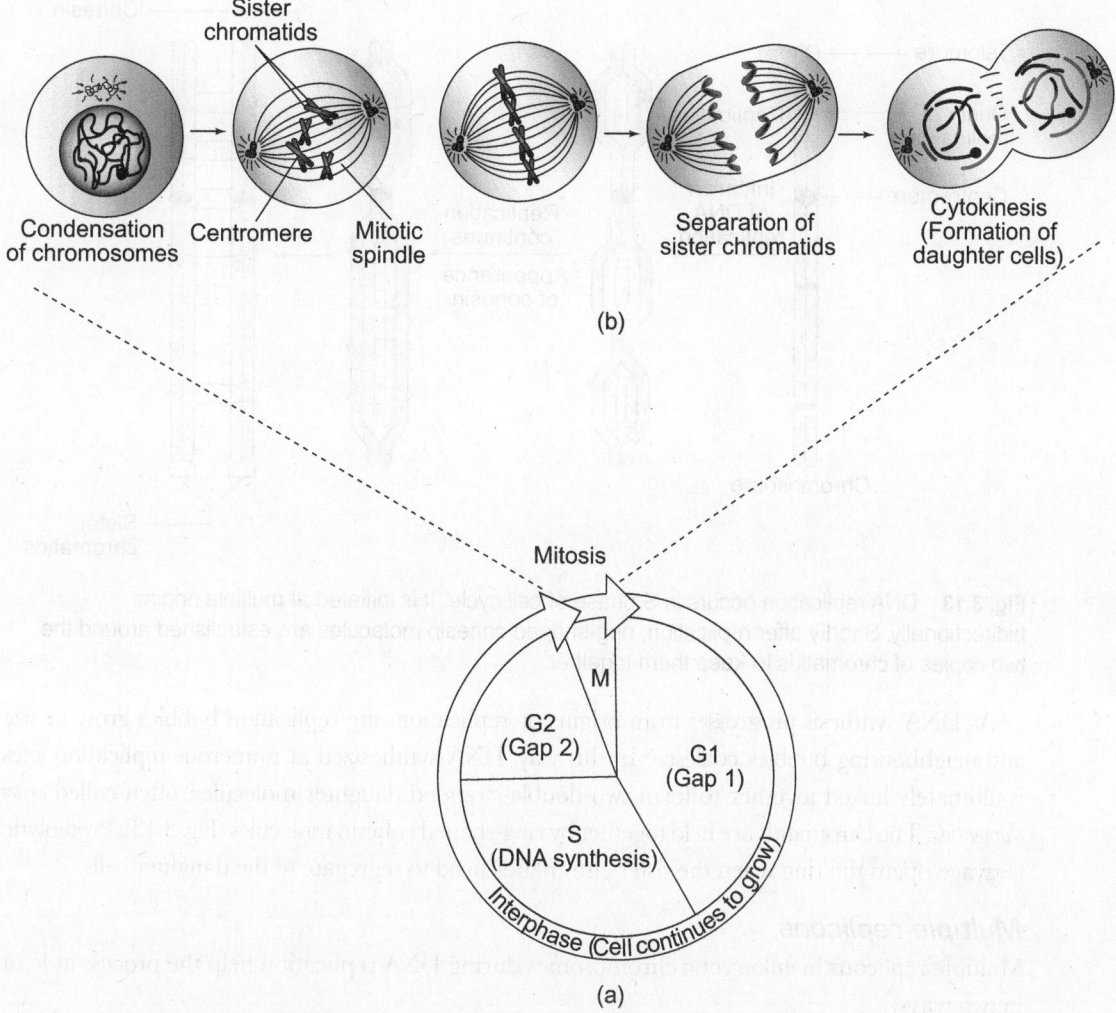

Fig. 3.12 (a) The four phases (G1, S, G2, and M) of eukaryotic mitotic cell cycle. (b) Events in M (mitotic) phase—chromosome condensation, orientation at the equator of mitotic spindle, and separation to two poles.

DNA replication in eukaryotes—Multiple replicons

In contrast to the single origin of replication in the circular chromosome of prokaryotes, the linear DNA molecule of a eukaryotic chromosome has multiple origins of replication. In general, origins of replication are located in non-coding regions of DNA. There could be thousands of origins of replication in a typical large eukaryotic chromosome.

Replication of the DNA molecule is initiated at multiple origins bidirectionally, forming *replication bubbles*, often called replicons (Fig. 3.13). Thus, an origin of replication remains at the central position of a replicon. Several groups of initiator proteins bind at the origin of replication to initiate DNA synthesis.

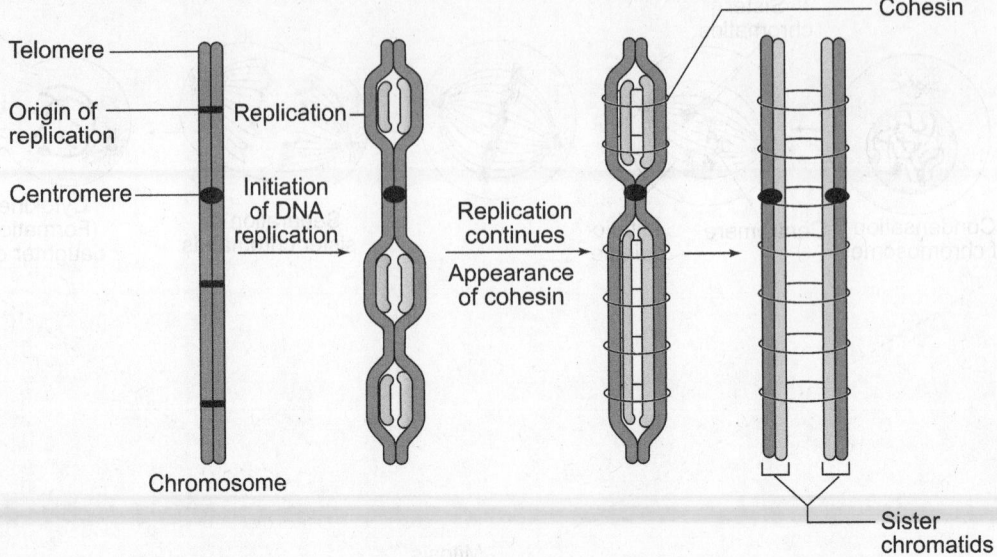

Fig. 3.13 DNA replication occurs in S phase of cell cycle. It is initiated at multiple origins bidirectionally. Shortly after replication, ring-shaped cohesin molecules are established around the two copies of chromatids to keep them together.

As DNA synthesis progresses from origins of replication, the replication bubbles grow in size and neighbouring bubbles coalesce. In this way, DNA synthesized at numerous replication sites is ultimately linked together to form two double-stranded daughter molecules, often called *sister chromatids*. The chromatids are held together by ring-shaped cohesin molecules (Fig. 3.13). Proteolytic cleavage opens the ring when the sister chromatids need to segregate to the daughter cells.

Multiple replicons

Multiple replicons in eukaryotic chromosomes during DNA replication help the process at least in two ways:

(i) DNA molecule in eukaryotic chromosome is much longer than in prokaryotic chromosomes; hence it would take much longer for replication to be completed if there was only a single replication origin.

(ii) Moreover, synthesis of DNA is notably slower in eukaryotes than in bacteria, presumably because of the presence of nucleosomes. (DNA replication in eukaryotes is at the rate of about 2,000 base pairs per minute compared to 50,000 base pairs per minute in bacteria.)

Interestingly, chromosomes in the embryonic cells of the fruit fly *Drosophila*, which divide much faster than adult cells, employ a larger number of simultaneously active replicons than the adult cells.

3.4.3 Chronological Events in the Replication Process

The process of replication of DNA is a finely tuned multi-step event, involving different enzymes with specific functions at different steps.

Replicator

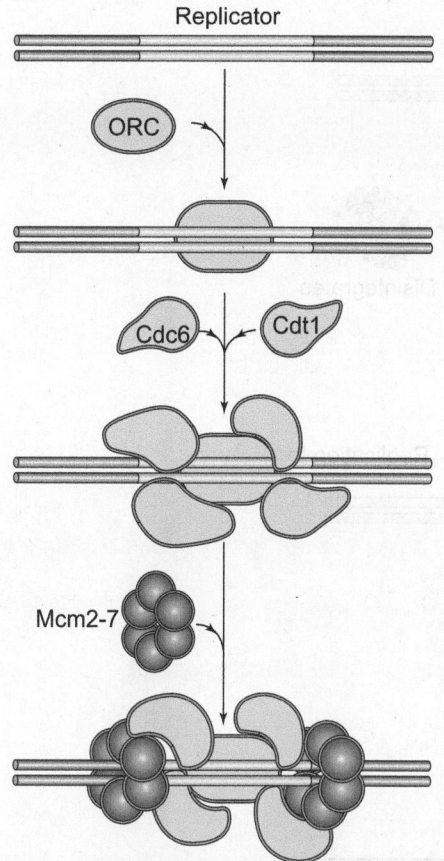

Fig. 3.14 Formation of the pre-replicative complex (pre-RC) with participation of four separate proteins.

Pre-replicative complex to initiate replication in eukaryotes

The pre-replicative complex (pre-RC) is formed at each replicator (origin of replication) with participation of four separate proteins (Fig. 3.14):

(i) The *origin recognition complex* (ORC) protein binds first at replicator.

(ii) This is followed by binding of *two helicase loading proteins* (Cdc 6 and Cdt 1).

(iii) Helicase [MCM2-7 (mini chromosome maintenance complex)], a hexamer protein, binds next to the ORC and helicase loader proteins to complete the formation of the pre-RC.

Helicase is often called a DNA helicase as it facilitates unwinding of the double helix.

With the formation of pre-RC, the eukaryotic DNA is 'licensed' for a single round of replication in one cell cycle. The pre-RC forms during the G1 phase of the cell cycle, but initiation of replication begins after cells pass from the G1 to the S phase, when the pre-RC is activated by protein kinases.

Pre-RC activation

Two protein kinases [cdk (cyclin-dependent kinase) and Ddk (Dbf4-dependent kinase)] phosphorylate the helicase loading proteins in the pre-RC, and the other replication proteins to be recruited at the pre-RC site (Fig. 3.15). Phosphorylation is the transfer of phosphate groups to target proteins. After phosphorylation, the helicase loading proteins are removed from the pre-RC site to make room for other phosphorylated proteins and DNA polymerase enzymes (δ and ε; Fig. 3.15b).

Unwinding of the DNA double helix by DNA helicases and topoisomerases

DNA helicases cause unwinding of the DNA double helix to expose the single strands to the enzymes involved in DNA replication. Helicases break the hydrogen bonds between nucleotide bases and unwind the DNA ahead of the replication fork, using energy derived from the hydrolysis of ATP.

The process of unwinding creates supercoiling of the DNA double helix ahead of the replication fork; under normal circumstances this would lead to tangling in the rest of the DNA molecule. *Topoisomerases* relieve this situation by creating swivel points in the DNA molecule by making and then quickly sealing single- or double-stranded breaks in the molecule (Fig. 3.16). In bacteria, topoisomerase is called gyrase and it opens up the double helix at the origin of replication and separates the linked circles of daughter DNA molecules at the end.

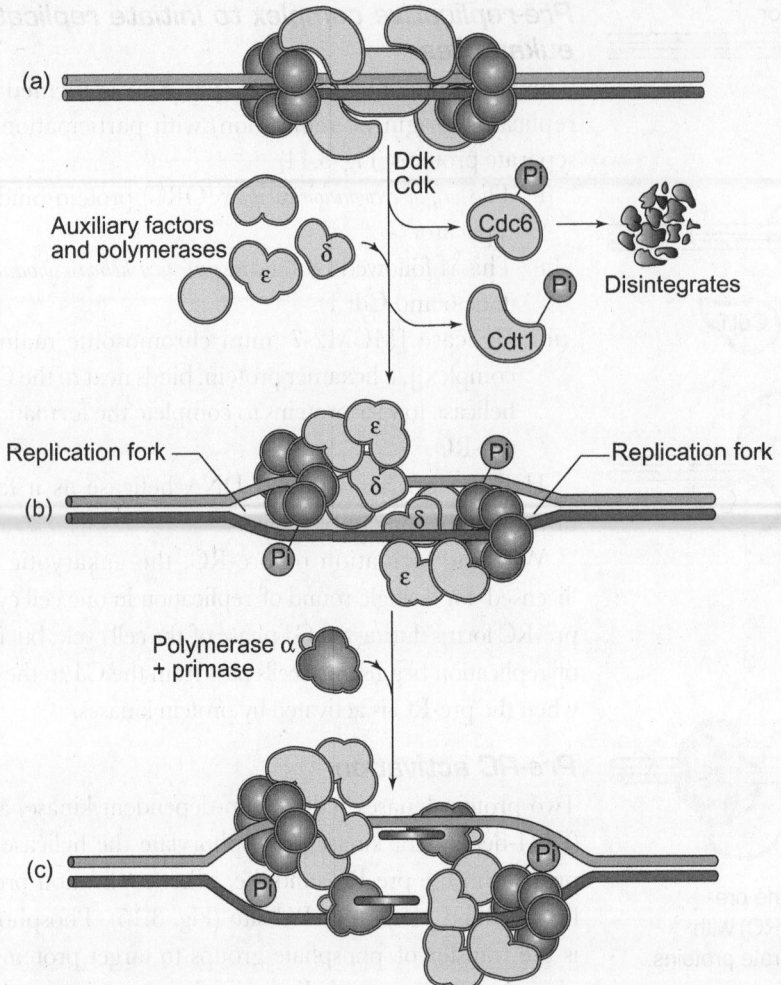

Fig. 3.15 Activation of the pre-RC leads to the recruitment (assembly) of other protein molecules at the replication fork.

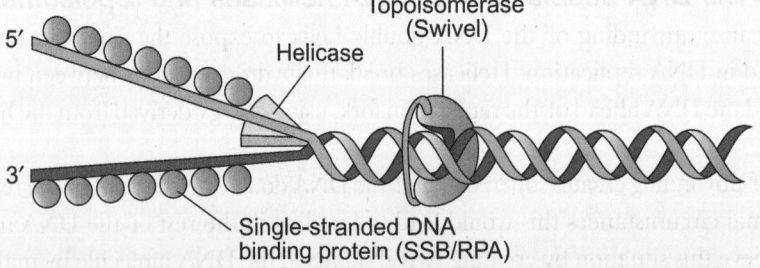

Fig. 3.16 Protein molecules involved in unwinding DNA at the replication fork. Topoisomerase relieves the supercoiling of the DNA double helix ahead of the replication fork.

Single-stranded DNA-binding protein (SSB in bacteria and RPA in eukaryotes)

When the DNA helix is unwound with the action of helicases and topoisomerase, RPA (replication protein A) or SSB (single-stranded DNA-binding protein) proteins quickly bind to the single strands to keep them in the unwound state, so that they are accessible to the DNA replicating enzymes. As soon as replication begins in a section of a single strand, the RPA or SSB molecules fall off from the section and are recycled for attaching to the newly opened single strands.

3.4.4 Directionality of DNA Synthesis

DNA polymerases catalyse the addition of nucleotides in the $5' \rightarrow 3'$ direction. The phosphate group on the $5'$ carbon (of the sugar) of the incoming nucleotide is covalently linked to the hydroxyl group on the $3'$ carbon of the nucleotide added in the previous step (Fig. 3.17). The covalent bond between the nucleotides catalysed by DNA polymerase is called a *phoshodiester* bond. DNA polymerase enzymes thus cause the growth of a DNA strand from the $5'$ end to the $3'$ end on a single-straned template DNA.

The two strands of the DNA double helix are in opposite orientations. One strand is in the $3' \rightarrow 5'$ orientation and the other one is in the $5' \rightarrow 3'$ direction. Synthesis of a new DNA strand in the $5' \rightarrow 3'$ direction on the template strand, which is in the $3' \rightarrow 5'$ orientation, by the polymerase produces a continuous new DNA strand and finally a complete DNA molecule. But how can the DNA polymerase, which is only capable of synthesizing DNA strands with $5' \rightarrow 3'$ orientation, function on the second $3' \rightarrow 5'$ DNA template strand. Reiji Okazaki's experiments with replication in bacteria, in 1968, provided the solution to this paradox.

On the $5' \rightarrow 3'$ template strand, the DNA polymerase acts in the reverse orientation synthesizing new DNA segments in the $5' \rightarrow 3'$ direction as usual, from the $3'$ end of the template DNA strand (Fig. 3.18). The synthesis of the fragments starts as soon as the double helix opens in a section. The short discontinuous DNA segments, thus produced, are called *Okazaki fragments*. ATP-dependent DNA ligase enzyme joins these fragments into a continuous $5' \rightarrow 3'$ DNA strand.

The two daughter strands formed after replication is completed can be distinguished on the basis of their mode of growth. The strand synthesized as a continuous chain, because it is growing in the $5' \rightarrow 3'$ direction, is called the *leading strand*. The other daughter strand formed by joining a series of short discontinuous Okazaki fragments is called the *lagging strand*.

Okazaki fragments are generally 1,000–2,000 nucleotides long in bacteria, but in eukaryotic cells they are only about one-tenth of this length.

RNA primers initiate the functioning of DNA polymerase

It has already been discussed that DNA polymerase can catalyse the addition of nucleotides to the $3'$ end of the previous nucleotide. But in the beginning how does the polymerase function when there is no previous nucleotide. A small RNA piece, about 3–10 nucleotides in length provides the $3'$ end for the DNA polymerase to initiate its function. This small sequence of RNA is known as the *RNA primer*.

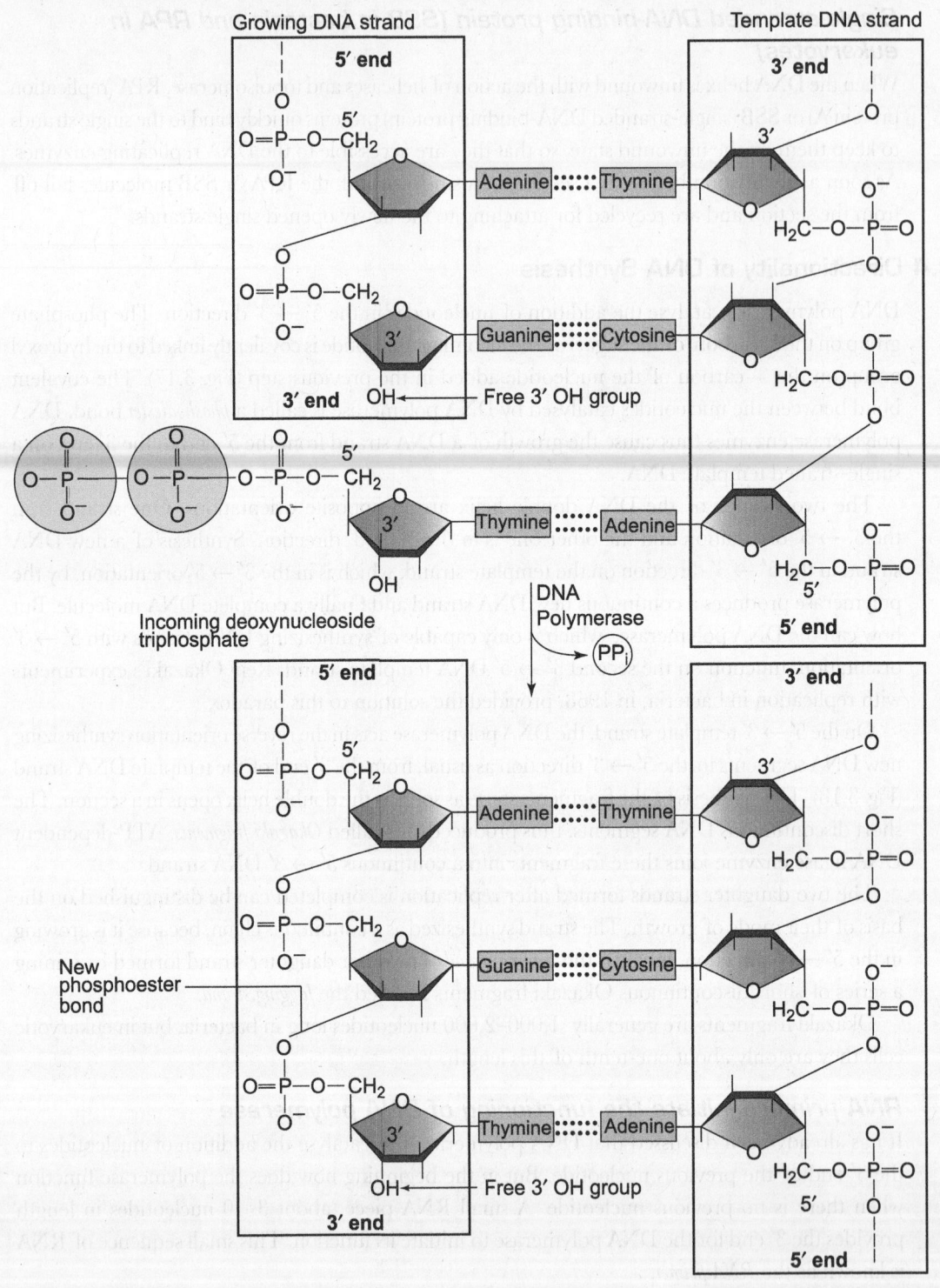

Fig. 3.17 The directionality of DNA synthesis.

Helicase enzyme in course of unwinding the helix in bacteria recruits an enzyme DNA *primase* that synthesizes the 3–10 nucleotide long RNA primer using DNA as the template. Unlike DNA polymerase, which adds nucleotides only to the ends of an existing nucleotide chain, the primase can initiate RNA synthesis from scratch by joining two nucleotides together. Primase is a specific kind of RNA polymerase that is involved only in the process of DNA replication. The eukaryotic primase on the other hand is tightly bound to DNA polymerase α, the main DNA polymerase involved in initiating DNA replication (Fig. 3.15) and is not so closely associated with helicase.

The synthesis of the leading strand is continuous and its initiation needs an RNA primer only once. In contrast, the lagging strand is synthesized as a series of discontinuous Okazaki fragments, each of which is initiated with a separate RNA primer. Thus, the lagging strand initially has multiple RNA primers.

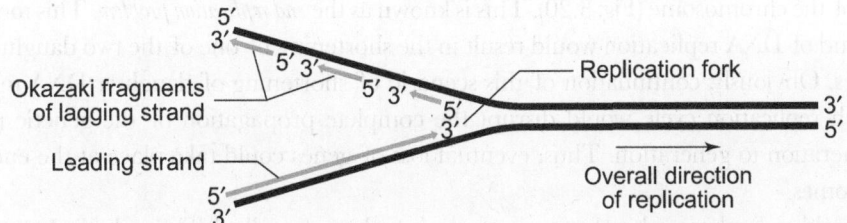

Fig. 3.18 Direction of DNA synthesis at a replication fork and formation of Okazaki fragments.

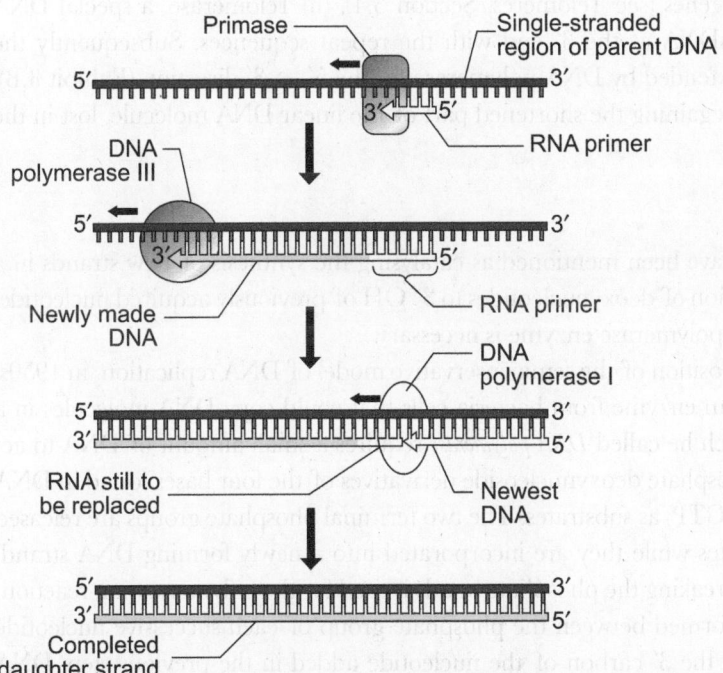

Fig. 3.19 The role of RNA primers in DNA replication. Primase, an RNA polymerase, synthesizes the RNA primer using a single strand of DNA as its template. Once a short stretch of RNA is available, DNA polymerase (here Pol III in *E. Coli*) initiates DNA synthesis by adding nucleotides in the 5′ → 3′ direction.

Removal of RNA primers and filling in the space with deoxyribonucleotides

DNA polymerase adds DNA nucleotides to each RNA primer until it reaches the adjacent Okazaki fragment (in the lagging strand). No longer needed at that point, the RNA primer is removed by 5′ → 3′ exonuclease activity. Then DNA polymerases fill in the space with deoxyribonucleotides.

In bacteria, DNA polymerase I removes the RNA primer, and at the same time, synthesizes DNA in the normal 5′ → 3′ direction filling in the resulting gaps. Adjacent fragments are subsequently joined together by DNA ligase (Fig. 3.19).

The leading strand in a replicating circular chromosome in bacteria continues to grow in the $5' \rightarrow 3'$ direction until its $3'$ end joins the $5'$ end of the lagging strand growing in the other direction. Actually the joining takes place with the last Okazaki fragment. In the course of this joining event, the RNA primer of the last Okazaki fragment is replaced by adding DNA nucleotides to the free $3'$ OH of the leading strand.

3.4.5 End Replication Problem for Linear DNA Molecules in Eukaryotes

For a linear DNA molecule such as that of a eukaryotic chromosome, the RNA primer of the last Okazaki fragment (in the lagging strand) is removed by a $5' \rightarrow 3'$ exonuclease. However, the gap cannot be filled because there is no $3'$ OH end for addition of deoxynucleotides by DNA polymerase. Thus, a short region of unreplicated single stranded DNA (ssDNA) will remain at the end of the chromosome (Fig. 3.20). This is known as the *end replication problem*. This means that each round of DNA replication would result in the shortening of one of the two daughter DNA molecules. Obviously, continuation of this scenario of shortening of daughter DNA molecules after each replication cycle would disrupt the complete propagation of the genetic material from generation to generation. Thus, eventual loss of genes could take place at the end of the chromosomes.

This problem has been solved by two strategies in eukaryotic cells. (i) The ends of chromosomes, called telomeres are made up of highly repetitive non-coding sequences of nucleotides, i.e., they do contain any functional genes (see Telomeres, Section 3.4). (ii) Telomerase, a special DNA polymerase, extends the ssDNA at the $3'$ end with the repeat sequences. Subsequently the complementary strand is extended by DNA polymerase in the $5' \rightarrow 3'$ direction (Exhibit 3.B). Thus, telomerase helps in regaining the shortened part of the linear DNA molecule, lost in the end replication process.

3.4.6 DNA Polymerases

So far DNA polymerases have been mentioned as catalysing the synthesis of new strands in a DNA molecule by the addition of deoxynucleotides to $3'$ OH of previously acquired nucleotide. Certain elaboration on the polymerase enzyme is necessary.

After a few years of proposition of the semiconservative model of DNA replication, in 1950s, Arthur Kornberg isolated an enzyme from bacteria cells that could copy DNA molecules in a test tube. This enzyme which he called *DNA polymerase* requires a small amount of DNA to act as a template and the triphosphate deoxynucleoside derivatives of the four bases found in DNA (dATP, dTTP, dGTP, and dCTP) as substrates. The two terminal phosphate groups are released from each of these substrates while they are incorporated into a newly forming DNA strand. The energy released from breaking the phosphate bonds is used for the polymerization reaction. A phosphodiester bond is formed between the phosphate group of each successive nucleotide and the hydroxyl group on the $3'$ carbon of the nucleotide added in the previous step. DNA polymerase precisely carries out this function in the $5' \rightarrow 3'$ direction.

Kornberg's discovery was later honoured with the Nobel Prize. His discovery encouraged other investigators to discover several different forms of DNA polymerase in prokaryotic and eukaryotic

① DNA replication from the origin forming the replication bubble with two replication forks moving in opposite directions (RNA primer in small black blocks in newly formed DNA strands).

② Only one primer remains finally on each daughter DNA molecule.

③ The primers removed by a 5′ → 3′ exonuclease, but no DNA polymerase can fill the resulting gaps as no 3′ OH available for adding a nucleotide; length of DNA gets shorter.

④ DNA molecules get shorter and shorter in each round of replication.

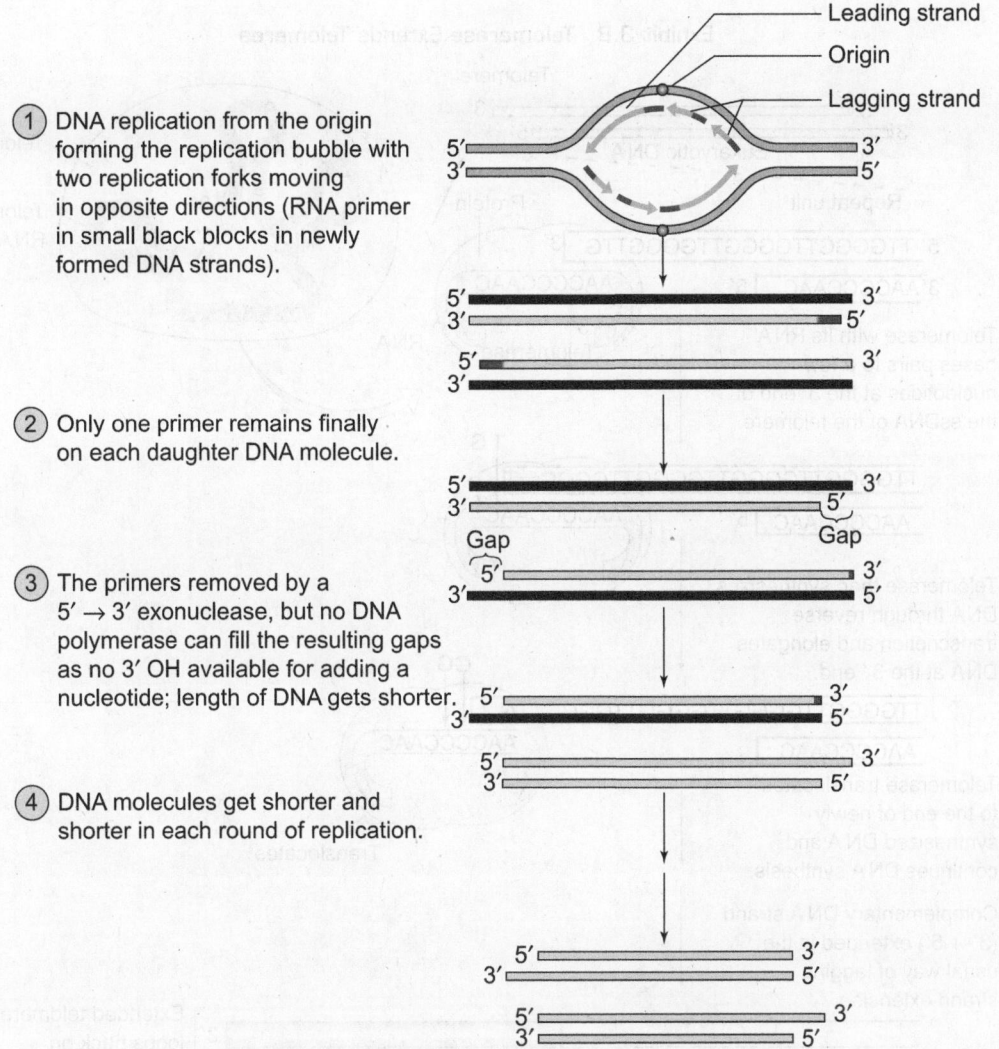

Fig. 3.20 How the end replication problem arises in a eukaryotic chromosome. A short region at the end of a linear DNA molecule, initially occupied by an RNA primer cannot be replicated and the chromosome gets shorter with each cycle of DNA replication.

cells. Five types of DNA polymerases were found in bacterial cells. They have been named using Roman numerals I to V. The original Kornberg enzyme is now called DNA polymerase I. The five different DNA polymerases from prokaryotes differ structurally, in the number of subunit components and some specific functions. The mechanism by which DNA polymerase I removes the RNA primer from Okazaki fragments by 5′ → 3′ exonuclease activity and then synthesizes small stretches of DNA to fill the gaps that are formed, has already been discussed. DNA polymerase I synthesizes DNA at a much slower rate than DNA polymerase III (Pol III) which averages about 50,000 base pairs per minute in bacteria. Pol III is the main enzyme

Exhibit 3.B Telomerase Extends Telomeres

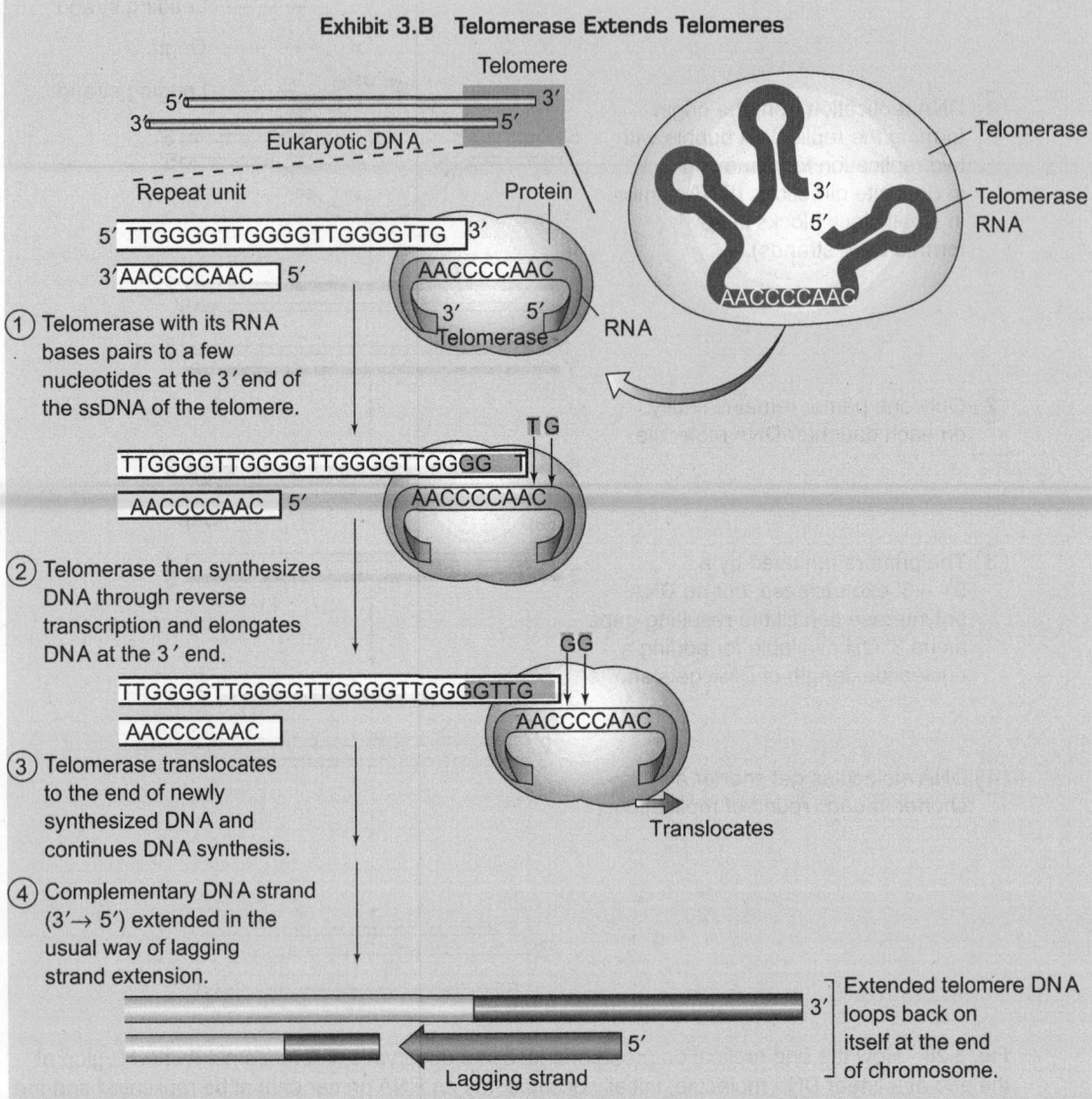

Fig. 3.B Telomerase extends telomeres

Telomeres consist of repeat sequences of DNA. In the human chromosome, the repeat unit on the 5′ → 3′ strand is TTAGGG. The telomerase enzyme is made up of a protein and RNA with complementary sequence for 1½ repeat unit of telomere. The telomerase catalyses the addition of nucleotides to the 3′ end. Then the regular DNA replication machinery synthesizes the lagging strand complementary to the strand elongated by telomerase. Ultimately, the telomere is lengthened to regain the loss during the earlier step of replication of linear DNA. Thus, the end replication problem for linear DNA molecules in eukaryotes is resolved.

to carry out DNA replication in bacterial cells. The other types of bacterial DNA polymerases, Pol II and IV, play specialized roles in DNA repair.

Eukaryotic cells also have more than a dozen DNA polymerases, with different structural entities and functional specialties, each named with a Greek letter. Of these, three are essential to duplicate the DNA molecule—DNA Pol α, DNA Pol β, and DNA Pol ε. DNA Pol α is attached to the primase; each is made up of two subunits.

The DNA Pol α–primase complex initiates synthesis of new DNA strands. The primase synthesizes an RNA primer and the resulting RNA primer–template junction is immediately taken over by DNA Pol α to initiate DNA synthesis.

Due to its relatively low processivity, DNA Pol α–primase is replaced by the highly processive DNA polymerases δ and ε. This process of replacing is called *polymerase switching*. These three DNA polymerases function at the eukaryotic replication fork.

DNA Pol γ is present in mitochondria and is involved in the replication of mitochondrial DNA. As in bacteria, the remaining eukaryotic DNA polymerases are involved in DNA repair.

Proofreading by the 3′ → 5′ exonuclease activity of DNA polymerase

DNA polymerase synthesizes new DNA strands in the 5′ → 3′ direction. The polymerase has an exonuclease active site to remove any incorrect nucleotide that may have been incorporated into the new strand; this is done in the reverse direction (3′ → 5′). Correction of such mistakes by the exonuclease activity (degradation of nucleic acid from one end) of DNA polymerase is known as *proofreading*. The incorporation of incorrect bases in DNA occurs in one out of every 100,000 nucleotides. Hence, the exonuclease activity of DNA polymerase removes improperly base-paired nucleotides to maintain the fidelity of DNA.

Processivity of DNA polymerases

Functional rate (processivity) of DNA polymerases from different sources is the point of discussion here. DNA polymerases, especially the ones that continuously add nucleotides to the primer strand for replication, catalyse quite rapidly. They are Pol III in bacteria, and Pol δ and ε in eukaryotes. The rate of DNA synthesis depends on the processivity of the polymerase enzyme, which has already been indicated as 50,000 nucleotides per minute, or approximately 1,000 nucleotides per second, in bacteria.

The initial binding of polymerase to the primer–template junction is the rate-limiting step. It takes approximately 1 second to locate a primer and form a primer–template junction. Once a primer has bound, the addition of nucleotides is very fast, in the range of milliseconds.

Three-dimensional structure of DNA polymerase

X-ray crystallography has played a major role in revealing the three-dimensional structure of DNA polymerase. DNA polymerase resembles a right hand that grips the primer–template junction (Fig. 3.21). The active synthesizing site for DNA is associated with the palm. DNA polymerase slides along the DNA template; electrostatic interactions between the phosphate backbone of DNA and the polymerase structure help in the process.

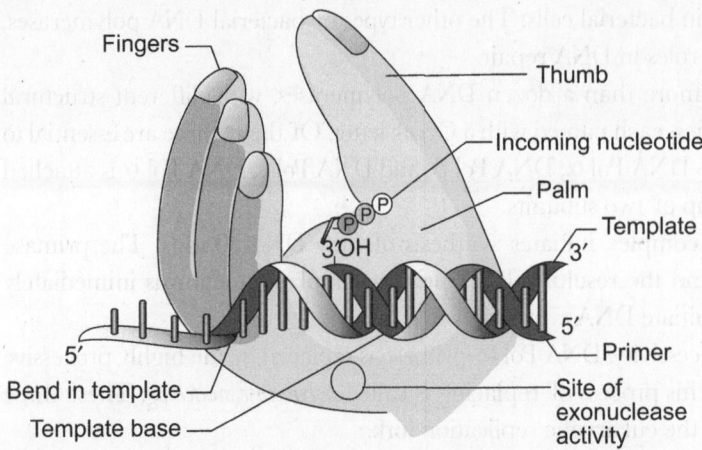

Fig. 3.21   Three-dimensional structure of DNA polymerase with its functional sites.

Sliding DNA clamp increases polymerase processivity

The high processivity of the DNA polymerases acting at replication forks is possible because of their close association with proteins called *sliding DNA clamps*. Each of these proteins is composed of multiple identical subunits that assemble in the shape of a circular ring with a central hole. The hole encircles the DNA double helix and enables the protein to slide easily along the newly synthesized double-stranded DNA. DNA polymerase, tightly bound to the sliding DNA clamp, catalyses at the primer–template junction (Fig. 3.22).

A special class of protein complexes, called sliding clamp loaders, catalyses the opening and placement of sliding clamps on the DNA in an ATP-dependent manner. In *E. coli*, the clamp loader is called the δ-complex, and in eukaryotic cells it is called replication factor C (RF-C). In eukaryotic cells, polymerase α bound to primase initiates DNA replication. But soon polymerase δ and ε replace Pol α during polymerase switching. Pol δ and ε get associated with sliding DNA clamps.

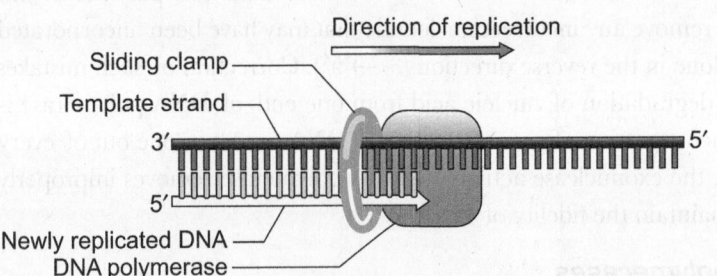

Fig. 3.22   A sliding DNA clamp in association with DNA polymerase.

Relative motion of DNA and the replication machinery—two hypotheses

There are two views regarding the relative motion of DNA and the replication machinery in the course of DNA replication.

All the different proteins involved in DNA replication (helicase, primase, polymerase, and sliding clamps) remain closely associated in a large complex often called a *replisome* that is about the size of a ribosome. The replisome's activity and movement are powered by the hydrolysis of the nucleoside triphosphates used for DNA and RNA primer synthesis.

One traditional view is that the replication machinery (replisome) moves from the replication fork along the DNA like a train advancing on its track. An alternate view suggests that the replisome remains static and the replicating DNA passes through it, similar to film moving through a movie projector. This hypothesis is known as 'replication factory'. The replisome remains static like a factory and the DNA is fed through it for replication. This hypothesis is being increasingly favoured, with more and more experimental evidence regarding the nature of DNA replication, both in prokaryotes and eukaryotes.

Table 3.1 Proteins that function at the replication fork at the beginning of replication in different organisms

Proteins	E. coli (Bacteria)	S. cerevisiae (Yeast)	Human
DNA helicase	DnaB	Mcm complex	Mcm complex
Topoisomerases	Gyrase, Topo I	Topo I, II	Topo I, II
SSB	SSB	RPA	RPA
Primase	DnaG	Primase	Primase

Recapitulation of DNA replication

- *DNA replication* is essentially the opening up of a double helix for synthesis of a new strand on an old template strand in a semiconservative replication process that results in the formation of two double helix DNA molecules.

- *Origin of replication* is the point of initiation of DNA replication. Usually, there is one point of origin of replication in prokaryotes and multiple points in eukaryotic cells (multiple replicons). DNA synthesis proceeds in both $3' \rightarrow 5'$ and $5' \rightarrow 3'$ directions from the origin of replication.

- *The molecular machinery* at the replication fork sequentially adds the different proteins. At the onset, helicase, topoisomerase, single strand DNA-binding proteins (SSB), and primase are required; helicase, to open up the turns of DNA helix, topoisomerase, to prevent supercoiling, SSB, to keep the single strand in the unwound state, and primase to synthesize RNA primer. The proteins are named differently in different organisms (Table 3.1), though the basic process of DNA replication is almost similar in these organisms.

- *RNA primer*, a small RNA piece of about 3–10 nucleotides, is synthesized by primase. This provides the $3'$ OH end for the DNA polymerase to initiate its function.

- *DNA polymerase* is next to be recruited in the replication machinery or replisome. It binds at the junction of the RNA primer and $3' \rightarrow 5'$ template strand and adds DNA nucleotides initially to the $3'$ OH of the primer and subsequently then to the $3'$ OH of the nucleotides recruited in the previous step. DNA polymerase thus synthesizes DNA in the $5' \rightarrow 3'$ direction on the $3' \rightarrow 5'$ template strand.

- *Multiple DNA polymerases* exist in prokaryotes and eukaryotes. Pol III is the main enzyme involved in DNA replication in bacterial cells. DNA Pol initiates DNA synthesis in eukaryotic cells; soon the highly processive DNA polymerases and take over the function of replication (polymerase switching).

- *Ring-like sliding DNA clamp protein* associate with the DNA polymerases to increase polymerase processivity.

- *Okazaki fragments* are formed on the template strand of DNA that is in $5' \rightarrow 3'$ orientation. As soon as the double helix is opened up, DNA polymerase acts in its functional orientation in the $5' \to 3'$ direction to produce the new DNA strand in segmented forms, which are joined later (after removal of RNA primer) by ligase into a continuous DNA strand. This strand is also called the lagging strand.

- *RNA primers are removed* by DNA polymerase I in bacteria and the polymerase adds nucleotides in that space. The gaps on two sides of the small DNA pieces are joined together by ligase. Exonuclease activity of DNA polymerase in eukaryotic cells removes the RNA primer, and DNA nucleotides are replaced by regular synthetic activity of the polymerase.

- After removal of the RNA primer, a short region at the ends of the linear DNA molecule in a eukaryote chromosome remains unreplicated (Fig. 3.20). This is called the *end replication problem*. Telomerase, a special DNA polymerase, extends the ssDNA at the $3'$ end and DNA polymerase synthesizes the complementary part on the other strand in the regular fashion, and thus, the loss of the terminal end is restored.

3.5 MALIGNANCY

The earlier pages have revealed that the whole process of cell division is finely regulated by the timely activities of different specific enzymes, factors, and structural elements in different phases. In addition, a vast number of activities are coordinated in the lifetime of every cell. Thus, it is not surprising that malfunctions occasionally arise in a cell. Sometimes malfunctions may lead to malignancy or cancer. The origin of the malfunction often lies in the defects of genes which are usually referred to as *mutations*. For example, protein kinases (enzymes) and cyclin proteins control progression of cells through the cell cycle; a mutation in the genes encoding these proteins could disrupt the cell cycle in such a way that control over the cell cycle is lost and rapid proliferation of cells lead to cancer. Mutation in genes for growth factors or receptors may lead to cancer.

Enormous progress towards an understanding of the molecular and genetic defects leading to cancer has been made in recent years. It is not yet complete, but has already helped in the diagnosis and curing of cancer patients, particularly when the patients are diagnosed in the early stage of the disease.

SUMMARY

- Chromosome in prokaryotes (bacteria) is a circular DNA molecule in the cytoplasm, without much association of proteins and without any nuclear envelope. Besides the chromosome, self-replicating small circular DNA molecules known as *plasmids* exist in the cytoplasm of prokaryotes.

- Usually several chromosomes are present within the nuclear membrane of eukaryote cells. Each chromosome consists of a DNA molecule and proteins (histones and non-histones). A given region of DNA with its associated proteins is called chromatin. An octamer of histone proteins wrapped by the DNA molecule constitute a *nucleosome*.

- Precise definition of a gene has changed over the years since Mendel's time. Currently, a gene is defined as a functional unit of DNA that either codes for the amino acid sequence of one polypeptide chain through mRNA or codes for one of the several types of RNA other than mRNA that have different functions.

- *Genome* of an organism is the DNA associated with one haploid set of chromosomes.

- The grand success of sequencing the *human genome* of 3.2 billion bases was achieved by April 2003 with the participation of scientists from different countries and a commercial company Celera Genomics.

- Replication of a chromosome is initiated at a point called the *origin of replication*. In contrast to a single origin of replication in the circular chromosome of prokaryotes, the linear DNA molecule of a eukaryotic chromosome has multiple origins of replication. In general, origins of replication are located in the non-coding regions of DNA.

- For initiation of the replication process, a specific group of initiator proteins binds to the origin and utilizing ATP-derived energy, unwinds the double helix to provide DNA polymerase and other proteins an access to single-stranded DNA.

- DNA replication proceeds from the origin bidirectionally. DNA polymerase can synthesize a new continuous strand of DNA (leading strand) in the $5' \rightarrow 3'$ direction on the template strand in the $3' \rightarrow 5'$ orientation (Fig. 3.18). On the $5' \rightarrow 3'$ template strand, the DNA polymerase enzymes act in reverse orientation synthesizing short segments, called *Okazaki fragments*, joined later by the enzyme ligase to produce the *lagging strand*.

- *End replication problem*, i.e., shortening of one newly synthesized DNA strand at the end of a replication cycle, takes place in eukaryotic linear chromosomes. This occurs at the telomere (end) region of a chromosome. Telomerase enzyme specifically helps in re-synthesizing the lost part of DNA.

- Malfunctions in replication or other cellular activitiess occasionally arise in a cell leading to mutations which in turn may cause malignancy.

Watson and Crick

James Dewey Watson (born 6 April 1928)
James D. Watson is an American zoologist, geneticist, and molecular biologist. He is the co-discoverer of the structure of the DNA molecule in 1953 with Francis Crick. Watson and Crick together with Maurice Wilkins shared the Nobel Prize in Physiology or Medicine in 1962 'for their discoveries concerning the molecular structure of nucleic acids and its significance for information transfer in living material'.

Watson was born in Chicago, Illionis, attended Horace Mann Grammar School for eight years and South Shore High School for two years. He was a precocious student and entered the University of Chicago when he was only 15. He obtained a BS degree in Zoology from the University of Chicago in 1947. After reading Erwin Schrödinger's book *What is Life?* in 1946, Watson changed his professional ambitions from the study of ornithology to genetics. He took admission as a graduate student in Indiana University, Bloomington.

At Indiana, he was deeply influenced by the geneticists H.J. Müller, Tracy M. Sonneborn, and Salvador E. Luria. Luria was an Italian born microbiologist in the bacteriology department at Indiana. Watson became the first PhD student under Luria from early 1948. In the summer of that year Watson made his first trip to Cold Spring Harbor Laboratory (CSHL) on Long Island, New York and met the German physicist-turned-geneticist Max Delbrück who had Luria as collaborator in his laboratory. Max Delbrück shared the Nobel Prize with Alfred D. Hershey and Luria in Physiology or Medicine in 1969 'for their discoveries concerning the replication mechanism and the genetic structure of viruses'.

Watson was awarded a PhD degree in 1950, at the age of 22. As a Merck Fellow of the National Research Council, USA, Watson joined Copenhagen University in September, 1950 for a year of postdoctoral research in the laboratory of the biochemist Herman Kalckar who was interested in the enzymatic synthesis of nucleic acid using phages as the experimental system. During the spring of 1951, he accompanied Kalckar to the Zoological Station at Naples, Italy; there, at a symposium Maurice Wilkins talked about his X-ray diffraction data for DNA. Watson now became certain that DNA had a definite molecular structure that could be deciphered. In 1951, the chemist Linus Pauling in California published the alpha helix model of protein structure. All this exposure motivated Watson to perform an X-ray diffraction experiment to determine the structure of DNA. Fortunately for him, that summer Luria met John Kendrew and arranged for a new postdoctoral research project for Watson at Cavendish Laboratory, Cambridge, where he started work in early October 1951. He soon met Francis Crick at Cavendish and both realized their common interest in solving the structure of DNA.

Watson and Crick talked endlessly and tried to guess the structure of DNA on the basis of X-ray diffraction results of Raymond Gosling and Rosalind Franklin of King's College, London. Rosalind Franklin shared the X-ray diffraction facilities in Kings College with Maurice Wilkins and produced some excellent X-ray diffraction photographs of DNA crystals and systematically analysed DNA's structural features. Her meticulous findings were useful in guiding Watson and Crick towards a correct molecular model of DNA.

At the same time Watson was also investigating the structure of tobacco mosaic virus (TMV), using the X-ray diffraction technique in order to determine whether its chemical subunits, earlier revealed by the elegant experiments of Schramm, were helically arranged. In late June 1952, the use of the Cavendish Laboratory's newly constructed rotating anode X-ray tube indeed revealed an unambiguous helical construction of the virus.

Watson and Crick deduced the double helix structure of DNA in 1953; their paper was published in *Nature* on 25 April 1953.

From 1953 to 1955, Watson was associated with California Institute of Technology as a Senior Research Fellow in Biology. He was back in Cavendish during between 1955 and 1956, again working with Crick and published several papers on the general principles of virus construction.

In the fall session of 1956, he joined Harvard University as assistant professor of biology and became associate professor in 1958, and professor in 1961. During this period, his main research interest was the role of RNA in protein synthesis. Watson became the Director of Cold Spring Harbor Laboratory (CSHL) in 1968, and served the Laboratory for the next 35 years. He successively became the President and the Chancellor of the Laboratory. He was invited by the National Institutes of Health (NIH) and appointed as the Head of the Human Genome Project in 1988. Watson left the Genome Project in 1992, having seen it off with a successful start and continued at CSHL.

He published numerous books, among which the *Molecular Biology of Gene*, *Molecular Biology of the Cell*, and *Recombinant* DNA are widely used as textbooks throughout the world; his bestselling book is *The Double Helix* (1968). His memoir is titled *Genes*, *Girls* and *Gamow* (2003).

Francis Harry Compton Crick
(June 1916–28 July 2004)

Francis H.C. Crick was an English physicist turned biophysicist, molecular biologist, and neuroscientist. The brilliant mind of Crick contributed significantly to formulating the model structure of DNA with J.D. Watson. This was even before he formally became a PhD holder. He was co-recipient of the Nobel Prize with Watson and Wilkins in 1962.

He was born in a small village near Northampton, England and went to Northampton Grammar School and Mill Hill School, London. He studied physics at the University College, London, obtained a BSc in 1937, and started research for a PhD under Professor E.N. da Costa Andrade on a project measuring the viscosity of water at high temperature. But his PhD work was interrupted by the outbreak of World War II in 1939. During the War he worked as a scientist for the British Admiralty, mainly in connection with magnetic and acoustic mines. He left the Admiralty in 1947 to study biology. Crick was interested in two fundamental unsolved problems of biology. One was, the transition of molecules from the non-living to the living, and the second was, how the brain makes a conscious mind.

Crick went to Cambridge and worked at the Strangeways Research Laboratory, supported by a studentship from the Medical Research Council and some financial support from his family. In 1949 he joined the Medical Research Council unit headed by Max F. Perutz, of which he became a member ever since. The unit was housed for many years in the Cavendish Laboratory of Cambridge, and in 1962 moved into a large new building—the Medical Research Council Laboratory of Molecular Biology. He became a research student for the second time in 1950, being accepted as a member of Caius College, Cambridge. Crick had to adjust from the 'elegance and deep simplicity' of physics to the 'elaborate chemical mechanisms that natural selection had evolved over billions of years'. Crick described this transition as 'almost as if one had to be born again'. He realized that X-ray crystallography offered the opportunity to reveal the molecular structure of large biological molecules like proteins and DNA. At this juncture his friendship with 23 year-old Watson began on his arrival at Cavendish in 1951. This had a critical influence on Crick's career. Crick studied the mathematical theory of X-ray crystallography in great detail. This helped him contribute enormously in building the DNA model. Watson and Crick were not officially working on DNA. They had separate problems to work on. Watson was supposed to obtain myoglobin crystals for X-ray diffraction experiments.

Crick obtained his PhD in 1954 on a thesis entitled 'X-ray Diffraction: Polypeptides and Proteins'. During the academic year 1953–54 Crick was on leave of absence at the Protein Structure Project of the Brooklyn Polytechnic, New York. During this stint he also delivered lectures at Harvard University, as a visiting professor, on two occasions. He was made an FRS (Fellow of Royal Society) in 1959. He left Cambridge in 1977 after a long association of 30 years. Eventually in the 1980s Crick devoted time to his second interest—consciousness—at the Salk Institute for Biological Studies in La Jolla, California. Two books by him are often mentioned, *Of Molecules and Men*, and his autobiography *What Mad Pursuit*.

Francis Crick and James Watson received many honorary awards and fellowships from universities and academic bodies and also from different governments for their *astounding discovery of molecular structure of DNA*, possibly the greatest discovery of the 20 th century in biology. For its far-reaching implications in biological world this discovery stands next to Darwin's Natural Selection.

The Francis Crick Institute is a planned £ 660,000,000 research centre to be located at London, with partnership among Cancer Research UK, Imperial College London, King's College London, the Medical Research Council, University College London, and Wellcome Trust. Once completed in 2015, it will be the biggest centre for biomedical research and innovation in Europe.

EXERCISES

Objective Questions

1. Fill in the blanks in the following sentences.
   (a) Nuclein from salmon sperms and human pus cells was discovered by _____.
   (b) Biochemical modification for histone takes place at _____.
   (c) 'One gene-one enzyme' concept was formulated by _____.
   (d) Gigabase refers to _____ base pairs.
2. Mark the following statements as True or False.
   (a) High number of chromosomes allow higher metabolic rate of a megakaryocyte.
   (b) Centromere comprises many repetitive DNA sequences.
   (c) Telomere is present in the chromosome of prokaryotes.
   (d) A dimer of each of the histones H1, H2, H3, and H4 constitutes the octamer histone core of nucleosomes.
   (e) At the time of replication of DNA molecules in eukaryotic cells, all the histones for new nucleosomes are newly synthesized.
   (f) Introns are present in DNA from all sources— virus, bacteria, and eukaryotes.
   (g) Introns are transcribed into the primary RNA transcript.
   (h) Rice genome is about forty times smaller than that of wheat.

(i)   Chromosomes of living organisms are contained within the membrane-bound nucleus.

(j)   Point of an origin of replication is AT base pair rich.

(k)   M phase is the shortest phase in the cell cycle.

(l)   Eukaryotic cells have only three types of DNA polymerase.

(m)   Synthesis of DNA is slower in eukaryotes than in bacteria.

(n)   Two copies of histone H1 interact with the linker DNA.

(o)   Telomerase is a special kind of DNA polymerase.

Review Questions

1. What are plasmids?

2. What is core DNA?

3. What are nucleosomes? Write a note on their contribution in DNA compaction in chromosomes.

4. Give the most recent definition of a gene.

5. Among viruses, prokaryotes, and eukaryotes, where do you think the density of genes per given length of DNA is highest?

6. Write short notes on intergenic sequences and introns.

7. What is a satellite DNA?

8. How do trasnsposons function?

9. Define a genome.

10. Mention at least two functions of the histone N-terminal tails.

11. What do you know about the Human Genome Project?

12. Differentiate centromere from kinetochore.

13. Elucidate the functions of telomerase.

14. What is semi-conservative replication of DNA? Who were the scientists to prove this mode of DNA replication? And what experiment did they perform to prove it?

15. What is the function of topoisomerase?

16. How does a replicon initiate DNA synthesis?

17. How does unwinding of the DNA double helix begin?

18. Name the single-stranded DNA-binding protein in bacteria and eukaryotes.

19. What are Okazaki fragments? Why are they formed?

20. Write brief notes on the following:
    (a)   RNA primer
    (b)   Removal of RNA primer
    (c)   DNA polymerase
    (d)   Proofreading activity of DNA polymerase

21. What is the end replication problem for linear DNA molecules in eukaryotes?

Biomolecules

LEARNING OBJECTIVES

♦ Composition of biomolecules
♦ Carbon backbone of biomolecules
♦ Monomeric subunits that constitute biological macromolecules
♦ Functional groups on biomolecules
♦ Biomolecules—carbohydrates, proteins, lipids, and nucleic acids

INTRODUCTION

Antoine Lavoisier (1743–1794) noted the chemical complexity of 'plant and animal worlds' in relation to the simplicity of the 'mineral world'. By the early 19th century, chemists realized the striking difference in the composition of living matter when compared to matter from the inanimate world. This realization ushered the development of organic chemistry, which in turn provided valuable insights for the development of *biochemistry*. During the 20th century, biochemistry developed fast, and gradually deciphered the structure and functions of molecules constituting living organisms; the blanket term biomolecules is used to describe these molecules.

4.1 FUNDAMENTAL CONCEPTS OF BIOMOLECULES

Only about 30 out of more than 100 naturally occurring chemical elements constitute biomolecules or living matter. Biologically used elements are of low atomic number and hence of low atomic weight (Table 4.1, also consult periodic table). Hydrogen, oxygen, nitrogen, and carbon together make up over 99 per cent of the total atoms (mass) of most cells. These elements form one, two, three, and four bonds, respectively (Fig. 4.1). Usually, the lightest elements are capable of forming the strongest covalent bonds. Essential and trace elements for living matter are listed in Table 4.1.

Atom of element (Electrons in outer shell shown as dots)	Number of unpaired electrons in outer shell	Number of electrons in complete outer shell	Formation of covalent bond (Example)
H·	1	2	H·+ H· → H:H = H—H Dihydrogen
:Ö·	2	8	2H·+ :Ö: → H:Ö: = H—O with H (Water)
:N̈·	3	8	:N̈· + 3H· → :N̈: H = N–H with H, H (Ammonia)
·C̈·	4	8	
:S̈·	2	8	
:P̈·	3	8	

Fig. 4.1 Covalent bond formation. Two atoms with unpaired electrons in their outer shell tend to achieve the required electron number in the outer shell by sharing electrons. The sharing of electrons forms covalent bonds between the atoms.

Table 4.1 Essential and trace elements, with atomic number in parenthesis, required for living matter

Essential elements (Atomic number)	Trace elements
H (1)	Mg (12)
C (6)	V (23)
N (7)	Mn (25)
O (8)	Fe (26)
Na (11)	Co (27)
P (15)	Ni (28)
S (16)	Cu (29)
Cl (17)	Zn (30)
K (19)	Se (34)
Ca (20)	Mo (42)
	I (53)

The ultimate sources of the elements for living matter are water, the earth's crust, and the atmosphere. Interestingly, the six most abundant elements in the human body are also among the nine most abundant elements in sea water, indicating the origin of life in the sea at a very early stage in evolution.

4.1.1 Carbon—The Backbone of Biomolecules

Carbon atoms share electron pairs with each other to form very stable carbon–carbon single covalent bonds. One carbon atom can join with another carbon atom by forming single, double or triple covalent bonds. Covalently linked carbon atoms can form linear chains, branched chains, and cyclic structures. Groups of other atoms known as *functional groups* are added to these different carbon skeletons. These functional groups attribute specific chemical properties to the molecule (Fig. 4.2). The bonding versatility of carbon is viewed as the main factor in its selection as the structural backbone of biomolecules. *Molecules with covalently bonded carbon backbones are called organic compounds*; they occur in an almost limitless variety.

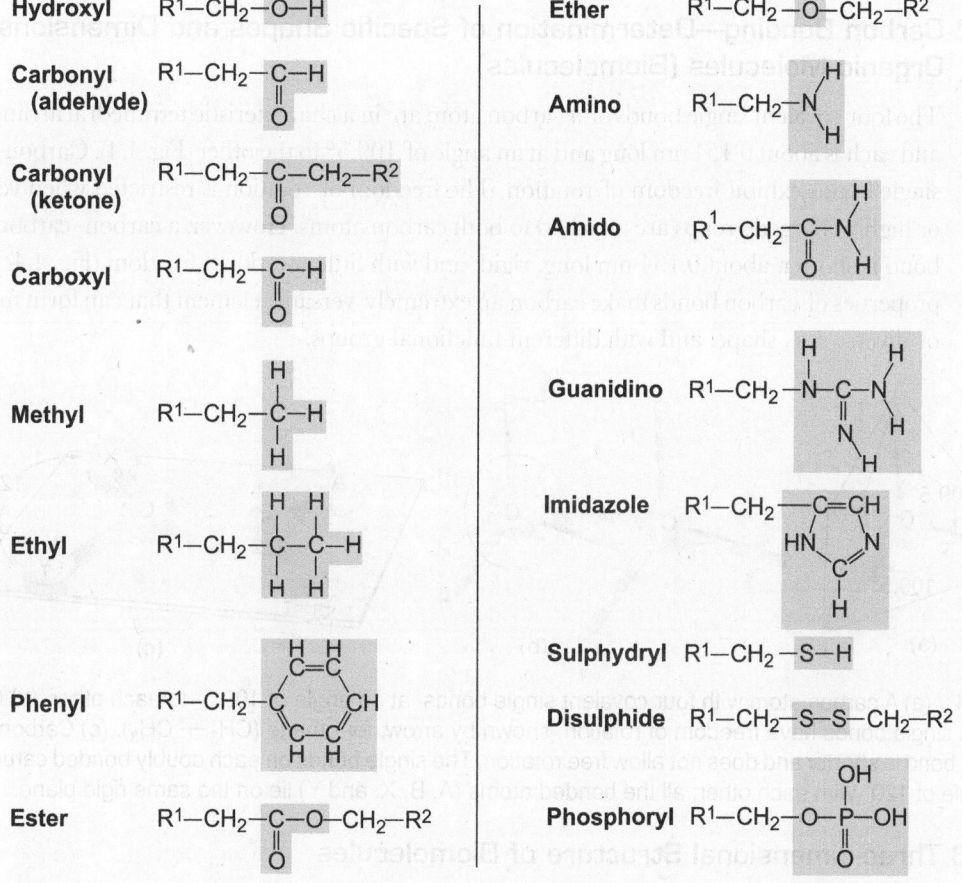

Fig. 4.2 Functional groups (shaded) of some biomolecules. All groups are in the uncharged (un-ionized) form.

To begin with, most biomolecules are derived from hydrocarbons, that is, molecules in which only hydrogen atoms are bonded to a backbone of covalently linked carbons. A variety of functional groups then replace the hydrogen atoms to produce different families of organic compounds (Fig. 4.2). Many biomolecules are polyfunctional, that is, they contain two or more different types of functional groups (Fig. 4.3).

COOH Carboxyl

H₂N—C—H
Amino |
        CH₂
        |
        C—NH
        ‖   \
        HC—N CH

Histidine
(Amino acid)

Methyl CH₃ H OH
        \ | |
         N—C—C—C
        / | | \
s-amino H   H   H

H   H
 \ /
  C=C   Phenyl
 /   \
C   C—OH
 \   /
  C—C   Hydroxyls
 /   \
H   OH

Epinephrine
(Hormone)

Fig. 4.3   Polyfunctional biomolecules with multiple functional groups (shaded).

4.1.2 Carbon Bonding—Determination of Specific Shapes and Dimensions of Organic Molecules (Biomolecules)

The four covalent single bonds of a carbon atom are in a characteristic tetrahedral arrangement, and each is about 0.154 nm long and at an angle of 109.5° to the other (Fig. 4.4). Carbon–carbon single bonds exhibit freedom of rotation. The freedom of rotation is restricted when very large or highly charged groups are attached to both carbon atoms. However, a carbon–carbon double bond is shorter, about 0.134 nm long, rigid, and with little rotational freedom (Fig. 4.4c). These properties of carbon bonds make carbon an extremely versatile element that can form molecules of diverse size, shape, and with different functional groups.

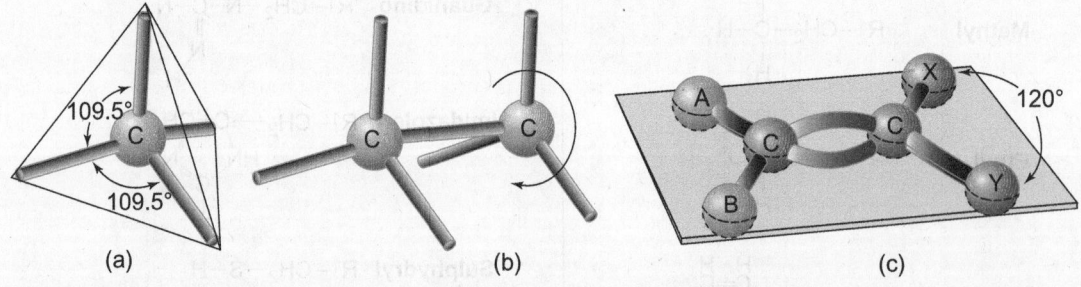

(a)   (b)   (c)

Fig. 4.4   (a) A carbon atom with four covalent single bonds, at an angle of 109.5° to each other. (b) Carbon–carbon single bonds have freedom of rotation, shown by arrow, for ethane (CH₃ — CH₃). (c) Carbon–carbon double bond is shorter and does not allow free rotation. The single bonds on each doubly bonded carbon make an angle of 120° with each other; all the bonded atoms (A, B, X, and Y) lie on the same rigid plane.

4.1.3 Three-dimensional Structure of Biomolecules

The tetrahedral covalent bonds of carbon atoms and the addition of other atoms and functional groups give a biomolecule a three-dimensional structure. The three-dimensional conformation is important when two biomolecules interact. They need to fit each other closely, in a complimentary fashion. Such complimentarity is necessary in binding such as enzyme–substrate, antigen–antibody, and hormone receptor–hormone. The technique of X-ray crystallography (diffraction) provides precise information regarding the arrangement of atoms in the three-dimensional structure of biomolecules.

When four different atoms or functional groups are bonded to a carbon atom in an organic molecule, the carbon atom is said to be asymmetric. It produces two different isomeric forms (stereoisomers) known as laevorotatory (L) and dextrorotatory (D) on the basis of their ability to rotate polarized light to the left (L) or right (D), respectively. Although both the isomers have similar chemical properties, only one form exists in living organisms. For example, amino acids occur in proteins only in L isomeric forms and glucose in starch occurs as the D isomer.

4.1.4 Chemical Reactivity of Organic Molecules

Carbon–carbon single bonds, without any double bond or substituent groups, constitute saturated hydrocarbon molecules. These hydrocarbons do not easily reacte with chemical reagents. On the other hand, biomolecules with their various functional groups attached to the carbon backbone (Figs 4.2 and 4.3) are much more chemically reactive. This is so because the functional groups alter the electron distribution and the geometry of atoms of neighbouring molecules, and thus, affect chemical reactivity. The breakage and formation of chemical bonds that takes place when biomolecules react with chemical reagents or with other biomolecules, releases energy, some of which is in the form of heat.

Biochemical reaction mechanisms are not very different from other chemical reactions. Biochemical reactions often involve interactions between *nucleophiles* and *electrophiles*. *Nucleophiles are electron-rich functional groups* and contribute electrons in a reaction, whereas *electrophiles are electron deficient functional groups that receive electrons* in a reaction. Functional groups containing oxygen, nitrogen, and sulphur are important biological nucleophiles, whereas positively charged hydrogen atoms (protons) and metal ions (cations) usually act as electrophiles in biochemical reactions. A carbon atom can act as either a nucleophilic or an electrophilic centre, depending on the type of bonds and functional groups that surround it.

4.1.5 Monomeric Subunits—Constituents of Biological Macromolecules

Several atoms of hydrogen, oxygen, carbon, and nitrogen make up simple biological molecules like glucose, amino acids, fatty acids, and nucleosides that have molecular weights in the range of hundreds of daltons. These simpler units are considered as monomers; several of these monomers of the same type are polymerized to larger molecules weighing in the range of tens of thousands to billions of Daltons (DNA). These larger molecules are called macromolecules, and basically, they are of four types—polysaccharides, proteins, lipids, and nucleic acids. These are the fundamental materials that make up cellular moieties.

The synthesis of macromolecules is a major energy-consuming activity of living cells. Macromolecules can further be assembled into supramolecular structures such as ribosomes, membranes, and other cell organelles.

4.2 CARBOHYDRATES

Carbohydrates can be described as the most abundant biomolecules on the earth and the principal source of energy for the living systems that provide many other essential functions including structural and protective elements in the cell walls of bacteria and plants as well as in the connective tissues and cell coats of animals. They are produced by the photosynthetic activity of plants and algae that involves more than 100 billion tonnes of CO_2 and H_2O, and sunlight. In this process, solar energy is trapped and used for the formation of bonds in carbohydrates.

Carbohydrate means *carbon hydrated*, that is, carbon with water. The ratio of C:H:O is 1:2:1 in carbohydrates. For example, glucose, a simple carbohydrate, has the empirical formula $C_6H_{12}O_6$ which can also be expressed as $(CH_2O)_6$ or $C_6(H_2O)_6$. Thus, carbohydrates can be represented by the general empirical formula $(CH_2O)n$.

4.2.1 Classes of Carbohydrates

Carbohydrates are grouped into three major classes on the basis of the number of basic unit structures constituting them.

Monosaccharides   These are simple sugars and consist of a single unit of polyhydroxy aldehyde or ketone (explained later). The most abundant monosaccharide in nature is the six-carbon sugar D-glucose (Fig. 4.5a).

> The word 'saccharide' is derived from the Greek word *Sakkharon*, meaning sugar. Interestingly, the very same word *sakkhar* means sugar in Sanskrit as well.

Oligosaccharides   These are monosaccharide units covalently joined by characteristic glycosidic bonds into a short chain. Two monosaccharides are joined to form a disaccharide, the most abundant form of oligosaccharides. A typical example is sucrose or cane sugar, consisting of the six-carbon sugars D-glucose and D-fructose (Fig. 4.5b). The other two common disaccharides are lactose and maltose. The names of monosaccharides and disaccharides end with the suffix 'ose'.

Oligosaccharides consisting three or more units do not exist as free entities. They are conjugated to lipids or proteins as glycolipids or glycoproteins, which are present in the cell membrane. Most proteins secreted by eukaryotic cells are glycoproteins.

Polysaccharides   These are long chains of hundreds or thousands of monosaccharide units. Cellulose is a linear chain polysaccharide, whereas glycogen is a branched chain polysaccharide. The most abundant polysaccharides, cellulose and starch in plants, have recurring units of D-glucose (they differ in the nature of glycosidic linkage, Fig. 4.5c).

Monosaccharides

Monosaccharides are colourless, crystalline solids, freely soluble in water but insoluble in non-polar solvents. Most of them are sweet to taste.

Monosaccharides have an unbranched carbon chain backbone, in which carbon atoms are linked by single bonds. One of the carbon atoms is linked to an oxygen by a double bond forming a *carbonyl group*; each of the other carbon atoms is linked to a hydroxyl group (OH). Depending on the position of the carbonyl group, there are two families of monosaccharides:

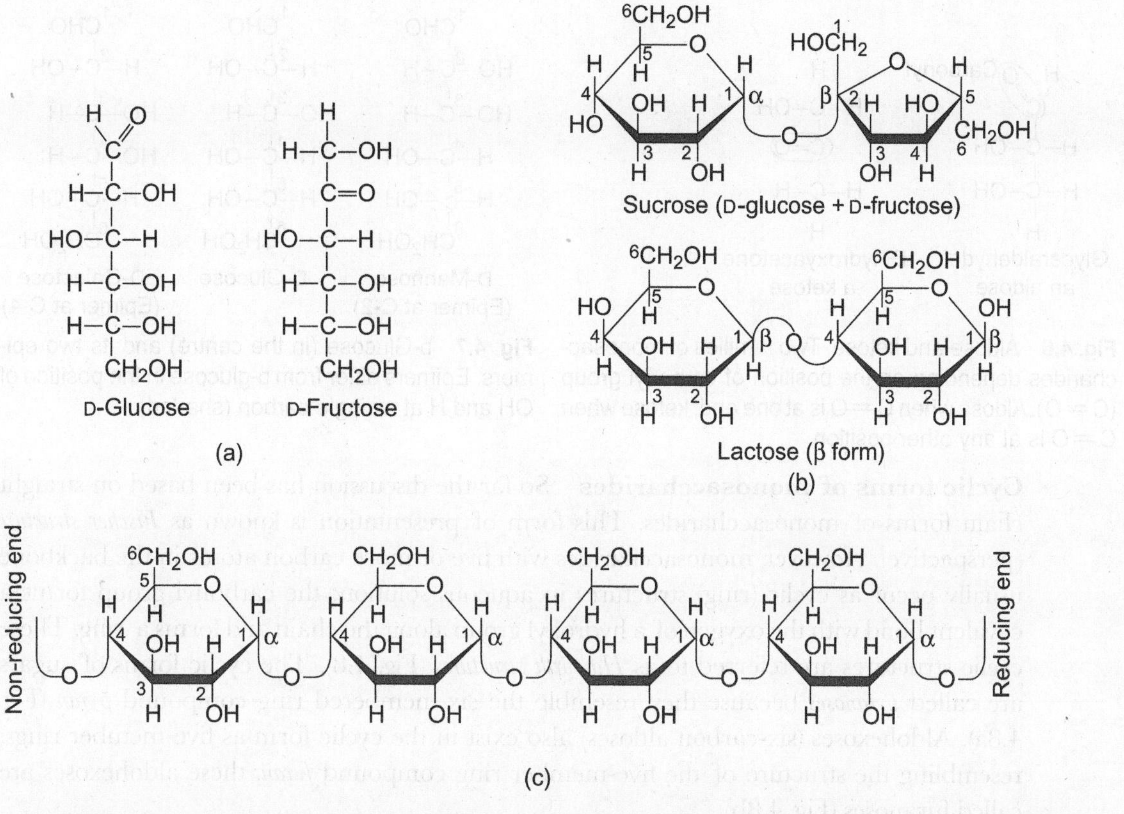

Sucrose (D-glucose + D-fructose)

Lactose (β form)

(b)

(a)

Non-reducing end

Reducing end

(c)

Fig. 4.5 (a) Monosaccharides: Six-carbon sugar molecules, D-glucose and D-fructose. **(b)** Disaccharides: Two monosaccharides joined by a *glycosidic bond*; examples—sucrose, lactose, and maltose. **(c)** Polysaccharide: Amylose (starch) is a linear polymer of D-glucose units formed by glycosidic bonds between C-1 and C-4.

Aldose   When the carbonyl group (C=O) is at one end of the carbon chain, the monosaccharide is an aldehyde and called an aldose (e.g., glucose).

Ketose   When the carbonyl group is at any other position, the monosaccharide becomes a ketone and is called a *ketose* (e.g., fructose).

The simplest monosaccharides are trioses comprising two or three carbon atoms; for example, *glyceraldehyde*, an aldose, and *dihydroxyacetone*, a ketose (Fig. 4.6).

Monosaccharides with four, five, six, and seven carbon atoms in their backbones are known as tetroses, pentoses, hexoses, and heptoses, respectively. Each of these varieties can be either aldoses or ketoses.

Epimers   When two monosaccharides differ only in the configuration around one carbon atom, they are called *epimers* of each other (the configurational change is due to the difference in the position of OH and H attached to a carbon.) In case of D-glucose and D-mannose, this stereochemical difference is at carbon-2 (C-2), and they are epimers. Similarly, the stereochemical difference for D-glucose and D-galactose, which are also epimers is at C-4 (Fig. 4.7).

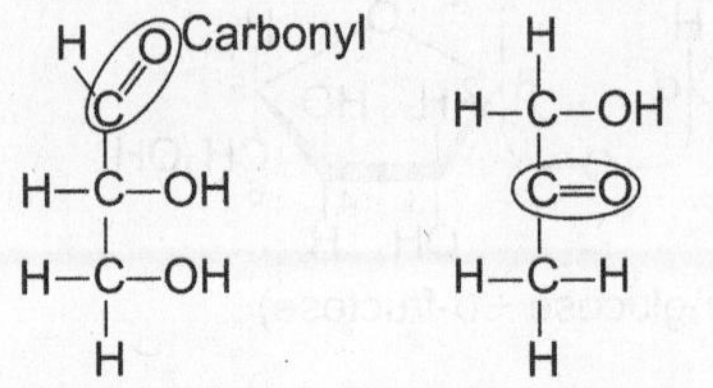

Glyceraldehyde, an aldose

Dihydroxyacetone, a ketose

Fig. 4.6 Aldose and ketose. Two families of monosaccharides depending on the position of carbonyl group (C=O). Aldose when C=O is at one end, ketose when C=O is at any other position.

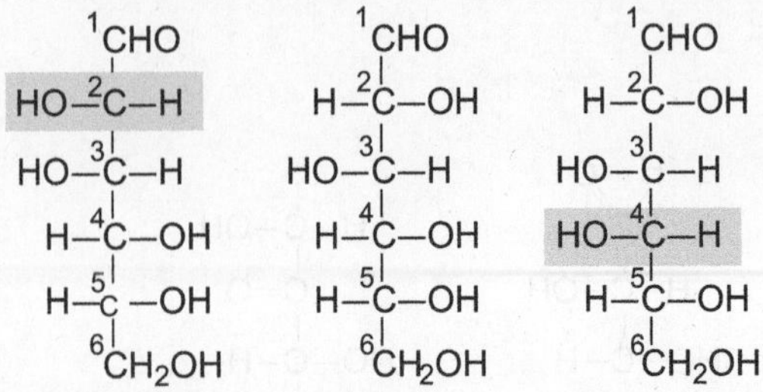

D-Mannose (Epimer at C-2)

D-Glucose

D-Galactose (Epimer at C-4)

Fig. 4.7 D-Glucose (in the centre) and its two epimers. Epimers differ from D-glucose in the position of OH and H at a single carbon (shaded).

Cyclic forms of monosaccharides So far the discussion has been based on straight chain forms of monosaccharides. This form of presentation is known as *Fischer structure* (perspective). However, monosaccharides with five or more carbon atoms in the backbone usually occur as cyclic (ring) structures in aqueous solution; the carbonyl group forms a covalent bond with the oxygen of a hydroxyl group along the chain and forms a ring. These cyclic structures are referred to as *Haworth structures* (Fig. 4.8). The cyclic forms of sugars are called *pyranoses* because they resemble the six-membered ring compound *pyran* (Fig. 4.8a). Aldohexoses (six-carbon aldoses) also exist in the cyclic form as five-member rings, resembling the structure of the five-member ring compound *furan;* these aldohexoses are called furanoses (Fig. 4.8b).

The carbon atom carrying oxygen (carbonyl group) in aldose and ketose is called the *anomeric carbon atom* because on formation of a ring structure, the carbon atom becomes asymmetric, leading to the formation of two configurationally anomeric forms α and β (Fig. 4.8a).

The biological properties and functions of some polysaccharides depend on the specific three-dimensional conformation of the monosaccharide units.

Variety of hexose derivatives

Several derivatives of the simple sugars, glucose, galactose, and mannose, are formed by replacement of a hydroxyl group with another substituent or by the oxidation of a carbon atom to carboxylic acid. The hydroxyl at carbon-2 (C-2) of glucose, galactose, and mannose is replaced with an amino group to produce glucosamine, galactosamine, and mannosamine, respectively (Fig. 4.9a). These are essential components of glycoproteins and glycolipids. The amino group may be condensed with acetic acid to produce N-acetylglucosamine. This glucosamine derivative is a constituent of many structural polymers, such as those found in bacterial cell walls. N-acetylglucosamine can be further modified when the three-carbon carboxylic acid, lactic acid, is ether-linked to the oxygen at C-3, to form N-acetylmuramic acid (Fig. 4.9b) which is also present in bacterial cell walls. The substitution of a hydrogen for the hydroxyl group at C-6 of galactose or mannose produces fucose or rhamnose, respectively (Fig. 4.9c).

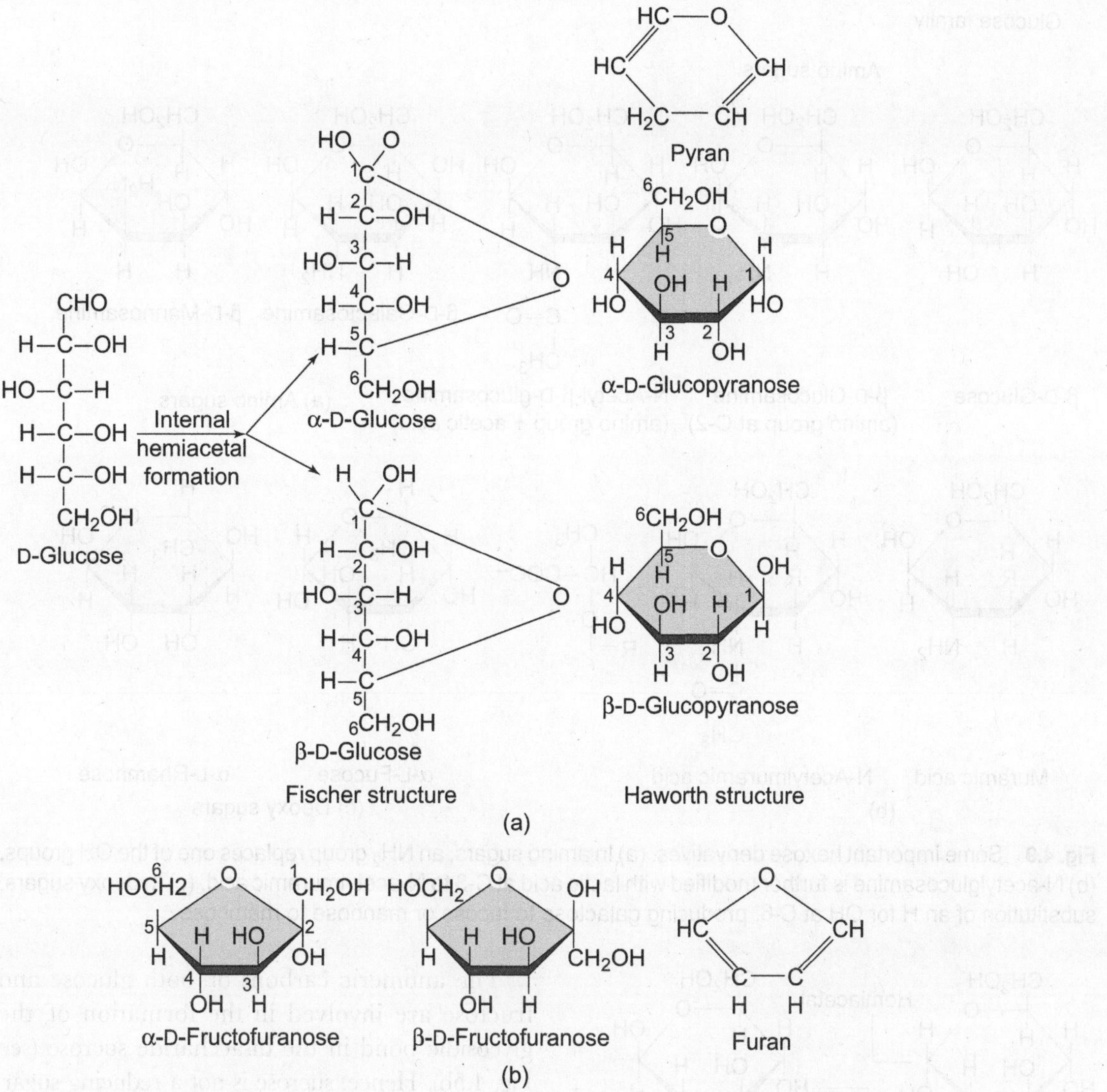

Fig. 4.8 (a) D-Glucose and its Fischer and Haworth structures. Haworth ring-like structure is the result of a reaction between aldehyde at C-1 and hydroxyl at C-5 to form a hemiacetal linkage. Haworth structure resembles a six-membered ring compound *pyran*; so the two anomeric forms of glucose are called glucopyranose. (b) Two anomeric forms of D-Fructofuranose resembling the five-membered ring structure of *furan*.

Monosaccharides as reducing agent

Monosaccharides with a free anomeric α carbon (bearing carbonyl group) are called reducing sugars because they can reduce alkaline solutions or copper salts (Cu^{2+} cupric) to yellow-red precipitates of cuprous oxide. This is the basic reaction in Fehling's and Benedict's test for the detection of reducing sugars which are routinely used for pathological detection of sugar in blood and urine samples of diabetic patients. Monosaccharides can also reduce Fe^{3+} salts.

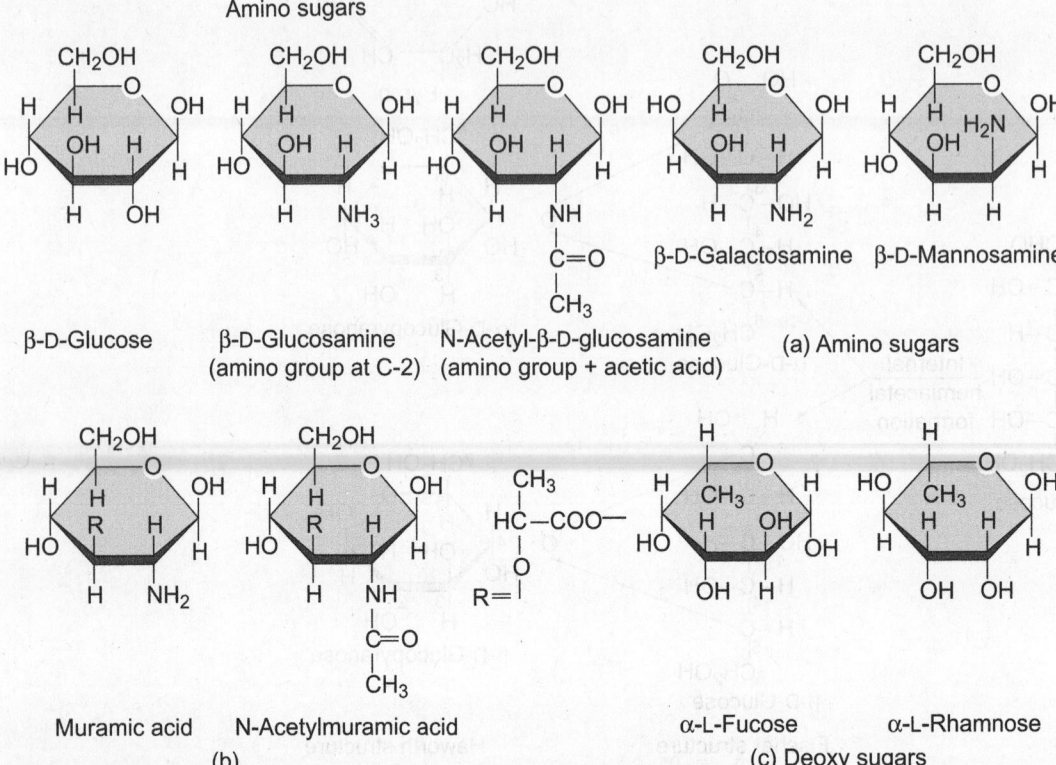

Fig. 4.9 Some important hexose derivatives. (a) In amino sugars, an NH_2 group replaces one of the OH groups. (b) N-acetylglucosamine is further modified with lactic acid at C-3 to N-acetylmuramic acid. (c) In deoxy sugars, substitution of an H for OH at C-6, producing galactose to fucose or mannose to rhamnose.

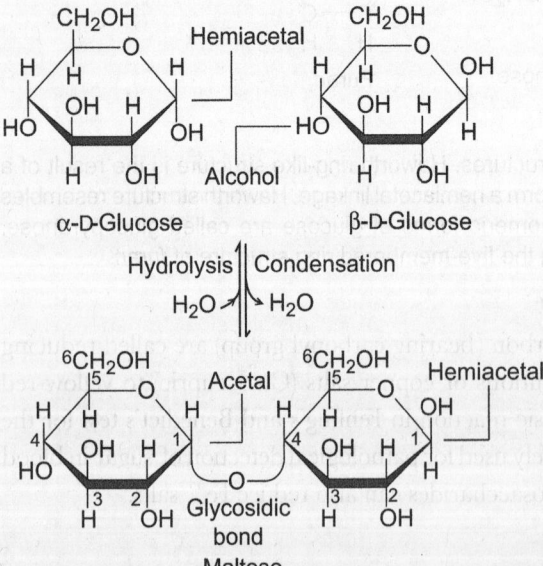

The anomeric carbons of both glucose and fructose are involved in the formation of the glycosidic bond in the disaccharide sucrose (see Fig. 4.5b). Hence, sucrose is not a reducing sugar. A free anomeric carbon at the end of the chain, not involved in the formation of a glycosidic bond in certain disaccharides like maltose and in polysaccharides, denotes the *reducing end* of the chain. The anomeric carbon end is indicated as *hemiacetal* in Fig. 4.10.

Fig. 4.10 Maltose, a disaccharide molecule is formed by joining two monosaccharides by a glycosidic bond. The glycosidic bond is formed by condensation (or removal of a water molecule), and hydrolysis breaks the bond. Maltose molecule retains a reducing hemiacetal at C-1, that is not involved in a glycosidic bond.

Glycosidic Bond

The anomeric carbon of one monosaccharide reacts with a hydroxyl group of another monosaccharide to form a covalent glycosidic bond to link the two monosaccharides (Fig. 4.10). In the process of formation of a glycosidic bond, a water molecule is eliminated. With the addition of more and more monosaccharides to a chain by glycosidic bonds, oligosaccharides (3–20 residues) and polysaccharides (more than 50 residues) are formed.

Glycosidic bonds are readily broken by hydrolysis in the presence of acid. The anomeric carbon of a sugar can form another type of glycosidic bond with a nitrogen atom which is called an N-glycosyl bond; this is present in all nucleotides.

Polysaccharides and proteoglycans

Polysaccharides are polymers of multiple monosaccharides. They have a high molecular weight, and are the most abundant form of carbohydrates in nature. Polysaccharides are often called *glycans*. They can be of two types:

Homopolysaccharides These contain a single type of monomeric unit or monosaccharide (Fig. 4.11a). For example:

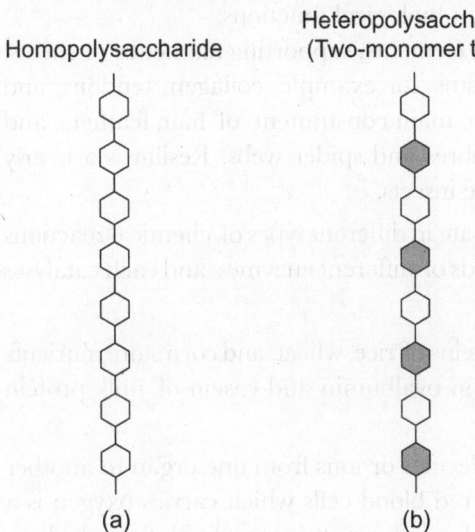

Fig. 4.11 (a) Homopolysaccharide contains a single type of monosaccharide. (b) Heteropoly-saccharide contains two or more different types of monosaccharides.

(a) Starch in plants and glycogen in animals are storage forms of sugars, to be used for energy resource.

(b) Cellulose and chitin serve as structural elements in plant cell walls and animal exoskeleton.

Heteropolysaccharides contain two or more different kinds of monomeric units (Fig. 4.11b). Heteropolysaccharides provide extracellular support and are found in organisms of all kingdoms. For example:

(a) Peptidoglycan which forms the rigid layers of bacterial cell envelopes

(b) Extracellular matrix in animal tissues

(c) Hyaluronic acid that accounts for the toughness and flexibility of cartilage and tendons in animals

(d) Proteoglycans which are large aggregates of heteropolysaccharides with proteins and serve as highly viscous and lubricating extracellular secretions

Synthesis of polysaccharides Each type of monosaccharide has a unique enzyme which adds it to the growing polysaccharide chain, and these enzymes perform this function only when the preceding monosaccharide has been inserted by its specific enzyme. Thus, the alternating action of several specific enzymes produces a polysaccharide with a precisely repeating sequence, but the exact length varies from molecule to molecule within a general range. This is the reason why polysaccharides do not have precise and definite molecular weights, unlike proteins which have defined sequences and lengths.

4.3 PROTEINS AND AMINO ACIDS

The word protein is derived from the Greek word *protos*, meaning 'first' or 'foremost'. There are thousands of varieties of proteins in a cell. Proteins make all kinds of structural elements of a cell, catalyse cellular reactions, and carry out a myriad of other tasks. Their significance in a cell can be appreciated by the fact that genetic information is ultimately expressed as protein.

In 1806, the first amino acid, Asparagine, was isolated from the edible plant asparagus. Threonine, in 1938 was the last of the 20 amino acids to be discovered.

The 20 amino acids of proteins are sometimes referred to as the *standard*, *primary*, or *normal amino acids* to distinguish them from the amino acids that are modified after the completion of protein synthesis, and from other types of amino acids present in living systems but not in proteins.

4.3.1 Diverse Functions of Proteins

Different kinds of proteins perform many different biological functions:

Structural proteins These provide structural elements, supporting filaments, or sheets, for strength, and protection of cells and organisms, for example, collagen, tendons, and cartilage. A tough insoluble protein, keratin, is the main constitutent of hair, feathers, and fingernails. Fibroin is a protein present in silk fibres and spider webs. Resilin is a nearly perfect elastic protein in the wing hinges of some insects.

Enzymes (catalytic proteins) These participate in different types of chemical reactions of organic biomolecules in cells; there are thousands of different enzymes, and each catalyses a specific reaction.

Nutrient and storage proteins The seed proteins of rice, wheat, and corn store nutrients for the germinating seedling. Egg white protein in ovalbumin and casein of milk protein are nutrient proteins.

Transport proteins These carry specific molecules or ions from one organ to another and across the cell membrane. Haemoglobin in red blood cells which carries oxygen is a prime example of such proteins. Lipoproteins of blood plasma transport lipids from the liver to the other organs. Other specialized proteins in the plasma membrane transport glucose, amino acids, and other substances across the membrane.

Motile or contractile proteins These are proteins which change shape and help in the movement of organisms; for example, actin and myosin. Tubulin protein in microtubules acts in concert with the dynein protein in flagella and cilia to propel cells.

Defensive proteins These are specialized proteins. Immunoglobulins or antibodies in vertebrates neutralize and kill invading bacteria, viruses, or foreign proteins. In invertebrates, lysins have a similar role. Blood-clotting proteins, fibrinogen, and thrombin prevent loss of blood and reduce the chance of being infected by pathogens. Snake venoms, bacterial toxins, and toxic plant proteins, such as ricin, are defensive protein devices of the biological world. Some of these proteins such as fibrinogen, thrombin, and some venoms act as enzymes.

Regulatory proteins Hormones (protein) and peptides coordinate and regulate diverse cellular functions. Hormonal signals which activate function-specific genes are mediated by other kinds of proteins, such as receptors, GTP-binding proteins, as well as kinases and other enzymes.

Other proteins for specific purposes There are many other proteins that cannot be categorized, but provide necessary support to the life process and some proteins that are yet to be assigned a function. For example, Monellin protein of an African plant is intensely sweet and used as sweetener by man. Antifreeze proteins are present in the blood plasma of some Antarctic fish that prevent freezing.

4.3.2 Protein Molecules

All proteins, from bacteria to higher organisms, are constructed from a set of 20 amino acids, covalently linked in characteristic linear sequences called polypeptides. Different combinations and linear sequences of these 20 amino acids are responsible for the diverse characteristics and functions of thousands of protein molecules in the living world. In general, protein molecules are large.

Bovine *insulin* is made up of 51 amino acids (residues) and has a molecular weight of approximately 5,733 Da. Human *cytochrome c* has 104 amino acid residues and a molecular weight of 13,000 Da. Human *apolipoprotein* B—a cholesterol-transport protein with 4,636 amino acid residues in a single polypeptide chain weighing 513,000 Da—lies at the upper end of the size limit. Most biological proteins have a maximum of 2000 amino acid residues.

Protein molecules can consist of single or multiple polypeptide chains. The enzyme ribonuclease has one polypeptide chain, whereas the haemoglobin molecule has four; two identical α chains and two identical β chains, all four held together by non-covalent interactions. When at least two polypeptides are identical in a multi-subunit protein, as in the case of haemoglobin, it is called an *oligomeric protein.*

The approximate number of amino acid residues in a simple protein containing no other chemical group can be estimated by dividing its molecular weight by 110. Considering the average molecular weight of an amino acid residue to be around 128 (actually about 137, but the smaller amino acids predominate in most proteins) and deducting the molecular weight of a water molecule (mol. wt 18), which is removed to create a peptide bond between two amino acids, one can arrive at the number 110 (128 – 18).

By the process of hydrolysis, with the help of water molecules, the peptide bonds between amino acid residues in a polypeptide chain can be broken down into a mixture of free amino acids.

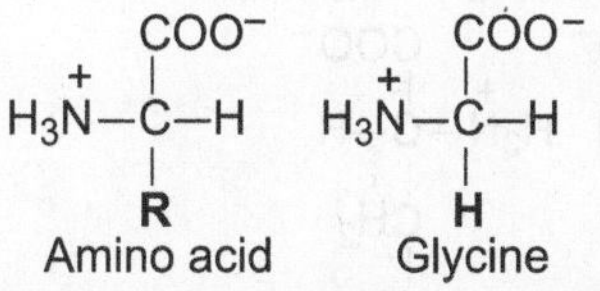

$$
\begin{array}{cc}
COO^- & COO^- \\
| & | \\
H_3\overset{+}{N}-C-H & H_3\overset{+}{N}-C-H \\
| & | \\
R & H \\
\text{Amino acid} & \text{Glycine}
\end{array}
$$

Fig. 4.12 Common structural features of amino acids. A carboxyl group and an amino group bonded to the same α carbon. Side chain or R group, attached to the same carbon, differs in different amino acids. R group is represented by H in glycine (the smallest amino acid).

4.3.3 Amino Acids

The following are some of the basic characteristic features of amino acids:

A common structural feature is that a carboxyl group and an amino group are bonded to the same carbon atom, the α carbon (Fig.4.12); The presence of these two groups is a basic feature of all amino acids.

The side chain or R-group differs in different amino acids. R-groups differ in structure, size, and electric charge (Fig. 4.13) and affect the solubility of amino acids in water. If the R-group contains additional carbons in a chain, they are named as β, γ, δ, ε, etc., in relation to the original α-carbon.

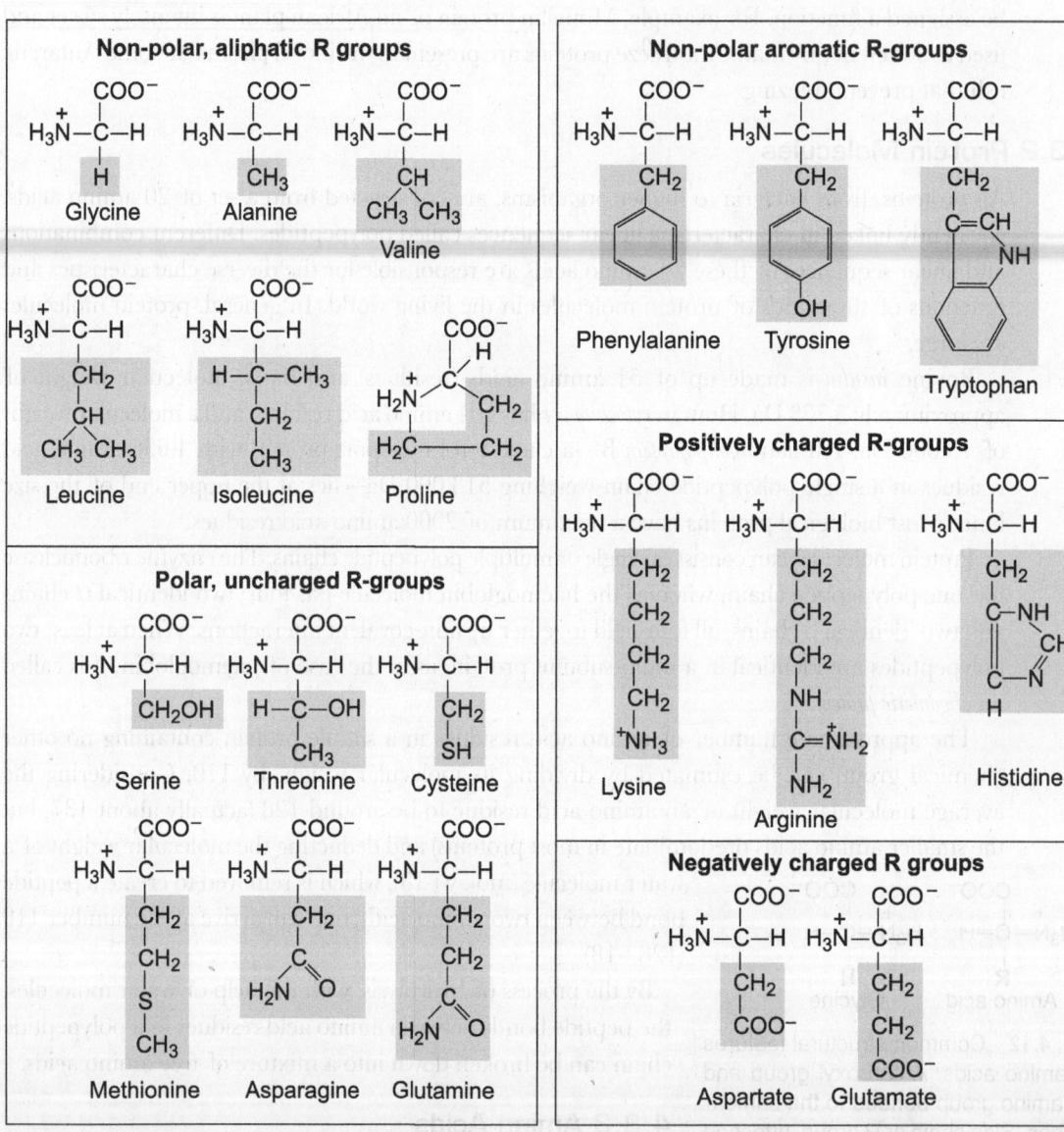

Fig. 4.13   The 20 standard amino acids with different R-groups (shaded). Their amino and carboxyl groups have been shown ionized, as they occur in this form at pH 7.0.

R-group allows grouping of the amino acids

The properties of the R-groups of different amino acids make their classification easy. To begin with, the polarity of the R-group, i.e., its tendency to interact with water at neutral or biological pH (near 7.0) classifies the amino acids into two groups—polar or hydrophilic (water soluble) and non-polar or hydrophobic (water insoluble).

Figure 4.13 represents the structure of the 20 standard amino acids with their R-groups (shaded) and the five classes they are divided into. The minor disagreement about these categories among various experts has not been taken into consideration here. The five classes with characteristics of the R-groups are (a) non-polar aliphatic (straight chain), (b) non-polar aromatic, (c) polar but uncharged, (d) positively charged, and (e) negatively charged. Within the members of each group or class, there are gradations in polarity, size, and shape of the R-groups.

Non-polar aliphatic R-groups   *Glycine* has the simplest amino acid structure and belongs to the non-polar (hydrophobic) class with only one hydrogen representing the R-group. Thus, there is minimal steric hindrance from the side chain, which is why glycine allows considerable structural flexibility to a protein molecule wherever it occurs.

The bulky hydrocarbon R-groups of *alanine, valine, leucine,* and *isoleucine* with their distinctive structures promote hydrophobic nature in protein molecules.

Proline with imino (NH_2^+, secondary amino) group in rigid conformation in the side chain reduces the structural flexibility of the protein at the point of its occurrence.

Non-polar aromatic R-groups   *Phenylalanine, tyrosine,* and *tryptophan* have aromatic (ring structure) non-polar (hydrophobic) side chains. They usually exist in the core area of protein molecules to avoid contact with water, and thus, are responsible for folding of newly synthesized polypeptide chains along with other hydrophobic amino acid residues. They also constitute the hydrophobic intra-membrane part of the protein molecule anchored in the lipid layer (also hydrophobic) of the plasma membrane. The hydroxyl group of tyrosine can form hydrogen bonds, and thus, serves as an important functional group in the activity of some enzymes. This hydroxyl group of tyrosine and the nitrogen of the indole ring in tryptophan make these two amino acids significantly more polar than phenylalanine.

Tryptophan, followed by tyrosine, and to a lesser extent, phenylalanine absorb ultraviolet light mostly at a wavelength of 280 nm. This is the reason why proteins exhibit strong absorbance of UV light at 280 nm. This property is utilized by scientists for the detection of proteins.

Polar uncharged R-groups   The functional groups present in the R-side chain form hydrogen bonds with water, thereby making the amino acids more soluble in water or hydrophilic. Serine, threonine, cysteine, methionine, asparagine, and glutamine have this type of polar uncharged R-group.

The polar nature (i.e., water solubility) of serine and threonine is due to the presence of a hydroxyl group in their side chain, and that of cysteine and methionine is due to their sulphur atom. The amide group in asparagine and glutamine makes them polar molecules.

Cysteine is readily oxidized to form a covalently linked dimeric amino acid called *cystine*, in which, two cysteine molecules are joined by a disulphide bridge. Such disulphide bonds stabilize the structure of many proteins.

Positively charged (basic) R-group The R-group of the amino acids lysine, arginine, and histidine has a net positive charge at pH 7. The charge is due to the presence of a second amino group at the ε position in the side chain of lysine, a positively charged guanidino group in the R-group of arginine, and an imidazole group in the side chain of histidine. (Some authorities prefer to include histidine in the category of non-polar aromatic group.)

Negatively charged (acidic) R-groups Aspartate and glutamate have R-groups with a net negative charge at pH 7; both these amino acids have a second carboxyl group (negatively charged) in the R-side chain. These amino acids are the parent compounds of asparagine and glutamine, respectively.

Ionization of amino acids in aqueous solution (acid–base behaviour) and zwitterions

The functional groups of α-amino acids, the amino group, and the carboxyl group attached to the α carbon are ionizable in aqueous solution. The net charge of the molecule depends on the pH of the solution. Both the groups are protonated at low pH with a net charge of +1 (Fig. 4.14) and the amino acid behaves as an acid. But at high pH, both the groups lose a proton and give the amino acid a net charge of −1 and the property of a base. The α-carboxyl group is a stronger acid than the amino group; it loses its proton first when the pH increases. This causes the formation of an intermediate form of an amino acid which has both a negative and a positive charge, at the middle range of pH. This intermediate form is known as *Zwitterion* (dipolar form, Fig. 4.14). The term Zwitterion is derived from German, and means 'hybrid ion'. Zwitterion is amphoteric (it has a dual nature) and its net charge is zero. As a zwitterion is electrically neutral, it remains stationary in an electric field.

Electrostatic forces between positively and negatively charged functional groups of neighbouring zwitterions strongly hold them in the stable ionic crystal lattice. That is why a much higher melting temperature is required for crystalline-amino acids which exist as zwitterions, as compared to other organic molecules of similar size.

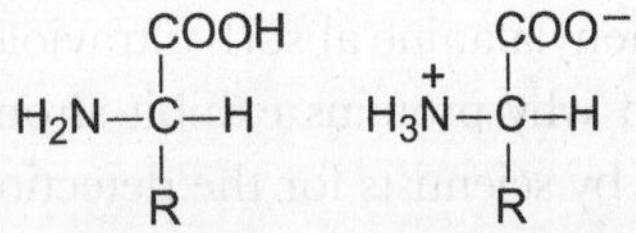

In addition to the α-amino and α-carboxyl groups, the R-groups of some amino acids can also be ionized. Therefore, the acid–base features of amino acids impart certain physical and biological properties to the proteins they constitute. The acid–base nature of amino acids is used for separating, identifying, and quantifying the amino acids in a protein and their sequence in the protein chain.

Fig. 4.14 Non-ionic and zwitterion forms of an amino acid. Amino and carboxyl groups ionize in aqueous solution at neutral pH (7.0) causing a dipole zwitterion with a negative and a positive charge that cancel each other so that the net charge is zero.

L-Amino acids in proteins

The α-carbon is asymmetric and a *chiral centre*. All molecules with a chiral centre can rotate plane-polarized light to the left or to the right, and accordingly become leavorotatory (L) or dextrorotatory (D) stereoisomers. *The amino acids in protein molecules are usually the L-stereoisomers.* D-Amino acids are present only in small peptides of bacterial cell walls and in some peptide antibiotics.

The ability of cells to specifically synthesize the L isomer of amino acids reflects the extraordinary specificity of the synthesizing enzymes—mainly their active sites.

Non-standard amino acids in cells

It has already been pointed out that the 20 standard amino acids constitute all proteins. After protein synthesis, in certain cases a modification of some standard amino acids that are already present in the protein chain may lead to the formation of non-standard amino acids. Thus, the non-standard amino acids are derivatives of the 20 standard amino acids. Examples of non-standard amino acids are given in Table 4.2.

Table 4.2 Examples of non-standard amino acids

4-Hydroxyproline COO^- $\|$ C–H H_2N^+ CH_2 H_2C —— C $\diagdown$ H OH	A derivative of proline due to modification an extra functional group OH (shaded) added; found in plant cell wall proteins and in collagen (of connective tissue in animals)
5-Hydroxylysine	A hydroxyl (OH) functional group added at C-4 of lysine (see Fig. 4.13); found in collagen
N-Methyllysine	A functional methyl (CH_3) group added to the nitrogen of terminal amino group of lysine; found in myosin, a contractile protein of muscle
γ-Carboxyglutamate	Derived from glutamine; found in the blood clotting protein prothrombin and in proteins that bind Ca^{2+} in their biological function
Desmosine	More complicated, a derivative of four separate lysine residues; found in elastin, a fibrous protein

[There are a few other such non-standard amino acids in biological systems.]

Characteristic chemical reactions of amino acids

The chemical reactions of amino acids are performed by their functional groups. The amino and carboxyl groups attached to the α-carbon undergo characteristic chemical reactions. The amino group can be acetylated or formylated and carboxyl group can be esterified (on reaction with alcohols). There are side chain (R) functional groups, such as the phenolic group in tyrosine, sulphydryl group in cystein, and alcoholic group in serine; each of these functions in a characteristic manner.

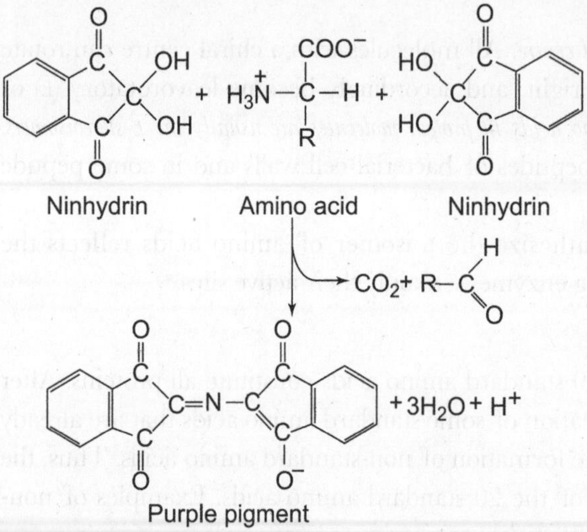

Fig. 4.15   Ninhydrin reaction for detection of amino acids.

Ninhydrin reaction

One of the distinguishing tests for all amino acids, except proline, is the reaction between the α-amino group of the free amino acid and ninhydrin. This is technically and historically a wonderful test. Heating with excess of ninhydrin yields a purple product.

Ninhydrin is a powerful oxidizing agent which brings about the oxidative deamination of amino acids, liberating ammonia, carbon dioxide, a corresponding aldehyde, and the reduced form of ninhydrin (Fig. 4.15). The ammonia formed from the α-amino group reacts with ninhydrin and reduces it giving rise to a blue or purple substance called *diketohydrin* (Ruhemann's purple).

The α-amino group in proline and hydroxyproline is substituted by an imino group which imparts an yellow colour to these amino acids. The intensity of colour produced by the ninhydrin reaction is proportional to the amino acid concentration. Amino acid concentration in a given sample can be accurately measured by comparing the optical absorbance of the solution with that of an appropriate standard solution.

4.3.4 Peptides

Biological peptides range in size from small molecules containing only two or three amino acids to large polypeptides or macromolecules containing thousands of amino acids. The amino acids in the linear chain of a peptide or polypeptide are joined by *peptide bonds*. A peptide bond is a covalent bond formed between the α-carboxyl group of one amino acid and the α-amino group of another amino acid with the liberation of one water molecule (Fig. 4.16). When a few amino acids are joined by a peptide bond, the structure is called an *oligopeptide*.

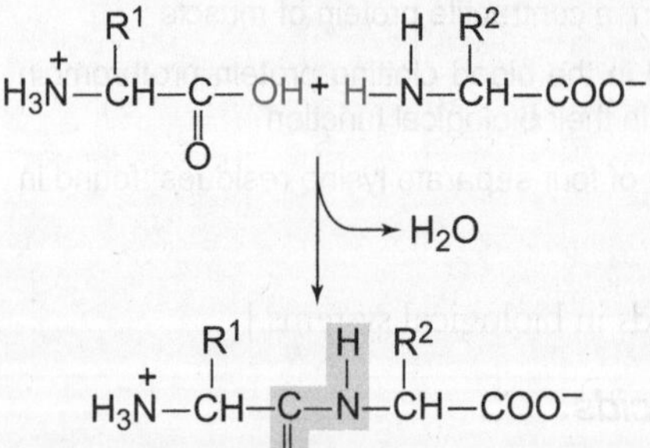

Peptide bonds can be hydrolysed (disintegrated) by boiling with either strong acid (typically 6M HCl) or base to yield the constituent amino acids of a protein. This is the reverse reaction of peptide bond formation (Fig. 4.16). Peptide bonds may also be hydrolysed by enzyme *proteases*.

Fig. 4.16   Formation of a peptide bond (shaded) between the α-carboxyl group of one amino acid and the α-amino group of another amino acid with the liberation of one water molecule (condensation reaction).

Significant biological activities of some small polypeptides

Much of the interest in biology is around the activities of large proteins with molecular weights in the range of tens, hundreds, or thousands of daltons. However, many oligopeptides and

Tyr—Gly—Gly—Phe—Met

Tyr—Gly—Gly—Phe—Leu

Fig. 4.17 Two enkephalins— small peptides produced in the brain that control pain and sleep.

small peptides have important biological functions, including the regulation of cellular activities.

There are several small *peptide hormones*, such as *oxytocin* (with nine amino acid residues), which is secreted by the posterior pituitary and stimulates contraction of the uterus; another nine-residue peptide *bradykinin* is involved in the inflammatory reaction of tissues; the three-residue long *thyrotropin-releasing factor* is synthesized in the hypothalamus and stimulates the release of another hormone, thyrotropin, from the anterior pituitary gland. The central nervous system produces *enkephalins*, which are five-amino acid peptides (Fig. 4.17) that bind to their receptors on the cells of certain areas of the brain to induce analgesia (inhibition of perception of pain). Enkephalins are components of the body's inherent mechanism to control pain and sleep. Interestingly, morphine, heroin, and other addicting opiate substances, can bind to the enkephalin receptors on the brain cells although they are not peptides. *Amanitin*, an extremely toxic mushroom poison, as well as many antibiotics are also examples of small peptides with diverse functions. The list of pharmacologically and commercially important small peptides is growing. Processes of synthetic organic chemistry enables their production.

Enzymatic removal of a portion of proinsulin produces *insulin* hormone which contains two peptides, one that has 30 amino acid residues and the other 21 amino acids, joined by two disulphide bonds (mol. wt 5,700 Da). Another pancreatic hormone *glucagon* that has 29 residues, increases the level of sugar in blood plasma, and thus, opposes the action of insulin. A 39-residue polypeptide hormone *corticotropin*, a product of the anterior pituitary gland, stimulates the adrenal cortex. Although these hormones are bigger than the small peptides just discussed, they actually represent polypeptides at the lower limit of molecular size and weight.

4.4 LIPIDS

Lipids are heterogeneous organic compounds consisting of molecules that contain at least eight carbon atoms in an aliphatic chain (of $-CH_2-$). They are sparingly soluble in water, but are soluble in non-polar solvents such as chloroform, ether, and benzene. However, the chain length and solubility criteria are not absolute; there are exceptions.

4.4.1 Fatty Acids

Fatty acids are the principal constituents of most lipids. A fatty acid is a long hydrocarbon chain with a terminal carboxyl group at one end (Fig. 4.18a). Most fatty acids have an even number of carbon atoms arranged in an unbranched chain.

Fatty acids are classified into saturated and unsaturated fatty acids. Saturated fatty acids do not have double bonds between the carbon atoms in the chain (and thus, do not have much scope to react chemically). They have the general formula $CH_3 (CH_2)_n COOH$, where n is an even number.

The most common saturated fatty acids from plant and animal fats are palmitic acid $[CH_3(CH_2)_{14}COOH]$ (Fig. 4.18b) and stearic acid $[CH_3(CH_2)_{16}COOH]$; palmitic acid is a

16-carbon compound and stearic acid has 18 carbons in its chain. Longer chain fatty acids containing up to 36 carbon atoms also exist. The carbon atoms are numbered with the terminal carboxyl as C-1, from the carboxyl to the CH_3 group.

Unsaturated fatty acids have one or more double bonds in the hydrocarbon chain. Unsaturated fatty acids can exist as geometric isomers, in *cis* or *trans* conformations. In all biological unsaturated fatty acids, the double bond has a *cis* configuration.

Unsaturated fatty acids are further classified depending on the number of sites with double bonds between carbon atoms into the following types.

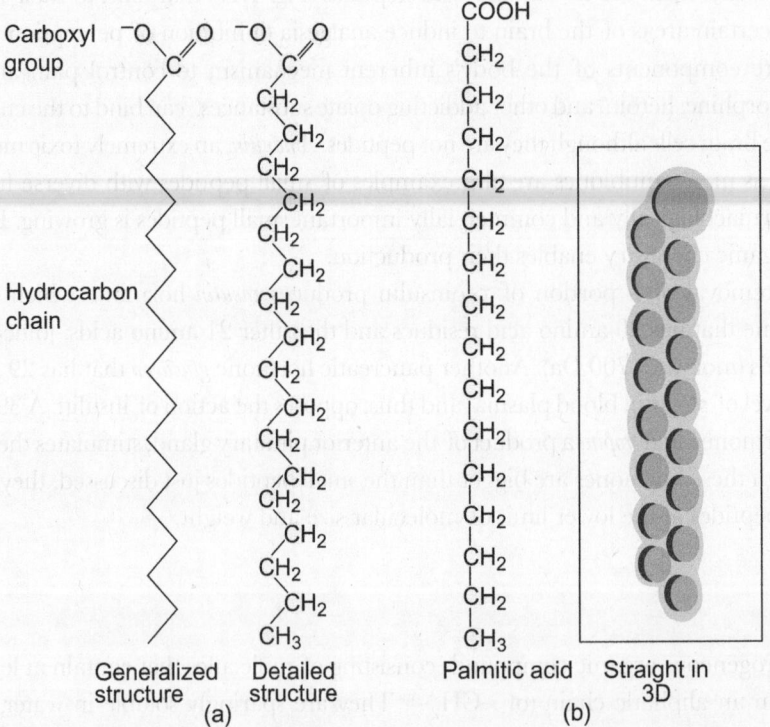

Fig. 4.18 (a) Generalized structure of a fatty acid: A long hydrocarbon chain with a terminal carboxyl group. (b) Palmitic acid (16-C) and stearic acid (18-C) are two of the most common saturated (no double bonds between carbons) fatty acids.

Monounsaturated fatty acids These have one site of unsaturation or double bond, e.g., palmitoleic and oleic acid. There is a notation that to specifies the number of carbon atoms (chain length), number of double bonds, and their site in the fatty acids (Table 4.3). The number of carbon atoms and number of double bonds are separated by a colon, then position(s) of double bond, if any, are specified by superscript numbers following delta or omega (Δ or ω) sign. In Table 4.3, palmitic acid is a 16-carbon saturated fatty acid and has been denoted as 16:0. Monounsaturated palmitoleic and oleic acid have been denoted as $16:1\omega^9$ (or $16:1\Delta^9$) and $18:1\Delta^9$, respectively. Palmitoleic acid has 16 carbons in the carbon chain, and one double bond between

C-9 and C-10 (the preceding number of carbon with double bond is mentioned). Similarly, oleic acid has 18 carbons with one double bond between C-9 and C-10.

Polyunsaturated fatty acids   These have two or more double bonds, which are separated by at least one methylene group. The examples are α-linoleic acid and linolenic acid. The abbreviated form of these acids indicating the numbers of carbons and position of double bonds have been presented in Table 4.3.

Table 4.3   Examples of saturated and unsaturated fatty acids

Fatty acid	Structural formula	Carbon skeleton abbreviated
Saturated Palmitic acid	$CH_3(CH_2)_{14}COOH$	C16:0
Monounsaturated		
Palmitoleic acid	$CH_3(CH_2)_5CH=CH(CH_2)_7\ COOH$	$C16:1\Delta^9$
Oleic acid	$CH_3(CH_2)_7CH=CH(CH_2)_7\ COOH$	$C18:1\Delta^9$
Polyunsaturated		
α-Linoleic acid	$CH_3(CH_2)_4CH=CHCH_2CH=CH(CH_2)_7\ COOH$	$C18:2\Delta^{9,12}$
Linolenic acid	$CHCH_2CH=CHCH_2CH=CHCH_2CH=CH(CH_2)$ COOH	$C18:3\Delta^{9,12,15}$

Chain length and unsaturation determine the physical characteristics of fatty acids

Chain length and the number of double bonds (i.e., the degree of unsaturation) determine certain characteristics of unsaturated fatty acids.

- Longer the chain, fewer the double bonds—lower solubility in water
- Shorter the chain length, more double bonds—lower melting point

For example, at room temperature (25°C), the saturated fatty acids of 12:0 to 24:0 show a waxy consistency, whereas unsaturated fatty acids of similar lengths are oily liquids.

The carboxyl group is polar and ionized at neutral pH, and accounts for the slight solubility of short-chain fatty acids in water. The non-polar (hydrophobic) hydrocarbon chain is responsible for poor solubility of fatty acids in water. In vertebrate animals, free fatty acids that have a free carboxylate group circulate in blood plasma bound to a protein carrier, the serum albumin.

In the fully saturated state of fatty acids, free rotation around each of the carbon–carbon bonds provides the hydrocarbon chain great flexibility. Thus, the most stable conformation is the fully extended form (Fig. 4.18a). In this state, the steric hindrance from neighbouring atoms is minimum.

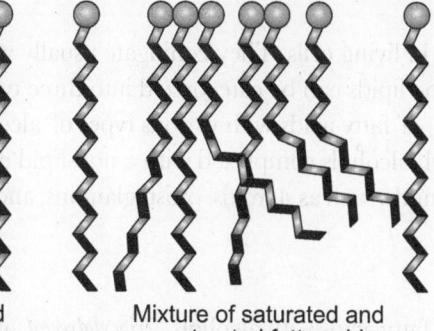

Saturated fatty acids (Stearic acid)

Mixture of saturated and unsaturated fatty acids

Fig. 4.19   The pack of fully saturated fatty acids are in stable and extended form. Unsaturated fatty acids with one or more twists at double bonded carbons, cannot pack together as tightly as fully saturated fatty acids.

The extended molecules can pack together tightly in nearly crystalline arrays, along their length with van der Waals bonds (Fig. 4.19). Hydrophobic interactions further stabilize this array. Therefore, it takes more thermal energy to disturb the array of saturated fatty acids; hence, saturated fatty acids have higher melting point in comparison to unsaturated ones.

In unsaturated fatty acids, a *cis* double bond causes a kink or twist in the hydrocarbon chains (Fig. 4.20). Unsaturated fatty acids with one or several such kinks cannot pack together as tightly as fully saturated fatty acids (Fig. 4.19); thus, their interactions with each other is weaker. Lesser amount of thermal energy is needed to disturb the poorly ordered array of unsaturated fatty acids; therefore, they have lower melting points than saturated fatty acids of the same chain length.

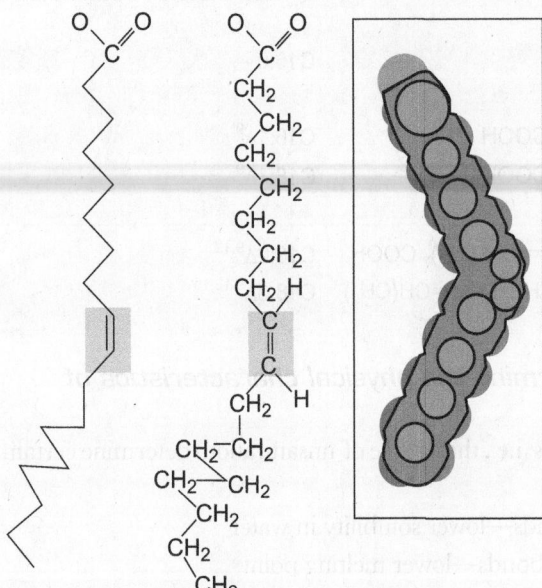

Fig. 4.20 In unsaturated fatty acids (oleic acid), the double bond (shaded) does not permit rotation and introduces a rigid bend (twist) in the hydrocarbon chain. All the other bonds are free to rotate.

4.4.2 Classification of Lipids

Fatty acids are rarely found in the free state in living cells. They conjugate usually with various alcohols through ester linkages and *lipids*. The lipids can be categorized into three types:

- (i) *Simple lipids*—Esters (with ester linkage) of fatty acids with various types of alcohols
- (ii) *Complex lipids*—Esters of fatty acids with alcohols complexed with a non-lipid moiety
- (iii) *Derived lipids*—Diverse group of compounds, such as steroids, prostaglandins, and fat soluble vitamins.

Simple lipids (storage lipids)

Simple lipids are formed by esterification of fatty acids with alcohols. *Triacylglycerol* (also referred as triglyceride) is the simplest lipid and is composed of three fatty acids in ester linkage with a single glycerol (Fig. 4.21). The polar hydroxyl (OH) groups of glycerol and the polar carboxylates of the fatty acids are involved in ester links, and thus, triacylglycerol is a non-polar, hydrophobic molecule, essentially insoluble in water. (That is why oil does not mix with water and floats on the aqueous phase as its specific gravity is lower.)

Glycerol

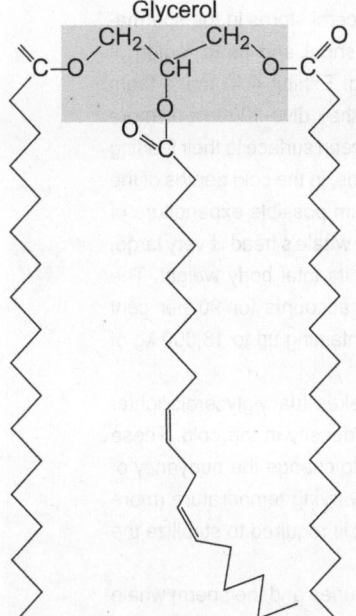

Fig. 4.21 Triacylglycerol (or triglyceride), the simplest lipid is composed of three fatty acids in ester linkage with a single glycerol.

Triacylglycerols containing the same kind of fatty acid in all three positions are called *simple triacylglycerols*, and are named after the fatty acids, for example, tripalmitin, tristearin, and triolein. Mixed triacylglycerols contain two or more different fatty acids.

When the alcohol is glycerol, the ester is called *neutral fat* or simply fat. If the alcohol has a long chain of hydrocarbons (C16–C30), instead of glycerol, the esters formed with long chain (C14–C36) saturated or unsaturated fatty acids are called *waxes* as produced by bees and other members of the living world. Waxes are chemically inert. They protect plants from water loss and abrasive damage. They act as a water barrier for insects, birds, and certain animals.

Functions of triacylglycerols

Tryacylglycerols remain in a separate phase as microscopic oily droplets in the aqueous cytosol of eukaryotic cells where they serve as a store of *metabolic fuel for energy* production.

Triacylglycerol is stored in large amounts in *adipocytes* or fat cells in vertebrates as a source of energy. It also has other functions (Exhibit 4.A). Triacylglycerols stored in the seeds of many varieties of plants provide energy and biochemical precursors during germination.

Polysaccharides, such as starch and glycogen, also serve as sources of energy in cells. Triacylglycerols surpass polysaccharides as a store of energy, on at least two counts: (i) the carbon atoms of fatty acids are more reduced than those of sugars; hence, oxidation of triglycerols produces more than double the energy that would be produced by an equal weight of carbohydrates. (ii) Being hydrophobic, triacylglycerols unlike polysaccharides are not hydrated and relieve the organisms of the extra burden of water of hydration which is needed for the storage of polysaccharides.

Exhibit 4.A Functional Spectrum of Triacylglycerols

The students are here requested to imagine the life processes of warm blooded animals, such as seals, walruses, penguins, and bears in polar regions at freezing temperature in the range of –30°C to –40°C. These processes have been made possible by the storage of triacylglycerol under the skin and other organs. Not only do triacylglycerols serve as energy stores, they also insulate the animals against very low temperatures. The blubber serves the same purpose when whales dive deep in cold water.

Besides insulation, subcutaneous stores of triacylglycerol protects all kinds of animals from mechanical impact and injuries.

The huge fat reserves accumulated before hibernation enable hibernating animals, including polar bears, to survive in holes for many months without food.

Triacylglycerols or fats deposited in the mammary glands allow mammalian females to nurse their young ones.

Fig. 4.A A sperm whale

Huge triacylglycerol stores in the sperma-ceti organ at the snout and head region of sperm whales (Fig. Exhibit 4.A) make them wonderful divers; they dive 1000 m or more from the tropical ocean surface to their feeding ground, full of squids, in the cold depths of the ocean with minimum possible expenditure of energy. The sperm whale's head is very large, about one third of its total body weight. The spermaceti organ accounts for 90 per cent of the weight of the head, containing up to 18,000 kg of spermaceti oil, a mixture of triacylglycerol and waxes.

An optimal combination of fatty acid chain length and unsaturation makes triacylglycerols lighter at warmer temperature and causes congealing (crystallizing) to higher density in the cold. These physical qualities of spermaceti oil are critical for the whales to be able to change the buoyancy of their bodies and match the density of sea water at different levels with varying temperature (more at the surface, less at increasing depth). Thus, no extra swimming effort is required to stabilize the body in water of varying density.

Unfortunately, spermaceti oil has fuelled man's greed for several centuries and the sperm whale population has been reduced to a limited number.

Vegetable oils, dairy products, and animal fat are complex mixtures of simple and mixed triacylglycerols and serve as a source of fatty acids for humans.

Lipase enzymes in the intestine catalyse the hydrolysis of triacylglycerols and help in the digestion and absorption of dietary fats. Lipases in adipocytes (cells of adipose tissue) and germinating seeds break down stored triacylglycerol to release fatty acids for export to other tissues in need of fuel (energy).

Hydrolysis of animal fats by heating with NaOH or KOH produces glycerol and Na$^+$ or K$^+$ salts of the fatty acids called *soap*.

Biological waxes are used in pharmaceutical, cosmetics, polishes, and other industries.

Hydroxylated fatty acids

A number of hydroxylated fatty acids can be synthesized by plants, e.g., *ricinoleic acid*, the structure of which is as follows:

$$CH_3 (CH_2)_5 \overset{\overset{\displaystyle OH}{\displaystyle |}}{CH} CH_2 CH = CH(CH_2)_7 COOH$$

Complex lipids (structural lipids)

Complex lipids comprise a lipid part and a non-lipid component that may be a small molecule or a macromolecule. This group includes mostly cell membrane lipids—glycerophospholipids, sphingophospholipids, glycolipids, lipoproteins, proteolipids, and sterols. Three main types, *glycerophospholipids*, *sphingophospholipids*, and *sterols*, are discussed here. One should remember the structure of triacylglycerol to appreciate the alteration of the structure in the case of complex lipids.

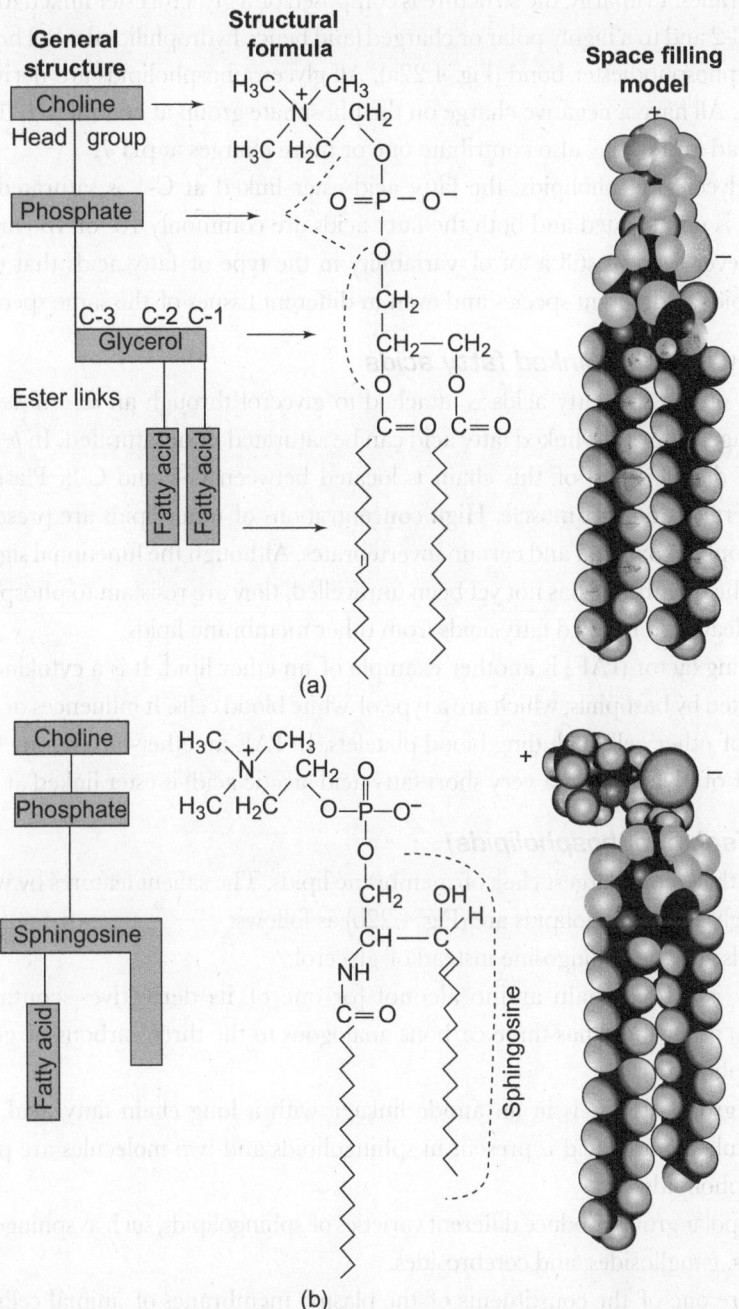

(a)

(b)

Fig. 4.22 Structure of phospholipids. (a) Phosphoglycerides (b) Sphingolipids are built on different backbones—glycerol for phosphoglycerides and sphingosine for sphingolipids. They resemble each other in three-dimensional structure and serve as cell membrane lipids.

Glycerophospholipids (phosphoglycerides)

Also called phosphoglycerides, glycerophospholipids are the most abundant lipids in cellular and subcellular membranes. Primarily, the structure is composed of a glycerol ester linked to two fatty acids at C-1 and C-2 and to a highly polar or charged (and hence hydrophilic) alcohol head group at C-3 through a phosphodiester bond (Fig. 4.22a). All glycerophospholipids are derivatives of phosphatidic acid. All have a negative charge on the phosphate group at neutral pH, that is, pH 7. The alcohol head group may also contribute one or more charges at pH 7.

Generally in glycerophospholipids, the fatty acid ester linked at C-1 is saturated and the one linked at C-2 is unsaturated and both the fatty acids are commonly 16- or 18-carbon long (Fig. 4.22a). However there is still a lot of variability in the type of fatty acids that constitute glycerophospholipids in different species and even in different tissues of the same species.

Phospholipids with ether-linked fatty acids

In *ether lipids*, one of the two fatty acids is attached to glycerol through an *ether linkage*, rather than an ester linkage. The ether-linked fatty acid can be saturated or unsaturated. In *plasmalogens* (unsaturated), the double bond of this chain is located between C-1 and C-2. Plasmalogens are enriched in vertebrate heart muscle. High concentrations of ether lipids are present in the membrane of halophilic bacteria and certain invertebrates. Although the functional significance of ether lipids in the membrane has not yet been unravelled, they are resistant to phospholipases which normally cleave ester-linked fatty acids from other membrane lipids.

Platelet-activating factor (PAF) is another example of an ether lipid. It is a cytokine (soluble cell product) secreted by basophils, which are a type of white blood cells. It influences or regulates various activities of other cells including blood platelets. In PAF, an ether-linked long fatty acid chain is at the C-1 of glycerol and a very short fatty acid (acetic acid) is ester linked at C-2.

Sphingolipids (sphingophospholipids)

Sphingolipids are the second largest class of membrane lipids. The salient features by which this class differs from glycerophospholipids are (Fig. 4.22b) as follows:

- Sphingolipids contain sphingosine instead of glycerol.
- Sphingosine is a long-chain amino alcohol [or one of its derivatives, containing an amino (NH_2) group] and has three carbons analogous to the three carbons of glycerol in glycerophospholipids.
- The amino group at C-2 is in an amide linkage with a long chain fatty acid. Thus, a single molecule of fatty acid is present in sphingolipids and two molecules are present in glycerophospholipids.

Changes in the polar group produce different varieties of sphingolipids, such as sphingomyelins, glycosphingolipids, gangliosides, and cerebrosides.

Sphingomyelins are one of the constituents of the plasma membranes of animal cells and the myelin sheath which surrounds and insulates the axons of myelinated neurons.

Glycosphingolipids have sugar moieties in the head group connected to C-1 of sphingosine (Fig. 4.22b). They are neutral in charge and occur largely in the outer face of the plasma membrane.

They are responsible for determination of human A, B, and O blood groups.

Derangements in the metabolism of cerebrosides and gangliosides are the underlying reasons for human genetic diseases such as Tay-Sachs and Niemann-Pick.

Although sphingomyelins differ significantly from glycerophospholipids in their chemical content, their three-dimensional structure and disposition in the cell membrane is quite similar to that of glycerophospholipids (Fig. 4.22).

Degradation of membrane phospholipids

Continuous degradation of membrane lipids and their replacement through synthesis are normal processes for most cells. Phospholipases of different specificities and some lysosomal enzymes are involved in degradation. Any defect (genetic) in these enzymes leads to severe diseases as indicated earlier.

Phospholipid breakdown initiates the cascade cell signalling process when a hormone or an antigen binds to its specific receptor on the external cell surface. For example, phospholipase C specifically cleaves phosphatidylinositols, releasing diacylglycerols and inositol phosphates, which in turn act as second messengers for the activation of specific genes or cytoplasmic proteins (bearing hormone or antigen receptors).

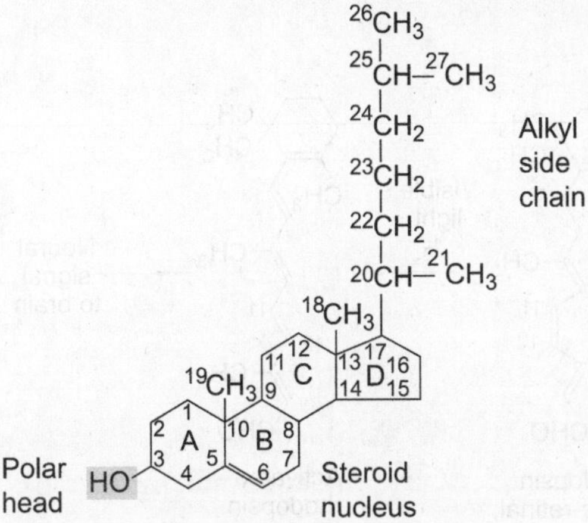

4.4.3 Sterols (Derived Lipids)

Sterols are structural lipids in the membranes of most eukaryotic cells. A sterol consists of a steroid nucleus characteristically made up of four fused rings, three with six carbons and one with five (Fig. 4.23). The sterol nucleus which is almost planar, is a rigid structure; the fused ring structure does not allow rotation around C–C bonds. A hydrocarbon side chain is attached to the nucleus at C17. The central ring structure and the side chain are non-polar. The side chain varies and can be hydrophilic as in the case of bile acid. A polar head consisting of a hydroxyl group (OH) is present at C3 position (Fig. 4.23). This polar head group may also form a sterol ester with a fatty acid for storage or transportation

Fig. 4.23 Cholesterol: the major sterol in animal tissues. The four fused rings of the steroid nucleus are marked A through D and the carbon atoms are numbered.

of the sterol. *Cholesterol* is a typical example of a sterol found in animal tissues; it is the major sterol and is amphipathic (it is both hydrophobic or non-polar and hydrophilic or polar). Similar sterols like stigmasterol and ergosterol are present in plants and fungi, respectively. Bacteria usually do not contain sterols.

Sterol derivatives for important biological functions

Besides their structural role, sterols serve as a mother (precursor) substance to many other chemicals which perform specific biological activities. For example, bile acids derived from cholesterol emulsify dietary fats in the intestine, in the same way as detergents do; emulsified fats are more accessible for digestion by lipases.

Other Notable Lipid Derivatives

The four fat soluble vitamins, A, D, E, and K are all isoprenoid compounds. Isoprenoids are synthesized by the condensation of isoprene units which are five-carbon precursors:

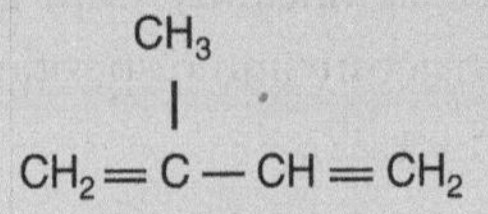

$$CH_2 = C - CH = CH_2$$

with CH_3 attached to the central C.

Isoprene

Exhibit 4.B

Vitamin A (retinol) is a lipid-derived pigment, essential for vision in human beings and animals. Plants do not contain vitamin A. However many plants contain carotenoids, which are light-absorbing pigments. Carotene provides the characteristic colour to carrots, sweet potatoes, and other yellow vegetables.

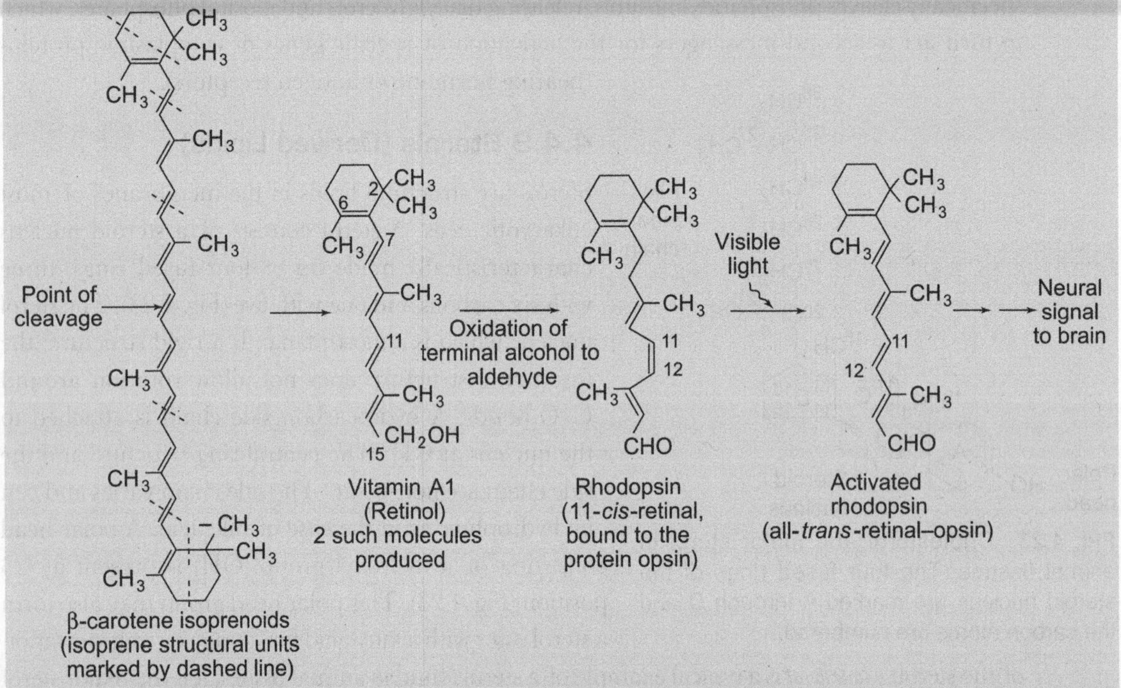

β-carotene isoprenoids
(isoprene structural units
marked by dashed line)

Point of cleavage → Vitamin A1 (Retinol) 2 such molecules produced → Oxidation of terminal alcohol to aldehyde → Rhodopsin (11-*cis*-retinal, bound to the protein opsin) → Visible light → Activated rhodopsin (all-*trans*-retinal–opsin) → Neural signal to brain

Fig. 4.B Vitamin A deficiency leads to many symptoms and disease conditions including xerophthalmia (dry eyes), night blindness, retarded growth, and sterility in males.

Enzymatic cleavage of a β-carotene yields two molecules of vitamin A (retinol). Oxidation at C-15 converts retinol to retinal which is complexed with the protein opsin to produce rhodopsin, the visual pigment in the retina of the eye. In the dark, the retinal part of rhodopsin is in the *cis* form at C-11. When exposed to visible light the rhodopsin molecule is excited; the 11-*cis* retinal undergoes a series of photochemical reactions that convert it to all-*trans*-retinal, causing a change in the shape of the entire rhodopsin molecule. This structural transformation occurring in the rod cells of the retina generates an electrical signal for transmission to the brain for visual perception.

All the major steroid sex hormones in males and females, and the cortisol and aldosterone hormones of the adrenal cortex are derivatives of cholesterols. These hormones retain the intact structure of the sterol nucleus (Fig. 4.24). They are carried as products from one tissue or organ by the blood to distant tissues for regulating the genetic functions of cells through specific cell surface hormone receptors. A very low amount of hormone is capable of producing the desired effect on target tissues.

Eicosanoids are derivatives of the 20-carbon (Gr. *eikosi*, 'twenty') polyunsaturated fatty acid arachidonic acid. They are extremely potent cytokines that have effects on various tissues of vertebrate animals. There are three classes of eicosanoids—prostaglandins, thromboxanes, and leukotrienes.

Lipid quinones, such as ubiquinone (in animals) and plastoquinone (in plants), are also isoprene derivatives and function as electron carriers in the production of ATP in mitochondria and chloroplasts. Ubiquinone is often called coenzyme Q in mammalian tissues.

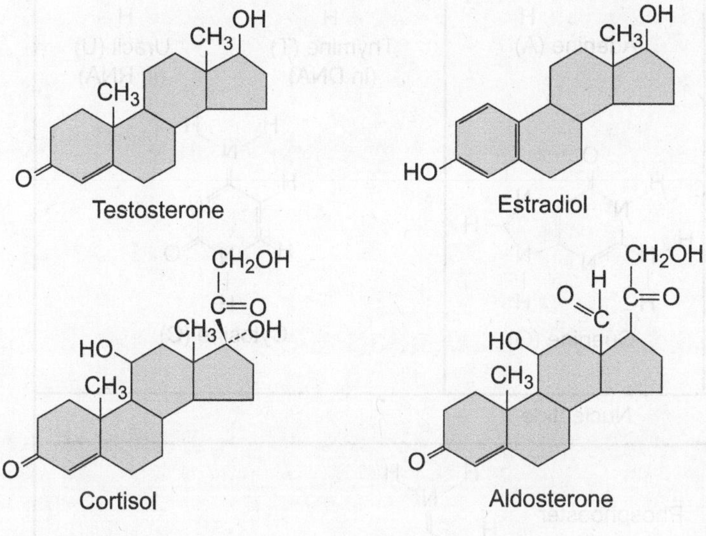

Fig. 4.24 Cholesterol derivatives include hormones with important biological functions—testosterone, the male sex hormone; estradiol, one of the female hormones; cortisol and aldosterone produced in the cortex of the adrenal gland regulate glucose metabolism and salt excretion, respectively.

4.5 NUCLEIC ACIDS AND NUCLEOTIDES

Nucleic acids are macromolecules made up of nucleotide monomers and are of paramount importance to cells and the biological world. They are a repository of the story of evolution and carry the process of life forward. They store genetic information and pass it from one generation to the next. Nucleic acids encode and direct the formation of proteins—some proteins in the form of enzymes help in the synthesis of biomolecules. Thus, nucleic acids are responsible for all the constituents of cells and the characteristics of life.

4.5.1 Types of Nucleic Acids

Nucleic acids are polymers of monomers called nucleotides. There are two major types of nucleic acids that occur in living organisms. They primarily differ in their chemical structure by the absence or presence of an oxygen atom in the pentose sugar component. They also differ in their functional roles in the cell. The two main types of nucleic acids are:

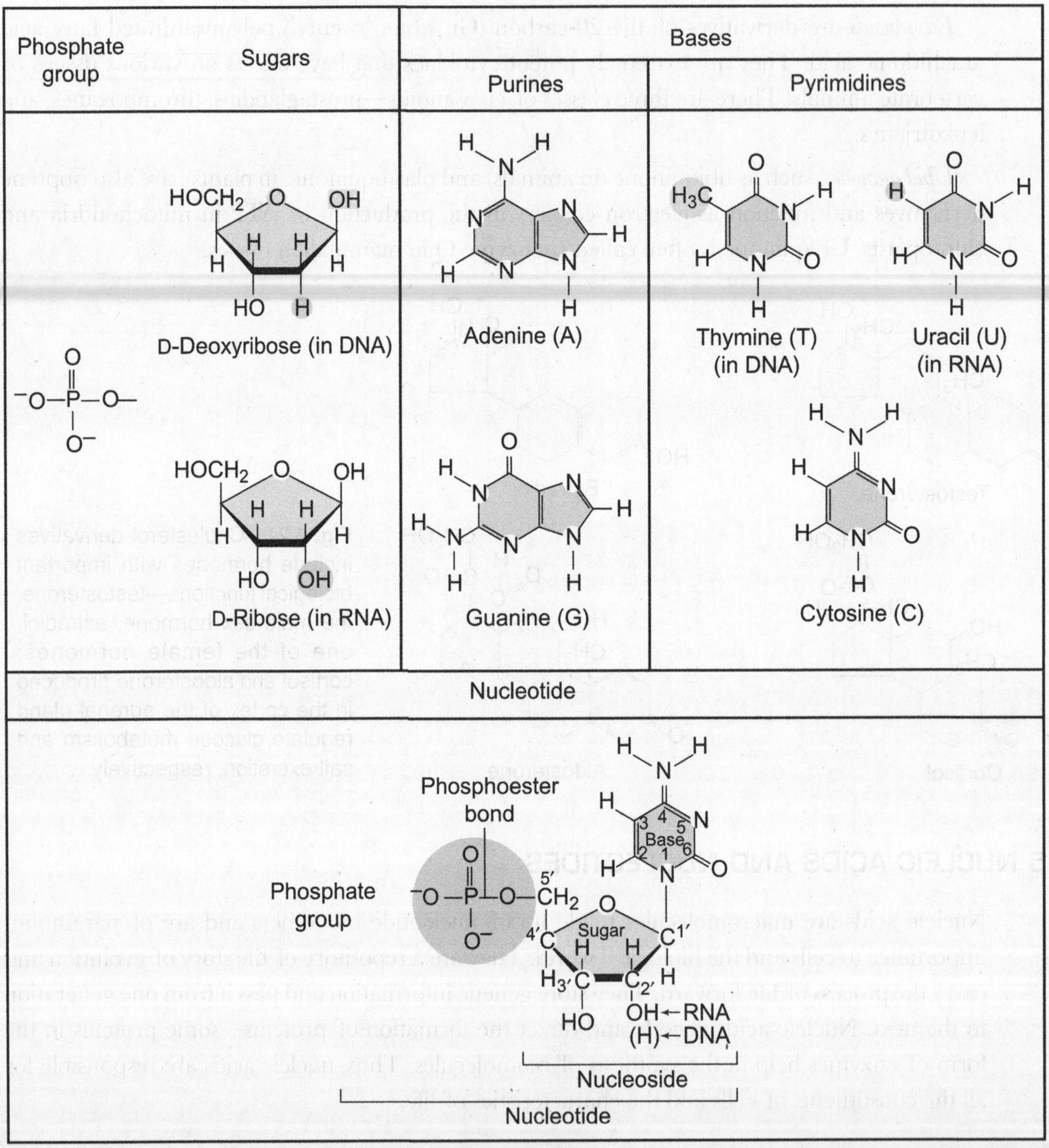

Fig. 4.25 The components and structure of a nucleotide. Base and sugar constitute nucleoside; nucleoside plus phosphate comprise a nucleotide.

Deoxyribonucleic acid (DNA)   The name itself suggests the absence of oxygen in the five-carbon (pentose) ribose sugar (at carbon-2) component of the nucleotides (Fig. 4.25). Thus the sugar of DNA is a *deoxyribose*. DNA carries genetic information (actually constitutes the genes) and directs the synthesis of messenger RNAs (mRNAs) and other RNAs.

Ribonucleic acid (RNA)   RNA contains a ribose sugar which has an oxygen atom in the sugar component of each of its nucleotides (Fig. 4.25). RNA molecules carryout multifarious activities in the cells:

Genetic material   RNA functions as the genetic material in RNA viruses (DNA is the genetic material for DNA viruses and all types of cells).

Messenger RNA (mRNA) synthesis   This is directed by DNA in the process of *transcription* inside the nucleus of a cell. The mRNA is processed within the nucleus and then transported through nuclear pores into the cytoplasm. The mRNA acts as *template for protein synthesis* in the cytoplasm. That is, the nucleotide sequence of mRNA directs the sequence of amino acids to be assembled in a protein molecule in the process of *translation*.

Ribosomal RNA (rRNA)   RNA complexed with certain proteins constitutes ribosomes on which protein synthesis (translation) takes place.

Transfer RNA (tRNA) molecules   These which are of at least 20 different types, each specific for a particular amino acid (20 kinds of amino acids make up protein molecules), and are responsible for recognizing the correct sequence on the mRNA fitted on the ribosome during protein synthesis and ensuring that the amino acid is incorporated at the right position.

Thus, three major species of RNA—mRNA, rRNA, and tRNA, are involved in the transcription of the genetic code of DNA in a cell.

Small interfering RNAs (siRNA) and micro RNAs (miRNA)   These are 21–22 nucleotide-long short RNAs processed in two different ways. They contain specific base sequences and can hybridize to a complimentary portion of target mRNA molecules and stop their function and thus inhibit the expression of genes. Before binding to mRNA, siRNAs bind to a group of proteins to form *RNA-induced silencing complex* (*RISC*), and miRNAs assemble with a group of proteins to form a *ribonucleoprotein complex* (miRNP). siRNA and miRNA have been recently discovered and have triggered extensive research interest especially because of their ability to regulate gene expression.

4.5.2 Nucleotides

The monomeric units of nucleic acids are called nucleotides. Each nucleotide has three characteristic structural components: (i) a nitrogenous aromatic base, (ii) a pentose sugar, and (iii) a phosphate (Fig. 4.25).

Nitrogenous base

Nitrogenous bases are derivatives of two parent compounds—*purine* and *pyrimidine*. Purine is a double-ring structure and is bigger than the single-ring pyrimidine. There are two purine bases—*adenine* (A) and *guanine* (G) and three pyrimidine bases—*thymine* (T), *uracil* (U), and *cytosine* (C), normally found in nucleic acids. Three bases, adenine (A), guanine (G), and cytosine (C) are common for both DNA and RNA, but thymine (T) is present only in DNA and uracil (U) is exclusively found in RNA in the place of thymine. These five aromatic bases together with amino acids, monosaccharides, and fatty acids are the most common small molecules of living systems.

Pyrimidine and purine bases are joined covalently to the 1′ carbon of the pentose sugar through an N-glycosyl linkage at N-1 and N-9 positions respectively.

The nitrogenous base plus pentose sugar linked structure (without the third component, the phosphate group is called a *nucleoside*.

Pentose sugar

The five carbons of the pentose sugar are denoted with a prime (′) following the number to distinguish them from the numbered atoms of the nitrogenous base. Thus, the pentose sugar has carbons marked from 1′ to 5′ (see nucleotide in Fig. 4.25). The nitrogenous base is connected at the 1′ carbon of the sugar by an N-glycosyl bond and the removal of a water molecule (a hydroxyl group from the pentose and a hydrogen from the base). The ribose form of the sugar has an OH (hydroxyl group) at the 2′ carbon and is present in RNA. In DNA, the hydroxyl group on the 2′ carbon is replaced by a hydrogen atom, so the sugar is deoxyribose.

Phosphate group

The pentose sugar is linked to a phosphate group at the 5′ carbon by a phosphodiester bond. Thus, a nitrogenous base, a pentose sugar, and a phosphate group together build a *nucleotide*, which is a unit or monomer of nucleic acids. A nucleotide is also called a *nucleoside monophosphate*, simply because it is a nucleoside with a single phosphate group attached to it. A nucleoside can have more than one phosphate group attached to it; the nucleotides thus formed participate in energy storing and releasing mechanisms in the cells, and are discussed in Section 4.5.3. It is the nucleoside monophosphate forms of the nitrogen bases (A, T, G, C, and U) that participate in the formation of nucleic acids (Table 4.4).

Table 4.4   Nomenclature of nucleosides and nucleotides (precursors) in RNA and DNA

Base (Nitrogenous)	Nucleoside (Base + pentose sugar)	Nucleotide (Nucleoside + phosphates[*])	Nucleic acid (Polymer of nucleotides)
		Purines	
Adenine (A)	Adenosine	Adenylate (ATP)	RNA
	Deoxyadenosine	Deoxyadenylate (dATP)	DNA
Guanine (G)	Guanosine	Guanylate (GTP)	RNA
	Deoxyguanosine	Deoxyguanylate (dGTP)	DNA
		Pyrimidines	
Cytosine (C)	Cytidine	Cytidylate (CTP)	RNA
	Deoxycytidine	Deoxycytidylate (dCTP)	DNA
Thymine (T)	Thymidine or deoxythymidine	Thymidylate or deoxythymidylate (dTTP)	DNA
Uracil (U)	Uridine	Uridylate (UTP)	RNA

Note: Ribonucleotide precursors for RNA are noted as ATP, GTP, CTP, and UTP; deoxyribonucleotides for DNA are noted with a prefix 'd' for deoxyribose sugar (dATP, dGTP, dCTP, and dTTP).

*Two phosphates from triphosphate nucleotide precursors are removed to release energy during synthesis of DNA/RNA; only monophosphate (single phosphate) nucleotides are incorporated into the chain of nucleic acids.

4.5.3 Linear Polymer of DNA and RNA Formed through the Linkage of Phosphate Groups

Phosphate groups connect the 5′ carbon of one nucleotide with the 3′ carbon of the next to form a linear chain of DNA or RNA (Fig. 4.26). The resulting linkage can be described as a 5′,3′ *phosphodiester bond*. This bond gives an intrinsic directionality to the growing polynucleotide chain, that is , 5′ to 3′ with a hydroxyl group at each end. The OH at the 5′ end belongs to the phosphate group and the one at the 3′ end belongs to the sugar (see Figs 4.9 and 4.27). The sugars and phosphate groups constitute the covalent backbone of the nucleic acid shown in light shade in Fig. 4.26. The purine and pyrimidine bases stick out of the backbone. In the case of DNA, a double helix structure (Fig. 4.27a), the bases of the two complimentary chains run in opposite directions, one 5′ to 3′ and the other 3′ to 5′, and join in a specific manner though hydrogen bonds (Fig. 4.27b). Adenine (A) pairs with thymine (T), and cytosine (C) pairs with guanine (G) in DNA. Adenine (A) combines with uracil (U) in RNA–DNA interactions or within the double chain region in RNA molecules as in tRNA.

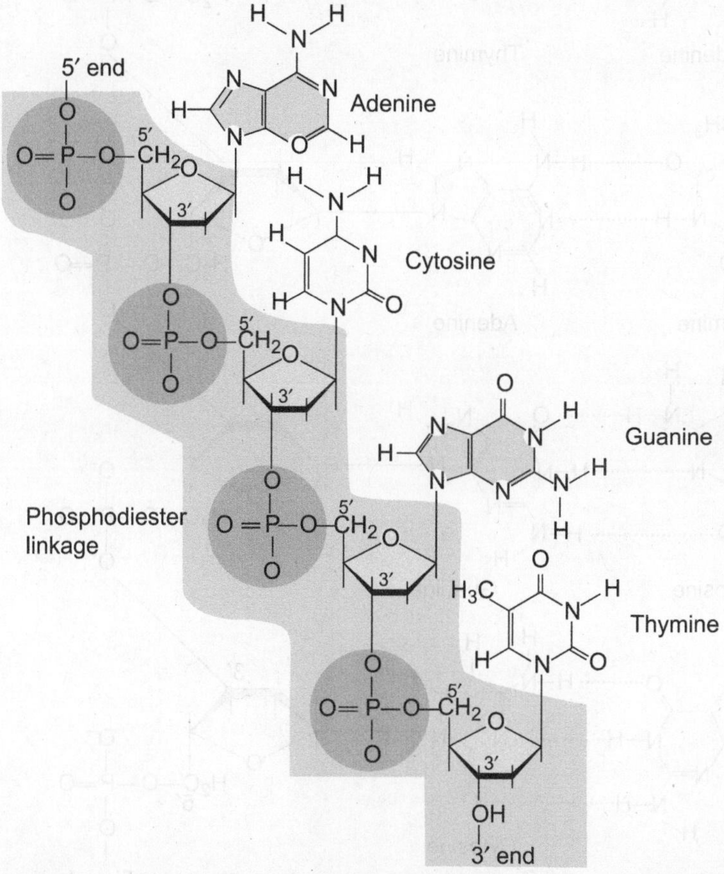

Fig. 4.26 A small portion of a DNA polynucleotide chain, showing the 3′ → 5′ phosphodiester linkage that connects the nucleotides. Phosphate groups actually connect the 3′ carbon of one nucleotide with the 5′ carbon of the next. The sugar and phosphate molecules constitute the outer backbone of DNA and the hydrophobic bases remain in the core of the molecule.

Complementarity for base pairing

Binding of A to T (or U) and C to G between two chains in a complementary fashion is possible due to the appropriate positioning of carbonyl groups and nitrogen atoms in the bases which permits the formation of hydrogen bonds. Two hydrogen bonds are involved in the binding of A and T (or U) and three hydrogen bonds are present between C and G (Fig. 4.27). This base pairing mechanism helps in maintaining the correct sequence for newly synthesized DNA molecules during replication and in synthesizing mRNA during transcription. Correct base pairing in both these processes is the basis of the fidelity of DNA as genetic material.

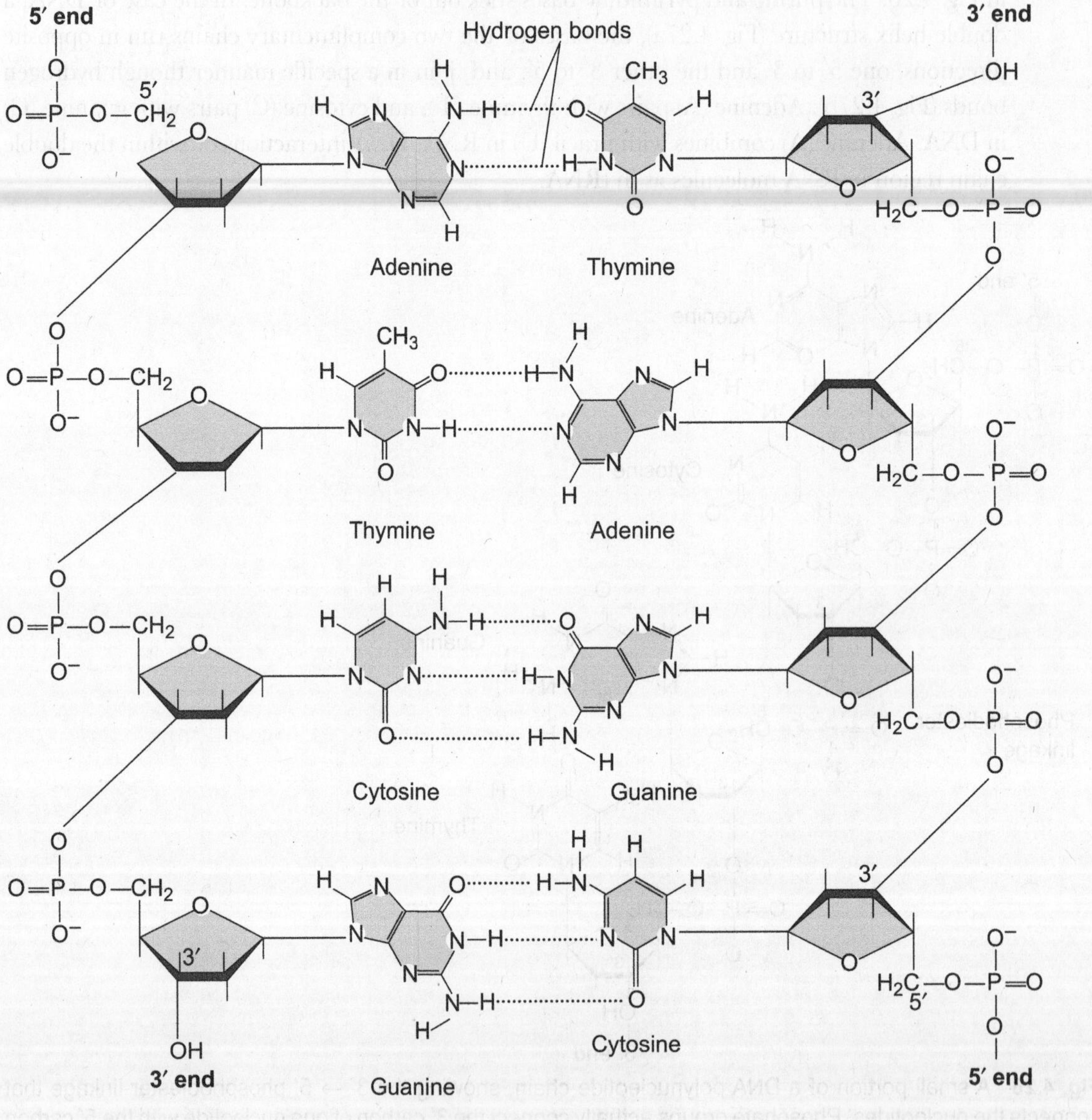

Fig. 4.27 DNA double helix and the antiparallel orientation of two strands and the complementary bases of two chains connected through hydrogen bonds.

Nucleotides—energy-rich compounds

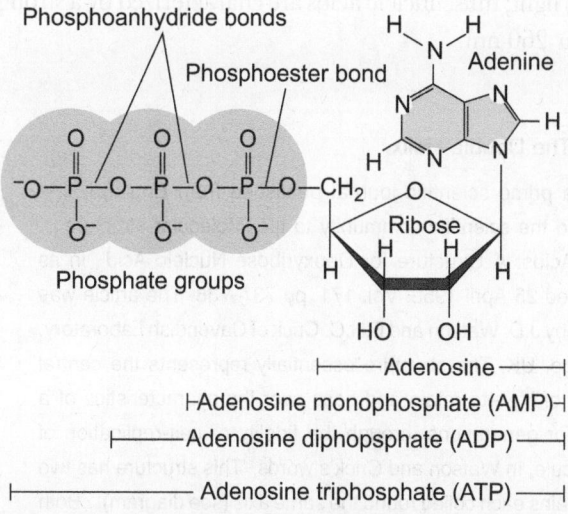

Fig. 4.28 Adenosine as nucleoside and its three phosphorylated forms—AMP, ADP, and ATP. The bond that links the first phosphate to the ribose (deoxyribose in DNA) of adenosine is a low-energy phosphodiester bond, whereas the bonds that link the second and third phosphate groups to the molecule are higher-energy phosphoanhydride bonds.

Nucleotides are energy-rich compounds that drive metabolic processes, primarily biosynthesis, in all cells. They also serve as chemical signals in biochemical cascade reactions in response to cell activation following extracellular stimuli from hormones, antigens, and other factors.

Nucleic acid synthesis requires energy. To provide energy for the formation of a new phosphodiester bond, each successive nucleotide enters as a high-energy nucleoside triphosphate (Fig. 4.28). The precursors for DNA synthesis are, therefore, dATP, dCTP, dGTP, and dTTP ('d' indicates deoxyribose). ATP, CTP, GTP, and UTP are required for RNA synthesis (see Table 4.4).

Figure 4.28 indicates the three forms of adenosine. Adenosine occurs as the free nucleoside, the monophosphate (AMP), the diphosphate (ADP), and the triphosphate (ATP). The triphosphate forms of adenine and other nucleotides are the precursors of DNA and RNA synthesis as indicated in the previous paragraph.

The triphosphate forms lose two phosphate groups to supply energy for the process of nucleic acid synthesis and enter the nucleic acid chain in the monophosphate form (AMP, GMP, CMP, TMP/UMP).

The bond that links the first phosphate to the ribose of adenosine is a low-energy *phosphodiester bond* in comparison to the bonds linking the second and third phosphate groups. The second and third bonds are higher-energy *phosphoanhydride bonds*; they liberate higher amount of free energy on hydrolysis (breaking of the bonds). Hydrolysis of a phosphoanhydride bond produces two to three times more energy than the hydrolysis of a phosphodiester bond.

4.5.4 Characteristics of Nucleic Acids

Acidic

Both pyrimidines and purines are weak bases and hydrophobic in nature. The hydroxyl groups of the sugar residues form hydrogen bonds with water. The phosphate groups in the polar backbone are completely ionized and negatively charged at pH 7; thus, *DNA and RNA are acids*. These negative charges are generally neutralized by ionic interactions with positive charges on proteins (as in eukaryotic chromosomes), metal ions, and polyamines.

The backbones (sugar + phosphate) of DNA and RNA are hydrophilic and remain at the outside, and the pyrimidine and purine bases being hydrophobic remain in the core of the nucleic acid.

UV absorption

The aromatic bases absorb ultraviolet (UV) light; thus, nucleic acids are characterized by a strong absorption of UV light at wavelengths near 260 nm.

EXHIBIT 4.C The Double Helix

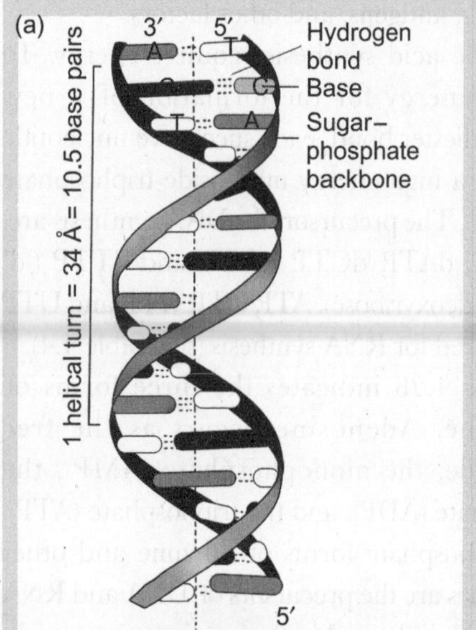

Fig. 4.C1

Nature, a prime scientific journal published from England, first introduced the scientific community to the 'Molecular structure of Nucleic Acids: A Structure for Deoxyribose Nucleic Acid', in its issue dated 25 April 1953, Vol. 171, pp 737–738. The article was authored by J.D. Watson and F.H.C. Crick of Cavendish Laboratory, Cambridge, UK. The structure essentially represents the central molecule of life—to retain and pass over the characteristics of a species for generations through the fidelity or self-replication of the molecule. In Watson and Crick's words, 'This structure has two helical chains each coiled round the same axis (see diagram)…Both chains follow right-handed helices, …' . Elsewhere, in Watson's words, the double helix structure – was too pretty not to be true. They merged data from chemistry, physics, and biology. From Columbia University, Erwin Chargaff's careful chemical analyses, establishing the pyrimidine to purine base ratio in DNA close to one, was one of the corner stones that helped build the double helix model. Watson and Crick also utilized the X-ray diffraction studies of Maurice H.E. Wilkins and Rosalind Franklin of King's College, London.

The work for building the model was done by Watson and Crick between the fall of 1951 and April 1953. The paper of some 900 words explaining the model was sent for publication on 2 April 1953 and was published within 23 days. The short publication had a profound impact on modern biology. In 1962, the Noble Prize for Medicine and Physiology was awarded to Francis H.C. Crick, James D. Watson, and Maurice H.F. Wilkins.

J.D. Watson's book, *The Double Helix* (Publisher Atheneum, New York, 1968) tells the 'inside story' of the interplay of ideas, temperaments of the scientists, and circumstances leading to the revolutionary discovery. It is a must read for the student who has the passion and romance for scientific research.

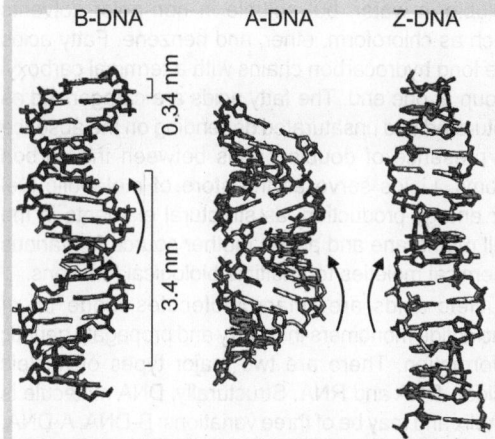

B-DNA   A-DNA   Z-DNA

0.34 nm

3.4 nm

Fig. 4.C2 Different conformations of the double helix.

The double helix exists in different conformations

The right-handed Watson–Crick helix in Fig. 4.C1 is actually an idealized version of a DNA molecule, called B-form DNA (B-DNA). B-DNA is the main form of DNA in cells. Two other structural variants of DNA—A-DNA and Z–DNA have been well characterized in crystal structures (Fig. 4.C2). They exist, perhaps in short segments interspersed within B-DNA molecules.

B-DNA	Usual form, that is, right-handed double helix, characterized by a helical turn consisting of 10 base pairs (3.4 nm); a base pair occupying 0.34 nm. Major and minor grooves distinguishable.
A- DNA	Right-handed double helix, shorter and thicker than B-DNA; 11 base pairs per helical turn. A large tilt of the base pairs with respect to the helix axis (Fig. 4C2). Major groove narrower and deeper than that of B-form, and minor groove broader and shallower. (Double helical RNA adopts A-form.)
Z-DNA	Left-handed double helix (radical feature). Name (Z) derives from zigzag pattern of its longer, thinner sugar–phosphate backbone. Its role in DNA replication, regulation, expression of genes, and genetic recombination is of current research interest.

SUMMARY

♦ Chemical elements present in organisms are of low atomic number and hence low atomic weight. Carbon, hydrogen, oxygen, and nitrogen together make up over 99 per cent of the total atoms (mass) of most cells. Covalently linked carbon atoms form the backbone of biomolecules; functional groups, formed by other atoms, are added to the backbone. Tetrahedral covalent bonds of carbon atoms, other atoms, and functional groups make the three-dimensional structure of a biomolecule.

♦ Biochemical reaction mechanisms are basically similar to chemical reactions. In reactions, electron-rich *nucleophiles* contribute electrons to electron-deficient *electrophiles.* Functional groups containing oxygen, nitrogen, and sulphur are important biological nucleophiles; positively charged hydrogen atoms

(protons) and positively charged metal ions (cations) act as electrophiles in biochemical reactions.

♦ Several atoms of H, O, C, and N make up simpler monomeric biological molecules like monosaccharides, amino acids, fatty acids, and nucleosides with molecular weights in the range of hundreds. These different types of monomers are polymerized into macromolecules of polysaccharides, proteins, lipids, and nucleic acids, respectively. The synthesis of macromolecules is a major energy-consuming activity of living cells. Macromolecules may be organized further into supramolecular structures, such as ribosomes, membranes, and other cellular organelles.

♦ Carbohydrates mean carbon with water. The empirical formula of carbohydrate is $(CH_2O)n$, where n indicates the number of monosaccharide units in a carbohydrate macromolecule. Carbohydrates are the most abundant biomolecules on the earth, serving as the principal source of energy for the living world and supporting structural elements in the cell walls of bacteria and plants, and in the connective tissues and cell coats of animals. Carbohydrates are categorized in three major classes on the basis of size—monosaccharides, oligosaccharides, and polysaccharides.

♦ There are thousands of different varieties of proteins in a cell. Proteins make all kinds of structural elements of a cell, catalyse cellular reactions as enzymes, and carryout a myriad of other functions in living systems. Twenty types of amino acids are covalently linked into linear sequences of polypeptides. Different combinations of these 20 amino acids are responsible for the specific folding, diverse nature, and functions of various protein molecules.

♦ Different types of fatty acids are the principal constituents of most lipids which are sparingly soluble in water, but soluble in non-polar solvents such as chloroform, ether, and benzene. Fatty acids are long hydrocarbon chains with a terminal carboxyl group at one end. The fatty acids are categorized as saturated and unsaturated depending on the absence or presence of double bonds between the carbon atoms. Lipids serve as the store of metabolic fuel for energy production, as structural elements in the cell membrane and as the mother source for various chemical moieties for multiple biological functions.

♦ Nucleic acids are macromolecules made up of nucleotide monomers that carry and propagate genetic information. There are two major types of nucleic acids—DNA and RNA. Structurally, DNA molecule is a helix and may be of three variations: B-DNA, A-DNA, and Z-DNA. RNA molecules can be of different types to carry out multifarious activities—genetic material for certain type of viruses, mRNA, rRNA, tRNA, siRNA, and miRNA.

Lehninger and Franklin

Albert Lester Lehninger
(17 February 1917 – 4 March 1986)

Albert L. Lehninger was an American biochemist in the field of bioenergetics. In 1948, along with Eugene P. Kennedy, he discovered that mitochondria are the site of oxidative phosphorylation in eukaryotes.

Lehninger was born in Bridgeport, Connecticut, USA. He earned a Bachelor of Arts degree in English from Wesleyan University, completing his courses between 1935 and 1939. Originally he intended to write stories and poetry. One of his teachers, Ross Fortner, Jr., introduced him to the emerging field of biochemistry and to the then recent discoveries of Otto Warburg and Hans Kreb on cellular metabolism. Inspired, Albert quickly shifted his interest and major to chemistry. He went on to pursue both MSc (1940) and PhD (1942) at the University of Wisconsin, Madison. His doctoral research was on the metabolism of acetoacetate and fatty acid oxidation by liver cells.

After earning his doctorate in biochemistry, Lehninger held various faculty positions at the UW, Madison and University of Chicago. At Chicago he was academically associated with Charles Huggins, who later won the Nobel Prize for work on cancer treatment. He was appointed DeLamar Professor of the Department of Biological Chemistry in Johns Hopkins University School of Medicine in 1952, and in 1978, he became University Professor of Medical Sciences. He held this title until his death in 1986. He obtained several honours and awards. He made major contributions in the field of bioenergetics and authored three lucid textbooks – *Biochemistry*, *The Mitochondrion*, and *Bioenergetics*. Besides these three classic texts, his collaborative text *Principles of Biochemistry* is a widely used textbook for biochemistry at the graduate and postgraduate levels throughout the world.

**Rosalind Elsie Franklin
(25 July 1920–16 April 1958)**

Rosalind Franklin was an X-ray crystallographer and a British biophysicist. Her data and X-ray diffraction photograph, according to Francis Crick, was 'the data we actually used' to formulate Crick and Watson's 1953 proposition for the structure of DNA.

Rosalind Franklin was born in Notting Hill, London to a merchant banker father and belonged to an affluent and influential British Jewish family. Her early education was at St Paul's Girls School and North London Collegiate School. In 1938, she went to Newnham College, Cambridge where she studied chemistry. Franklin received second class honours in her final examinations in 1941 which was accepted as a bachelor's degree. She earned a PhD from Cambridge University in 1945; her study was on the porosity of coal by X-ray diffraction. Then Franklin took an appointment with Jacques Mering at the *Laboratoire Central des Services Chimiques de l'Etat* in Paris. Mering was a crystallographer and applied X-ray diffraction to the study of rayon and other amorphous substances. This training proved to be useful later when she began working with biological material.

In January 1951, Franklin was appointed as research associate at King's College, London in the Medical Research Council's (MRC) biophysics unit directed by John Randall. Originally she was supposed to work on X-ray diffraction of proteins and lipids in solution; Randall redirected her work to DNA fibres before she started working at King's. This was because she was the only experienced experimental diffraction researcher at King's in 1951, and the pioneering work of Maurice Wilkins and Raymond Gosling had produced an outstanding diffraction picture of DNA using crude equipment, and had fuelled further interest in this molecule. They were working since May 1950. Randall's lack of communication to Wilkins about this

reassignment of DNA diffraction work and guidance of Gosling's PhD thesis to Franklin significantly contributed to the friction that developed between Wilkins and Franklin.

She applied her expertise in X-ray diffraction techniques to decipher the structure of DNA. She used a new fine focus X-ray tube and microcamera ordered by Wilkins, which she refined, adjusted, and focussed carefully. Drawing upon her physical chemistry background, she skillfully manipulated the critical hydration (humid condition) of her specimens. Soon Franklin and Gosling discovered two forms of DNA—'B' form (wet long and thin DNA) and 'A' form (dried, short, and fat DNA).

Randall divided the work on DNA—Franklin chose the data rich 'A' form and Wilkins selected 'B' form because his preliminary pictures had hinted it might be helical. He showed tremendous insight in this assessment of preliminary data. The X-ray diffraction pictures produced by Franklin at this time were described by J.D. Bernal as 'amongst the most beautiful X-ray photographs of any substance ever taken'.

By the end of 1951 it was generally accepted at King's that the 'B' form of DNA was a helix, but Franklin was unconvinced about helical structure of 'A' form and undertook intensive analysis with Gosling by applying the Patterson function to the X-ray pictures of DNA they had produced. By January, 1953 Franklin reconciled her view about the helical structure of DNA and started to write a series of three-draft manuscripts, two of which included a double helical DNA backbone. Her two 'A' form manuscripts reached *Acta Crystallographica* journal in Copenhagen on 6 March 1953, one day before Crick and Watson had completed their model. Franklin definitely had mailed them while the Cambridge team was in the process of building their model, and had written a lot earlier without any knowledge of their work. On 8 July 1953 she modified one of these Acta articles in proof 'in light of recent work' by the King's and Cambridge research teams. The third manuscript on the 'B' form of DNA, dated 17 March 1953 was located years later in her papers by Aaron Klug, a colleague at Birbeck College. On the other hand, Watson and Crick's article on DNA model was published in *Nature* on 25 April 1953. It was submitted to the journal on 2 April 1953; this was a remarkably rapid publication even for that time.

Watson vividly described in his book *The Double Helix* that on 30 January 1953, he travelled to King's carrying a preprint of Linus Pauling and Corey's incorrect proposal for DNA structure and an urgent message that they should all collaborate before Pauling found his error. The unimpressed Franklin became angry when Watson suggested she did not know how to interpret her own data. On his hasty retreat Watson met Wilkins; Wilkins showed him Franklin's famous diffraction photograph 51, without Franklin's permission or knowledge. This photograph provided the Cambridge pair critical insights into the DNA structure which changed the course of DNA history.

Rosalind moved to Birkbeck College in March, 1953. John Randall asked her to leave her DNA data with King's College. With her critical contribution to DNA structure building, the question often arose as to why she was not considered for the Nobel Prize along with Watson, Crick, and Wilkins in 1962. It was indicated that the Nobel Prize rules forbid posthumous nomination. She died of ovarian cancer at the age of 37 on 16 April 1958 at Chelsea, London.

EXERCISES

Objective Questions

1. Multiple Choice Questions
   (i) Which of the following is a correct statement?
   (a) Hydrogen (H), oxygen (O), nitrogen (N), and carbon (C) together contribute 66 per cent of the total atomic mass of most cells.
   (b) H, O, N, C, and water together make 99 per cent of the atom mass of the most cells.
   (c) Usually H, O, N, and C together constitute 99 per cent of the total atom mass of a cell.
   (d) All the above statements are not true.

2. Mark the following statements as True or False.
   (a) Usually the lightest elements are capable of forming the strongest covalent bonds.
   (b) Organic compounds can only be produced by living organisms.
   (c) Molecules with covalently bonded carbon backbones are called organic compounds.
   (d) The freedom of rotation of carbon–carbon single bond is restricted when only highly charged groups are attached to both carbon atoms.
   (e) The freedom of rotation of carbon–carbon single bond is restricted when very large or highly charged groups are attached to both the carbon atoms.
   (f) Amino acids in proteins are all in the leavorotatory (L) isomeric form in the eukaryotic cells.
   (g) Amino acids in the proteins of eukaryotic cells can be both L and D (dextrorotatory) isomeric forms.

   (h) Lysine and arginine are the only basic amino acids.
   (i) Lysine, arginine, and histidine are the basic amino acids.
   (j) Zwitterion has both a negative and a positive charge at the middle range of pH.
   (k) Zwitterion's dipolar nature appears at the high range of pH.
   (l) Crystalline amino acids have zwitterions and require a much higher melting temperature compared to other molecules of the same size.
   (m) Non-standard amino acids are derivatives of standard amino acids.
   (n) Ninhydrin reaction identifies all amino acids.
   (o) Ninhydrin reaction identifies all amino acids except proline.
   (p) Glycoproteins are polysaccharides conjugated to proteins.

3. Match the statements in Column 2 with the items in Column 1. There could be more than one match for an item in Column 1.

Column 1	Column 2
(i) Amino acids	(a) Hydrophobic
(ii) Fatty acids	(b) Triacylglycerol
(iii) Glycans	(c) Saturated and unsaturated
state	
(v) Protein	(d) Act as both acid and base
(v) Adipocyte	(e) Ganglioside
(vi) Grey matter of	(f) Most of them are with
human brain	even number of carbon
atoms	(g) Polymer

Review Questions

1. Differentiate between the following:
   (a) Nucleophiles and electrophiles
   (b) Aldose and ketose sugars
   (c) DNA and RNA

2. Define carbohydrates.

3. What are the ultimate sources of the elements for biomolecules?

4. What are functional groups? Give a few examples.

5. What are epimers?

6. Why are monosaccharides called reducing sugars? What are the routine tests employed to detect them?

7. How are polysaccharides formed? Explain the basis of classification of polysaccharides.

8. Give some examples of transport proteins.

9. What are the basic characteristics of amino acids?

10. What is the basis of classification of amino acids?

11. What are the biological activities that small peptides participate in?

12. Write a short note on triacylglycerol.

13. What are structural lipids and why have they been given this name?

14. Define nucleoside.

15. What is RISC?

16. Where does the energy required to form phosphodiester bonds during nucleic acid synthesis come from?

5

Fundamentals of Biochemical Engineering

LEARNING OBJECTIVES

♦ Optimum biological conditions for cell growth such as pH (acidity and basicity), and buffering
♦ Different systems of physical units, dimensions, and physical variables
♦ Stoichiometry for material and energy balance in biochemical reactions and cell growth
♦ Statistical analysis and probability
♦ Components in a bioprocess
♦ Bioreactors (fermenters) for producing biomass or metabolites in biotechnology

---

INTRODUCTION

Biochemical engineering is the branch of life sciences that deals with the design and construction of unit processes that involve biological organisms or molecules. Living cells or their components, such as enzymes, are allowed to catalyse different types of biochemical reactions in large vessels (bioreactors) to produce useful compounds. Microorganisms and eukaryotic cells are cultured in these facilities for the large-scale production of antibiotics, vitamins, vaccines, monoclonal antibodies, hormones, and many other beneficial compounds. Genetically manipulated microorganisms are employed for the production of a targeted compound, which is normally synthesized by other strains or species or higher organisms, or by human beings. *Biochemical engineering* is often called *bioprocess engineering and technology*, and is the core of modern biotechnology that involves mass-scale industrial production.

Biochemical engineering involves an understanding of the optimum biological conditions for cell growth and catalytic reactions and the principles of process technology, which is a branch of chemical engineering. Use of a quantitative approach for different biological aspects helps in the development of processing instruments, and the regulation and monitoring of key variables in large-scale industrial production.

The parameters such as pH, dissolved oxygen, agitation and aeration rates, pressure, heat transfer, and homogeneity of the reaction are measured in the course of production of biomass and certain products in a fermenter. The key variables involved are cell density, substrate concentration, and product concentration. The quantitative study of all the above-mentioned parameters provides the information necessary for reactor design. Modern reactors (fermenters) are equipped with instrumentation to measure these parameters online (in the course of running) very accurately. Optimization of input versus product is another fundamental aspect in biochemical engineering.

Biotechnological processes require quantification in terms of physical units, dimensions, and physical variables which are also dealt with in this chapter.

5.1 CONCEPT OF PH

pH is a measure of the acidity or basicity of a solution. It indicates whether a solution is acidic, alkaline, or neutral. Measurement of pH is an important criterion for biochemical reactions. It affects the structure and activity of enzymes, and the well-being of organisms and their cells. Usually, cells thrive better in neutral pH (pH 7). Figure 5.1 shows the pH range (0–14) and the positions of fluids in the range, including human blood and tears around neutral pH. Acidic solutions have a pH in the range of 0 –7 and the pH of basic solutions ranges between 7–14.

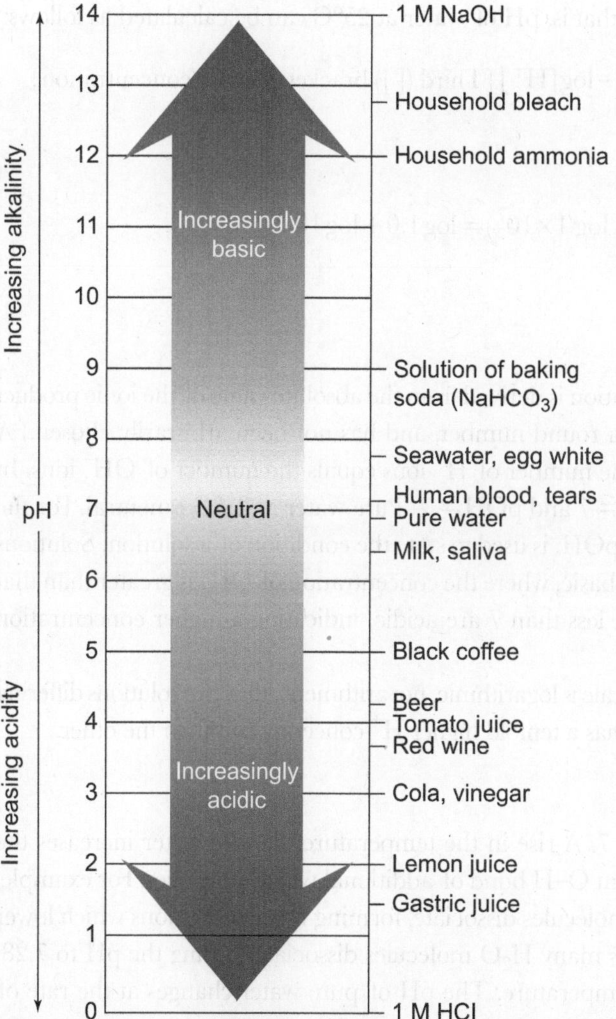

Increasing alkalinity
Increasing acidity
pH
14
13
12
11
10
9
8
7
6
5
4
3
2
1
0
Increasingly basic
Neutral
Increasingly acidic
1 M NaOH
Household bleach
Household ammonia
Solution of baking soda (NaHCO3)
Seawater, egg white
Human blood, tears
Pure water
Milk, saliva
Black coffee
Beer
Tomato juice
Red wine
Cola, vinegar
Lemon juice
Gastric juice
1 M HCl

Fig. 5.1 pH values of some aqueous fluids.

A small proportion of water molecules in any solution dissociates into ions by breaking one of the covalent bonds between the oxygen and hydrogen atoms. When the bond breaks, one of the hydrogen atoms loses its electron to the oxygen atom. With the extra electron, the OH part becomes negatively charged and by losing the electron, H turns into a positively charged ion.

$$\underset{\text{Water molecule}}{\text{H :O: H}} \xrightarrow{\text{Dissociates}} \underset{\text{Hydroxyl (OH}^-\text{) ion}}{\text{H :O:}^-} + \underset{\text{Hydrogen (H}^+\text{) ion}}{\text{H}^+}$$

Thus, the dissociation of a water molecule produces one hydroxyl ion (OH^-) and one proton or hydrogen ion (H^+). By convention, the hydrogen ion is shown as the end product, but in reality it combines with a water molecule, forming a hydronium ion (H_3O^+).

Definition of pH

pH is the measure of the hydrogen ion concentration in any aqueous solution. The symbol p denotes *negative logarithm* of H, that is, hydrogen ion (H^+).

The dissociation of water into ions is a reversible process. The dissociation of water is always accompanied with the re-association of ions and formation of water molecules; both events occurring simultaneously. In due course, only a very small proportion of water molecules remains dissociated at any given time, approximately 1×10^{-7} at room temperature (25°C). This value is determined on the basis of the molarity of pure water and its electrical conductivity due to ions arising from the dissociation of H_2O molecules. Thus, the concentration of protons (H^+) arising from the dissociation of water molecules at 25°C is 10^{-7} M (so the number for OH^-). The concentration of hydrogen ions, that is, pH of water at 25°C can be calculated as follows:

$$pH = \frac{\log 1}{[H^+]} = -\log[H^+] \{\text{Third ([ ]) brackets denote concentration}\}$$

At 25°C,

$$pH = \frac{\log 1}{1 \times 10^{-7}} = \log(1 \times 10^7) = \log 1.0 + \log 10^7$$

$$= 0 + 7.0$$

$$= 7$$

The value 7 for pH of a neutral solution is derived from the absolute value of the ionic product of water at 25°C. (It happens to be a round number, and has not been arbitrarily chosen.) A solution is considered neutral when the number of H^+ ions equals the number of OH^- ions. In other words, pH = pOH, that is, pH = 7 and pOH = 7. Pure water at 25°C is neutral. For the sake of convenience, pH, rather than pOH, is used to state the condition of a solution. Solutions with pH greater than 7 are alkaline or basic, where the concentration of OH^- is greater than that of H^+. Conversely, solutions with pH less than 7 are acidic, indicating a higher concentration of H^+ ions.

It may be stressed here that the pH scale is logarithmic, not arithmetic; thus two solutions differing by 1 pH unit means that one solution has a tenfold higher H^+ concentration than the other.

5.1.1 Effect of Temperature on pH

At 25°C, pure water has neutral pH 7. A rise in the temperature of pure water increases the thermal energy that breaks the covalent O–H bond of additional water molecules. For example, at 45°C, almost twice as many H_2O molecules dissociate, forming more of H^+ ions which lower the pH to 6.72. At 5°C, about half as many H_2O molecules dissociate, raising the pH to 7.28. Thus, the pH varies inversely with temperature. The pH of pure water changes at the rate of 0.017 units per degree Celsius.

Changes in pH or temperature affect the ionization states of amino acids which, in turn, affect the way in which a protein folds or functions.

5.1.2 Acids and Bases Alter the pH of Water

An acid may be defined as a proton donor and a base as a proton acceptor. An acid has a tendency to lose its proton in an aqueous solution. For example, hydrochloric acid (HCl) dissociates in water into H^+ and Cl^-, thereby increasing the hydrogen ion concentration, and lowering the pH value of the solution. When sodium hydroxide (NaOH), a base, is dissolved in water, it rapidly dissociates into Na^+ and OH^-. The OH^- ions arising out of the NaOH dissociation rapidly interact with H^+ ions to form H_2O, reducing the $[H^+]$ and increasing the pH.

The degree to which acids and bases change the pH of a solution depends on the ease with which the molecules dissociate (ionize) under normal conditions. The stronger the acid, the greater is its tendency to lose its proton. Inorganic acids such as hydrochloric acid (HCl), sulphuric acid (H_2SO_4), and nitric acid (HNO_3), are considered strong acids; they are completely ionized in dilute aqueous solutions. The strong bases NaOH and KOH are also completely ionized to release OH^-.

Many biological molecules are weak acids or weak bases, which are only partially ionized under physiological conditions.

5.1.3 Dissociation Constant

Dissociation of any acid (HA) produces H^+ and an anion, A^-. A reversible chemical reaction can be represented by the following equation

$$HA \rightleftharpoons H^+ + A^-$$

The relationship between the substrate (HA) and products (H^+ and A^-) in a reversible reaction can be defined by *mass action ratio*, and expressed in terms of the equilibrium constant (K).

$$K = \frac{[H^+] \ [A^-]}{[HA]}$$

An increase in both $[H^+]$ and $[A^-]$ results in a decrease in $[HA]$; this also causes an increase in the mass action ratio. At some point, the reaction slows down with $[HA]$ reaching a minimum and $[H^+]$ and $[A^-]$ reaching a maximum. When this occurs, the reaction is at equilibrium. At equilibrium, the mass action ratio or K attains a specific value for a particular acid. It is called *equilibrium constant* or *dissociation constant*. The dissociation constant of acids is often designated as Ka. In most cases, the dissociation constant is converted to its negative log ($-\log Ka$), analogous to the way $[H^+]$ is converted to pH, following the equation

$$pKa = \log \frac{1}{Ka} = -\log Ka$$

The Ka and pKa values for some acids are given in Table 5.1.

Table 5.1   Dissociation constant (Ka) and pKa values of some common weak acids at 25°C

Acid	Dissociation	Ka (M)	pKa
Acetic acid	$CH_3COOH \rightarrow CH_3COO^- + H^+$	1.74×10^{-5}	4.76
Carbonic acid	$H_2CO_3 \rightarrow HCO_3^- + H^+$	1.70×10^{-4}	3.77
Bicarbonate	$HCO_3^- \rightarrow CO_3^{2-} + H^+$	6.31×10^{-11}	10.2
Formic acid	$HCOOH \rightarrow HCOO^- + H^+$	1.78×10^{-4}	3.75
Phosphoric acid	$H_3PO_4 \rightarrow H_2PO_4^- + H^+$	7.25×10^{-3}	2.14
Dihydrogen phosphate	$H_2PO_4^- \rightarrow HPO_4^{2-} + H^+$	1.38×10^{-7}	6.86
Monohydrogen phosphate	$HPO_4^{2-} \rightarrow PO_4^{3-} + H^+$	3.98×10^{-13}	12.4
Ammonium	$NH_4^+ \rightarrow NH_3 + H^+$	5.62×10^{-10}	9.25

The more strongly dissociated the acid, the lower its pKa.

5.1.4 Effects of Changes in pH on Enzyme and Cellular Activities

It has already been mentioned that changes in pH affect the ionization states of certain amino acids, and in turn, the folding and function of protein molecules. It is the catalytic activity of enzymes that is especially sensitive to changes in pH. Enzymes typically show maximal catalytic activity at a characteristic pH, called the *optimum* pH (Fig. 5.2). A small change in pH can lead to a significant change in the catalytic rate of some crucial enzymes. Ionic interactions are among the forces that stabilize a protein molecule and allow an enzyme to recognize and bind to its substrate.

Human blood plasma normally maintains a pH close to 7.4. When pH-regulating mechanisms fail, as may happen in severe uncontrolled diabetes, metabolically produced acids cause *acidosis*. The pH of the blood can fall to 6.8 or below, leading to irreparable cell damage and death. In some other diseases, pH may rise to a lethal level, causing the condition known as *alkalosis*.

Thus, biological control of the pH of cells and body fluids is of central importance in all aspects of metabolism and cellular activities. This control is achieved through *buffer systems*.

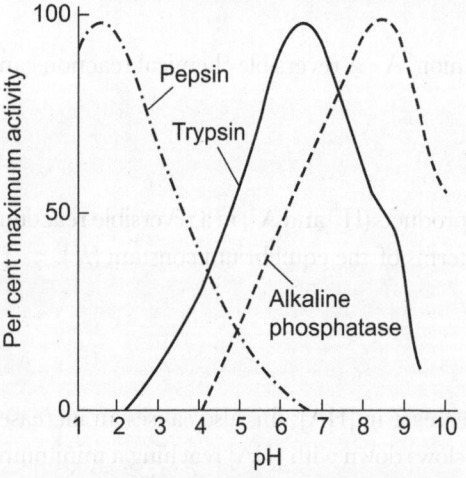

Fig. 5.2 Variation in the pH optima (at which enzymes act best) of some enzymes. Pepsin, a digestive enzyme secreted into acidic gastric juice; trypsin, a digestive enzyme, functions in the small intestine at more or less neutral pH; alkaline phosphatase of bone tissue.

5.1.5 Buffering to Maintain Constant pH in Biological Systems

A specific and constant cytoplasmic pH, usually near pH 7, is maintained in cells and organisms to keep biomolecules in their optimal ionic state that stabilizes their structure and function. In multicellular organisms, the pH of extracellular fluids such as the blood is also tightly regulated.

A constant pH is achieved primarily by biological buffers—mixtures of weak acids and their conjugate bases. The protonated acid liberates protons when the pH increases. Conversely, the deprotonated base accepts protons when a higher concentration of H^+ ions lowers the pH.

5.1.6 Buffer

A *buffer system* is an aqueous system, consisting of a weak acid (the proton donor) and its conjugate base (the proton acceptor). It resists changes in its pH when small amounts of acid (H^+) or base (OH^-) are added, for a particular range of pH values.

Acetic acid and acetate buffer—an example of a buffer system

Acetic acid is a weak acid and proton donor, and acetate as a base is a proton acceptor. Their combination can be used as a buffer system for a pH range from 4.26 to 5.26 (Fig. 5.3). In this range of pH, small addition of OH^- or H^+ ions does not change the pH significantly (shown in the figure in grey). Figure 5.3 represents the titration curve of acetic acid, after increasing amounts of NaOH (X-axis) have been added to the acetic acid solution for neutralizing it (that is, to maintain the pH at 7). Changes in the pH values are plotted on the Y-axis.

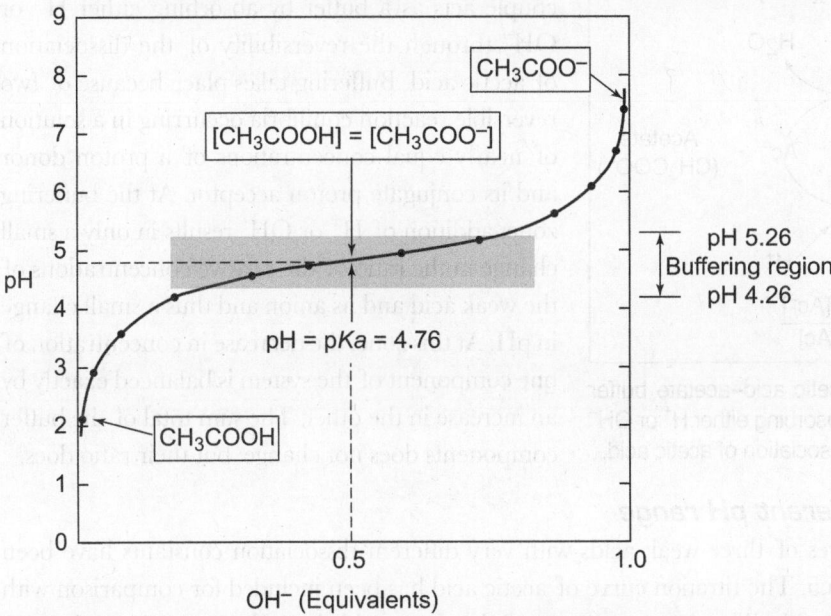

Fig. 5.3 The titration curve of a 0.1 M solution of acetic acid with 0.1 M NaOH at 25°C. The predominant ionic forms at designated points in the titration are mentioned. At the midpoint of the titration, the concentration of the proton donor (acetic acid) and proton acceptor (acetate) are equal. The pH at this point is numerically equal to the pKa of acetic acid. The shaded zone is the buffering region.

At the beginning of the titration (before adding any NaOH), acetic acid is already slightly ionized. This can be estimated from its dissociation constant or Ka (see Table 5.1). As NaOH is gradually added, the released OH^- combines with the free H^+ in the solution to form H_2O. As free

H^+ is removed, HAc (acetic acid) dissociates further to satisfy its own dissociation or equilibrium constant. As the titration (adding of NaOH) proceeds, more and more HAc ionizes, forming Ac^-. At the midpoint of titration (Fig. 5.3), at which exactly 0.5 equivalent of NaOH has been added, one half of the original acetic acid has undergone dissociation, and the concentration of proton donor (HAc) now equals that of the proton acceptor Ac^- (acetate). At this midpoint, the pH of the equimolar solution of acetic acid and acetate is exactly equal to the pKa of acetic acid, that is, 4.76 (See Table 5.1).

As the titration proceeds with the continuous addition of NaOH, the remaining undissociated acetic acid dissociates gradually to transform into acetate. The end point of the titration is reached at about pH 7, when all the acetic acid has dissociated and lost its protons to OH^- to form H_2O and acetate. Throughout the titration, two equilibria coexist, each conforming to its dissociation constant:

$$H_2O \rightleftharpoons H^+ + OH^-$$

$$HAc \rightleftharpoons H^+ + Ac^-$$

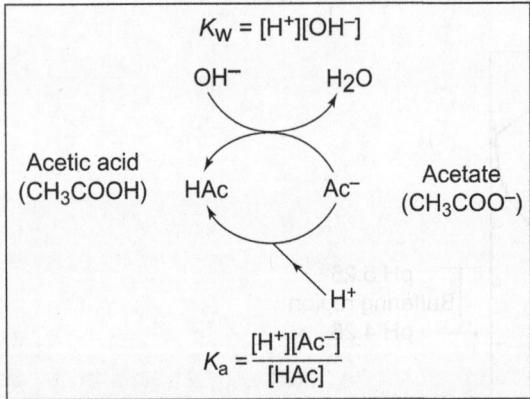

Fig. 5.4 Principle of the acetic acid–acetate buffer system which is capable of absorbing either H^+ or OH^- through reversibility of the dissociation of acetic acid.

Figure 5.4 summarizes how acetic acid–acetate couple acts as a buffer by absorbing either H^+ or OH^- through the reversibility of the dissociation of acetic acid. Buffering takes place because of two reversible reaction equilibria occurring in a solution of nearly equal concentrations of a proton donor and its conjugate proton acceptor. At the buffering zone, addition of H^+ or OH^- results in only a small change in the ratio of the relative concentrations of the weak acid and its anion and thus a small change in pH. At this zone, the decrease in concentration of one component of the system is balanced exactly by an increase in the other. The sum total of the buffer components does not change, but their ratio does.

Buffers of different pH range

The titration curves of three weak acids with very different dissociation constants have been presented in Fig. 5.5. The titration curve of acetic acid has been included for comparison with that of two other acids. The titration curves of the three acids have the same shape, but are placed at different points on the pH scale depending on the strength of the acids. Acetic acid is the strongest, and loses its proton most readily, thus its Ka is the highest and pKa is the lowest of the three. Acetic acid gets half dissociated at pH 4.76. Dihydrogen phosphate dissociates less readily and gets half dissociated at pH 6.86. Ammonium is the weakest acid among the three and reaches the half dissociated condition at pH 9.25. Each conjugate acid–base pair serves as buffer at different pH zones. The $H_2PO_4^-$ and HPO_4^{2-} pair that has a pKa of 6.86 serves as a buffer system around pH 6.86, precisely from pH 6.36 to pH 7.36. Similarly the NH_4^+ and NH_3

pair with a pKa of 9.25 acts as buffer near pH 9.25, that is, from pH 8.75 to 9.75. Thus, as per the need for buffering at different zones of pH scale, the right combination of weak acid–base can be chosen.

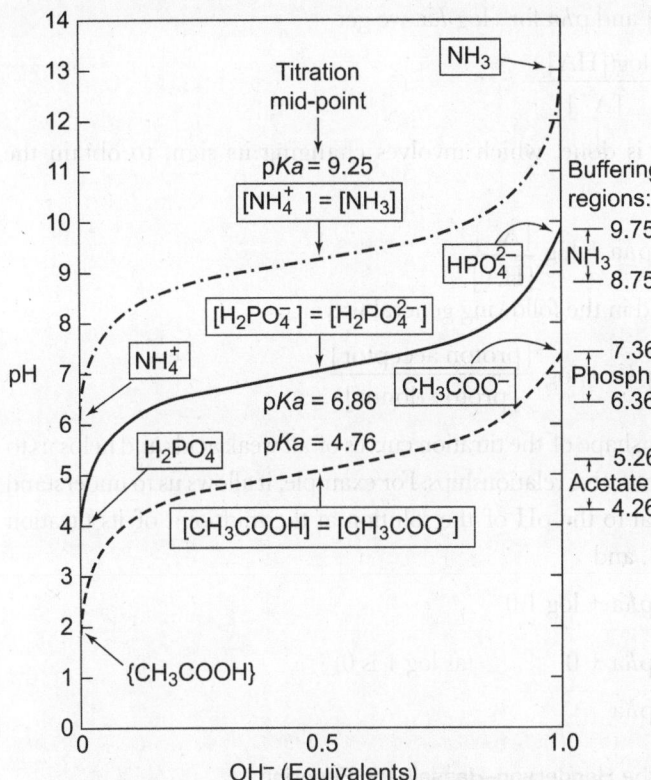

Fig. 5.5 The titration curves of three weak acids: acetic acid (CH₃COOH), dihydrogen phosphate(H₂PO₄), and ammonium (NH₄⁺) are compared. They have different dissociation values (pKa). The predominant ionic forms at designated points in the titration are given in the boxes.

5.1.7 Henderson–Hasselbach Equation Relates pH, p*K*, and Buffer Concentration

The similar shape of the titration curves of acetic acid, $H_2PO_4^-$, and NH_4^+ (Fig. 5.5) suggests some fundamental correlation among pH, the buffering action of a mixture of weak acid and its conjugate base, and the pKa of the weak acid. Henderson and Hasselbach introduced the equation for understanding acid–base balance and buffer action in the blood and tissues of the vertebrate organisms. In fact, they rearranged the equation for the dissociation constant of an acid.

The original equation for the dissociation constant of a weak acid HA into H⁺ and A⁻ is (as stated earlier)

$$Ka = \frac{[H^+][A^-]}{[HA]}$$

Henderson and Hasselbach changed it to

$$[H^+] = \frac{Ka[HA]}{[A^-]}$$

Taking the negative logarithm of both sides, we get

$$-\log[H^+] = -\log Ka - \frac{\log([HA])}{[A^-]}$$

Substitution of pH for $-\log[H^+]$ and pKa for $-\log Ka$, we get

$$pH = pKa - \cdot\frac{\log([HA])}{[A^-]}$$

Inversion of $-\log([HA]/[A^-])$ is done, which involves changing its sign, to obtain the Henderson–Hasselbach equation:

$$pH = pKa + \log\frac{[A^-]}{[HA]}$$

The above equation can be stated in the following general way:

$$pH = pKa + \log\frac{[\text{proton acceptor}]}{[\text{proton donor}]}$$

This equation explains the similar shape of the titration curves of all weak acids and helps us to deduce a number of important quantitative relationships. For example, it allows us to understand why the pKa of a weak acid is equal to the pH of the solution at the midpoint of its titration (Fig. 5.5). At this point [HA] = [A$^-$], and

$$pH = pKa + \log 1.0$$
$$= pKa + 0 \qquad (\text{as log 1 is 0})$$
$$= pKa$$

Exhibit 5.A   Use of the Henderson–Hasselbach Equation

The Henderson–Hasselbach equation makes it possible to find out the pKa of any acid or the molar ratio of proton acceptor and proton donor or the pH of a given acid–base conjugate, when three of these values are given.

Example

The pH of a mixture of 0.1 M acetic acid and 0.2 M sodium acetate, when pKa of acetic acid is 4.76, can be easily calculated by placing these values in the Henderson–Hasselbach equation:

$$pH = pKa + \log\frac{[\text{acetate}]}{[\text{acetic acid}]}$$
$$= 4.76 + \log\frac{0.2}{0.1}$$
$$= 4.76 + 0.301$$
$$= 5.06 \ (\text{pH of the mixture})$$

5.2 PHYSICAL VARIABLES

Bioprocess engineering is planned and executed on the basis of quantitative physical measures and variables. Examples of such measures and variables are mass, length, time, velocity, area, temperature, viscosity, and density. It is difficult to assign a numerical value to qualitative realizations such as odour, and hence they cannot be considered for bioprocess engineering.

Physical quantities are often classified into two groups: — substantial variables and natural variables.

Substantial variables   The magnitude of substantial variables is expressed in terms of precise physical standards, called *units*, such as metre, seconds, and gram.

Natural variables   The expression of natural variables does not require units or any other standard of measurement. That is why these variables are referred to as *dimensionless constants* or *dimensionless numbers*. For example, the ratio of the length of a cylinder to its diameter is a dimensionless number.

Dimensionless numbers are used by engineers to represent physical phenomena and for use in designing. For example, in fluid mechanics, transition from laminar to turbulent flow depends not only on the velocity of the fluid, but also on its viscosity and density, and the geometry of the flow conduit (diameter of the pipes). Reynolds number (R_θ) is used as a parameter for fluid flow:

$$R_\theta = \frac{Du\rho}{\mu}$$

Here D is the pipe diameter, u is the fluid velocity, ρ is the fluid density, and μ is the fluid viscosity.

5.3 UNITS AND DIMENSIONS

Bioprocess engineering is executed on the basis of physical quantities. Units are used to measure physical quantities, and there are several systems of units. Again, a class of physical quantities is called a dimension. Mass, length, and time are the three basic dimensions of physical quantities. Depending on these base dimensional quantities, other dimensions of physical quantities such as area, density, acceleration, efficiency, and energy can be calculated.

5.3.1 Unit

To improve our understanding about units, let us answer a few fundamental questions.

What is meant by a unit?

It is a standard which is used for the measurement of a physical quantity.

What is the meaning of a measurement?

A measurement is a combination of a numerical value (n) and the name of the unit (x). For example, 102 kilometres, here 102 is n, kilometres is the unit.

What are the basic essentials which a unit must possess?

- The unit should have a convenient magnitude.
- It should not vary due to a variation in external factors.
- It should be easily accessible.
- It should be non-perishable.

How can one show that the numerical value of a measurement and the size of the unit used for a measurement are inversely proportional to each other?

Since, nx = constant

$$n \propto \frac{1}{x}$$

For example 1 km = 1000 m = 1000 × 100 cm

As the unit decreases from km to m to cm, its numerical value increases from 1 to 1000 to 1000 × 100.

5.3.2 Systems of Units

For measurements, the following unit systems are used (Table 5.2).

CGS system (also known as *French system*)   It is a system of measurement in which the fundamental units of measurement of length, mass, and time are taken as 1 centimetre (cm), 1 gram (gm) and 1 second (s), respectively. It is also known as *Gaussian system*. Many of the derived units of this system are sometimes inconveniently small.

FPS system (also known as *British system*)   In this system, the fundamental units of length, mass, and time are foot (ft), pound (lb), and second (s), respectively. Pound can also be used as a unit of force in the English gravitational system.

MKS system (also known as *Metric system*)   It is the system of measurement in which the fundamental units of length, mass, and time are 1metre (m), 1 kilogram (kg), and 1 second (s), respectively.

Table 5.2   The three systems of units

Measurement of	CGS	FPS	MKS
Length	1 cm	1 ft	1 m
Mass	1 gm	1 lb	1 kg
Time	1 s	1 s	1 s

SI units (International systems of units)   The three fundamental units of length, mass, and time serve the purpose of basic measurements. But these units are not sufficient for measurements of the physical world (or so to say Physics). For example, these three units cannot be used to measure temperature, luminosity in light, and electric current. Thus, four more units have been introduced to measure some other fundamental qualities, and thereby, increasing their number from three to seven. The SI system is an extension and modification of the MKS system.

The fundamental SI units are:
1. Unit of mass   1 kilogram   (kg)
2. Unit of length   1 metre   (m)
3. Unit of time   1 second   (s)

4. Unit of electric current   1 ampere   (A)
5. Unit of temperature   1 Kelvin   (K)
6. Unit of luminosity   1 candela   (cd)
7. Unit amount of a substance   1 mole   (mol)

Supplementary SI units

Two more fundamental units for the measurement of angle and solid angle have been added and are called *'supplementary SI units'*.

Unit of angle (radian, rad)   Radian is the angle subtended, at the centre of a circle, by an arc whose length is equal to the radius of the circle.

$$2\pi \text{ radian} = 360°$$

Unit of solid angle (steradian, sr)   Steradian is the solid angle subtended at the centre of a sphere by a surface area of the sphere whose magnitude is equal to the square of the radius of the sphere.

Coherent system of units (SI system)   SI system is also described as a *coherent system of units*. A coherent system of units is a system in which the units of all the quantities can be obtained from the fundamental units by simple multiplication or division without introducing any numerical value.

Rational system of units (SI system)   In a rational system of units all the physical quantities, when qualitatively similar, are expressed in the same unit. A coherent system of units can be considered as a rational system of units.

For example, in the MKS system, energy is expressed in *Joule* (in mechanics), *calorie* (in heat), and *watt-hour* (in electricity). This system is not rational. On the other hand, the SI system expresses all types of energy in *Joule*, and hence it is a rational system of units.

5.3.3 Basis of Fundamental Units

The units of fundamental quantities are called fundamental units. The fundamental quantities for different units are as follows:

Kilogram   It is the mass of a cylinder of platinum–iridium alloy, kept at the International Bureau of Weights and Measures in Sevres, France.

Metre   It is defined as 1650763.73 times the orange red wavelength of Kr-86 (isotope of krypton), emitted due to decay or electronic transitions between two particular energy levels $5d_5 \rightarrow 2p_{10}$.

Second   It is the time interval which corresponds to 9,19,26,31,770 time periods of a selected transition between two hyperfine levels of the ground state of Cs-133 (isotope of caesium).

Ampere   It is the strength of current which when flowing through two parallel straight conductors, spaced 1 metre apart produce a force of 2×10^{-7} N/m length of the conductors.

Kelvin   It is (1/273.16) of the thermodynamic temperature of the triple point of water.

Candela   It is the luminous intensity of a black body with a surface area of $(1/6,00,000)$ m^2, in the direction perpendicular to the surface, at the temperature of freezing platinum and at a pressure of $1,01,325$ N/m^2 (N = Newton, unit of force).

Mole   It is the amount of substance which contains as many elementary constituents as there are atoms in 0.012 kg of carbon-12.

5.3.4 Metric Prefixes to Standard Units

To make precise expressions of large quantities with huge numbers or very small quantities with minute fractions of the 'standard unit', the metric system of multiples of 10 or 1/10 is used with a definitive prefix to the unit, for the sake of convenience (Table 5.3).

Table 5.3   Prefixes for multiple and submultiple units in powers of 10

Multiples			Submultiples		
Prefix	Abbreviation	Value	Prefix	Abbreviation	Value
Deca-	D	10^1	deci-	d	10^{-1}
Hecta-	H	10^2	centi-	c	10^{-2}
Kilo-	k	10^3	milli-	m	10^{-3}
Mega-	M	10^6	micro-	μ	10^{-6}
Giga-	G	10^9	nano-	n	10^{-9}
Tera-	T	10^{12}	pico-	p	10^{-12}
Peta-	P	10^{15}	femto-	f	10^{-15}
Exa-	E	10^{18}	atto-	a	10^{-18}

5.3.5 Symbols for SI Units

In addition to the seven fundamental units in the SI system, there are other derived units. Table 5.4 represents symbols of all the units.

Table 5.4   Symbols of fundamental units in the SI system and other derived units

Unit	Symbol	Unit	Symbol
Metre	m	Joule	J
Kilogram	kg	Watt	W
Second	s	Coulomb	C
Ampere	A	Weber	Wb
Kelvin	K	Ohm	Ω
Candela	cd	Volt	V
Radian	rad	Farad	F
Steradian	sr	Henry	H

(Contd)

Table 5.4   (*Contd*)

Unit	Symbol	Unit	Symbol
Newton	N	Siemen	S
Hertz	Hz	Tesla	T

Exhibit 5.B   Range of Variation of Fundamental Physical Quantities and Some Practical Units

There are tremendous variations in mass and length (or distance between two bodies) in the objects of the physical world. The magnitude of variation in the physical quantities is often referred to as *atomic to astronomical* or *microscopic to macroscopic*.

Very small objects, such as ions, atoms, and molecules are referred to as microscopic objects. Very small distances are known as microscopic or atomic distances, for example, radius of an atom or an electron, or size of a dust particle.

Heavy bodies of millions and billions tonnes of mass, such as the sun and earth are known as macroscopic entities. Similarly very large distances like the distance between planets and distance of the sun from the earth are known as macroscopic or astronomical distances.

The usual SI units are used for measurements of microscopic and macroscopic objects and distances, with multiples of 1/10 or 10 of the metric system as shown in Tables 5.B1 and 5.B2. The notation with multiples of 10 is often called *exponential notation*.

Table 5.B1   Range of variation of mass

Microscopic entities		Macroscopic entities	
Entity	**Mass**	**Entity**	**Mass**
Electron	9.11×10^{-31} kg	Man	7.0×10^{1} kg
Proton	1.67×10^{-27} kg	Elephant	5.0×10^{3} kg
Hydrogen atom	1.70×10^{-27} kg	Ship	8.0×10^{7} kg
Uranium atom	4.00×10^{-26} kg	Moon	7.33×10^{22} kg
Oxygen molecule	5.30×10^{-26} kg	Earth	5.98×10^{24} kg
Penicillin molecule	5.00×10^{-17} kg	Sun	1.99×10^{30} kg
Pencil mark	1.00×10^{-15} kg	Our Galaxy	2.23×10^{38} kg
Dust grain	2.30×10^{-13} kg		
Grape	3.00×10^{-3} kg		

Practical units for macroscopic and microscopic lengths

There are some practical units to measure very big macroscopic lengths and very small microscopic lengths.

Astronomical unit (AU) is used for measuring the distance (length) of the earth from the sun. It is equal to about 1.496×10^{11} m.

Light-year—Often distances between planets and galaxies are measured in light-years. It is the distance travelled by light, in vacuum, in one year.

Velocity of light in vacuum = 3×10^8 ms^{-1} (metre per second)

Time in a year = 365 ¼ days = 365 ¼ × 24 × 60 × 60 seconds

Thus, 1 light-year = $3 \times 10^8 \times 365$ ¼ × 24 × 60 × 60 = 9.467×10^{15} m (approximately)

Important practical units for measuring microscopic lengths, such as wavelength of light and size of nuclei

1 micrometre (micron)(μm) = 10^{-6} metre = 10^{-4} centimetre = 10^{-3} mm

1 nanometre (millimicron)(nm) = 10^{-9} metre = 10^{-7} cm

1 angstrom (Å) = 10^{-10} metre = 10^{-8} cm

1 fermi(fm) = 10^{-15} metre = 10^{-13} cm

Table 5.B2 Range of variation of length

Atomic entities		Astronomical entities	
Entity	Size	Entity	Size
Radius of electron	10^{-16} m	Size of human head	6.0×10^{-1} m
Radius of proton	1.2×10^{-15} m	Height (mean) of a person	1.8×10^0 m
Size of nucleus	10^{-14} m	Height of Mt Everest	8.9×10^3 m
		Radius of earth	6.4×10^6 m
		Radius of moon	10^8 cm
		Radius of sun	6.9×10^8 m
Distance between the atoms in a solid	10^{-13} m	Distance (mean) of moon from earth	3.8×10^8 m
Radius of hydrogen atom	5.0×10^{-11} m	Distance (mean) of sun from earth	1.49×10^{11} m
Distance between air molecules in a room	10^{-8} m	Distance (mean) of nearest star (Alpha centauri) from earth	4.3×10^{16} m
Size of a polio virus	1.2×10^{-8} m	Radius of our Galaxy (Milky Way)	6×10^{19}m
Thickness of gold foil	1.0×10^{-7} m		
Thickness of a book leaf	10^{-4} m	Distance of nearest nebula	2×10^{22} m

Similarly, there is an extensive variation in the range of short times and long times. Only two examples are presented here.

Example of a short time: Time for an electron to revolve around the hydrogen nucleus = 10^{-15} s

Example of a long time: Time of earth's revolution around the sun (1 year) = 3.1×10^7 s

5.4 DIMENSIONS

The characteristic attribute that is common to a class of physical quantities is called their dimensionality. For example, quantities like distance between the earth and moon, radius of the earth, height of a mountain, thickness of a sheet of paper, width of a finger and all those listed in Table 5.B2 belong to one particular class of physical quantities. The common attribute to all these quantities is that they are all measured only in the units of length. Therefore, they are all said to possess the *dimension of length.* Similarly all the examples in Table 5.B1 possess *dimension of mass.* In mechanics, *mass, length,* and *time* provide three basic dimensions, which are denoted by letters [M], [L], and [T], respectively. Each class of these physical quantities represents one dimension.

All other quantities are derived from fundamental quantities according to their defining relations, and they may have mixed multiple dimensions $[M^a L^b T^c]$. The defining relations of other quantities tell the exact manner of their dependence on the fundamental quantities.

Determination of dimensions of a quantity

For example, surface area cannot be expressed by a single dimensional fundamental quantity. It is equal to the product of its length and breadth. However both length and breadth are measured with the help of units of length [L], and area is represented as $[L \times L]$ or $[L^2]$. The unit of mass [M] and unit of time [T] have not been used for the purpose, hence the unit of area can be represented by $[M^0 L^2 T^0]$, which is called the *dimensional formula* of area. The third bracket [] only indicates the nature of the fundamental quantity, but not its magnitude. Power 0, 2, and 0 of the fundamental units are called the dimensions of area in mass, length, and time, respectively.

Dimensions of a particular quantity are the powers to which the fundamental units have to be raised in order to represent that quantity.

Dimensions of physical quantities   Now, we know how to work out the dimension of area. Some other examples are as follows:

Volume

$$\text{Volume} = \text{length} \times \text{breadth} \times \text{thickness}$$

$$= [L \times L \times L] = [L^3] = [M^0 L^3 T^0]$$

Velocity

$$\text{Velocity} = \text{distance} / \text{time}$$

$$= [L]/[T] = [LT^{-1}] = [M^0 L^1 T^{-1}]$$

Acceleration

$$\text{Acceleration} = \text{velocity} / \text{time}$$

$$= [LT^{-1}] / [T] = [LT^{-2}] = [M^0 L^1 T^{-2}]$$

Momentum

$$\text{Momentum} = \text{mass} \times \text{velocity}$$

$$= [M] \, [LT^{-1}] = [M^1 L^1 T^{-1}]$$

Force

$$\text{Force} = \text{mass} \times \text{acceleration}$$
$$= [M] \, [LT^{-2}] = [M^1L^1T^{-2}]$$

Moment of a force

$$\text{Moment} = \text{force} \times \text{distance}$$
$$= [MLT^{-2}] \, [L] = [M^1L^2T^{-2}]$$

Work

$$\text{Work} = \text{force} \times \text{distance}$$
$$= [MLT^{-2}] \, [L] = [M^1L^2T^{-2}]$$

Energy   Using the dimensional formula, it can be shown that the dimensions of energy are the same as those of work. Furthermore, it can also be shown that although kinetic energy and potential energy are different categories of energy, they possess the same dimensions.

Kinetic energy $= \frac{1}{2} \, mv^2$ ($\frac{1}{2} \times \text{mass} \times \text{velocity}^2$)

Therefore, kinetic energy $= [M^1] \, [LT^{-1}]^2 = [M^1L^2T^{-2}]$

(As $\frac{1}{2}$ is dimensionless, it is ignored).

Thus, the dimensions of kinetic energy are 1, 2, and –2 in mass, length, and time, respectively.

Potential energy $= mgh$ (where 'g' is the acceleration due to gravity)

In this manner, dimensions of other physical quantities, such as specific heat capacity, specific latent heat, pressure, power, velocity of light, gravitation constant, Planck's constant, angular velocity, temperature, and resistance in electricity can be determined.

Such dimensional analysis can be used to (i) derive a relation between various physical quantities and (ii) convert values of a physical quantity from one system of units to another.

5.5 MEASUREMENT CONVENTIONS

Biotechnological processes require quantification of common physical variables with their magnitude. For this purpose, some (engineering) conventions used are:

(a) *Density* is a substantial variable defined as mass per unit volume at a fixed temperature.

(b) *Specific gravity* also known as relative density is a dimensionless variable.

To mention specific gravity, the temperature of the substance and its reference material are specified. For example, the specific gravity of ethanol is given as $0.789_{4°C}^{20°C}$; this means that the specific gravity is 0.789 for ethanol at 20°C actually referenced against water at 4°C. The density of water at 4°C is almost exactly 1.000 g cm^{-3}, so one can say that the density of ethanol at 20°C is 0.789 g cm^{-3}.

Engineering calculations often require information about physical and chemical properties of materials used and produced in a bioreactor. Since the discussion on these topics is beyond the scope of this book, interested readers may refer to books on related areas for further information.

5.6 STOICHIOMETRY

Stoichiometry deals with material and energy balances in biochemical reactions involving bacterial growth during fermentation to produce necessary chemicals, or during biodegradation for pollution control, or in biogeological processes.

The nature of the substrates to be converted by bacterial action, the type of micro organisms, the biochemical pathway, and the reaction kinetics of the process involved comprise the basic information used in stoichiometry. Stoichiometry and reaction kinetics together provide the quantitative approach essential for the design of bioreactors and other design-related calculations.

5.6.1 Stoichiometry of Microbial Growth

Microbial growth systems are generally characterized (Fig. 5.6) by (i) their catabolism using the available electron donor and electron acceptor, (ii) their anabolism using the available C source and N source, and (iii) involvement of HCO_3^-, H_2O, and H^+.

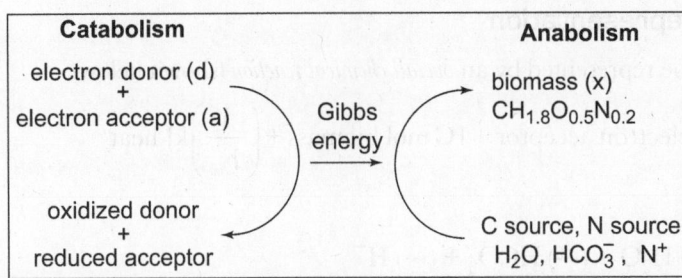

In chemical and biochemical reactions (including catabolism and anabolism), atoms and molecules rearrange to form new groups.

In the process of growth, the substrate is biochemically converted to biomass (Fig. 5.6). Bacterial growth is quantified using well-known parameters such as the following:

Fig. 5.6 Growth can be presented as coupled anabolism and catabolism.

(1)   Maximum biomass yield (x) on substrate (s)

$$Y_{sx}^{\,max} = \text{maximal growth yield of biomass } (x) \text{ on substrate (s). } (x \text{ can be expressed as } Y_{sx})$$

or

$$Y_{dx}^{\,max} = \text{when substrate is called electron donor (d or } Y_{dx})$$

$$= \text{C-mol } x \text{ per C-mol s}$$

(2)   Maintenance requirements for substrate (s) or electron donor (d) (ms or md)
[Consumption of substrate for maintenance of cellular functions, such as (i) electrical gradients across cellular membranes by maintaining large concentration gradients of protons, minerals and other ions, (ii) phosphorylation, hydrolysis, and other biochemical functions to maintain the living state, and (iii) continuous turnover of macromolecules, for example, continuous synthesis and degradation of mRNA (with short half life). All these processes consume Gibbs free energy (in the form of ATP) without formation of cell mass.]

(3) μ_{max} = maximum specific growth rate of bacteria per hour (h^{-1})
[determined by biomass produced per hour; it is a complex function of many parameters such as substrate concentration (both carbon and nitrogen sources), pH, temperature, by-products and oxygen availability]

(4) K_s = affinity constant $(\text{mol } l^{-1})$

(measure of the affinity of the cells towards the substrate)

(The values of these parameters may vary by one to two orders of magnitude, depending on the growth systems.)

Such a growth system uses the available electron donor and electron acceptor for catabolism, and the available C and N sources for anabolism. In addition, HCO_3^-, H_2O, and H^+ are involved in each growth system. Many microorganisms have similar elemental compositions with one C-mol formula for biomass, that is, $C_1H_{1.8}O_{0.5}N_{0.2}$. For practical purposes, this is considered as a standard biomass composition, unless specific information is available.

In fermentation processes, information on the biomass yield from the substrate (Y_{sx} or Y_{dx}), O_2 requirement, CO_2 production, and heat (J, Joules) production are important for the purpose of designing an optimal process. Stoichiometric calculations play an important role in the design of an optimal process.

5.6.2 General Stoichiometric Representation

A microbial growth system can be represented by an *overall chemical reaction* (shown below).

$$\left(\frac{1}{Y_{dx}}\right)\text{electron donor} - \left(\frac{1}{Y_{ax}}\right)\text{electron acceptor} + 1\text{C mol biomass} + \left(\frac{1}{Y_{hx}}\right)\text{kJ heat}$$

$$+ \left(\frac{1}{Y_{Gx}}\right)\text{kJ Gibbs energy} + (\cdots)\,H_2O + (\cdots)\,HCO_3^- + (\cdots)\,H^+$$

where one C mol of biomass is formed which accounts for the role of electron donor and electron acceptor, the N source, HCO_3^-, H_2O, H^+, Gibbs energy, and heat. Dots in the parenthesis,$(...)$, indicate unknown stoichiometric coefficients. Y_{dx}, Y_{ax}, Y_{hx}, and Y_{Gx} are known as stoichiometric yield coefficients and are involved in the growth of C mol biomass. A minus sign in the equation signifies consumption.

One C mol of biomass contains 12 grams of carbon. This is equal to about 25 gm of dry matter as the carbon content of biomass is typically about 45 per cent. The energy is generated in a *redox reaction* between the electron donor and the electron acceptor. Gibbs energy ΔG combines the heat-related enthalpic (ΔH) and entropic (ΔS) contributions ($\Delta G = \Delta H - T\,\Delta S$) responsible for the formation of the complex molecules in the biomass from simple carbon compounds.

Mass and molar relationship between the reactants consumed and products formed are deduced by stoichiometric calculations. Balanced chemical equations and atomic weights of the elements help in this deduction. This is exemplified below in reference to alcoholic fermentation, and then in mass balance during the degradation of casein.

$$C_6H_{12}O_6 \rightarrow 2\,C_2H_5OH + 2\,CO_2 \ (\textit{Balanced equation})$$

$$\text{(Carbohydrate)} \qquad \text{(Alcohol)}$$

During this (or other) chemical or biochemical reactions, the following two quantities are conserved.

1. *Total mass*

   Total mass of reactant(s) = Total mass of product(s)

   (here carbohydrate)   (here alcohol and CO_2)

2. *Number of atoms of the elements*

   Number of C, H, and O atoms in the reactant(s) = Number of C, H, and O atoms in the products

   In the above equation, complete conversion of the reactant has occurred. In industrial reactions, this does not happen, for several reasons; when the reactants used are not in exact proportions indicated by the reaction equation, or when the reactants are consumed in the side reaction to some other side products not indicated by the principal equation for the reaction. Making appropriate adjustments using this additional information can only contribute to better stoichiometric calculations and control of chemical reactions in reactors.

Mass balance

Oxidative degradation of casein in waste water   In waste water, casein can be oxidized by microorganisms in the presence of oxygen, for their growth. In the process, there is reduction of the BOD (biological oxygen demand) level in the waste water and its polluting effect. The equation for this oxidation reaction is given below. The values in parentheses given below the equation are the corresponding molecular weights.

Mass balance equation

$$C_8H_{12}O_3N_2 + 3\ O_2 \rightarrow C_5H_7O_2N^* + NH_3 + 3\ CO_2 + H_2O$$
$$\text{(Casein)} \qquad\qquad \text{(Bacterial cell)}$$
$$(184) + (96) \qquad\quad = (113) + (17) + (132) + (18)$$
$$(280) = (280)$$

[*The empirical formula, obtained from the literature, for calculating cell mass, on an average (with a few exceptions) is as follows:

$$N:O:C:H :: 1:2:5:7]$$

The *mass balance equation*, as mentioned above, simply states that *the input of mass* in biochemical reactions carried by bacteria *is equal to the output*. To develop this equation, the chemical formulae of the bacterial cells are needed. Since, many microorganisms have similar composition an empirical, if not exact, generalized formula, noted above, is used.

Often, the generalized mass balanced equation is stated as

$$\text{Input} - \text{Output} = \text{Accumulation}$$

From the mass balance equation, the quantity of oxygen to be supplied in dissolved form by aerating the waste water, for the biological degradation of a known quantity of casein in waste water, and the quantity of biomass solids (that is, dry sludge) that will be produced in the reactor can easily be calculated. From kinetic data, the retention time required for digestion or degradation can be determined. The basic volume of the digester is calculated by multiplying the volume of waste water to be treated per day by the retention time. The size and shape of the reactor is decided on the basis of the volume needed to carry out digestion for the total volume of inputs and outputs. The aeration system should be appropriately designed to allow adequate transfer of oxygen into the system. The outlets for gaseous and solid outputs should also be properly designed.

It may be mentioned here that although the major constituents for microbial growth are C, H, O, and N, elements such as phosphorus, sulphur, and iron are required in trace amounts in the waste water. For example, the phosphorus requirement is about 2 per cent of the cell mass on a dry weight basis.

5.7 DATA AND CALCULATIONS

Various types of data are generated in the course of experimentation in field laboratories and during operation of bioreactors and through products analysis. Thus, proper analysis of data, determination of range of variability, elimination of errors, correlations between different sets of observations or data, and so on, help in proper running and future projections of laboratory experiments as well as biotechnology plants.

5.7.1 Errors in Data

Experimentally determined quantities have some inaccuracies due to errors in measurement. Such errors should be taken into consideration to determine the significance or reliability of observations. Estimation of error and principles of error propagation in calculations prevent misleading representation of data and are important in engineering analysis.

Absolute and relative uncertainty

On the basis of several observations, if the temperature of a place or a reaction is expressed as $32 \pm 0.4°C$, then $\pm 0.4°C$ represents the actual temperature range, due to which the reading is uncertain. It is known as *absolute error*. An alternate expression can be $32°C \pm 1.25\%$, then $\pm 1.25\%$ is called the *relative error* or relative uncertainty.

5.7.2 Types of Errors

Two broad classes of measurement errors are systemic errors and random errors.

Systemic error   It is often associated with measuring equipment and affects all measurements of all variables in the same way. When a systemic error is detected, the readings should be accounted for using the correction factor. A simple example is an imperfectly calibrated analytical balance or a scale.

Random error   It is revealed when repeated measurements of an unchanging quantity cause scatter of different results. The causes for this variation are unknown, and hence, cannot be corrected by a factor. Statistical analysis is used to quantify random error.

The term *precision* indicates the reliability or reproducibility of data, and tells us to what extent the measurement is free from random error. *Accuracy* requires both random and systemic errors to be as small as possible and ideally zero.

5.8 STATISTICAL ANALYSIS

The first step in statistics is to measure or count and then to connect data and reality. A set of data represents the reality based on numerical or measurable scales. Data is *primary type* when the analyst collects the data relevant to his or her investigation. Data collected from other sources is *secondary type* data.

Decision-making process under uncertainty invites application of statistical data analysis for probabilistic risk assessment; in other words, for better decisions about uncertain situations.

In a nutshell, *statistics* is a set of methods that are used to collect, analyse, present, and interpret data. Statistical methods are used in diverse fields, such as education, economics, business, agriculture, breweries, and different branches of biology including biotechnology and biochemical process engineering, and medicine. More recently, the use of statistical methods has commenced in the arena of astronomy, geology, and physics.

5.8.1 Data Analysis

The statistical methods for analysing data may be divided into two categories:

Exploratory methods   These are used to find out what the data seems to project by using simple arithmetic, schematic drawings to summarize the data.

Confirmatory methods   These use the probability theory. It provides a mechanism for measuring, expressing, and analysing the uncertainties associated with similar events in the future. Thus, probability helps in decision-making.

5.8.2 Collection of Data

A *population* refers to the entire set of individuals, objects, or measurements that a researcher is interested in studying. Populations are characterized by certain properties referred to as parameters. To get information about the *parameters* of a population, a small portion of the population is studied. This small portion is usually referred to as a *sample*. For several reasons, it is not practical to pursue the collection of data for an entire population. However, there should be no bias in choosing the individuals for a sample. The sample should be a *random sample*, that is, every member of the population should have an equal chance of being a member of the sample. Thus, all types of individuals—say super tall, tall, medium, short in a population for height—need to be included in sample. It should be remembered that any statistical inference is dependent on sample size and the method of sampling employed. Statistical inference refers to extending the knowledge obtained from a random sample of a population to the whole population. It is inductive reasoning from a mathematical point of view, that is, the knowledge of the whole from a part.

5.8.3 Measures of Central Tendency

The readings for individuals in a random sample (of man/animal/objects/events), expressed in numbers, differ from each other. The readings could be for height or weight of people of a region, size of a particular species of animals, pod size of a variety of legume, diameter of some objects, and so on. Noting individual readings is the first step for statistical calculations. Then an average of all these varying readings is made to get a representative or a single figure to describe a finding about the random sample or by logic the population. Since an *average* value usually occupies the central position of the varying readings, with some readings larger and some others smaller than the average value, the method of calculating the average is known as 'measure of central tendency'. The other two measures of central tendency are *median* and *mode*.

Usually, the average is determined by calculating the *arithmetic mean*. (There are two other means—harmonic and geometric.)

Arithmetic mean   The arithmetic mean is obtained by adding the values of readings for individuals (say $X_1, X_2, X_3 \ldots X_N$ in a sample together and then dividing the total sum by the number of individuals (N) in the sample. The mean (X) for the values $x_1, x_2, x_3 \ldots x_N$ can be indicated as

$$\text{Mean} = \bar{X} = \frac{X_1 + X_2 + X_3 \ldots + X_N}{N}$$

This equation can be abbreviated by using the symbol Σ, which means summation, and X_i which means the value of an individual (X) at ith (last) position.

$$\bar{X} = \frac{\sum X_i}{N}$$

Example 5.1   The weights of five individuals are noted in Table 5.5; $i = 1, 2, \ldots 5$. On the basis of these readings, calculate the arithmetic mean ($\bar{X}$) using the formula.

Table 5.5   Individuals versus body weights

Individual	X_1	X_2	X_3	X_4	X_5
Weight in kg	46	43	37	44	65

Solution

$$\bar{X} = \frac{\sum X_i}{N} = \frac{(46 + 43 + 37 + 44 + 65)}{5}$$

$$= \frac{235}{5} = 47$$

Sometimes $\sum X_i$ is also denoted by $\sum_{1}^{i} X_i$ indicating summation of i individuals increasing in position by 1.

Median   An alternative to the mean as a measure of central tendency is the *median*. Median is the middle or central value of a set of values arranged in ascending order. This arrangement of numbers is referred to as an array. The set of numbers in Table 5.5 when arranged in ascending order would be 37, 43, 44, 46, and 65. The median of this set of numbers is 44. The median divides the set of numbers into two equal parts on the left and the right. The median number 44 may not be equal to the mean value of 47, as shown above, but it provides a fair idea about the central tendency of the numbers. The set of numbers in Table 5.5 is an odd one, consisting of 5 numbers. If the numbers in a set is even, the median is the average of the two middle values. For a set of numbers 3, 6, 8, and 11, the median would be (6+8)/2 = 7.

When data is distributed more or less symmetrically, the mean and median will differ but only slightly. (However, when the distribution is not symmetrical and most of the numbers are on the lower side, the mean becomes considerably larger than the median, and in such cases the mean is quite misleading.)

Mode   Mode of a set of numbers is the value of the most frequent occurrence. This is also another measure of central tendency.

***Example* 5.2**   For a set of numbers 6, 7, 7, 7, 8, 10, 10, 11, 14, the mode is 7. This set is also referred to as a *unimodal set*, as there is only one mode.

When a set consists of numbers like 3, 7, 9, 4, 8, 9, 13, 7, 6, 2, 18, there are two modes—7 and 9. This is referred to as a *bimodal set*.

A set of numbers 4, 1, 2, 10, 12, 13, 54, 34, 77 does not have any mode, as no number occurs more often than the others.

5.8.4 Measures of Dispersion

Mean, median, and mode provide an idea about the average of a group of data, such as weight, and height of a group of people to compare with those of another group. In the following two sets of numbers,

$$10, 30, 40, 50, 70 \text{ (set } a\text{)}$$

$$34, 38, 40, 42, 46 \text{ (set } b\text{)}$$

the mean of the two sets are the same, that is, 40. But the second set appears to be more bunched about the mean than the first set. This aspect of the sets is not reflected by mean numbers; but would be revealed in measures of dispersion or variation (deviation from the mean).

Measures of dispersion are *range, variance,* and *standard deviation*.

Range   It is defined as the difference between the largest and the smallest values in a set of observations. For the above-mentioned set a, the range is $70 - 10 = 60$ and for set b, the range is $46 - 34 = 12$. Range is not a good measure of dispersion (spread of data), since it is based only on two observations.

Variance   When $X_1, X_2, X_3 \dots X_N$ are the N values (number of individual values) of a variable X, the arithmetic mean is $\bar{X}$. Then *deviation* of any value (X_i) from mean X can be indicated by $d = X_i - \bar{X}$. This d is called the *deviation* of $\bar{X}_i$ (an individual value) from mean $\bar{X}$. It indicates how far X_i lies from the mean $\bar{X}$.

The *average deviation* of all the values, called *mean deviation* (MD) from the mean is calculated as

$$\text{MD} = \frac{\sum |d|}{N}$$

$$= \frac{\sum_{i=1}^{N}(X_i - \bar{X})^2}{N} \text{ (assigning value of } d\text{)}$$

where negative signs are removed from the deviations by using the absolute value (modulus).

Another way of removing the negative signs is by squaring the deviations and then finding the average.

The mean of squared deviations is called *variance*, devoted by *var(X)* or σ^2. Then

$$\sigma^2 = \frac{\sum\limits_{i=1}^{N}(X_i - \bar{X})^2}{N}$$

This can be simplified to

$$= \frac{\sum\limits_{i=1}^{N}(X_i - \bar{X})^2}{N}$$

Here X indicates an individual value, $\bar{X}$ the arithmetic mean, and N the total number of values. Let us see the variance (σ^2) for the numbers in *set a*, presented above.

$$\sigma^2 = \frac{\sum(X - \bar{X})^2}{N}$$

$$= \frac{(10-40)^2 + (30-40)^2 + (40-40)^2 + (50-40)^2 + (70-40)^2}{5}$$

$$= \frac{900 + 100 + 0 + 100 + 900}{5} = 400$$

For the numbers in *set b*

$$\sigma^2 = 18$$

Although the mean of numbers in *set a* and *set b* is equal, their variances are quite far apart. This is due to large deviation or dispersion of values from the mean in *set a* and the proximity of the numbers to the mean in *set b*.

One defect with variance is that its units are different from the units of variable X and their mean $\bar{X}$. Variance is expressed in terms of square value. Thus, when the heights of a group of students are measured in centimetres, the variance will be in *square centimetres*; it is difficult to visualize or comprehend. *Standard deviation* (SD) or *root mean squared deviation*, which is the *square root of variance, is often preferred*. It is denoted by σ.

Therefore,

$$\text{Standard deviation} = \sigma = \sqrt{\frac{\sum(X - \bar{X})^2}{N}} \text{ (i.e., } \sqrt{\text{variance}})$$

It can also be written as follows:

$$\sigma = \sqrt{\frac{\sum X^2}{N} - \bar{X}^2}$$

When values of $(X - \bar{X})$ are small, the first formula is used; when values of X are convenient, the second formula is used.

Sample variance (σ^2) and standard deviation (σ)

σ^2 and σ are considered as variance and standard deviation, respectively, for a population and s^2 and s for a sample (part of a population). The method of calculation remains the same. Only in the case of a small sample size, the denominator is taken as $N-1$ (instead of N). The quantity represented by $N-1$ is termed as the *degree of freedom*. This is supposed to make the SD calculation unbiased. It is considered that for a large sample size (greater than 30), whether N or $N-1$ is used in the denominator makes little differences in the value of s or σ.

5.8.5 Data Presentation

Data of a study can be presented in many ways, such as tables, charts, diagrams, and graphs. The values, thus presented usually intend to summarize and characterize the data. The tables present the actual numbers or percentages. They can also club together the same values in a class and thus, several classes in a study. The number of occurrences of the same value in a class is the frequency of the value; this can also be included in the table.

Graphic presentations are more precise in summarization and characterization of data, and thus more impressive than frequency tables which list the numbers. Figure 5.7 shows the presentation of data with vertical lines and bars. Relative frequencies or proportions are essentially probabilities, implied by a glance at the height presented by lines and bars or columns.

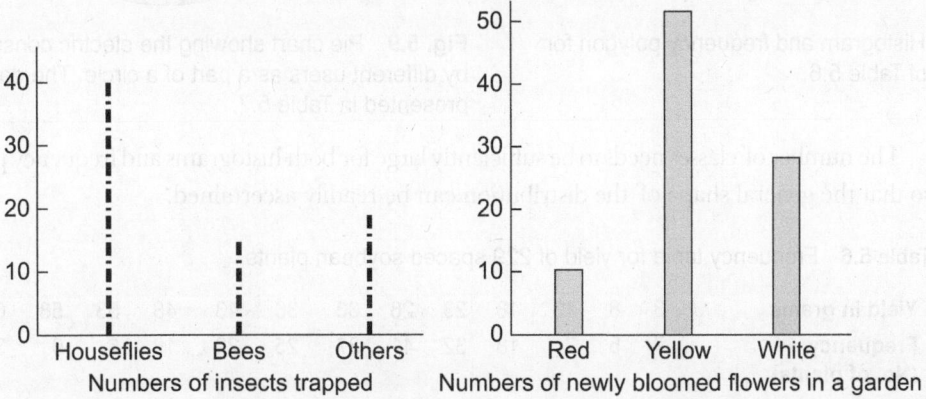

Fig. 5.7   Presentation of discrete data obtained on a particular day, by line and bar diagrams.

Histogram and frequency polygon

The histogram and frequency polygon of Fig. 5.8 are common methods of presenting a considerable amount of data. The *histogram* pictures the data with *class values* (midpoints of class intervals) along the horizontal (or X) axis and with rectangles (or bars) above the class intervals to represent frequencies along the Y-axis. The histogram presents data in a readily understandable form so that at a glance, the general nature of the distribution is revealed. If necessary, an observed distribution can be compared with the theoretical one by superimposing it on the histogram and discrepancies can easily be ascertained. A graph paper can be used to draw a histogram to scale.

The *frequency polygon* is drawn by locating the midpoint of each class interval at the top of the frequency bar and then connecting the midpoints by straight lines. The frequency polygon tends to imply the smooth curve of the population from which the sample was drawn. The histogram and frequency polygon in Fig. 5.8 are representations of the data in Table 5.6.

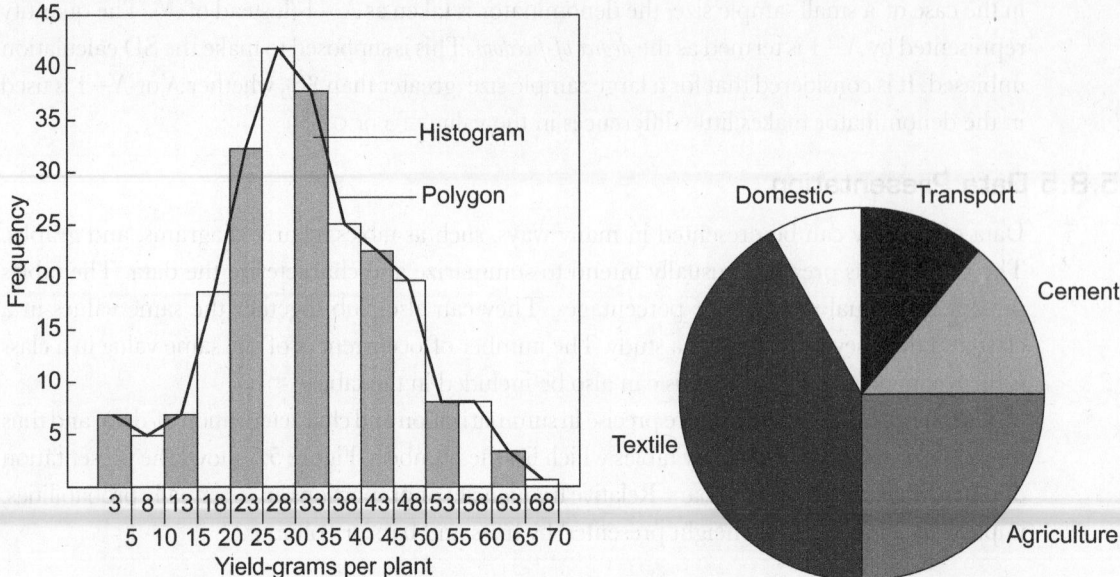

Fig. 5.8   Histogram and frequency polygon for the data of Table 5.6.

Fig. 5.9   Pie chart showing the electric consumption by different users as a part of a circle. The data is presented in Table 5.7.

The number of classes needs to be sufficiently large for both histograms and frequency polygons, so that the general shape of the distribution can be readily ascertained.

Table 5.6   Frequency table for yield of 229 spaced soybean plants

Yield in grams	3	8	13	18	23	28	33	38	43	48	53	58	63	68
Frequency (No. of plants)	7	5	7	18	32	41	37	25	22	19	6	6	3	1

Source: Adapted from Steel, R.G.D. and J.H. Torrie, 1960, Principles and Procedures of Statistics, McGraw-Hill Book Co., Inc., New York, p.12

Pie chart   A pie chart presents the data in a way that is easy to understand. The data in Table 5.7 have been presented in a pie chart (Fig. 5.9). In a pie chart, the data is presented as a part of a circle whose area is proportional to the data. The pie chart in Fig. 5.9 displays data on the consumption of electricity by different types of users in a particular city.

Table 5.7   Consumption of electricity by different types of users

Type of user	Percentage of electricity consumed
Domestic	8
Transport	11.2
Cement industry	14.3
Agriculture	25.5
Textile industry	41

To present a value, say consumption by transport (11.2%), an arc of a circle is drawn which subtends an angle (11.2/100) × 360 degrees at the centre and each end of the arc is joined by a line to the centre. This provides a piece of the circle whose area is proportional to that used for transport. Similarly, the proportions of other users are drawn to complete the circle as in Fig. 5.9.

The type of graphical presentation depends on the nature of the data and the audience for which the data is intended.

Frequency distribution   An orderly arrangement of data is described as an *array*. The array of frequencies (repeated occurrence) of different quantitative or qualitative classes in a sample population is called frequency distribution. This is often referred to as *probability distribution*. The distribution pattern can be graphically presented (Fig 5.10). The figure indicates three types of frequency distribution curves—normal (Gaussian) distribution, asymmetric distribution (skewed), and bimodal distribution.

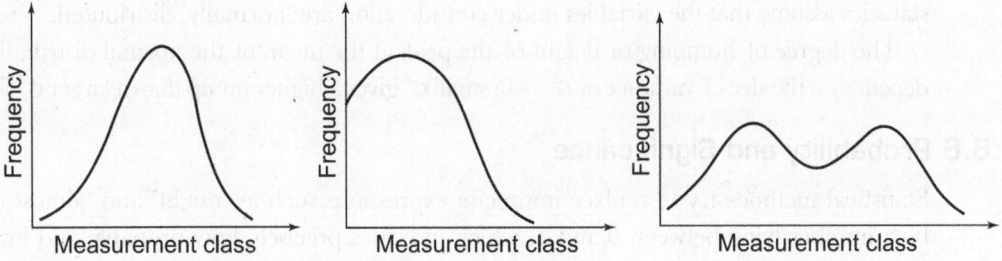

Fig. 5.10 Graphical presentation of (a) normal, (b) symmetrical, and (c) bimodal distribution of data.

Normal distribution   When a qualitative parameter is studied in a large population, it reveals that a very few individuals possess extreme values of the parameter, and that progressively more individuals are found nearer the average (mean) value of that population. The graph of such a parameter, the normal curve, is symmetrical and characteristically bell shaped. This normal distribution curve is also called Laplacian or Gaussian curve.

The mean of the distributions is exactly at the centre of the normal curve which also marks the point where the number of observations is maximum. The curve is symmetrical as one half of the area lies on each side of the mean. Median and mode coincide with the mean.

The fact that a large number of observations in a normal curve are towards the centre is obvious from Fig. 5.11. The area between $-\sigma$ and σ is 68 per cent of the total area of the curve. This indicates that in a normal distribution 68 per cent of the observations lies within a distance equal to the standard deviation (σ) on each side of the mean. Similarly, the distance between -2σ and 2σ includes 95 per cent, and -3σ to 3σ region includes 99.7 per cent of observations. Therefore, in a normal distribution, practically all the observations lie within the range of -3σ to 3σ, that is, within the range of six times the standard deviation. In other words, when the number of observations is large, the standard deviation (σ) should be approximately one sixth of the total range of distribution of data.

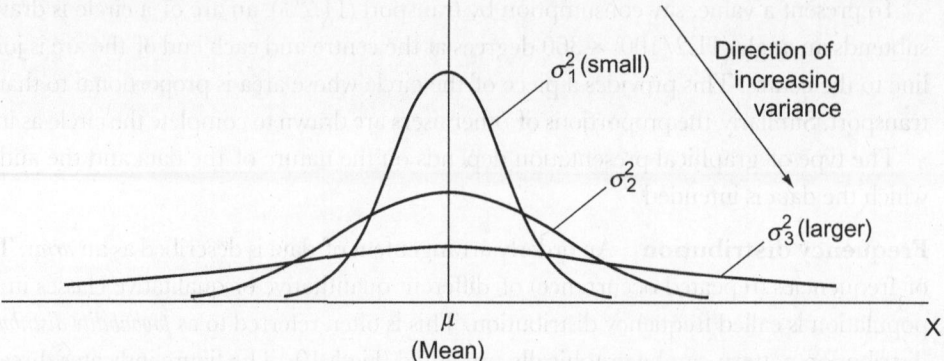

Fig. 5.11 The height of the normal distribution curve depends on the size of the variance (σ^2).

Normal distribution is important in theory as well as practice of statistics. Most methods in statistics assume that the variables under consideration are 'normally' distributed.

The degree of humping or height of the peak at the mean of the normal distribution curve depends on the size of variance or σ^2—a small σ^2 gives a higher hump than a larger σ^2 (Fig. 5.11).

5.8.6 Probability and Significance

Statistical methods try to replace imprecise expressions such as 'might' and 'almost certainly' by a number lying between 0 and 1, which indicates precisely how probable and improbable an event is. The ratio of the number of successful events to the total number of events under certain conditions is known as *probability*; in other words, probability (p) of the successful events.

$$p = \frac{\text{number of successes}}{\text{total number of events (successes + failures)}}$$

Statistically, it can be expressed as follows.

If m is the number of times a successful event E occurs in N independent repetitions, then the probability of the event E occurring at all times

$$p(E) = \lim_{N \to \infty} \frac{m}{N}$$

For example, if a coin is tossed 100 times, and it falls as 'head' 50 times, then the probability of the event ('head', in this case) is

$$p(\text{'H'}) = \frac{50}{100} = \frac{1}{2} = 0.5$$

On the other hand, the probability of the coin falling as 'tail' will also be 0.5 since there are only two events that can occur when a coin is tossed and one excludes the other. If the probability lies between 0 and 1, and if q denotes the non-occurrence of the event, then $p + q = 1$ or $q = (1 - p)$.

If events A and B are *mutually exclusive* (as in the case of coin), the probability of occurrence of A or B is given by the sum of the probabilities of A and B. Thus,

$$p(A + B) = p(A) + p(B)$$

This first rule of probability is referred to as the addition theorem or the sum rule.

On the other hand, if the two events A and B are independent, then the probability of occurrence of the events A and B simultaneously is given by

$$p(AB) = p(A) \times p(B)$$

This is referred to as the multiplication theorem or the product rule.

The term *significance* is used in statistics to indicate the deviation of a particular estimate from its expected value and to explain the difference occurring by chance as a result of random sampling or due to something unusual. In practice, odds of 19 to 1 against an occurrence by chance are usually considered as indicating the significance of that occurrence. This can be expressed in terms of probability (p) as a decimal fraction so that odds of 19 to 1 becomes $p = 0.05$ (or 5% level of significance). In other words, if two samples were from the same population, one would expect to observe deviations by chance done in less than 5 per cent cases.

Testing hypothesis on the difference between the means (by t-test)

William Sealy Gosset (1876–1937), a British statistician in a brewery firm, wrote many papers under the pseudonym of 'Student' and introduced the *t*-test to find out the significance of the difference between the means of separate samples. The samples could be from a large population.

$$t = \frac{\bar{X} - \mu}{s\bar{X}}$$

$$= \frac{\bar{X} - \mu}{s / \sqrt{n-1}}$$

where $\bar{X}$ is the sample mean and μ is the population mean, $s\bar{x}$ is the standard deviation of the mean, and n is the number in the testing sample ($n - 1$ = degree of freedom).

(The *standard deviation of the mean*, $s\bar{X} = s / \sqrt{n-1}$, is also known as *standard error*. Often, histograms of data are presented with standard error as a measure of variance instead of standard deviation.)

From the *t*-test equation, t is the deviation of the sample mean from the population mean, in units of sample standard deviations of means (used as denominator). This unit of measurement is commonly used to judge the usualness or unusualness of a deviation.

t-test for two independent samples

So far the discussion was on testing the significance of difference between samples drawn from the same population. Sometimes, samples are collected from two independent or separate populations, for example, one group from an experimental population, and the other from a control population, to test whether the experimental index significantly differs from that of the control population or not. The *t*-test can also be applied for the purpose with the following necessary modifications in the formula.

$$t = \frac{\overline{X}_1 - \overline{X}_2}{\sqrt{(SE_1^2 + SE_2^2)}}$$

Here, $\overline{X}$ is the mean, SE is the standard error, and subscripts 1 and 2 represent the different samples.

The value of t is then judged from the necessary table in reference to the degree of freedom. Usually the t value ≤ 0.5 is considered as significant. This means that the difference between the means of two populations will hold true in 95 per cent or more cases.

Chi square (χ^2) test (test of goodness of fit)

The statistical tests described so far are based on the assumption of a normal distribution in the population under study. These tests are often referred to as *parametric tests*. Chi square tests are considered as non-parametric or distribution-free tests.

In some experiments, the experimental results sometimes differ from the expected results on the basis of some theoretical considerations. In such cases it is necessary to know whether the departure (*discrepancy*) is due to chance alone or not. If the discrepancy between observed and expected/predicted results is very large, the discrepancy may not be due to chance alone; the assumption may not be a tenable one.

The *chi square test* is a statistical tool to determine the *goodness of fit* of expected and observed values. The objective of the chi square test is to determine whether or not the difference is due to chance alone. The formula for calculating chi square is

$$\chi^2 = \sum \frac{(o-e)^2}{e}$$

Here o is the observed frequency, e is the expected (calculated) frequency, and Σ is summed for all classes. Both o and e are calculated in actual numbers and not in percentages or proportion.

The above formula may also be expressed as

$$\chi^2 = \sum \frac{d^2}{e}$$

where $d = o - e$

***Example* 5.3**   *A test cross (or back cross) of a monohybrid grey mouse and an albino strain results in 66 grey and 50 albino progeny. Test the goodness of fit of this data to a 1:1 ratio, using the chi square test.*

Solution

	Classes	
	Grey	Albino
Observed (o)	66	50
Expected (e)	58	58
Observed–expected = $o - e = d$	+8	−8
$(o - e)^2 = d^2$	64	64
$\dfrac{d^2}{e}$	1.1	1.1

Therefore,

$$\chi^2 = \sum \frac{d^2}{e} = 1.10 + 1.10 = 2.20$$

Degree of freedom (df) is one less than the number of classes; here $n - 1 = 2 - 1 = 1$.

Interpretation   The chi square value is judged in reference to the degree of freedom to determine the probability (p) that the deviation of the observed results from the expected values is due to chance alone.

For the above case, that is, for the chi square value of 2.20 with one degree of freedom, the p-value is between 0.10 and 0.20. This means that in 10–20 cases out of every 100 cases, one can expect χ^2 value of this magnitude or even larger due to chance alone, and the hypothesis (expectation) is valid.

When the χ^2 value is greater than 5 out of 100 (i.e., in 5% of the cases, $p > 0.05$), the deviation of observed data and theoretically expected values is statistically insignificant, and the proposed hypothesis is accepted as a valid explanation of the experimental results. When the probability of χ^2 is less than 5 per cent ($p < 0.05$), the deviation between observed and expected values is considered not due to chance alone and is statistically significant enough to consider the hypothesis as invalid.]

5.8.7 Use of Graph Paper with Logarithmic Coordinates for Plotting Data

For graphical presentation of data spread over a wide range, a logarithmic graph paper is used. For example, values from 30 to 1000 or from 0.01 to 10 can be easily presented in logarithmic paper on any or both the axes (X and Y). The axis is divided as a unit of the power of 10, 1 means 10^1, 2 means 10^2, 3 means 10^3, and so on. In other words, it is a logarithmic scale to the base ten (log to the base ten). (It is difficult to present data which spans such a wide range on a graph paper with a unitary scale; to get a better idea, compare a standard linear graph paper with a log graph paper.) The log values for the data in between 2 units can easily be found from the log table. A datum of 16 has a logarithm value of 1.20 and is plotted at 1.20. Negative numbers and zero cannot be plotted in logarithmic coordinates. The logarithm of zero or a negative number is not suitably defined.

5.9 COMPONENTS IN A 'BIOPROCESS' (PROCESS FLOW DIAGRAM)

A 'bioprocess' is a large-scale operation for transformation of raw materials (biological or non-biological) into a product by means of microorganisms, animal or plant cell cultures, or materials derived from them, for example, enzymes, organelles, and so on. Such processes produce commercial biotechnological 'products', such as insulin, penicillin, enzymes, specific proteins, and growth hormones, and help in waste treatment.

A typical bioprocess has three necessary components—the *upstream part*, the *bioprocess* or fermentation itself, and the *downstream part* (Fig. 5.12). Bioprocess engineering treats the raw material to make it suitable for biological reactions in the upstream part. It also includes the isolation,

Process stages   **Operations**

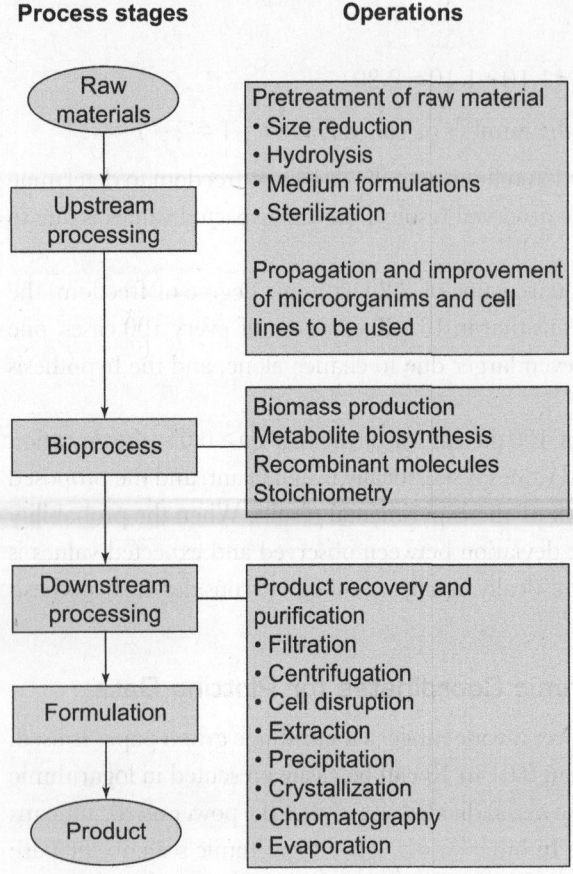

Fig. 5.12 A flow diagram to depict the components in a bioprocess.

preservation, propagation, and improvement of microorganisms and cell lines to be used for biological reactions and media formulation. Transformation of raw materials into biomass and metabolites takes place in the bioprocess part. The optimization of stoichiometry and the transfer of energy, heat and mass improves the bioprocess as well as the end quality and quantity of products. Downstream processing involves product recovery, purification, and packaging. These are mostly physical operations.

An integrated design of the whole process is essential for optimization of its functional efficiency. Knowledge about whether the product is extracellular or intracellular is essential for the design of methodology and equipment for product recovery and purification. Simple parameters such as the substrate to be used or the temperature for growth can affect heat and mass transfer calculations, and thereby necessitate major changes in fermenter (bioreactor) design and, in turn, in the overall process economics.

In most bioreaction processes, a carbon source (carbohydrates such as starch, molasses, sucrose, and glucose) provides both the raw material and energy to the growing cells to carry out the desired biochemical reaction (Table 5.8).

Table 5.8   Carbohydrates used for bioconversion to produce useful compounds

Substrate/Carbon source	Microorganisms/Cell types	Product
Molasses	Yeast	Methyl alcohol
Molasses	*Aspergillus niger*	Citric acid
Starch hydrolysate	*Streptomyces*	Streptomycin
Glucose	*Acetobacter suboxidans*	Vitamin C
Whey	*Lactobacillus*	Lactic acid
Glucose	Animal cells	Erythropoietin
Sucrose	Plant cells	Shikonin

5.10 FACTORS CONSIDERED IN THE DESIGN OF A BIOREACTOR

A *bioreactor* or fermenter is a specialized container which can maintain optimal conditions for the growth of microorganisms and is used to carry out biochemical conversion for large-scale industrial production of desired chemicals.

Careful consideration of some physical and chemical factors is required for maintaining a suitable biological environment for maximum bioconversion in a cost-effective manner. The important factors are discussed in the following paragraphs.

Mass balance

Substrate is consumed and the biomass of microbial cells (or other cell lines) and product is generated in a bioreactor. The mass (or material) balance of cell growth gives an idea of substrate consumption and product formation. Information about mass balance is useful to keep track of input and output amounts and the overall cost involved in the manufacture of the biochemicals. Mass balance analysis during fermentation also indicates the shortage or excess build up of nutrients that may adversely affect the bioconversion process. Mass balance has already been discussed in detail under Section 5.6.2.

Fluid flow and mixing

A bioprocess operates in and around physiological temperature and pH in a suitable fluid medium containing nutrients. Physical mixing of the medium to provide *uniform concentration of nutrients and microorganisms* in the reactor vessel, maintenance of *uniform temperature*, and *supply of oxygen* to the growing cells are the three most important physicochemical variables that are necessary for cell growth and proper product yield. The bioreactor is fitted with a motorized stirrer for mixing of the fluid content to maintain homogeneous conditions.

Fluid behaviour, particularly the viscosity of the fluid, is an important factor for mixing.

- Water, honey, and bacterial fermentation broths behave like *Newtonian fluids*, which continue to flow, regardless of the forces acting on them and their viscosity depends on temperature and pressure.
- *Non-Newtonian fluids* are mostly highly viscous and stirring them can leave a long lasting 'hole'. Thus, these types of fluids impose serious mass transfer limitations in terms of nutrient mixing and oxygen supply to the growing cells. Fungal fermentation broths for the production of penicillin or streptomycin, are examples of non-Newtonian fluids. Different stirring arrangements are made in the bioreactor, depending on the viscosity of the fermentation broth, to create uniformity in the physiological environment which is important for the growth of microorganisms.

Special care needs to be taken to ensure proper mixing of nutrients and gaseous oxygen during aerobic fermentation, keeping in mind the low solubility of oxygen in water. Different types of reactors, such as stirred tank reactor, bubble column, and airlift reactor are designed to carry out aerobic fermentation.

Mass transfer

Mass transfer deals with diffusive transfer of oxygen from the gas phase and of nutrients to the fermentation broth and to the cell surface and finally to the site of biochemical reaction inside the cells. Usually, mass transfer involves a chain of mass transfer steps.

The gas–liquid interface in the fermentation broth is one of the most critical factors that influence the availability of oxygen to the cells. Mixing with proper impellers (propeller blades), creates smaller bubbles of incoming gas, and baffles inside the reactor reduce the mass transfer resistance to oxygen and thus, help in better aeration of the culture. A homogeneous environment of nutrients and oxygen inside the reactor helps in better growth of the microorganisms or cells.

Diffusion is controlled by concentration gradients and is given by Fick's first law:

$$N_A = D\left(\frac{dC_A}{dx}\right)$$

where N_A is the flux of component A per unit area, D is the diffusivity, and (dC_A/dx) is the concentration gradient.

This is similar in form to the equation governing heat transfer.

Effects of transfer limitations

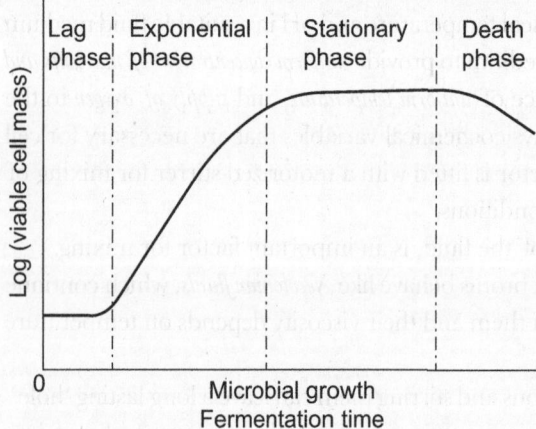

Fig. 5.13   Growth kinetics in fermentation process.

The slowest of the steps shown in Fig. 5.13 will determine the overall mass transfer rate, and its value needs to be compared with the slowest kinetic reaction step to find out whether or not mass transfer will affect the overall process performance. (In enzymatic biotransformations, that is, transformations that do not use cells, there are fewer mass transfer steps. However, the same concept is applied.)

If one mass transfer step is slower than the key kinetic reaction step, it will limit the metabolic activity of the microorganisms, and reduce the quantity of desired product that is formed from a selected substrate.

Sometimes, a slower mass transfer step may alter a biochemical reaction. For example, in the formation of baker's yeast with glucose as substrate, oxygen serves as electron acceptor. In the absence (or lowering the supply) of oxygen, the electrons will be directed to pyruvate resulting in the formation of ethanol and CO_2 instead of yeast.

Heat transfer

Most fermentation processes need careful temperature control for proper growth of the microorganisms and biochemical reactions.

Heat is generated in a bioreactor by (i) metabolism of the organisms and (ii) stirring and aeration (agitation). Heat from the reactor is lost due to (i) evaporation, (ii) the heat content of the input and output flows, and (iii) the heat exchange with the surroundings.

In a large reactor, the metabolic heat often contributes more than 80 per cent of the heat load to the system. This load can be calculated on the basis of the rate of biomass formation, which is determined by growth kinetics and the enthalpy efficiency of the process.

Once the heat generation in a bioreactor due to metabolism (fermentation), agitation, and aeration is known, the desired temperature for the heat transfer area (the mass within the bioreactor) is calculated. Accordingly, heating and cooling coils are designed by chemical engineers to provide suitable optimal temperature during the fermentation process. Steam for heating or chilled water for cooling is circulated in well-designed jackets around the bioreactor to control the temperature. The same heating and cooling facility in a bioreactor can also be used for sterilization before starting the fermentation process and for autoclaving and cleaning of the bioreactor after the batch is over.

5.11 UNIT OPERATION

Biotechnological processes entail bioconversion of raw materials to generate useful products. A process involves several individual operations or steps to convert and separate components; these steps are often called unit operations. The following is a list of examples of unit operations:

(i) Heating   (ix) Filtration
(ii) Humidification   (x) Evaporation
(iii) Cooling   (xi) Drying
(iv) Milling   (xii) Membrane separation
(v) Mixing   (xiii) Dialysis
(vi) Precipitation   (xiv) Crystallization
(vii) Centrifugation   (xv) Chromatography
(viii) Distillation   (xvi) Solvent extraction

5.11.1 Growth Kinetics in the Fermentation Process

Growth and metabolism of the microorganisms (or cells) are the central issues in the fermentation process. Following inoculation, the growth of microorganisms proceeds in a typical pattern as illustrated in Fig. 5.13. The growth pattern can be divided into three phases—lag, log, and stationary. In the lag phase, soon after introduction of microorganisms in the fermenter (bioreactor), there is no apparent increase in biomass. This does not mean there is no metabolic activity in the cells. Rather the lag phase is a preparatory phase which preceeds rapid exponential growth and is an adaptation period.

In the exponential (log) phase, the cell number (and therefore the mass) increases exponentially with time. Cell mass undergoes several doublings during the log phase and the specific growth rate of the culture remains constant. During the log phase, the increase in cell mass with time, dX/dt, is the product of the *specific growth rate* μ and the biomass concentration (X):

$$\frac{dX}{dt} = \mu X$$

This increase can also be denoted in cell number, dN/dt, which is the product of specific growth rate and the cell number N:

$$\frac{dN}{dt} = \mu N$$

The specific growth rate (μ) is a function of limiting nutrients and environmental parameters, such as temperature, pH, composition of the medium, and dissolved oxygen levels.

For cell mass concentration X_0 at the beginning of the phase of exponential growth, the time required for doubling the biomass can be calculated:

$$\ln[X/X_o] = \mu t$$

Thus,

$$t = \frac{\ln(X/X_o)}{\mu}$$

The generation time of a culture is the time required for the biomass to double under defined conditions. The value of μ for microorganisms used in biotechnological production usually varies from 2.1 to 0.08 per hour, which corresponds to a doubling time of 20 minutes to 8 hours. Specific growth rate and doubling time of the organisms determine the medium requirements and fermentation batch time for the production of biochemicals in a fermenter.

Exponential growth is followed by a stationary phase during which the growth and death rates are equal. Exhaustion of the growth-limiting nutrients in the medium or accumulation of toxic products is possibly a reason for the onset of the stationary phase. Eventually, the cell culture enters the death phase where loss of cell viability overtakes growth.

5.11.2 Effect of Temperature on Cell Growth

Table 5.9 Three types of microorganisms, depending on optimum temperature for growth

Types	°C
Psychrophiles	~15
Mesophiles	~37
Thermophiles	~55

Temperature essentially controls the rate of chemical reaction and thus the overall growth rate of the microorganisms. On the basis of the optimum temperature required for growth, microorganisms can be categorized into three types as shown in Table 5.9.

5.11.3 Effect of Substrate Concentration on Cell Growth

The concentration of a growth-limiting substrate, such as a carbon source, affects the growth of microorganisms. The relation between the substrate (nutrient) concentration and growth of the organisms has been shown by the equation:

$$\mu = \frac{\mu_m S}{K_s + S}$$

Here, μ is the specific growth rate, S is the concentration of the limiting nutrient, K_s is the substrate-specific constant (affinity of the organism for the substrate), and μ_m is the maximum specific growth rate.

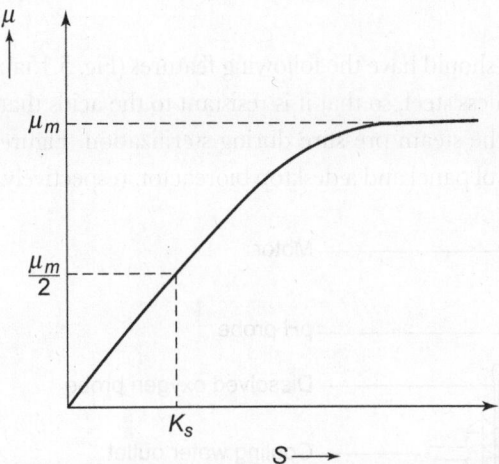

Fig. 5.14 Specific growth rate (μ) increases with substrate concentration (S).

The specific growth rate μ increases with substrate concentration until the concentration is no longer growth limiting. This can be realized from Fig. 5.14. High substrate concentrations inhibit the growth rate. It can be seen from Fig. 5.14 that numerically K_s is the substrate concentration at $\mu_m/2$. Thus, growth on a particular substrate may be described by two constants—μ_m and K_s. Evaluation of μ_m, K_s, and μ of an organism in a substrate is essential to predict the suitability of a substrate for growth.

Formation of a biochemical product may or may not be associated with growth. When primary energy metabolism and cell growth simultaneously result in product formation, the product concentration is found to be directly proportional to cell growth or cell concentration. Alcohol production during fermentation belongs to the growth-associated class. In non-growth associated fermentation processes, product concentration is not directly proportional to the biomass concentration. Antibiotic fermentations are non-growth associated processes. The metabolites accumulate after the active growth phase is over. Understanding the relation between cell growth kinetics and product kinetics is essential for maximization of product formation.

5.11.4 Design of Growth Media

Careful production of growth and production media is a very important aspect of the biotechnological fermentation process. It should take care of the following factors:
- Product yield should be at minimum medium cost.
- Medium must provide sufficient carbon, nitrogen, minerals, and other nutrients to yield requisite cell mass and product.
- Minimum requirements are estimated from the stoichiometry of growth and product formation. It is customary to supply most of the nutrients at levels well above the minimum concentrations required.
- Media should be developed keeping in mind the overall process—types of microorganisms, precise nutritional requirement, environmental needs, and so on. For example, using glucose as a carbon source instead of molasses may increase the cost initially but it simplifies pollution control and waste treatment.

5.12 BIOREACTORS

Bioreactors (fermenters) play a central role in biotechnology-based production processes—producing biomass or metabolites, transforming one compound into another, or degrading unwanted wastes.

The reactions occurring in a bioreactor are carried out by biocatalysts enzymes, microorganisms, cells of plants and animals, or subcellular organelles, such as chloroplasts and mitochondria. A bioreactor provides an environment conducive to the optimal functioning of a biocatalyst.

5.12.1 Stirred Tank Bioreactor Design

A stirred tank bioreactor, which is a closed vessel, should have the following features (Fig. 5.15a):

(i) It should be made up of 4–5 mm thick stainless steel, so that it is resistant to the acids that are produced during fermentation and to the steam pressure during sterilization. Figure 5.15(b) and (c) shows a bioreactor with control panel and a desktop bioreactor, respectively.

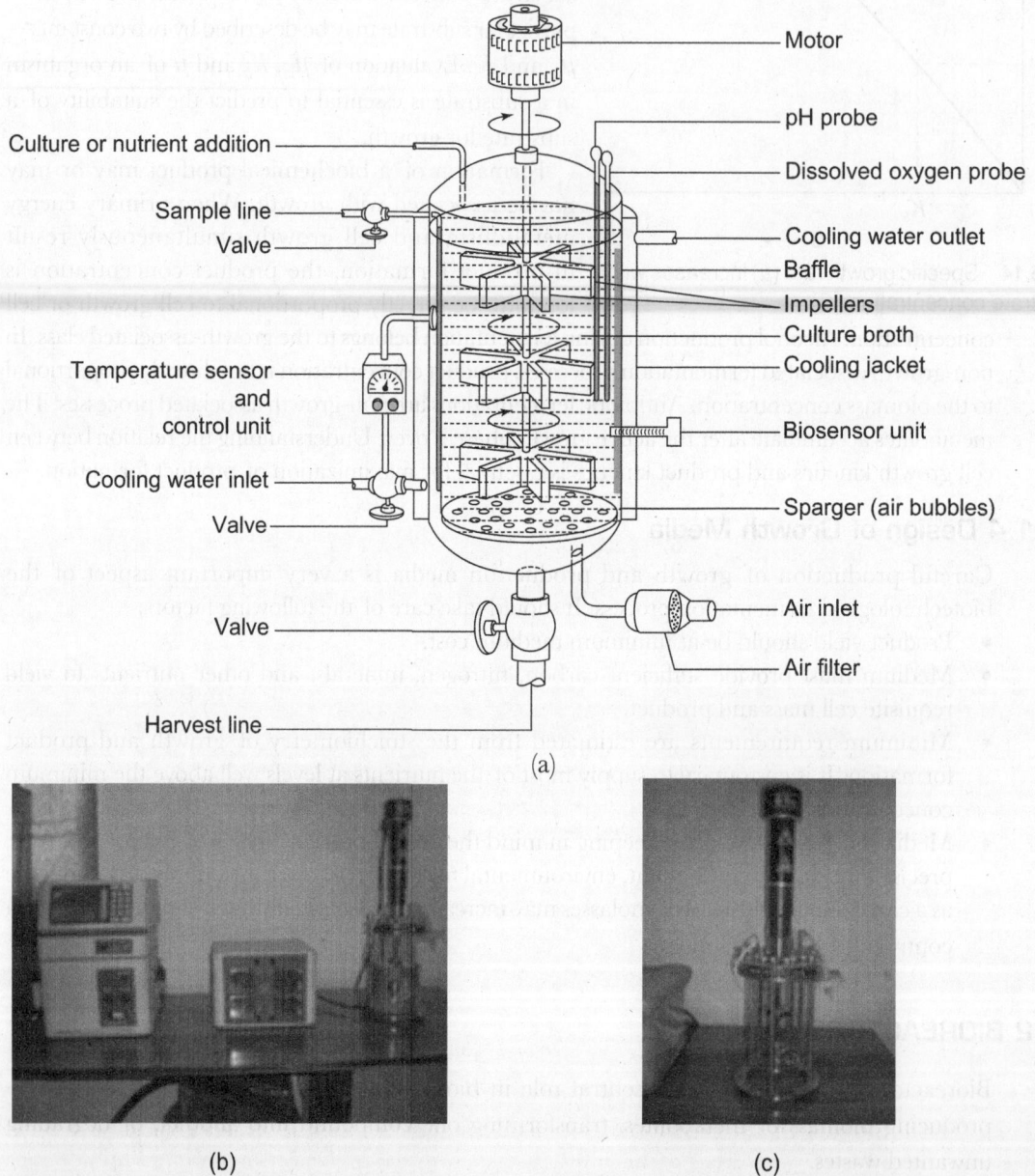

(a)

(b)   (c)

Fig. 5.15 (a) Basic design of a stirred tank bioreactor. (b) A bioreactor with control panel (c) A desktop bioreactor. (Also see colour plate 1.)

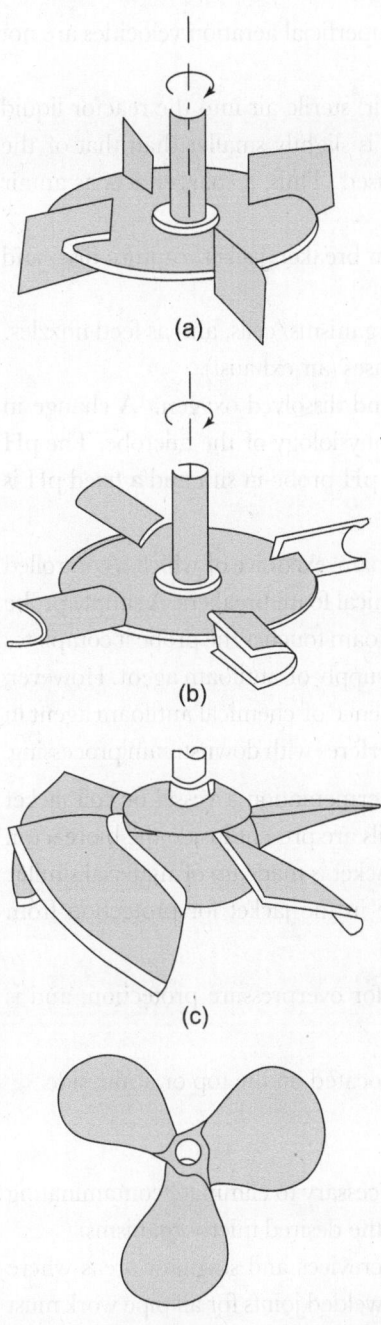

Fig. 5.16 Design of some impellers. (a) Rushton stirrer, (b) Scaba agitator, (c) Prochem maxflow agitator, and (d) marine propeller.

(ii) It should have a *height-to-diameter ratio* (H/D) that varies from 1 (for small reactors) to 3 (for larger vessels). A high (H/D) ratio saves space (smaller footprint) and provides ease of construction and better mixing since vessels with larger diameters require proportionally larger agitator blades on the rotor. However, a limitation of tall tubular reactors with high (H/D) ratios is frequent oxygen starvation and reduced oxygen transfer rates.

(iii) It must have *baffles*, rectangular metal strips which project from the inner wall of the vessel and prevent swirling and vortexing of the fluid and improve the efficiency of oxygen transfer by increasing the turbulence of the agitated culture medium. Microbial culture vessels are generally provided with four baffles, and animal cell culture vessels are without baffles so that turbulence is minimized.

(iv) *A vertical motor-driven shaft* with agitator blades (impellers) may be fitted from the top or bottom of the reactor vessel. The number of impellers for a reactor depends on the volume of the vessel and the type of culture. The impeller diameter is about one-third of the vessel diameter. The bottom impeller is located at a distance of about one-third of the tank diameter from the bottom of the tank. Additional impellers are spaced one to two impeller diameter distances apart.

The *design of the blades of the impellers* can influence oxygen transfer and mixing. The blades of impellers (Fig. 5.16) may be of the following types:

- Rushton stirrer—for good radial mixing
- Scaba agitator—with concave blades for better bulk mixing
- Prochem maxflow (hydrofoil) agitator—also for bulk mixing, used in fermenters with highly viscous fungal broths (non-Newtonian fluids)
- Marine propeller—a single low-shear impeller used in animal cell culture vessels

For animal or plant cell culture, the speed of the impeller does not exceed 120 rpm in vessels larger than 50 L. Higher stirring rates are employed for microbial cultures. For mycelia and filamentous cultures, the impeller speed is lower.

In a stirred bioreactor, the superficial aeration velocity, that is the volumetric gas flow rate divided by the cross-sectional area of the vessel, must remain below the value needed to flood the impeller. An impeller is flooded when it receives more air/gas than it can effectively disperse. A flooded impeller is a poor mixer. Usually, superficial aeration velocities are not allowed to exceed 0.05 m s^{-1}.

(v)   *Sparger* is a perforated pipe ring to sparge (introduce air) sterile air into the reactor liquid below the bottom impeller. The diameter of the ring is slightly smaller than that of the impeller. Alternatively, a single-hole sparger may be used. Thus, a sparger acts as an air inlet.

(vi)   *Other features* — these include inlets, outlets, ports, foam breaker, jacket, rupture disc, and sight glass.

Inlets   These are used for medium, inoculums of microorganisms/cells, and as feed nozzles.

Outlets   These are used for harvest and for removal of gases (air exhaust).

Ports   These are used for sensors of pH, temperature, and dissolved oxygen. A change in the pH of the culture medium occurs with a change in the physiology of the microbe. The pH of the fermentation broth is continuously monitored with a pH probe in situ and a fixed pH is maintained by addition of acid or alkali.

Foam breaker   Aeration and agitation inevitably produce foam, excessive of which is controlled with a combination of chemical antifoam agents and mechanical foam breakers. A simple probe placed above the culture medium detects foaming. When the foam touches the probe it completes an electrical circuit and this activates a pump connected to a supply of antifoam agent. However, mechanical foam breakers are used exclusively when the presence of chemical antifoam agent in the product is not acceptable or when the antifoam agent interferes with downstream processing.

Jacket   To counter the metabolic heat generated during fermentation a vessel or coil jacket around the bioreactor is used. Sometimes, heat exchange coils are present inside the bioreactor. Cooling water is circulated through the coils or jacket. The jacket is made up of material similar to that of the vessel. There is a provision of a relief valve in the jacket for protection from excessive pressure.

Rupture disc   This is located on top of the bioreactor for overpressure protection, and is made of graphite.

Sight glass   Often called the view port, the sight glass is located on the top or at the side.

5.12.2 Sterilization

Before starting the fermentation of a batch, sterilization is necessary to eliminate contaminating microbes from the system so that it can be used for growing the desired microorganisms.

To begin with, the vessel (bioreactor) should be free of crevices and stagnant areas where pockets of liquid or solid may accumulate. As far as possible, welded joints for all pipe work must be used instead of couplings, as they are better for maintaining sterile conditions.

All inputs to the system, such as air, media, feed, inoculum, and regulators of acid/base concentration, need to be sterile.

Inlets for the above items and exhaust pipes are installed with in-situ stem sterilizable gas filters. Typically, hydrophobic membrane cartridge filters are used.

The reactor vessel can be sterilized with clean steam at high pressure (212 kPa) and at 121°C for 20 minutes in situ.

5.12.3 Inoculum

After the bioreactor is properly sterilized and the right type of medium has been added, the temperature, pH, and aeration are brought to the optimum level for the microorganisms to be used in bioreactor.

The microbial cells are grown from a stock culture initially in small volumes (5–10 mL). Subsequently, they are grown in a shaker flask (200–1000 mL) and then in a seed fermenter (10–100 L). This constitutes the inoculum for the bioreactor. The inoculum consists of microorganisms in a phase of rapid exponential growth, added at a concentration of 5–10 per cent by volume to the bioreactor. A large volume of slower growing organisms can be used as inoculum to avoid long fermentation time in the production vessel; this reduces the cost involved in the process.

The outline of the steps involved in the production of the inoculum and fermentation in a bioreactor and harvesting of two types of cell products are presented in Fig. 5.17.

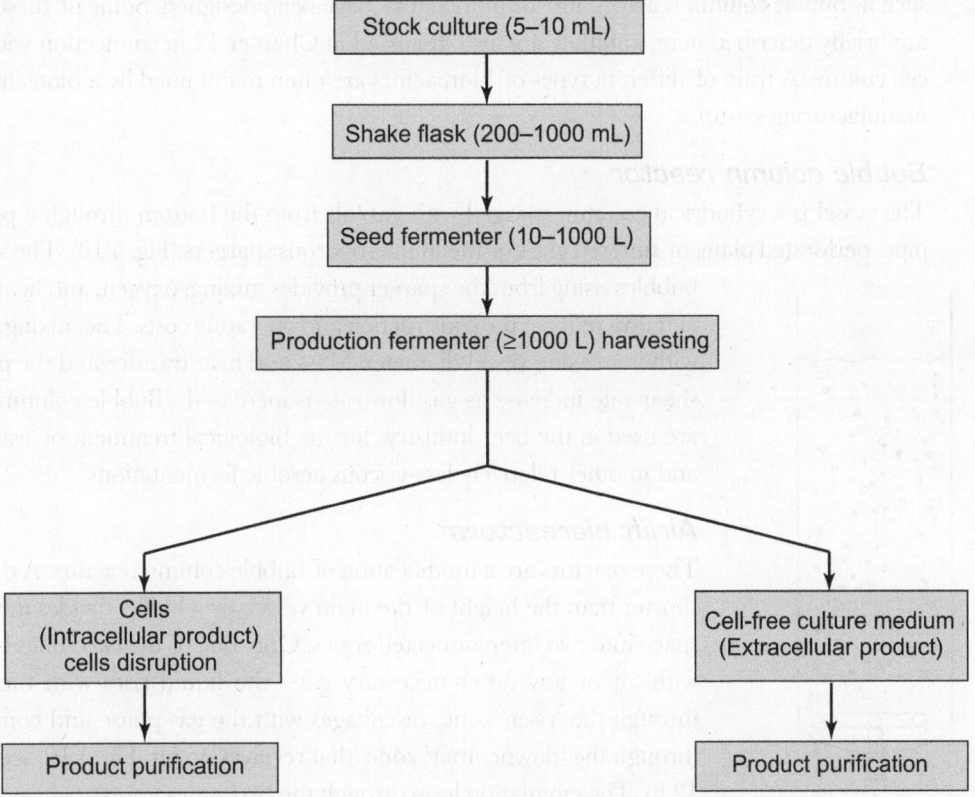

Fig. 5.17   Steps to produce inoculum for large-scale fermentation process.

When the fermentation is completed, the microbial cells are separated from the fermentation liquid broth either by filtration or centrifugation. Extracellular products, such as vitamins and most of the antibiotics are purified from the cell-free culture supernatant during downstream processing. Intracellular products like recombinant proteins produced in *E. coli* or other bacteria are collected by disrupting the cells and removing cell debris. The purification of the desired protein (or other products) is then carried out during downstream processing by employing proper physical and biochemical procedures.

For optimal recovery of the fermented product, it is necessary to integrate both the fermentation and downstream processes on the basis of prior analysis, so that the number of steps in purification can be limited as much as possible.

5.13 OTHER TYPES OF BIOREACTORS

Stirred tank is the most widely used type of bioreactor, especially in production of antibiotics and organic acids. Often the problems of heat and oxygen transfer are not well addressed with the conventional stirred tank design. To take care of these problems, other types of bioreactors, such as bubble column reactors and airlift reactors have been designed. Some of these reactors are briefly described here, and they are also discussed in Chapter 12 in connection with animal cell culture. A train of different types of bioreactors are often maintained by a biotechnological manufacturing set-up.

Bubble column reactor

The vessel is a cylindrical column, sparged with gas/air from the bottom through a perforated pipe, perforated plate, or sintered glass or metal microporous spargers (Fig. 5.18). The stream of bubbles rising from the sparger provides mixing, oxygen, and heat transfer, and thus reduces the construction and operating costs. The mixing improves with increasing vessel diameter. Mass and heat transfer and the prevailing shear rate increase as gas flow rate is increased. Bubble column reactors are used in the beer industry for the biological treatment of waste water and in other relatively less-viscous aerobic fermentations.

Airlift bioreactors

These reactors are a modification of bubble column reactors. A draft tube, shorter than the height of the main vessel, or a baffle, divides the internal space into two interconnected zones. Only one of the two zones is sparged with air or any other necessary gas—the liquid rises with the bubbles through the 'riser' zone, disengages with the gas phase and comes down through the 'downcomer' zone (that receives no air; Fig. 5.19, see also Fig. 12.8). The circulation loop through the two zones improves heat and mass transfer rates. Airlift bioreactors are highly energy efficient relative to stirred fermenters, and the productivities of both types are comparable.

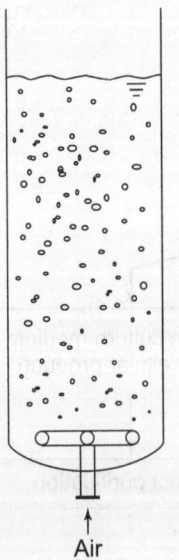

Fig. 5.18 A bubble column bioreactor.

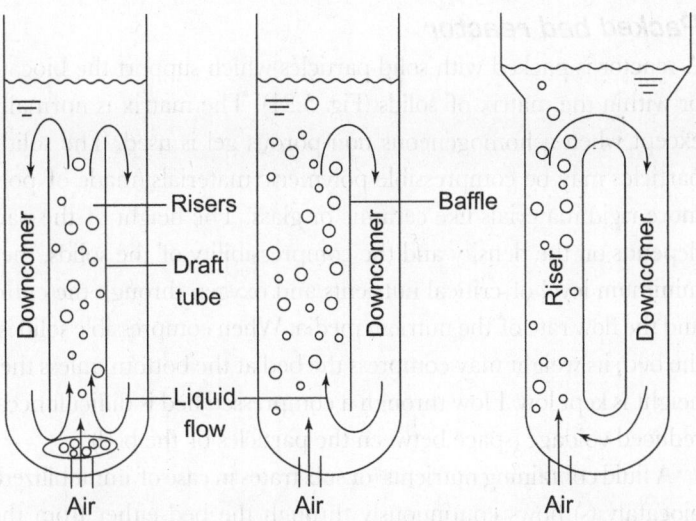

Fig. 5.19 Airlift bioreactors.

Airlift bioreactors are suited to shear-sensitive cultures like animal cells. Thus, they are often employed in large-scale manufacture of biopharmaceutical proteins produced by fragile animal cells in culture. Airlift devices are also used in high rate biotreatment of waste water, production of insecticidal nematode worms and other low-viscosity fermentations.

Fluidized bed reactor

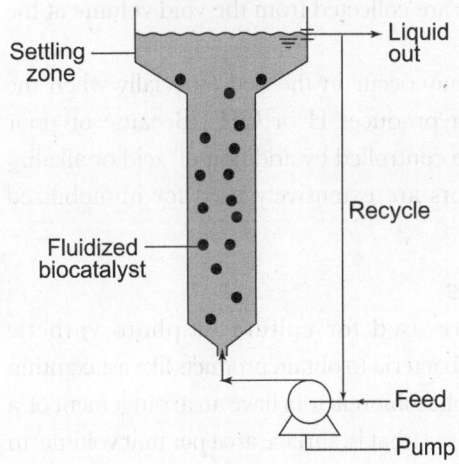

Fig. 5.20 A fluidized bed bioreactor.

Particulate biocatalysts such as immobilized enzymes and cell or microbial particles continue reactions in the suspended condition in a fluid, in a fluidized bed bioreactor. An upflowing stream of liquid which is pumped in is used to suspend or fluidize the particles (solids). The vessel is similar to a bubble column with a modification at the top; the top section is expanded to reduce the superficial velocity of the fluidizing liquid to a level below the level that is needed to keep the particles in suspension (Fig. 5.20). Consequently, the solids sediment in the expanded zone and drop back into the narrower reactor column below; the solids are retained in the reactor and the liquid which flows out may be recycled especially when the solid–liquid contact time in the column is insufficient for the reaction. The liquid fluidized bed may be sparged with air to produce a gas–liquid–solid fluid bed.

In case, the solid particles are too light, they may be artificially weighted by embedding stainless steel balls. A high density of solids improves solid–liquid mass transfer by increasing the relative velocity between the phases. The density of the particles should not be too high when compared to that of the liquid, otherwise fluidization will be difficult.

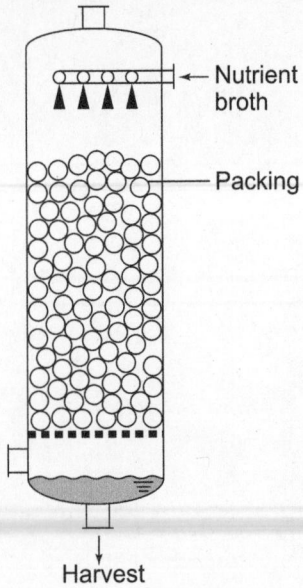

Fig. 5.21 A packed bed bioreactor.

Packed bed reactor

A reactor is packed with solid particles which support the biocatalysts on or within the matrix of solids (Fig. 5.21). The matrix is normally porous except when a homogeneous non-porous gel is used. The solid packing particles may be compressible polymeric materials (made of polymer) or more rigid materials like ceramic or glass. The height of the packed bed depends on the density and the compressibility of the solids, the need of minimum level of critical nutrients and oxygen through the entire depth, and the flow rate of the nutrient media. When compressible solid is used for the bed, its weight may compress the bed at the bottom unless the packing height is kept low. Flow through a compressed bed is difficult because of a reduced voidage (space between the particles of the bed).

A fluid containing nutrients (or substrates in case of immobilized enzyme biocatalysts) flows continuously through the bed either from the top or from the bottom. Downward flow under gravity is the norm. In case of upward flow, the maximum flow velocity needs to be limited; the velocity cannot exceed the minimum fluidization velocity or the bed will fluidize (will be disturbed).

The concentration of the nutrients decreases as the fluid moves down the bed and concentrations of metabolites and products increase. The concentration variations of nutrients and metabolites near the bottom of the bed can be overcome to some extent by increasing the flow rate. The metabolites or products are collected from the void volume at the bottom.

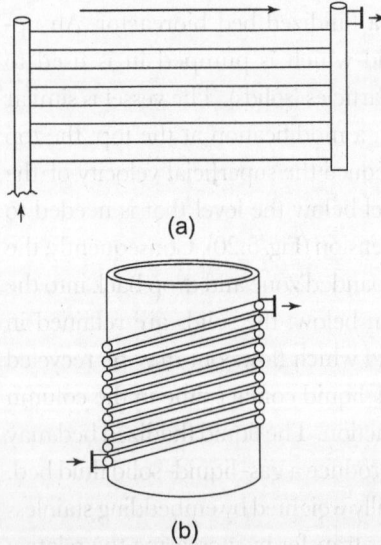

Fig. 5.22 Photobioreactors. (a) Solar receiver of multiple parallel tubes. (b) Helically wound tube.

Gradient of pH may occur in the bed especially when the reaction consumes or produces H or OH⁻. Because of poor mixing, pH cannot be controlled by addition of acid or alkali.

Packed bed reactors are extensively used for immobilized enzymes.

Photobioreactors

Photobioreactors are used for culture of photosynthetic microalgae and cyanobacteria to obtain products like astaxanthin and β-carotene. The photobioreactors have an arrangement of a large specific surface area (that is, surface area per unit volume) to provide a large amount of incident sunlight. Artificial illumination for the purpose is impractically expensive.

Closed photobioreactors for monoculture are constructed using arrays of transparent tubes of glass or more commonly clear plexiglass for sunlight penetration. The tubes may be arranged horizontally, or as long rungs on an upright ladder (Fig. 5.22). A continuous tube may be wound helically around a vertical cylindrical support.

The medium enters with a small inoculum, flows through the tubes, and provides a high cell density at the outlet. Often, a small fraction of the outlet cells is recycled to provide a continuous inoculum. The culture is circulated through the solar receiver variously—by centrifugal pump, positive displacement mono pumps, or airlift devices. The flow in a solar receiver panel needs to be turbulent enough for movement of cells from the deeper poorly illuminated interior to the regions nearer the wall. The velocity of the flow should be sufficient to prevent sedimentation of cells. Typical linear velocity through receiver tubes needs to be $0.3 - 0.5$ m s^{-1}.

Light penetration in a photobioreactor also depends on biomass density, cellular morphology and pigmentation, and the adsorption characteristics of a cell-free culture medium.

Based on the principle of outdoor photobioreactors, open ponds and raceways (narrow channels) are often used to culture microalgae especially to treat waste water.

SUMMARY

- Biochemical engineering is concerned with all the aspects of large-scale production of materials of cellular origin. Microorganisms, eukaryotic cells, and genetically engineered cells are mainly used for production of commercially important and pharmaceutical products.

- Biochemical engineering involves the knowledge of biological optimum conditions for cell growth and catalytic reactions, and the principles of process technology from chemical engineering.

- Information about pH, dissolved oxygen, agitation rate, aeration rate (mass transfer), heat transfer, pressure, cell density, substrate requirement, and so on allows design of a proper bioreactor, a large vessel for cell growth, and fermentation. pH is the hydrogen ion concentration in any aqueous solution. p denotes negative logarithm of, H refers to hydrogen ion (H$^+$).

- pH value of a fluid indicates whether it is acidic, alkaline, or neutral. Neutral pH value is 7. pH affects the structure and activity of enzymes, and the well-being of organisms and their cells. Usually, cells thrive better in neutral pH. pH varies inversely with temperature.

- Constancy of pH is achieved by using biological buffers—mixtures of weak acids and their conjugate bases.

- Bioprocess engineering is planned and executed on the basis of quantitative physical measures and variables, such as mass, length, time, velocity, area, temperature, viscosity, and density. A unit is used to measure a physical quantity. There are several systems of units—CGS, FPS, MKS, and SI.

- Mass, length, and time provide three basic dimensions. Using dimensional formula it can be shown that the dimensions of energy are the same as those of work.

- Stoichiometry deals with material and energy balance in biochemical reactions involving bacterial growth during fermentation. Stoichiometry and reaction kinetics together provide the quantitative approach for the design of bioreactors.

- The total input of mass and the oxygen requirement for the biochemical reaction during fermentation by bacteria and the amount of output (products) can be calculated apriori on the basis of the mass balance equation.

- Data generated in the course of experimentation in the field and in the laboratory and during operation of bioreactors needs to be properly analysed to understand the significance or reliability of observations. This helps in proper running of experiments, as well as biotechnology plants and for future projections.

- Decision-making process under uncertainty requires statistical data analysis for probabilistic risk assessment. Statistical methods are used to collect, analyse, present, and interpret data.

- The two fundamental steps in statistical analysis are measures of central tendency of individual readings by calculating mean or median or mode, and then measuring dispersion in terms of range, variance, and standard deviation (or standard error).

- Data of a study can be presented in many ways—using tables, charts, diagrams, graphs, and so on. Graphic presentations are more precise for summarizing and

- characterization of data than frequency tables which list the numbers.
- Statistical significance indicates whether the deviation of a particular estimate from its expected value is a chance occurrence as a result of random sampling or is something unusual.
- A 'bioprocess' is a large-scale operation for transformation of raw materials (biological or non-biological) into a commercial product by means of microorganisms, animal or plant cell cultures, or by materials derived from them, such as enzymes.
- A typical bioprocess has three necessary components—upstream part, bioprocess or fermentation, and downstream part.
- A bioprocess engineer designs a bioreactor, a specialized large container for microbial growth and fermentation, on the basis of physical and chemical consideration of mass balance, fluid flow mixing, mass transfer, effects of transfer limitations, and heat transfer.
- There are several types of bioreactors for various industrial processes. Stirred tank bioreactors are the most commonly used.

J.B.S. Haldane

**John Burdon Sanderson Haldane
(5 November 1892–1 December 1964)**

J.B.S Haldane was born in Oxford to a renowned physiologist father John Scott Haldane from an aristocratic intellectual Scottish family. Notably, Haldane began his science career as his father's assistant at an early age of eight. He received his education from Eton and New College, Oxford, and served the British army during World War I. From 1919 to 1922, he served New College, Oxford, as a Fellow and later taught as a Reader in Biochemistry at the Trinity College at Cambridge University. While at Cambridge, Haldane devoted his time to studying enzymes and genetics, with particular emphasis on the mathematical aspect of genetics. He was a pioneer in developing Population Genetics. Later, he joined the University College, London as Professor of Genetics.

Haldane along with G. E. Briggs formulated a new interpretation of the law of enzyme kinetics.

During the early 1930s, when the molecular nature of enzymes was not yet fully understood, J.B.S. Haldane wrote a treatise entitled *Enzymes*. This book put forward the remarkable suggestion that weak-bonding interactions between an enzyme and its substrate might be used to distort the substrate and catalyse the reaction. Interestingly this insight is the basis of our current understanding of enzymatic catalysis. He was an outstanding scientist who contributed significantly to physiology, genetics, biochemistry, statistics, and biometry. He was talented and had interest in many subjects including subjects outside the arena of science. Haldane wrote on varied subjects including science fiction, stories for children, and popular science.

For his outstanding contributions, Haldane received much recognition. Haldane was elected a Fellow of the Royal Society in 1932. The French Government awarded him the Legion of Honour in 1937. He was also a recipient of the Weldon Memorial Prize from Oxford University (1938), the Darwin Wallace Medal of the Linnean Society (1958). He was the President of the Genetical Society (1932–1936).

In 1957, Haldane moved to India, as a mark of protest against the Anglo-French invasion of Suez. He joined the Indian Statistical Institute (ISI), Calcutta. In 1962, he moved to Bhubaneswar, Orissa to set up a Genetics and Biometry Laboratory. Haldane had a deep appreciation of Indian culture. In April 1961, he became an Indian citizen.

In Kolkata (Calcutta), the busy connecting road from Eastern Metropolitan Bypass to Park Circus area on which the Science City is located, is named after J.B.S. Haldane.

EXERCISES

Objective Questions

1. Multiple Choice Questions
   (a) Bioprocess engineering involves only
       (i) bacteria
       (ii) eukaryotic cells
       (iii) enzymes
       (iv) all of the above
   (b) Catabolism in microbial growth systems uses
       (i) electron donor
       (ii) C and N sources
       (iii) electron
       (iv) electron donor and electron acceptor
   (c) A buffer system normally uses a
       (i) strong acid
       (ii) weak acid
       (iii) both of these
       (iv) none of these

2. State whether the following statements are True or False.
   (a) The pH affects the structure and activity of enzymes.
   (b) The pH of an equimolar solution of acetic acid and acetate is exactly equal to the pKa of acetic acid.
   (c) The same acid-conjugate base pair can serve as buffer at different pH zones.
   (d) Natural variables require units.
   (e) The numerical value of a measurement and the size of the unit used for a measurement are inversely proportional to each other.
   (f) Potential energy and kinetic energy possess the same dimensions.

3. Fill in the blanks in the following statements.
   (a) Dissociation or equilibrium constant of an acid is expressed as _____.
   (b) Histogram presents the data of _____ values.
   (c) In a pie chart, the data is presented as a part of a _____.
   (d) The centre of the curve for normal distribution marks the point where the number of observations is _____.
   (e) The chi-square (χ^2) test is a statistical tool to determine the *goodness of fit* of _____ and _____ values.

Review Questions

1. Define pH.
2. How does a hydronium ion (H_3O^+) form in water?
3. Pure water at 25°C shows neutral pH 7. Why is this not so at other temperatures?
4. Briefly explain why addition of NaOH to water increases the pH.
5. Mention one property of a strong acid.
6. How is a specific cytoplasmic pH maintained in a cell?
7. Show how Hendersson–Hasselbach's equation is derived from the original equation for the dissociation constant of a weak acid.
8. What are the basic properties which a unit for measurement of a physical quantity must have?
9. What are supplementary SI units?
10. What is the unit *Candela* used for?
11. What does dimension signify for a physical quantity?
12. Why is a knowledge of stoichiometry essential to design a bioreactor and to run an industrial fermentation?
13. What are the essential deductions from mass balance equation during fermentation or conversion of a substrate by biochemical reaction?
14. Bacterial growth at the expense of a substrate can be expressed as Y_{sx}^{max}. Explain this expression.
15. Briefly explain the difference between *systemic error* and *random error*.
16. What does a sample mean statistically?
17. Define arithmetic mean, median, and mode. What is their importance in statistics?
18. Why is the range not considered as a good measure of dispersion?
19. What is variance?
20. What does lead ti device standard deviation?
21. How can one draw a frequency polygon curve?
22. Explain probability and significance as used in statistics.
23. What is *t*-test?

24. Why does one need to plot data on a graph paper with a log scale?

25. Mention and characterize the components of bioprocess engineering.

26. Write a brief note on the following with respect to the fermentation process:
    (a) Mass balance
    (b) Effect of substrate concentration on cell growth
    (c) Mass transfer
    (d) Impellers in stirred tank bioreactor
    (e) Heat transfer

27. What are the reasons for the generation of heat in a bioreactor during operation?

28. Explain the growth kinetics of microorganisms in the fermentation process.

29. What are the essential features required for a growth medium of a bioreactor?

30. Write about the essential features of a stirred tank bioreactor which is commonly used in the biotechnology industry.

31. Discuss the steps involved in producing the proper inoculum required to initiate processing in a bioreactor.

32. What are the advantages of an airlift bioreactor as compared to a stirred tank bioreactor?

Genetic Engineering

LEARNING OBJECTIVES

♦ The discovery of a factor (gene) by Gregor Johann Mendel (1822–1884) that gave birth to genetics and its most modern application, genetic engineering

♦ Recombinant DNA (rDNA) technology which is the basis of genetic engineering

♦ The purification of DNA, cleavage by restriction endonucleases, construction of various vectors, cDNA synthesis, gene library, transfer of rDNA into host cells, and selection of clones

♦ The polymerase chain reaction technique for amplification of specific DNA sequences into millions of copies which is a rapid and important method used in rDNA technology and site-directed mutagenesis

♦ The role of transgenic organisms in the mass production of desired proteins, hormones, and biopharmaceuticals

INTRODUCTION

Genetic engineering has now become a major subject by itself and an integral part of other disciplines of biological science and commercial technologies. *Genetic engineering, by definition, is the technique of modifying the genome of an organism by using recombinant DNA technology. Recombinant DNA (rDNA) technology isolates a specific DNA segment (gene / genes) from one organism to insert it in another DNA molecule of a different organism at a desired position.* Thus, the process of genetic engineering makes it possible to transfer just one or a few desirable genes between species that are distantly related or not related at all. This is not possible by the traditional breeding processes that are used for the improvement of the stock. Genetic engineering has made it possible for scientists to fish out a desirable gene from virtually any living organism and insert it into any other organism. For example, it is possible to put a gene from a rat into a lettuce plant to make it a producer of vitamin C or insert a microbial toxin gene into a cotton plant to make it insect pest-resistant. All these genetic manipulations became possible with the discovery of the techniques of gene splicing and recombinant DNA technology.

Genetic engineering includes diverse techniques: (i) isolation, purification, and quantitation of cellular DNA; (ii) use of restriction endonucleases; (iii) construction of recombinant (rDNA) molecules; (iv) use of cloning vectors; (v) introduction of vectors into host cells to produce clones; (vi) construction of gene libraries; (vii) selection of specific clones; (viii) screening gene libraries for the gene of interest; (ix) site-directed mutagenesis; (x) polymerase chain reaction (PCR) to amplify specific DNA sequences or genes in vitro; and (xi) transgenic technology. rDNA technology is important for commercial and pharmaceutical products. For gene transfer, genetic engineering has broken the barriers of genus, order, class, phylum, and kingdom generated through the process of evolution over millions of years. This technology, thus, demands due recognition and at the same time lots of caution.

Genetic engineering is based on classical knowledge and modern concepts related to the behaviour of genes, starting from Gregor Mendel's concept of *factors*.

Here, Mendel and his laws elucidating the inheritance of traits (genes) are mentioned briefly to remind the students about Mendel's profound insight and his experiments that laid the foundation of genetics, which in a way culminated in genetic engineering.

6.1 MENDEL AND THE TWO LAWS OF GENETICS

Gregor Johann Mendel (1822–1884), a contemporary of Charles Darwin, laid the foundation for genetics, another revolutionary milestone in the annals of biology as was Darwin's proposition of evolution through natural selection. The impacts of these contributions were as phenomenal as those of Nicolaus Copernicus on astronomy and Isaac Newton on physics.

Mendel was from a family of farmers in Moravia, then a part of the Hapsburg Empire in Central Europe. Thus, he had a natural inclination towards plant and animal husbandry. At the age of 21, Mendel left the farm and entered a Catholic monastery in the city of Brünn (today, Brno in the Czech Republic). In another four years, in 1847, he was ordained a priest, adopting the clerical name Gregor. He taught at the local high school, and during the period 1851–1853 he studied in the University of Vienna, Austria. After returning, he became a teaching monk and within a few years began carrying out experiments of crossing between pea plants with different traits in the garden adjoining the monastery. These experiments eventually allowed Mendel to propose his hypothesis in the form of two laws of inheritance, which brought him fame 34 years after the publication of his results and 16 years after his demise at the age of 62.

Exhibit 6.A   Mendel's Study of Heredity with Peas

Ronald A. Fisher, a British statistician and geneticist concluded (*Annals of Science*, 1:115–137, 1936) that Mendel began cultivation of stocks of peas (obtained from local farmers) in 1857 and hybridization experiments in 1858. Mendel followed the progeny of the crosses for as many as six generations. Fisher conjectured that Mendel began his dihybrid and trihybrid crosses in 1861. In the same year, he apparently initiated test crosses to determine gametic ratios from heterozygous plants. Mendel's experiments with peas spanned eight years, from 1857–1864. During this period, Mendel grew more than 5000 pea plants in the monastery garden, for progeny analysis.

After completion of his work in 1864, Mendel presented the results before the local Natural History Society of Brönn in 1865. In 1866, he published a detailed report in the Society's proceedings. The paper which technically launched the science of genetics and remained in obscurity until 1900, was entitled 'Versuche Über Pflanzenhybriden'—translating from German it becomes 'Experiments with Plant Hybrids'. The paper is lucid and presents the results in a systematic manner. This reflects that Mendel was an experienced and successful teacher. His ingenuity was in successful approximation in calculating ratios of different genotypes of progeny plants after a cross.

Initially, Mendel used several species of garden plants as experimental material, and even tried some experiments with honey bees. His real success was with the garden pea, *Pisum sativum*. *P. sativum* is a dicot (with two cotyledons in the germinating seed) plant and grows easily in pots or gardens. It carries bisexual flowers.

One interesting feature of pea flowers is that the petals close tightly, preventing pollen grains from entering or leaving. The flowers get self-fertilized, that is, the male and female gametes from the same flower unite to produce seeds. Thus, individual pea strains are highly inbred, showing negligible variation from generation to generation. This uniformity allows such strains to become true-breeding.

Mendel obtained many different true-breeding varieties of peas, each distinguished by a particular true-breeding characteristic. A strain of tall plants was about 2 m high, and a short variety was only half a metre in height. One variety produced green seeds, whereas another strain produced yellow seeds.

Mendel crossed pea plants with contrasting traits and analysed the progeny to understand how the distinctive characteristics are inherited. He focussed on one particular trait at a time, for example, plant height. Other biologists of that time studying the inheritance in plants and animals followed many traits simultaneously, and ended up with complex results without arriving at any fundamental principles of heredity. Much of Mendel's success lay in his selection of contrasting differences between plants—tall versus short, green seeds versus yellow seeds, round versus wrinkled seeds, and so on. His calculations of the ratios of the inherited traits in a large number of progeny in different generations collected meticulously, allowed him to propose the laws of inheritance. He assigned the term 'factor' to the entity that determined trait, which was later renamed gene.

Fig. 6.A   Mendel at work

Mendel's discovery was ahead of his time; his 1866 paper suffered obscurity until 1900, when it was rediscovered by three botanists—Hugo de Vries in Holland, Carl Correns in Germany, and Erich von Tschermak-Seysenegg in Austria. In the course of searching the scientific literature on hereditary studies, they were astonished to find Mendel's careful analysis that had been carried out 34 years earlier. Mendel's findings quickly got recognition among biologists. A British biologist, William Bateson's promotional efforts spread Mendel's idea far and wide in European countries.

Bateson also coined a new term to describe the study of heredity—*genetics*, from the Greek word meaning 'to generate'.

6.1.1 Experiment of Monohybrid Crosses—The Law of Dominance and Segregation

A cross between two organisms bearing contrasting traits for a single character is known as monohybrid cross. In one monohybrid experiment, Mendel cross fertilized tall and dwarf pea plants to investigate how the traits for height were inherited.

At the beginning of each experiment, Mendel carefully removed the anthers (male reproductive structure) from one variety (tall or short plant) before the pollen matured. Then on maturity of the flower, pollen grains from the other variety were transferred to the stigma of the emasculated (male organ removed) flower. The stigma is the sticky part on top of the pistil leading to the ovules in the ovary. Thus, he carried out cross fertilization between flowers of a tall female plant and a short male plant or vice versa.

The seeds obtained from these cross fertilizations were sown the next year, yielding hybrids that were uniformly tall. Mendel observed that the dwarf characteristic seemed to have disappeared in the progeny of plants in the first filial generation (denoted as F_1; Fig. 6.1). Mendel allowed the tall hybrids to undergo self-fertilization (a normal event in peas). Interestingly, in the progeny of second filial generation (F_2), some dwarf plants appeared besides the vast number of tall plants. In fact, among 1064 plants of the F_2 generation, there were 787 tall and 277 dwarf progeny. Mendel calculated the ratio of tall to short plants to be approximately 3:1.

Mendel was struck by the reappearance of the dwarf characteristic in the F_2 generation. Clearly, the hybrids of the F_1 generation had the dwarf characteristic, though they were themselves tall. Mendel inferred that the F_1 hybrids carried a latent *factor* (gene) for dwarfness, which remained masked by the expression of the factor for tallness. He considered the latent factor (for dwarfness) as *recessive* and the expressed factor (tall) as *dominant*. He also inferred that these dominant and recessive factors did not mix with each other; rather they maintained their identity in the F_1 hybrids, and separated from each other when the hybrid plants reproduced after self-fertilization for the F_2 generation. This explained the reappearance of the dwarf characteristic in the next generation.

Mendel repeated the experiment of monohybrid crosses with plants bearing contrasting characters for six other traits—seed coat texture, seed colour, pod shape, pod colour, flower colour, and flower position; he found that the results corroborated the findings with tall and dwarf plants.

The 'factors' responsible for different characteristics are now called *genes*; the word 'gene' was coined by the Danish plant breeder Wilhelm Johannsen in 1909.

Although chromosomes and homologues had not been discovered at that time, Mendel realized from his monohybrid cross experiments that factors (genes) for a trait come in pairs (alleles). The parental strains (true breeding) that he used in his experiments carried two identical copies of a gene, either TT for tall or tt for dwarf; in modern terminology, they were *diploid* and *homozygous*. *Alleles* are two genes that occupy the same locus (or site) in two homologous chromosomes (a pair of similar chromosomes, one from the mother and the other from the father). During the production of gametes, the chromosomes of a diploid organism are reduced to half or to a *haploid* condition through meiotic cell division—so a pair of genes (alleles) is reduced to one allele in a gamete.

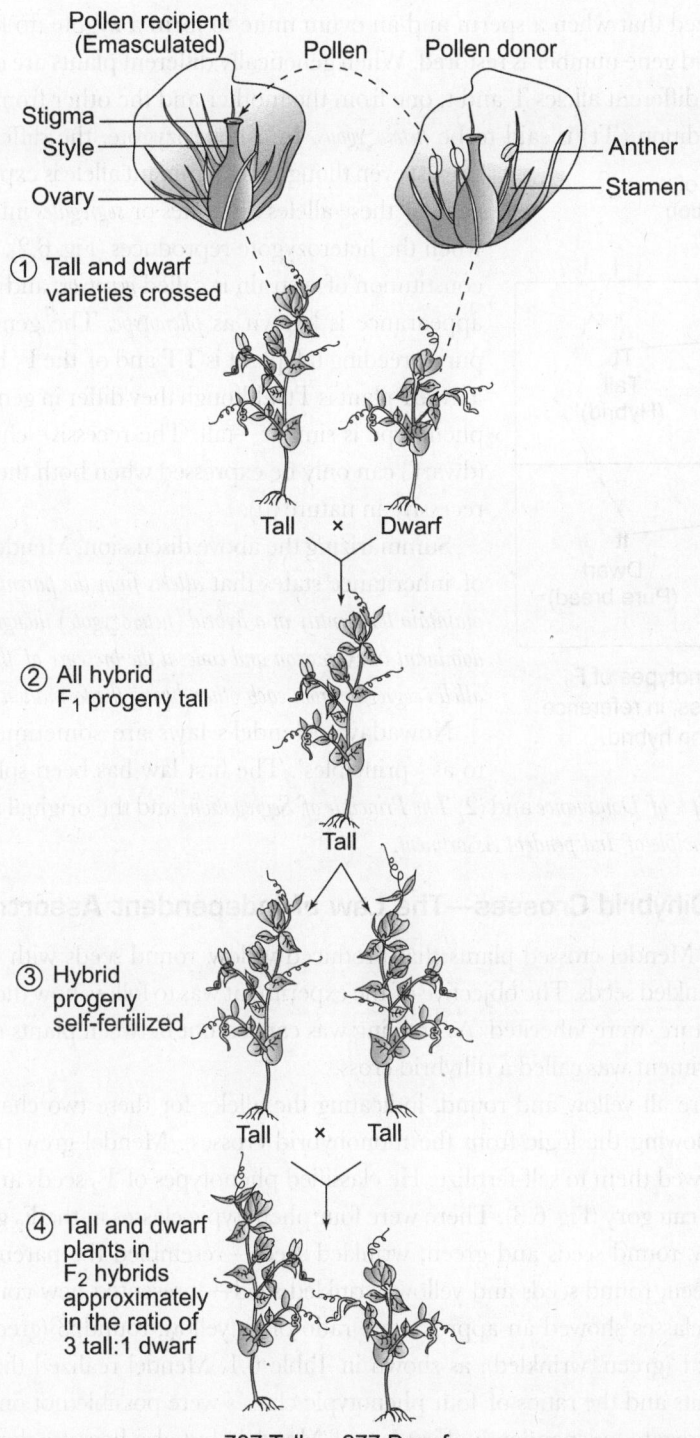

Pollen recipient
(Emasculated)   Pollen   Pollen donor

Stigma

Style

Ovary   Anther

                                                   Stamen

① Tall and dwarf
   varieties crossed

Tall   ×   Dwarf

② All hybrid
   F₁ progeny tall

Tall

③ Hybrid
   progeny
   self-fertilized

Tall   ×   Tall

④ Tall and dwarf
   plants in
   F₂ hybrids
   approximately
   in the ratio of
   3 tall:1 dwarf

787 Tall   277 Dwarf

Fig. 6.1 Mendel's monohybrid cross experiment involving tall and dwarf varieties of pea plants.

Mendel recognized that when a sperm and an ovum unite to form a zygote (to form a new offspring), the diploid gene number is restored. When genetically different plants are crossed, the hybrid inherits two different alleles T and t, one from the mother and the other from the father and its genetic condition (Tt) is said to be *heterozygous*. In a heterozygote, the different alleles coexist even though the dominant allele is expressed, and each of these alleles separates or *segregates* into a gamete when the heterozygote reproduces (Fig. 6.2). The allelic constitution of a strain is called *genotype*, and its physical appearance is known as *phenotype*. The genotype of a pure-breeding tall plant is TT and of the F_1 hybrid with a dwarf plant is Tt; although they differ in genotype their phenotype is similar—tall. The recessive characteristic (dwarf) can only be expressed when both the alleles are recessive in nature (tt).

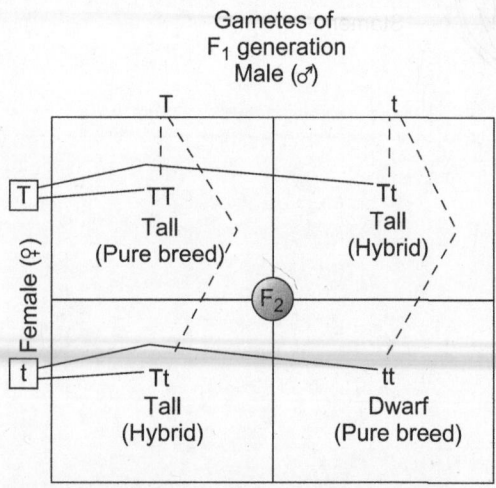

Fig. 6.2 Genotypes and phenotypes of F_2 generation of monohybrid cross, in reference to the gametes of F_1 generation hybrid.

Summarizing the above discussion, Mendel's first law of inheritance states that *alleles from the parental generation maintain their entity in a hybrid (heterozygote) though one may be dominant in expression and conceal the presence of the other. Two alleles segregate from each other during the formation of gametes.*

Nowadays, Mendel's laws are sometimes referred to as 'principles'. The first law has been split into two parts—(1) *The Principle of Dominance* and (2) *The Principle of Segregation*, and the original second law becomes (3) *The Principle of Independent Assortment*.

6.1.2 Experiment of Dihybrid Crosses—The Law of Independent Assortment

In his next project, Mendel crossed plants that produced yellow, round seeds with plants that produced green, wrinkled seeds. The objective of the experiment was to follow how the two traits, seed colour and texture, were inherited. As crossing was carried out between plants differing in two traits, the experiment was called a dihybrid cross.

The F_1 seeds were all yellow and round, indicating the alleles for these two characteristics were dominant (following the logic from the monohybrid crosses). Mendel grew plants from these seeds and allowed them to self-fertilize. He classified phenotypes of F_2 seeds and counted the number in each category (Fig. 6.3). There were four phenotypic classes in the F_2 generation. Two classes—yellow, round seeds and green, wrinkled seeds—resembled the parental strains. The other two—green, round seeds and yellow, wrinkled seeds—presented new combinations of traits. The four classes showed an approximate ratio of 9 (yellow, round):3 (green, round): 3 (yellow, wrinkled): 1 (green, wrinkled), as shown in Table 6.1. Mendel realized that the new combinations of traits and the ratios of four phenotypic classes were possible not only because alleles for a trait segregate independently (First law of Mendel), but also because the two genes were inherited independently.

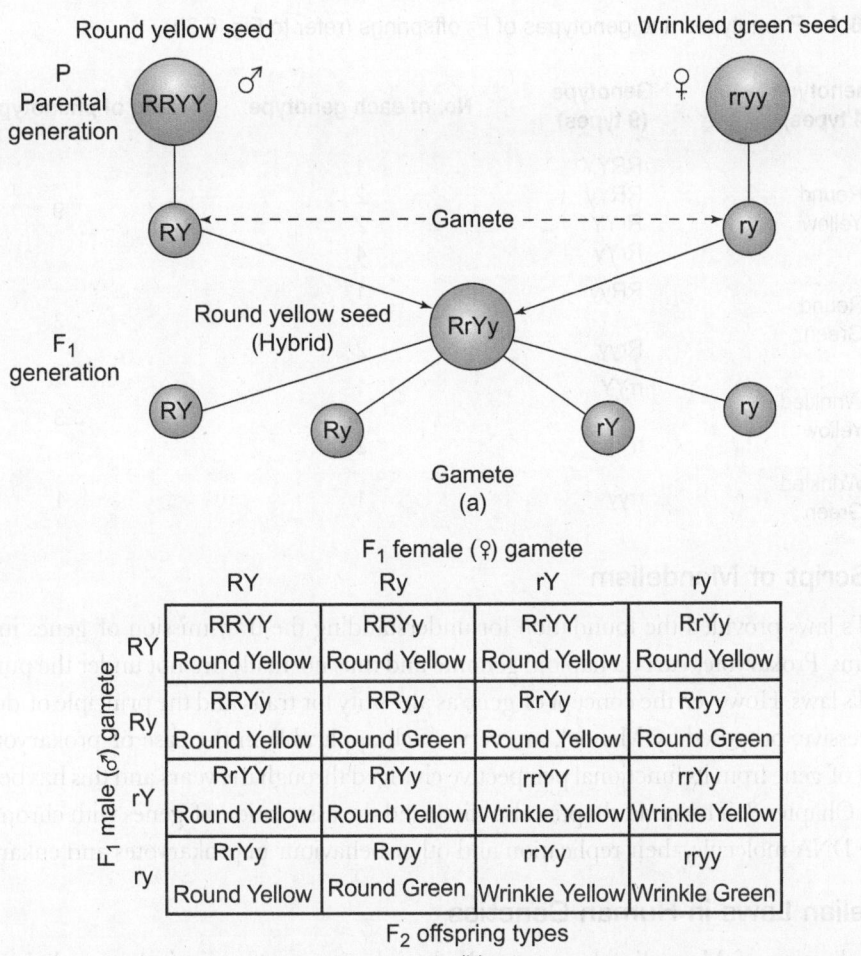

F₁ female (♀) gamete

	RY	Ry	rY	ry
RY	RRYY Round Yellow	RRYy Round Yellow	RrYY Round Yellow	RrYy Round Yellow
Ry	RRYy Round Yellow	RRyy Round Green	RrYy Round Yellow	Rryy Round Green
rY	RrYY Round Yellow	RrYy Round Yellow	rrYY Wrinkle Yellow	rrYy Wrinkle Yellow
ry	RrYy Round Yellow	Rryy Round Green	rrYy Wrinkle Yellow	rryy Wrinkle Green

F₁ male (♂) gamete

F₂ offspring types

(b)

Fig. 6.3 Mendel's dihybrid cross experiment involving pea plants of two types—bearing (i) round and yellow seed, and (ii) wrinkled and green seed. (a) Generation of gametes (b) Genotypes and phenotypes of offspring of F₂ generation in reference to the gametes presented in a checkerboard fashion.

Thus, the *second law of inheritance*, proposed by Mendel is the *law of independent assortment: The alleles of different genes segregate or assort independently of each other.*

Mendel then performed dihybrid crosses with other combinations of traits and found corroboration for his first experiment as described above.

Mendel formulated the two laws of inheritance using pea plants. Later, other investigators found that these laws were applicable to other plants and animals including human beings. However, not all genes abide by the law of independent assortment especially when they are closely linked on a chromosome. Interestingly, the seven pairs of characteristics chosen by Mendel in pea are located on different chromosomes and are not linked. This possibly worked in Mendel's favour and helped him formulate the laws without complication in the results, due to linkage.

Table 6.1 Phenotypes and genotypes of F_2 offsprings (refer to Fig. 6.3b)

Phenotype (4 types)	Genotype (9 types)	No. of each genotype	Ratio of phenotype
Round Yellow	RRYY RRYy RrYY RrYy	1 2 2 4	9
Round Green	RRyy Rryy	1 2	3
Wrinkled Yellow	rrYY rrYy	1 2	3
Wrinkled Green	rryy	1	1

6.1.3 Post Script of Mendelism

Mendel's laws provided the foundation for understanding the transmission of genes in diploid organisms. Prokaryotes carry a haploid genome and thus normally are not under the purview of Mendel's laws. However, the concept of gene as an entity for traits and the principle of dominant and recessive, proposed by Mendel, are very much applicable in the case of prokaryotes. The concept of gene from its functional perspective changed through the years and this has been dealt with in Chapter 3. The same chapter also discussed the association of genes with chromosomes and the DNA molecule, their replication and other behaviour in prokaryotes and eukaryotes.

6.1.4 Mendelian Laws in Human Genetics

The application of Mendelian laws to study the inheritance of traits in human beings began soon after the rediscovery of Mendel's paper in 1900. Initially, the progress of study was slow for several reasons, and some important ones are mentioned below:

(i)   Controlled crosses with human beings are not possible.
(ii)   The analysis depends on family records, which are often incomplete.
(iii) Human beings, unlike experimental organisms, do not produce many offspring; this makes it difficult to calculate Mendelian ratios properly.

In spite of all these difficulties, the drive to understand human genes in reference to their functions and inheritance has been very strong. Today, thousands of human genes are known.

Table 6.2 lists the phenotypic expression of some genes as examples. Completion of the Human Genome Project in 2003 made us aware about detailed genetic, cytogenetic, and physical maps of all human chromosomes and about 3×10^4 genes. The project sequenced more than three billion (3.2×10^9) nucleotide pairs in the human genome (see Exhibit 3.A in Chapter 3). The Human Genome Project has also served as an umbrella for similar mapping and sequencing projects on the genomes of several other organisms, including the bacterium *Escherichia coli*, yeast *Saccharomyces cerevisiae*, fruit fly *Drosophila melanogaster*, plant *Arabidopsis thaliana*, and worm *Caenorhabditis elegans*.

Table 6.2 Some examples of dominant and recessive traits in man and other organisms

Organism	Dominant traits	Recessive traits
Man	Woolly hair Brown eyes Thick lips Colour sensitiveness 'A' and 'B' blood group	Smooth hair Blue or grey eyes Thin lips Colour-blindness 'O' blood group
Guinea pig	Black hair Rough hair	White hair Smooth hair
Drosophila	Striped body	Black body
Tomato	Red fruit Round fruit Brown stem	Yellow fruit Oval fruit Green stem
Maize	Coloured seed coat Smooth seed	Colourless seed coat Wrinkled seed

Some Other Dominant and Recessive Traits in Humans

Dominant traits: Achondroplasia (dwarfism), brachydactyly (short finger), congenital night blindness, Huntington's disease (a neurological disorder)

Recessive traits: Albinism (lack of pigment), alkaptonuria (a disorder of amino acid metabolism), ataxia telangiectasia (a neurological disorder), cystic fibrosis (a respiratory disorder), Duchenne muscular dystrophy, phenylketonuria (a disorder of amino acid metabolism), sickle-cell anaemia (a haemoglobin disorder), Tay-Sachs disease (a lipid storage disorder)

The detailed nucleotide mapping of human genes through the human genome project made scientists hopeful that a defective gene could be repaired by genetic engineering early in ontogeny (gene therapy), so that a child inheriting a bad gene can have a normal life.

The structure of the DNA molecule constituting the genes, and DNA replication have been discussed in Chapters 3 and 4.

DNA being the physical basis of genes, isolation and purification of DNA from organisms and viruses are the first steps in the manipulation of genes for genetic engineering.

6.2 ISOLATION AND PURIFICATION OF CELLULAR DNA

There are some (minor) differences in the DNA isolation procedure depending on whether DNA is being isolated from bacterial cells or eukaryotic plant and animal cells.

6.2.1 Isolation and Purification of Total DNA from Bacterial Cells

The procedure involves the following four steps (Fig. 6.4).

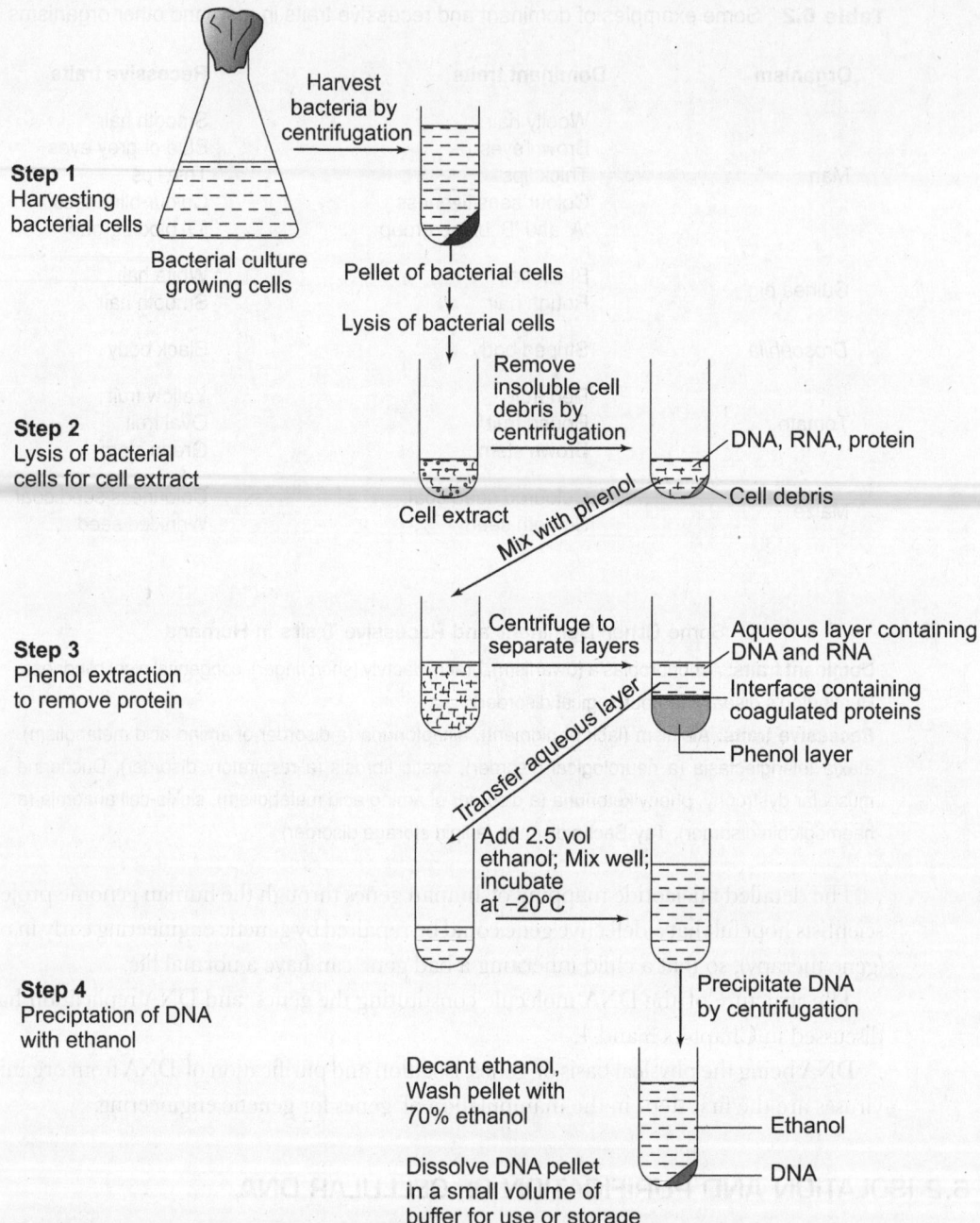

Step 1
Harvesting
bacterial cells

Bacterial culture
growing cells

Harvest
bacteria by
centrifugation

Pellet of bacterial cells

Lysis of bacterial cells

Step 2
Lysis of bacterial
cells for cell extract

Cell extract

Remove
insoluble cell
debris by
centrifugation

DNA, RNA, protein

Cell debris

Mix with phenol

Step 3
Phenol extraction
to remove protein

Centrifuge to
separate layers

Aqueous layer containing
DNA and RNA

Interface containing
coagulated proteins

Phenol layer

Transfer aqueous layer

Add 2.5 vol
ethanol; Mix well;
incubate
at –20ºC

Step 4
Precipitation of DNA
with ethanol

Precipitate DNA
by centrifugation

Decant ethanol,
Wash pellet with
70% ethanol

Ethanol

Dissolve DNA pellet
in a small volume of
buffer for use or storage

DNA

Fig. 6.4 The steps for isolation and purification of total DNA from bacterial cells.

Growing and harvesting a bacterial culture

Bacteria are grown in defined liquid medium (broth culture) until the appropriate cell density
is reached. The medium contains a mixture of inorganic nutrients to provide essential elements

such as nitrogen, magnesium, and calcium, as well as glucose to supply carbon and energy. When bacteria grow to an optimum level, they are harvested by centrifugation at 3,000–8,000g for approximately 5 minutes.

Lysis of bacterial cells for preparation of cell extract

Physical and chemical methods alone or in combination are employed to break open the cell wall and membrane of bacteria.

Physical methods　These involve mechanical forces, for example, vortexing, and sonication.

Chemical methods　These involve the use of lysozyme and EDTA (ethylenediaminetetraacetic acid, disodium salt) as disruptive chemical agents. Lysozyme causes hydrolytic cleavage in the components of peptidoglycan which normally provides rigidity to the bacterial cell wall. The breakdown of peptidoglycan leads to osmotic lysis of bacterial cells. On the other hand, Na_2EDTA (disodium EDTA) being a chelating agent removes divalent metal cations Mg^{2+} and Ca^{2+} and hence disrupts the overall structure of the cell envelope. Removal of Mg^{2+} leads to inactivation of cellular DNases and thus prevents degradation of DNA in the cell extract. A detergent such as sodium dodecyl sulphate (SDS) is also used for chemical disruption of cells. The detergent dissolves the lipids in the cell membrane leading to lysis of the cells.

Removal of proteins and RNA from the cell extract

A bacterial cell extract contains large amounts of proteins and RNA in addition to DNA. These contaminating biomolecules are systematically removed to obtain pure DNA for cloning and other purposes.

Deproteination that is removal of proteins can be achieved in two ways—phenol:chloroform extraction and treatment with proteases.

Phenol : chloroform extraction　Organic solvents, such as phenol, chloroform, and isoamyl alcohol, denature and precipitate proteins efficiently, leaving the nucleic acids (DNA and RNA) in the aqueous cell extract. To the cell extract, an equal volume of phenol:chloroform (1:1) is added. Phenol should be of high quality and without impurities and at pH of 7.8–8.0. Normally, phenol is very unstable and oxidizes rapidly into quinones which form free radicals that break the phosphodiester linkage of nucleic acids. Thus, phenol needs to be distilled before use. Furthermore, nucleic acids tend to partition into the organic phase if phenol has not been adequately equilibrated to a pH around 8 with 1M Tris–HCl. Chloroform facilitates partitioning of the aqueous and organic materials.

Following proper mixing of the cell extract and phenol:chloroform to form an emulsion, the mixture is centrifuged in a centrifuge tube at high speed (12,000g for 10 min). At the end of centrifugation, the aqueous phase occupies the top while the organic phase settles at the bottom of the tube. The protein molecules in the form of a white coagulated mass, precipitate at the interface between the organic and aqueous layers. Nucleic acids (DNA and RNA) have a highly charged phosphate backbone, which is polar and hydrophilic. Therefore, the nucleic acids are water soluble and remain partitioned in the aqueous layer along with salts.

The aqueous phase containing nucleic acids and salts is removed with the help of a pipette and transferred to a fresh tube; the organic phase and the interface containing denatured proteins are discarded. For further purification from contaminating proteins, phenol:chloroform extractions may be repeated several times, until no protein precipitation is visible at the interface. After the first extraction step, isoamyl alcohol may be included in the organic solvents in the ratio of phenol:chloroform:isoamyl alcohol–25:24:1. Isoamyl alcohol helps in deproteination and reduces foaming in the course of the extraction procedure.

At the end of organic extraction(s), an equal volume of chloroform is added to the aqueous solution (phase). Phenol is extensively soluble is chloroform. Thus, this step of extraction removes the traces of phenol so that quinones from phenol do not disturb the integrity of the nucleic acid molecules.

One should remember that phenol, chloroform, and their combinations are caustic and carcinogenic, and phenol is highly corrosive; in no case should they come in contact with the skin, eyes, or any other part of the body. In case of accidental contact, the affected area needs to be washed immediately with abundant amount of water. A lab coat, gloves, and protective eye glass should be worn while working with these chemicals.

Treatment with proteolytic enzymes   A certain amount of breakage of DNA occurs in the course of several rounds of phenol:chloroform extractions, resulting in a decrease in the DNA yield. Thus, prior to phenol extraction the cell extract may be treated with proteases, such as *pronase* and *proteinase K*. These enzymes break down a large spectrum of proteins. Occasionally, the protein digestion is carried out in the presence of Na_2EDTA to inhibit labile Mg^{2+}-dependent nucleases. Proteinase K is obtained from the mold *Tritiachium album* and can be used at concentrations up to 50 μg/mL. The recommended working concentration for pronase is about 1 mg/mL.

Removal of RNA   Phenol extraction discussed in the earlier step removes some RNA molecules, especially mRNAs. The aqueous layer from phenol extraction is treated with RNAase (RNase) for degradation of RNA into ribonucleotide subunits, which are then easily removed by phenol extraction. RNase A and RNase T1 in Tris-HCl buffer are normally used. RNase preparations should be made free of all intrinsic DNase. This is routinely done by heating RNase stock solutions to near boiling (90°C) for 10 minutes, followed by cooling on ice. At 90°C, DNase activity can be quickly eliminated without compromising on the RNase activity. The DNase-free RNase thus obtained is stored frozen in aliquots. DNase-free RNase is also commercially available.

Precipitation of DNA with ethyl alcohol (EtOH)

EtOH being less polar than water, and with a much lower dielectric constant, disrupts the hydration (water) shell on negatively charged phosphate groups of nucleic acids and allows positively charged ions to interact with phosphates easily. Positively charged ions (cations) such as sodium acetate, sodium chloride, and ammonium acetate neutralize the negative charge on the nucleic acid molecule, making it far less hydrophilic and less soluble in water. Thus, the DNA precipitates out of the solution.

With a concentrated solution of nucleic acid, two to three volumes of absolute EtOH is layered on top of the sample, allowing DNA molecules to precipitate at the interface. Usually, a thin glass rod is inserted through the EtOH into the nucleic acid solution. Nucleic acid molecules adhere to the glass rod and are pulled out (*spooled out*) of the solution in the form of long fibres. When nucleic acid concentration is low, the solution is centrifuged to obtain DNA as a pellet. The supernatant is discarded and the pellet (nucleic acid) is washed with 70% EtOH.

Washing with 70% EtOH removes much of the salt present in the supernatant or bound to the DNA pellet and also loosens the precipitates from the wall of the tube. The pellet is then air dried by leaving the tube open on the laboratory bench for a few minutes; it should not be allowed to dry out completely as that leads to denaturation of DNA and makes it difficult to resuspend. The pellet of DNA is redissolved in Tris-HCl, EDTA buffer at pH 8.0 and stored at 4°C. For long-term storage, DNA as a suspension in EtOH is maintained at −20°C.

Other methods of precipitation of DNA   Instead of ethanol (EtOH), *isopropanol* can be used for DNA precipitation. Isopropanol precipitation is more effective than EtOH in separating primers from PCR products.

Extraction with a secondary *butyl alcohol*–isobutanol or *n*-butyl alcohol (*n*-butanol) results in a reduction in the volume of the DNA preparation; DNA can then be recovered easily by precipitation with EtOH.

Ultracentrifugation   The nucleic acid solution is forced under centrifugal force through a semipermeable membranous disk. The aqueous medium and small solute molecules pass through the semipermeable membrane, leaving the large molecules on the membrane. Thus, efficient desalting and concentration of nucleic acid samples can be achieved.

Dialysis   Another alternative to EtOH precipitation is dialysis. Dialysis tubing or a dialysis bag is made up of semipermeable cellophane (cellulose acetate) membrane with a specific pore size. The DNA solution is poured into the bag, which is kept suspended in a flask containing a buffer solution. Small molecules, such as salt, small biochemicals, and water, diffuse across the membrane (through pores) driven by the concentration differential between the solutions on either side of the dialysis membrane, but large molecules like DNA are retained within the bag. Dialysis desalts and concentrates DNA preparations.

6.2.2 Isolation and Purification of Genomic (Nuclear) DNA from Eukaryotes

Genomic DNA from plant and animal cells is required to clone genes for genetic engineering experiments. The basic steps in DNA isolation and purification are the same as those outlined for prokaryotic DNA. There are some modifications depending on the nature of eukaryotic cells.

The major modification is in the procedure of cell lysis. Lysozymes, which are used for disrupting bacterial cells, do not lyse plant and animal cells.

Cell lysis
Plant cells   10–15 g of fresh plant tissue is collected from young plants; the plant is kept in a dark place one to two days prior to harvest in order to reduce the starch content. The tissue is

rinsed with cold, sterile water, blotted dry, and frozen with liquid nitrogen. Then the frozen tissue is ground to a fine powder with a mortar and pestle or a homogenizer (Fig. 6.5).

For chemical disruption, plant cell wall-degradative enzymes specific for cell types are used.

Animal cells Animal tissue is disrupted in a homogenizer (see Fig. 6.5) or the cultured cells are centrifuged into a pellet. Animal cells do not have a cell wall and can be lysed simply by treating with a detergent.

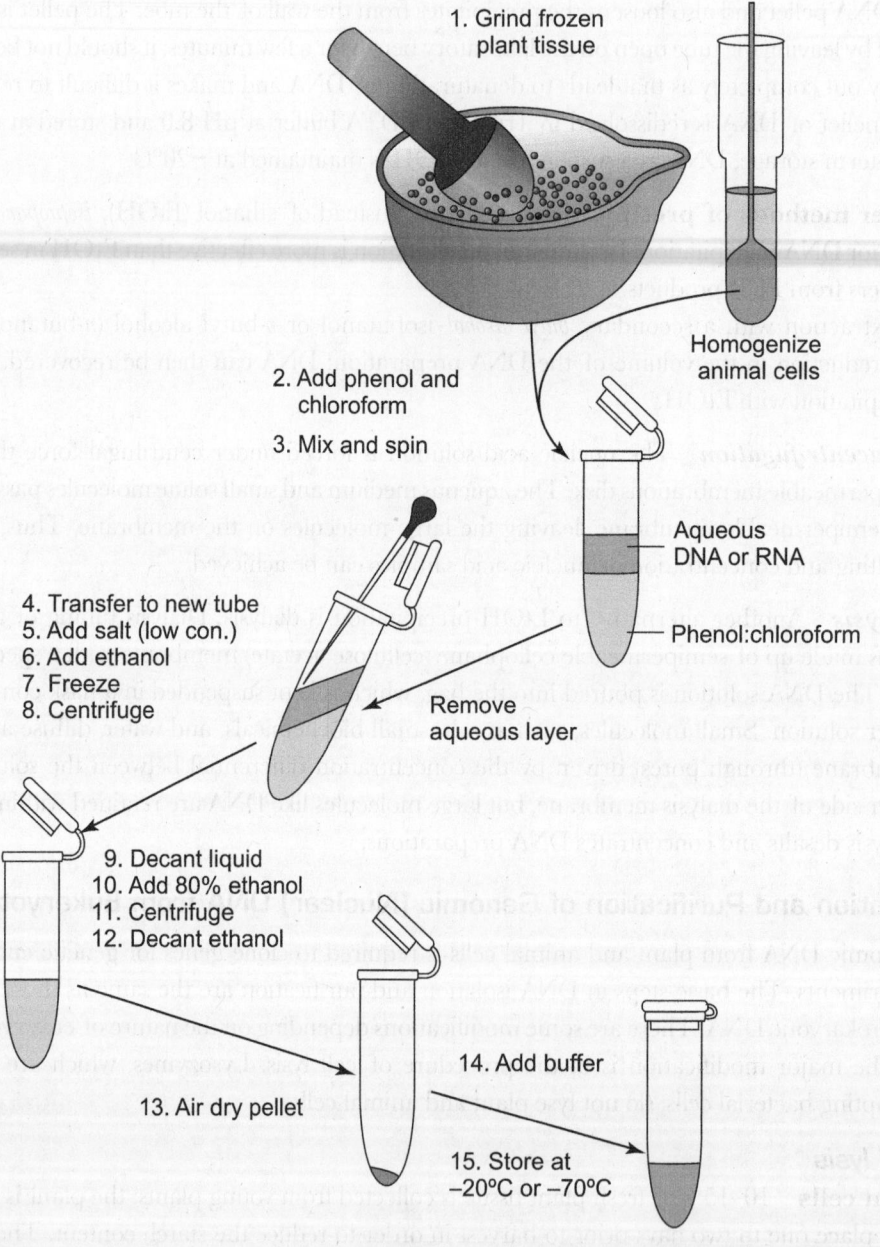

1. Grind frozen plant tissue

Homogenize animal cells

2. Add phenol and chloroform

3. Mix and spin

Aqueous DNA or RNA

Phenol:chloroform

4. Transfer to new tube
5. Add salt (low con.)
6. Add ethanol
7. Freeze
8. Centrifuge

Remove aqueous layer

9. Decant liquid
10. Add 80% ethanol
11. Centrifuge
12. Decant ethanol

14. Add buffer

13. Air dry pellet

15. Store at −20°C or −70°C

(a)

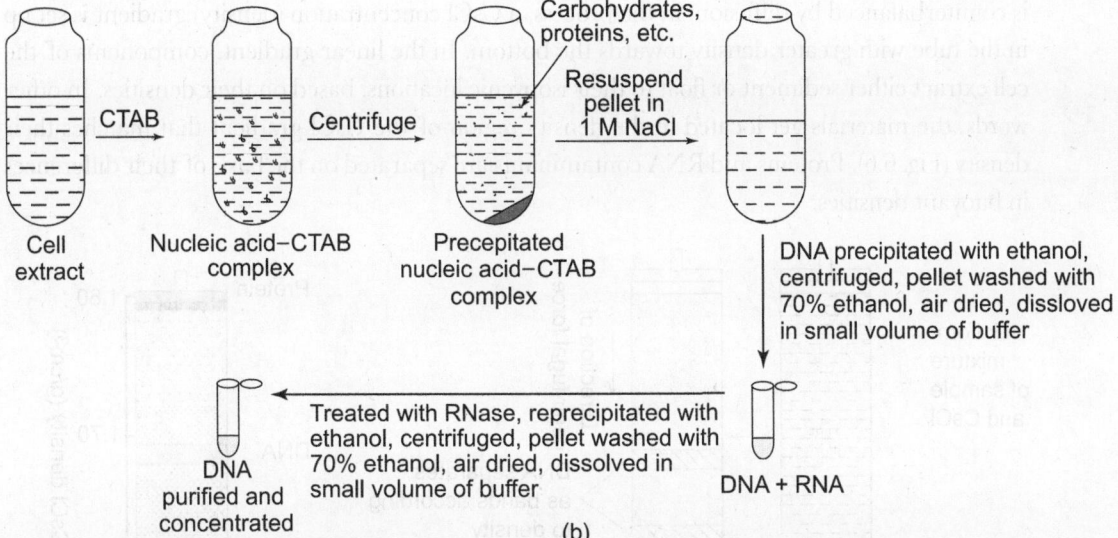

Fig. 6.5 (a) Extraction and precipitation of nucleic acid from plant source and animal cell source (b) CTAB method for purification of DNA from plant source.

Removal of contaminating biomolecules to recover DNA

Plant cells contain significant quantities of polysaccharides, polyphenolics, and secondary metabolites. Polysaccharides are not removed by phenol extraction. A detergent *cetyl trimethyl ammonium bromide* (CTAB) forms an insoluble complex (precipitate) with nucleic acids; following centrifugation, carbohydrates, proteins, and other contaminants are left out in the supernatant to be discarded (Fig. 6.5). The precipitate is then resuspended in 1 M NaCl which also allows breakdown of the nucleic acid–CTAB complex. Subsequent precipitation with EtOH concentrates the nucleic acids, and RNase treatment removes RNA.

The EtOH precipitation technique is rapid, virtually foolproof, efficient, and it can precipitate even minute quantities of DNA. Even faster and simpler recovery of DNA is possible nowadays by following certain variants of this technique, such as silica-based spin column chromatography, magnetic bead capture technique, and anion exchange chromatography. The alcohol precipitation technique can be miniaturized in Eppendorf micro centrifuge tubes to obtain genomic DNA for PCR analysis.

Separation of DNA in a caesium chloride (CsCl) density gradient—a classical and traditional method

Historically, Matthew Meselson and Franklin Stahl used a CsCl density gradient in 1958 to establish the semiconservative mechanism of DNA replication.

Solid CsCl salt is added to an extract containing the nucleic acid mixture which is then subjected to ultracentrifugation (above 30,000 rpm for several hours). The gradient develops on the basis of differences in the isopycnic point or buoyant density; the high centrifugal force pulls the cesium and chloride ions towards the bottom of the tube. Downward migration of Cs and Cl is counterbalanced by diffusion; in the process, a CsCl concentration (density) gradient is set up in the tube with greater density towards the bottom. In the linear gradient, components of the cell extract either sediment or float to their isopycnic locations, based on their densities. In other words, the materials get located in the density region of the CsCl gradient that matches their density (Fig. 6.6). Proteins and RNA contaminants are separated on the basis of their differences in buoyant densities.

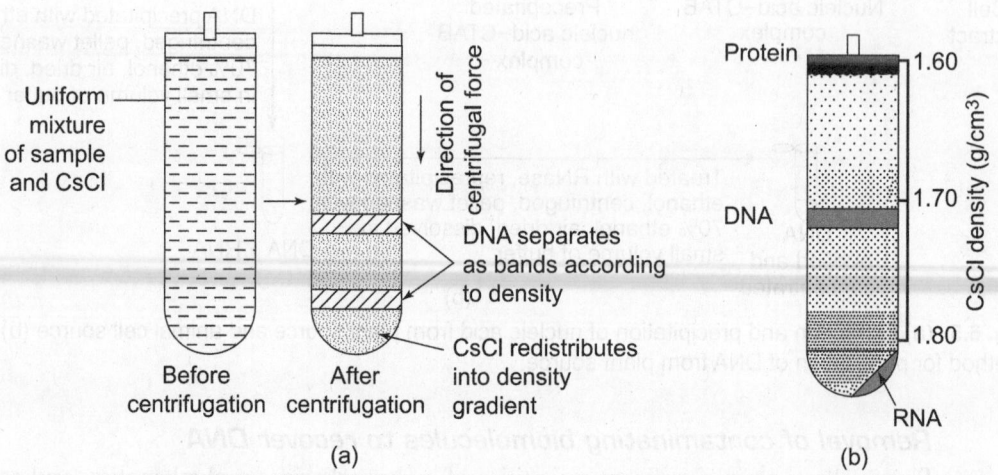

Fig. 6.6 (a) Separation of DNA in a CsCl density gradient. The upper band of DNA includes linear and open circular DNA (slow sedimenting); the lower fast sedimenting band includes supercoiled DNA. (b) Separation of DNA from proteins and RNA through CsCl gradient centrifugation.

The CsCl density gradient separation can also be carried out with a microultracentrifuge and the required centrifugation time can be significantly reduced.

DNA isolation kits   These commercial kits are based on the methods described above and make the process of DNA isolation and purification rapid and simple.

Isolation and purification of plasmid DNA

Plasmid DNA is often used as a vector to transfer a gene into a host cell. The relatively small size and covalently closed circular nature of plasmid DNAs help in the separation of the plasmid DNA from the genomic DNA (from the chromosome) of bacteria.

For plasmid DNA, selective propagation of bacterial cells in culture is followed by lysis of the bacteria as outlined earlier. Equilibrium centrifugation in CsCl–ethidium bromide gradients has been the method of choice for many years to prepare large amounts of plasmid DNA. Many alternative methods have been developed in recent times to surpass the expensive and time-consuming CsCl technique. The majority of modern methods use ion-exchange or gel filtration chromatography or differential precipitation to separate plasmid and chromosomal DNAs. Differential precipitation with polyethylene glycol (PEG) has been improved to yield plasmid DNA of high purity in large quantities. However, the three most popular procedures for the small-scale isolation of plasmid DNA from bacteria are the alkaline lysis method, the boiling method, and precipitation with CTAB. The general scheme for plasmid DNA preparations is depicted in Fig. 6.7. After removal of bacterial cell walls with lysozyme treatment, in the boiling procedure, chromosomal DNA forms the debris, leaving the soluble plasmid DNA in the supernatant.

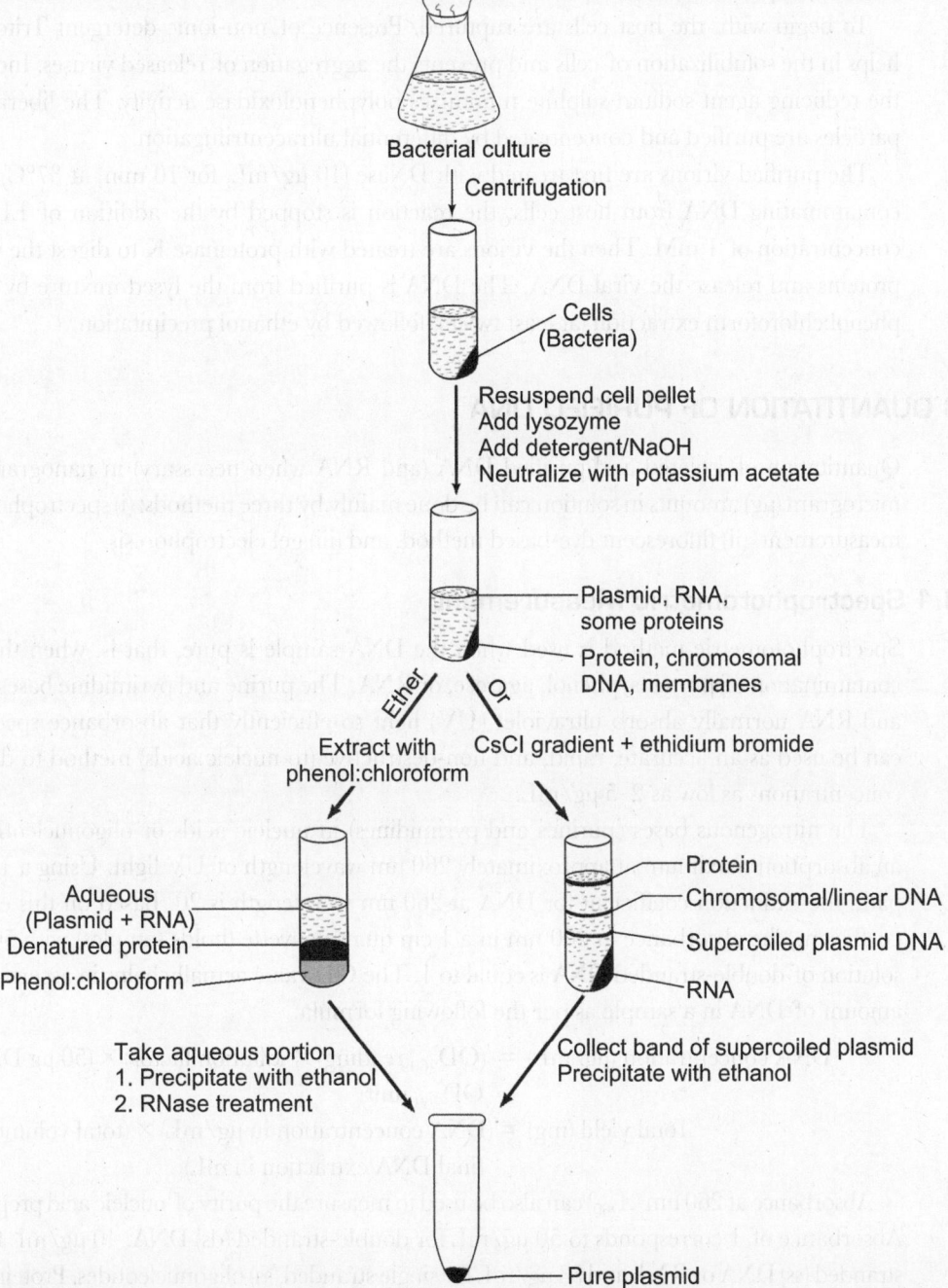

Bacterial culture

Centrifugation

Cells
(Bacteria)

Resuspend cell pellet
Add lysozyme
Add detergent/NaOH
Neutralize with potassium acetate

Plasmid, RNA,
some proteins

Protein, chromosomal
DNA, membranes

Ether   Or

Extract with
phenol:chloroform

CsCl gradient + ethidium bromide

Aqueous
(Plasmid + RNA)
Denatured protein
Phenol:chloroform

Protein
Chromosomal/linear DNA
Supercoiled plasmid DNA
RNA

Take aqueous portion
1. Precipitate with ethanol
2. RNase treatment

Collect band of supercoiled plasmid
Precipitate with ethanol

Pure plasmid

Fig. 6.7 Outline scheme for preparation of plasmid DNA.

6.2.3 Outline of Viral Nucleic Acid Precipitation

The genome of viruses varies between virus families. Unlike the genome of true organisms, the virus genome may consist of DNA or RNA, which may be double- or single-stranded, circular or linear, depending on the type of virus.

To begin with, the host cells are ruptured. Presence of non-ionic detergent Triton X-100 helps in the solubilization of cells and prevents the aggregation of released viruses. Inclusion of the reducing agent sodium sulphite minimizes polyphenoloxidase activity. The liberated virus particles are purified and concentrated by differential ultracentrifugation.

The purified virions are first treated with DNase (10 µg/mL, for 10 min, at 37°C) to digest contaminating DNA from host cells; the reaction is stopped by the addition of EDTA to a concentration of 1 mM. Then the virions are treated with proteinase K to digest the viral coat proteins and release the viral DNA. The DNA is purified from the lysed mixture by standard phenol:chloroform extraction (at least twice), followed by ethanol precipitation.

6.3 QUANTITATION OF PURIFIED DNA

Quantitation of isolated and purified DNA (and RNA when necessary) in nanogram (ng) or microgram (µg) amounts in solution can be done mainly by three methods: (i) spectrophotometric measurement, (ii) fluorescent dye-based method, and (iii) gel electrophoresis.

6.3.1 Spectrophotometric Measurement

Spectrophotometric method is used when the DNA sample is pure, that is, when there is no contamination of proteins, phenol, agarose, or RNA. The purine and pyrimidine bases of DNA and RNA normally absorb ultraviolet (UV) light so efficiently that absorbance spectroscopy can be used as an accurate, rapid, and non-destructive (to nucleic acids) method to determine concentrations as low as 2–5 µg/mL.

The nitrogenous bases (purines and pyrimidines) in nucleic acids or oligonucleotides have an absorption maximum at approximately 260 nm wavelength of UV light. Using a 1 cm light path, the extinction coefficient for DNA at 260 nm wavelength is 20. Based on this extinction coefficient, the absorbance at 260 nm in a 1 cm quartz cuvette (holds samples) with 50 µg/mL solution of double-stranded DNA is equal to 1. The OD value actually helps in ascertaining the amount of DNA in a sample as per the following formula:

$$\text{DNA concentration (µg/mL)} = (\text{OD}_{260} \text{ reading}) \times (\text{dilution factor}) \times (50 \text{ µg DNA/mL OD}_{260} \text{ unit})$$

$$\text{Total yield (mg)} = (\text{DNA concentration in µg/mL}) \times (\text{total volume of final DNA extraction in mL})$$

Absorbance at 260 nm (A_{260}) can also be used to measure the purity of nucleic acid preparations. Absorbance of 1 corresponds to 50 µg/mL for double-stranded (ds) DNA, 40 µg/mL for single stranded (ss) DNA or RNA and 33 µg/mL for single stranded (ss) oligonucleotides. Proteins absorb maximally at approximately 280 nm, mainly due to tryptophan residues. Therefore, the ratio of readings at 260 and 280 nm (A_{260}/A_{280}) provides an estimate of the purity of the DNA preparation and it should fall at 1.8. A lower value suggests protein contamination of the DNA preparation.

6.3.2 Fluorescent Dye-based Method

Fluorescence by EtBr

UV absorption spectroscopic measurement cannot be carried out if there is not sufficient DNA (<250 ng/mL) or the DNA sample is heavily contaminated with other substances that also absorb UV irradiation. To overcome these problems, quantitation of ds DNA is done by spectrophotometric analysis of DNA intercalated with ethidium bromide (EtBr). EtBr binds (intercalated into) DNA and emits fluorescence when excited with UV light. The amount of DNA in a sample is determined by estimating the level of fluorescence emitted by the EtBr molecules intercalated into the DNA at 590 nm. The fluorescence is proportional to the total mass of the DNA. The quantity of DNA in a sample can be estimated by comparing the fluorescence of the sample with that of a series of standards of known value of DNA; as little as 1–5 ng of DNA can be detected by this method.

The fluorescence of EtBr is increased about 50 fold when it is intercalated into the DNA. In contrast to the UV absorbance technique described earlier, this test destroys the DNA, that is, after the test the DNA cannot be used. To work with EtBr one needs to avoid contact with the compound as it is considered carcinogenic (cancer inducing). Gloves should be worn while working with EtBr.

Fluorometric quantitation of DNA using Hoechst 33258

Hoechst 33258 is a type of bis-benzimidazole fluorescent dye that binds (does not intercalate) with high specificity to the minor groove of ds DNA. Like other non-intercalative dyes, Hoechst 33258 binds preferentially to A–T rich regions of the DNA helix. When bound to DNA, Hoechst 33258 absorbs maximally at 356 nm and emits maximally at 458 nm. The fluorescence of the dye is approximately three fold lower with ssDNA. Thus, it helps in better quantitation of dsDNA in solution as compared to EtBr.

The assay can only be used to measure the concentration of DNA fragments that are longer than one kilo base pairs (kbp), as Hoechst 33258 binds poorly to smaller DNA fragments. The assay should be carried out rapidly to minimize photobleaching of the dye. pH extremes adversely influence the binding of the dye.

6.3.3 Gel Electrophoresis

Conventional agarose gel electrophoresis is used to determine the size of DNA fragments in a sample. It can be used to analyse fragments ranging in size from 50 bp to ~20 kbp. This method can be used to select DNA fragments for a gene cloning experiment. Pulse field gel electrophoresis (PFGE) is used for DNA molecules greater than 20 kbp in length. Polyacrylamide gels are effective for separating and visualizing fragments of DNA between 5 and 500 bp.

A horizontal agarose (2%) gel slab of about 3 mm thickness is cast in a gel tray and sample wells are made at one end using an acrylic comb, and it is allowed to polymerize for some time.

After polymerization, the comb is removed and the gel slab is placed in a horizontal electrophoresis apparatus filled with Tris buffer. The DNA samples are mixed with the tracking dye bromophenol blue and carefully loaded with a micropipette in separate sample wells of the gel slab. The electrophoresis apparatus is run for 3 hours at 100 V (Fig. 6.8). The DNA fragments from the wells separate into different bands along the lanes beneath the wells according to their sizes. The slab is then taken out of the apparatus and stained with 1 μg/mL ethidium bromide. Pink DNA bands can be visualized on illumination with UV light and can be photographed through a red filter (Fig. 6.8). Necessary precautions should be taken not to expose the eyes directly to the UV light.

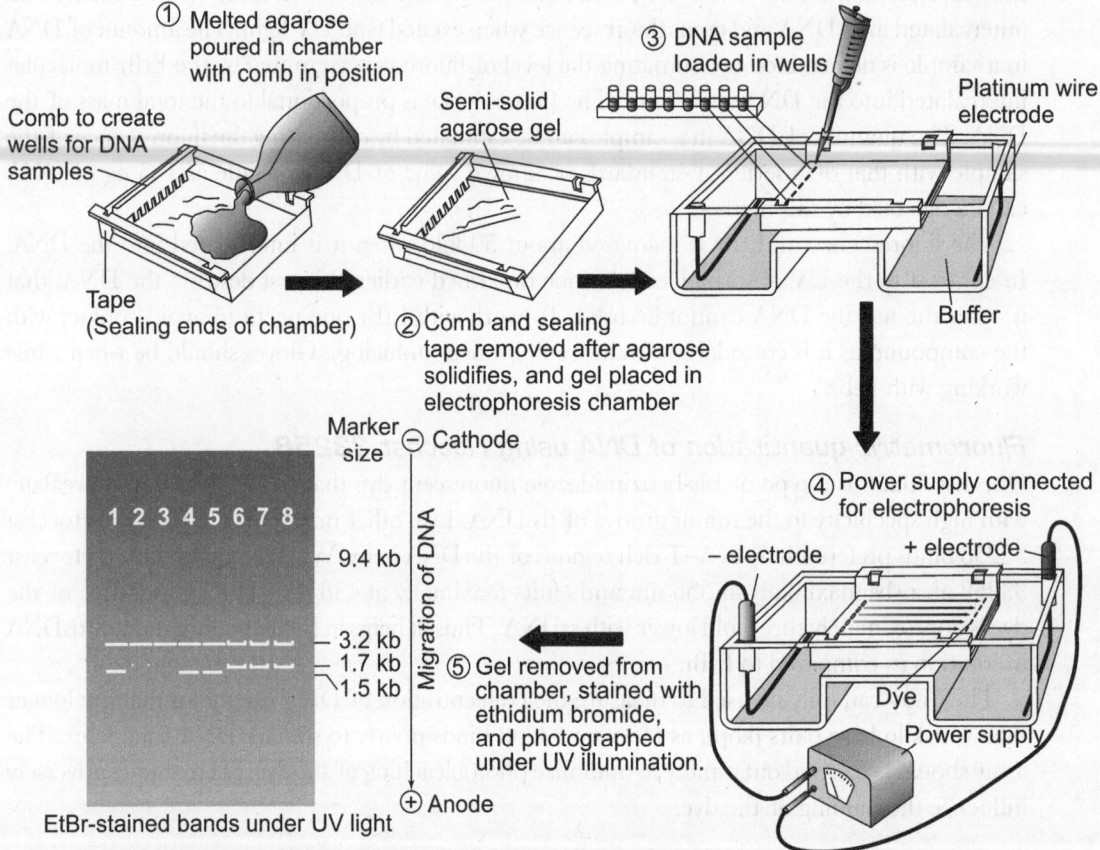

Fig. 6.8 The separation of DNA fragments of different sizes in bands by agarose gel electrophoresis. The DNA samples are dissolved in loading buffer with density higher than that of the electrophoresis buffer so that DNA samples do not diffuse out of the wells into the electrophoresis buffer. Tracking dye is added to the loading buffer to monitor the rate of migration of samples through the gel. Ethidium bromide (EtBr) binds to DNA and the bands fluoresce when illuminated with UV light.

Besides viewing the bands of different sizes of DNA on an agarose gel, digitizing image analysing software allows quantitation of DNA concentration in bands. Size and concentration of

DNA in different bands on an agarose gel are measured in reference to a known mass standard, for example, ΦX174 or λ phage DNA (see Fig. 6.8). This method is also used to determine the quantity of synthesized cDNA.

6.4 ISOLATION AND PURIFICATION OF RNA

RNA isolation is required for purification of mRNA which is used for the preparation of complementary DNA (cDNA). cDNA is then used to make a gene library.

In general, the extraction methods produce cytoplasmic RNA, nuclear RNA, or a mixture of both, commonly known as cellular RNA. RNA isolation involves the same basic steps as DNA isolation—cell lysis, removal of contaminating biomolecules, and precipitation of RNA.

The first step in RNA isolation is cell disruption by cell lysis buffer. The cell lysis buffer includes a chaotropic and denaturing agent, deproteination agent, detergent, chelating agent, and an RNase inhibitor.

Complete inhibition of RNase activity is absolutely essential for mRNA extraction, otherwise the family of ribonucleases (RNase) degrades RNA easily and quickly during extraction, purification, and storage. Moreover, RNA exhibits a short half-life.

RNase is normally sequestered within the cell and liberated on cellular lysis. To neutralize RNase, the cell lysis buffer is designed to have RNase inhibitors [for example, heparin, iodoacetate, dextran sulphate, polyvinyl sulphate, macaloid, vanadyl ribonucleoside (VDR), and cataionic surfactant]. Extrinsic sources of RNase can be almost anything ranging from laboratory wares, chemicals, water, gel box, bacteria, molds on airborne dust particles to the person performing the isolation.

There are many prophylactic measures that need to be taken to restrict the entry of RNase from extrinsic sources. In general, proper microbiological sterile techniques need to be observed for washing laboratory ware and equipment, and for sterilization, autoclaving, handling, and preparation of reagents. Glassware and plasticware can be made RNase free by washing with DEPC (diethyl pyrocarbonate) water. DEPC water is prepared by treating water with 0.1% DEPC for at least 1 hour at 37°C and autoclaving it for 15 minutes at 15 psi (pressure). RNase-free solutions of different reagents are also commercially available. The fingertips of humans are notoriously rich in RNases. Hence, wearing rubber gloves throughout the RNA isolation and purification process is a must. Changing of the gloves several times is recommended so as not to take any chance of extrinsic contamination of RNase.

Finally, RNA is efficiently precipitated with 2.5–3 volumes of EtOH from solutions containing various salt ions, such as Na^+, K^+, Li^+, or NH_4^+. The salt ions in combination with nucleic acid facilitate precipitation.

6.5 RECOMBINANT DNA (rDNA) TECHNOLOGY

Pure, isolated DNA becomes the source of genes to be inserted in the DNA molecule of the vector for transfer to host cells. Thus, recombinant DNA (rDNA) is constructed in a test tube by covalently joining (recombining) DNA molecules from different origins; for example, one from an eukaryote and another from a bacterium or virus. The DNA molecule from bacterial or viral source is called a *vector* as it can enter an appropriate cell to transfer the desired gene. The objective of the insertion of recombinant DNA in a bacterium is to amplify the transferred gene through multiplication of the *transformed bacterium*. After the bacteria have multiplied, the desired product of the transferred gene can be recovered in large amounts. A *clone* refers to the progeny derived from a single cell. In the course of multiplication of the transformed bacterium, the inserted genes also multiply to form a gene clone. The process is known as cell-based (in-vivo) *molecular cloning*.

The gene to be inserted into a vector needs to be cut out from a full length DNA molecule and a corresponding nick has to be made in the vector DNA molecule for insertion of the gene. These steps are fundamental for rDNA technology. *Restriction endonucleases* (enzymes) act as molecular scissors to provide cuts in intact DNA molecules. The enzymes cut at sites that contain a specific nucleotide sequence.

The crucial aspects of recombinant DNA technology, as indicated below, are discussed in the following sections.

- Restriction endonucleases
- Vectors
- Gene library
- Selection and screening of clones
- cDNA insert
- Introduction of rDNA vectors into bacteria
- PCR to amplify specific DNA
- Site-directed mutagenesis

6.6 RESTRICTION ENDONUCLEASES

Restriction endonucleases, commonly known as *restriction enzymes*, are prokaryotic (bacterial) nucleases that recognize specific nucleotide sequences in the DNA, and cleave the phosphodiester bond in the sugar–phosphate backbone between nucleotides, either within or close to a recognition site. The restriction enzymes make two incisions at a site, one each through of the two strands of the intact dsDNA, without any damage to the nitrogenous bases. The incisions create a gap in the intact DNA molecule (Fig. 6.9). Two such successive gaps on an intact DNA molecule created by an endonuclease allow separation of a DNA piece containing gene(s) to be used as a graft in rDNA techniques. A single gap in the vector DNA molecule provides the site for insertion of the graft.

The presence of a methyl group at the endonuclease-binding site on DNA does not allow the enzyme to function and hence, no incision is made.

6.6.1 Discovery

The discovery of restriction endonucleases was a breakthrough in molecular biology. In the 1950s, Werner Arber and his colleagues reported a restriction in the growth of some phages (viruses) in certain strains of bacteria. In 1970, Daniel Nathans and Hamilton Smith of John Hopkin's

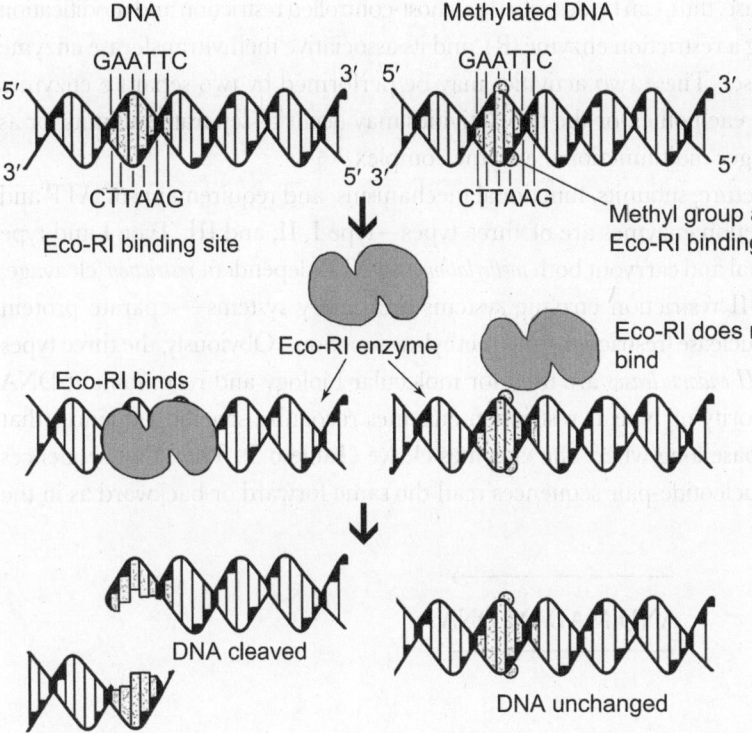

DNA   Methylated DNA

GAATTC   GAATTC

Eco-RI binding site

Eco-RI enzyme

Eco-RI binds   Eco-RI does not
                       bind

DNA cleaved

DNA unchanged

Methyl group at
Eco-RI binding site

Fig. 6.9 Restriction endonuclease Eco RI binds to the specific site (a palindrome) and cleaves the DNA with overhanging sticky ends on the left side. The enzyme does not bind to and will not cut methylated DNA shown on the right side.

School of Medicine discovered restriction endonucleases as the cause behind the restriction in the growth of some phages. Accordingly, the enzymes were named as restriction enzymes. The enzymes cleave the DNA molecule entering from phages into a bacterial host cell. Consequently, the phage DNA is destroyed and cannot multiply in bacteria. In other words, restriction enzymes have apparently evolved as a primitive immune measure in bacteria—to protect the host cell from invasion by foreign DNA, such as virulent and temperate phages, and conjugative plasmids from other species of bacteria. Nearly 1000 restriction enzymes and 130 different nucleotide recognition sequences for the enzymes have been identified from several hundred bacterial strains.

The pioneering discovery of Arber, Nathans, and Smith was honoured with the Nobel Prize in Physiology and Medicine in 1986.

6.6.2 Characteristics

A restriction endonuclease has dual functions—(i) to cleave and degrade foreign DNA (restriction function) and (ii) to methylate, that is to add a methyl group (CH_3) to the host cell DNA. Presence of a CH_3 group does not allow cleavage by the enzyme, that is, methylated DNA is rendered insensitive to degradation. Thus, the host DNA is modified by methylation and remains protected from autorestriction, that is, cleavage by its own enzyme(s). This phenomenon is defined as 'host-controlled modification'. Even a single modified (methylated) strand in the DNA duplex is sufficient to prevent cleavage.

A restriction endonuclease, thus, can be described as a host-controlled restriction and modification (R–M) system, comprising a restriction enzyme (R), and its associative methyltransferase enzyme (M or methylase or MTase). These two activities may be performed by two separate enzymes that act independently of each other, or the two activities may occur as separate subunits, or as separate domains of a larger multifunctional enzyme complex.

Depending on the structure, subunits, functional mechanisms, and requirements of ATP and Mg^{2+}, the bacterial restriction enzymes are of three types—type I, II, and III. Type I and type III enzymes are bifunctional and carryout both *methylation* and ATP-dependent *restriction* (cleavage) activities. Whereas, type II restriction enzyme systems are binary systems—separate protein molecules perform endonuclease (restriction) and methylase functions. Obviously, the three types differ in size. Often, *type II endonucleases* are used for molecular biology and recombinant DNA technology. The vast majority of type II restriction enzymes recognize specific sequences that are mostly 4-base and 6-base long where the enzymes cleave (Table 6.3). The DNA sequences are *palindromes*—that is, nucleotide-pair sequences read the same forward or backward as in the case of a nonsense phrase

AND MADAM DNA

6.6.3 Nomenclature

A system of nomenclature for a large number of endonucleases was proposed by Smith and Nathans, the pioneers in the discovery of endonucleases, in 1973, which is still in use today. Restriction endonucleases are named by using the first letter of the genus and the first two letters of the species that produces the enzyme. If the enzyme is produced only by a particular strain of the species, a fourth letter designating the strain is added. The first three letters of the name are italicized as they are abbreviations of genus and species. The first restriction enzyme identified from a bacterial strain is designated I, the second – II, and so on. For example, the restriction endonuclease *Eco*RI is produced by *Escherichia coli* strain RY13, and is the first (I) such enzyme identified from the organism (Table 6.3).

6.6.4 Recognition Sites

The restriction endonucleases recognize specific nucleotide sequences on dsDNA. These recognition sequences are called variously—recognition site/binding site/cognate DNA/DNA site/host specificity site/restriction site, and so on. In case of type II endonucleases, the recognition and restriction sites are the same. It has already been mentioned that the DNA sequences at the sites are palindromes. All type II binding sites have a high GC content (68% GC and 32% AT). The high GC content is favoured for better interaction between DNA and enzyme protein.

6.6.5 Types of Ends Generated after Cleavage with Type II Endonucleases

The arrow marks in the third column of Table 6.3 indicate the phosphodiester bonds cleaved by the endonuclease at the recognition site. In some cases, the endonuclease cuts on the two strands

Table 6.3 Some restriction endonucleases

Name of bacteria (Source)	Name of enzyme	Nucleotide sequence (Palindrome) arrow (↓) at cleavage site 5′ → 3′ 3′ ← 5′
Bacillus amyloliquefaciens H	*Bam* HI	G↓G A T C C C C T A G↑G
Brevibacterium albidum	*Bal* I	T G G↓C C A A C C↑G G T
Escherichia coli RY 13	*Eco* RI	G↓A A T T C C T T A A↑G
Haemophilus aegyptius	*Hae* II	Pu G C G C↓Py Py ↑C G C G Pu
Haemophilus aegyptius	*Hae* III	G G↓C C C C↑G G
Haemophilus haemolyticus	*Hha* I	G C G↓C C↑G C G
Haemophilus influenza Rd	*Hind* II	C T Py↓Pu A C C A Pu↑Py T G
Haemophilus influenza Rd	*Hind* III	A↓A G C T T T T C G A↑A
Haemophilus parainfluenzae	*Hpa* I	G T T↓A A C C A A↑T T G
Haemophilus influenza	*Hpa* II	C↓C G G G G C↑C
Providencia stuartii 164	*Pst* I	C T G C A↓G G↑A C G T C
Streptomyces albus G	*Sal* I	G↓T C G A C C A G C T↑G
Xanthomonas oryzae	*Xor* II	C G A T C↓G G↑C T A G C

of DNA are asymmetric, that is, they do not lie exactly opposite each other. In other cases, the cuts are symmetric, running through both strands at the same place. Thus, restriction cuts by endonucleases produce two types of ends—*staggered* and *blunt ends*.

Staggered ends Asymmetric cuts made by endonucleases at the restriction site produce staggered ends. The staggered ends are usually with two or four nucleotide bases in single stranded form overhanging on each strand (Fig. 6.10). Such ends are also called sticky or cohesive ends. They are complementary to each other and can be held together only by weak hydrogen (H) bonds between complementary bases of two strands. The weakness of these bonds, however, eventually allows the DNA fragments to separate. Again, fragments produced by the same endonuclease enzyme from DNA of different organisms bear similar sticky ends with the same nucleotide composition; this helps in easy alignment of nucleic acid fragments from two different sources, with H bonding, for efficient annealling of the two DNA fragments in the presence of the enzyme ligase. Ligase restores the phosphodiester bonds in the sugar–phosphate backbone of the two DNA fragments to be ligated. In other words, the ligase restores the bonds that were originally broken (hydrdrolysed) by the endonuclease. Ligases are obtained from two sources, *E. coli* and T4 bacteriophages.

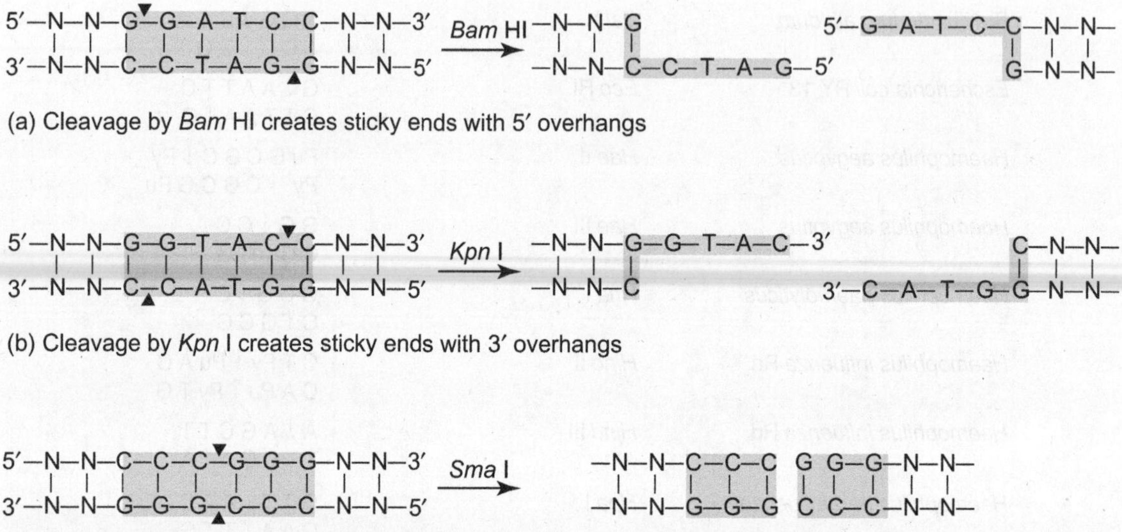

```
5′—N—N—G▼G—A—T—C—C—N—N—3′   Bam HI  —N—N—G            5′—G—A—T—C—C—N—N—
    |  |  |  |  |  |  |  |  |  |                  |  |  |              |  |  |
3′—N—N—C—C—T—A—G▲G—N—N—5′          —N—N—C—C—T—A—G—5′         G—N—N—
```

(a) Cleavage by *Bam* HI creates sticky ends with 5′ overhangs

```
5′—N—N—G—G—T—A—C▼C—N—N—3′   Kpn I  —N—N—G—G—T—A—C—3′          C—N—N—
    |  |  |  |  |  |  |  |  |  |                  |  |  |          |  |  |
3′—N—N—C▲C—A—T—G—G—N—N—5′          —N—N—C             3′—C—A—T—G—G—N—N—
```

(b) Cleavage by *Kpn* I creates sticky ends with 3′ overhangs

```
5′—N—N—C—C—C▼G—G—G—N—N—3′   Sma I   —N—N—C—C—C G G—G—N—N—
    |  |  |  |  |  |  |  |  |  |                |  |  |  |  |  |  |  |
3′—N—N—G—G—G▲C—C—C—N—N—5′          —N—N—G—G—G C C—C—N—N—
```

(c) Cleavage by *Sma* I generates blunt ends

Fig. 6.10 Cleavage by different endonucleases produces different types of ends. Palindrome sequences (recognition site by the enzymes) are shaded. ▼ Indicates cleavage site. (a) and (b) sticky ends, (c) blunt ends.

There can be two types of staggered ends depending on the position of the cuts in the two strands. When the restriction enzyme cuts asymmetrically at the 5′ ends of both the strands, it produces 5′ *overhanging* ends (Fig. 6.10a). When the cuts are at 3′ ends of both the strands, 3′ *overhanging* termini are produced (Fig. 6.10b).

Blunt ends Symmetric cuts through two strands by certain endonucleases produce blunt or flat-ended DNA (Fig. 6.10c). In the absence of sticky ends, the blunt ends of DNA fragments cannot be easily annealled or joined. Blunt ends can be joined using bacteriophage T4 DNA ligase, although the efficiency of the joining reaction is somewhat lower than for ligation of sticky or cohesive ends. Sometimes, using specific terminal transferase enzymes, homopolymer (CCCCC or GGGGG) sticky ends may be generated at the end of a strand of blunt ended DNA and two blunt ends with complementary homopolymer ends (tails) can be annealled.

6.6.6 Production of Recombinant DNA (rDNA) Molecules in Vitro

A restriction endonuclease cleaves an intact DNA molecule at a specific nucleotide sequence regardless of the source of the DNA. It can cleave phage or viral DNA, bacterial DNA, corn

DNA, human or other DNA, as long as the DNA contains the specific nucleotide sequences that it recognizes. Two cleaved fragments of different DNA bearing the same nucleotide sequence at the cut ends can be covalently fused regardless of their origin. The construction of a recombinant DNA molecule has been depicted in Fig. 6.11, where *Eco*RI produced DNA fragments from two different sources have been fused or recombined into one molecule.

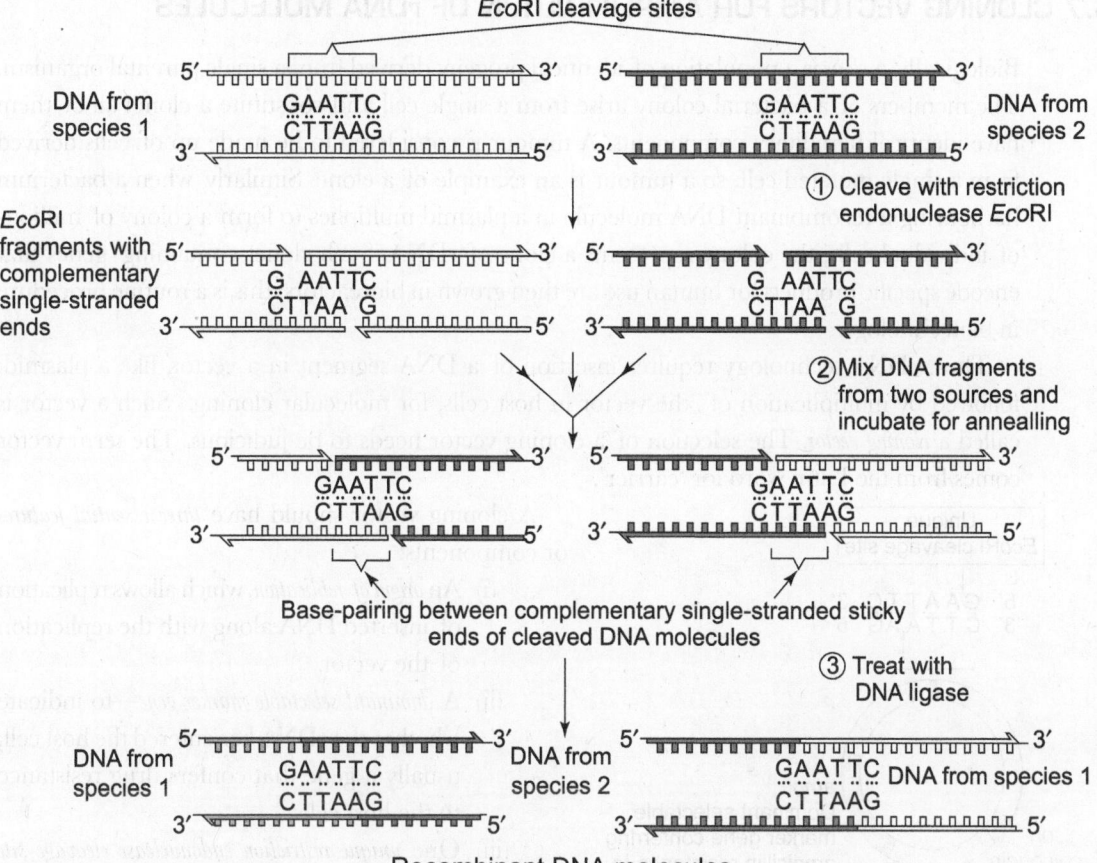

Fig. 6.11 The construction of recombinant DNA molecules in vitro. Initially DNA molecules isolated from different species (prokaryote, plant, animal) are cleaved with a particular restriction enzyme which produces the same complementary nucleotide sequences at the cut ends. The DNA fragments are mixed and allowed to anneal, DNA ligase joins the fragments covalently to produce the recombinant DNA molecules.

The first recombinant DNA molecules were produced in Paul Berg's laboratory at Stanford University in 1972. Berg's research team recombined lambda phage DNA with a DNA molecule of simian virus 40 (SV40). For this pioneering work, Berg was a corecipient of the Nobel Prize in Chemistry in 1980.

Shortly after Berg's discovery, Stanley Cohen and colleagues, also at Stanford, inserted an *Eco*RI restriction DNA fragment from one source into the cleaved unique *Eco*RI restriction site of a self-replicating plasmid (circular extrachromosomal DNA in a bacterium). When this recombinant plasmid was introduced into *E. coli* cells by the process of transformation it exhibited autonomous replication, just like the original plasmid. The recombinant plasmid is considered as a *vector* or vehicle that is used to transport the gene of interest into a host cell and is responsible for its replication.

6.7 CLONING VECTORS FOR AMPLIFICATION OF rDNA MOLECULES

Biologically, a *clone* is a population of identical progeny derived from a single parental organism. The members of a bacterial colony arise from a single cell and constitute a clone; all of them have identical hereditary components. A tumour is considered to be made up of cells derived from a single mutated cell, so a tumour is an example of a clone. Similarly, when a bacterium harbouring a recombinant DNA molecule in a plasmid multiplies to form a colony of millions of individual cells, the colony represents a clone of rDNA. Such clones containing genes that encode specific products for human use are then grown in bioreactors; this is a routine procedure in biotechnology.

Thus, rDNA technology requires insertion of a DNA segment in a vector, like a plasmid, followed by multiplication of the vector in host cells, for molecular cloning. Such a vector is called a *cloning vector*. The selection of a cloning vector needs to be judicious. The term vector comes from the Latin word for 'carrier'.

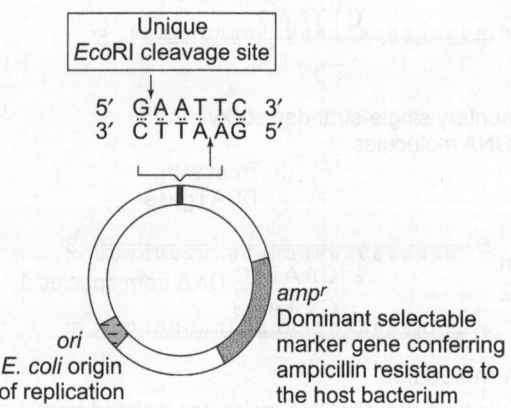

A cloning vector should have *three essential features* or components:

(i) An *origin of replication*, which allows replication of inserted DNA along with the replication of the vector.

(ii) A *dominant selectable marker gene*—to indicate whether the rDNA has entered the host cell, usually a gene that confers drug resistance to the host cells.

(iii) One *unique restriction endonuclease cleavage site (insertion site)*—far away from the origin of replication site and the selectable marker gene. This is the site for introducing the piece of foreign DNA (Fig. 6.12).

Fig. 6.12 A cloning vector with three essential features— an origin of replication, a restriction endonuclease cleavage site, and a dominant selectable marker.

Currently used cloning vectors carry a cluster of unique restriction sites called a *polylinker* or a *polycloning site* (Fig. 6.13). The site bears nucleotide sequences for multiple restriction endonucleases.

Although efficiency of ligation and transformation is greater with small vectors that have the three essential features mentioned above, two other genetic characteristics can be considered as *desirable features* for vectors. They are mentioned in the following text:

(i) Insertion site (unique restriction endonuclease site) right within a vector gene whose function would be disrupted as soon as a foreign DNA piece is inserted at the site. Inactivation of

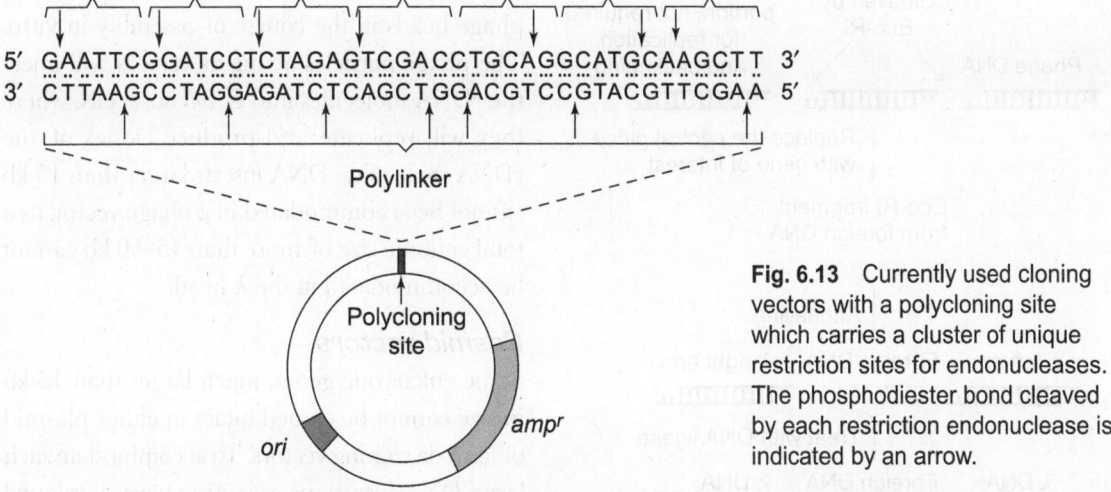

Fig. 6.13 Currently used cloning vectors with a polycloning site which carries a cluster of unique restriction sites for endonucleases. The phosphodiester bond cleaved by each restriction endonuclease is indicated by an arrow.

an easily identifiable function of the gene would indicate insertion of foreign DNA. Then clones containing the recombinant plasmid can easily be selected.

(ii) Presence of a promoter for initiating transcription of the inserted DNA fragment. The promoter on the vector can undertake transcription of the inserted DNA, in case normal promoters of the inserted genes are not recognized by the transcription machinery of the host cells.

6.7.1 Types of Vectors

Here, three different types of vectors which initially contributed to the development of recombinant DNA technology are described first. Next, a limited discussion on some other vectors that serve special purposes, especially to accommodate large segments of eukaryotic genome is considered.

Plasmid vectors

Plasmids are extrachromosomal, double-stranded circular DNA molecules present in the cytoplasm of bacteria. Their sizes range from about 1 kb (1000 base pairs) to over 200 kb, and most of them can replicate autonomously. Antibiotic resistant genes are often present in plasmids and found to be ideal selectable markers. Plasmid pBR322 from *E. coli* is one of the first widely used cloning vectors, especially since it has both ampicillin- and tetracycline-resistance genes and unique restriction enzyme cleavage sites within these genes. Many of the cloning vectors currently in use are derivatives, at least partly, of plasmid pBR322.

Bacteriophage (virus) vectors

Most of the bacteriophage cloning vectors are derived from the λ (lambda) phage chromosome. The chromosome of wild-type λ phage is 48,502 (48.5 kb) nucleotide base pairs long. The central one-third portion of the chromosome (about 15 kb) contains genes that are required for lysogeny (integration with bacterial host chromosome). This part of the λ chromosome is excised with restriction enzymes and foreign DNA is accommodated in its place (Fig. 6.14). The

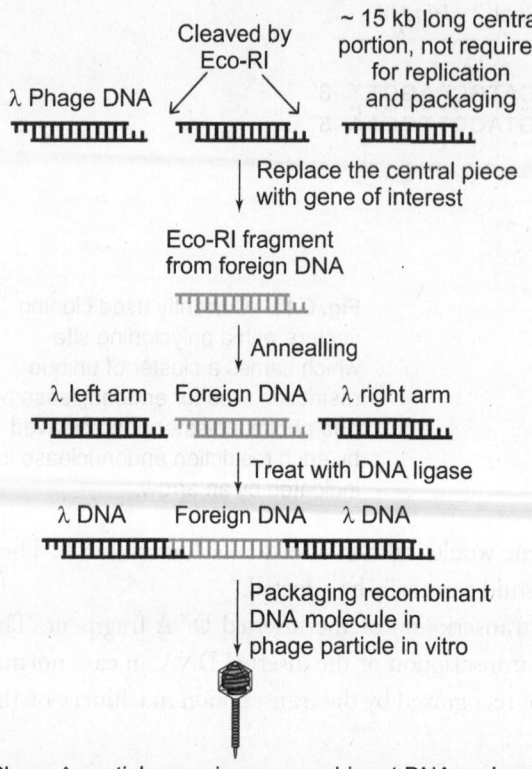

λ Phage DNA

Cleaved by
Eco-RI

~ 15 kb long central
portion, not required
for replication
and packaging

Replace the central piece
with gene of interest

Eco-RI fragment
from foreign DNA

Annealling

λ left arm   Foreign DNA   λ right arm

Treat with DNA ligase

λ DNA   Foreign DNA   λ DNA

Packaging recombinant
DNA molecule in
phage particle in vitro

Phage λ particle carrying a recombinant DNA molecule

Fig. 6.14   Construction of a bacteriophage
cloning vector.

resulting rDNA molecules can be packaged in phage heads in the course of assembly in vitro. The phage particles acting as vectors will inject the rDNA molecules into *E. coli* host cells, where they will replicate and produce clones of the rDNA molecules. DNA inserts larger than 15 kb cannot be accommodated in a phage vector, as a total genome size of more than 45–50 kb cannot be accommodated in the λ head.

Cosmid vectors

Some eukaryotic genes, much larger than 15 kb in size cannot be cloned intact in either plasmid or lambda cloning vectors. To accommodate such large DNA insertions, scientists have developed *cosmids*, hybrids between the λ phage chromosome and plasmids. *Cos* indicates cohesive site of λ chromosome and *mid* is derived from plasmid. The cos site contains the 12 complementary base pairs recognized by the phage λ DNA packaging apparatus, which causes staggered cuts at this site during packaging to produce single stranded complementary cohesive ends of the mature lambda chromosome.

Cosmids have key advantages of both the plasmid and the λ phage vector. The cosmid vector includes the origin of replication and a marker, an antibiotic-resistance gene of the plasmid and the λ cos site for packaging of recombinant DNA in the λ head. In the absence of other genes of λ phage, cosmid vectors can accept foreign insert DNA as large as 35–45 kb and still be accommodated in λ heads.

Other vectors

Phagemid vectors   These vectors are constructed using a combination of both *phage* chromosomes and *plasmids*. On certain counts, they are different from cosmids and serve some special purposes. Phagemids replicate in *E. coli* as double-stranded plasmids. When a *helper phage* is provided, they switch to the phage mode of replication and pack ssDNA in phage particles. Actually, the helper phage is a mutant and replicates its own DNA inefficiently, but it makes the provision of viral replication enzymes and structural proteins for the production of phagemid DNA molecules packed in phage coat proteins. These particles do not lyse the host cells at the time of release as is the case with phages such as T4. Instead, the progeny phagemids are extruded through the cell membrane and cell wall of the host cells which keep producing more phagemids.

The phagemid vectors *pUC118* and *pUC19* were designed by the University of California; that is why the name includes the abbreviated form of the University (UC). These phagemids contain the origin of replication from phage 13 which has a single stranded DNA genome.

Use of the pUC allows a simple colour test to differentiate the host cells harbouring phagemids with foreign DNA inserts from those containing no insert.

Eukaryotic and shuttle vectors   In the early years of rDNA technology, scientists found that plasmid, phage λ, and cosmid cloning vectors were very useful and all of them replicated in *E. coli* cells. However, these vectors did not prove to be as useful for various other taxonomic groups with their distinct origins of replication and regulatory signals. Different cloning vectors for different species became the need of the hour. Thus, different cloning vectors have been developed which can replicate in prokaryotes other than *E. coli* and in eukaryotic organisms. Currently, many unique cloning vectors are available for use in *S. cerevisiae* (yeast), *D. melanogaster* (fruit fly), plants, mammals, and other organisms.

Shuttle vectors can replicate in both *E. coli* and eukaryotic species. Such shuttle vectors designed for use in the yeast *S. cerevisiae* contain origins of replication of both *E. coli* and *S. cerevisiae* and selectable marker genes, along with a polycloning site to insert the gene of choice. Shuttle vectors are very useful in certain genetic studies. For example, site-specific mutagenesis (directed changes in nucleotide sequence) can be induced in a yeast gene cloned in a shuttle vector in *E. coli*, and the vector can then be transferred to yeast to examine the effect of induced mutations. Similarly, shuttle vectors can be used in *E. coli* and animal systems.

Artificial chromosomes   Researchers tried to develop vectors that would accept DNA sequences of eukaryotic genes much bigger than the 35–45 kb insert normally accommodated by a cosmid vector. They ended up with *yeast artificial chromosomes* (YACs). YAC vectors are genetically engineered yeast minichromosomes that contain (i) a yeast origin of replication, (ii) a yeast centromere, (iii) two yeast telomeres at the terminals, (iv) a selectable marker, and (iv) a polycloning site for accomodation of DNA inserts of 200–500 kb.

YAC cloning vectors are very useful for investigating sequences of large eukaryotic genomes. One YAC clone can perform the role of several cosmid clones. YAC cloning was especially valuable in the Human Genome Project.

Bacterial artificial chromosomes (BACs) and bacteriophage *P1 artificial chromosomes (PACs)* have been constructed from bacterial fertility (F) factor, circular DNAs, and bacteriophage P1 chromosomes. Like YACs, BACs and PACs can have large inserts up to 150–300 kb in size. BACs and PACs are less complex and easier to construct than YACs. Furthermore, BACs and PACs replicate in *E. coli* just like plasmid, lambda, and cosmid vectors. For all these advantages, BACs and PACs currently have replaced the need of YAC vectors, for the studies with large genomes from humans and other mammals.

In the past few years, BAC and PAC vectors have been further modified to enable them to replicate both in *E. coli* and in mammalian cells.

So far, genomic DNA, cleaved by restriction endonuclease has been considered as source material for DNA inserts in the cloning vectors. Now *complementary DNA (cDNA)* as another source for DNA insert and *polymerase chain reaction (PCR)* which can produce a huge number of copies of a selected region (gene) of DNA in vitro (without using cloning vectors) are discussed in the following text.

6.7.2 Synthesis of Complementary DNA (cDNA) for Cloning Vectors

Major parts of the large genomes of higher plants and animals do not encode proteins. Again a particular stretch of DNA for a gene includes both exons and introns; introns are non-functional

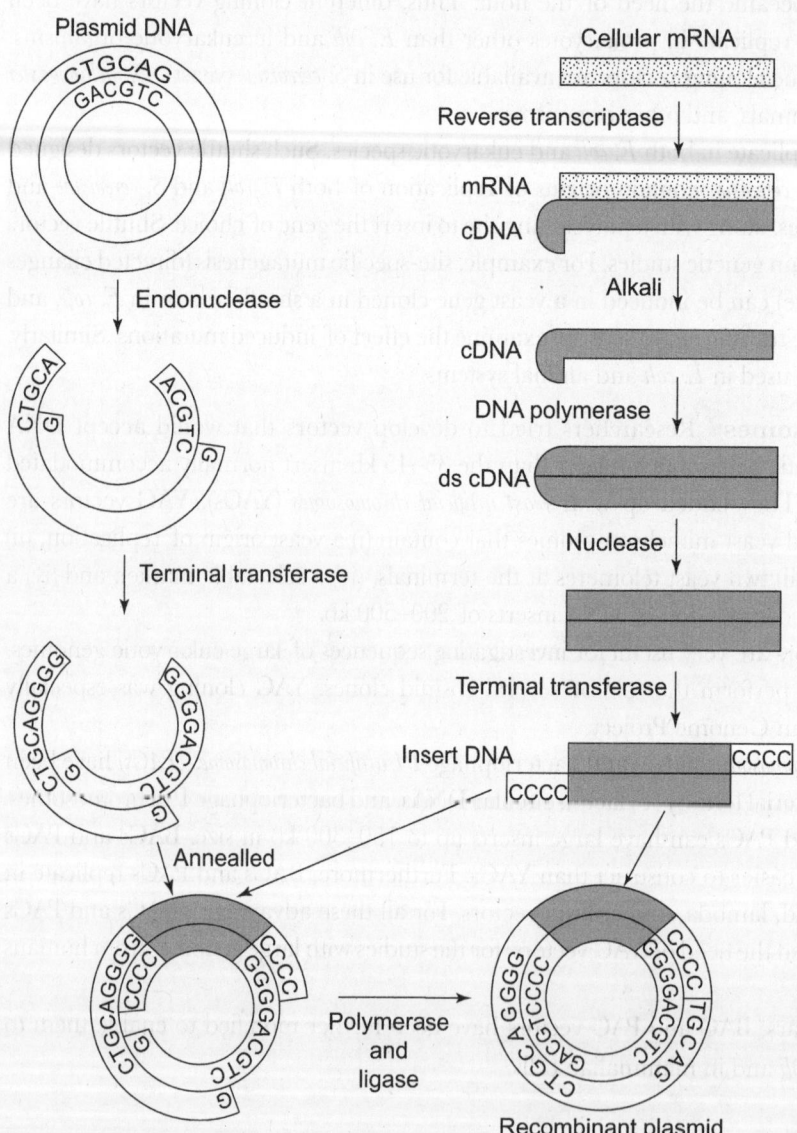

Fig. 6.15 Construction of a recombinant plasmid vector. Synthesis of cDNA with reverse transcriptase shown on the right.

regions. A stretch of DNA that codes for a gene is transcribed into pre-mRNA, which is then edited to get rid of introns by splicing and thereby converted to a functional mRNA in which the exons are placed together. Thus mRNA represents a template of ribose nucleotides without the non-coding parts or introns. Therefore if mRNA is used as the starting material then the size of the inserts for cloning vectors can be considerably reduced. *Reverse transcriptase* (RNA-dependent DNA polymerase) is used to synthesize the complementary DNA strand. Then the RNA–DNA duplex is converted to double-stranded DNA molecule by the combined activities of ribonuclease H, DNA polymerase I, and DNA ligase (Fig. 6.15). Ribonuclease H degrades the mRNA template strand, and short RNA fragments produced during the degradation serve as primers for DNA synthesis (see Chapter 3). DNA polymerase I catalyses the synthesis of the second DNA strand and removes the RNA primers from the DNA strand. DNA ligase joins the single-strand breaks and a double-stranded DNA molecule is formed. These double-stranded cDNAs can be easily used as inserts for plasmid or phage λ cloning vectors by enzymatically adding complementary single-stranded tails to the cDNAs and vectors as in the case of blunt-ended restriction fragments.

In the classical method of ds cDNA synthesis, first strand synthesis on mRNA is initiated by oligo (dT) primers, and mRNA is removed from the mRNA:ss cDNA hybrid by alkali hydrolysis or boiling. Then the ss cDNA loops back on itself due to the hydrophobicity of the bases at the 3′ end. The few nucleotides at the hairpin like loop act as a self-primer for synthesis of the second cDNA strand. The hairpin loop is then cleaved with S1 nuclease (Fig. 6.15).

mRNA which is the starting material is highly labile. RNA isolation and special care to protect RNA from ribonuclease have been discussed earlier. mRNA has to be isolated from the cells types where it is expressed abundantly if it is required in large quantities. Some examples of the preferable sources of specific abundant mRNAs include chick oviduct cells for ovalbumin mRNA, reticulocytes for β-globin mRNA, pancreatic β-cells for insulin mRNA, and wheat seeds for gliadin mRNA.

6.8 CONSTRUCTION OF DNA LIBRARIES

The entire genome of an organism is cleaved with a restriction endonuclease into many fragments which are then accommodated into many appropriate cloning vectors. The collection of such recombinant DNA clones constitutes the genomic or DNA library of an organism.

6.8.1 Genomic Libraries

A genomic (DNA) library is a collection of clones that is supposed to contain at least one copy of every DNA sequence in the genome of an organism. In other words, a set of clones collectively contains the entire genome of an organism. The DNA library can be screened for genes and other sequences for further analysis or to amplify them for use in the rDNA technique.

Genomic libraries are constructed by isolating total genomic DNA from an organism, digesting the DNA with a restriction endonuclease, and then inserting or recombining the restriction fragments into appropriate cloning vectors, as outlined earlier. The cloning vectors are then introduced into host cells for replication along with the new recombinant DNA. There are several techniques for introduction of vectors or recombinant DNA in host cells, which are outlined later.

The transformed host cells are allowed to grow into colonies in culture.

6.8.2 cDNA Libraries

The procedure of synthesis of cDNAs and their insertion into plasmid or phage cloning vectors have been discussed in Section 6.7.2. These vectors can be introduced into bacterial host cells for replication and production of colonies containing particular cDNA inserts. The clones that together contain all the cDNA inserts from an organism constitute the cDNA library for the organism. The total number of cDNA clones of an organism is smaller than the number of clones constituting the genomic DNA library of the same organism, as the cDNA clones do not include introns and intergenic DNA (non-functional DNA stretches between the genes). It is easy to accommodate an entire gene in a single clone in a cDNA library, which is sometimes difficult to accomplish for a genomic library.

Since splicing does not occur in bacteria due to lack of proper enzymes, eukaryotic gene sequences containing introns cannot be processed and expressed in prokaryotes. In such cases, cDNA clones find applications where bacterial expression of foreign eukaryotic DNA is necessary, especially in the production of proteins for industrial purposes.

6.9 INTRODUCTION OF rDNA VECTORS INTO BACTERIAL CELLS

In classical genetics of prokaryotes, the process of introducing a piece of foreign DNA into a bacterium is known as *transformation*. The methods devised for transformation of *E. coli* cells with rDNA vectors can be classified into chemical and physical methods.

6.9.1 Chemical Method—CaCl₂ and Heat Shock Technique

The physiological state of bacteria when they can take up DNA from an external source is known as the *competent* state. *E. coli*, the most commonly used host cell for the propagation of recombinant molecules can be made competent for DNA uptake by treating a rapidly growing culture with high concentrations of calcium chloride (50–100 mM $CaCl_2$) in a hypotonic buffer. Water from the hypotonic buffer enters into the bacterial cells and makes the cells spheroblasts. This treatment is combined with the heat shock. The heat shock is generated by raising temperature of the fluid containing bacteria and added DNA (recombinant vector molecules) from 37°C to 42°C for a brief period of 1–2 minutes. During the heat shock, the spheroblasts contract and the bacteria take up any bound DNA.

After the heat shock, bacterial cells are incubated with non-selective general growth medium for the recovery of the bacterial cells. Then the cells are subjected to the selective medium to select the (transformed) cells containing the plasmid vector. Non-transformed cells do not survive in the selective medium (Fig. 6.16).

The $CaCl_2$ and heat shock method, originally devised by Cohen, was modified by others to induce a much higher frequency of competence in bacteria and hence a higher number of transformants.

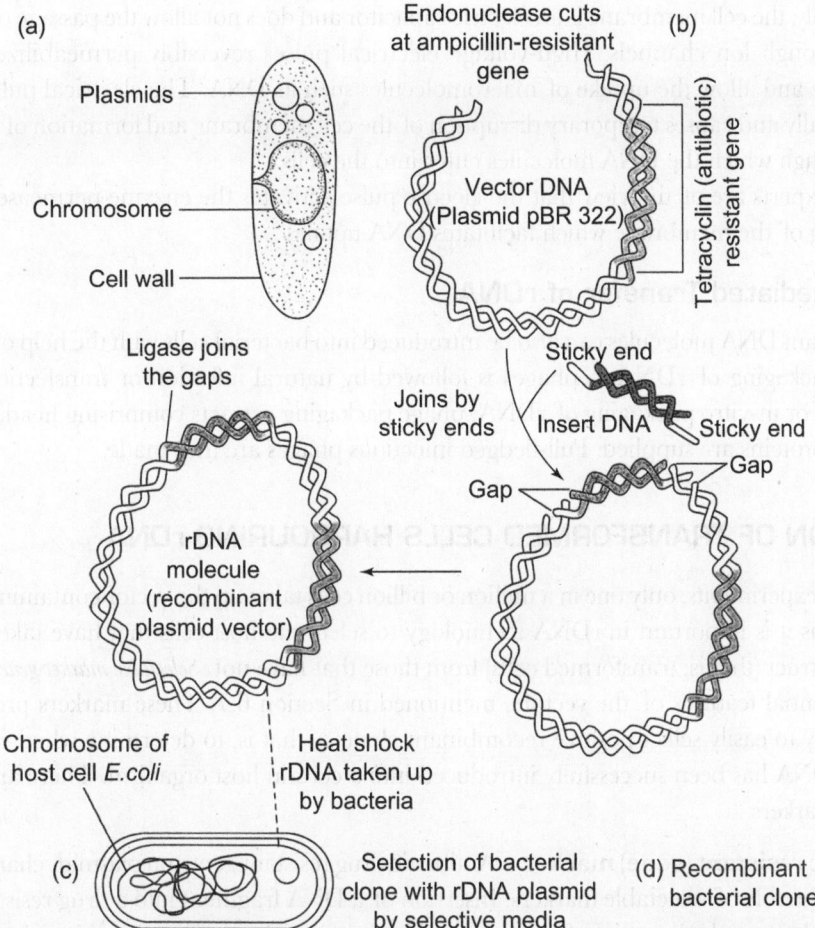

Fig. 6.16 (a) A bacterial cell with a circular chromosome and plasmids as additional circular DNA molecules in the cytoplasm. (b) Steps of producing rDNA with insert DNA in *E. coli* plasmid pBR322. (c) Inserting the rDNA plasmid in *E. coli* with heat shock. (d) Selection of bacterial clone with rDNA plasmid: Bacteria survive in medium with tetracycline antibiotic only when they have been transformed with a plasmid with the tetracyclin resistant gene. The same bacterial clones will die in medium with ampicillin (antibiotic) indicating disruption of ampicillin resistant gene for insertion of foreign DNA at this site. Thus, the entry of the plasmid with recombinant DNA is confirmed.

Transformation efficiency is defined as the number of transformed colony forming units (cfu) produced by 1 µg of plasmid DNA. Transformation efficiency decreases rapidly as the size of the DNA molecule increases and is almost insignificant when the size exceeds 50 kbp.

6.9.2 Physical Method—Transformation by Electroporation

Electroporation is a direct physical method for gene transfer. It is a fast and about 100-fold more efficient method of transformation than the chemical technique. However, electroporation requires costly electrical equipment and specially designed expensive disposable cuvettes for the bacteria cells and plasmid DNA.

Normally, the cell membrane is an electric capacitor and does not allow the passage of current except through ion channels. High-voltage electrical pulses reversibly permeabilize the cell membrane and allow the uptake of macromolecules such as DNA. The electrical pulse decays exponentially and causes temporary disruption of the cell membrane and formation of transient pores through which the DNA molecules enter into the cells.

Some experts are of the view that the electric pulse activates the enzyme permease leading to thinning of the membrane which facilitates DNA uptake.

6.9.3 Phage-mediated Transfer of rDNA

Recombinant DNA molecules can also be introduced into bacterial cells with the help of phages. In-vitro packaging of rDNA in phages is followed by natural infection or transfection of the host cells. For in-vitro packaging of rDNA, phage packaging extracts comprising head, tail, and assembly proteins are supplied. Full-fledged infectious phages are thus made.

6.10 SELECTION OF TRANSFORMED CELLS HARBOURING rDNA

In cloning experiments, only one in a million or billion cells takes up the vector containing foreign DNA. Thus it is important in rDNA technology to select the host cells that have taken up the DNA construct (that is, transformed cells) from those that have not. *Selectable marker genes* are one of the essential features of the vectors, mentioned in Section 6.7. These markers provide the opportunity to easily select positive recombinant clone(s), that is, to determine whether a piece of insert DNA has been successfully introduced into a certain host organism. There are several types of markers.

Antibiotic resistant (gene) marker   Antibiotic drug resistant genes on plasmids characteristically play the role of selectable markers. Insertion of a DNA fragment into a drug resistant gene (Fig. 6.16) destroys the integrity of the gene and its function (that is, insertional inactivation of a gene) and the recombinants can be identified on the basis of sensitivity of the clone to the drug (Fig. 6.17). Here the DNA fragment has been originally inserted in the ampicillin-resistance gene. The sensitivity of the clone to the drug is tested on a replica plate; the original clone with the DNA insert remains intact in the master plate. The transformant clone on master plate is identified with reference to the position of the missing clone on the replica plate treated with ampicillin.

In bacteria, resistance to antibiotics, such as ampicillin, chloramphenicol, streptomycin, tetracycline, and kanamycin are almost exclusively used as selectable markers.

Reversal of auxotrophy (complementation) marker   Reversal of auxotrophy is another type of positive selection technique for transformants. The host cells are auxotrophic mutants, and survive in a growth medium supplemented with a specific nutrient which the mutant (host) cells cannot synthesize. The vector contains a wild type functional version of the mutated gene of the host, as selectable marker. When the vector is properly inserted in the host cell, its marker gene can complement the function of the mutated gene in the host cells which will grow in non-supplemented medium, and thus the auxotroph is reverted to the prototroph condition.

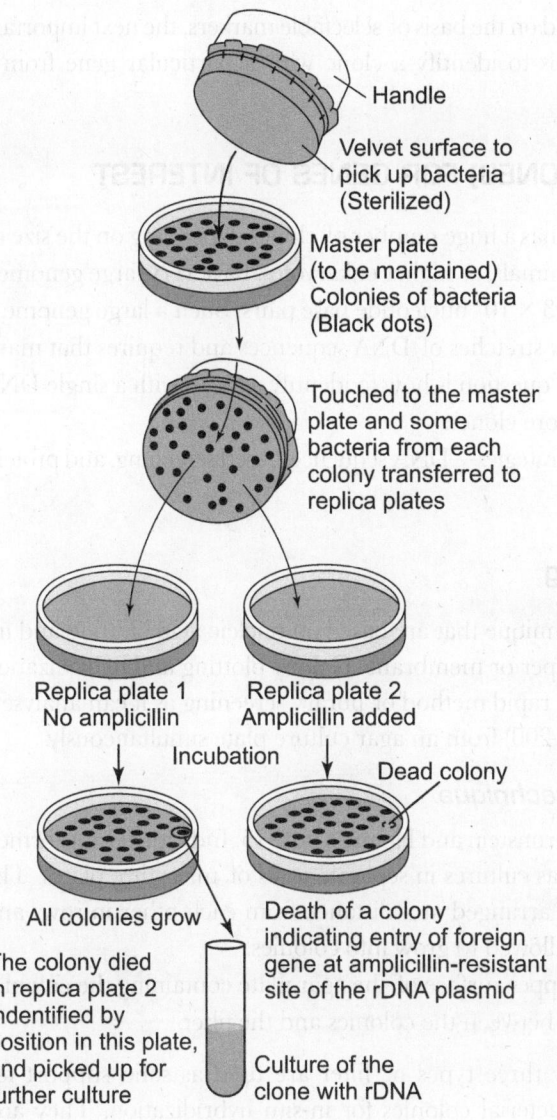

Handle

Velvet surface to pick up bacteria (Sterilized)

Master plate (to be maintained) Colonies of bacteria (Black dots)

Touched to the master plate and some bacteria from each colony transferred to replica plates

Replica plate 1
No amplicillin

Replica plate 2
Amplicillin added

Incubation

Dead colony

All colonies grow

Death of a colony indicating entry of foreign gene at amplicillin-resistant site of the rDNA plasmid

The colony died in replica plate 2 indentified by position in this plate, and picked up for further culture

Culture of the clone with rDNA

Fig. 6.17 Replica plating technique to select transformed bacteria harbouring rDNA.

Reversal of auxotrophy markers are frequently used to select transformants in yeast, as the drug resistance genes are not as frequent in yeast as they are in bacteria.

LacZ gene marker The *lacZ* gene in *E. coli* encodes β-galactosidase, the enzyme that cleaves lactose into glucose and galactose. This first step in the catabolism of lactose in *E. coli* can occur only if *lacZ* remains functional. Thus the *lacZ* fulfils the condition of a selectable marker for entry of a vector into host cells. This gene with some initial modification is incorporated in phagemid pUC vectors (in Section 6.7.1). Initially, only a small part of the *lacZ* gene, called *lacZ'* gene segment, is incorporated in the vector (the whole of lacZ gene sequence is difficult to accommodate in a phagemid vector). The *lacZ'* gene segment encodes only a portion of the amino terminal of β-galactosidase and can complement the defective or mutant *lacZ* gene of the host cells to come up with functional β-galactosidase.

The presence of the vector with foreign DNA insert can be monitored on the basis of the formation of β-galactosidase out of complementation. As substrate, 5-bromo-4-chloro-3-indolyl-β-D-galactoside (in short, X-gal) is added to the bacteria culture plate. X-gal is cleaved by β-galactosidase to 5-bromo-4-chloro-indigo and galactose. X-gal is colourless and 5-bromo-4-chloro-indigo is blue. Thus, cells containing active β-galactosidase produce blue colonies on agar medium in the presence of X-gal. Cells lacking β-galactosidase activity grow as white colonies on X-gal plates.

Other selectable gene markers Scientists have already discovered a long list of other selectable marker genes to select transformed bacteria, plant, and animal cells. Some of them are of bacterial origin.

For example, antibiotic resistance genes are used as selectable markers in plants. Although plants are eukaryote, antibiotics efficiently inhibit protein synthesis in the chloroplasts. Neomycin phosphotransferase (*npt II*) gene confers resistance to the antibiotic kanamycin. The *npt II* marker gene provides a direct means to obtain only transformed cells in culture.

Once the recombinant clones are selected on the basis of selectable markers, the next important step in genetic engineering experiments is to identify a clone with a particular gene from a genomic or cDNA library.

6.11 SCREENING DNA LIBRARIES (CLONES) FOR GENES OF INTEREST

Essentially, a DNA library of an organism has a huge number of clones depending on the size of the genome of the organism. Plants and animals, as higher eukaryotes, have very large genomes; for example, the human genome contains 3×10^9 nucleotide base pairs. Such a large genome is distributed over a million or more different stretches of DNA sequences and requires that many number of clones in gene library. Now the question is how to identify a clone with a single DNA sequence of interest out of a million or more clones.

There are two basic types of screening strategies—DNA sequence-based screening, and protein structure/function-dependent screening.

6.11.1 DNA Sequence-based Screening

There are several modifications of the technique that are based on nucleic acid blotting and in-situ hybridization on a supportive filter paper or membrane. Colony blotting and hybridization technique is the most commonly used and rapid method of library screening as it can analyse a large number of transformed clones (100–200) from an agar culture plate simultaneously.

Colony blotting and hybridization technique

Colony blot hybridization was devised by Grunstein and Hogness in 1975. Individual transformed clones in libraries are arrayed and stored as cultures in separate wells of microtitre plates. The different clones from the gene library are arranged at a distance from each other in rows and columns like a grid in an agar plate, and allowed to grow into colonies.

- A filter membrane is placed on the upper surface of the agar plate containing the colonies of the library, making direct contact between the colonies and the filter.

Choice of filter/membrane   Usually three types of filter are used as solid support for transferred colonies and for lysing the bacterial colonies for in-situ hybridization. They are, Whatman 541 filter papers, nitrocellulose filters, and nylon filters. Previously, Whatman 541 filter paper was used. DNA and RNA attach to nitrocellulose membrane by hydrophobic interactions and leach slowly from the matrix during hybridization and washing at higher temperatures. Nitrocellulose filter with 0.45 μm pore is normally used; 0.22 μm filter may also be useful in the study of smaller nucleic acid molecules. Nitrocellulose membrane exhibits certain drawbacks—comparatively low binding capacity for nucleic acids ($\sim$50–100 μg/cm^2) and slight negative charge.

Nylon membrane has been devised to avoid the shortcomings in nitrocellulose and has largely replaced nitrocellulose in blotting experiments. Nylon membranes are the most durable among the three and easily withstand repeated hybridization (probing) on the same membrane. The membrane shows higher nucleic acid binding capacity (400–500 μg/cm^2).

- By coming in direct contact with the colonies arranged on the agar plate, the filter picks up some bacteria from each colony. Then the filter is removed with forceps (Fig. 6.18). The bacteria on the filter are lysed either by alkali or by a combined treatment with alkali and heat at 80°C. SDS, denaturing solution, and neutralizing solution are added sequentially on the filter to denature DNA from lysed cloned colonies on the filter. All these treatments may be carried out by placing the filter in a sealed plastic bag.

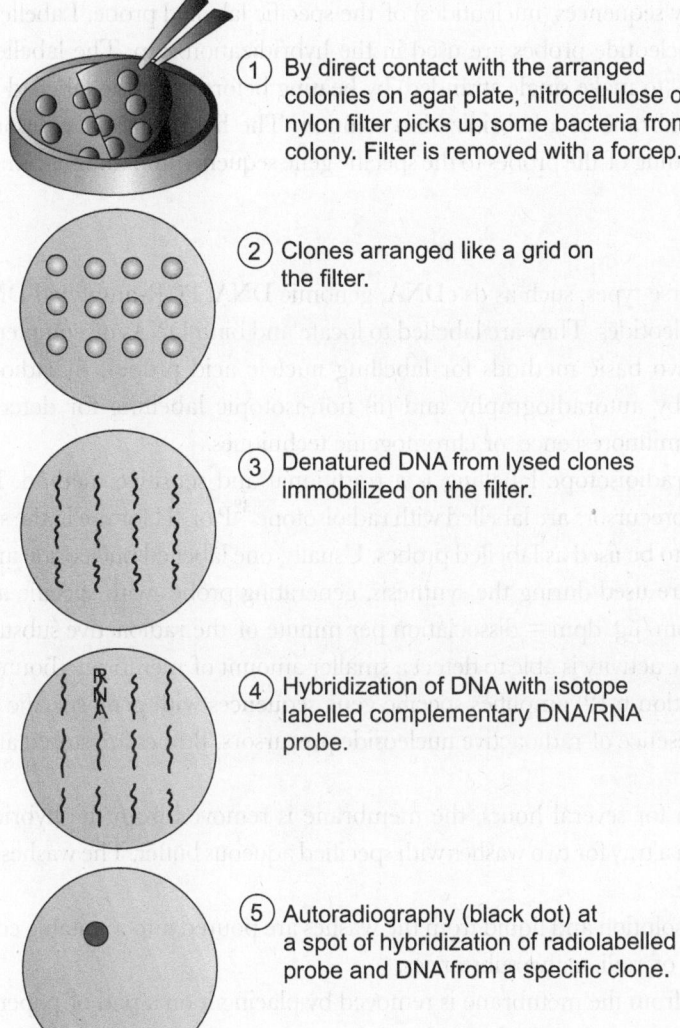

1. By direct contact with the arranged colonies on agar plate, nitrocellulose or nylon filter picks up some bacteria from each colony. Filter is removed with a forcep.

2. Clones arranged like a grid on the filter.

3. Denatured DNA from lysed clones immobilized on the filter.

4. Hybridization of DNA with isotope labelled complementary DNA/RNA probe.

5. Autoradiography (black dot) at a spot of hybridization of radiolabelled probe and DNA from a specific clone.

Fig. 6.18 Screening of a genomic library for a particular gene by colony hybridization.

The bacterial colonies in agar plate are stored as a *master plate* at 4°C until the results of the screening procedure is obtained.

- The denatured DNA is fixed firmly either by baking the filter at 80°C or by treatment with UV light.

A brief exposure to a calibrated UV light source effectively cross-links nucleic acids to the nylon membrane. This is a dependable method for immobilization of DNA on the membrane to carryout further steps of hybridization with probes.

- The DNA molecules become attached to the membrane through their sugar–phosphate backbone, so the bases are free to pair with the complementary nucleic acid molecules or *probes*.

- Denatured DNAs attached to the membrane are single stranded and allow easy binding of the complementary sequences (nucleotides) of the specific labelled probe. Labelled DNA, RNA, and oligonucleotide probes are used in the hybridization step. The labelled DNA probe is denatured (to make single stranded) by heating before use. The labelled probe is applied to the membrane in a hybridization solution. The hybridization solution brings better clarity in binding of the probes to the specific gene sequence and inhibits non-specific binding.

DNA probes

DNA probes are of diverse types, such as ds cDNA, genomic DNA, PCR-amplified DNA, and single-stranded oligonucleotides. They are labelled to locate and bind DNAs of complementary sequences. There are two basic methods for labelling nucleic acid probes: (i) radioisotopic labelling for detection by autoradiography and (ii) non-isotopic labelling for detection by chemiluminescence, chemifluorescence, or chromogenic techniques.

Use of probes with radioisotope labelling is a traditional and sensitive method. Initially, nucleoside triphosphate precursors are labelled with radioisotope ^{32}P or ^{3}H for use in the synthesis of nucleic acid stretches to be used as labelled probes. Usually, one labelled nucleoside and three unlabelled nucleosides are used during the synthesis, generating probes with specific activities of 5×10^8 to 5×10^9 dpm/μg (dpm = dissociation per minute of the radioactive substance). A probe with higher specific activity is able to detect a smaller amount of membrane-bound DNA. PCR technique (see Section 6.12) amplifies specific gene sequences with gene specific primers or from cDNA in the presence of radioactive nucleoside precursors. Probes are stored at −20°C and −80°C until use.

- After hybridization for several hours, the membrane is removed from the hybridization container, placed in a tray for two washes with specified aqueous buffer. The washes remove the unbound probe.

- The hybridization solution and liquid from the washes are poured into a suitable container for proper disposal of radioactive substances.

- Most of the liquid from the membrane is removed by placing it on a pad of paper towels. Then the damp membrane is placed on a sheet of transparent Saran wrap and subjected to autoradiography to find out the colonies that have been radiolabelled, that is, those colonies that harbour the specific DNA sequences (gene) to which radiolabelled DNA probe has hybridized. The hybridized spot can be detected on the X-ray film.

Autoradiography is a simple and sensitive photochemical technique to record the spatial (positional) distribution of radiolabelled compounds, within a specimen or an object. An X-ray film with radio-and light-sensitive emulsion, is kept in close proximity to the radiolabelled specimen for 16–24 hours at –70°C. During exposure to the film, energy emitted from the radiolabelled substance with the release of electrons is absorbed by silver halide grains within the film emulsion. The resulting negatively charged halides (sensitivity specks) attract positively charged silver ions, thus forming atoms of metallic silver. When the film is developed through chemical processes (of developer and fixative), the metallic silver spots indicate the sites of emission of radioactivity and represent the autoradiography of the original radiolabelled material.

Nowadays, there are many modifications to increase the sensitivity of autoradiography method and to obtain better autoradiographic image.

- A colony whose hybridized DNA print in autoradiography gives a positive result is then identified in the *master plate* kept at 4°C which is set aside right after transfer of the colonies on the membrane.

The positive clone is then expanded for the particular gene or its product for technological purposes.

Plaque lift and hybridization technique

Plaque lift and hybridization technique is used for mass screening to isolate specific recombinants from libraries of bacteriophage λ (λ phage containing eukaryotic gene sequences).

Aliquots of the diluted bacteriophage stock (or library) are mixed with appropriate amount of freshly prepared bacteria host cells and incubated. Plaques develop due to multiplication of phages and lysis of host cells. The plaques are then lifted on filter or membrane by making direct contact with plaques on the agar plate. Then the membrane is treated in a step-by-step procedure as followed in the colony blotting and hybridization technique with radioactive DNA probe. The plaque containing the specific gene can be identified from the autoradiography plate.

6.11.2 Other Techniques Developed on the Basis of Hybridization with Radioactive Probe

A radioactive probe can be used to detect a specific gene sequence or its transcriptional (expression) product in the following techniques. These techniques do not have much to do with screening of transformed colonies, but bear certain procedural similarities with it, especially in the process of hybridization.

Southern blotting and hybridization

This technique is named after its inventor British biologist Edwin Southern (1975) and is used to detect specific DNA sequences. The whole genome is cleaved with a restriction endonuclease and subjected to electrophoresis in an agarose gel. The DNA sequences disperse in a column in the lane under each well according to their size. The separated DNA bands are transferred to a membrane by the capillary action of fluid from the gel induced by a stack of blotting paper placed on top of the membrane (Fig. 6.19). Hybridization with specific DNA probe is then carried out on the membrane. A hybridized band is revealed by autoradiography and corresponds to a specific sequence of DNA that is complimentary to the probe.

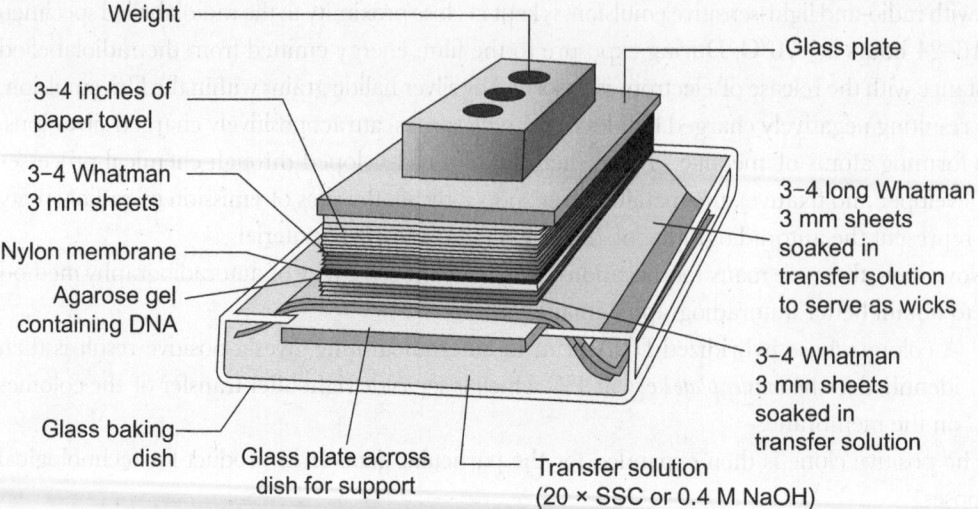

Weight

Glass plate

3–4 inches of
paper towel

3–4 Whatman
3 mm sheets

Nylon membrane

Agarose gel
containing DNA

3–4 long Whatman
3 mm sheets
soaked in
transfer solution
to serve as wicks

3–4 Whatman
3 mm sheets
soaked in
transfer solution

Glass baking
dish

Glass plate across
dish for support

Transfer solution
(20 × SSC or 0.4 M NaOH)

Fig. 6.19 Southern blotting technique. The bands of fragments of genomic DNA in different sizes cleaved with restriction endonuclease and separated in gel electrophoresis. The capillary action of fluid from the gel induced by the stack of blotting papers, facilitates the transfer of DNA bands to the nylon membrane. The DNA on contact binds to the membrane which is then dried and baked under vacuum to affix the DNA for hybridization with labelled probe. SSC is a solution of sodium chloride and sodium citrate.

Northern blotting and hybridization

Northern blotting is used to study gene expression, that is, the amount and size of RNAs transcribed from eukaryotic genes. Technique wise it is similar to Southern blotting with a key difference that it analyses RNA, rather than DNA. The name of the technique was assigned due to its similarity with Southern blotting. This technique was developed in 1977 by Alwine and coworkers at Stanford University.

Western blotting and immunological screening of gene products (proteins)

The western blotting technique is used for detection of proteins. It uses antibodies that specifically recognize antigenic determinants on the polypeptide synthesized by a target clone. This technique is often described as '*protein structure/function dependent screening (expression screening)*', especially when it is employetd to screen a clone with a specific gene sequence. It has already been mentioned that this technique follows the DNA sequence-based method for screening gene libraries.

Dot blot for DNA and RNA

Dot blot (or slot blot, Fig. 6.20) only confirms the presence or absence of a biomolecule, which can be detected by nucleic acid probes or the antibody. Dot blot analysis of RNA is slightly trickier than dot blotting of DNA. Purified preparations of RNA denatured with glyoxal or formaldehyde produce better results. This technique is less time-consuming than the Southern or the northern blotting methods, as gel electrophoresis and complex transfer procedure to the membrane are not required.

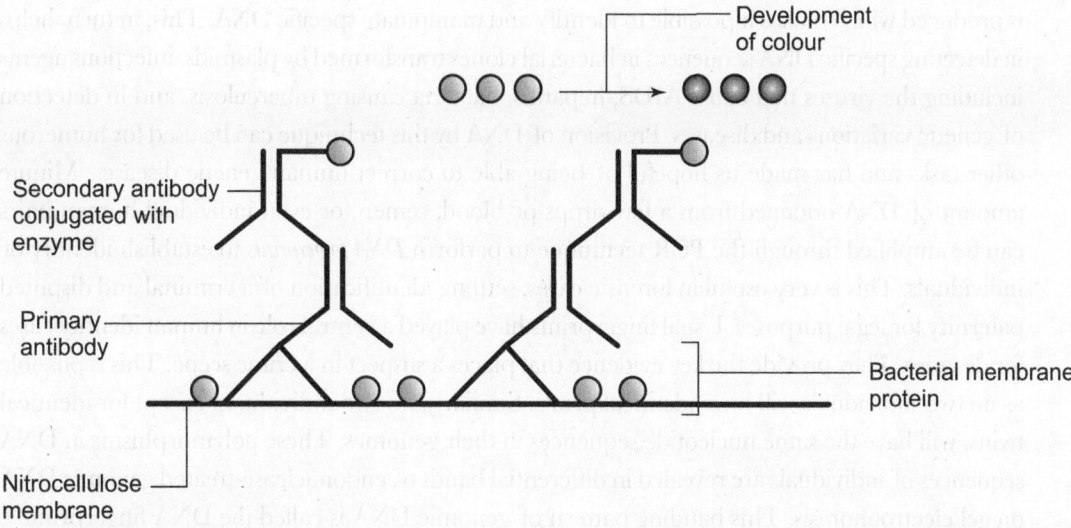

Development
of colour

Secondary antibody
conjugated with
enzyme

Primary
antibody

Bacterial membrane
protein

Nitrocellulose
membrane

Fig. 6.20   Dot blot analysis.

DNA sequencing

The genome of a eukaryote is quite large. It is a tedious task to decipher the nucleotide sequences of the genomic DNA of individual species. Clones maintaining the gene (or DNA) sequences of an organism in a library have contributed in a big way to the effort of DNA sequencing of its genome. DNA sequencing consequently (in reference to the codons) helps in derivation of amino acid sequences in proteins. DNA sequencing with the help of gene libraries also suggested the way to initiate site-directed mutagenesis to alter the properties of proteins as needed in industry.

DNA sequencing in an automatic analyser also uses a gene library. Human genome sequencing has been discussed in Chapter 3.

6.12 Polymerase Chain Reaction to Amplify Specific DNA Sequences In Vitro

At present, nearly complete or complete nucleotide sequences of many genomes, including the human genome are available in Gene Bank and other databases. Now, researchers can isolate genes or other DNA sequences of interest without using cloning vectors or host cells. *Polymerase chain reaction (PCR)* is a technique that results in the exponential amplification of a selected region of a DNA molecule in vitro in short time. The sequence can be amplified to produce a million copies in just a few hours. Knowledge of short nucleotide sequences flanking the sequence (gene) of interest is essential to initiate the process. A short (18–30 nucleotides) length of oligonucleotides complementary to the flanking sequences is synthesized to use as primers for the two strands of the DNA segment of interest.

6.12.1 Applications of PCR

PCR technique is often held as one of the most important inventions of the 20th century in molecular biology. With this technique, an enormous number of copies of a given stretch of DNA is produced which makes it possible to identify and manipulate specific DNA. This, in turn, helps in detecting specific DNA sequences in bacterial clones transformed by plasmids, infectious agents including the viruses that cause AIDS, hepatitis, bacteria causing tuberculosis, and in detection of genetic variations and diseases. Provision of DNA by this technique can be used for numerous other tasks and has made us hopeful of being able to correct human genetic diseases. Minute amount of DNA obtained from a few drops of blood, semen, or even individual human hairs can be amplified through the PCR technique to perform *DNA fingerprints* to establish identity of individuals. This is very useful in forensic cases, settling identification of a criminal and disputed paternity for legal purposes. Usual fingerprints have played a central role in human identity cases for decades. They provide the key evidence that places a suspect in a crime scene. This is possible as no two individuals will have identical prints. Similarly, no two individuals, except for identical twins, will have the same nucleotide sequences in their genomes. These polymorphisms in DNA sequences of individuals are revealed in differential bands of endonuclease-treated genomic DNA on gel electrophoresis. This banding pattern of genomic DNA is called the DNA fingerprint.

The PCR procedure was developed by Kary Mullis, a research scientist at Cetus, a biotech company in California, in 1983. Mullis received the Nobel Prize in chemistry in 1993 jointly with Michael Smith.

6.12.2 Components of PCR Reaction

The basic constituents of a PCR reaction are—(i) one or more molecules of target DNA, (ii) two synthesized oligonucleotide primers (upstream and downstream primers) in excess, (iii) all the deoxyribonucleotide triphosphates (dNTPs, where N is A/T/G/C, and (iv) thermostable DNA polymerase (*Taq* polymerase). These components are added in a small PCR tube and mixed well before placing them in a PCR machine, known as a thermocycler. Mg^{2+} in the buffer forms a soluble complex with dNTP that is essential for dNTP incorporation, especially when Taq polymerase is used.

6.12.3 Steps in PCR Reaction

The PCR reaction involves three steps and this set of three steps are repeated many times (Fig. 6.21). *In the first step*, the genomic DNA containing the target sequence to be amplified is denatured by heating to 94–95°C for about 30 seconds. Denaturation means breakage of hydrogen bonds between the nucleotides of two strands of DNA at high temperature and separation of the two strands. *In step 2*, the denatured DNA strands are allowed to anneal to an excess of the synthetic oligonucleotide primers at 50–60°C for 30 seconds. The ideal annealling temperature depends on the base composition of the primers. The oligonucleotide primers provide the free 3′-OH for covalent extension or synthesis of new strands of DNA on the template of denatured separated strands. *In step 3*, DNA polymerase precisely does this extension function or synthesis starting from the primer. This step of polymerization is usually carried out at 70–72°C for about 1.5 minutes. Starting with a single DNA molecule, the end products of the first cycle are two DNA molecules.

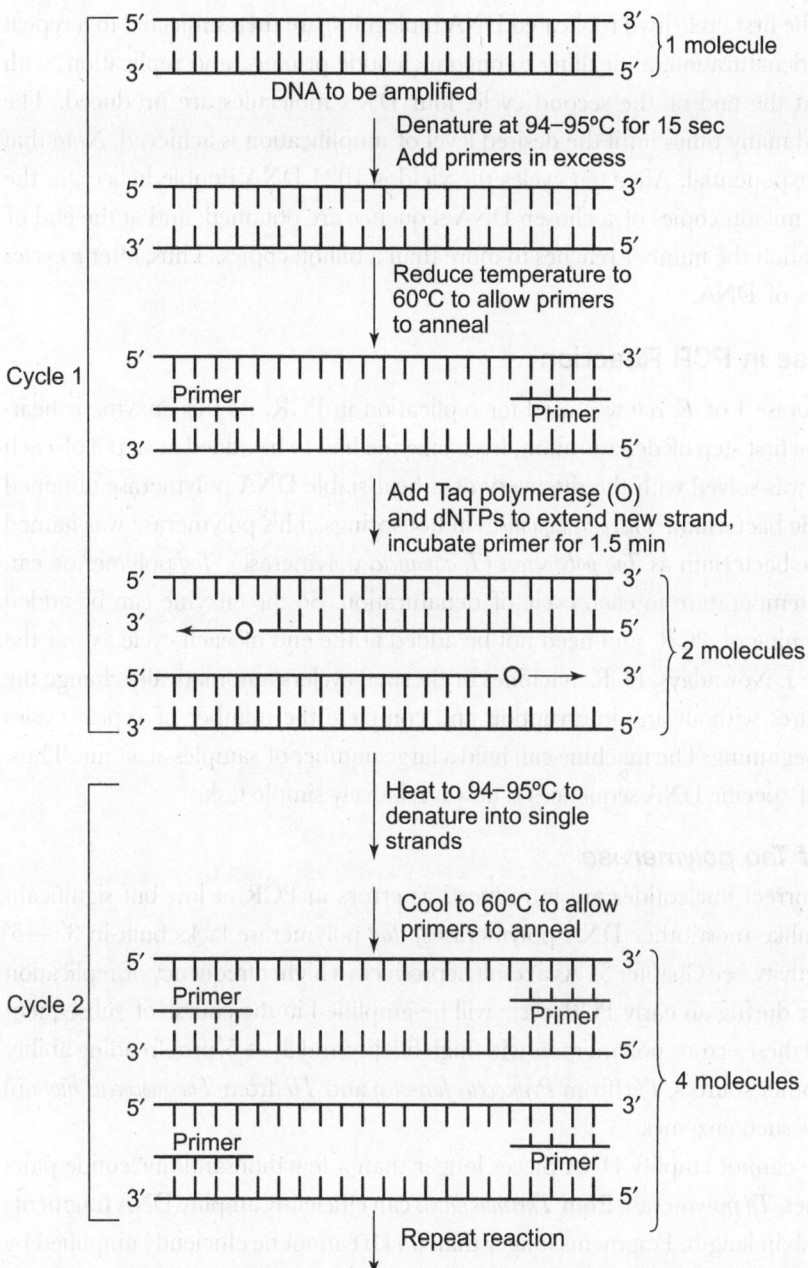

25 repeat cycles produce a million copies of a chosen DNA sequence

Fig. 6.21 The polymerase chain reaction (PCR). The DNA sample is denatured into single strands by a brief heat treatment and then cooled in the presence of an excess of oligonucleotide primers complementary to the DNA sequences flanking the desired DNA segment (gene). The DNA strand is extended from the primer with the help of a heat-resistant DNA polymerase (Taq polymerase) in the presence of four deoxyribonucleotide triphosphates (dNTPs). The heating and cooling cycle can be repeated many times, resulting in alternate DNA melting and synthesis and thus the rapid amplification of a chosen sequence.

The products of the first cycle (two replicated DNA molecules) are then subjected to a repeat of the cycle, that is, denaturation, annealling to oligonucleotide primers, and replication with DNA polymerase. At the end of the second cycle, four DNA molecules are produced. The procedure is repeated many times until the desired level of amplification is achieved. Note that the amplification is exponential. After ten cycles the yield is 1024 DNA double helices, at the end of 20 cycles one million copies of a chosen DNA sequence are obtained, and at the end of 30 cycles of amplification the number reaches to more than a billion copies. Thus, after n cycles there will be 2^n copies of DNA.

6.12.4 DNA Polymerase in PCR Reaction

Earlier, DNA polymerase I of *E. coli* was used for replication in PCR. As this enzyme is heat-inactivated during the first step of denaturation, fresh enzyme had to be added at step 3 of each cycle. This problem was solved with the discovery of a heat-stable DNA polymerase obtained from the thermophilic bacterium, *Thermus aquaticus* in hot springs. This polymerase was named after the name of the bacterium as *Taq polymerase* (*T. aquaticus* polymerase). *Taq* polymerase can withstand the 95°C temperature in each cycle of denaturation. So the enzyme can be added in excess at the beginning of PCR, and need not be added at the end of each cycle as was the case with polymerase I. Nowadays, PCR machines or thermal cyclers automatically change the sequential temperatures without any interruption and complete the number of repeat cycles programmed at the beginning. The machine can hold a large number of samples at a time. Thus, PCR amplification of specific DNA sequences is now a relatively simple task.

Disadvantages of Taq polymerase

(i) In general, incorrect nucleotides are introduced as errors in PCR at low but significant frequencies. Unlike most other DNA polymerases, *Taq* polymerase lacks built-in $3' \rightarrow 5'$ proofreading activity (see Chapter 3). As a result, it produces a higher frequency of replication errors. An error during an early PCR cycle will be amplified in the course of subsequent cycles. To avoid these errors, polymerases with high fidelity and $3' \rightarrow 5'$ proofreading ability are used from other sources. *Pfu* (from *Pyrococcus furiosus*) and *Tli* (from *Thermococcus litoralis*) are examples of such enzymes.

(ii) *Taq* polymerase cannot amplify DNA pieces longer than a few thousand nucleotide pairs efficiently. Rather, *Tfl* polymerase from *Thermus flavus* can efficiently amplify DNA fragments upto about 35 kb in length. Fragments longer than 35 kb cannot be efficiently amplified by PCR.

6.12.5 Variants of PCR

PCR is a highly versatile technique which can be modified in diverse ways to suit the need of researchers for different conditions and applications. Some of the modified PCR techniques, for example, are as follows:

(i) *Inverse PCR* is used when flanking sequences of the target DNA are not known, and primers cannot be designed, and only one internal sequence of the target DNA is known. The restriction enzyme digested DNA fragments are allowed to self-ligate to form circular DNA; the primers used correspond to a known sequence that faces outward from the target DNA and they carryout the polymerization reaction outwardly.

(ii) *Colony PCR* is used for screening recombinant DNA products in bacteria. Bacteria colonies are the starting material.

(iii) *Hot start PCR* prevents polymerase activity until the PCR reaction conditions are set at high temperature. This type of PCR reduces non-specific amplification and dimer formation between primers, and is thus more efficient. This is best suited for multiplex PCR.

(iv) *Multiplex PCR* amplifies several sequences of interest simultaneously. It saves template, time, and cost. Generally, upto eight primer pairs can be used in a standard multiplex reaction. This type of PCR is used for forensic purposes, prenatal diagnosis, and for clinical applications in which tissue/DNA samples are limited.

(v) *Nested PCR* needs two or more pairs instead of one pair of PCR primers to amplify a DNA fragment. The first pair of primers amplifies a fragment similar to a standard PCR. The next pair(s) of primers are called *nested primers* as they hybridize to sites nested within the first fragment produced by the first pair of primers. The second fragments produced are obviously shorter than the first one.

In case a wrong fragment is amplified in the first PCR cycle, the second pair of nested primers will not find their binding sites in the first product, and the reaction will stop. Thus, the nested primers increase the specificity and magnitude of PCR amplification.

(vi) *Reverse transcriptase PCR (RT - PCR)* is a technique to amplify cDNA copies of mRNA, in two steps. The first step of the reaction produces cDNA from mRNA by reverse transcriptase, and in the second step, cDNA is amplified with gene specific forward and reverse primers in a regular PCR process.

(vii) *Real-time PCR* is also known as kinetic PCR, qPCR, and QRT–PCR. It quantitates the copy number of amplifying DNA or cDNA segments in the course of the PCR reaction, that is, in 'real time', on the basis of the fluorescence emitted. This is in contrast to standard PCR where detection is possible only at the end when the entire process of amplification is completed .

6.12.6 Site-directed Mutagenesis

With the development of recombinant DNA technologies, researchers can now manipulate a gene at any level—it is now possible to alter a single nucleotide. They can induce a mutation at a nucleotide (point mutation) or at several sites of a gene to alter the function of a gene including its protein product. Such defined alteration at a site in a gene sequence of a clone in vitro is known as *site-directed mutagenesis* (or site-specific mutagenesis).

The site-specific mutagenesis is accomplished by using a mutagenic primer (Fig. 6.22) in course of in-vitro amplification of a DNA sequence by PCR. The mutagenic primer is an oligonucleotide, 12–15 nucleotides in length, which is complementary to one strand of the DNA sequence of interest for most of its length, but has one or more non-complementary or 'mismatched' bases. The mismatched bases will generate the desired mutant sequence. Another regular primer complementary to the other strand of the DNA sequence of interest is used for elongation of the strand in the usual manner. DNA polymerase (*Taq* polymerase) actually participates in elongation or adding bases for both the newly forming strands. After many cycles of amplification, the PCR products will consist mostly of the mutant DNA fragments.

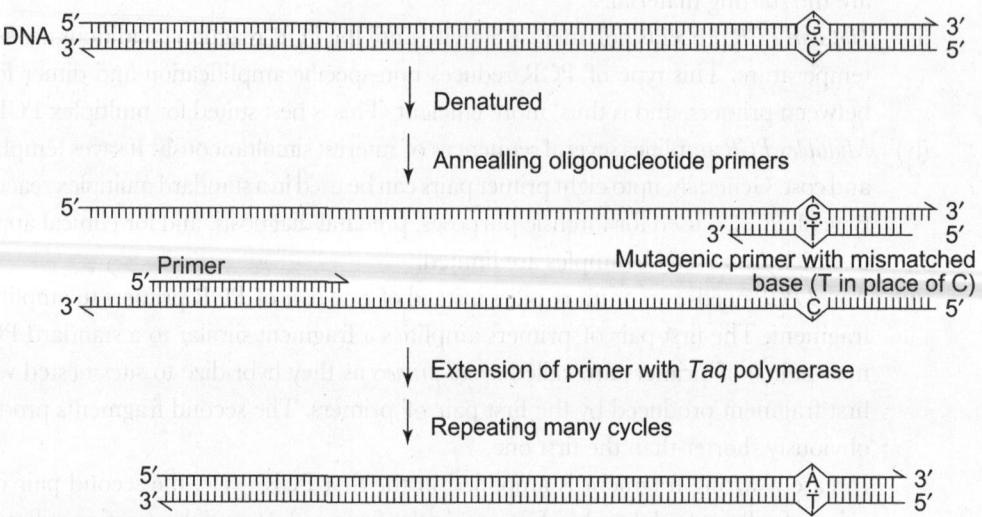

Fig. 6.22 Induction of site-specific mutagenesis with mutagenic PCR primer.

The mutated DNA is then reintroduced into the same or different species for further study. Site-directed mutations help in (i) understanding gene function, (ii) creation of new restriction enzyme cleavage sites in DNA molecules, and (iii) expression of new proteins with different properties, such as enhanced thermal tolerance, pH stability of enzymes, increased resistance to proteases, or increased catalytic efficiency of enzymes.

Thus, techniques of site-directed mutagenesis find applications in protein engineering where they are used to change the structure and activity of proteins. The oligonucleotide-directed mutagenesis is central to all the available techniques of site-directed mutagenesis.

British born Canadian biochemist Michael Smith developed oligonucleotide-directed mutagenesis in 1978, and was the co-winner of the Nobel Prize in 1993 for chemistry with Kary Mullis, who invented the PCR technology.

6.13 TRANSGENIC ORGANISMS (PLANTS AND ANIMALS)

So far, we have discussed the transformation of prokaryotes (bacteria) with foreign genes inserted in vectors using rDNA technology. When one or more, synthetic, modified, and foreign genes are introduced into eukaryotes—plants and animals—the resulting hosts are called *transgenic* organisms. In eukaryotes, 'transformation' means change towards malignancy (cancer); that is why 'transgenic' is the equivalent term used for eukaryotes to refer to the technique of 'transformation' in prokaryotes. Usually a transgenic plant or animal is a fertile organism that carries an introduced gene(s) in its germ line so that the new gene(s) becomes part of the genome of subsequent generations. Transgenic conversion of somatic (body) cells of an organism are not carried to the next generation.

Transgenic organisms are used to study the functions and regulation of genes, for mass production of desired proteins, hormones, and pharmaceuticals. They help in creating novel products. Transgenic plants and animals are changing our food habits and the world we live in. The field of transgenic engineering holds promise, and also generates debates. The mass production of rare plants with new traits, flowers with long shelf life, tomatoes, and broccoli producing pharmaceutical compounds, vaccine-producing bananas, better growth of animals, animals with novel characters, and production of industrial chemicals, have all been made possible through transgenic engineering.

Transfection of the eukaryotic organisms with external genes can be achieved by several different methods. The commonly used procedures carryout transfection with:

(a) Plasmids
(b) P element (transposon)
(c) Viruses (as in case of prokaryotes)
(d) Forced introduction of DNA sequences with the help of gene gun
(e) Electroporation
(f) Microinjection of DNA
(g) A few other techniques for direct gene transfer
(h) Transfer of a nucleus to an enucleated ovum or a stem cell

The specific discussion on transgenic plants and animals begins with transformation of yeast cells, which are simple eukaryotes.

6.13.1 Transformation of Yeast Cells

Yeast is a single cell, a simple eukaryote that shares the complex internal cell structure of plants and animals. It is an excellent organism for genetic engineering studies.

The following features of yeast allow its usage in genetic engineering studies and in production of biotechnological items.

(i)   The life cycle of yeast is short, only about 90 minutes and it is easy to grow and study and inexpensive to maintain in cultures.
(ii) It has a small and haploid genome, which is genetically well characterized. The yeast genome was the first eukaryotic genome to be completely sequenced; detailed genetic maps of *Saccharomyces cerevisiae* and *Schizosaccharomyces pombe* are available. (It is estimated that yeast shares approximately 23 per cent of its genome with that of human beings.)
(iii) Several strains of yeast have naturally occurring plasmids that can be used as part of an endogenous yeast expression vector. Very large pieces of DNA can be cloned in yeast artificial chromosomes (YACs), already discussed under vectors (Section 6.7.1).

(iv) Many yeast genes are functionally expressed in *E. coli*. Shuttle vectors are designed containing *E. coli* and *S. cerevisiae* origins of replication, selectable marker genes, and sites for inserting the desired gene. These shuttle vectors can replicate and express in both *E. coli* and eukaryotic species.

(v) A relatively high rate of recombination between homologous DNA sequences in yeast allows precise insertion of DNA sequences at specific locations within the yeast genome.

(vi) Normally, yeast secretes very few proteins. Thus, when yeast is engineered for extracellular release of a heterologous (recombinant) protein, the protein is easily purified. A number of vaccines, pharmaceuticals, and diagnostic agents produced in yeast are commercially available.

(vii) Naked DNA can be introduced in yeast with ease following some standard transformation techniques as in the case of bacteria.

Transformation techniques for yeast

Spheroplasts technique The cell wall is removed enzymatically by lyticase or zymolyase, and the resulting spheroplasts (protoplasts) are fused with polyethylene glycol (PEG) in the presence of DNA sequences and $CaCl_2$ (Ca^{2+} ions). The spheroplasts then generate new cell walls in a medium containing 3 per cent agar. Till this stage the technique is efficient, but the subsequent retrieval of transformed cells is a laborious process. This is why other techniques have been developed.

Lithium acetate treatment The cells are first suspended in 0.1 M lithium acetate, and then DNA and PEG are added. Following this the cells are subjected to a brief heat shock.

Electroporation Yeast cells are subjected to a short pulse of electric current which creates transient pores in the membrane, through which DNA enters the cell.

6.13.2 Transgenic Plants with the Help of Ti Plasmid of *Agrobacterium tumefaciens*

Agrobacterium tumefaciens is a soil bacterium and the causative agent of *crown gall* tumour of dicotyledonous plants (which have two cotyledons in the early state). Gall is a tumour that often forms at the crown—junction between the root and the stem of a plant. Usually, a wound in the plant body facilitates the infection of *A. tumefaciens*, even in parts other than the crown. The *Ti plasmid* present in *A. tumefaciens* is mainly responsible for induction of crown galls in plants.

Ti plasmid Ti plasmid is named after its *tumour-inducing capacity*. It is a large circular plasmid and remains as extrachromosomal DNA in *A. tumefaciens*. It is made up of about 200,000 nucleotide base pairs. Two components of the Ti plasmid, *T-DNA* and *vir region*, are essential for the malignant transformation of plant cells (Fig. 6.23).

T-DNA is a 23,000 nucleotide pair segment that carries 13 known genes. There are two strains of *A. tumefaciens*—nopaline and octopine, depending on the type of arginine derivative compounds induced by the strains in the host cells. These compounds provide sustenance to the specific strains of bacterium in the host cells. The Ti plasmid of the *nopaline strain* carries only one T-DNA segment and that of the *octopine type* carries two T-DNA segments. This discussion is based on the nopaline Ti plasmid.

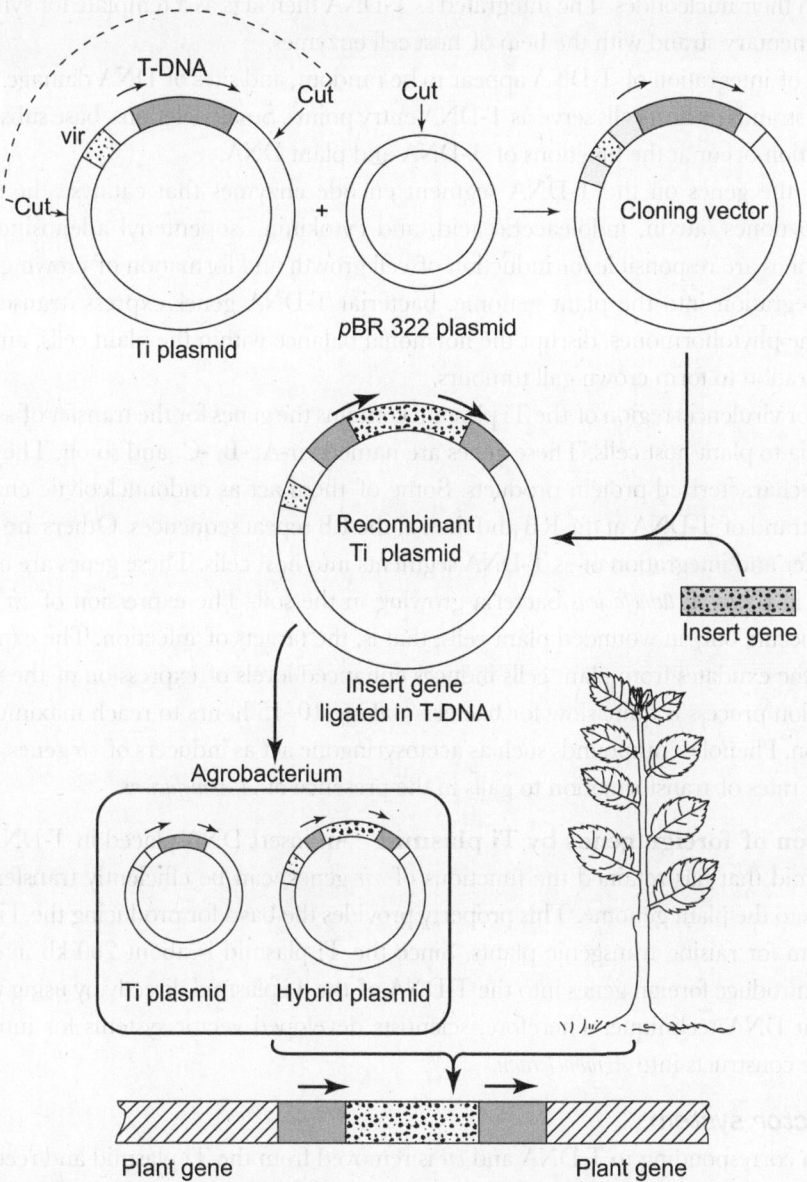

Fig. 6.23 Gene transfer in plant with the help of Ti plasmid.

The T-DNA is flanked on both the left and right side by 25-nucleotide-pair repeats, which are often called left and right border (LB and RB) sequences. Excision at these repeat sequences produces the single-stranded (ss) T-DNA for transport to the plant host cells.

T-DNA continuously produces ss DNAs which pass through the membrane and cell wall of the bacterium and through the cell membrane and the nuclear membrane of the infected plant cells. The ss T-DNA integrates to one strand of DNA at different sites of the plant genome through illegitimate recombination, that is, at sites where the host and T-DNA do not share much homology in their nucleotides. The integrated ss T-DNA then acts as a template for synthesis of the complementary strand with the help of host cell enzymes.

The sites of integration of T-DNA appear to be random, and sites of DNA damage (nicks) at one or both strands of host cells serve as T-DNA entry points. Small deletions, base substitutions, and duplication occur at the junctions of T-DNA and plant DNA.

Some of the genes on the T-DNA segment encode enzymes that catalyse the synthesis of phytohormones (auxin, indoleacetic acid, and cytokinin, isopentenyl adenosine). These phytohormones are responsible for induction of cell growth and formation of crown galls.

After integration into the plant genome, bacterial T-DNA genes express (transcribe and translate), the phytohormones, disrupt the hormonal balance within the plant cells, and induce their proliferation to form crown gall tumours.

The *vir* (for virulence) region of the Ti plasmid contains the genes for the transfer of ss T-DNA from bacteria to plant host cells. These genes are named vir-A, -B, -C, and so on. They encode several well-characterized protein products. Some of them act as endonucleolytic enzymes to cleave one strand of T-DNA at the RB and then at the LB repeat sequences. Others are involved in the transfer and integration of ss T-DNA segments into host cells. These genes are expressed at very low levels in *A. tumefaciens* bacteria growing in the soil. The expression of *vir* genes at high levels occurs only in wounded plant cells, that is, the targets of infection. The exposure of bacteria to the exudates from plant cells induces enhanced levels of expression of the *vir* genes. This induction process is quite slow for bacteria, taking 10–15 hours to reach maximum levels of expression. Phenolic compounds such as acetosyringone act as inducers of *vir* genes, and also increase the rates of transformation to galls in the presence of *A. tumefaciens*.

Introduction of foreign genes by Ti plasmid   An insert DNA placed in T-DNA within the Ti plasmid that has retained the functions of *vir* genes, can be efficiently transferred and integrated into the plant genome. This property provides the basis for producing the Ti plasmid vector system for raising transgenic plants. Since the Ti plasmid is about 200 kb in size, it is difficult to introduce foreign genes into the T-DNA of the Ti plasmid directly by using the usual recombinant DNA technique. Therefore, scientists developed vector systems for introducing foreign gene constructs into *Agrobacterium*.

Hybrid vector system

The portion corresponding to T-DNA and *vir* is removed from the Ti plasmid and recombined with the common cloning vector pBR322 of *E. coli*. Foreign DNA is inserted in the T-DNA region between the left and right border sequences through recombination between homologous base sequences. The combination of part of the Ti plasmid and pBR322 creates a co-integrative vector in *Agrobacterium* (Fig. 6.23). Presence of the co-integrative vector in *Agrobacterium* can be easily ascertained by selection for antibiotic (kanamycin) resistance marker(s) present in the pBR322 derivative.

Agrobacterium containing the hybrid vector and inserted gene is co-cultured with plant cells or protoplasts or plant tissues in vitro.

On infection of plant cells, the inserted foreign gene will be transferred and integrated in the plant genome in the same way as T-DNA. (In the process of insertion of a foreign gene at the T-DNA site, T-DNA loses its ability to induce galls.) The plant cells containing foreign genes become transgenic cells which can be grown in cell culture to a callus and then into transgenic plants.

Extensive research has been carried out to find ways of increasing the efficiency of Ti plasmid-mediated generation of transgenic plants.

6.13.3 Examples of Transgenic Plants and their Utility

The powerful tool of Ti plasmid-mediated gene transfer primarily in dicotyledonous plants allowed researchers to produce transgenic plants with features never heard of before. The discussion in the following text starts with the often quoted example of *Bt* crops, which are made by transformation of plants with an in-built insecticide gene from bacteria.

Bacillus thuringienesis-based biopesticides

Bacillus thuringienesis, a Gram-positive soil bacterium, harbours a *cry* gene which encodes a protein that is toxic to many types of insects. This endotoxin is named *Bt* toxin (after the bacterium) or Cry protein. Transfer of *cry* gene to different plant species gives them an endogenous insecticidal property. *Bt* toxin disrupts the gut epithelium of insects and kills them. The toxin is active against lepidopteran insects that normally damage corn and cotton (Bollworm larva), dipteran, and coleopteran insects.

The plants that carry *cry* gene for *Bt* toxin are naturally resistant to attacks by pest insects. Farmers who cultivate these plants can avoid using chemical pesticides which usually cause land and water pollution and also kill the non-targeted insects. Big multinational companies are involved with this rDNA or transgenic technology to produce multiple food and agri crops.

Plants and animals that have been altered by the introduction of foreign genes are called GMOs—*genetically modified organisms*. The development and marketing of GMOs has stirred up controversy worldwide. Concern about marketing *Bt* crops in India has been discussed in Chapter 15. Several issues need to be settled before introducing all kinds of GMOs for public consumption: (i) the conflicting interest of small farmers—they need to buy GM seeds to sow every year from big business houses as the seeds from cultivated plants do not produce plants with vigour; (ii) concerns about the safety of consuming GM food; (iii) concern that *Bt* crop might kill non-pest species of insects such as butterflies and honey bees, which normally act as pollinators; and (iv) concerns about field reports that *Bt* crops may be infested by new varieties of pests and compel the use of chemical pesticides.

Herbicide resistance in agronomic crops

The most widely used transgenes are those that produce herbicide resistance in agronomic crops. Nearly 10 per cent of the global crop produce is lost through weed infestation every year, despite the expenditure of more than $10 billion on 100 different chemical herbicides. Herbicides are used primarily to kill undesirable plants or weeds that compete with desirable crop plants for space, water, and nutrients. Herbicides typically disable target enzymes in metabolic pathways unique to plants, for example, those involved in photosynthesis and biosynthesis of essential amino acids. It is difficult to develop herbicides which selectively affect weeds without causing harm to crop plants. An alternative approach would be to modify crop plants to become resistant to broad-spectrum herbicides.

Genetic engineers developed resistant plants for bromoxynil (3,5-dibromo-4-hydroxybenzonitrile), a herbicide that acts by inhibiting photosynthesis. They transferred a gene that encodes the enzyme *nitrilase,* from the soil bacterium *Klebsiella ozaenae.* Nitrilase enzyme inactivates bromoxynil by removing nitrogen before the herbicide can act. The removed nitrogen may be utilized by the plant for its own growth. The transgenic tobacco plants expressing nitrilase gene are resistant to the toxicity of bromoxynil. Thus, weeds in a field of transgenic tobacco plants can be killed by spraying bromoxynil without affecting the tobacco plants.

Transfer of animal gene to plant

Flashing or glowing of fireflies is known as bioluminescence. This is possible when luciferin, present in the specialized cells of a firefly, comes in contact with the enzyme luciferase. Researchers introduced the luciferase gene of firefly into the cells of tobacco plant with the help of recombinant Ti plasmid. Luciferase is produced in the tobacco plant containing the transgenic cells and can glow when sprayed with water containing luciferin. This kind of plant has tremendous horticultural value.

Only a few examples of transgenic plants are mentioned here. Other products from transgenic plants have been enlisted at the end where broader outline of applications of genetic engineering will be discussed.

6.13.4 P Element-based Transformation in *Drosophila melanogaster*

For nearly 40 years after the discovery of transformation of bacteria with isolated DNA by Oswald Avery and colleagues, it was not possible to carryout similar transformations in eukaryotic organisms. G.M. Rubin and A.C. Spradling published a paper, entitled 'Genetic transformation of *Drosophila* with transposable element vectors' in *Science* in 1982 (Vol. 218, p. 348) showing the possibility of transformation in eukaryotes. What is a *transposable element vector?* It is simply called a *transposon* and more specifically a *P element* in *Drosophila melanogaster.*

Transposon was originally discovered by Barbara McClintock in maize and she received the Nobel Prize in 1983, 35 years after her first publication on transposable elements.

Transposons These are DNA sequences that can move from one position to another in the genome. Transposons are present in organisms from bacteria to human, and constitute an appreciable fraction of the genome. They clearly have roles in shaping the structure of chromosomes and in modulating the expressions of genes. Transposons are flanked at both ends with identical or nearly identical nucleotide sequences. These terminal sequences are always in

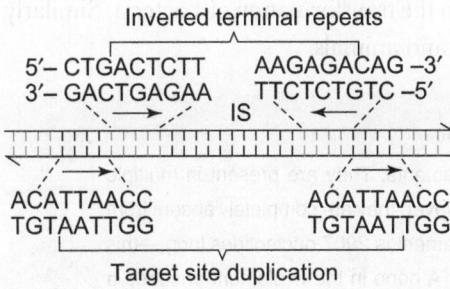

Inverted terminal repeats

5'– CTGACTCTT   AAGAGACAG –3'
3'– GACTGAGAA   TTCTCTGTC –5'

IS

ACATTAACC   ACATTAACC
TGTAATTGG   TGTAATTGG

Target site duplication

Fig. 6.24 Transposon flanked at both ends with identical or nearly identical terminal inverted repeats and target site duplication. IS is insertion sequence (transposon).

inverted orientation with respect to each other and they are called *inverted terminal repeats* (ITRs; Fig. 6.24). The excision of a transposon from its position in a chromosome is made at these inverted terminal repeats and it is then inserted into another position of the same or a different chromosome. The process of excision and insertion are catalysed by an enzyme called *transposase*, which is usually encoded by the transposon itself. This mechanism is referred to as *cut-and-paste transposition* because the transposon element is physically cut out from one site in a chromosome and pasted into a new site; the transposons are called *cut-and-paste transposons*. P element of *Drosophila* belongs to this type.

(The second category of transposons is referred to as *replicative transposon*, because the transposon is replicated at the site and one copy of it is inserted at the new site, while one copy remains at the original site.)

When a transposon inserts into a site of a chromosome, a duplication of a part of the DNA sequence of the chromosome occurs at the site of the insertion following staggered cleavage of the double-stranded chromosomal DNA (Fig. 6.25). There are two such duplication copies, one on each side of the transposon (see also Fig. 6.24). These short (2 to 13 nucleotide pairs) directly repeated sequences are called *target site duplications*.

Two strands of target DNA cleaved (arrows).

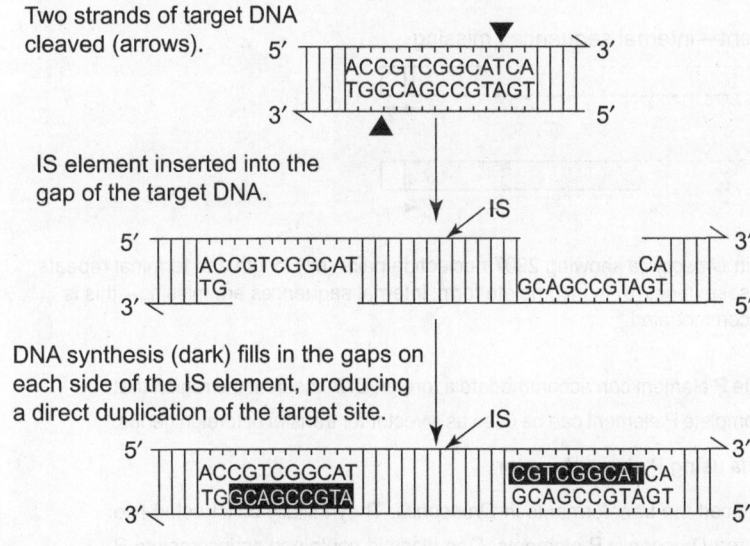

5' ACCGTCGGCATCA 3'
3' TGGCAGCCGTAGT 5'

IS element inserted into the gap of the target DNA.

IS

5' ACCGTCGGCAT ... CA 3'
3' TG ... GCAGCCGTAGT 5'

DNA synthesis (dark) fills in the gaps on each side of the IS element, producing a direct duplication of the target site.

IS

5' ACCGTCGGCAT ... CGTCGGCATCA 3'
3' TGGCAGCCGTA ... GCAGCCGTAGT 5'

Fig. 6.25 A staggered cleavage of the chromosomal DNA allows insertion of a transposon, then target site duplications take place.

6.13.5 Transfection of Plants and Animals with Viral Vectors

Viral infection introduces new genetic material to be expressed in host cells; this can be considered as a natural example of genetic engineering. The use of bacteriophages (virus) for transfer of rDNA in bacteria has been discussed in connection with the transformation of bacteria. Similarly, viruses play important roles as vectors for both plants and animals.

Exhibit 6.B   P Element

Drosophila P strain-specific transposons are known as P elements. They are present in multiple copies and at different locations in the genome of P strains; however, they are completely absent from the genomes of M strains. They vary in size. The largest P element is 2907 nucleotides long;—this includes the terminal inverted repeats of 31 nucleotide pairs. A gene in the P element encodes a transposase which binds near the ends of a P element and moves the P element to a new location in the genome to paste. When some nucleotide sequences are missing, the P element is called an incomplete P element; that is unable to produce the transposase enzyme (Fig. 6.B). However, the incomplete P element possesses the terminal and subterminal sequences at the two ends that can bind to the transposase synthesized by a complete P element in some other location of the genome. Thus the incomplete P element can also be mobilized by transposase, supplied by a complete P element.

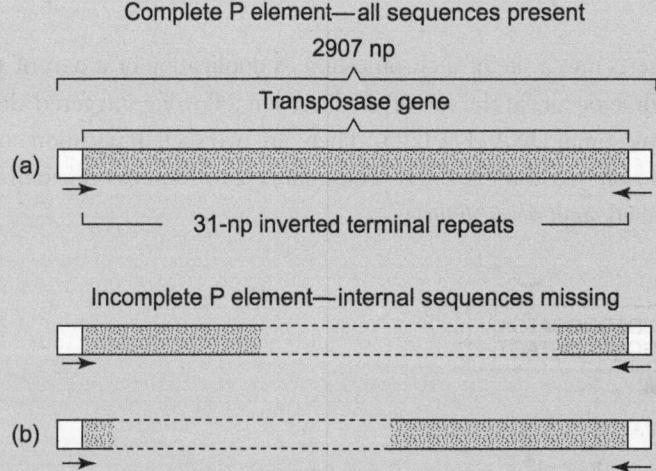

Complete P element—all sequences present

2907 np

Transposase gene

(a)

31-np inverted terminal repeats

Incomplete P element—internal sequences missing

(b)

Fig. 6.B   (a) Structure of P elements in *Drosophila* showing 2907 nucleotide pairs (np) length and terminal repeats in complete form representing transposase gene. (b) In incomplete form, internal sequences are missing—this is where a foreign DNA insert may be accommodated.

At the same time the incomplete P element can accommodate a foreign DNA insert at the region that has been deleted. Thus an incomplete P element can be used as a vector for transfer of foreign genes.

Transformation of *Drosophila* using P element vector

Rubin and Spradling first produced the transformants in *Drosophila*. They initially constructed two bacterial plasmids that harboured *Drosophila* P elements. One plasmid contained an incomplete P element into which a gene for wild-type red eye colour had been inserted. The other plasmid contained a complete P element capable of producing the P transposase in vivo, to compensate for the missing function of the incomplete P element. The mixture of the two plasmids was injected into *Drosophila* embryos that were homozygous for a recessive mutation of the eye colour gene [brown (rosy) eyes].

Rubin and Spradling hoped that transposase produced by the complete P element would catalyse the transfer of the incomplete P element with wild eye colour gene from its plasmid into the chromosomes of the injected embryos of the fly. (The complete P element, here, could not transpose itself as it had mutated ITRs). When the embryos matured, they were mated with flies homozygous for mutated eye colour and looked for progeny that had wild type eye colour. Indeed, some of the progeny had red wild type eyes, indicating that the wild type gene carried by the incomplete P element had been successfully incorporated into the genomes of some of the injected embryos. In effect, this event of transformation corrected a mutated gene in a eukaryotic organism by inserting a copy of the wild type gene with the help of an incomplete P element, a transformation vector.

The transformation technique of Rubin and Spradling has been successfully used for the transfer of genes from diverse organisms—ranging from bacteria to humans into *Drosophila* chromosomes. Similar techniques are in the process of development for the genetic transformation of other organisms including humans.

Since then, the discovery of other transposons such as *Hermes*, *Hobor' Minos*, *MosI*, and piggyBac (TTAA-specific element, discovered from insects of the order, *Lepidoptera*) and more advanced DNA delivery devices like microinjection, electroporation, and particle bombardment have facilitated the genetic transformation of various insects.

Additional genetic material incorporated in the genome of a plant or an animal virus is likely to replicate and express in the plant or animal cells. Different viral vectors have been engineered for their specific applications. In the course of the development of viral vectors, the viral genomes (especially when pathogenic) are also modified to curb their virulence so that they cannot adversely affect the physiology of the host cells. Most viral vectors are engineered to infect a wide range of cell types. However, sometimes a receptor on the target cell for virus is genetically modified to direct the virus to a specific cell type.

Essential points for construction of viral vectors Modification of viral genomes is a prerequisite for construction of viral vectors and needs to take into account the following points:

(a) There is a strict packaging limitation on the amount of nucleic acid (DNA/RNA) that can be accommodated in virus particles.

(b) Hence, non-essential genes including pathogenicity-conferring genes are deleted, to make room for the accommodation of exogenous DNA.

(c) There should not be any loss of infectivity of viruses, otherwise viral vectors would be incapable of transmitting the genes to the target cells.

Examples of viral vectors

Plant viral vectors can be DNA as well RNA viruses. Cauliflower mosaic virus (CaMV) and Gemini virus (GV) are two groups of *DNA viruses*. CaMV is the first plant virus to be discovered that contains DNA and is the first plant viral genome to be sequenced. It is also the first plant virus to be manipulated by rDNA technology. It is spherical, isometric, and ~50 nm in diameter, and harbours a ds circular DNA, 8,024 bp long. Virion DNA or cloned CaMV DNA is infectious when simply rubbed on the surface of susceptible leaves. Gemini virus is characterized by its two partially fused icosahedral capsids and has a comparatively smaller genome.

There are two basic types of *ss RNA viruses*–monopartite and multipartite. The monopartite viruses have an undivided genome of fairly large size. Tobacco mosaic virus (TMV) and potato virus X (PVX) are two prominent examples of monopartite viruses. The multipartite viruses, as the name suggests, have their genome divided among small RNAs.

Besides the usual expression vectors in plants, certain plant viruses have been designed to present short antigenic peptides (epitopes) on their surface. The epitope display systems on alfalfa mosaic virus, cowpea mosaic virus, PVX, and tomato bushy stunt virus have become potential sources of vaccines, particularly against animal viruses.

Animal viral vectors are used for gene transfer to cultured cells as well as to living animals. Various types of viral vectors have been devised for use in human gene therapy, such as retrovirus, adenovirus, herpes virus, and adeno-associated virus (AAV). Transgenes (foreign genes) are inserted into the intact genome of viral vectors or by replacement of one or more viral genes, through the process of ligation at unique restriction sites or through homologous recombination. The choice of a wide variety of animal virus vectors provides the opportunity to express a foreign gene in different cell types at different levels and with different consequences to the host cells. Small DNA viruses may have limited capacity for foreign DNA due to packaging constraints of the icosahedral virus capsid. On the other hand, rod-shaped viruses such as the baculoviruses, have no such size constraint as the capsid is formed around the whole genome.

None of the animal viruses is an ideal vector for gene transfer due to certain associated disadvantages. Attempts have been made to develop several hybrid vectors by combining the favourable features of two or more viruses.

Epstein-Barr virus and Herpes virus as vector

The herpes simplex viruses (HSVs) are large ds DNA viruses that include Epstein–Barr virus (EBV). EBV has a large ds DNA genome ~170 kb in length and mostly infects primate and canine (dog family) cells. It is naturally lymphotrophic, infecting B lymphocytes in human beings.

HSV has a linear ds DNA molecule, 152 kbp in length. By deleting non-essential genes, ~40–50 kbp of foreign DNA can be accommodated within the virus. HSV vectors are predominantly suitable for gene therapy in the nervous system as the virus efficiently infects neuronal cells.

Vaccinia virus and other pox viruses as vectors

Vaccinia virus belongs to the family *Poxviridae*, and has a complex structure and a large linear ds DNA genome ~300 kbp in size. It can accommodate a large piece of foreign DNA. This virus encodes its own DNA replication, transcription, and packaging machinery (enzymes), which is a unique feature as most DNA viruses use the machinery of the host cell nucleus. Furthermore, replication takes place in the cytoplasm of the infected cell rather than in the nucleus unlike other DNA viruses. The extraordinary replication strategy and large size of the *vaccinia* genome make the design and construction of expression vectors more complex as compared to other viruses.

Simian virus (SV40) as vector

The DNA tumour virus SV40 (simian vacuolating virus 40) was first isolated from monkeys and was the first animal virus to be molecularly characterized in detail. It was also the first animal virus

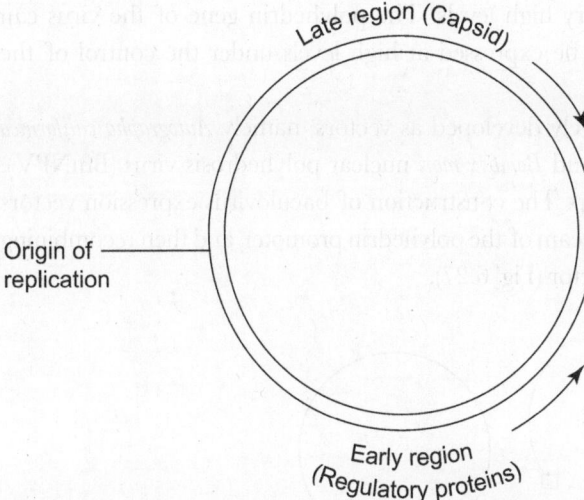

Late region (Capsid)

Origin of replication

Early region (Regulatory proteins)

Fig. 6.26 SV40 genome consists of a circular ds DNA of ~5 kbp, with two regions facing in opposite directions. Transcripts from the early region produce regulatory proteins and transcripts from the late region produce components of the viral capsid during maturation or formation of viral particles. The two regions are under two separate promoters. Foreign genes can be inserted in either of the regions.

to be exploited as a vector for higher animals, such as mouse, rabbit, and hamster cells. SV40 has a small icosahedral capsid and a circular ds DNA genome of ~5 kbp. The circular genome has two transcripts, known as the early and late regions, under two separate promoters. The regulatory proteins are produced by the early region, whereas the late region produces components of the viral capsid. The small size of the genome has made in vitro manipulation easy. Either the early region or the late region can be replaced with foreign insert DNA. The foreign gene is inserted into the genome of animal cells by transduction (Fig. 6.26).

Normally, only about 2.5 kbp of foreign DNA could be incorporated in the engineered SV40 genome. To overcome this size constraint, shuttle vectors containing a small SV40 DNA fragment that included the origin of replication, in an *E. coli* plasmid were developed. Such hybrid vectors can accommodate a large size of foreign DNA in the plasmid part and can be easily grown in mammalian (monkey) cell lines to transcribe the product of the foreign DNA at a high rate. The SV40 origin of replication actually directs the synthesis of a high copy number of DNA. The product of foreign DNA from a cell line is often used for commercial purposes.

Similar vectors which have been designed by incorporating the murine (mouse) polyomavirus origin, are functional in mouse cells and can be used to produce high levels of recombinant proteins.

Invitrogen currently markets a series of more versatile vectors, that have both the SV40 and the murine polyomavirus origins, and allow high level of protein expression in cells that are permissive for either virus.

Baculoviruses as vector

Baculoviruses are ds DNA viruses with rod-shaped capsids. They infect diverse species of insects and are used as biocontrol agents. Many of the viruses of the family *Baculoviridae* are produced by the midgut cells and other cells of infected insects in two forms—(i) viruses embedded within the polyhedral crystals known as polyhedral inclusion bodies (PIBs) and (ii) budded viruses (BVs) responsible for systemic infection of the host. PIBs are proteinaceous particles in which viruses are embedded for survival in harsh environmental conditions. The viruses can be released from PIBs by treatment with alkaline solutions similar to those found in the midgut of the insect host. A single type of protein, called *polyhedrin* is the main constituent of the proteinaceous embedding material of PIBs, and is expressed at very high levels. The polyhedrin gene of the virus can be replaced with foreign DNA that can be expressed at high levels under the control of the endogenous polyhedrin promoter.

Two baculoviruses have been extensively developed as vectors, namely *Autographa californica* multiple polyhedrosis virus (AcMNPV) and *Bombyx mori* nuclear polyhedrosis virus (BmNPV). The genome size of AcMNPV is 133 kbp. The construction of baculovirus expression vectors involves insertion of the transgene downstream of the polyhedrin promoter, and then recombining the engineered portion to the transfer vector (Fig. 6.27).

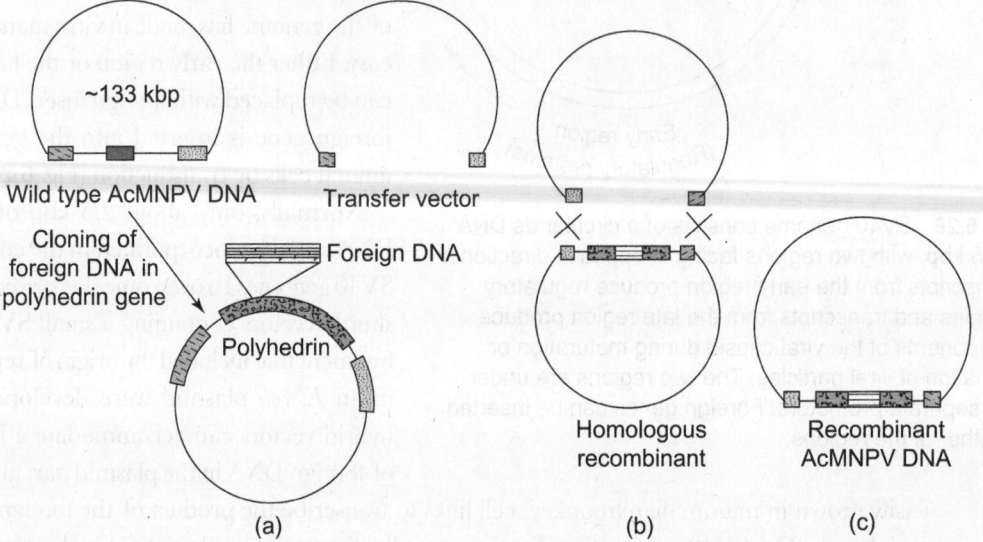

Fig. 6.27 Procedure for the construction of recombinant baculovirus expression vector. (a) Foreign gene is inserted next to the promoter of polyhedrin in the multiple cloning site. Transfer vector (on right hand top) with a large deletion to accommodate a foreign gene and the adjoining promoter. (b) Recombination occurs at homologous sequences of cloning vector and transfer vector. (c) Recombinant baculovirus to transfect insect cells.

Recently transduction of mammalian cell lines with baculoviruses has been reported. This encourages scientists to try and develop recombinant baculoviruses as vectors for gene therapy.

6.13.6 Forced Introduction of DNA Sequences using Gene Gun

Scientists found it difficult to induce transformation with *Agrobacterium tumefaciens* in monocot plants and conifers. They were compelled to devise other methods. Bombardment of microprojectiles coated with DNA was then developed to introduce transgenes into plant cell suspensions, callus cultures, meristematic tissues, immature embryos, coleoptiles, and pollens of a wide range of plants including dicots, monocots, and conifers. This method also enables the delivery of genes into organelles like chloroplasts and mitochondria. This is not possible with *A. tumefaciens* as it targets the transgenic DNA only into the nucleus of plant cell.

Transformed cells are grown into a callus, leading to the development of plants in culture.

Exhibit 6.C   Gene Gun

Small high-density microparticles (microprojectiles), made up of gold or tungsten, are initially coated with the DNA to be transferred and placed in the central hole of a plastic cartridge or macrocarrier. Helium pressure drives the macrocarrier alongwith the DNA-coated microparticles with high velocity towards the bottom of the gene gun (Fig. 6.C.1). A metallic plate with a small opening at the centre, known as stopping screen stops the moving macrocarrier, but the DNA-coated gold particles are released from the carrier and travel through the central hole of the stopping screen with kinetic energy sufficient to penetrate the target cells or tissue kept at the bottom. Thus, the DNA on the microparticles enters the target cells and is released in the cytoplasm or nucleus and subsequently integrates in their genome.

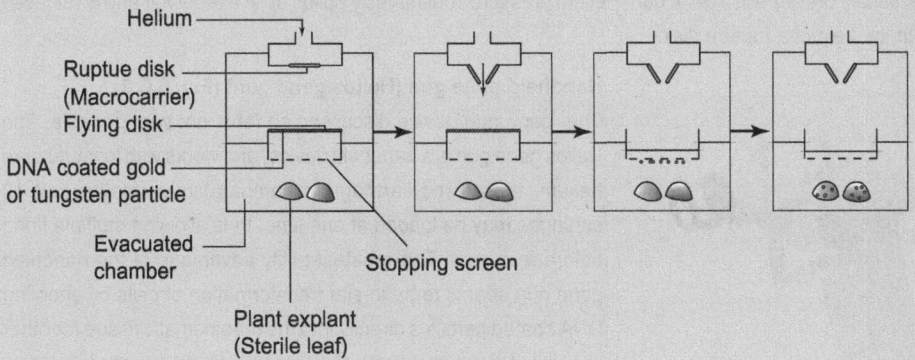

Fig. 6.C.1   Schematic representation of a gene gun.

Microparticles

Microparticles are spherical in shape, ranging in diameter from 0.5–2 µm (about the size of some bacterial cells). As a rule of the thumb, the gold or tungsten particles should be approximately one-tenth the diameter of the target cell. These microparticles are small enough to enter plant cells without causing too much damage (can be repaired later), yet large enough to have the mass and kinetic energy to penetrate the cell wall and membrane and to carry an appropriate amount of DNA. Gold microparticles offer certain advantages in comparison to tungsten microparticles. Gold microparticles are biologically inert and do not produce cytotoxic effects; furthermore, uniform particles can be made out of gold as it is more malleable.

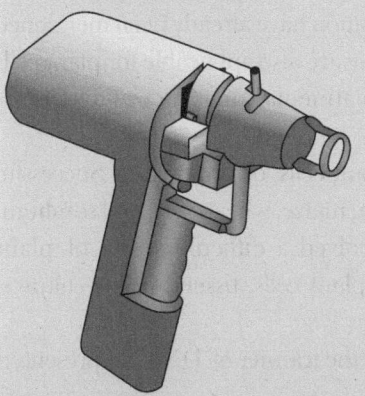

Plastic rupture disk

Some types of gene guns include a plastic rupture disk, fitted in a disk holder (Fig. 6.C.2). This rupture disk bursts after sufficient high pressure (~1000 psi) of helium gas is generated, so that the macrocarrier loaded with DNA coated microparticles can be driven at the required velocity.

Vacuum pump

The lower chamber of the gene gun, where the target cells are placed, is partially evacuated with a vacuum pump. The DNA coated microparticles are accelerated after their release from the macrocarrier

Fig.6.C.2   A handheld gene gun.

under partial vaccum due to reduction in drag through the gas phase. At the same time the partial vacuum reduces the damaging shock wave of helium pressure.

Development of gene gun design

The gene gun technique is also referred to as the 'biolistic' (a combination of biological and ballistics – bombardment of microprojectiles), 'particle gun', 'particle accelerator', 'particle bombardment' or 'bioblaster' technique. Since its first development in 1987, sophistication and much more developments have been brought into the gene gun, the basics remaining same. Initially the propelling force for the macrocarrier used to be generated by chemical explosion with gun powder, and later by discharge of compressed air. Nowadays, helium is used. Helium is a light gas and expands faster to create the necessary pressure in a gene gun. Helium pressure continuously builds up in a reservoir and is released with the help of a rupture disk.

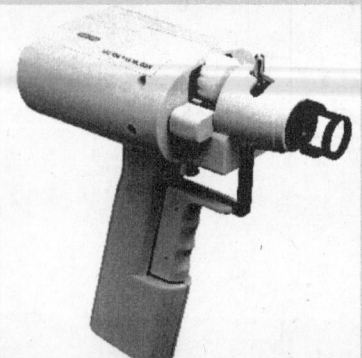

Fig. 6.C.3 Helios Gene Gun (Biorad). Also see colour plate 2.

Handheld gene gun (Helios gene gun) (Fig. 6.C.3)

The gene gun device discussed so far is not very portable. The Helios gene gun is a handheld device, and works with low-pressure helium; the plastic cartridge is comparatively smaller and 12 cartidges may be loaded at one time, thus allowing multiple firing before reloading. The greatest utility advantage of the handheld gene gun seems to be in-situ transformation of cells by shooting DNA coated particles directly into the meristematic tissue (or other tissues) of plants growing in experimental fields or in greenhouses.

Transformation of *Drosophila* using P element vector

Rubin and Spradling first produced the transformants in *Drosophila*. They initially constructed two bacterial plasmids that harboured *Drosophila* P elements. One plasmid contained an incomplete P element into which a gene for wild-type red eye colour had been inserted. The other plasmid contained a complete P element capable of producing the P transposase in vivo, to compensate for the missing function of the incomplete P element. The mixture of the two plasmids was injected into *Drosophila* embryos that were homozygous for a recessive mutation of the eye colour gene [brown (rosy) eyes].

6.13.7 Plant Transformation by Electroporation

The principle and technique for transformation by electroporation have already been mentioned in connection with the introduction of rDNA in bacteria; these are also applicable for plant cells (Fig. 6.28). However, plant cells may require specific treatments, such as pre-and post-electroporation incubation in high osmotic buffers.

Electroporation is suitable for both monocot and dicot plant cells, or protoplasts. Successful gene transfer by electroporation is possible in tobacco, petunia, maize, wheat, rice, and sorghum. Initially, protoplasts were used for transformation and involved a difficult course of plant regeneration through calli. Now it is possible to use intact plant cells, tissues, callus cultures, immature embryos, and inflorescence.

Other physical methods comparable to electroporation for the transfer of DNA are presented here.

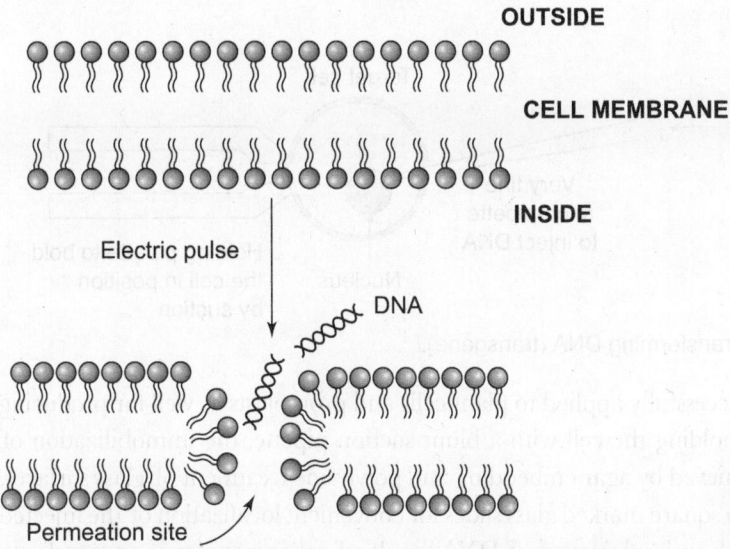

OUTSIDE

CELL MEMBRANE

INSIDE

Electric pulse

DNA

Permeation site
(Hole)

Fig. 6.28 Principle of electroporation and entry of DNA through permeation hole.

6.13.8 Laser Microbeam Irradiation and DNA Uptake

A high-energy laser beam can be finely focussed through a microscope phase contrast objective on target tissue which is less than 1 μm in diameter. It creates perforations in the plant cell wall and membrane, which allow DNA uptake through passive diffusion. The perforations are self-healing. This technique can be used for genetic transformation of plant tissue, individual cells, protoplasts, and organelles, particularly chloroplasts.

Sonoporation is ultrasonic frequency mediated acoustic cavitation or disruption of the cell membrane to allow DNA to move into cells. In *Magnetofection*, DNA-coated magnetic particles are brought into contact with cells in monolayer culture by placing a magnet underneath the tissue culture dish.

6.13.9 Microinjection of DNA

The transforming DNA (transgene) is directly injected into the nucleus or cytoplasm of a target cell with the help of a glass capillary needle (micropipette). The tip of the capillary needle ranges from 0.5 to 10 micrometre in diameter. Microinjection is performed under a phase contrast microscope, applying gentle air pressure to the capillary through a manual or electronic air pulse system of a micromanipulator. The recipient target cell or plant cell protoplast is immobilized on a solid support such as a cover slip or a slide during micromanipulation by the suction created by a blunt holding pipette (Fig. 6.29).

The DNA microinjection technique was originally developed for animal cells. The DNA is delivered to the male pronucleus just before fertilization of an ovum. Then the transformed ovum is placed in the uterus of a surrogate mouse mother to produce transgenic offspring.

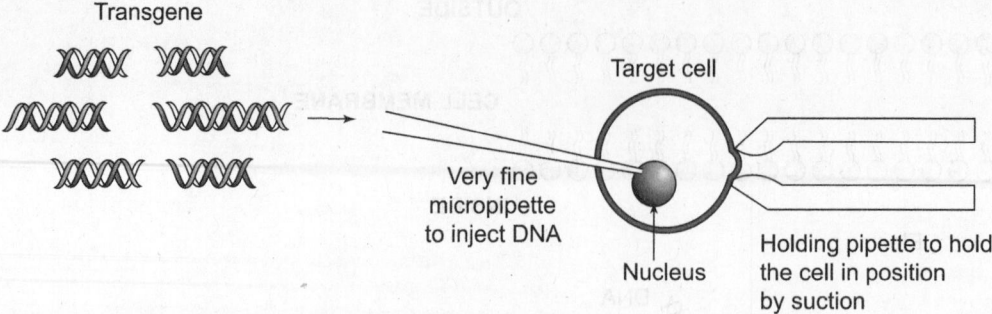

Transgene

Target cell

Very fine
micropipette
to inject DNA

Nucleus

Holding pipette to hold
the cell in position
by suction

Fig. 6.29   Microinjection of transforming DNA (transgene).

Microinjection is now successfully applied to plant cells and protoplasts as well for producing transgenic plants. Besides holding the cell with a blunt suction pipette, the immobilization of the recipient cells is also achieved by agar embedding and polylysine treatment of glass surfaces. Recipient cells are grown on square marked glass slides for convenient localization of the injected cells. The cells that survive the microinjection of DNA develop further into clones on transfer to optimal cultures, and give rise to transgenic cell lines and plants.

6.13.10 Other Techniques for Direct Gene Transfer in Plant Cells

Silicon carbide fibre-mediated transfer

This is a very simple technique and does not involve any sophisticated equipment. Silicon carbide forms microscopically small needles or fibres (0.3 – 0.6 µm diameter and 10–100 µm length), which are also called *silicon whiskers*. The mixture of silicon carbide fibres, plant tissues or cells, and transforming DNA, in a suitable buffer, is vortexed in a microfuge tube. The silicon needles perforate plant cell walls and plasmalemma by forceful intercellular collisions during vortexing. The DNA adsorbed on the cells finds entry into the cell by diffusion through the perforations. Thus, the silicon carbide fibres act as microinjection needles facilitating DNA delivery inside the cells.

At present, this technique is used to transform cells in culture, embryos, and embryo-derived calli of only a few plant species. It is a promising technique for wide use.

Lipofection

Liposomes are closed, self-sealing, solvent filled vesicles bound by a lipid bilayer similar to the cell membrane (Fig. 6.30a). A wide variety of molecules including DNA can be encapsulated within their aqueous interior and delivered to cells via endocytosis or membrane–membrane fusion (Fig. 6.30b). The use of liposomes for transfection of cells with DNA is called *lipofection*. This technique is mostly used in genetic engineering to deliver foreign DNA molecules into plant protoplasts.

Brief sonication of phospholipids (phosphatidyl serine or phosphatidyl choline or phosphatidyl glycerol) and an aqueous buffer containing DNA to be packaged in a large volume of organic solvent (diethyl ether or isopropyl ether), in a screw capped glass tube, produces a stable water-in-oil emulsion. The organic solvent is then evaporated in a rotary evaporator till a viscous gel is obtained. The gel is vortexed briefly in vacuum to produce DNA-encapsulated liposomes which are stable for months in an inert atmosphere at 4°C. Incubation of liposomes with plant protoplasts for about 30 minutes is necessary for transfer of DNA into the cells. Use of PEG or dimethyl sulphoxide (DMSO) during incubation facilitates DNA delivery. At the end of incubation, the cells are washed with buffer and plated in growth medium.

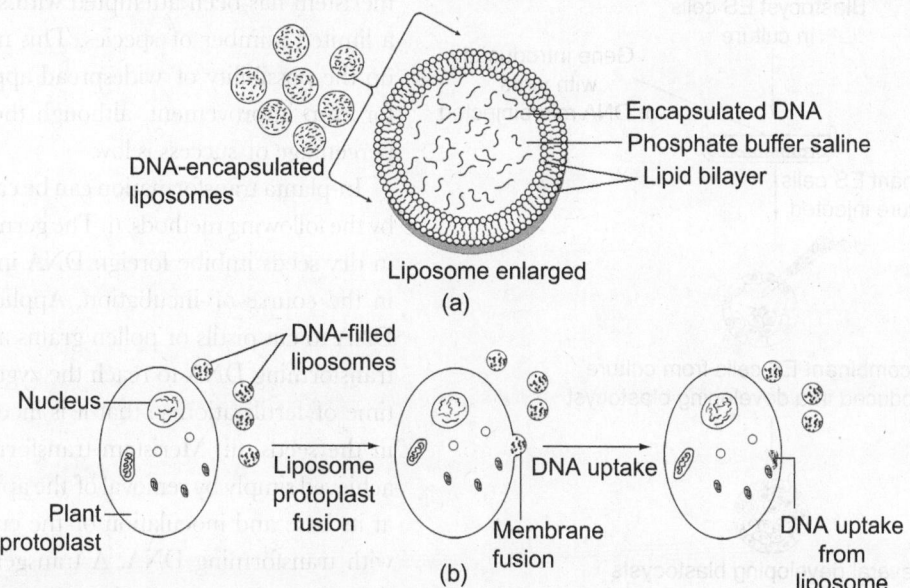

Fig. 6.30 (a) Liposome enclosed by lipid bilayer, similar to cell membrane. (b) Fusion of DNA-filled liposomes with protoplast and delivery of DNA.

DNA–calcium phosphate co-precipitation

DNA is mixed with CaCl$_2$ solution and isotonic phosphate buffer to form a DNA–calcium phosphate co-precipitate (also called Ca-DNA co-precipitate). The co-precipitate enters the cells or protoplasts by endocytosis and a part of it escapes from endosomes or lysosomes to the cytoplasm and finally reaches the nucleus. Several hours of incubation allows entry of the Ca-DNA co-precipitate in a calcium-requiring process. The protoplasts are washed and grown in fresh culture medium. The tobacco leaf protoplast was first transformed by this process.

There are some facilitators for uptake of Ca-DNA co-precipitate and expression of DNA in host cells. For example, treatment of the co-precipitate with polyvinyl alcohol and physiological shock with DMSO increase the uptake. Sometimes, the level of expression increases when the transfected cells are exposed to hormones, heavy metals, or other substances, that activate cellular transcription factors.

In-planta transformation

Gene transfer in plant explants, cells, or protoplasts that has been discussed so far require an obligatory time-consuming tissue culture step for the regeneration of whole fertile transgenic plants. The extensive period of tissue culture for many months involves loss of time and a

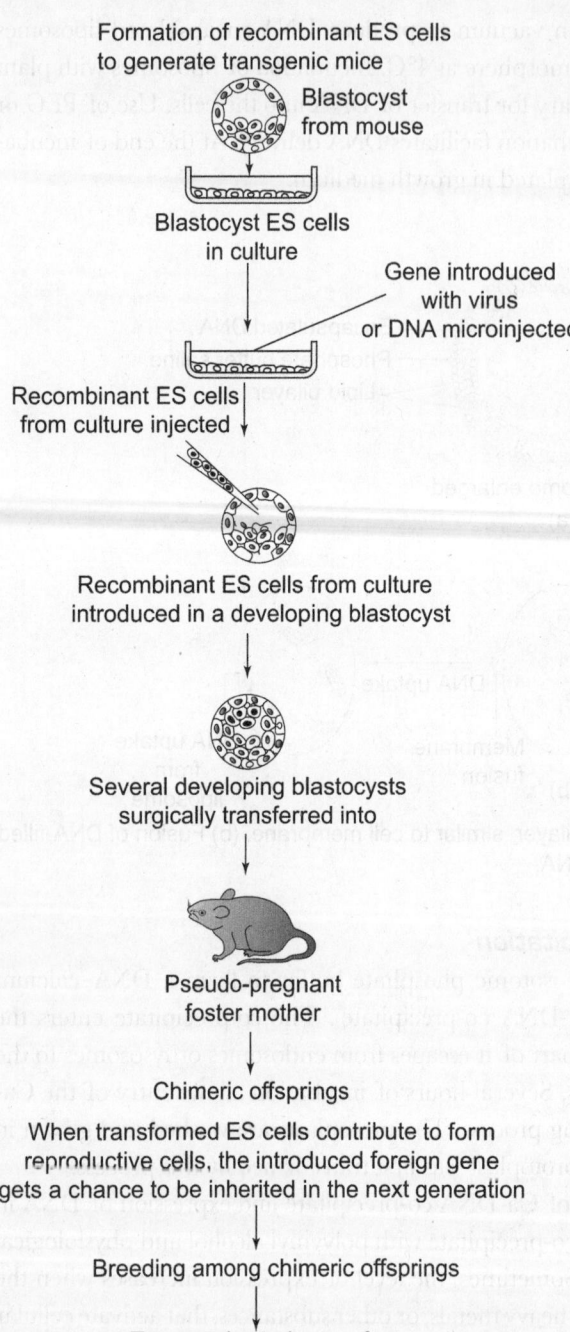

Formation of recombinant ES cells
to generate transgenic mice

Blastocyst
from mouse

Blastocyst ES cells
in culture

Gene introduced
with virus
or DNA microinjected

Recombinant ES cells
from culture injected

Recombinant ES cells from culture
introduced in a developing blastocyst

Several developing blastocysts
surgically transferred into

Pseudo-pregnant
foster mother

Chimeric offsprings

When transformed ES cells contribute to form
reproductive cells, the introduced foreign gene
gets a chance to be inherited in the next generation

Breeding among chimeric offsprings

Transgenic strain may form

Fig. 6.31 Embryonic stem (ES) cell technology to raise transgenic animals.

requirement of space, labour, and a monetary resource. At the same time, the extensive period leads to the appearance of somatoclonal variations. Therefore, recently, in-situ or in-planta transformation of germ cells and meristem has been attempted with success in a limited number of species. This may open up the possibility of widespread applications for crop improvement, although the present percentage of success is low.

In-planta transformation can be carried out by the following methods. (i) The germinal cells in dry seeds imbibe foreign DNA in solution in the course of incubation. Application of DNA to cut pistils or pollen grains allows the transforming DNA to reach the zygote at the time of fertilization so that it is incorporated in the seeds. (ii) Meristem transformation is achieved simply by removal of the apical shoot at its base and inoculation of the cut surface with transforming DNA. A transgenic shoot then grows from this meristem.

6.13.11 Overview of Genetic Transformation of Animals

The methodologies for producing transgenic organisms, both in plants and animals, have been already discussed in a general fashion.

Usually to produce transgenic animals, the transgene or DNA is introduced by virus vectors or direct microinjection in an early embryonic stage, preferably at the stage of the freshly fertilized egg.

Microinjection directly in the nucleus

Immediately after fertilization, the egg nucleus (female pronucleus) and the sperm nucleus (male pronucleus) are discrete. The egg is immobilized with a suction pipette (Fig. 6.29) and DNA solution (~2 pL) is injected into the male pronucleus with the help of a fine glass needle. The male pronucleus is larger and thus facilitates the microinjection. This technique is reliable with transfection efficiency in the range of 5–40 per cent. The site of DNA integration appears random and may depend on the occurrence of natural chromosome breaks.

In case of insects, the microinjection of DNA is done at the micropyle end of the egg as the germinal cells are located at this area.

Embryonic stem cell technology to raise transgenic animals

Manipulation of animal cells in culture is considerably easier than introducing genes into a fertilized egg in vitro.

Embryonic stem cells (ES cells) are derived from the inner cell mass (ICM) found in the blastula stage of mouse or mammalian embryos. The cells are *pluripotent*; that is, each of them can differentiate into any of the different types of cells in the body. These cells are in experimental use for therapy of damaged tissue in heart and brain. The ES cells can be cultured in vitro, transfected or injected with DNA, and then introduced into other developing blastocysts (Fig. 6.31). Some of the transfected ES cells may participate in the formation of certain tissues, so that the newborn mice become *chimeras*—mixtures of two types of cells—cells of the developing blastocyst and those derived from transfected ES cells. When ES cells contribute to the formation of the chimera's germ line (reproductive) tissue, the introduced foreign DNA has a chance of being transmitted to the next generation. Breeding of such a chimeric mouse is likely to establish a transgenic strain. One of the first experiments showed that the transgenic mouse carrying human growth hormone gene became double the size of a normal mouse.

Usually the transfected fertilized eggs and blastocysts with genetically manipulated embryonic stem cells are placed in the uterus of a surrogate mother in mammals like mouse or cattle for full development into live offspring. The newborn offsprings are screened for the presence of the transgene by Southern hybridization and PCR. The transgenic siblings are mated to obtain homozygous transgenic animals. A similar procedure is also employed to raise transgenic birds like chicken and transgenic fish like salmon.

Besides microinjection of DNA, *physical methods* of transfection of animal cells also include particle bombardment (gene gun) and electroporation.

Chemical and biochemical methods of transfection of animal cells include calcium phosphate–DNA coprecipitation, lipofection, and cell membrane permeation by polycations (poly-L-ornithine, polybrene) in the presence of DMSO.

6.13.12 Somatic Cell Nuclear Transfer (SCNT) in an Enucleated Ovum

So far we have discussed how to raise transgenic organisms by transfer of foreign DNA (gene) in the genome of an organism by using various DNA delivery systems.

Now we shall discuss the transfer of a somatic nucleus, that is, whole genome, to an enucleated ovum (nucleus of ovum removed beforehand). This is known as somatic cell nuclear transfer (SCNT) technique. This was experimented in amphibia a long time ago. Resurgence of interest about the technique arose with the successful production of a lamb, named Dolly at the Roslin Institute near Edinburgh, Scotland in early 1997. Dolly was produced by transfer of the nucleus from an epithelial cell of the udder (mammary gland) of a Finn Dorset ewe (white faced) to an

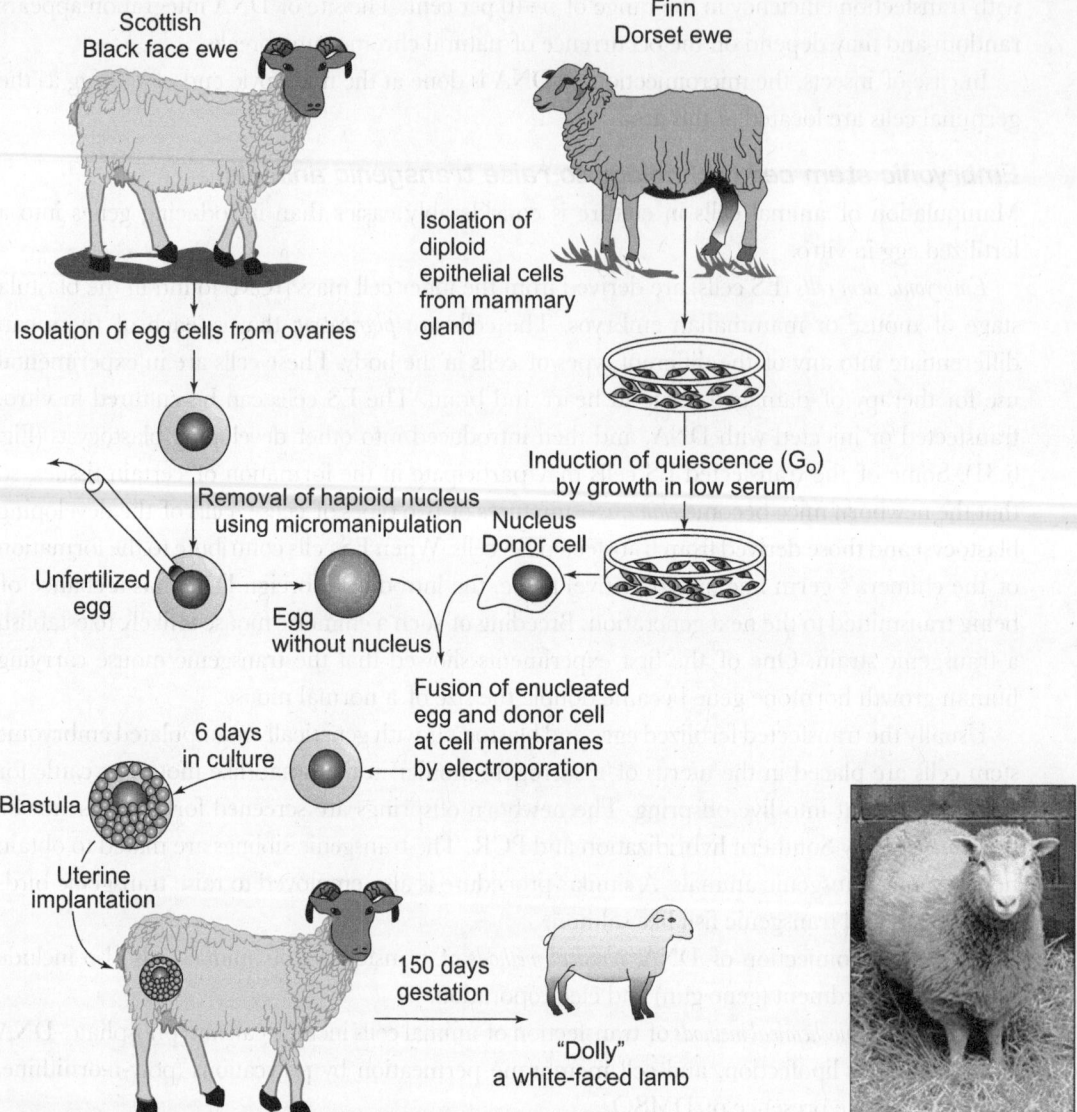

Fig. 6.32(a) Somatic cell nuclear transfer (SCNT) in an enucleated **Fig. 6.32(b)** Dolly—the sheep
ovum produced 'Dolly' (a lamb) and demonstrated the possibility of
cloning of animals.

enucleated egg from a Blackface ewe; this manipulated egg was stimulated to divide in vitro and
then the blastocyst was implanted in the uterus of another Blackface ewe for gestation and birth
(Fig. 6.32a). Dolly was white faced like its genetic (nucleus contributing) mother.

Interestingly, Dolly did not have a father, but had three mothers—egg contributing mother,
nucleus (gene) contributing mother, and surrogate or gestating mother. Credit of creation of Dolly
(Fig. 6.32b) goes to the group of scientists guided by I. Wilmut and K.H. Campbell.

Creation of Dolly clearly establishes the following facts:

A differentiated nucleus in a somatic (body) cell has the potential to develop into an organism in the same way as a zygotic nucleus with the provision of the cytoplasmic environment of an ovum. In other words, the nuclei of most somatic cells retain all the genetic information required for the entire developmental programme; the nuclei can be reprogrammed for development by the cytoplasm of the egg.

This technique (SCNT) can be used for the production of clones of animals with the same genotype by transplanting somatic nuclei from several cells of the same individual into a series of enucleated eggs.

Since the cloning of Dolly, somatic cell nuclei have been used to clone cattle, goats, and pigs.

Molly and Polly

SCNT technique was developed further to include *transgene(s) in the* transplanted *nucleus.* In December 1997, the Roslin Institute reported the birth of two transgenic lambs, named Molly and Polly. They were generated from the nucleus of foetal donor cells that had been stably transfected with a mammary-specific β-lactoglobulin expression vector. The vector carried the human Factor IX gene. Molly and Polly were genetically identical and they also possessed the potential to produce large quantities of the human blood clotting protein Factor IX in their milk. One of the transgenic experiments used pigs to produce human haemoglobin.

Experiments are being carried out to genetically transform sheep with two bacterial genes so that they can synthesize their own cystein (an amino acid that has an influence on the growth of wool). These transgenic sheep have produced two times more wool than normal sheep. Transgenic technology in sheep has also attempted the expression of insecticidal proteins in the wool to avoid the need of chemical spray to reduce the infestation by insects. Chitinase genes from plants are used for the purpose. If sheep could be induced to secrete chitinase in the skin, insect larvae would be killed by ingesting the protein before they could harm the sheep. Chitinase is known to be non-toxic to mammals. This is analogous to using *Bt* toxin as an in-built insecticide in plants.

6.14 APPLICATIONS OF GENETIC ENGINEERING

Recombinant DNA (rDNA) technology or genetic engineering is a major aspect of biotechnology. Genetic engineering revolutionized the concept and practice of breeding in organisms. By transferring new DNA sequences into microbes, plants, and animals, or by removing or altering DNA sequences in their genomes, completely new strains or varieties can be created to perform specific tasks of human interest. A *genetically engineered organism* (GEO), or *genetically modified organism* (GMO) is also called a *transgenic organism* or *transformant.* The DNA introduced is called a *transgene* and the process of insertion of a gene or part of a gene from one organism into the genome of another organism is called *transgenesis* or *transformation* or *transgenic technology*.

Transgenic organisms exhibit at least one new and useful trait which is transmitted through the germ line as a simple Mendelian trait. The transgene must be inserted into the germ line (cells involved in producing ova or sperm) in a multicellular organism, so that it can be propagated in the subsequent generations. On the other hand, genes can also be inserted into somatic (body) cells for correction of a defective gene; this process is known as *gene therapy*.

Applications of genetic engineering are innumerable and some of these have been mentioned in this chapter and are listed below. Commercial implementation of genetic engineering has opened a multi-billion dollar market throughout the world and has changed the lifestyle of human beings in certain ways.

6.14.1 Applications of Recombinant Microorganisms

1. Transgenic bacteria help in *understanding the unknown function(s) of some genes and in regulation* of the expression of some other genes.
2. *Commercial production of hundreds of therapeutic proteins and biopharmaceuticals* are carried out with recombinant microorganisms; some of the examples are presented in Table 6.4. Most of these are used to treat human beings. Sometimes, growth hormones produced by transgenic bacteria help in high milk and meat production in cattle.
3. *Production of novel antibiotics* with increased activities against selected targets and decreased side effects is possible by rDNA technology. Furthermore, genetic manipulation of the producer strain can enhance production of antibiotics and lower the cost of production.
4. *Cloning antibiotic biosynthesis genes* has been made possible by constructing bacterial artificial chromosome (BAC) vectors with large DNA fragments, encoding enzymes for entire antibiotic biosynthetic pathways and transferring them into *Streptomyces* host cells.
5. *Commercial production of enzymes* such as restriction endonucleases for genetic engineering, and lipase and proteinase for use in detergents is carried out by genetically manipulated microorganisms in bioreactors.
6. *Production of recombinant vaccines* consisting of antigenic proteins present in the virus coat is possible and can be used against many dreaded viruses (Tables 6.4 and 6.5). The gene encoding viral surface antigen is ligated into the *Vaccinia* genome under the control of a *Vaccinia* promoter. After injection of this live vaccine in the host, the recombinant *Vaccinia* virus replicates and expresses its own surface antigen (vaccine for small pox) as well the antigen encoded by the ligated gene (for example, *Hepatitis B surface antigen*; Fig. 6.33). Thus, the recombinant *Vaccinia* viruses as live vaccines can induce immunity against small pox and other diseases at the same time. Recombinant anthrax, cholera, malarial (*Plasmodium falciparum*) vaccines are under development.

Table 6.4 Some examples of biopharmaceuticals produced commercially in recombinant microorganisms

Type	Product	Use
Blood proteins	Erythropoietin	Treats anaemia
	Granulocyte-macrophage colony stimulating factor (G-MCSF)	Stimulates leucocyte production from bone marrow
	Factor VIII, Factor IX	Coagulation factors; treat haemophilia
	Urokinase	Thrombolytic agent
	Tissue plasminogen activator	Anticoagulant, dissolves clots
	Serum albumin	Supplements plasma

(Contd)

Table 6.4 *(Contd)*

Hormones	Insulin	Treats diabetes
	Growth hormone	Promotes growth
	Adrenocorticotrophic hormone	Treats rheumatic diseases
	Follicle stimulating hormone (FSH)	Treats reproductive disorders
	Calcitonin	Treats osteomalacia
	Relaxin	Facilitates childbirth
	Epidermal growth factor	Promotes wound healing
	Nerve growth factor	Promotes nerve damage repair
	Thyroid stimulating hormone	Treats metabolic dysfunctions
Cytokines	Interleukins (IL-1, IL-2, etc.)	Cancer therapy and immune disorders
	Interferons (IFN-γ)	Antiviral, anti-tumour
	Tumour necrosis factor(TNF-α)	Anti-tumour agent
Vaccines (Engineered *Vaccinia* viruses)	Hepatitis B antigen	For vaccination against hepatitis B
	Measles antigen	Prevention of measles
	Rabies antigen	Prevention of rabies
Regulatory substance	Endorphins and enkephalins	Analgesic agent

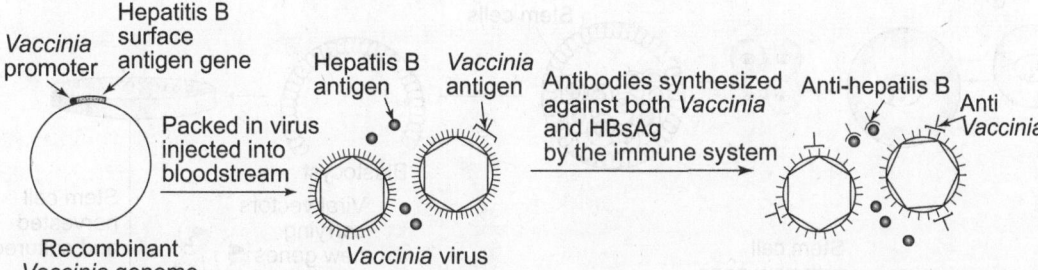

Fig. 6.33 A recombinant *Vaccinia* virus with gene for hepatitis B surface antigen (HBsAg) for use as vaccine.

Table 6.5 Some examples of viral genes expressed in recombinant *Vaccinia* viruses, to be used as vaccines

Viral Genes	Vaccines
Hepatitis B surface antigen (HBsAg)	Hepatitis B
Influenza virus haemagglutinin	Influenza
Rabies virus G protein	Rabies
Herpes simplex glycoproteins	Herpes
Vesicular stomatitis virus coat proteins	VSV
Human immunodeficiency virus (HIV) envelope proteins (gp 120, gp 41)	AIDS

7. *Gene therapy* by viruses that carry normal genes and deliver them to human cells to cure diseases caused by defective genes is a relatively new method in medicine. It has been used

to treat genetic disorders such as severe combined immunodeficiency (deficiency in both B and T cells) and incurable diseases such as cystic fibrosis, sickle cell anaemia, and so on. A futuristic projection of stem cell gene therapy is depicted in Fig. 6.34.

8. *Production of biopolymers* by genetically engineered microorganisms helps in developing new biopolymers to replace synthetic polymers like plastics. The biopolymers are biodegradable and thus, eco-friendly.

9. *Synthesis of textile-quality indigo* is possible by genetically engineered *E. coli* which produces indoxyl. On oxidation, indoxyl produces indigo. In earlier times, it was extracted from plants.

10. *Improvement in many other commercial products and fermented products* including alcoholic beverages are made by genetically engineered microorganisms.

11. *The use of living organisms to degrade the contaminants or wastes to clean up the environment is termed as bioremediation.* Bioremediation of contaminated land, water bodies, and air by using genetically manipulated microorganisms is currently being developed. Contaminants can be both inorganic and organic chemicals. The genetic manipulation of the microorganisms for bioremediation can be performed by the following two strategies: (i) transfer of plasmids (Exhibit 6.D) and (ii) specific gene alteration.

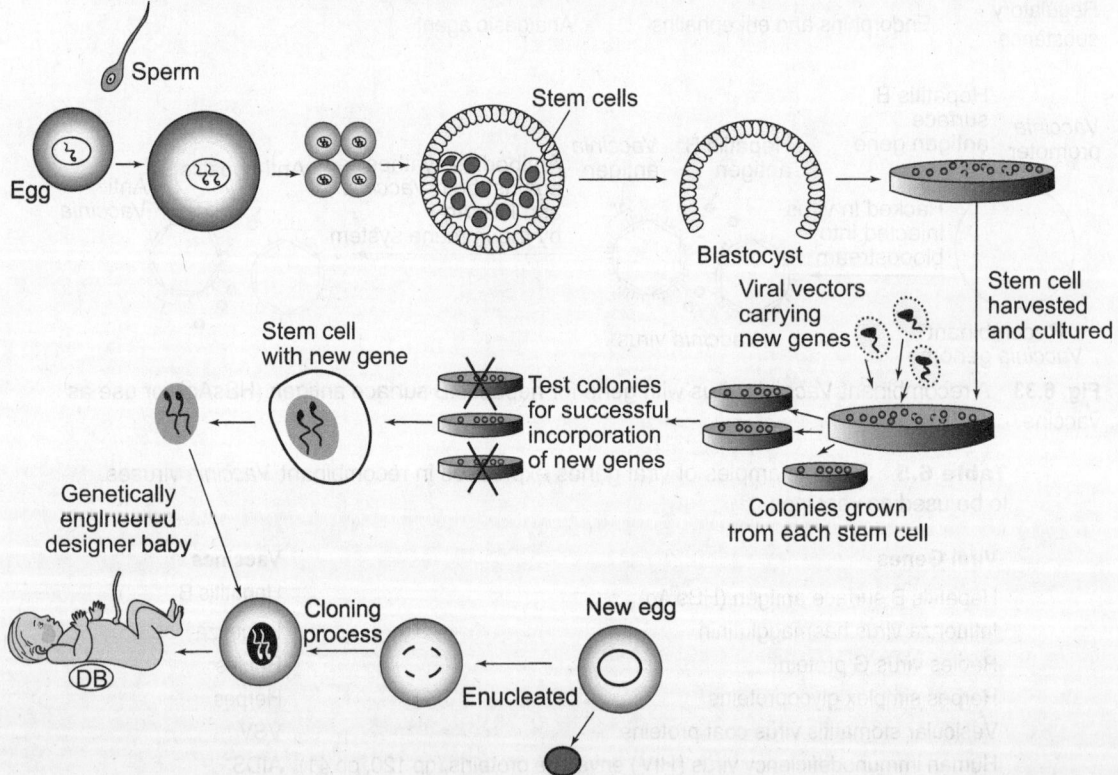

Fig. 6.34 Stem cell gene therapy. Genetically correcting the hereditary defects in human stem cells and then implanting the corrected nucleus into an enucleated ovum to produce a genetically engineered or designed baby (DB). This is a futuristic projection.

Exhibit 6.D Superbug for Cleaning Oil Spills

Petroleum oil is a combination of variety of hydrocarbons, the major types being xylenes, naphthalenes, octanes, and camphors. Different strains of *Pseudomonas putida* can consume one or two of these hydrocarbons, depending on the presence of genes encoding catalysing enzymes for a type of hydrocarbon. These genes are present on plasmids, and they are referred to as xylene-degrading (XYL), naphthalene-degrading (NAH), octane-degrading (OCT), and camphor-degrading (CAM) plasmids.

In 1979, Ananda Mohan Chakrabarty, an India-born US scientist and his colleagues introduced the four plasmids from different strains of *Pseudomonas* through the process of conjugation (mating) into a single cell, and the resulting bacterium could consume the four types of hydrocarbons present in oil. This unique bacterial strain which had the ability to degrade the hydrocarbons in petroleum oil was named *superbug*. To create the superbug, first the CAM plasmid was transferred by conjugation into a strain carrying the OCT plasmid. These CAM and OCT plasmids are incompatible and cannot exist in the same bacterium independently. That is why, the relevant genes from these two plasmids were brought together into a single plasmid by recombination. At the same time, the NAH plasmid was transferred by conjugation into a strain that had the XYL plasmid. The NAH and XYL plasmids are compatible and coexist within the same host cell. At the end, the CAM/OCT fusion plasmid was transferred by conjugation into the strain carrying NAH and XYL plasmids (Fig. Exhibit 6D1). The final strain, the superbug containing the genes of all four plasmids in one cell could convert the four major components of oil into harmless non-polluting products.

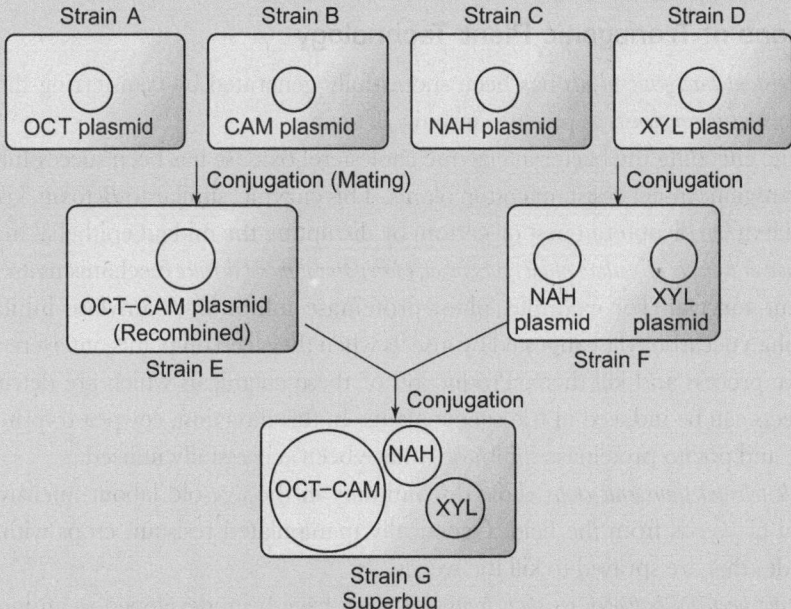

Fig. 6.D Planned mating for creation of superbug to degrade petroleum oil.

The objective behind the creation of the superbug was to first mix them with straw and then dry and store them. The superbug-laden straw would be scattered over the oil spills from tankers in the sea, where the straw would soak up the oil and the superbug would act on oil to convert it into harmless products. Thus, the sea surface would be cleaned and marine ecosystems would not be affected.

Although the superbug has not been used for the practical purpose of cleaning oil spills, it has played a significant role in formulating patent rights for biotechnology research and industry. In a landmark judgement, the US Supreme Court in March 1981, granted a US patent to the inventor of this superbug describing its construction and use. This was the first patent in the world granted ever for genetically engineered microorganism. This enabled the biotechnology companies to protect their inventions in the same way as chemical and pharmaceutical industries.

Genetically engineered soil and other bacteria can detoxify organophosphate pesticides, trichloroethylene, and toxic heavy metals like mercury. Some bacteria have been genetically modified to remediate the effects of organic pollutants present in radioactive environments.

12. *Plant growth promotion* through recombinant soil microorganisms may be achieved by (i) increasing the level of nitrogen fixation, by expressing *nif* genes, and as a consequence reducing the dependence on chemical fertilizers; (ii) increasing the efficiency of root nodulation by expressing *nod* genes; (iii) microbial synthesis of a variety of substances that limit the damage caused to plants by pathogens, which include siderophores, antibiotics, other small molecules, and various enzymes; (iv) cloning enzymes for phytohormones to stimulate plant proliferation.

Applications of recombinant microorganisms have been summarized in Fig. 6.35.

6.14.2 Applications of Transgenic Plant Technology

1. *Insect-resistant transgenic plants* has been successfully generated by transferring the *cry* gene from *Bacillus thuringiensis* to produce in-built *Bt* toxin.

   The gene encoding the bacterial enzyme cholesterol oxidase has been successfully used to raise transgenic insect-resistant cotton plants. This enzyme, similar to *Bt* toxin, kills the boll weevil larva (a coleopteran pest of cotton) by disrupting the midgut epithelial membrane.

2. *Expression or transfer of certain plant genes involved in natural insect defence* mechanisms are sufficient for plant survival. For example, plant proteinase inhibitor, α-amylase inhibitor, and tryptophan decarboxylase ingested by insects when they feed on plants, interfere with their digestive process and kill them. Production of these chemicals which are detrimental to the insects can be induced in transgenic plants. In this direction, cowpea trypsin inhibitor (CpTI) and potato proteinase inhibitor II have been successfully utilized.

3. *Herbicide tolerant transgenic crops* allow discontinuity in the age-old labour-intensive regular removal of weeds from the field. Genetically manipulated resistant crops withstand the herbicides that are sprayed to kill the weeds.

4. *Fungal and bacterial pathogen resistant transgenic plants* have been developed as an inexpensive, effective, and environment-friendly measure to avoid the use of chemical agents like fungicides and bactericides. The chemical agents usually have detrimental effects on the environment and living organisms.

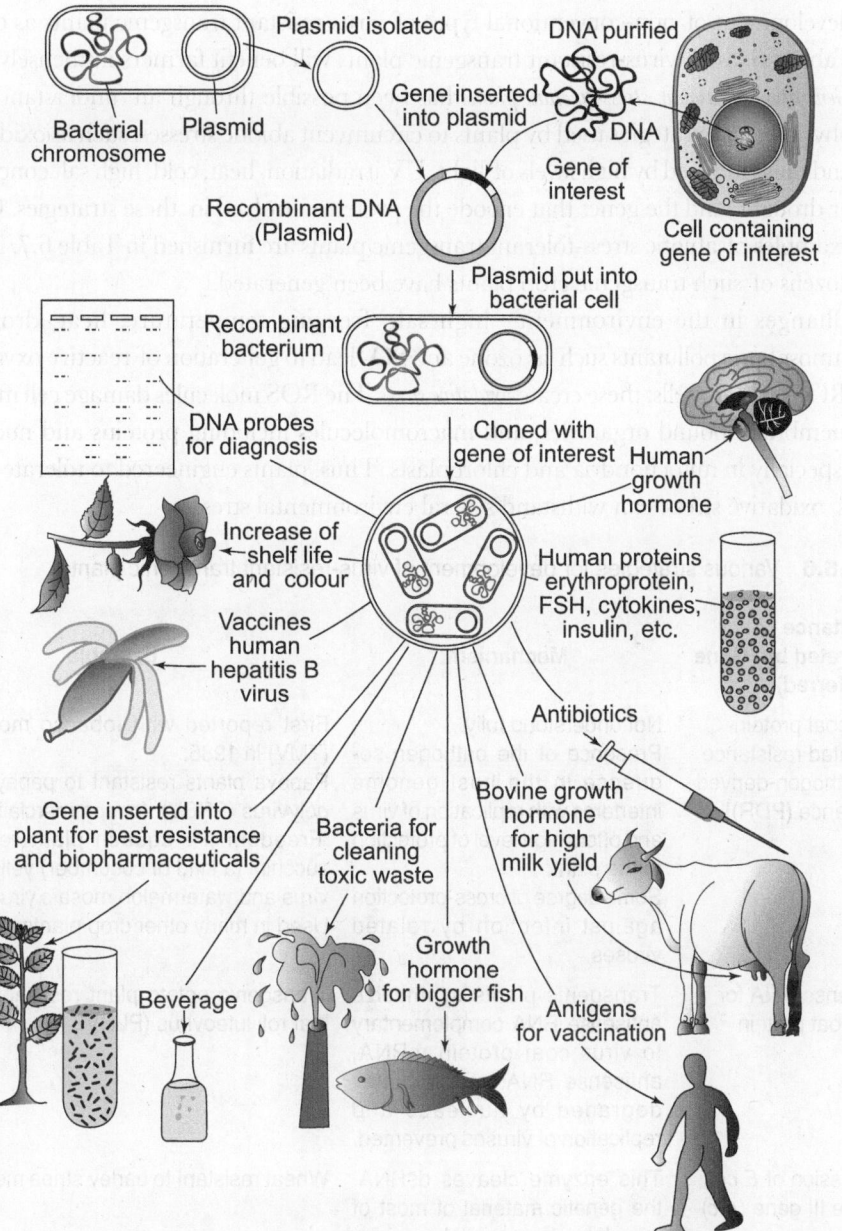

Fig. 6.35 Summary of applications of recombinant microorganisms for various products.

5. *Development of virus-resistant plants* has been attempted by plant breeders by the transfer of *naturally occurring virus resistance genes* from one plant strain (cultivar) to another. However, the resistance is not permanent and resistance to one virus does not necessarily confer resistance against other similar viruses. Genetic manipulation for expression of antiviral plant proteins in transgenic plants is under development. Genetic engineering has brought success in the

development of non-conventional types of virus-resistant transgenic plants as depicted in Table 6.6. Such virus-resistant transgenic plants will benefit farmers immensely.

6. *Generation of abiotic stress-tolerant plants* has been possible through an understanding of the physiological strategies used by plants to circumvent abiotic stresses, such as oxidative stress and those induced by high levels of light, UV irradiation, heat, cold, high salt concentrations, or drought, and the genes that encode the proteins involved in these strategies. Only a few examples of abiotic stress-tolerant transgenic plants are furnished in Table 6.7. Nowadays, dozens of such transgenic crop plants have been generated.

Changes in the environment—high salt, freezing temperature, heat, drought, and atmospheric pollutants such as ozone and SO_2 lead to generation of reactive oxygen species (ROS) in plant cells; these create *oxidative stress*. The ROS molecules damage cell membranes, membrane-bound organelles, and macromolecules including proteins and nucleic acids, especially in mitochondria and chloroplasts. Thus, plants engineered to tolerate ROS, that is, oxidative stress, can withstand several environmental stresses.

Table 6.6 Various strategies for development of virus-resistant transgenic plants

Resistance generated by (Gene transferred)	Mechanism	Example
Viral coat protein-mediated resistance [or pathogen-derived resistance (PDR)]	Not understood fully. Presence of the pathogen sequence in the host genome interferes with replication of virus and offers high level of protection to the plant. Some degree of cross-protection against infection by related viruses.	First reported with tobacco mosaic virus (TMV) in 1986. Papaya plants resistant to papaya-ringspot polyvirus (PRSV), for commercialization. 'Freedom II'—squash plant resistant to zucchini (a kind of cucumber) yellow mosaic virus and watermelon mosaic virus. Used in many other crop plants.
Antisense RNA for viral coat protein	Transgenic plants synthesize antisense RNA complementary to virus coat protein mRNA, antisense RNA:mRNA duplex degraded by nuclease and replication of viruses prevented.	Transgenic potato plant resistant to potato leaf roll luteovirus (PLRV).
Expression of *E.coli* RNase III gene (*rnc*)	This enzyme cleaves dsRNA, the genetic material of most of the plant viruses, and provides resistance to a broad spectrum of viruses.	Wheat resistant to barley stripe mosaic virus.
Modification: use of a mutant gene (*rnc* 70)	As the normal enzyme interacts with the plant RNA to cause stunted growth of the plant, the mutant gene (rnc 70) product is unable to cleave the plant cell RNA, but cleaves viral dsRNA.	

For example, the *enzyme SOD detoxifies* the superoxide anion, a potentially damaging oxygen radical belonging to ROS. Different isoforms of SOD are present in chloroplasts, mitochondria, and the cytosol. The enzyme SOD converts the anion to hydrogen peroxide, which in turn is broken down to water by different cellular peroxidases or catalases. Tobacco plants transformed with a SOD gene had reduced superoxide anion damage under stress conditions in comparison to control plants.

Oxidative stress is also reduced when the level of oxidized glutathione is increased within a plant cell by the activity of glutathione-S-transferase and glutathione peroxidase. Introduction of the genes for both these enzymes and SOD in a transgenic plant would be a right strategy to lower oxidative stress in plants.

7. *Improvement of nutritional status of plants* can be achieved through transgenic plants capable of producing seed grains with higher content of certain amino acids (specifically methionine and lysine), synthesis of vitamin E and β-carotene (pro-vitamin A) and increasing the levels of available iron, and so on. After ingestion by mammals, β-carotene is converted to vitamin A. Such transgenic modifications increase the nutritional status of the food and hence improve human health and the growth of farm animals.

Table 6.7 Some examples of abiotic stress tolerant transgenic plants

Transgene	Encoded product	Host crop plant	Tolerance to stress
Arthobacter globiformis cod A	Choline oxidase	*Arabidopsis* chloroplast, rice	Salt, freezing, heat, strong light
Annexine	Annexine like protein	*Brassica*	Salinity, water deficit
Antifreeze gene from cold adapted fish and plants	Antifreeze protein	Tobacco, potato	Cold
Anti-ProDH	Proline dehydrogenase (osmoprotectant)	*Arabidopsis*	Salt, freezing
Bacillus subtilis sac B	Fructans osmoprotectant	Sugar beet	Drought
E. coli bet A	Choline dehydrogenase (osmoprotectant)	Rice, tobacco	Drought, salt
Cytosolic Cu/Zn-SOD	Cu/Zn-SOD	Tobacco cytosol	Acute ozone exposure
Glycerol-3-phosphate acyl-transferase	Glycerol-3-phosphate acyl-transferase	Tomato	Cold
GST/GPX	Glutathione S-transferase with glutathione peroxidase	Tobacco	Oxidative stress
Hsp 17.7	SmHSP (small heat-shock protein family)	Carrot	Heat
Mothbean P5 CS	Pyrroline carboxylate synthetase (proline osmoprotectant)	Rice	Drought, salt, oxidative stress
		Soybean	Osmotic stress, heat

Source: Based on Rastogi, S. and N. Pathak, 2009, *Genetic Engineering*, Oxford University Press, New Delhi, pp. 641–643.

For example, lupine, a grain legume used as feed for cattle, pigs, and chicken in Australia, lacks methionine and cysteine. The gene encoding *sunflower seed albumin,* a rich source of the amino acids methionine and cystein, has been successfully engineered into lupine. The technology can also be used to modify the fatty acid or lipid composition of both edible and non-edible oil-producing crops to make the oil better suited for intended uses. The technology is also being used for expression of sweet tasting proteins in plants as low-calorie sugar substitutes.

8. *Improvement of flower colour* through manipulation of genes encoding enzymes involved in *anthocyanin and carotenoid pigment biosynthesis* pathway has been done. Using this technology, the first blue rose of the world was produced and approved for commercial cultivation in Australia in 2009. (Also see colour plate 2.)

9. *Increase in shelf life (long-term storage) of agri-horticultural products,* such as fruits, vegetables, and flowers are possible by gene manipulation in transgenic plants. Post-harvest softening and discolouration of fruits, as well as wilting, discolouration, and ageing of flowers can be delayed by inhibiting expression of some genes producing enzymes responsible for all these events, which are detrimental when these products are used for commercial purposes.

For example, *antisense polyphenol oxidase (PPO) gene* inhibits function of the PPO gene and reduces discolouration of potato. Similarly, by interfering with the expression of cellulose and polygalacturonase (PG) using genes for antisense RNA, the ripening process can be delayed. Transgenic tomato plants constitutively expressing *antisense PG gene* have been produced and marketed under the trade name 'Flavr Savr' (flavour saver). The Flavr Savr tomato was the first transgenic fruit developed by antisense RNA technology by Calgene in the United States and released in the market in 1994. The idea behind production of Flavr Savr tomato was to produce healthy, vine-ripened tomatoes that could withstand the hazards of transportation and shipping. By 1997 it was found that the plants lacked consistent production qualities and these transgenic tomatoes did not have much market appeal.

Ripening of fruits including tomatoes can also be delayed by antisense *inhibition of the ethylene* biosynthesis pathway, and thus the shelf life of the fruits can be increased. Normally the plant growth regulator ethylene induces expression of a number of genes encoding enzymes involved in fruit ripening. Inhibition of ethylene also delays wilting in flowers, thereby extends their storage life, even without refrigeration or application of expensive chemicals.

10. *Detoxification of toxic metals (phytoremediation),* such as mercury, lead, cadmium, and selenium polluting the environment can be achieved by transgenic plants harbouring genes for *phytochelatin synthase* (PC synthase) or *glutathione synthetase* enzymes. These enzymes help in the synthesis of phytochelatin (PC), a metal chelator (binder), which improves metal uptake and sequestration by the transgenic plants.

11. *Increase in the yield of plants* may be achieved by increasing the availability of iron or oxygen or increasing the rate of photosynthesis.

The compounds mugineic acid and avenic acid secreted by the roots of graminaceous plants are natural iron chelators (bind to iron in the soil) and are also called *phytosiderophores*. The deficiency of iron and stunted growth of rice can be remedied by transforming it with a barley *naat* gene; *naat* A and *naat* B genes encode subunits of nicotinamine aminotransferase (NAAT) enzyme, catalysing the biosynthesis of the mugineic acid family of phytosiderophores to acquire iron from the soil.

Increase in oxygen concentration   Increased O_2 concentration often helps in plant respiratory metabolism and better growth. The gene encoding a *dimeric Hb* capable of binding oxygen has been inserted from a Gram-negative bacterium *Vitreoscilla* into tobacco plant.

Enhancement of photosynthetic activity   The photosynthetic process fixes atmospheric carbon and increases the overall size of the organic carbon pool responsible for plant growth. For increasing the photosynthetic activity, the genes encoding better light-utilizing *phytochrome enzymes* are included. Attempts have been made to introduce the gene components of the energy efficient C4 photosynthetic pathway from maize into C3 plants.

12. *Increase in cellulose production and nutritional value of cattle feed by decreasing lignin content of plants* also come under the purview of increasing yield of plants by transgenic technology.

    The presence of higher amount of lignin in timber-producing large trees is a major obstacle in obtaining quality cellulose for pulp in the paper industry. Removal of lignin from cellulose involves harsh physical and chemical treatments, which are energy-intensive, harmful to the environment, and expensive. Lignin is intimately associated with cellulose and hemicelluloses in forage crops and limits their digestibility by cattle and thus energy production. Thus, reduction of lignin content for optimal utilization of plant biomass in pulp and paper industry and cattle feed has been a goal for plant genetic engineers. *Antisense RNA* or RNAi technology successfully inhibits certain enzymes related to lignin biosynthesis. Several genes have been targeted, for example, *omt* encoding O-methyltransferase, *f5h* for ferulate 5-hydroxylase, *cad* encoding cinnamyl aldehyde dehydrogenase, and so on.

    Conversion of bioenergy crops into a liquid fuel, bioethanol, can be significantly enhanced by reducing lignin biosynthesis and over-expression of *cellulases, endoglucanase,* and other enzymes in transgenic plants. These manipulations of genes cause higher saccharification of the bioenergy crops and save the expenses of chemical treatments.

13. *Large-scale production of special chemicals and pharmaceuticals* by transgenic plants is possible as in the case of transformed bacteria in bioreactors. In fact, there are several advantages of transgenic plants over recombinant microbial systems:

    (i) Transgenic plants are relatively easy to grow and maintain on a large scale. Highly trained personnel or expensive equipment are not required for the purpose. Scale-up can be easily done by simply planting seeds. The product is obtained in high quantity. Bulk production also allows the products to be cheaper.

    (ii) Cross breeding between transgenic plants producing different chemicals allows the generation of plants that can produce multiple products.

(iii) Integration of foreign DNA into plant genomes is more stable than transforming microorganisms with plasmids; the latter may get lost during prolonged culture or large-scale fermentation.

(iv) Post-transcriptional processing (say glycosylation) and assembly of proteins encoded by genes of animal origin occur in a similar fashion in plant cells and animal cells. This is because plant and animal cells share similar types of enzyme systems, both being eukaryotes; the enzyme systems in bacterial cells are different.

Transgenic materials in the form of fruits or seeds can be stably stored for a longer period and transported easily.

(vi) Transgenic plant products, particularly therapeutics, do not face the problem of contaminating pathogens and thus get more public acceptability.

Purification of high-value products from an enormous quantity of harvested transgenic plants, sometimes pose problems. Scientists are modifying the technique to obtain the products as exudates from the plants or targeting foreign genes to chloroplasts.

Advantages of transgenic plant systems as bioreactors allow the large-scale production of (i) biopolymers, (ii) therapeutic proteins, (iii) vaccines including edible ones, (iv) nutraceuticals, (v) hormones, and (vi) industrial enzymes.

The production of biopharmaceuticals in transgenic plants is especially free of the risk of mammalian viral contamination which occurs when the products are raised in animal cell cultures.

Antibodies and antibody fragments against various antigens can be produced in transgenic plants into which mammalian or human antibody genes have been introduced. These antibodies produced in plants are called *plantibodies* and they are mostly used for diagnosis of pathogens and for therapeutic purposes. The cost of production of antibodies in transgenic plants is almost 100–500 times cheaper than other conventional techniques of production, such as hybridoma cell culture and transformed bacteria.

Transgenic plant technology is bringing in more and more items of commercial interest under its aegis. Some of the examples of applications of transgenic plant technology involving particular genes are furnished in Table 6.8. Table 6.9 presents a list of therapeutic recombinant proteins produced in plants and used as biopharmaceuticals. Transformation of choloroplasts with foreign genes and products of transplastomic plants have been elucidated in Exhibit 6.E.

Table 6.8 Applications of transgenic plant technology

Objective	Genes transferred (Encoding)	Application
Insect resistance	*Cry* gene (Bt toxin)	Target cotton bollworm
	Bacterial gene (Cholesterol oxidase)	Boll weevil larva
	Cp T1 from cowpea (Trypsin inhibitor)	Coleoptera, Lepidoptera
Herbicide (Glyphospate) resistance	*Agrobacterium CP4-epsps* [Enzyme 5-enol pyruvyl shikimate-3-phosphate synthase (EPSPS)]	Herbicide affects EPSPS enzyme of weeds, but EPSPS increases the enzyme in resistant plants–soybean, tomato, oilseed, rapeseed

(Contd)

Table 6.8 *(Contd)*

Objective	Genes transferred (Encoding)	Application
Herbicide [Bromoxynil (nitriles)] resistance	Nitrilase gene from soil bacteria *Klebsiella ozaenae* (Nitrilase enzyme)	Bromoxynil inhibits photosynthesis; in resistant plants nitrilase detoxifies the herbicide in cotton, potato, rapeseed, tomato, tobacco
Fungus resistance	Chitinase and/or glucanase gene from bean or other plants (chitinase and glucanase-hydrolytic enzymes)	Antifungal
β-Carotene (pro-vitamin A) biosynthesis in 'Golden rice' and 'Beta-sweet carrot' (Food improvement)	Three genes for β-carotene synthesis *psy*, *lcy* from daffodil and *crt* from bacteria *Erwinia uredovora* (3 catalysing enzymes)	Genes express in endosperm (major part) of rice grain and carotene pigment of carrot; food improvement
α-Tocopherol (Vitamin E) enrichment in oil (Food improvement)	δ-TMT (δ-Tocopherol methyltransferase) gene from *Arabidopsis* under a seed-specific promoter DC3 from carrot (δ-TMT enzyme for methylation of δ-tocopherol to α–tocopherol)	Several-fold increase in production of vitamin E in oil seeds
Manipulation of flower pigmentation	Delphinidin gene from pansy (primary blue pigment)	First blue rose produced
Increase in shelf life of agri-horticulture products	Antisense RNA gene to S-adenosyl synthetase (SAM synthetase), or 1-amino cyclopropane-1-carboxylic acid synthase (ACC synthase) or ACC oxidase, the enzymes block ethylene biosynthetic pathway from methionine (antisense RNA)	Delay in ripening of fruits and wilting of flowers
Production of biopolymer PHB (poly-3-hydroxy butyric acid)	Three genes in a single operon in plasmid of bacteria (3-ketothiolase, acetyl-CoA reductase, and PHB synthase catalyse synthesis of PHB from acetyl-CoA)	Biodegradable plastics produced by plants
Plantibodies (antibodies from plant)	*sIgA* gene in tobacco plant (Secretory immunoglobulin A)	To treat oral pathogen *Streptococcus mutans*
	IgG gene in soybean plant (Immunoglobulin)	To treat against herpes simplex virus containing glycoprotein B
Production of specific antigens	*Chitinase* gene in tobacco plant (Chitinase enzyme)	To raise antibodies
	Gene for human cancer cell surface antigen in pea (Tumour cell antigen)	To diagnose and treat cancer patients
Edible vaccine	HBsAg gene in tobacco and potato (trying in banana) (Hepatitis B surface antigen)	Vaccination against Hepatitis B virus

Table 6.9 Some biopharmaceuticals produced in transgenic plants

Recombinant protein	Application	Plant
Human erythropoietin	To treat anaemia	Tobacco

(Contd)

Table 6.9 (*Contd*)

Recombinant protein	Application	Plant
Human enkephalins	Antihyperanalgesic by opiate activity (for sleep)	Canola
Human epidermal growth factor	Controlled cell proliferation for wound repair	Tobacco
Human α-interferon	For hepatitis B and C	Rice, turnip
Human protein C	Anticoagulant	Tobacco
Human haemoglobin	Blood substitute	Tobacco
Human serum albumin	To treat liver cirrhosis	Potato, tobacco
Human growth hormone	To treat dwarfism, promote wound healing	Tobacco
Angiotensin-1-converting enzyme	Hypertension	Tobacco, tomato
α-Tricosanthin	HIV therapy	Tobacco
Interleukins	Antiviral, anti-cancer	Potato, tobacco
Lysozyme	Antimicrobial	Rice
Human collagen	Tissue repair	Tobacco
Factor XIII (A-domain)	To stop bleeding	Tobacco
Antigens for edible vaccines		
Rabies Virus glycoprotein	Anti-rabies vaccine	Tomato, spinach, tobacco
Malarial antigens	Anti-malarial vaccine	Tobacco
Haemagglutinin	Anti-influenza vaccine	Tobacco
gp41 peptide	Anti-HIV-1 vaccine	Cowpea
C-Myc	Anti-cancer vaccine	Tobacco

Source: Based on Rastogi, S. and N. Pathak, 2009. *Genetic Engineering*, Oxford University Press, New Delhi, pp. 654 & 656.

Exhibit 6.E　Chloroplast Engineering

Chloroplast Genome and Transformation

The chloroplast or plastid genome is a small, circular DNA that resembles the prokaryotic genome. The protein synthesis machinery of chloroplasts is also like that of prokaryotes and uses 70s ribosomes; therefore genes of prokaryotic origin can be appropriately expressed when transferred to a chloroplast. **Genes of eukaryotic origin can also be transferred** Chloroplast engineering apparently means change(s) in the chloroplast genome or plastid proteins. In reality, the chloroplast genome encodes only a portion of the plastid proteins; the vast majority of the plastid proteins are encoded by nuclear genes. Thus, manipulation of plastid proteins or synthesis of a novel plastid protein can be achieved by either targeting chloroplast genome or in some cases the nuclear genome. For targeting the nuclear genome, a chimeric gene construct includes the gene of interest and sequences coding for a plastid transit peptide plus appropriate transcriptional control sequences; the gene product after its synthesis in the cytoplasm would then reach the chloroplast.

First transplastomic plant The plants carrying transgenic plastid genome (plastome) are often called transplastomic plants. Boynton and Gillham in 1988, first successfully transformed *Chlamydomonas reinhardtii* containing a single large chloroplast.

Development of vectors for chloroplast transformation   A number of DNA sequences are carefully included in the vector to facilitate transfer of genes of interest to the plastid genome, and not to the nucleus. The sequences (Fig. 6.E.1) are as follows: (i) flanking sequences from the chosen insertion sites in the plastid genome for easy entry by homologous recombination, (ii) promoter sequence for the marker genes, (iii) selective marker gene(s) for selection of cells with transformed plastids, (iv) 3′ untranslated region (UTR), (v) promoter sequence for the gene of interest (GOI), (vi) 5′regulatory elements (RE), (vii) GOI, (viii) 5′ UTR. 5′ RE and 3′ UTR help in the stability of mRNA transcribed from the marker gene. The most commonly used marker genes confer resistance to streptomycin, spectinomycin, and kanamycin.

Insertion sites for vectors in a chloroplast genome   A circular chloroplast genome with three common insertion sites, indicated by arrow, has been presented in Fig. 6.E.2. Vectors containing a foreign gene get integrated in these transcriptionally active sites. There are altogether 14 such insertion sites in the genome.

Development of vectors through two decades (since 1988)   The first successful chloroplast transformation in a higher plant, tobacco was made in 1989 using a vector of the ZS series. This vector design was based on pUC vector which has been discussed in Section 6.7.1. Since then, many vectors are being developed for improved efficiency and to serve different purposes. Different vectors insert transgenes at different sites of the chloroplast genome. Polycistronic vectors such as pPRV11OL with no separate promoter for the marker gene have been developed, in which the same promoter allows transcription of the marker gene and several other genes (GOIs).

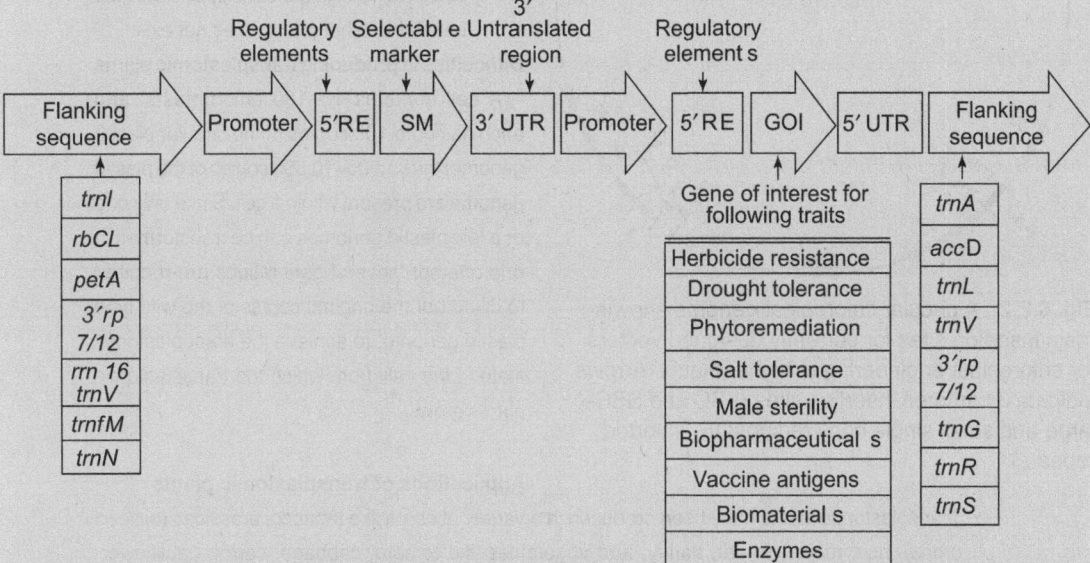

Fig. 6.E.1   Construction of a chloroplast specific expression cassette (vector) showing (i) the integrating sites (flanking sequences), (ii) promoters, (iii) 5′ regulatory elements (RE), (iv) selectable marker genes (SM) for selection of cells with transformed plastids, (v) 3′ and 5′ UTRs, and (vi) genes of interest (GOI) to be inserted. A few specific examples of flanking sequences and genes of interest have been mentioned in the boxes.

Scientists also tried to develop a universal vector for the transformation of the chloroplast genome, mainly by including flanking sequences from the highly conserved intergenic spacer of the IR region, containing genes *trnA* and *trnI* (Fig. 6.E.2). However, the experts concluded that a higher level of transformation efficiency is achieved with species–specific vectors as compared to the universal one.

Techniques for insertion of foreign genes   In general, four methods are followed for the introduction of foreign genes into the chloroplast genome—(i) biolistic or gene gun process to bombard leaves with tungsten particles coated with DNA, (ii) *Agrobacterium tumefaciens* (Ti plasmid) mediated, (iii) polyethylene glycol (PEG) mediated direct DNA uptake by protoplasts, (iv) microinjection of DNA directly into the plastid by a femtosyringe.

Advantages of chloroplast transformation

(i) It has already been mentioned that expression of genes from prokaryotes is easier when they are inserted in the chloroplast genome due to the similarity of the protein synthesizing machinery with that of prokaryotes.

(ii) As multiple copies of chloroplast genome exist in a cell, the level of expression and accumulation of foreign protein is high in the chloroplast.

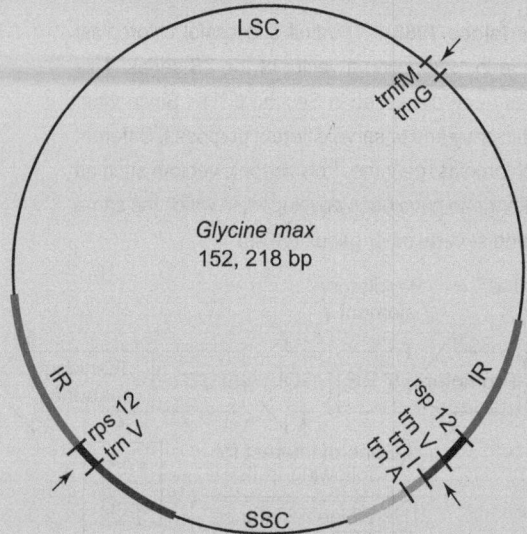

(iii) Multiple genes may be transferred to the chloroplast genome in a single attempt ('transgene stacking') and they may co-transcribe polycistronic mRNA.

(iv) The most important advantage of transplastomic plants over transgenic plants is biosafety, that is, the problem of escape of foreign genes as pollen and their transfer to other plants does not exist.

Difficulties in producing transplastomic plants

A cell contains 10–100 chloroplasts, and each has upto 100 copies of the circular plastid genome; thus 1,000–10,000 copies of the plastid genome are present within a cell. Since only one or a few plastid genomes can be transformed in one attempt, several generations are required to dilute out the original copies of the wild type plastid genome, to achieve the homoplastomic state in the calli from which the transplastomic plants grow.

Fig. 6.E.2   A circular chloroplast genome showing main insertion sites for currently designed vectors for chloroplast engineering in higher plants. Arrows indicate 3 common insertion sites (LSC and SSC–large and small single copy regions; IR–inverted repeats).

Applications of transplastomic plants

- Transplastomic plants have been produced in a variety of crops like tobacco, brassicas (oilseed crops), rice, maize, wheat, barley, and vegetables like tomato, cabbage, carrot, cauliflower, potato, and so on.
- Transplastomic plants with *rbcS* gene encoding 'rubisco' enzyme, *ndh* gene complex for improving photosynthetic efficiency, and lycopene β-cyclase (*lcy*) gene for improving the level of pro-vitamin A (β-carotene) in tomato fruits are some salient examples of chloroplast transformation.
- Chloroplast transformation also acquires other traits for the plants, such as insect resistance (with *cry* gene), herbicide resistance, salt and drought tolerance, and phytoremediation.
- Transformed plants can produce several biopharmaceutical proteins such as insulin-like growth

factor, interferons, monoclonal antibodies, vaccines for dreaded diseases—tetanus, plague, hepatitis, cholera, and so on.
- Useful biomaterials like elastin-derived polymer, polyhydroxybutyrate, xylanase, and other compounds may also be produced from chloroplast transformed plants.

A note may be added here; the genome of plastids other than chloroplasts, for example, leucoplasts, can also be targeted for production of proteins.

Exhibit 6.F   Terminator Technology

Crop plants with the transgenic terminator gene can bear seeds with all the normal characteristic features; however these seeds do not germinate. The terminator gene produces a protein that is toxic to the embryo and therefore does not allow the seed to germinate. A seed thus produced carries the endosperm, but not the embryo, and it can be sold as grain, but cannot be sowed.

The technology involves two gene systems.

Gene System I Terminator gene encodes ribosome inactivating protein (RIP) under the control of transiently active (during embryogenesis) *Lea* promoter—so that RIP does not allow protein synthesis and embryo development.

Expression of terminator gene is blocked by incorporation of a blocking sequence between promoter and the terminator gene (Fig. 6.F.).

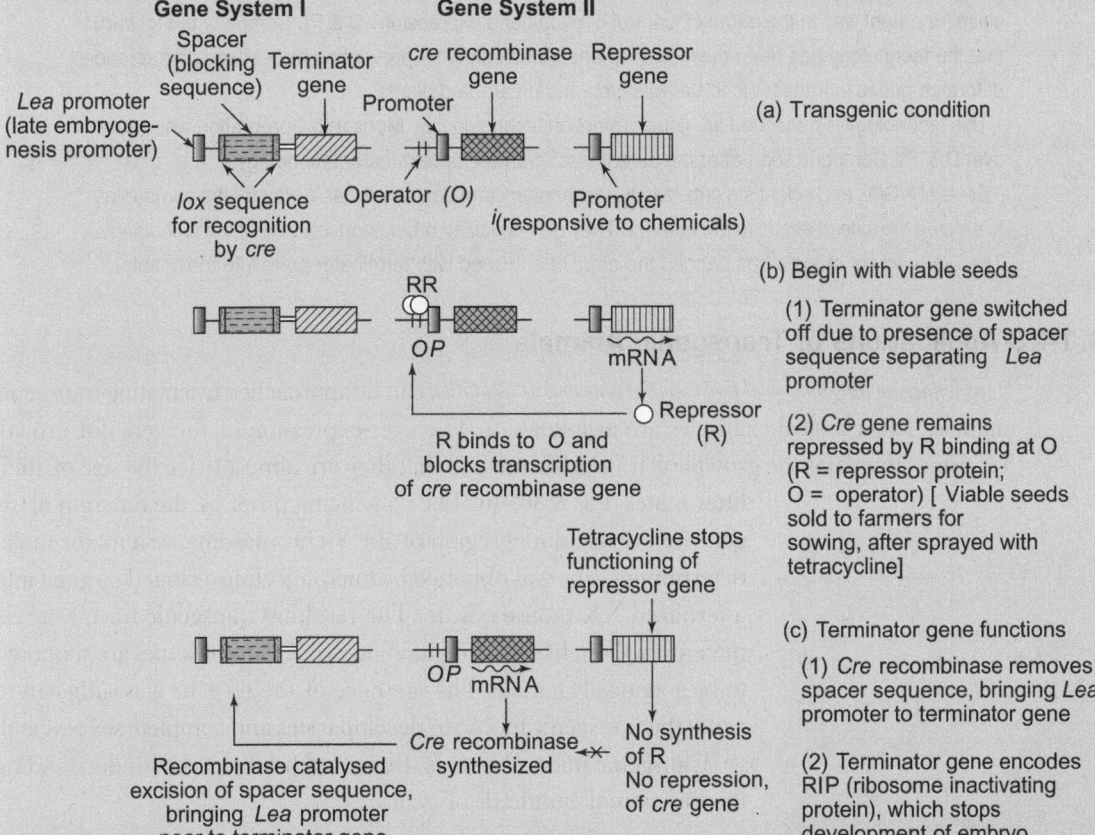

Fig. 6.F.   Genetic basis of terminator technology.

Source: Based on Rastogi, S. and N. Pathak, 2009, *Genetic Engineeering*, Oxford University Press, New Delhi, P. 694.

This blocking sequence can be removed by recombinase enzyme encoded by *cre* gene in gene system II. The *cre* recombinase acts at lox sequences on both sides of blocking sequence. The lox sequence is obtained from P1 bacteriophage.

Gene System II A *cre* gene encodes *cre* recombinase that acts at the *lox* sequence to remove the blocking (spacer) sequence in between the promoter and terminator gene. But *cre* cannot function due to a repressor protein (R) bound to its operator region (O). R is produced by a third gene. The *cre* gene can be derepressed (activated) by exogenous application of tetracycline. Tetracycline is absorbed by the seedling tissue.

Gene systems I and II are developed separately and then through hybridization are recombined into one transgenic variety of crop with functional hybrid seeds containing a viable embryo. Such viable seeds are sold to the farmers after they are sprayed with tetracycline. As the seed already contains an embryo, it germinates and produces the crop. At the time of seed formation as the tetracycline has already derepressed the *cre* gene, the blocking sequence would be removed, *Lea* promoter comes close to the terminator gene leading to its transcription and production of RIP that destroys the embryo. The seeds will not germinate but can be sold as grain.

Melvin Oliver, a scientist engaged in research with the US Department of Agriculture (USDA), Lubbock, Texas and the Delta and Pine Land (D&PL) Company, Mississippi mainly carried out the work on terminator technology for which patent was granted jointly to both the institutes in 1998. The claim for patent was in the name of 'control of plant gene expression'. D & PL Company put forward that the technology has been developed to ensure biosafety by preventing the escape and spread of foreign genes in transgenic to the wild type and untargeted plants.

This technology is described as 'gene protection technology' by Monsanto Corporation which took over D & PL Company soon after the patent was granted for terminator technology.

Several NGOs and scientists criticized the technology and described it as a way for the companies to enforce the sale of seeds to the farmers. They also criticized other shortcomings of the technology. The Government of India has banned the entry of any seed with terminator gene into the country.

6.14.3 Applications of Transgenic Animals

1. (a) *Understanding biological function of a gene and its regulation* can be approached by creating transgenic animals. Two classical examples are as follows: (i) The over-expression of the gene for growth hormone stimulates the growth of transgenic mice so that they are almost twice the size of their litter mates (Fig. 6.36). (ii) The convincing proof of the function of *sry* gene (sex-determining region of the Y chromosome) as a major male-determining gene was obtained by microinjecting a cloned *sry* gene into a fertilized XX mouse oocyte. The resulting transgenic mice were all male, even though oocytes containing two X chromosomes are supposed to be genetically female. The presence of the *sry* gene was sufficient to cause the transgenic mouse to develop testes and complete sex reversal.

Transgenic mice, *Drosophila*, frogs, and fish helped in understanding the function of hundreds of genes.

(b) *Gene knock-out technology* disrupts the function of a particular gene and the resulting effects provide clues to the gene's function(s). This

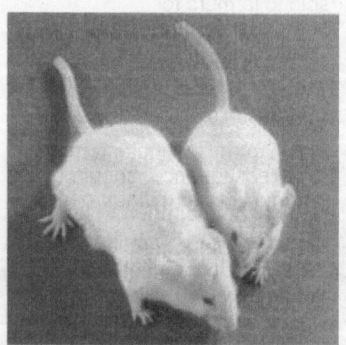

Fig. 6.36 Giant and normal mice. (Also see colour plate 2.)

technique relies on the process of homologous recombination of a constructed transgene with a particular gene in situ. In other words, regions of strong sequence similarity need to be present between the in-situ gene and the transgene for recombination to occur. Sometimes, the mutated form of a gene to be probed for function is used in a transgene construct. Knock-out mice with inactivated tumour suppressor genes, which normally suppress the spontaneous development of cancers, become prone to the development of different kinds of cancer.

(c) *Gene knock-in technology* can replace an endogenous gene with another functional gene following the homologous recombination procedure used for knock-out technology.

(d) *Specific deletion of a gene in a tissue* by using specific recombinase allows the functional assessment of a gene in a particular tissue. The transgene construct for the purpose includes *cre* (for recombinase) and *loxP* (for recognition site of *cre* recombinase) just as it is used for terminator technology in plants (see Exhibit 6.F).

2. *Animal modelling of human disease* by transfer of mutated or disease causing genes from human beings helps understand the development of diseases and their treatment with drugs. Some common examples of disease models in mice are for cystic fibrosis and Alzheimer's. Scientists from Harvard University created a genetically engineered mouse incorporating exogenous oncogenes (cancer genes) and named it 'OncoMouse' or the Harvard mouse, to test potential cancer drugs. The scientists obtained a US patent for the mouse and the company DuPont holds exclusive rights to its use.

In the course of studies on breast cancer, the transgenic mouse model revealed a regulatory element in the mouse mammary tumour virus (MMTV) that acts specifically in cells of the mammary gland, and indicated the involvement of *myc* and *ras* oncogenes.

A gene, *brac* 1 responsible for a certain form of brain cancer, was detected and cloned by AnyGene, a pharma company in Manchester. This gene was inserted in mice and the resulting transgenic mice as models of brain cancer provided opportunities to study the effects of potential anti-cancer medicines.

The above models used single gene defects in human beings; recently the modelling is being improved to include more complex diseases that involve multiple genes. In some cases, cross breeding between different modified mouse lines is needed.

3. *Improvement of production traits of animals* can be manipulated by transgenic technology. This leads to an increase in growth rate, milk and wool production, and improvement of the quality of milk and meat (Table 6.10). Thus, transgenic animal technology has great potential in agriculture to improve livestock to meet the demand of food for the growing population. Besides the examples in Table 6.10, there are many other items on the bench, and genetic engineers are trying to find different ways to improve farm animals and their products. Some of them are mentioned here: (i) expression of β-galactosidase in cow milk to reduce lactose content of milk, as 70 per cent of the world's population lacks lactase (lactose-hydrolyzing enzyme); (ii) alteration of milk *casein* content to increase its nutritive value, cheese yield, and ability to withstand processing; (iii) production of human lysozymes (antimicrobial enzymes) in cow milk to reduce mammary gland infection and

the chance of infection of the gut following consumption of milk; (iv) production of superior quality cystein-rich wool; (v) disease resistant animals; (vi) creation of inherited immunological protection by transgenesis with genes contributing to the immune system (MHC molecules, T cell receptor, lymphokines, antibody H and L chains); (vii) cold tolerance in fish by transferring the full length of antifreeze protein (*AFP*) gene, and so on. Many of these items are under development and trials in laboratory mice, and problems related to their commercial use are being resolved.

Table 6.10 Transgenic animals and improvement of production traits

Transgene	Animal	Product
Growth hormone (GH) (somatotropin) gene (inserted in embryo under metallothionein promoter)	Pig, sheep, cattle, goat, fish	Increase in body weight (meat)
rtGH cDNA	Catfish	More protein, less fat, and less moisture in edible muscle
c-ski oncogene	Pig, cattle	Increase in growth of skeletal muscle
Insulin-like growth factor I (IGF-I)	Pig	Reduction in body fat, increase in muscle fibres
Fatty acid desaturase gene from spinach (plant)	Pig	20% increase in linoleic fatty acid in adipose tissue (example of modification of fatty acid composition of animal with a plant gene)
Introduction of human lactoferrin gene and elimination of β-lactoglobulin gene	Cow	Cow milk with lactoferrin protein involved in iron transport and inhibits bacterial growth, absence of β-lactoglobulin removes a major allergen in cow milk (the quality of cow milk becomes near to human milk, better for consumption by infants)
Anti-freeze promoter of a fish (ocean pout) cloned with the growth hormone cDNA from salmon	Salmon (fish)	Fast growth (meat) and survive cold water (extending cultivation in cold climate)

4. Xenograft for human organ transplants can be found from genetically conditioned animals, particularly pigs. Xenograft is the tissue or organ from one species grafted to another species; for example, transplanting a pig organ into a human. The demand for donated organs like heart, liver, kidney, and lung is steadily increasing throughout the world and has far exceeded the availability. Xenotransplants from animals to humans are a way of solving the organ shortage problem.

Pig's organs are similar in size and physiological function to those of human beings and natural porcine (pig's) pathogens (carried along with the grafts) are mostly non-infecting for human beings. Pigs reach reproductive maturity within a year, have a short gestation period of only four months, and produce a large litter size of about a dozen piglets. All these features of pigs have been considered as suitable for their use as a source of xenotransplants.

The problem with organ transplantation is not only the availability of organs, but also the immunological reactions of the host leading to hyperacute rejection of the foreign graft. 'Knock-out' pigs with a deletion of the gene encoding the protein against which the host's immunity works would solve the rejection problem to a good extent. Incorporation of one or more genes for human complement inhibiting protein in pigs would also protect a transplanted organ from the inflammatory response that initiates graft rejection. In fact, transgenic pigs with human complement inhibitor genes have been produced.

5. *Transgenic 'pharm' animals producing pharmaceuticals* and nutritional supplements for human use have been designed. Once genes are incorporated in the transgenic animals, they pass the cloned genes to their offspring in a Mendelian fashion. Several genes can be incorporated at a time.

(i) *As milk protein* The foreign genes encoding pharmaceutically important proteins are placed under a promoter active only in mammary tissue, for example, the β-lactoglobulin promoter and the casein promoter. Thus, the expression of foreign proteins is restricted only to the mammary glands; production of foreign proteins in all cells is not desirable and may sometimes be fatal for the animals. To prevent the entry into the bloodstream and the functioning of pharmaceutical proteins in the host animals, the transgene is usually modified such that the protein produced in the mammary gland cannot work unless some proteolysis occurs. Proteolysis is normally done during processing and purification of the products from the milk (Table 6.11). Thus, a drug protein remains confined in the mammary gland and does not affect the normal physiology of the transgenic animals, and it can easily be purified. This transgenic technology can be used to enrich the milk itself with specific and necessary proteins, and also allows the use of the mammary glands of dairy cows, goats, sheep, and pigs as biological bioreactors for producing large amounts of pharmaceuticals and other proteins. So these farm animals are often called 'pharm' animals.

Table 6.11 Some human recombinant proteins (pharmaceuticals) obtained from the milk of transgenic[1] animals

Gene for the product	Transgenic animal
α₁–Antirypsin[2]	Sheep
α–Lactablumin	Rat, cow
Antithrombin III[3]	Goat
β-Interferon	Mouse
γ-Interferon[4]	Mouse
Factor VIII	Sheep
Factor IX[5]	Mouse, sheep
Fibrinogen	Mouse
Growth hormone	Mouse
Hepatitis B surface antigen[6]	Goat
Human antibodies	Mouse
Interleukin–2	Rabbit
Insulin	Cow
Lactoferrin[7]	Mouse, cow
Lysozyme	Mouse
Protein C[8]	Mouse
Serum albumin	Mouse
Surfactant protein B	Mouse
tPA (tissue plasminogen activator)	Goat

Note: Transgenic experiments that are initially successful with mouse are often transferred to higher animals.

1. Gene constructs include β-lactoglobulin or casein promoter from other animals.
2. Protease inhibitor, treatment of emphysema and cystic fibrosis.
3. Blood clotting protein, treatment of ATIII deficiency disease and in open heart surgery.
4. Anti-cancer, antiviral.
5. Blood clotting protein, treatment of haemophilia B.
6. Vaccine.
7. Iron transport protein.
8. Anticoagulant, treatment of haemophilia and used for surgery.

Milk is a renewable secreted body fluid, produced in large quantities, collected frequently without any harm to the animal, and consumed throughout the world by all age groups. Several grams of transgenic protein may be obtained from a litre of milk.

Animal transgenics have certain advantages over the use of recombinant bacteria for generating pharmaceuticals; maintaining them is less expensive than running industrial-scale fermenters and purifying of the product is comparatively easy. There is no problem of post-translational modifications of protein as is the case in bacterial cells.

(ii) *As egg protein* A large amount of ovalbumin is secreted by the cells of the reproductive tract of a hen. Transgenes expressed in these cells can cause accumulation of the transgenic proteins to be encased in the egg shell. Neutraceuticals, monoclonal antibodies (McAbs), growth hormone, human serum albumin, insulin, and α-interferon can be derived from this source.

6. *Mammalian cell lines produce many recombinant therapeutic products.* Mammalian cells have been the focus of biopharmaceutical industrial research since the mid 1990s, mainly because these cells can glycosylate (add a carbohydrate molecule) human proteins in the correct manner. Some of the cell lines in use are Chinese hamster ovary (CHO) cells, some murine myeloma cell lines, and the human retinal line PER-C6.

7. *Gene therapy* is used to replace or repair inherited non-functional or mutated genes causing diseases in humans. An abnormal gene may be replaced with a normal gene through homologous recombination. The abnormal gene may also be repaired through selective reverse mutation, to bring back its normal function. The most common vector used for the purpose is a virus that has been genetically modified to carry normal human DNA as an insert (gene). Adenoviruses, adeno-associated viruses (AAV) retroviruses, and herpes simplex viruses are usually used as vectors for gene therapy.

Non-viral options for gene delivery are bombardment of DNA coated particles by a gene gun, lipofection of correcting or therapeutic DNA, and chemically linking the DNA to a molecule capable of binding to special cell surface receptors. The last option is less effective.

Gene therapy using adenovirus as vector   Initially, the normal gene of interest is isolated and copied by PCR, and then a construct containing all the genetic elements for correct expression is recombined with the viral genome. The gene is introduced by the viral vector into the host cell (Fig. 6.37). In most cases, the introduced gene gets inserted at an unspecified location in the host genome.

Adenovirus is responsible for respiratory, intestinal, and eye infections and common cold in human beings. It is a double-stranded DNA virus. The genetic material of adenoviruses is not incorporated into the genetic material of the host cells, unlike other viruses. The DNA molecule from adenovirus exists as an independent molecule in the nucleus and transcribes just like any other gene. However it cannot replicate along with the genetic material of the host cells and thus cannot continue in the descendant cells. Overcoming this transient nature of adenovirus, requires re-administration of the genetically engineered viruses in a dividing cell population. The absence of integration in the host cell genome makes adenovirus a better candidate for gene therapy; the integration of some other types of viruses, in an inadvertent region of the host cell genome can lead to cancer. The adenovirus-mediated gene therapy product 'Gendicine' has been licensed to treat cancer. Gendicine is composed of adenoviruses bearing *p53* gene known to down-regulate the development of malignancy.

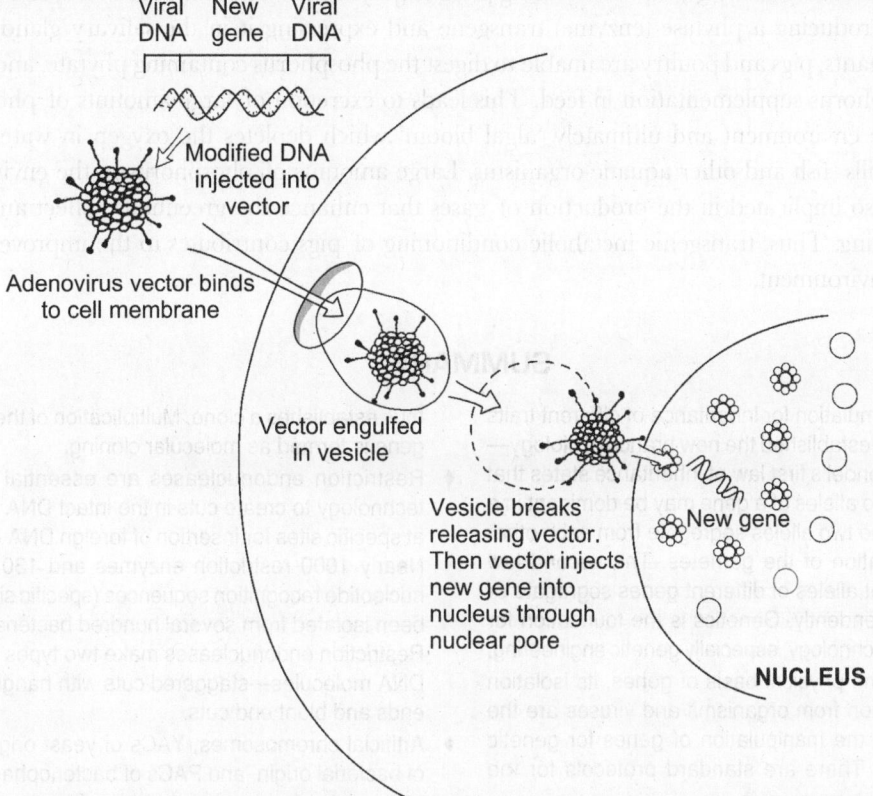

Viral New Viral
DNA gene DNA

Modified DNA injected into vector

Adenovirus vector binds to cell membrane

Vector engulfed in vesicle

Vesicle breaks releasing vector. Then vector injects new gene into nucleus through nuclear pore

New gene

NUCLEUS

Fig. 6.37   Gene therapy using Adenovirus vector.

Gene therapy can be performed either in germ cells (sperm or egg) or in somatic cells.

8. *Production of spider silk (biosteel)* has been initiated in lactating Nigerian dwarf goats by Nexia Biotechnologies in Canada. A single gene for spider web was implanted into egg cells of the goat and was designed for expression in mammary glands. Cloning produced the first 'web kids' or 'silk milk goats'. These GM goats secreted spider silk along with their milk. Basically both the proteins are secretory products. The spider silk can be isolated and purified from the milk into polymer strands which can be spun into thread for the production of light, tough, flexible material, known as 'biosteel', to be used for a long list of utility items such as ultra-light bullet-proof vests, military uniforms, surgical thread, fishing line, artificial ligament to repair broken limbs, coating of space stations, aircraft, racing cars, and sometimes bridges.

The spider silk is so strong and its tensile strength is $1,36,000$ kg/in.^{2}; in other words a square inch of spider silk material can hang that much weight. Thus this flexible lightweight substance is comparable to steel in its capability.

9. *Ornamental fish production* has got a commercial boost from transgenic technology. Transfer of red fluorescent coral gene to common Zebra Danio fishes made them red fluorescent Glow Fish™, the first commercialized transgenic pet in the US.

10. *Pro-environmental pigs*, capable of using phytate (phytic acid) of plant-based feed, can be raised by introducing a phytase (enzyme) transgene and expressing it in the salivary gland. Unlike ruminants, pigs and poultry are unable to digest the phosphorus containing phytate, and require phosphorus supplementation in feed. This leads to excretion of large amounts of phosphorus to the environment and ultimately 'algal bloom' which depletes the oxygen in water bodies and kills fish and other aquatic organisms. Large amounts of phosphorus in the environment are also implicated in the production of gases that enhance the greenhouse effect and global warming. Thus, transgenic metabolic conditioning of pigs contributes to the improvement of the environment.

SUMMARY

♦ Mendel's formulation for inheritance of different traits in pea plants established the new branch of biology—Genetics. Mendel's first law of inheritance states that one of the two alleles of a gene may be dominant in a hybrid, but the two alleles segregate from each other during formation of the gametes. The second law proposes that alleles of different genes segregate or assort independently. Genetics is the foundation for modern biotechnology, especially genetic engineering.

♦ DNA being the physical basis of genes, its isolation and purification from organisms and viruses are the first steps in the manipulation of genes for genetic engineering. There are standard protocols for the purpose.

♦ Quantitation of isolated and purified DNA (or in some cases RNA) in nanogram (ng) or microgram (µg) amounts in solution is performed mainly by three methods—spectrophotometrically, fluorescent dye-based method, and gel electrophoresis.

♦ Recombinant DNA (rDNA) is constructed by recombining DNA molecules from different origins. The DNA molecule from the bacterial or viral source is called a vector as it can carry a foreign gene and transfer it to a target cell. Different types of vectors can be constructed as per requirement. Often, a recombinant DNA containing a gene of interest is introduced into a bacterium to amplify the transferred gene in course of multiplication of the transformed bacterium into millions of copies and then to get the desired product encoded by the transferred gene in large amounts. The transformed original bacterium thus establishes a clone. Multiplication of the inserted gene is termed as molecular cloning.

♦ Restriction endonucleases are essential in rDNA technology to create cuts in the intact DNA molecule at specific sites for insertion of foreign DNA or genes. Nearly 1000 restriction enzymes and 130 different nucleotide recognition sequences (specific sites) have been isolated from several hundred bacterial strains. Restriction endonucleases make two types of cuts in DNA molecules—staggered cuts with hanging sticky ends and blunt end cuts.

♦ Artificial chromosomes, YACs of yeast origin, BACs of bacterial origin, and PACs of bacteriophage origin, are constructed to play the role of vectors with the ability to accommodate much bigger piece of DNA containing large eukaryotic genes.

♦ Complementary DNA (cDNA) is synthesized from mRNA in vitro using reverse transcriptase, and used as a gene insert for cloning vectors. mRNAs are without intergenic region and introns, and thus represent genes in a more precise or shortened form.

♦ A genomic (DNA) library is a collection of clones representing the entire genome of an organism. The DNA library is useful for studying individual genes or as a source of genes for amplification and for biotechnological products.

The library can also be constructed with cDNA. As cDNA clones do not include non-functional DNA stretches, it is easy to accommodate an entire gene in a clone in a cDNA library, which is difficult for a genomic library. Furthermore, since splicing does not occur in bacteria due to lack of proper enzymes, eukaryotic gene sequences in cDNA form can easily be expressed in bacteria for technological products.

♦ rDNA vectors are introduced into bacteria or other host cells by different chemical and physical methods.

♦ Selectable marker genes are one of the essential features of vectors. These markers help in selecting the recombinant (transformed) clones harbouring rDNA.

♦ For screening DNA libraries (clones) for genes of interest, two basic types of screening strategies are DNA sequence-based screening, and protein structure/function dependent screening. Several specific techniques have been developed for these purposes.

♦ Polymerase chain reaction (PCR) is an in-vitro technique to amplify exponentially a selected region of a DNA molecule (gene) in a short time. The sequence can be amplified to a million copies in just a few hours. Primers, 18–30 oligonucleotides in length, complementary to the flanking sequences of the DNA piece to be amplified, DNA polymerase (*Taq*) and other reaction components are provided in small PCR tubes which are then placed in a PCR machine (often called a thermocycler).

♦ PCR is a highly versatile technique which can be varied to suit the need of researchers.

♦ PCR technique is one of the most important inventions of the 20th century in molecular biology. It makes it possible to identify and manipulate specific genes. This in turn helps detect various infectious agents including viruses. It has made us hopeful of being able to correct genetic diseases. It helps in the process of getting DNA fingerprints which is a molecular way of identifying individuals beyond doubt.

♦ Defined alteration in a gene sequence of a clone in vitro is made by site-directed mutagenesis. It helps in understanding gene function and expression of new proteins, and protein engineering.

♦ A transgenic plant or animal is a fertile organism that carries an introduced gene(s) in its germ line so that the new gene(s) becomes part of the genome of the next generation. Transgenic plants and animals are used for mass production of desired proteins, hormones, and pharmaceuticals.

Transfection of eukaryotic organisms with external genes can be done following several methods.

♦ Ti plasmid from *Agrobacterium tumefaciens* is a widely used vector for producing transgenic plants. Transgenic plants with *Bt* toxin (*Cry* protein) have an in-built pesticide. Similarly herbicide resistant transgenic agronomic crops can be raised. Plants may be made transgenic with animal genes with the help of the Ti plasmid.

♦ Animal viral vectors are used for gene transfer to cultured cells as well as to living animals. Retrovirus, adenovirus, herpesvirus, and adeno-associated virus (AAV) are used for producing transgenic animals and for human gene therapy.

♦ P-element of *Drosophila* is a transposon (wandering gene) class of genetic material and is used as a vector for transformation of flies.

♦ Monocot plants and conifers cannot be transformed by Ti plasmid. The gene gun helps in transformation of plants by bombardment of microparticles coated with foreign DNA.

♦ Electroporation, microinjection of DNA, lipofection (using liposomes) are other usual methodologies to produce transgenic plants.

♦ To produce transgenic animals, the transgene or DNA is introduced by virus vectors or direct microinjection in an early embryonic stage, preferably after fertilization.

♦ The technique of somatic cell nuclear transfer (SCNT) to the enucleated ovum came into practice to produce transgenic animals and led to the creation of the sheep Dolly. Since the cloning of Dolly, somatic cell nuclei have been used to clone cattle, goats, and pigs.

♦ Embryonic stem (ES) cells are derived from the inner cell mass in the mammalian blastula. These cells are pluripotent and can develop into any types of tissue. These cells have been experimentally used for therapy of damaged tissue in heart and brain. Introduction of foreign genes through ES cells produces transgenic animals.

♦ By transferring new DNA sequences into microbes, plants and animals, or by removing or altering DNA sequences in their genomes, completely new strains or varieties are created to perform specific tasks of human interest.

♦ Applications of genetic engineering are innumerable. Commercial implementations of genetic engineering opened a multi-billion dollar market throughout the world and has changed certain life styles of human beings. This chapter presented an enormous list of the applications of genetic engineering.

Sanger

Frederick Sanger
(born 13 August 1918)

Frederick Sanger, born on 13 August 1918 in Rend-comb, England has the unique distinction of being the only person to win the Nobel Prize in chemistry twice. His seminal works led to the development of techniques to sequence proteins and nucleic acids.

Sanger obtained his BA degree in natural sciences from St John's College, Cambridge in 1939. After an initial stint in N.W. Pirie's lab where he investigated whether edible protein could be obtained from grass,

he moved on to Albert Neuberger's lab to study the metabolism of lysine. He was awarded a PhD degree in 1943.

In 1951, he determined the complete amino acid sequence of the two polypeptide chains of bovine insulin. For this work he was awarded his first Nobel Prize in 1958. Subsequently, his laboratory developed interest in solving the problem of sequencing DNA. It should be noted that around this time, the central dogma of molecular biology was being formulated. In 1977, Sanger and colleagues introduced the 'Sanger method' or the dideoxy chain-termination method for sequencing DNA molecules. This breakthrough led to the second Nobel Prize in 1980, which he shared with Walter Gilbert and Paul Berg. Sanger's sequencing technique is still in use in most molecular biology and biochemistry laboratories around the world, more than three decades after its discovery. The technique also lies at the heart of global sequencing projects like the Human Genome Project and can be considered responsible for the birth of the field of genomics. A modest person at heart, Sanger still terms his path-breaking discoveries as methods he developed to solve the problems at hand, not necessarily keeping in mind their larger implications.

EXERCISES

Objective Questions

1. Multiple Choice Questions
   (i) The approximate total number of genes in human beings is
   (a) 10×10^4
   (b) 3×10^3
   (c) 3×10^4
   (d) 3.2×10^5
   (ii) The lacZ gene in *E. coli* encodes
   (a) lactose
   (b) galactose
   (c) β-galactosidase
   (d) none of these
   (iii) Neomycin physphotransferase (*npt II*) gene confers resistance to the antibiotic
   (a) neomycin
   (b) kanamycin
   (c) streptomycin
   (d) tetracycline
   (iv) Nylon membrane is the recent choice for filter/membrane in blotting experiments because
   (a) it has 0.45 μm porosity

   (b) it holds smaller nucleic acid molecules it is most durable and withstands repeated hybridization
   (c) it shows higher nucleic acid binding
   (v) For successful xenograft transplantation from pig to man, the solution is
   (a) histocompatibility gene 'knock-out' pig
   (b) immunosuppression of human host
   (c) pig with histocompatibility gene 'knock out' and transgenic for human complement inhibition is the right solution
   (d) none of these
   (vi) Human insulin gene can be expressed in large quantity in
   (a) plant cells
   (b) mouse cells
   (c) bacteria
   (d) human cells
   (vii) Lipase and proteinases are used as detergents
   (a) for genetically manipulated microorganisms in bioreactors

(b) in genetically altered animal cells in in vitro cultures

(c) for transgenic plant cells in bioreactors

(d) none of these

(viii) Genetically manipulated *Vaccinia* viruses as vaccine can provide immunity only against

(a) small pox

(b) small pox and hepatitis B

(c) hepatitis B, rabies, herpes simplex and small pox

(d) none of these

(xi) In 1979, Ananda M. Chakrabarty produced the oil cleaning superbug out of *Pseudomonas* by

(a) rDNA technology

(b) electroporation

(c) conjugation

(d) transduction

(x) Abiotic stress-tolerant transgenic plants need to be engineered

(a) with a separate gene for each kind of stress

(b) to tolerate ROS to withstand wide range of environmental stress

(c) both of these

(d) none of these

(xi) The gene(s) responsible for imparting insect resistance in different transgenic plants are

(a) only *Cry*

(b) cholesterol oxidase only

(c) both *Cry* and cholesterol oxidase

(d) *Cry*, cholesterol oxidase, and trypsin inhibitor

(xii) To restrict the expression of pharmaceutically important proteins of transgenic 'farm' animals to the mammary gland, the foreign genes are

(a) controlled by steroid hormones

(b) kept under casein promoter

(c) kept under control of growth hormone

(d) none of these

2. Fill in the blanks in the following sentences.

(a) The technique for exponential amplification of a selected region of a DNA molecule is known as ....

(b) The technique of exponential amplification of DNA was developed by ....

(c) Primers for synthesis of DNA is made up of ....

Review Questions

1. What are the findings of Mendel's monohybrid cross experiments?

2. What is the proposition in the 'second law of inheritance' formulated by Mendel?

(d) DNA fingerprinting is often used for ....

(e) dNTPs are required for ....

(f) Defined alteration at a site in a gene sequence of a clone in vitro is known as ....

3. Mark the following statements as True or False.

(a) Mendel performed his experiments with *Pisum sativum*.

(b) Mendel presented his results in the Science Society of Vienna, Austria.

(c) Mendel assigned the term 'gene'.

(d) Bateson coined the term 'genetics'.

(e) Blue eye of man is a dominant trait.

(f) Brown stem of tomato plant is a dominant trait.

(g) Isopropanol can be used instead of ethanol for precipitation of DNA in aqueous solution.

(h) Animal cells are disrupted with lysozyme to obtain DNA.

(i) The non-ionic detergent Triton X-100 helps in the precipitation of viral nucleic acid.

(j) Absorption maximum of nucleic acids is at 260 nm wavelength of UV light.

(k) Hoechst 33258 binds into the major groove of DNA.

(l) Yeast is a prokaryote.

(m) Yeast has a haploid genome.

(n) Life cycle of yeast is 90 minutes.

(o) It is estimated that yeast shares approximately 23% of its genome with the human genome.

(p) Yeast genome has been sequenced after the human genome project.

(q) Several strains of yeast have naturally occurring plasmids.

(r) Yeast profusely secretes different types of protein.

(s) Yeast chromosome can be used to create shuttle vectors.

4. Match the statements in Column 2 with the items in Column 1.

Column 1	Column 2
(a) EtBr	(i) binds preferentially to A-T rich regions of DNA
(b) Hoechst 33258	(ii) is used as tracking dye
(c) Bromophenol blue	(iii) interacalates into DNA

3. Define allele.

4. Mention two chemical agents necessary for disruption of bacterial cells for DNA extraction.

5. Explain the procedure of deproteination of the cell extract to obtain nucleic acid.

6. Write down the total number of base pairs in the human genome.

7. What treatment destroys all intrinsic DNase activity in RNase preparations?

8. What is the mechanism for precipitation of DNA in solution using ethyl alcohol?

9. Provide a brief note on using 'dialysis' for precipitation of DNA in solution instead of ethyl alcohol?

10. Why is a plant kept in a dark place one to two days prior to harvest of tissue for DNA extraction?

11. What is the reason for using the detergent cetyl trimethyl ammonium bromide (CTAB) in the course of DNA extraction from eukayotic cells?

12. Why is ethidium bromide used for detection and quantitation of DNA?

13. How is RNase activity inhibited during mRNA extraction?

14. What do you understand by 'molecular cloning'?

15. Characterize 'restriction endonucleases'.

16. Explain the process of assigning name to a restriction enzyme.

17. Provide a brief note on the types of ends generated after cleavage with type II endonucleases.

18. Define vector and mention three essential features of a cloning vector.

19. Differentiate the following vectors and mention their specific utilities in rDNA technology: (i) Plasmid vector, (ii) Bacteriophage vector, and (iii) Cosmid vector.

20. Write briefly on YACs and BACs.

21. What are the advantages of using a cDNA insert rather than genomic sequences of DNA for rDNA?

22. How is the RNA-DNA duplex converted to a double -stranded cDNA molecule?

23. What are the differences between genomic library and cDNA library?

24. What are the methods used for introduction of rDNA-vectors into bacteria?

25. How do the 'selectable marker genes' help in the selection of transformed cells?

26. What is the colony blotting and hybridization technique used for?

27. What is a probe? Name a few DNA probes.

28. Discuss the technique of Southern blotting and hybridization.

29. Provide a brief note on *Taq* enzyme.

30. Name at least four variants of PCR.

31. Elucidate the molecular mechanism behind the use of Ti plasmid to generate transgenic plants.

32. Discuss GMOs with reference to *Bt* toxin and the ethical concern.

33. How did genetic engineers develop plants resistant to bromoxynil, a herbicide?

34. Mention two essential points for construction of viral vectors for transfection of plants and animals.

35. What is P element in *Drosophila*? What is the significance of 'inverted terminal repeat sequences' of the element? What does 'incomplete P element' mean?

36. Write a note on SV40 virus as vector.

37. What is the basic principle of gene gun to transfect cells?

38. What is the meaning of 'Electroporation'?

39. Mention the name of three techniques for direct gene transfer in plant cells, other than microinjection, gene gun and electroporation.

40. Describe point-wise 'Somatic Cell Nuclear Transfer' (SCNT) technique for the production of clones of animals.

41. Provide your arguments for the following statement:

'Genetic engineering revolutionized the concept and practice of beeding in organisms'.

7

Genomics, Proteomics, and Bioinformatics

LEARNING OBJECTIVES

♦ Genomics, which deals with mapping, sequencing, and comparative analyses of genomes of different organisms
♦ Importance of single nucleotide polymorphisms (SNPs), that is, variation in single nucleotides throughout the genome
♦ The notion that functional genomics initiated proteomics
♦ Structures and functions of all proteins encoded by a genome
♦ Bioinformatics, which aided by computers, systematically maintains sequence data of DNA, protein, mRNA, genome content and arrangement in various organisms
♦ Bioinformatics allows retrieval of data by researchers for various purposes including prediction of structure and function of newly discovered macromolecules

INTRODUCTION

The term *genome* refers to one complete copy of the genetic information or one complete set of chromosomes (haploid or monoploid) of an organism.

In the first decade of the 20th century, geneticists began making efforts to identify and map genes of organisms of interest, such as *Drosophila*, maize, mice, algae, yeast, and bacteria with the help of mutants. The geneticists identified spontaneous mutations or mutants induced by chemical and physical agents as the first step. Then they worked out the linkage of the genes and their linkage maps. They even identified the location of genes on particular chromosomes. The whole endeavour was difficult, labour intensive, and time-consuming.

From the mid-1980s, geneticists used recombinant DNA technology for genetic analysis. The collection of DNA sequences in clones of a genomic library of an organism (see Section 6.6) were pieced together into overlapping sets to construct the final genetic and physical maps for the entire genome. This was the beginning of genomics.

In a way, functional genomics initiated proteomics which deals with the structure and functions of all the proteins encoded by a genome.

Knowledge of the nucleotide sequences of entire genomes provided an enormous wealth of information. Storing, retrieving, and classifying the recorded information from a genome represented by billions of nucleotide base pairs and using it for human utilities became a challenging task. This challenge virtually gave birth to another discipline—*bioinformatics* (biology + informatics). Bioinformatics also provides support to proteomics as it does to genomics.

7.1 GENOMICS

Thomas H. Roderick in 1986 coined the term *genomics*. *Genomics* is a sub-discipline of genetics devoted to the mapping, sequencing, and functional and comparative analyses of genomes of different organisms.

More detailed maps and sequences of genomes are now available and have given impetus to the enormous growth of the sub-discipline of genomics in recent times. Now, genomics can be divided into the following categories:

(i)   Structural genomics—study of genome structure
(ii) Functional genomics—study of genome function
(iii) Comparative genomics—study of genome evolution

Structural genomics forms the base of all sub-disciplines and is concerned with accumulating data that has been generated since the inception of genomics. Functional genomics has just entered an explosive growth phase. It will allow researchers to monitor the expression of entire genomes, that is, all the genes in an organism throughout its growth and development or in response to environmental changes.

Functional genomics includes studies of the *transcriptome*, the complete set of RNAs transcribed from a genome, and the *proteome*, the complete set of proteins encoded by a genome. In fact, functional genomics created another new discipline, *proteomics*, which deals with the structure and functions of all the proteins in an organism.

7.1.1 Three Types of Genomes in Eukaryotic Cells

It may be mentioned here that eukaryotic cells have a nuclear genome, a mitochondrial genome, and additionally a chloroplast genome in plants and algae. Mitochondrial and chloroplast genomes consist of circular DNA molecules, similar to those of bacteria. The nuclear genomes in most cases are multiple DNA molecules in the form of a haploid set of chromosomes. Nuclear genome is discussed in this chapter.

7.2 GENOME SEQUENCING

The ultimate structure of a genome is the map of specific genes with exact nucleotide-pair sequence (complementary nucleotide pairs on two strands of a DNA molecule) in spatial relation to the chromosome(s). The complete report of a genome may also include a chart of all nucleotide-pair changes that alter the function of the genes or chromosome(s).

Before 1975, to sequence entire chromosomes was a matter of conjecture and considered a laborious task that could extend for years. However, by late 1976, the entire 5386 nucleotide long chromosome of phage ΦX174 had been sequenced. The information content of DNA lies in the arrangement of four nucleotides (A, C, G, and T) and the process of determining the sequence of these nucleotides in a given DNA molecule is referred to as *DNA sequencing*. Thus, DNA sequencing forms the basis of genome sequencing.

7.2.1 Development of Techniques for Sequencing DNA

Four major advances in molecular biology essentially contributed to the development of the routine procedure of DNA sequencing:

(i)   Discovery of *restriction enzymes* that are nucleotide site specific and cut DNA molecules to produce specific segments or fragments.

(ii) Improvement of *gel electrophoresis* procedures to the point where DNA fragments differing in length by a single nucleotide could be resolved.

(iii) *Gene cloning* technique facilitated the preparation of large quantities of a particular DNA fragment and construction of gene libraries.

(iv) The techniques of *DNA sequencing*, developed in 1977 by Allan Maxam and Walter Gilbert at Harvard University, and by Fred Sanger and colleagues at Cambridge, England, helped in determining the nucleotide sequences of DNA molecules.

The first three techniques have been elaborated in Chapter 6.

7.3 DNA SEQUENCING TECHNIQUES

There are two major techniques that are used for sequencing DNA: (i) Maxam–Gilbert's method and (ii) Sanger's method. Although these methods are based on different principles, they both generate four populations of oligonucleotides (short DNA molecules) terminating at A/C/G/T residues. When these oligonucleotides are subjected to polyacrylamide gel electrophoresis (PAGE), the individual DNA fragments differing in length (and thus the mol. wt) resolve into different bands (Fig. 7.1). The porosity of the gel is such that individual DNA fragments differing in length even by a single nucleotide would form separate bands. The four populations of oligonucleotides ending at four different nucleotides are loaded into adjacent lanes of a sequencing polyacrylamide gel, and the order of nucleotides at the end of the fragments can be determined directly from the image or autoradiogram of the gel. For autoadiography, the double-stranded DNA fragment is first radio-labelled at the 5′ end with α-^{32}P in Maxam and Gilbert's method.

The principle underlying the generation of oligonucleotide fragments of varying size is different in Maxam–Gilbert's and Sanger's techniques.

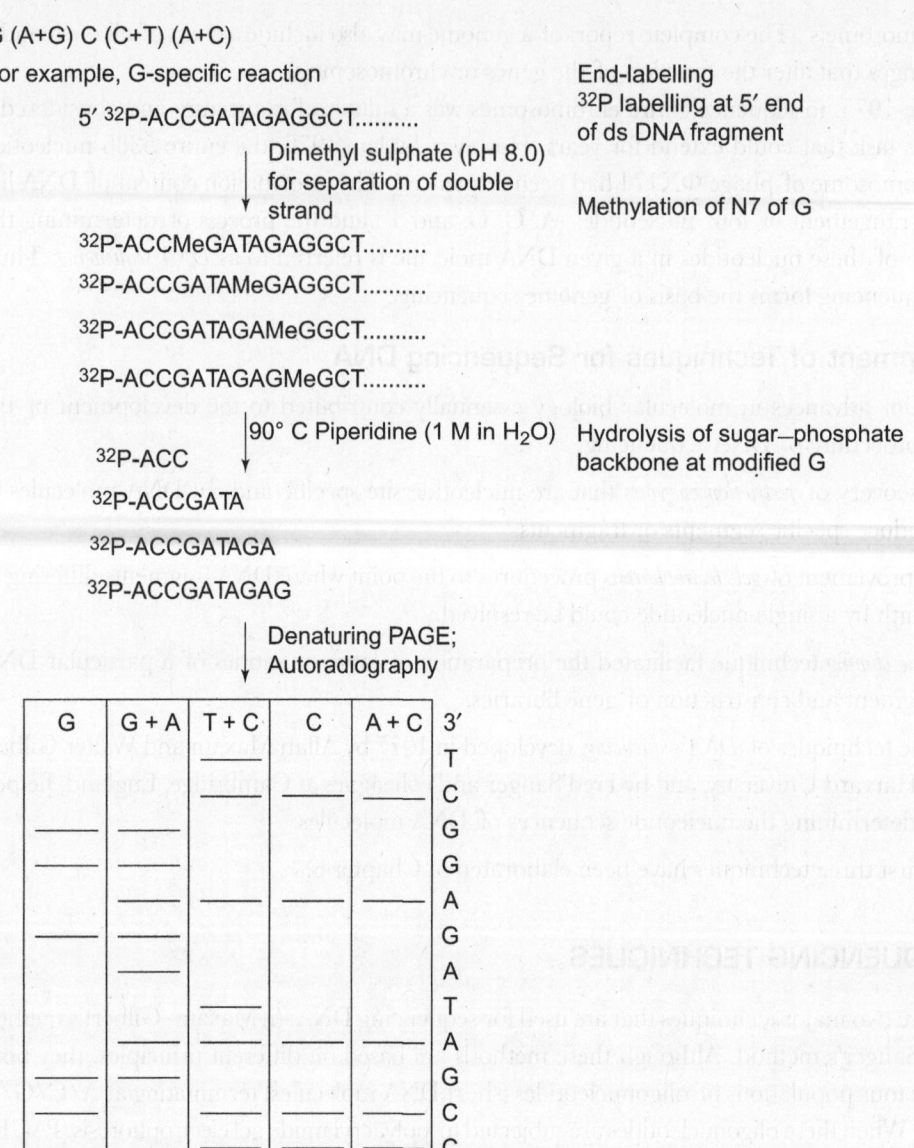

G (A+G) C (C+T) (A+C)

For example, G-specific reaction

5′ ³²P-ACCGATAGAGGCT..........

↓ Dimethyl sulphate (pH 8.0) for separation of double strand

³²P-ACCMeGATAGAGGCT..........

³²P-ACCGATAMeGAGGCT..........

³²P-ACCGATAGAMeGGCT..........

³²P-ACCGATAGAGMeGCT..........

End-labelling
³²P labelling at 5′ end of ds DNA fragment

Methylation of N7 of G

90° C Piperidine (1 M in H₂O)   Hydrolysis of sugar–phosphate backbone at modified G

³²P-ACC
³²P-ACCGATA
³²P-ACCGATAGA
³²P-ACCGATAGAG

↓ Denaturing PAGE; Autoradiography

G	G+A	T+C	C	A+C

3′
T
C
G
G
A
G
A
T
A
G
C
C
A
5′

Fig. 7.1 Maxam–Gilbert method of chemical degradation for DNA sequencing after base-specific modification.

7.3.1 Maxam and Gilbert's Technique (Chemical Degradation)

In Maxam–Gilbert's technique, a specific base is modified by a chemical agent and the subsequent chemical treatment (hot piperidine 1 M in water, 90°C) hydrolyses the sugar–phosphate backbone of the modified base and terminates the chain length at that point (Fig. 7.1). The base-specific modifications for the four nucleotides are done by different chemicals.

Base-specific modification for subsequent cleavage

Dimethyl sulphate at pH 8 causes methylation of guanine residues so that they become susceptible to cleavage; piperidine formate at pH 2 weakens the glycosidic bonds of adenine and guanine residues; hydrazine makes C+T vulnerable; hydrazine along with 1.5 M NaCl reacts with cytosine residues; and 1.2 M NaOH at 90°C allows strong cleavage at adenine residues and weaker cleavage at cytosine residues. These base-specific cleavage reactions are carefully controlled to modify on an average only one of the target bases in each DNA molecule.

Thus, base-specific cleavage reactions or chemical degradations generate the different lengths of oligonucleotides. Often, Maxam and Gilbert's technique is designated as the chemical degradation method. At the end, the hydrolytic agent piperdine is removed either by ethanol precipitation or vacuum drying or a combination of both.

This procedure is no longer used as Sanger's method is comparatively easier for sequencing DNA. Sanger's Chain Termination Method is now widely accepted for sequencing DNA.

7.3.2 Sanger's Chain Termination Method (Dideoxy Method)

A single-stranded DNA fragment is initially employed as a template to guide the synthesis of a new complementary DNA strand with the help of a primer and a DNA polymerase. DNA synthesis is performed in the presence of the deoxynucleotides, dATP, dCTP, dGTP and dTTP, the normal precursors for A, C, G, and T bases, respectively, which are incorporated in growing DNA chains. Also included are four dye-labelled dideoxynucleotides (ddATP, ddCTP, ddGTP, and ddTTP) at much lower concentrations than the normal deoxynucleotides; the ratio of dXTP : ddXTP, where X represents any of the four bases, is about 100 : 1. Dideoxynucleotides are analogs of regular deoxynucleotides and lack the hydroxyl group present at the 3′ carbon of normal deoxynucleotides (Fig. 7.2a).

Incorporation of a dideoxynucleotide in a growing chain terminates the elongation of the DNA chain as a ddXTP lacks an OH group at the 3′ position and thus does not allow the DNA polymerase to add any further nucleotide (Fig. 7.2b). Absence of the 3′ OH group makes it impossible to form a bond with the next nucleotide. The prematurely terminated DNA strand thus ends up with a dye-coloured dideoxy base (Fig. 7.3, Step 2). Normal deoxyribonucleotide bases which are incorporated in the growing chain are proportionately higher in amount ~100 times more than the dideoxyribonucleotides.

However, every now and then, at random, a coloured dideoxynucleotide is incorporated instead of its normal analog and halts the elongation of the DNA strand. Thus, DNA chains of different lengths, each with a specific dideoxynucleotide at the 3′ end, which can be identified by its colour code, gradually accumulate (Fig. 7.3).

Next, the newly synthesized DNA fragments, terminated at different lengths, are released from the template strands by denaturation and subjected to polyacrylamide gel electrophoresis. The shorter fragments migrate through the gel faster than the longer fragments due to their lighter molecular weight (Fig. 7.3, Step 3). As they move through the gel, a special camera detects the colours of the various fragments due to the dye at the terminal dideoxynucleotide. The camera

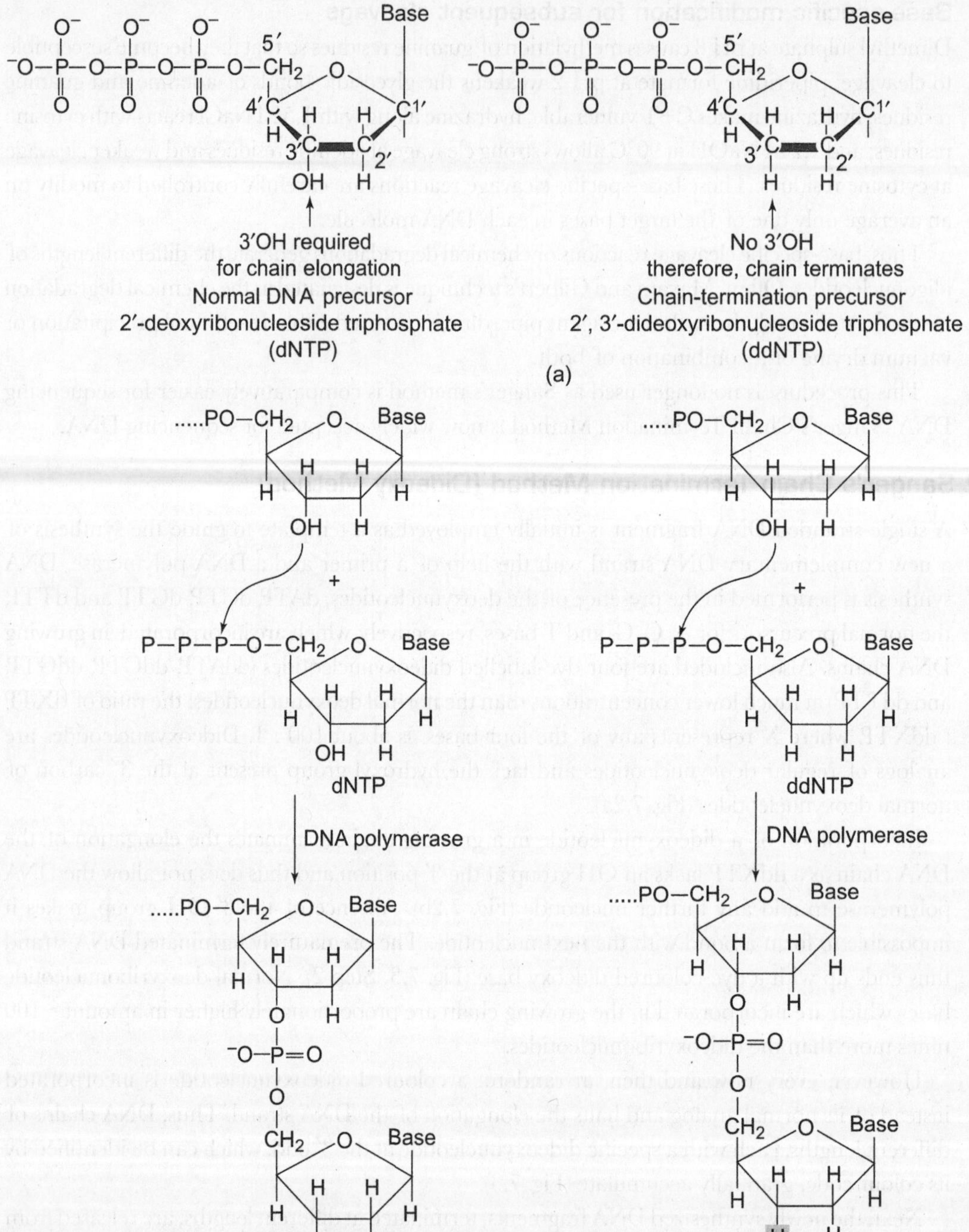

Fig. 7.2 (a) Structures of the normal DNA precursor 2′ deoxyribonucleoside triphosphate and the chain-terminator 2′, 3′ dideoxyribonucleoside triphosphate used in DNA sequencing reactions. (b) Normal process of elongation of the DNA chain with dNTP in the presence of DNA polymerase on the left, and termination of elongation of the DNA chain due to incorporation of ddNTP on the right.

is hooked to a computer to show the exact colour sequence of the dideoxynucleotides. This information allows the DNA base sequence to be determined (Fig. 7.3, Step 4). In the example shown in Fig. 7.3, the shortest DNA fragment is blue and the next shortest fragment is green. Since blue and green are the colours of ddCTP and ddGTP, respectively, the first two bases added to the primer must have been cytosine followed by guanosine.

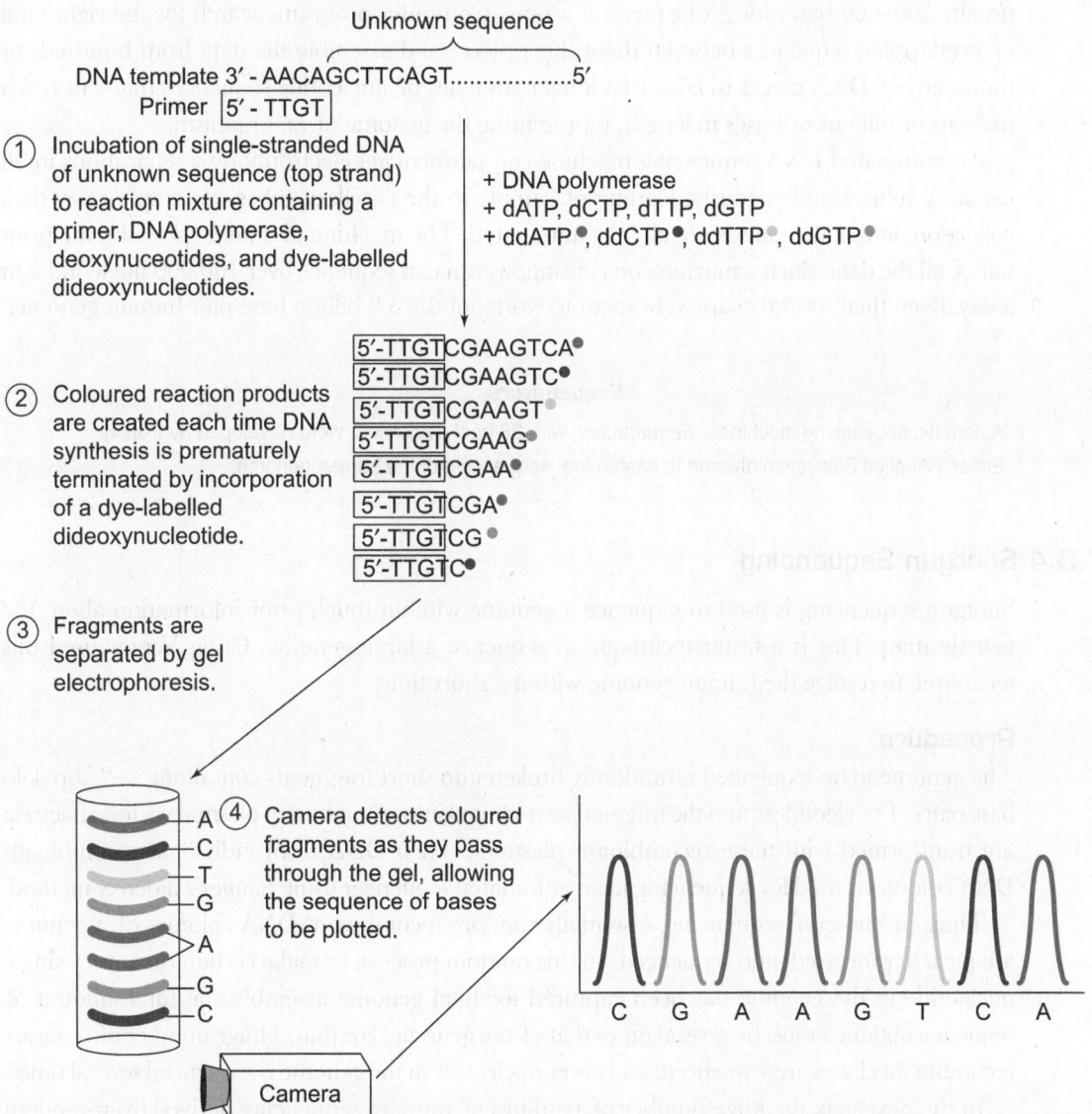

Fig. 7.3 DNA sequencing by Sanger's chain termination technique, which employs dye-labelled dideoxynucleotides for the high-speed, automated sequencing machine. In the figure, only the first eight bases of a DNA sequence have been analysed. Experiments of this type typically determine the sequence of 500–800 bases in DNA fragments. There are four main steps involved in the procedure.

7.3.3 Automatic Sequencing Machine

Sanger's technique was adapted in automatic sequencing machines, which collect information spanning hundreds of bases to provide the complete sequence of nucleotides in DNA fragments in a short period of time.

DNA sequencing machines can only determine the sequence of short pieces of DNA, usually 500–600 bases long, one piece at a time. Computer programs search for the right kind of overlapping sequences between these short pieces and assemble the data from hundreds or thousands of DNA pieces to construct longer stretches of nucleotide sequences that can reach millions or billions of bases in length, representing the genome of an organism.

An automated DNA sequencing machine can perform gel electrophoresis separations in 96 capillary tubes simultaneously; loading of sample in the capillary tubes, electrophoresis, data collection, and data analysis are all fully automated. The machine ultimately provides the print out of all the data. Such a machine on continuous run can sequence over 100,000 nucleotides in a day. Even then several years were spent to work out the 3.2 billion base pair human genome.

> **Sequenators**
>
> Automatic sequencing machines, *Sequenators*, with 384 gel capillaries, were developed by Perkin Elmer's Applied Biosystem division to hasten the sequencing of the human genome.

7.3.4 Shotgun Sequencing

Shotgun sequencing is used to sequence a genome without much prior information about the genetic map. This is a faster technique to sequence a large genome. Craig Venter used this technique to resolve the human genome within a short time.

Procedure

The genome to be sequenced is randomly broken into short fragments containing 1–2 kbp (kilo base pairs of nucleotides), and the fragments are cloned into recombinant plasmid vectors. Bacteria are transformed with these recombinant plasmids. Then DNA from individual recombinant DNA colonies is used for sequencing in an automated sequencer using Sanger's dideoxy method.

Thus, in 'shotgun' sequencing, essentially random recombinant DNA colonies of a genome are picked, processed, and sequenced. In this random process, to make certain that every single nucleotide in the genome has been captured for final genome assembly, the total amount of sequence obtained must be several times that of the genome. For that, a huge number of separate recombinant clones are sequenced; and every nucleotide in the genome is sequenced several times.

In the next step, the huge number of readings of random sequencing derived from random genomic DNA fragments is loaded into the computer with special programs that assemble the DNA fragments on the basis of overlapping DNA sequences. Thus, random DNA fragments are assembled based on matching sequences at their ends. This matching process is conceptually similar to the assembly of the pieces in a jigsaw puzzle.

The sequential assembly of short DNA sequences leads to a continuous longer assembly of sequences called a *contig* (Fig. 7.4). Individual contigs are usually composed of 50,000–200,000 bp.

The contigs are then linked by sequencing the ends of large DNA fragments. For example, if one end of a random large (100 kb) genomic DNA fragment contains sequence matches within contig 1, and the other end matches sequences in contig 2, then obviously these two contigs would be placed adjacent to one another as in Fig. 7.4. There are certain other modifications to close the gap between the contigs.

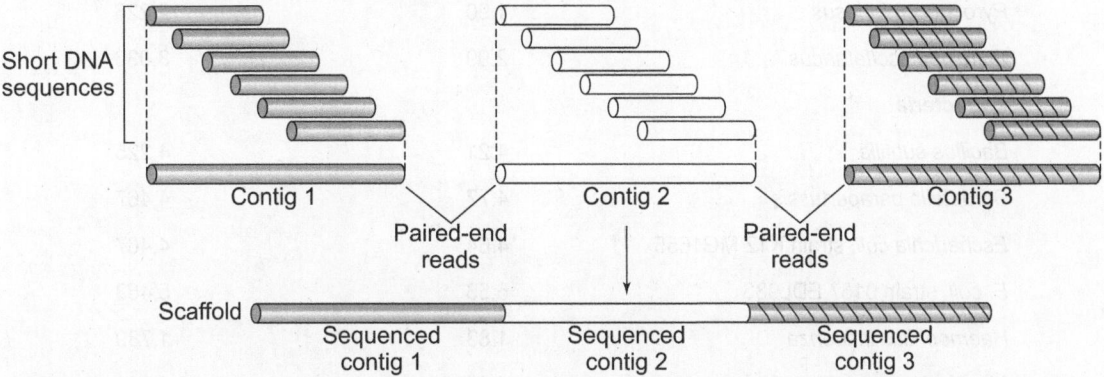

Fig. 7.4   Contigs are assemblies of short sequences from random shotgun DNA fragments. The ordered arrangement of short fragments is done by matching the overlapping sequences using a computer program. (A contig may be composed of 50,000–200,000 bp.) Two contigs are put into sequential continuity in the scaffold when a fragment has sequence match at one end within contig 1 and at the other end with contig 2.

The shotgun method might seem tedious, but it is considerably faster and less expensive than the earlier technique of systematically sequencing every defined restriction DNA fragment (generated by restriction endonucleases) on the physical map. Craig Venter's Celera Genomics, a private company armed with the shotgun technique, competed with the public-funded International Human Genome Sequencing Consortium and completed the Human Genome Project very much ahead of schedule. The Human Genome Project was discussed briefly in Chapter 3.

7.3.5 Sequencing of Genome of Various Organisms

Today, sequencing is a routine laboratory procedure. Nucleotide sequences of the genomes of over 2000 viruses, more than 1000 plasmids, about 1500 chloroplasts and mitochondria, over 700 bacteria and archaea, and about 30 eukaryotes have been completed. The sequencing of another 200 eukaryotic genomes is underway. The size and gene content of genomes of some selected organisms are presented in Table 7.1. Table 7.2 presents similar information for selected mitochondrial and chloroplast genomes.

7.3.6 Gains from Genome Sequencing

- Although there were other earlier methods such as determining the recombination frequencies between genes, cytogenetic microscopic observations, and so on, to specify the location and linkage order of genes in the chromosomes of different organisms, the

Table 7.1   Size and gene content of genomes of some select organisms

Species (Common name)	Genome size in nucleotide base pairs (Mb)[1]	Number of genes predicted
Prokaryote		
Archaea		
Archaeoglobus fulgidus	2.18	2,486
Pyrococcus furiosus	1.90	2,228
Sulfolobus solfataricus	2.99	3,033
Eubacteria		
Bacillus subtilis	4.21	4,225
Bordetella parapertussis	4.77	4,467
Escherichia coli, strain K12 MG1655	4.64	4,467
E. coli, strain 0157 EDL933	5.53	5,463
Haemophilus influenza	1.83	1,789
Mycobacterium tuberculosis	4.40	4,293
Mycobacterium genitalium	0.58	525
Salmonella typhimurium	4.85	4622
Eukaryote		
Saccharomyces cerevisiae (Yeast)	12.05	6,268
Arabidopsis thaliana (Mustard plant)	119.18	28,152
Plasmodium falciparum (Malarial protozoa)	22.82	5,361
Caenorhabditis elegans (Roundworm)	100.29	20,516
Anopheles gambiae (Mosquito)	278.25	14,707
Drosophila melanogaster (Fruit fly)	131.00	13,792
Canis familiaris (Dog)	2359.82	18,201
Danio rerio (Zebra fish)	1571.02	23,524
Gallus gallus (Chicken)	1054.18	17,709
Mus musculus (Mouse)	2932.36	25,396
Pan troglodytes (Chimpanzee)	2928.56	21,098
Homo sapiens (Man)	2851.33	22,287

[1]Mb = megabase pairs (10^6 bp)
Source: The NCBI website (http://www.ncbi.nih.gov/Genomes/)

nucleotide *sequences of the genomes provide the ultimate physical maps of genes and chromosomes.* These maps have been constructed with the finest precision at the nucleotide base pair level.

- The complete genetic information present in the genomes of various organisms equips researchers to work in the field of *comparative genomics and evolution, and enables a better understanding of phylogenetic relationships.*

- The availability of the nucleotide sequences of entire genomes has led to the development of microarray, gene-chip, and reporter gene technologies that permit the study of expression of all the genes of an organism simultaneously. This kind of study *provides the basis of 'functional genomics', and gives insights into the molecular mechanisms for development of organisms.*

Table 7.2 Size and gene content of selected mitochondrial and chloroplast genomes

Species (Common name)	Genome size in nucleotide base pairs	Number of genes predicted
Mitochondrial genomes		
Saccharomyces cerevisae (Yeast)	85,779	43
Chlamydomonas reinhardtii (Green alga)	15,758	25
Arabidopsis thaliana (Mustard plant)	366,924	57
Oryza sativa (Indica rice)	491,515	96
Zea mays mays (Corn)	569,630	218
Caenorhabditis elegans (Roundworm)	13,794	12
Drosophila melanogaster (Fruit fly)	19,517	37
Danio rerio (Zebra fish)	16,596	37
Mus musculus (Mouse)	16,299	37
Homo sapiens (Man)	16,571	37
Chloroplast genomes		
Arabidopsis thaliana (Mustard plant)	154,478	129
Chlamydomonas reinhardtii (Green alga)	203,828	109
Oryza sativa (Japonica rice)	134,525	159
Zea mays mays (Corn)	140,384	158

- Genome projects inspired scientists to develop *excellent fast technologies and equipment* for sequencing and paved the way for the *development of new branches in science—bioinformatics and proteomics.*

- The existing information about the genome inspired Craig Venter and his colleagues to synthesize chemically in vitro the entire genome of the smallest bacterium *Mycoplasma genitalium*, which was 582,970 bp long, and twenty times longer than any DNA molecule synthesized earlier. They reported this in 2008; this indicated that genome information will allow creation of an organism in a test tube.

- The study of 'single nucleotide polymorphisms' (SNP, Section 7.3) is helping researchers identify genes that are involved in susceptibility to various diseases. This will help *predict the risk of an individual for a disease and help find ways to control it.*
- Genomics will also help understand why different individuals respond differently to the same drug and will *establish 'pharmacogenomics'*.
- The *pathogenicity of microorganisms and their management would be better understood.*

Important Findings of the Human Genome Project
- The number of genes in the human genome is between 20,000 and 25,000, much lesser than earlier predictions.
- Half of the human genome is made up of repeating sequences.
- Approximately, 28% of the genome is transcribed to RNA.
- Approximately, 1.1% to 1.4% of the genome encodes protein.
- The difference in the genomic content between human beings is less than 0.1%.
- The chimpanzee and human genomes vary by an average of just 2%—on an average, in a stretch of 100 bp there are only two nucleotide substitutions.

7.3.7 Gene Prediction and Counting

The next task that follows genome sequencing is annotation. *Annotation* is an analytical process to identify encoding genes that encode both RNAs and proteins. Annotation identifies encoding genes, their regulatory sequences, and their functions. In the process, mobile genetic elements (transposons) and repetitive sequence families in the genome are also identified and characterized.

Protein encoding genes are identified by open reading frames (ORFs). An ORF has a series of codons that specify an amino acid sequence. The series begins with an initiation codon, usually ATG and ends with a termination codon, such as TAA, TAG, or TGA. These codons are identified with the aid of a computer. This method is effective for bacterial genomes.

This method as such is difficult to use for eukaryotic genomes including the human genome, mainly due to the presence of introns (noncoding DNA regions) in between exons (coding regions), and large intergenic distances (wide space between genes). These two factors increase the chances of identifying false genes. To resolve this difficulty, new versions of ORF scanning software have been developed.

7.3.8 Identifying the Functional Groups of Genes

After the prediction of functional genes from the genomic sequence, the function of the encoded product of each gene is determined. Several techniques are employed for the purpose. Databases from GenBank are searched for similar sequences or genes isolated from other organisms, the predicted ORFs are compared with those from known, well-characterized bacterial genes, or a search is carried out for similar sequences of functional motifs encoding proteins with specific functions. Thus, genome analysis leads to determining the function of all the genes and their interactive functions in the development of organisms.

7.4 SINGLE NUCLEOTIDE POLYMORPHISMS (SNPS)

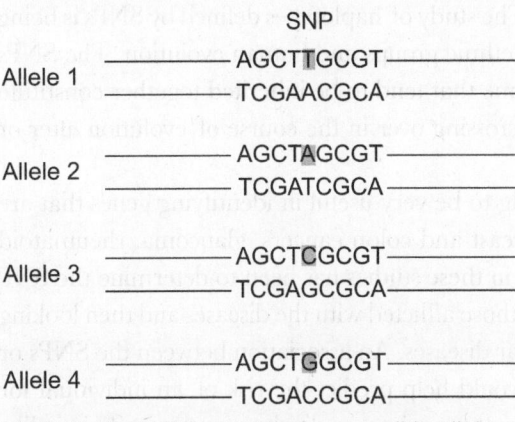

Allele 1 ——————— AGCTTGCGT———————
           ——————— TCGAACGCA———————

Allele 2 ——————— AGCTAGCGT———————
           ——————— TCGATCGCA———————

Allele 3 ——————— AGCTCGCGT———————
           ——————— TCGAGCGCA———————

Allele 4 ——————— AGCTGGCGT———————
           ——————— TCGACCGCA———————

Fig. 7.5 An SNP locus with four possible alleles. The base pair T.A (allele 1) is changed to A.T (allele 2), and C.G (allele 3) to G.C (allele 4). The remaining base sequences are identical in the four alleles.

The human genome project revealed the nucleotide base pair wise detailing of genetic content and functional genes in *Homo sapiens*. It also contributed to the fact that two individuals differ by a minimal number of the bases. The difference is due to single nucleotide substitutions at places, for example, A:T to G:C or G:C to A:T substitution. The four possible alleles at an SNP locus are shown in Fig. 7.5. Base-pair substitutions of this type throughout the genome have produced a large number of *single nucleotide polymorphisms* (*SNPs*, pronounced 'snips'). Scientists have already created a database of most of the common SNPs in the human genome. A comprehensive account of SNPs can be obtained from HGBASE database.

In a few cases, SNPs located in coding regions of genes are associated with altered phenotypes. For example:

SNP	Gene	Phenotype
A → T	β-globin	Sickle cell anaemia; RFLP (abolishes *Mst* II recognition site)
C → T	α₁-antitrypsin	Emphysema

Most of the SNPs are not located in the coding regions of genes and do not result in mutant phenotypes. Even when SNPs are present within the coding sequence they do not necessarily change the amino acid sequence of the encoded protein. This is because of the degeneracy of the genetic code, that is, there is more than one code for an amino acid. Such a change in the code which does not cause any change in the function of the protein is designated as a silent mutation. However, SNPs at non-coding regions may have consequences for gene splicing, transcription factor binding, or the sequence of non-coding RNA. Some SNPs are located in the recognition sequence of an endonuclease. They change the base sequence of the recognition sequence and generate RFLP (restriction fragment length polymorphism).

When the nucleotide sequences of the same chromosome of two individuals are compared, one SNP is present on an average in every 1200 nucleotide pairs. As many as 10 million SNPs may be present in the human genome and they provide a valuable resource for the study of DNA polymorphism. An SNP originates in a genome by the process of point mutation. SNPs can be used as the finest molecular markers on the chromosome for individuals.

7.4.1 SNPs in Studying Ancestry and Evolution in Human Population and Disease Associations (HapMap Project)

The frequency and distribution of SNPs throughout the human genome are proving to be valuable genetic markers for various purposes. The study of haplotypes defined by SNPs is being used to trace the relationships among different ethnic groups and human evolution. The SNPs on a chromosome or a segment of a chromosome that tend to be inherited together constitute a genetic unit of a 'haplotype'. Mutation and crossing over in the course of evolution alter or modify haplotypes as usual.

The study of SNPs and haplotypes is proving to be very useful in identifying genes that are involved in susceptibility to diseases such as breast and colon cancers, glaucoma, rheumatoid arthritis, and Alzheimer's disease. The strategy in these studies has been to determine the SNP genotypes of large samples of people including those afflicted with the diseases and then looking for associations between the SNPs and particular diseases. An association between the SNPs or haplotype and a particular disease, if found, would help predict the risk of an individual for developing the disease. Then one can take the appropriate medical precautions. This is like genetic counselling. Furthermore, the SNP study may also identify the actual disease-causing gene.

Recognizing the importance of the study of SNP haplotypes, researchers around the world have initiated the 'International HapMap Project'. The goal of this international enterprise is to map SNPs using DNA samples from many different human populations.

7.4.2 Pharmacogenomics

The response of different patients suffering from the same disease to a particular drug may vary widely, by 25 per cent to 80 per cent. This is thought to be due to the genetic differences between patients. Genetic difference as the basis of differential drug response is described as *pharmacogenetics*. The new field of study concerned with disease susceptibility and drug response among individuals is known as *pharmacogenomics*.

Scientists are working with SNPs to find out genetic variations at the personal level. Genetic differences may affect drug response in many ways; for example, metabolism of drug, transport of the drug, action of the drug on its target molecules, and so on. SNPs are likely to provide clues for genetic differences of two individuals affected differentially by the same drug. Thus, pharmacogenomics is expected to deliver personalized treatment that will be safer and more effective. In the near future, the genome of an individual may be sequenced at an affordable cost of $100 to $1000 within a few hours or in lesser time.

7.4.3 Detection of SNPs—Microarray Hybridization and 'Gene-chip' Technology

SNPs can be detected in the human genome by microarray hybridization or gene-chip technology.

Oligonucleotides (short chain of DNA) or DNAs representing different genes in a genome are synthesized or multiplied by PCR. They are arranged and immobilized in an array on a solid support (Fig. 7.6). Then test DNA is added to the array to allow hybridization between the arranged and test DNA molecules. The diagnostic oligonucleotides or test DNAs are tagged with a fluorescent compound, and on hybridization they produce fluorescence which can be detected and measured by a scanner and densitometry, aided by a programmed computer.

If there is a single-nucleotide difference between the diagnostic oligonucleotide or DNA molecule and the test DNA, hybridization will not take place. For example, a segment of DNA from one individual having an A:T base pair at a specific position may hybridize to a diagnostic DNA (in a spot of the array), while the corresponding segment of DNA from another individual having a C:G base pair at the same position will not hybridize. Thus an SNP can be detected.

Oligonucleotide or cDNA bearing sequences of genomic DNA which are used as diagnostic probes are immobilized in microarrays on a glass slide or a silicon wafer in a gene chip. They are produced in several ways: (i) microsynthesis of oligonucleotides in situ on the chip, (ii) spotting prefabricated oligonucleotides on the support, and spotting DNA fragments or cDNA on the support. In case of spotted DNA arrays, each array or spot contains 10^6–10^9 ds DNA molecules that are 100–300 bp long. Up to 5000 such spots can be accommodated per cm^2 of the chip. The DNA molecules are obtained from genomic libraries, cDNA clones, or by PCR amplification. Spotting is done by robots. The *printed oligonucleotide chips* are produced by a UV light-guided printing technology called *photolithography*.

Types of screening

There are two ways in which screening can be done.

Hybridization with labelled oligonucleotide   One end of the oligonucleotide probe is labelled with fluorescent dye and the other end with a quencher (stops fluorescence). The oligonucleotide is designed in such a way that the base pairs at both its ends are complimentary, and join to form a hairpin structure, placing the quencher next to the fluorophore. Therefore, normally there is no fluorescence. Hybridization between the oligonucleotide and a test DNA disrupts the base pairing of the two arms of the oligonucleotide, and causes the quencher to move away from the fluorophore and a fluorescent signal is generated (Fig. 7.6). Thus, fluorescence indicates hybridization of the labelled oligonucleotide and the test DNA. As mentioned earlier, a difference of a single nucleotide between the oligonucleotide and the test DNA can prevent hybridization.

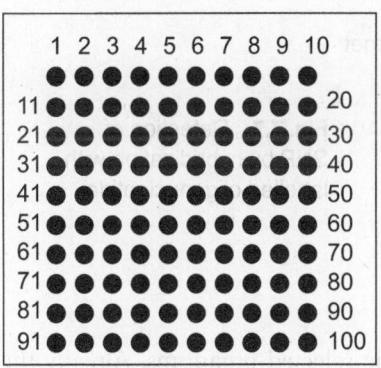

An array of 100 probes

Fig. 7.6 Array hybridization with gene-specific probes. After hybridization, the fluorescent-labelled test DNA bound to probes is quantitated by densitometry or scanners that measure the amount of fluorescence.

Hybridization is performed under highly stringent conditions. The temperature for incubation is maintained just below the melting point (T_m) of the oligonucleotide. At this temperature, a stable hybrid is formed. If there is a single mismatch of nucleotide, no hybrid is formed. Thus, oligonucleotide hybridization easily discriminates between the two alleles of an SNP.

DNA chips, oligonucleotides without fluorescent dye   Different oligonucleotides in a high-density array on a DNA chip can also be used for screening SNPs. Oligonucleotides are not labelled here. On the contrary, the test DNA is labelled with a fluorescent marker and pipetted onto the surface of the chip. Hybridization on the chip is examined under a fluorescent microscope. The fluorescent signal is emitted wherever the immobilized oligonucleotide has hybridized with the test DNA. A DNA chip with 2,50,000 oligonucleotides per cm^2 can type as many as 1,25,000 SNPs in a single run, considering that oligonucleotides for both alleles of each SNP are used.

7.5 FUNCTIONAL GENOMICS

Fine discrimination of a huge number of SNPs with DNA microarray inspired scientists to study the expression of all the genes in an organism at the same time or during development. Expression of a gene means transcription of mRNA. Thus, cDNA synthesized on the mRNA is used as test DNA to hybridize with oligonucleotides or DNA on chips. The test DNAs are labelled with fluorescent dye. On the chip, oligonucleotides or DNA represent all kinds of genes in an organism. Therefore, the hybridization of test cDNA and fluorescence indicate that the gene(s) is expressed in the organism (Fig. 7.7). Such studies of the expression of multiple genes in an organism simultaneously has created the new sub-discipline *functional genomics*. This has made it possible to study the expression of close to 6000 genes of budding yeast, approximately 14,000 genes of *Drosophila*, and approximately 23,000 human genes at a time if required.

Pharmacogenomics, sometimes, is viewed as a topic under functional genomics.

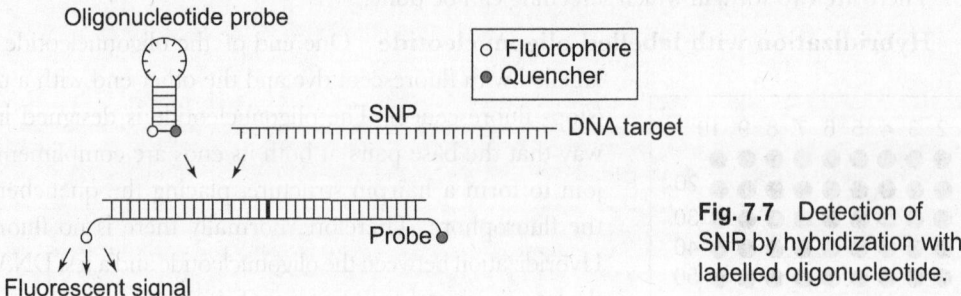

Fig. 7.7 Detection of SNP by hybridization with labelled oligonucleotide.

7.6 COMPARATIVE GENOMICS

Table 7.1 presents information about the genome of some selected organisms. Already the complete nucleotide sequences of the genomes from thousands of viruses, and eubacteria to hundreds of eukaryotes have been compiled—in terms of total number of sequences it might be over 90 billion nucleotide pairs. This is an immensely large information base with which to study similarities and differences in genome structure and organization of different organisms. This is the precise reason behind establishing *comparative genomics*. Comparative genomics helps in understanding evolutionary relationships between various species and other taxonomic groups.

Comparative genomics converts the DNA sequence data into proteins of known function for the purpose of comparison. Homologous genes present in different organisms encoding proteins with the same function are called *orthologues*. Orthologues are considered to be present in different organisms by vertical or evolutionary descent; they diverged by accumulating mutations in the course of time. In contrast, homologous genes within an individual may diverge for different functions, sometimes giving rise to different protein families. These genes are called *paralogues*. They can be used to trace the evolution of some genes in a species. Paralogues seem to originate by gene duplication, followed by accumulation of mutations. Genes of the globin family can be referred as paralogues.

Comparative genomics, thus, provides a new powerful tool to study evolution and phylogenetic relationships between species and other taxonomic groups.

7.7 PROTEOMICS

Proteome is the complete set of proteins encoded by a genome. Functional genomics initiated an entirely new discipline, *proteomics* with the objective of determining the structures and functions of all the proteins encoded by a genome.

Identification of the complete set of proteins in an organism is one of the big concerns of proteomics to begin with.

There are thousands of different proteins in a cell. To identify them, they need to be separated by several means systematically.

Isoelectric focussing The proteins separate according to their isoelectric point. A gradient of pH is generated in a gel. The isoelectric point is the pH at which a protein exhibits no net charge and thus becomes stationary (focussed) in the pH gradient.

SDS gel electrophoresis The proteins separate according to size. Two-dimensional gel electrophoresis separates proteins in two dimensions and thus brings a better resolution of protein spots.

Mass spectroscopy Each protein separated by two-dimensional gel electrophoresis is subjected to mass spectrometry to determine its exact molecular weight.

A slice of gel containing a spot with a large protein molecule is treated with trypsin, which cleaves the polypeptide chain after each of its positively charged amino acids. These peptides are separated on a gel and then eluted from the gel for mass spectroscopy (MS). MS analysis determines the molecular weight of the proteolytic fragments and their precise sequence in the protein molecule. Mass spectrometry is a high-speed, extremely sensitive technique that utilizes magnetic and electrical fields to separate proteins or protein fragments based on differences in mass and charge.

Finally, the peptide sequences from the proteins are assigned to a particular protein-coding sequence in the genome, with the tools of bioinformatics.

The three-dimensional structures of proteins are experimentally determined by X-ray crystallography and nuclear magnetic resonance (NMR) spectroscopy.

7.7.1 Contributions of Proteomics

Proteomics contributes to multiple aspects of the molecular biology of proteins, starting from predicting the probable structure and functions of a protein to identifying disease-specific proteins and studying the generation of diversity in the structures of the huge number of proteins that are more numerous than the number of encoding genes in a genome. These are systematically stated below.

- Structural proteomics is concerned with building a body of structural information about known proteins that will help *predict the probable structure and potential function of almost any protein* from the knowledge of its coding sequence and analogy with known genes.

- Proteomics assembles information about protein–protein interactions and the interactions of proteins with other compounds. This will help in understanding the exact contributions of a type of protein molecule in building a superstructure like an organelle or a cell, and in identifying particular proteins as targets or receptors for drugs and other compounds.

- To study the interactions and functional properties of the vast number of protein molecules in a proteome, *protein microarrays* (or protein 'chips') can be used. Tiny spots of thousand different proteins are immobilized on a microscope slide or a support to test their ability to react with other proteins or compounds, in a process similar to 'DNA chip' technology described earlier in this chapter. This approach will immensely help in drug discovery.

- Expression proteomics is useful in identifying disease-specific proteins. For example, over-expression or under-expression of certain proteins in cancerous or diseased cells in comparison to that in normal cells can be analysed.

- *E. coli* and *S. cerevisiae* are two intensively studied organisms, in which more than half the genes in their genomes do not have any known function. Such genes are designated as *orphans* or *ORFans* as they do not belong to any known gene family. Proteomics tries to find the function of orphan genes by determining the three-dimensional (3D) structure of the proteins encoded by these genes. Such proteins are called *hypothetical proteins*. The 3D structure of a hypothetical protein is compared with the database of protein structure and function. The 3D structure of proteins is much more conserved than their primary structures, and the function of a protein depends a great deal on its 3D structure. When a hypothetical protein is structurally similar to a known protein, there is 66 per cent chance that it is functionally similar to the known protein.

  Scanning the surface of the protein may reveal clefts (as in enzyme molecules) and domains most likely involved in interactions with other proteins.

- Proteomics helps in understanding the complexity of an organism's proteome that is considerably greater than that of its genome. For example, about 23,000 genes in human genome produce 200,000 or more proteins. This is possible as an individual gene can be 'read' in multiple ways to produce multiple versions of proteins from the same gene. Furthermore, there are post-translational modifications of proteins. Alternate splicing of mRNA is post-transcriptional modification of gene function. In post-translational modifications, end groups are cleaved, and chemical groups, such as methyl-, acetyl-, phosphoryl or sugar and lipids are added. All these cause a high level of diversity in protein molecules and their functions.

7.8 BIOINFORMATICS

The 21st century began with a glorious achievement in modern biology—sequencing the human genome. Its historical foundation was laid in the last century and was supported by all types of disciplines: biology, chemistry, physics, mathematics, and above all the development of computer science. The handling of the enormous data of the genome project would have been almost impossible without the aid of computers and the establishment of the discipline of bioinformatics. First, through bioinformatics the information generated by more than 3 billion repetitions of the four letters A, T, G, C, the basic notations of life—the nucleotides, was computed and stored in a proper fashion. In the second step, bioinformatics took charge of proteomics—a subject that has rapidly developed along with genome research in this century.

7.8.1 What is Bioinformatics?

Bioinformatics has been developed as an interdisciplinary field involving biology, computer science, mathematics, and statistics to analyse biological sequence data, genome content and arrangement, and to predict the structure and function of macromolecules, especially DNA and proteins.

Bioinformatics organizes biological data related to genomes in order to apply this information in pharmacology, for the identification of new proteins as targets for drug therapy, in agriculture and in other commercial applications.

Informatics has been developed by mathematicians, computer scientists, statisticians, and engineers, towards information management which is extremely critical for the organization of the increasing volume of data produced everyday by a number of publications in the fields of gene sequencing, genetics, molecular biology, biochemistry, cell biology, and health care.

To summarize, *bioinformatics* is used for the following purposes:
- The development of computer hardware and software programs for collection, storage, analysis, and visualization particularly of DNA and protein sequence data
- The use of algorithms and statistics to find relationships among large sets of sequence data
- Management of storage and access of the large volume of information

7.8.2 History of Bioinformatics—Establishment of Databases

Bioinformatics started its journey with protein sequencing data. Sanger's development of protein sequencing methods in 1951 led to the sequencing of representatives of several common protein families, such as cytochromes from different organisms. Margaret O. Dayhoff and her collaborators at the National Biomedical Research Foundation (NBRF), Washington, D.C., were the first to assemble databases of these sequences in the *Atlas of Protein Sequence and Structure*. She edited this collection from 1965 to 1978. The collection centre eventually became known as the Protein Information Resource (PIR, formerly Protein Identification Resource). Dayhoff and coworkers also developed computer programs for comparing distantly related sequences of amino acids to understand phylogenetic relationships. In 1988, the PIR-International Protein Sequence Database (http://www.nbrf.georgetown.edu/pir) was established as a collaboration of NBRF, the Munich Centre for Protein Sequences (MIPS), and the Japan International Protein Information Database (JIPID).

7.8.3 DNA Sequence Database

Assembly of information on nucleotide sequences of DNA obtained from different organisms began after the protein sequence database was established.

GenBank

George I. Bell founded the theoretical biology and Biophysics Group in 1974 at Los Alamos National Laboratory (LANL) in New Mexico, USA and initiated the entry of DNA sequences into the *GenBank database*, as Dayhoff did with the protein database. (Los Alamos National Laboratory was famous for great nuclear physicists and for the first detonation of an early small device of a nuclear bomb). The physicists of the newly formed Theoretical Biology and Biophysics group initially were interested in providing a theoretical background to experimental work in biology, primarily in immunology. Under the leadership of Walter Goad and colleagues, the first version of the GenBank was developed and maintained at LANL between 1982 and 1992.

Today the GenBank database (http://www.ncbi.nlm.nih.gov/) is maintained by the National Center for Biotechnology Information (NCBI), which is part of the National Library of Medicine (NLM) at the National Institute of Health (NIH) in Bethesda, Maryland. The content of the database has grown enormously since the time of Goad. GenBank had 680,338 nucleotide pairs of sequenced DNA at the end of 1982, which had grown to 90 billion nucleotide pairs by April, 2008.

Other databases

Databases equivalent to GenBank were also established in Europe and Japan. The European Molecular Biology Laboratory (EMBL) Data Library was founded in Germany in 1980, and in 1984 the DNA Databank of Japan (DDBJ) came into existence. GenBank, EMBL, and DDBJ have now formed the *International Nucleotide Sequence Database Collaboration* (http:/www.ncbi.nlm.nih.gov/collab), which facilitates search of all three databases simultaneously.

Search and retrieval programs have developed to an extent that one can screen databases for sequences similar to input sequences from various scientists. NCBI's *ENTREZ* retrieval system has become particularly invaluable. This system is now available free on the Internet (http://www.ncbi.nlm.nih.gov/entrez).

The volume of searchable and retrievable information available at the ENTREZ website is increasing rapidly. In addition to DNA and protein sequence databases, it now includes a huge bibliographic database called *PubMed* that covers most of the journals in medicine and biology.

The *items available from the ENTREZ databases* are (i) whole genome sequences, (ii) protein sequences, (iii) three-dimensional macromolecular structures, (iv) expressed sequences (gene function), (v) RNA secondary structures, (vi) SNPs, (vii) cancer chromosomes and genes, (viii) PubMed, and many more.

7.8.4 FASTA and BLAST Programs for Rapid Database Searches

FASTA is pronounced 'fast A', and stands for 'Fast–All', because it works with any alphabet, an extension of 'FAST–P' (protein) and 'FAST–N' (nucleotide) alignment. The current FASTA package has programs for database similarity in protein : protein, DNA : DNA, protein : translated DNA (with frameshift), and ordered or unordered peptide searches.

W. Pearson and D. Lipman developed FASTA to scan database similarity in a short time and to find homologues for short stretches of newly found sequences with sequences in the database. Now, one of the quickest ways to obtain this information is to run a BLAST (Basic Local Alignment Search Tool) program. This program was developed by S. Altschul and coworkers in early 1990.

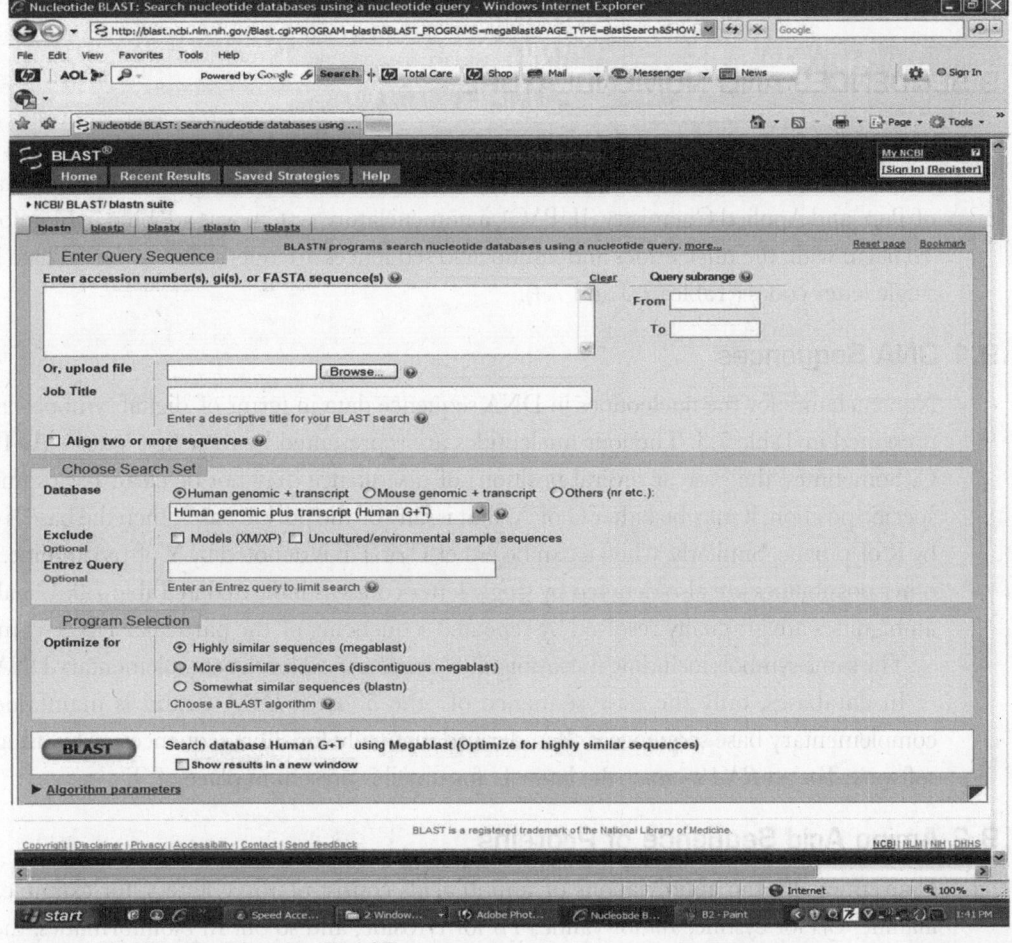

Fig. 7.8 NCBI page showing BLAST.

To work with BLAST, one needs to start at the NCBI home page—http://www.ncbi.nlm. nih.gov/. After opening the site, select 'BLAST' along the top toolbar. Then, under 'Nucleotide' at the top left, click on 'Quickly search for highly similar sequences (megablast)', and a window (similar to that indicated in Fig. 7.8) appears. Type in the new sequence to be compared and the operation is continued. After the search is complete, the number of similarities that exist and the similar sequences are displayed on the screen. Motif and domains in the sequence will also be displayed. Score points are assigned to each sequence and the higher the points the more closely related are the sequences.

Further operation will allow access to the PubMed literature and original paper(s) giving an account of similar sequences reported earlier.

So far our discussion was to illustrate the power and convenience of the software and databases that are now available in bioinformatics to analyse DNA sequences. Similar tools are also used for analysing protein sequence data. These tools of bioinformatics are helping geneticists and molecular biologists make sense of the torrent of data on DNA sequences that is currently available.

7.9 SEQUENCES AND NOMENCLATURE

The enormous databases created by different research groups has to be stored in a way that it can be retrieved easily in an understandable format. On the recommendation of the International Union of Pure and Applied Chemistry (IUPAC), a nomenclature system was adopted in bioinformatics. To begin with, the nucleotides and amino acid sequences are referred to in digital data by using single letter codes (Tables 7.3 and 7.4).

7.9.1 DNA Sequences

Nomenclature for the nucleotides in DNA sequence data in terms of digital symbols have been presented in Table 7.3. The four nucleotides are represented by their first letter—A, T, G, and C. Sometimes, the base at several positions of a sequence may not be clear. For example, at a specific position, it may be either G or A, that is, any of the purine bases; then the base is denoted by R of purine. Similarly, when it can be either C or T it is denoted by Y of pyrimidine. Certain other possibilities are also denoted by single letter codes as indicated in Table 7.3. Usually, these ambiguities are gradually resolved by repeated sequencing of the particular DNA segments.

The same symbols including the ambiguities are also used for the complementary DNA strand.

In databases, only the base sequence of the 5′ to 3′ DNA strand is maintained. The complementary base sequence is then derived manually for short sequences or by using proper software. For an *RNA sequence*, the letter U for uracil is present in place of T.

7.9.2 Amino Acid Sequence of Proteins

Conventionally, the abbreviations of amino acids consist of three letters; for example, *Ala* for alanine, *Cys* for cystine, *Val* for valine, *Tyr* for tyrosine, and so on. In bioinformatics, the amino acids are represented by single letters, such as A for alanine, C for cystine, V for valine, and Y for tyrosine (Table 7.4). Ambiguities are denoted by an exclusive letter as it was done for DNA. For example, Z symbolizes glutamine or glutamic acid; B denotes either asparagine or aspartic acid. When a position can be occupied by any of the amino acids it is indicated by X.

In a protein database, the amino acid sequence is presented from the N (amino)—terminus at the left to the C (carboxyl)—terminus at the right. Amino acids are assembled in this direction during protein synthesis.

7.9.3 Types of Sequences in Nucleotide Sequence Databases

Different types of sequences are available in DNA sequence databases. They are mentioned below.

Genomic DNA sequences   These represent the complete genome of organisms; in eukaryotes it represents the nuclear DNA in different chromosomes.

cDNA sequences   These represent the functional parts of the genome that are transcribed into RNA. cDNA synthesized on the mRNA template, represents only the exon sequences of the genes, that is, the sequences which are translated into proteins.

Expressed sequence Tag (EST) sequences   These comprise only a part of the cDNA sequences produced on an mRNA template and are used for hybridization with the corresponding gene in the genomic DNA. ESTs were used by J. Craig Venter and coworkers for isolating the functional or expressed sequences of the human genome.

An EST database (dbEST) contains millions of EST sequences. Duplication of ESTs is a problem; there may be two or more ESTs for a long gene. A short EST sequence, representing a single exon or its part may provide some solution.

Organellar DNA sequences   These represent databases for mitochondrial (mtDNA) and chloroplast (cpDNA) DNA.

Table 7.3   Single-letter codes for bases in DNA sequences, used in bioinformatics

Code	Meaning	Basic code	Code for the complementary sequences
A	Adenine	Adenine	T
C	Cytosine	Cytosine	G
G	Guanine	Guanine	C
T	Thymine	Thymine	A
R	G or A	Purine	Y
Y	T or C	Pyrimidine	R
M	A or C	Amino (bases having)	K
K	G or T	Keto (bases having)	M
S	G or C	Strong (nature of base pairing)	S[1]
W	A or T	Weak (nature of base pairing)	W[1]
H	A or C or T	Not G[2]	D
B	C or G or T	Not A[2]	V
V	A or C or G	Not U[2]	B
D	A or G or T	Not C[2]	H
N	A or C or G or T	Nucleotide	N

[1]Same code for the complementary strands since G pairs with C (S denotes both), and A pairs with T (W denotes both).
[2]The next letter of the alphabet to not G → H; not A → B; not U → V (in DNA, T equivalent to U); not C → D.

Table 7.4 Amino acids in protein sequences represented by single-letter codes in bioinformatics and by three-letter codes for conventional use

Single-letter code	Amino acid	Three-letter code
A	Alanine	Ala
B	Asparagine or Aspartic acid	Asx
C	Cystine	Cys
D	Aspartic acid	Asp
E	Glutamic acid	Glu
F	Phenylalanine	Phe
G	Glycine	Gly
H	Histidine	His
I	Isoleucine	Ile
K	Lysine	Lys
L	Leucine	Leu
M	Methionine	Met
N	Asparagine	Asn
P	Proline	Pro
Q	Glutamine	Gln
R	Arginine	Arg
S	Serine	Ser
T	Threonine	Thr
V	Valine	Val
W	Tryptophan	Trp
Y	Tyrosine	Tyr
Z	Glutamine or Glutamic acid	Glx
X	Any amino acid	Xaa

7.10 DATABASES

Databases are at the core of bioinformatics. *A database is a large collection of data on a specific topic,* electronically collected, analysed, and used. DNA sequence databases have already been discussed and protein sequence databases have been mentioned. Besides these, there are many other databases, numbering several hundred. Databases on sequences of tRNA and other categories of RNAs including small RNAs have also been created. Some of the important databases and their sources of availability are mentioned in Table 7.5. Three sources—National Center for Biotechnology Information (NCBI), USA; European Molecular Biology Laboratory (EMBL), Germany; DNA Databank of Japan (DDBJ)—helped in the initial development of databases, and are still the major players in the world of bioinformatics.

Table 7.5 Some important databases and their source

Databases	Information	Source
GenBank	Nucleotide sequences of genomic DNA (non or less redundant sequences)	NCBI, USA
EMBL nucleotide	Nucleotide (DNA and RNA) sequences	EMBL
DDBJ (DNA DataBank of Japan)	Nucleotide sequences	GenomeNet, Japan

(Contd)

Table 7.5 (*Contd*)

Databases	Information	Source
dbEST	EST sequences	NCBI
HomoloGene	Homologous genes in human, mouse, rat, zebrafish, cow	Unigene
KEGG (Kyoto Encyclopedia of Genes and Genomes)	Complete genomes and genes for metabolic pathways in microorganisms	GenomeNet, Japan
IMGT Database	Nucleotide sequences of immunologically important genes	EBI, UK
Mito	Mitochondrial DNA sequences	NCBI
NR	Non-redundant entries from GenBank, EMBL, DDBJ, and PDB databases	
RefSeq	mRNAs and proteins of human, mouse, and rat	
Ribosomal RNA database	Ribosomal RNA sequences	NCBI, EBI
PDB (Protein DataBank)	Sequences of proteins of 3D structures known	NCBI, EBI
PIRs (Protein Identification Resource)	Redundant protein sequences	NCBI
SWISS-PROT	Non-redundant annotated protein sequences	NCBI, EBI
TrEMBL	Sequences of proteins translated from all coding sequences in EMBL database	NCBI
PALI	Phylogenetic analysis and alignment of proteins	
Kabat	Nucleotide sequences immunologically relevant	NCBI
E. coli	Nucleotide sequences of *E. coli* genome	NCBI
Vector	Vector sequences from GenBank	NCBI, EBI
Yeast	Genome sequences of yeast	NCBI

7.10.1 Utilization of Databases and Analysis Tools

Suitable search engines and analysis tools help in the utilization of the databases described earlier. These tools are often described as *database mining tools* because the use of database is described as *database mining*. A list of some commonly used tools is provided in Table 7.6.

Table 7.6 Some commonly used database search tools

Search tool (Supported through the Web)	Function performed
BLAST [Basic Local Alignment Search Tool] (NCBI, USA)	Search for sequence similarity information and detect homologous sequences in both DNA and protein sequences
ENTREZ (NCBI)	Search for bibliographic material and biological data from databases like GenBank, EMBL, SWISS-PROT, PDB, and PubMed.
DNAPLOT (EBI, UK)	Sequence alignment
LOCUS LINK (NCBI)	Provide information on homologous genes in different species, derived by common ancestry

(Contd)

Table 7.6 (*Contd*)

Search tool (Supported through the Web)	Function performed
LIGAND (GenomeNet, Japan)	Access through a chemical database and other regular databases for combinations of enzymes and metabolic enzymes
BRITE (GenomeNet)	Information for biomolecular relations, transmission and expression
TAXONOMY BROWSER (NCBI)	Provide genetic information and taxonomic relationships of the species
PROSITE (Linked with SWISS-PROT)	Search for functional sites and sequence patterns in many proteins, and match a sequence submitted
STRUCTURE	Analysis of molecular structure and support Molecular Modelling Database (MMDB)

Many other sophisticated and user-friendly programs for bioinformatics tools have been developed.

7.10.2 Types of BLAST Programs

The utility and basics of operation of BLAST program for the quickest way of identifying homologues for a given sequence has been discussed earlier in connection with DNA and other databases.

BLAST is a group of user-friendly sequence similarity search tools on the Web. There are several BLAST programs, each for a specific purpose. For example:

BLASTn It compares a submitted nucleotide sequence with nucleotide databases.

BLASTp It compares the submitted protein sequences against protein databases.

BLASTx It translates the submitted nucleotide sequence into amino acid sequence for comparing with a protein database.

tBLASTn It converts the submitted protein sequence into nucleotide sequence for comparing the latter with nucleotide sequence databases.

tBLASTx It translates the submitted nucleotide sequence and nucleotide sequence database into amino acid sequences and searches for homology between the two.

Databases Developed in India

The National Research Centre on Plant Biotechnology, New Delhi has developed two databases.
GM Crops database stores information on biosafety of transgenic plants released in India. 139 transgenic lines using four genes (3 *cry* genes for *Bt* toxin and 1 *vip3A*) and a promoter (CaMV 35S) are the focal theme.
Vanshanudhan database has been initiated to maintain information about India rice genome; information on 56,298 rice genes is available.

7.11 BIOINFORMATICS TOOLS IN THE DETECTION OF GENES AND FUNCTION OF A NEW GENE

The significance of bioinformatics is essentially summarized here. Databases were developed to store enormous nucleotides by nucleotides data of genomes, protein sequences, and sequences of other macromolecules from thousands of organisms. The tools of bioinformatics analyse the databases to detect new genes and their functions from newly discovered sequences from various organisms. Different strategies are adopted for detection or prediction of genes.

Homology search tools like BLAST are employed to compare the newly submitted genome with sequences of known genes, cDNAs, ESTs, and proteins present in the databases. In another strategy, gene-specific features, such as promoters, splice sites, polyadenylation sites, and open reading frames (ORF, long sequence frame uninterrupted by a stop codon) are analysed in the newly submitted genome with the help of bioinformatic tools.

The gene prediction programs in eukaryotes utilize the output of several algorithms to generate a *whole gene model*. This model defines a gene as a series of exons that are coordinately transcribed.

To identify the function of a new gene, the gene sequence is translated into the amino acid sequence it is supposed to encode. Then this translated protein sequence is compared with protein databases of known characteristics.

Bioinformatic studies contribute in a big way to *molecular phylogeny*. Bioinformatics tools can establish phylogenetic relationships of species in the quickest possible way.

Bioinformatics—Today and Tomorrow

Knowledge of the nucleotide sequences of entire genomes has provided a wealth of information and a new challenge for managing them. Bioinformatics which uses computer programming, statistics, and algorithms has met the challenge. It has changed the research methodology in certain areas of biological science.

Although earlier experiments of cytogenetics, breeding, and protein chemistry paved the path for development of genomics, proteomics, and bioinformatics, these new subdisciplines have already replaced the time-consuming and laborious experimental approaches.

Bioinformatics equipped with huge databases and tools can now detect a new gene or assign definitive function to a new protein sequence in a few hours or a few days; the same operation would take years to complete with the experimental approaches that were used earlier. Bioinformatics allows scientists and biotechnologists to alter the function of a protein in a desired fashion for research and commercial purposes. Bioinformatics in the long run will have a profound effect on the lifestyle, health, and thinking of human beings and on society at large.

SUMMARY

♦ Genomics is devoted to the mapping, sequencing, and functional and comparative analyses of genomes of different organisms.
Earlier, the DNA sequences in the clones of a genomic library of an organism (Chapter 6) were pieced together into overlapping sets to construct the final

genetic and physical maps for the entire genome.

♦ Proteomics deals with structures and functions of all the proteins in an organism.

♦ Bioinformatics has been developed as an interdisciplinary field involving biology, computer science, mathematics, and statistics to analyse enormous biological sequences of data, genome content, and arrangement.

♦ Genomics, proteomics, and bioinformatics, all depend on the successful development of techniques for DNA sequencing. Maxam–Gilbert's technique of chemical degradation and Sanger's chain termination method are the two prime techniques that have helped in sequencing DNA of different organisms. Sanger's method has been widely accepted for sequencing DNA and for automatic sequencing machines.

♦ Craig Venter of Celera Genomics, a private company, introduced shotgun sequencing, a fast technique to sequence a large genome. In this method, essentially random DNA colonies are picked up and sequenced using Sanger's technique. The shotgun technique helped resolve the human genome within a short time.

♦ Today, nucleotide sequences of the genomes of over 2000 viruses, over 1000 plasmids, about 1500 chloroplasts and mitochondria, over 700 bacteria and archaea, and about 30 eukaryotes have been completed; and many more are being sequenced.

♦ Gains from genome sequencing are many. Genome analysis provides physical maps of genes and chromosomes, a better understanding of phylogenetic relationships, the basis of 'functional genomics'— proteomics, and so on. Above all, genome projects inspired the scientists to develop excellent fast technologies and equipment for sequencing and a new branch in science—bioinformatics.

♦ The human genome project revealed the large number of single nucleotide polymorphisms (SNPs), their importance in studying ancestry and evolution in human populations, and the basis of pharmacogenetics. Microarray hybridization and 'gene chip' technology have been developed for the detection of SNPs.

♦ Structural proteomics builds a body of structural information about known proteins that will help predict the probable structure and potential function of almost any protein.

♦ Proteomics also assembles information about protein–protein interactions and the interactions of proteins with other compounds that will help in understanding particular proteins as targets or receptors for drugs and other compounds.

♦ Expression proteomics is useful in identifying disease specific proteins and their over- or under-expression during disease conditions.

♦ Proteomics has revealed that an organism's proteome is far more complex than its genome. For example, about 23,000 genes in the human genome produce 200,000 or more proteins. This is possible as an individual gene can be 'read' in multiple ways to produce multiple versions of proteins from the same gene, and due to post-translational modifications of proteins.

♦ In short bioinformatics has taken charge of the development of computer hardware and software programs for collection, storage, analysis, and visualization of DNA and protein sequence data. It also participates in the management of storage and access of the large volume of information. There are a good number of databases managed by different research institutes.

John Craig Venter

John Craig Venter (born 14 October 1946)

John Craig Venter is an American molecular biologist and entrepreneur, most famous for his role in being one of the first to sequence the human genome with private funding and for his effort in creating the first cell with a synthetic genome in 2010.

John Craig Venter was born in Salt Lake City, Utah, graduated from Mills High School and entered the College of San Mateo in California. In his biography, *A Life Decoded: My Genome: My Life* (2007) he admitted not being serious in studies and had C and D grades in his school days. In his youth he was more interested in boating or surfing, than in studies.

Although he had reservations against the Vietnam War, Venter was drafted and enlisted in the US Navy where he worked as a medic in the intensive-care ward of a field hospital. While in Vietnam, he attempted suicide by swimming out to sea, but changed his mind more than a mile out in sea. His experience with wounded, maimed, and dying soldiers influenced his decision to study medicine, but he later switched to biomedical research.

He obtained his BS degree in biochemistry in 1972 and PhD degree in physiology and pharmacology in 1975 from the University of California, San Diego. He studied under the biochemist Nathan O. Kaplan. He worked as an associate professor and later as professor at the State University of New York, Buffalo, and then joined the National Institute of Health (NIH) in 1984. At NIH, Venter learned a technique for rapidly identifying all the mRNAs present in a cell and began to use it to identify human brain genes. The short cDNA sequence fragments discovered by this method are called *expressed sequence tags* (ESTs); they are used as tags to identify unknown genes.

Venter was passionate about the power of genomics to radically transform health care. He introduced *shotgun sequencing* which was the fastest and most effective technique for genome sequencing that allowed him to resolve the human genome three years earlier than the original programme of the public funded Human Genome Project. In Celera, he used DNA from five demographically different individuals including himself. In 2000, Craig Venter of Celera Genomics and Francis Collins of the NIH and US Public Genome Project jointly made the announcement of mapping of the human

genome. Celera published the first Human Genome in the *Journal of Science* (vol. 291, 1304–1351, 2001). The Human Genome Project was published in *Nature* (vol. 409, 860–921, 2001).

The ambitious competition of Celera Genomics with the government funded Human Genome Project has been elucidated in the book, *The Genome War* by James Shreeve.

Craig Venter founded Celera Genomics, the Institute for Genomic Research and the J. Craig Venter Institute; he now works at the latter to create synthetic biological organisms. He created a small bacterial cell, in 2010, by synthesizing the genome of *Mycoplasma* by arranging the nucleotides A, T, G, C, in a test tube, with the knowledge of its genomics. Venter would like to patent the first life form created by a human being and will probably name it *Mycoplasma laboratorium*. He believes that synthetic microorganisms will help produce clean next generation biofuels and pharmaceuticals, and can be used for bioremediation in a more effective way. Craig Venter has another ambitious project to document genetic diversity in the world's oceans with the hope to expand the universe of protein families.

It is no wonder that a flamboyant and active personality like Craig Venter would receive awards and honours from different scientific bodies, and the media attention. He was listed on *Time* magazine's 2007 and 2008 *Time 100 list* of the most influential people in the world.

EXERCISES

Objective Questions

1. Multiple Choice Questions
   (a) Genome refers to the
      (i) complete set of chromosomes in a somatic cell of an organism
      (ii) total DNA in a cell
      (iii) genetic information in a haploid set of chromosome of an organism
      (iv) total number of genes of an organism
   (b) Transcriptome is
      (i) a particular nucleotide sequence present in the mRNA transcript
      (ii) the complete set of RNAs transcribed from a genome
      (iii) the other name of DNA dependent RNA polymerase

   (iv) a factor that helps in the process of transcription
   (c) 'Contigs' are usually composed of
      (i) 20,000 bp
      (ii) 50,000 to 200,000 bp
      (iii) 200,0000 bp
      (iv) much above 200,0000 bp
   (d) Most of the SNPs
      (i) are located in the coding regions of genes
      (ii) lead to mutant phenotypes
      (iii) are never located in the recognition sequence of an endonuclease
      (iv) are located in the non-coding regions of genes
   (e) ENTREZ databases include

(i) whole genome sequences
(ii) protein sequences
(iii) similarities in DNA:DNA
(iv) SNPs
(v) PubMed

(vi) all of the above

2. State whether the following statement is True or False.
   (a) Genome size in general increases with the advancement of evolution.

Review Questions

1. Compare and contrast genomic libraries with cDNA libraries. What type of library would you choose if you have to clone regulatory gene elements such as promoters and enhancers?

2. Why is Maxam and Gilbert's method for sequencing DNA often described as the chemical degradation technique?

3. What is the structural difference between a deoxyribonucleotide (dNTP) and a dideoxyribonucleotide (ddNTP) used for DNA sequencing?

4. Briefly state the basic principle of the shotgun method of genome sequencing.

5. What are the gains from genome sequencing?

6. State four exclusive findings of the 'Human Genome Project'.

7. What is SNP? How can SNP be useful in human health?

8. How do protein structure studies help proteomic research?

9. Describe the procedures used for the detection of SNPs.

10. Define and contrast 'orthologues' and 'paralogues'.

11. What are ORFan genes?

12. What are the biological items handled by bioinformatics?

13. What is a DNA sequence database?

14. Write a short note on GenBank.

15. What will be your approach to utilize databases?

8

Enzyme Biotechnology

LEARNING OBJECTIVES

- ♦ Classification system of enzymes
- ♦ Working mechanism of enzymes
- ♦ Multiple uses of enzymes
- ♦ Impact of biotechnology on the enzyme industry
- ♦ Enzyme immobilization technology

INTRODUCTION

In 1878, Kuhne coined the term 'enzyme' from the Greek word *enzumos*, which refers to the leavening of bread by yeast. Enzymes are usually globular proteins and biological catalysts that facilitate chemical reactions in living cells under relatively mild range of biological temperature. *Enzymes catalyse reactions in a highly substrate-specific manner, with accelerated rates, without being destroyed in the process.*

An enzyme may be made up of a single or many polypeptide chain(s). The active site of an enzyme is a restricted region that has catalytic activity and binds the substrate (the molecule on which the enzyme acts). Sometimes, when a molecule binds at places other than the active site, the conformation of the tertiary structure of the enzyme may change. This distortion can affect the activity of the enzyme. Tertiary structure and enzyme activity can also be influenced by factors, such as temperature, pH, and ionic strength of the environment of the enzyme.

Chemical conversion of a substance by an enzyme is an energy-efficient process; the consumption of raw materials required for the conversion is low. CO_2 emission is significantly reduced, production costs are usually reduced, and such processes often replace toxic chemical processes. The production and application of enzymes, nature's own toolset, for various industrial processes are termed as *'white biotechnology'*. Implementation of white biotechnology can be differentiated into 'red' (medical) and 'green' (agricultural) biotechnology.

8.1 NOMENCLATURE AND CLASSIFICATION OF ENZYMES

Enzymes are recognized by the type of specific reaction they catalyse. The Enzyme Commission (EC), established by the International Union of Biochemistry (IUB), suggested some rules for naming enzymes and also assigned a definite (EC) number for each enzyme. On the basis of the catalytic reactions carried out, the enzymes are grouped into the following six classes (see also Table 8.1).

Class 1 Oxidoreductases   These carry out redox reactions, that is, transfer of hydrogen or oxygen atoms between molecules, for example, dehydrogenases (hydride transfer), oxidases (e^- transfer to O_2), oxygenases (oxygen atom transfer from O_2), and peroxidases (e^- transfer to peroxides).

Class 2 Transferases   These transfer an atom or a group of atoms, such as acyl-, alkyl- e^- and glycosyl- groups between two molecules, for example, oxidoreductases.

Class 3 Hydrolases   These catalyse hydrolytic reactions; for example, esterases, glycosidases, proteases, and lipases.

Class 4 Lyases   These remove a group of atoms from the substrate molecule; for example, aldolases, decarboxylases, dehydratases, and pectinases.

Class 5 Isomerases   These catalyse the formation of isomers of molecules, for example, epimerases, racemases, and xylose isomerases.

Class 6 Ligases (synthetases)   These catalyse covalent bond formation between two molecules using the energy released by hydrolysis of a nucleoside triphosphate (ATP/GTP), for example, glutathione synthase.

Table 8.1 Enzyme classes and reactions catalysed by them

Enzyme class	Type of reaction	Reaction
Oxidoreductase	Catalyses oxidation/reduction	$X + e^- + H^+ \leftrightarrow XH$
Transferase	Transfers functional groups like phosphate, amino, acetyl, etc.	$X\text{-}Y + Z \leftrightarrow X\text{-}Z + Y$
Hydrolase	Catalyses hydrolysis of substrate	$X\text{-}Y + H_2O \leftrightarrow X\text{-}H + Y\text{-}OH$
Lyase	Breaks covalent bonds	$X\text{-}Y \leftrightarrow X + Y$
Isomerase	Changes into isomer of the substrate	$X \leftrightarrow X^*$
Ligase (Synthetase)	Joins two molecules	$X + Y + ATP \leftrightarrow X\text{-}Y + ADP + Pi$

8.2 NON-TRADITIONAL ENZYMES

Two forms of biological catalysts, (i) ribozymes and (ii) abzymes, were discovered in the 1980s and can be discussed separately from the traditional enzymes classified earlier.

8.2.1 Ribozymes

In the course of unveiling the structural and functional complexity of RNA, in the 1960s, Carl Woese, Francis Crick, and Leslie Orgel proposed informational as well as catalytic functions of RNA molecules. The investigation of post-transcriptional processing of RNA molecules, indeed showed that *rRNA can act as an enzyme and is called ribozyme.*

- Ribozymes function in a substrate-specific manner by transesterification (for splicing) and phosphodiester bond hydrolysis (cleavage). In case of self-splicing, the substrate is itself a part of the ribozyme.

- Ribozymes align the RNA substrate by base pair matching and binding.

- Similar to enzymes, ribozymes enhance the rate of chemical reactions without being destructed, and follow Michaelis–Menten kinetics.

- Ribozymes can catalyse synthesis of the peptide bonds of proteins.

> The best characterized ribozymes are self-splicing group I intron from *Tetrahymena* and RNase P from *E. coli.* RNase P is also present in other organisms, including human and plant RNA viruses.

Discovery of the ribozyme function of rRNA allowed the scientists to envisage that RNA was the primary macromolecule in an early biological world to function as both a hereditary and a catalytic substance; later protein took charge of efficient enzyme function, and DNA as a more fidel molecule for heredity.

Ribozymes can be very useful for *pharmaceutical purposes* in the selective destruction of pathogenic mRNAs (from viruses). Ribozymes are more effective than antisense RNAs as a single ribozyme molecule can specifically inactivate multiple target mRNAs.

Improvement of the catalytic function of existing ribozymes and isolation of entirely new ribozymes are few of the research areas of molecular biotechnology.

In recent years, a number of clinical trials in patients with diseases, such as HIV and cancer, have been initiated with trans-cleaving ribozymes. For such therapies, either a ribozyme expression cassette in a viral vector is used or the synthetic ribozyme is directly injected.

8.2.2 Abzymes

Abzymes are antibodies capable of catalysing specific chemical reactions. By its virtue, an antibody recognizes specific molecules (antigen) and binds to them at their ground state of energy. On the other hand, enzymes have binding sites that preferentially bind to the transition states of their substrate. Thus, normal antibodies do not function as abzymes. A catalytic abzyme is produced against the expected transition state of a specific substrate. Abzymes have been produced for catalysing hydrolysis of esters and carbonates, cleavage of amide bonds, and other biochemical reactions. Sometimes metal ions act as cofactors with abzymes like conventional enzymes. Research is underway to come up with specific abzymes which recognize specific amino acid residues to provide a cut at a specific peptide bond, much as restriction endonucleases cut DNA at specific sites.

8.3 MECHANISM OF ENZYME ACTION

Enzymes catalyse reactions by lowering the *activation energy* of the molecules. *Activation energy* of a substance is the energy that is required for breaking the covalent bonds of the substance in a molecular transformation. Activation energy constitutes the *energy barrier* which prevents a biomolecule from transforming spontaneously. A higher activation energy corresponds to a slower reaction. At the same time, constant specific conversion or transformation of biomolecules is necessary to maintain the life process.

The binding of an enzyme to a substrate (Fig. 8.1) distorts the structure of the substrate, forming an intermediate state. This lowers the level of activation energy (Fig. 8.2), allowing

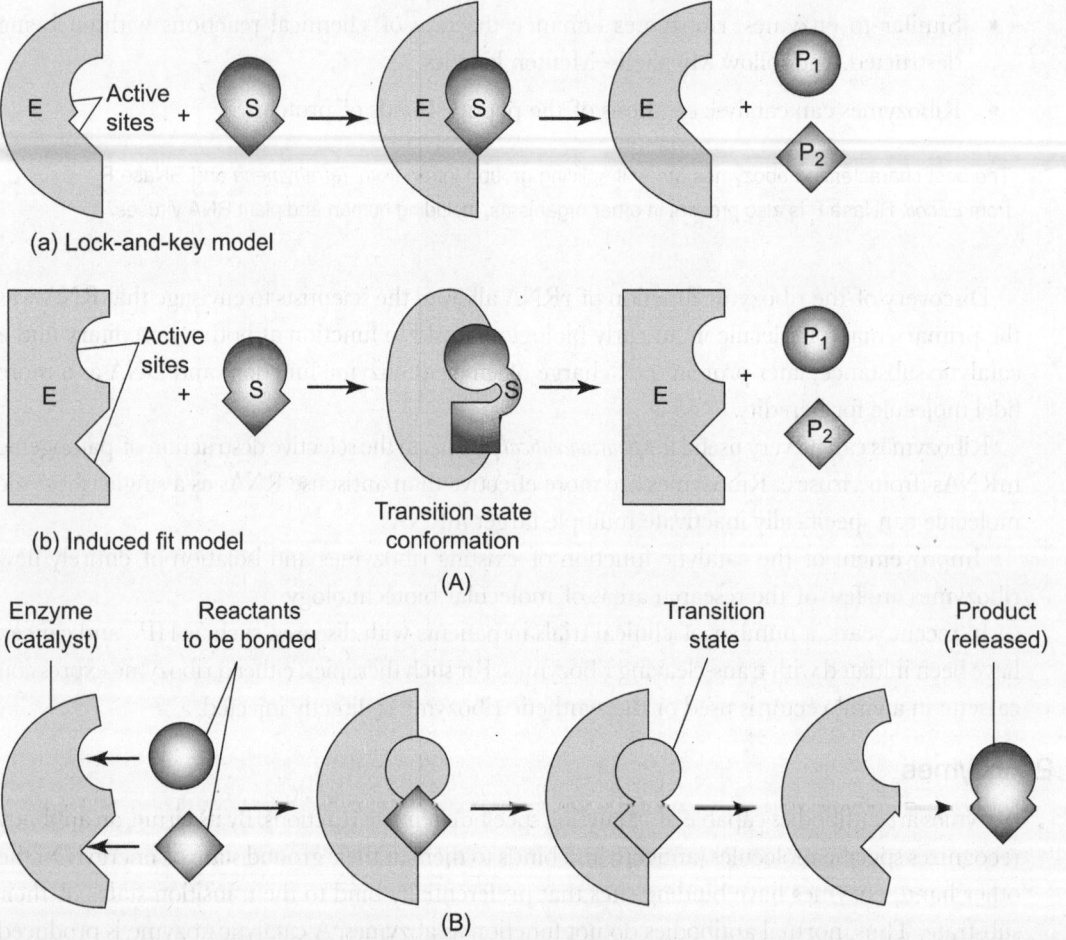

Fig. 8.1 Enzyme–substrate interaction: (A) In cleavage interaction (as in hydrolysis or endonuclease function) – Model (a) represents the early lock-and-key model, the active site of the enzyme fits the substrate as a lock does a key. Model (b) represents the induced fit model as elaboration of the lock-and-key model. Both enzyme and substrate are distorted on binding. The substrate is forced into a conformation approximating the transition state by the enzyme to lower the level of activation energy (lower energy requirement, see Fig. 8.2). (B) To remain in synthetic interaction (as in joining amino acids into a protein or nucleotides into a RNA or DNA strand), two reactants are bound to an enzyme, which ensures correct mutual orientation and proximity and binds them strongly to force them into the transition state.

the transformation of the original molecule(s) with lesser energy involvement. Thus, an enzyme plays the role of a catalyst to allow biological reactions (transformation) to occur at physiological temperature of the biological system, without itself being destroyed. An enzyme molecule can, therefore, function repeatedly.

> An enzyme, as a catalyst, increases the rate of a reaction by lowering the activation energy requirement, thereby allowing a thermodynamically feasible reaction to occur at a reasonable rate without thermal (high temperature) activation.

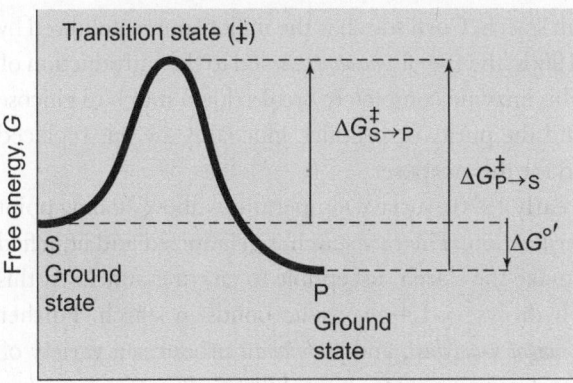

Fig.8.2 Diagram depicting energetics of the enzyme reaction. Reaction coordinate (horizontal axis) indicates progression of chemical changes (bond breakage or formation) as S is converted to P. S and P, here, also symbolize free energies of the substrate and product at ground state, the stable state. Symbol ‡ indicates the transition state, $\Delta G^\ddagger$ activation energies for the S → P and P → S reactions. $\Delta G'$ indicates the overall standard free energy change in going from S to P.

8.4 TRADITIONAL USES OF ENZYMES

Some examples of enzymes with traditional uses are *papain* of plant origin used as a meat tenderizer, *rennin* of animal origin used to clot milk during cheese manufacture, and so on. However, extraction of enzymes on an industrial scale from plant and animal origins to meet the industrial demand has proved difficult. So for large-scale applications, use of enzymes from microbial sources is favoured, as microorganisms can be produced in large-scale industrial fermentation units.

Microbial enzymes are used in food and beverage industries. Amylases liquefy and hydrolyse starch for manufacturing alcoholic beverages, starch paste, and glucose syrups; proteases hydrolyse proteins for tenderizing meat and clot milk for manufacturing cheese; pectins improve fruit juice extracts and prevent haze formation in fermented fruit juices; glucose isomerase produces fructose syrup as a 'high sweetener'. Microbial proteases and lipases (fat/lipid-digesting enzymes) are produced on a large scale for inclusion in 'biological detergents' to remove protein and oily stains from clothing.

8.5 MULTIPLE USES OF ENZYMES

Enzymes are used for many different purposes. Some of them are discussed in the following text.

8.5.1 Detergent Enzymes

By far, the widest application of enzymes today is in detergents. *Proteases*, *lipases*, and *amylases* are used in detergents to remove stains of proteins, lipids (fats), and polysaccharides, respectively. *Cellulases* clean and maintain the look of a fabric. Addition of two or more of these enzymes in the detergent allows better efficiency in cleaning, cuts down the washing time, and water consumption, reduces energy consumption by lowering wash temperatures and time, increases the longevity of the fabrics, and diminishes the environmental impact since enzymes are biodegradable.

8.5.2 Enzymes for Processing Starch and Production of Fuel

Several necessary products are derived from starch. Corn starch is the major source, followed by wheat, tapioca, and potatoes. In the early 1960s, the use of *glucoamylase* led to the introduction of enzyme technology in the food industry. This enzyme completely breaks down starch to glucose hydrolytically. Owing to its higher yield and the purity of product, glucoamylase has replaced acid hydrolysis, the traditional process used for this purpose.

Heat stable *x-amylases*, discovered in the early 1970s, survive temperatures above boiling point and revolutionized industrial starch saccharification. The raw starch is gelatinized and liquefied by heating with steam in tank reactors to make the starch susceptible to enzyme attack. At this temperature, the heat stable x-amylases hydrolyse x-1,4-glycosidic bonds in starch. Further hydrolysis by the addition of *glucoamylase*, *fungal x-amylase*, and *pullulanase* produces a variety of syrups—maltose syrups, glucose syrups, and mixed syrups. Immobilized *glucose isomerase* can then be used to isomerize glucose into fructose for fructose syrup. Fructose syrups are used as sweeteners in many food products.

For production of fuel ethanol, starch is hydrolysed to glucose, which is then fermented by yeast to ethanol. Ethanol is a renewable source of energy and burns cleaner than gasoline, producing fewer harmful emissions. Cellulose, the most abundant organic polymer on the earth, has also been targeted to produce ethanol fuel by *cellulase* and *hemicellulase* enzyme technologies. Cellulosic waste materials, such as rice straw, corn stalks, and wood chips, are primary sources.

Futuristic projection to replace petroleum-based carbon with glucose derived from starch and cellulose for the synthesis of organic molecules and polymers has already been made. Starch and cellulose conversion technologies to produce fuels, power, and chemicals from renewable raw material will give birth to *biorefineries*. These will be analogous to today's petroleum refineries, producing petroleum-based multiple fuels and other products.

8.5.3 Improving Fermented Alcoholic Drinks

The natural enzymes *pectinesterase* and *polygalacturonase* present in grapes are unable to completely hydrolyse pectin and other complex polysaccharides in the cell wall. The use of the enzyme *pectinase* for this purpose has improved the quality and production of wine since the 1970s.

Glycosidase acts on aroma precursors and enhances the wine aroma of Muscat or similar varieties of grapes that contain bound terpenes.

Yeast glucans and grape pectins interact during fermentation to from colloids. Blends of *pectinase* and *b-glucanase* hydrolyse the colloids, improving the quality of wine. This enzyme mixture also accelerates the ageing process of wine, reducing the storage time.

β-Glucanase also improves the quality of wines produced from late harvest of grapes affected by the fungus *Botrytis cinera*.

Industrial enzymes, like microbial amylases, have replaced the use of malt or *koji* (fermented rice) for fermentation of alcoholic drinks from starch-based raw materials, such as sugarcane, potatoes, cereals, fruits, and wine.

For producing beer of a consistently high quality, modern breweries use enzymes in the regular processes of production and fermentation.

8.5.4 Enzymes Used in Baking

Fungal *X-amylase* is used to degrade starch to maltodextrins as it helps the yeast in its function. A special *variety of amylase* alters a specific portion of the starch so that it does not get stale easily and thus prolongs the shelf life of bread. Some *oxidative enzymes* have replaced potassium bromate, a chemical additive that has been banned in many countries.

Xylanases (type of hemicellulases), *lipases*, and *oxidases* increase the strength of the network of proteins (gluten in flour) and improve the quality of bread.

8.5.5 Fruit and Vegetable Processing

Pectin (see Chapter 2) binds adjacent plant cell walls together and can retain water. Thus, its presence makes it difficult to release the juice from the crushed fruit. *Pectinase* or pectinolytic enzymes solve this problem.

Pectin esterase is used to maintain the firmness of canned and frozen fruits and vegetables.

8.5.6 Forest Product and Paper Industry

To de-colourize wood pulp, *xylanases* (hemicellulases) are gradually replacing the use of harsh chemicals, such as chlorine and chlorine dioxide. Environmental regulations discourage the use of chlorine-based harmful chemicals.

Enzymes also help in the smooth running of high-speed paper manufacturing machines by removing pitch (resinous substance from the pulp) and slime (deposits of microbial origin).

8.5.7 Dairy Products

Earlier, *chymosin*, collected from calf stomach extract, was used to coagulate milk for cheese production. Bovine chymosin is nowadays produced in recombinant microorganisms.

Ripened cheese with distinctive flavours is now produced faster by progressive enzymatic hydrolysis of proteins and lipids in cheese by endo- and exo-peptidases and lipases.

8.5.8 Enzymes Used in Animal Feed

Adding enzymes to livestock feed effectively improves the digestibility of food and thus the growth of the animals. A good range of enzymes is available to degrade substances, such as cellulose, hemicelluloses, glucan, starch, protein, pectin, phytic acid, xylan, raffinose, and stachyose present in animal feed. Supplementing feed with enzymes improves the feed conversion ratio, that is, better utilization and gut absorption of feed, leading to better health and faster growth of livestock.

Pigs and poultry are incapable of digesting phytic acid (inositol hexakisphosphate), the phosphorus storage compound in legume plants like soya bean. *Phytase* supplementation in feed is beneficial in two ways. It hydrolyses phytic acid, removing several phosphate groups, making them available as a nutrient. By degrading phytic acid it diminishes the level of organic phosphorus in animal manure—phosphorus is harmful to the environment.

8.6 RECENT EXPANSION OF THE ENZYME INDUSTRY

Recently, the enzyme industry has shown tremendous growth with the development of genetic engineering techniques. Biotechnological methods have led to the production of recombinant microorganisms, which are the major sources of enzymes for a wide variety of industries. Large-scale production of recombinant microorganisms is carried out in modern and sophisticated versions of large fermenters for bulk production of enzymes. The volume of enzymes manufactured has increased by 12 per cent annually over the last decade. Approximately, 400 companies are currently involved in the manufacture of enzymes. Novozymes of Denmark, Danisco/Genencor (Denmark and USA), BASF of Germany and DSM of the Netherlands together earn three-fourths of the total revenue on account of manufacturing enzymes. 60 per cent of enzyme production takes place in Europe, with 15 per cent in USA and 15 per cent in Japan. However, USA and Europe each consume 30 per cent of the world output. The current market of enzyme production is around US$ 3 billion and is expanding quite rapidly.

Enzymes are Consumed Broadly in Three Sectors

(i) Technical applications in detergents, textiles, leather, pulp and paper, and fuels consume about 60% of all enzymes.
(ii) Food manufacture uses close to 30% of the total enzyme production.
(iii) Animal feed consumes about 6% of enzymes, but is a rapidly growing sector.

8.7 ENZYME PRODUCTION

There are certain specific ways to develop recombinant microorganisms capable of producing high yields of the desired enzymes. The techniques involved are rDNA technology, traditional mutation and breeding, protein engineering, and directed evolution.

8.7.1 Selection and Development of the Producer Strains of Microorganisms

Selection of microorganisms or genes producing particular enzymes perfectly suited for a specific purpose and their *development* through genetic manipulations are two fundamental sequential steps for modern enzyme biotechnology.

Selection

For selection of microorganisms, the following methods are used:

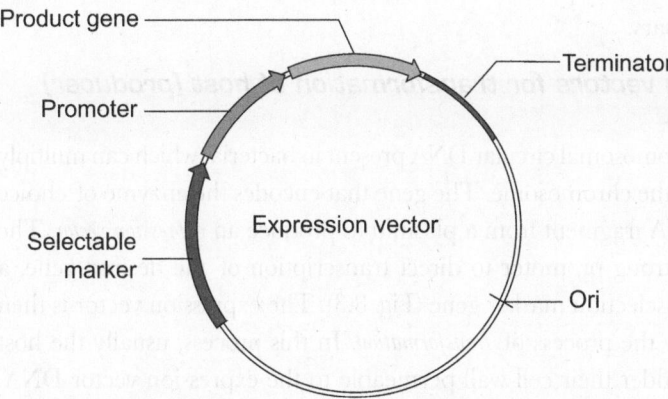

Fig. 8.3 Construction of expression vector (plasmid) for production of industrial enzymes: Strong promoter to direct transcription of product gene; Product gene producing specific enzyme; Terminator to terminate transcription; Selectable marker gene for selection of transformed cells into which the expression vector has entered, Ori indicates the point of origin of replication of the plasmid.

Screening   Microorganisms that produce the desired enzymes are obtained either from nature or from large collections of cultures that are maintained by enzyme companies. These are then cultured in nutrients at optimum temperature and pH which are adjusted to favour the growth of one particular variety, that is, the best enzyme-producing candidate. For example, to obtain an ideal protease for cold water laundry detergents, the microorganisms are cultured under alkaline conditions at temperatures of 5°C in the presence of chemical additives in various detergents. The best growing variety of microorganisms under the desired conditions is selected and further cultured for enrichment. The specific gene encoding the desired enzyme is then selected from the organisms in enriched culture for rDNA technology to construct an expression vector (Fig. 8.3).

Gene libraries and screening through expression cloning   Nowadays, enzyme companies maintain diverse collections of ready-to-screen DNA and cDNA libraries from microorganisms. The genes are introduced into surrogate hosts such as *Saccharomyces cerevisiae* (yeast) and *Escherichia coli* that are easy to manipulate in the laboratory, and screened for expression clones producing the desired enzymes.

Selection of the right kind of expression clones involves an enormous amount of repetitive work; therefore sophisticated computer programming and automation through robotics are used.

Genome sequencing and bioinformatics   Many bacterial and fungal genomes are being sequenced. The expanding sequence databases help find DNA sequences responsible for novel enzymes.

Development of producer strains

Once the genomic sequence for a desired enzyme is screened, several types of genetic technology are employed for its higher yield at high purity and with better economic prospect. The gene encoding a specific enzyme is incorporated into genomes of safe and environmental friendly bacteria (e.g., *Bacillus* species), yeast, and filamentous fungi (e.g., *Aspergillus* species) for stable transcription and translation. Such host organisms for the gene should not produce any substance or protease detrimental to the desired product.

At the same time the host should be capable of carrying out post-translational modifications of the required enzyme, if necessary.

Construction of expression vectors for transformation of host (producer) strains (rDNA technology)

As we know plasmids are extrachromosomal circular DNA present in bacteria, which can multiply and transcribe independently of the chromosome. The gene that encodes the enzyme of choice is introduced in a plasmid or DNA fragment from a plasmid to produce an *expression vector*. The expression vector consists of a strong promoter to direct transcription of the desired gene, a transcriptional terminator, and a selection marker gene (Fig. 8.3). The expression vector is then allowed to enter the host cells by the process of *transformation*. In this process, usually the host cells are chemically treated to render their cell wall permeable to the expression vector DNA. Then the host cells are placed in selective growth medium that allows the growth of only those host cells that have incorporated the expression vector with the desired gene and the selection marker. These host cells are called transformants and are tested for their ability to produce the desired (recombinant) enzyme.

Further improvement of the transformant strain through mutagenesis

The selected transformant is subjected to mutation and then screened for individuals with a high yield of the desired enzyme. High-yielding cells cut down the production cost of the enzyme and thus are desirable.

Traditional improvement through mutation and breeding of producer strains (alternative to rDNA technology)   When the performance of the product depends on several enzymes, it is not easy to obtain high-yielding strains through recombinant DNA technology. Sometimes the use of genetically modified organisms (GMOs), especially for food industry, arouses much public concern. In such cases, traditional strain improvement through mutation and breeding is sought, to obtain high-yielding strains (Fig. 8.4).

Protein engineering and molecular-directed evolution for improvement of enzymes to meet industrial requirements   Sometimes the conditions of industrial processes, such as high temperatures, extreme pH values, and presence of harsh chemicals may be too extreme for naturally occurring enzymes. In such situations, academic laboratories make use of protein engineering where some specific amino acids are selectively replaced within an enzyme by altering the genes (more particularly codons for amino acids) to allow the enzyme to withstand harsh conditions. This is done by site-directed mutagenesis of the gene encoding the enzyme. Site-directed mutagenesis is carried out in a test tube rather than in a living microbe. The mutated genes are then cloned into plasmids and introduced into a unicellular expression host (as shown in Fig. 8.3) to judge and select the strains producing the required enzyme.

Years of research has established a relationship between enzyme structure and function, and this data has helped to develop *predictive protein modelling software*. This software package is essential for protein engineers to identify specific amino acid substitutions that can be introduced. (Readers may consult Chapter 7.)

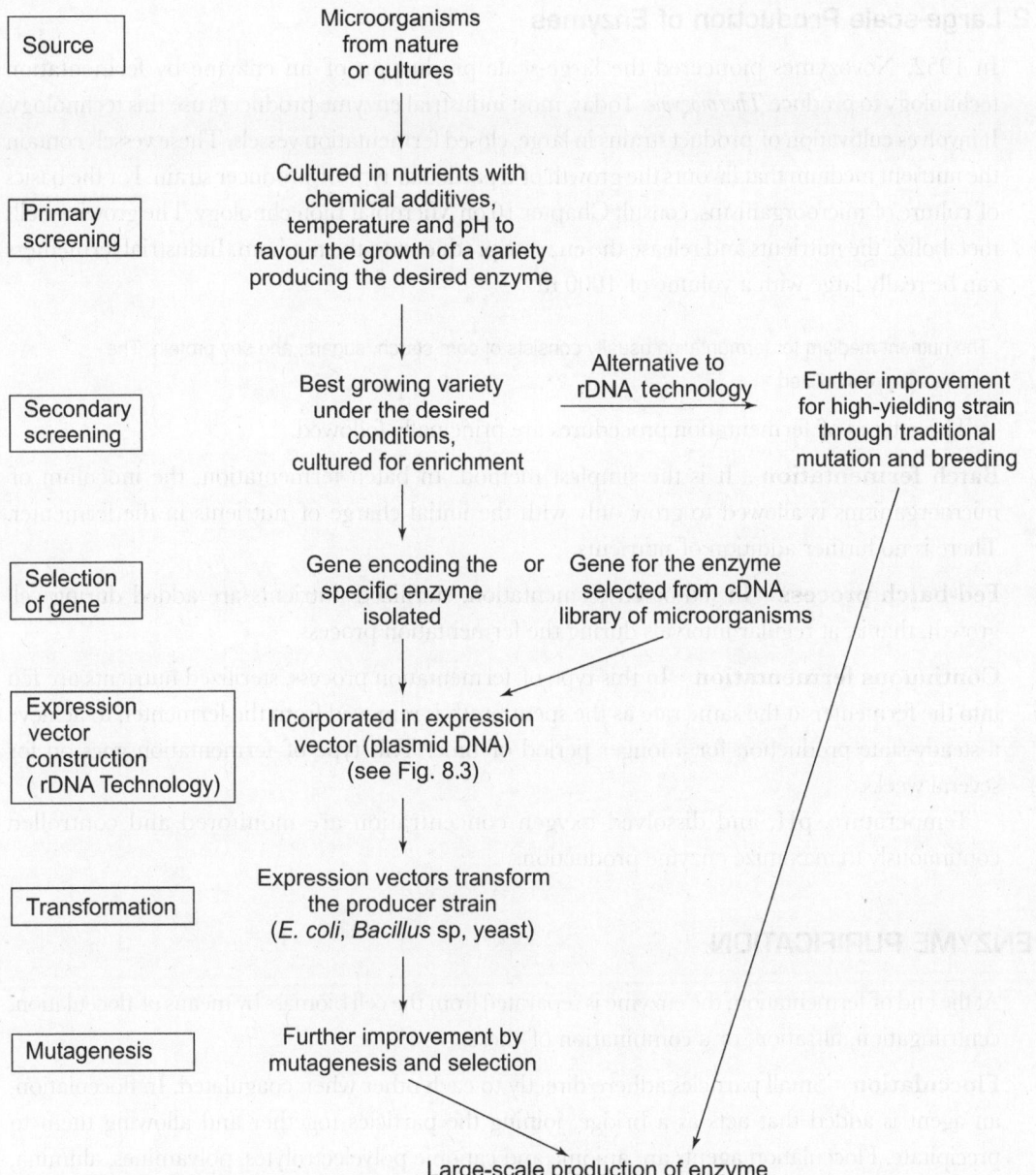

| Source | Microorganisms from nature or cultures |

| Primary screening | Cultured in nutrients with chemical additives, temperature and pH to favour the growth of a variety producing the desired enzyme |

| Secondary screening | Best growing variety under the desired conditions, cultured for enrichment |
Alternative to rDNA technology →
Further improvement for high-yielding strain through traditional mutation and breeding

| Selection of gene | Gene encoding the specific enzyme isolated   or   Gene for the enzyme selected from cDNA library of microorganisms |

| Expression vector construction (rDNA Technology) | Incorporated in expression vector (plasmid DNA) (see Fig. 8.3) |

| Transformation | Expression vectors transform the producer strain (*E. coli*, *Bacillus* sp, yeast) |

| Mutagenesis | Further improvement by mutagenesis and selection |

Large-scale production of enzyme in industrial fermenter

Fig. 8.4   Scheme for selection and development of a microbial strain to produce a desired enzyme.

At times, it is difficult to obtain the desired level of improvement using only random mutagenesis. To solve this problem, biotechnologists perform recombination between already mutated and improved enzyme genes. This recombination can be carried out either in vitro by using a variety of techniques called *DNA shuffling* techniques or directly in yeast using its natural (homologous) recombination system.

8.7.2 Large-scale Production of Enzymes

In 1952, Novozymes pioneered the large-scale production of an enzyme by fermentation technology to produce *Thermozyme*. Today, most industrial enzyme producers use this technology. It involves cultivation of product strains in large, closed fermentation vessels. These vessels contain the nutrient medium that favours the growth of a particular type of producer strain. For the basics of culture of microorganisms, consult Chapter 10 on Microbial Biotechnology. The growing cells metabolize the nutrients and release the enzyme product into the medium. Industrial fermenters can be really large with a volume of 1000 m^3.

> The nutrient medium for fermentation usually consists of corn starch, sugars, and soy protein. The medium is pre-sterilized.

Three types of fermentation procedures are principally followed.

Batch fermentation　It is the simplest method. In batch fermentation, the inoculum of microorganisms is allowed to grow only with the initial charge of nutrients in the fermenter. There is no further addition of nutrients.

Fed-batch process　In fed-batch fermentation, sterilized nutrients are added during cell growth, that is, at regular intervals during the fermentation process.

Continuous fermentation　In this type of fermentation process, sterilized nutrients are fed into the fermenter at the same rate as the spent broth is removed from the fermenter, to achieve a steady-state production for a longer period of time. This type of fermentation goes on for several weeks.

Temperature, pH, and dissolved oxygen concentration are monitored and controlled continuously to maximize enzyme production.

8.8 ENZYME PURIFICATION

At the end of fermentation, the enzyme is separated from the cell biomass by means of flocculation, centrifugation, filtration, or a combination of these.

Flocculation　Small particles adhere directly to each other when coagulated. In flocculation, an agent is added that acts as a bridge, joining the particles together and allowing them to precipitate. Flocculation agents are anionic and cationic polyelectrolytes, polyamines, alumina, and synthetic polymers. These agents effectively flocculate the cell biomass and debris in broth culture, leading to their easy precipitation.

Centrifugation　Centrifugation separates particles in a solution based on density. Cells and cell debris from a culture broth can be precipitated by centrifugation. Different types of centrifuge machines are used for the purpose. Tubular bowl centrifuge is the most preferred type.

Filtration　This process works under pressure applied by a pump. Elements of a solution are separated on the basis of size with the help of a filter cloth or some other porous materials. As the solution passes through the filter, filter cake builds up from accumulated particles on the filter and resists broth flow. To solve this problem, a rotary vacuum filter is used, and the filter cake is removed after each rotation. In industrial modifications, the filtration of broth through filter can also be done under suction pressure. The enzyme is concentrated from liquid media by semipermeable membranes or evaporation.

Precipitation   Addition of an ionic solution to an ionic fermentation broth helps in formation of insoluble particles of enzymes which precipitate out. Similarly, addition of organic solvent to an aqueous fermentation broth decreases the dielectric constant and therefore reduces the solubility of the enzyme.

Enzymes can also be precipitated out from solution like other proteins by adding ammonium sulphate $[(NH_4)_2SO_4]$ salt up to saturation. This is known as *salt precipitation*. Subsequently, $(NH_4)_2SO_4$ is removed by dialysis. Salt precipitation is an inexpensive procedure.

Chromatography   Chromatography is the major technique for high resolution purification of proteins and can be employed for enzyme purification (enzyme being protein). Different types of chromatography, such as gel filtration, ion-exchange chromatography, hydrophobic interaction chromatography, hydroxyapatite chromatography, and immunoaffinity chromatography, are used depending on the nature of enzyme products and other associated factors.

Selective precipitation, adsorption of the impurities, or crystallization can produce extremely pure enzymes. The final enzyme products can be supplied in dry (lyophilized), liquid, or immobilized form, as per the requirement of the industries.

8.9 IMMOBILIZATION OF ENZYMES

After performing their catalytic function, enzymes remain structurally and functionally intact. Thus, enzymes should be reused and not discarded with the reaction fluid after each cycle of use. This is also more economical. This has been achieved by immobilizing the enzymes on solid supports that can easily be separated from the reaction products and reused. Many support matrices, also called carriers, are available including glass beads, resins, modified cellulose polymers, agarose, gelatin, chitosans, and other polysaccharides.

8.9.1 Methods of Enzyme Immobilization

There are several techniques to immobilize an enzyme on a solid support (Fig. 8.5).

Adsorption   It is the simplest method that does not involve any chemical reagent which might denature the enzyme. Adsorption of enzyme molecules onto the surface of the support is mediated by low-energy bonds, such as ionic interactions, hydrogen bonds, and van der Waals forces. Since the bonds formed between enzymes and support matrices are weak, the enzymes can be easily desorbed (removed) by small changes in ionic strength, pH, or temperature.

Entrapment of enzymes   This method includes entrapment of the enzyme molecules inside a polymer matrix or gel. The size of the pores in the matrix is adjusted to prevent the loss of enzymes.

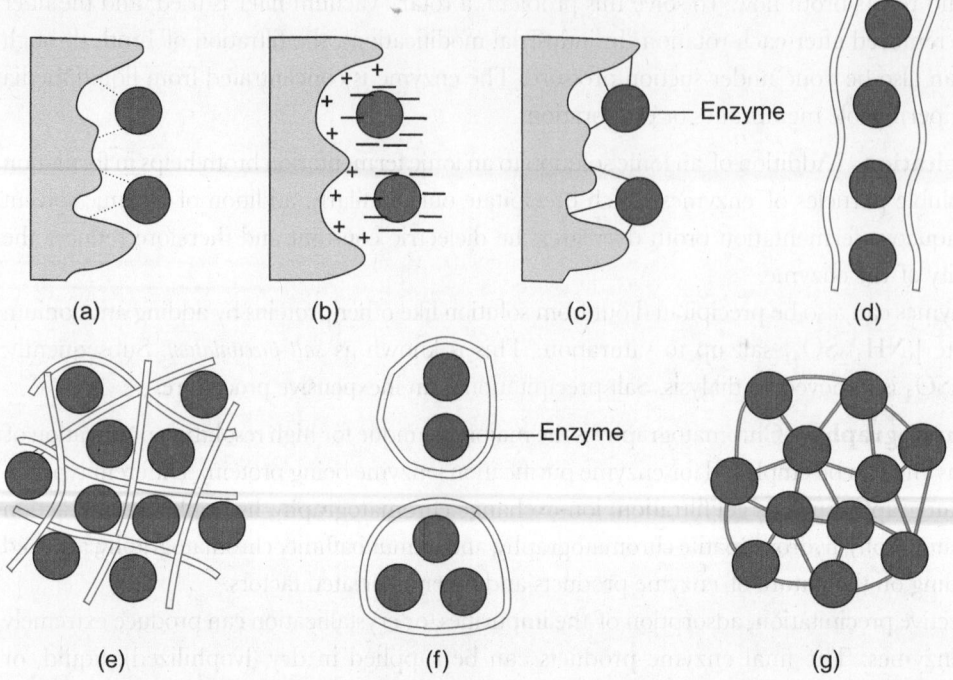

Fig. 8.5 Different techniques of immobilization of the enzyme on a solid support. (a) Adsorption by van der Waals forces (b) Adsorption by hydrogen bonding (c) Direct covalent bonding (d, e, and f) Immobilization by entrapment (g) Cross-linking of enzyme molecules.

Covalent linkage This method involves formation of covalent bonds between chemical groups of the enzyme and the support. The supports require chemical treatment (activation) prior to enzyme coupling. Loss of enzyme during the process is least in this method as the bond formation is quite strong. This kind of bonding requires use of purified enzymes of high quality. Covalent linkage can be done by the following methods:

(a) Diazotation, that is, bond formation between the amino group of the support (such as amino benzyl cellulose, amino derivatives of polystyrene, and aminosilane-treated porous glass) and a tyrosyl or histidyl group of the enzyme.

(b) Formation of peptide bond, that is, the amino/carboxyl groups of the support and the carboxyl/amino groups of the enzymes get linked via a peptide bond.

(c) Group activation, that is, cyanogen bromide activates glycol groups of a support (such as cellulose, sephadex, and sepharose) for binding of the enzymes.

(d) Polyfunctional reagents, that is, a bi- or multifunctional reagent, such as glutaraldehyde, creates bonding between the support and the amino groups of the enzyme.

Cross-linking of enzyme molecules This is characterized by the absence of any support; covalent links between multiple enzyme molecules create a matrix via a polyfunctional reagent. Cross-linking of enzyme molecules with proteins, such as serum albumin, gelatin, or haemoglobin, reduces the cost of immobilization.

8.9.2 Advantages of Immobilization

The process of immobilization has the following advantages:

(a) The process of immobilization often increases the activity and stability of the enzyme molecules as compared to that in free solution.

(b) It allows continuous industrial conversion processes, keeping non-productive downtime on account of cleaning in between batches to a minimum.

(c) Overall, the immobilization of enzymes is economically advantageous to the industry.

8.9.3 Applications of Immobilized Enzymes

In this section, a few applications of immobilized enzymes are discussed.

Bioreactors with immobilized enzymes for continuous conversions

The enzyme, immobilized on/in support material or crosslinked into particles, is packed into a reactor (usually columnar), and the support material is held in place by gravity or by a retaining mesh. The substrate is then pumped in to percolate through the packed enzyme that progressively converts the substrate to product. While the product passes out of the reactor in the product stream, the enzyme is retained in the reactor in an immobilized form. Thus, an efficient continuous system for enzyme conversion can be operated in *packed bed reactors*.

Many original papers on the application of immobilized enzymes have been published in recent times. Today, Japan leads other countries in immobilized enzyme technology.

Use of immobilized enzymes in bioelectrodes and biosensors   Immobilized enzyme technology enabled the designing of bioelectrodes and biosensors for detection of some substances in an extremely sensitive manner. An enzyme immobilized on an electrode can generate electrons, that is, a current or voltage, in the course of its reaction with its substrate. The electrode picks up that current or voltage for measuring. Thus, the substrate from an unknown source or its amount can be easily detected.

SUMMARY

♦ Enzymes are usually globular proteins. They work as biological catalysts that facilitate chemical reactions in living cells under a relatively mild biological temperature range.

♦ Enzyme-mediated chemical conversion of a substance is an energy efficient process; raw material consumption is decreased, CO_2 emission is reduced, production costs are reduced, and often replace toxic chemical processes.

♦ Some enzymes have traditional uses; for example, papain from papaya is traditionally used for tenderizing meat, and rennin of animal origin is used to clot milk during cheese manufacture.

♦ In recent times, the industrial demand for enzymes in the manufacture of detergents, textiles, leather, paper, human food, and animal feed has been increasing every year.

♦ Genetic engineering has made it possible to use microorganisms in bioreactors for the bulk production of enzymes in various industries.

♦ Certain specific techniques are used to develop recombinant microorganisms for high yield of the desired enzymes. These include rDNA technology, traditional mutation and breeding, and protein engineering and directed evolution.

♦ Enzymes remain structurally and functionally intact after a catalytic cycle. Thus, reutilization of enzymes instead of discarding with the reaction fluid after each cycle is economical. Immobilization of enzymes on solid supports precisely achieves this goal.

♦ There are several techniques for immobilization of enzymes on various solid supports for industrial purposes.

David Baltimore, Renato Dulbecco, and Howard Martin Temin

David Baltimore (Born 7 March 1938)

David Baltimore was born in New York City, USA. He graduated in 1956 from Great Neck High School and then obtained his bachelor's degree from Swarthmore College in 1960. In 1964, David received his PhD from Rockefeller University and did his postdoctoral work from Massachusetts Institute of Technology (MIT) and Albert Einstein College of Medicine. He joined MIT as a faculty in 1968 and later was elected a fellow of the American Academy of Arts and Sciences in 1974.

**Renato Dulbecco
(22 February 1914 – 19 February 2012)**

Renato Dulbecco was born in Catanzaro, South Italy. He graduated in anatomy and pathology from the

University of Turin, Italy. Dulbecco was sent to France and Russia during World War II. After the War, he moved to Bloomington, Indiana where he worked on bacteriophages. In 1949, he began working on animal oncoviruses. He joined the Salk Institute in 1962 and then in 1972 moved to The Imperial Cancer Research Fund. He was also among the scientists who launched the Human Genome Project. He last served as the President of the Institute of Biomedical Technologies at C.N.R. in Milan.

**Howard Martin Temin
(10 December 1934 – 9 February 1994)**

Howard Temin was born in Philadelphia, Pennsylvania, USA. He majored in biology from Swarthmore College in 1955 and graduated in animal virology. Howard received his Ph D in 1959 from the California Institute of Technology. In 1960, he joined the McArdle Laboratory for Cancer Research at the University of Wisconsin-Madison as assistant professor. The author of this text had privilege to be a student in a course on oncology taught by Dr Temin at UW-Madison.

David Baltimore and Howard Temin received the Nobel Prize in Physiology and Medicine along with Renato Dulbecco in 1975 for the discovery of reverse transcriptase.

The existence of reverse transcriptase enzyme in RNA

viruses of the Retroviridae (L. *retro,* 'backward') family infecting animal tissues was predicted by Howard Temin in 1962, and the enzyme was finally demonstrated to occur in such viruses by Temin and independently by David Baltimore in 1970. Their discovery aroused much attention, particularly as it provided molecular proof that genetic information can sometimes flow 'backward' from RNA to DNA. Although, Dulbecco was not directly involved in the experiments of Baltimore and Temin, he taught them the methods that they used for their discovery.

On infection, the single-stranded RNA viral genome and the reverse transcriptase enzyme (also called RNA-directed DNA polymerase) enter the host cell, and the reverse transcriptase catalyses the synthesis of a DNA strand complementary to the viral RNA. The same enzyme then degrades the RNA strand in the resulting RNA–DNA hybrid and replaces it with DNA. The double helix DNA so formed often becomes incorporated into the genome of the eukaryotic host cell. Under certain conditions such integrated (and dormant) viral genes become activated and are transcribed to generate new RNA viruses, or they may sometimes interfere with the functions of other regular genes of the host cells leading to malignant condition of the cells.

EXERCISES

Objective Questions

1. Multiple Choice Questions
   (i) Which of the following methods is not used for the separation of enzyme from the biomass at the end of production (or fermentation)?
      (a) semipermeable filtering membranes
      (b) general filtration
      (c) flocculation
      (d) centrifugation
      (e) evaporation

2. Mention the name of enzyme needed in industries for the following functions:
   (a) To clot milk during cheese manufacture
   (b) To manufacture glucose syrup
   (c) To prevent haze formation in fermented fruit juice
   (d) To increase the efficiency of detergents

Review Questions

1. Normally enzymes function at the physiological range of temperature, say 37°C in a cell; do you think in an in-vitro industrial set-up increased temperature will accelerate the enzyme reaction? Justify your answer.

2. Mention four characteristic features of an enzyme.

3. What is *white biotechnology*?

4. What environmental factors can influence enzyme technology?

5. Explain how activation energy is responsible for maintenance of the structure of a biomolecule.

6. What is the correlation between activation energy and enzyme function?

7. Elucidate the concept of 'selection and development of the producer strains of microorganisms' for industrial purposes.

8. When do the scientists opt for protein engineering to improve enzyme functions for industrial purposes? How do they achieve it?

9. How does 'continuous fermentation' differ from the 'batch fermentation' procedure in the large-scale industrial production of enzymes?

10. What technique would you employ to obtain the purest form of an enzyme?

11. What are the merits and demerits of the 'adsorption' technique for immobilization of enzyme?

12. What is diazotation?

13. Provide a short note on covalent linkage as a technique for enzyme immobilization.

14. What is the working principle of bioelectrodes coated with a specific enzyme?

9

Protein Structure and Engineering

LEARNING OBJECTIVES

♦ Three-dimensional (3D) structure of proteins
♦ Structure and function relationship in proteins
♦ Various methods used for purification of proteins
♦ Different protein-based products
♦ Basics of protein engineering

INTRODUCTION

Proteins are more varied in structure and more versatile in their function than the other macromolecules. Their functional diversities and polymeric residues (amino acids) have been discussed in Chapter 4. The twenty amino acids can be arranged in an almost infinite number of sequences to make an almost infinite number of different proteins. A specific sequence of amino acids folds up into a unique three-dimensional structure that determines the function of the protein.

It is important to note here that certain variations in amino acid sequence of a particular protein may occur within a population. Such proteins are known as *polymorphic proteins*. Many of these variations in sequence have little or no effect on the function of the protein. Changes in the amino acid sequence in the functional regions of a protein can have critical effects.

Advances in protein science, molecular biology, and genetic engineering have given rise to protein engineering. Protein engineering is concerned with altering the function of an existing protein in a predictable way by chemical and genetic means. In other words, it is concerned with altering the structure of a protein in order to make it function 'better' for human benefit.

9.1 THREE-DIMENSIONAL (3D) STRUCTURE OF PROTEINS

The three-dimensional (3D) arrangement of atoms in a protein is called its *conformation*. Of the innumerable conformations that are theoretically possible in a protein containing many amino acids linked by peptide bonds, one generally predominates. This is usually thermodynaamically most stable conformation, that is, having the lowest Gibb's free energy (G). Proteins in their functional conformation are called *native* proteins.

The amino group of one amino acid combines with the carboxyl group of the neighbouring amino acid to form a peptide bond, a rigid covalent bond. A number of covalent bonds arranged in series join the amino acids in a chain to form the primary structure. Let us now focus on the peptide bonds and the structural arrangement of atoms of amino acids. This would help in understanding how a polypeptide chain assumes its secondary structure and then its final conformation—the tertiary or quaternary structure.

Three-dimensional Structure

A protein with a given amino acid sequence has a specific three-dimensional structure over a range of conditions such as temperature, pH, ionic strength of the medium, and so on. It is the three-dimensional structure of the protein that confers biological activity to it. Function of a protein may thus be said to arise from its conformation, that is, the three-dimensional arrangement of atoms in the structure. The simple fact that proteins could be crystallized provided strong evidence in favour of their three-dimensional structure.

9.1.1 Peptide Bond, and phi (φ) and psi (ψ) Angles of Rotation

In an amino acid, a carboxyl group and an amino group remain bonded to the α-carbon, to which the side group R is also attached (refer to basic features of amino acids in Chapter 4 and Figs 4.12 and 4.13). The α-carbons of adjacent amino acids are arranged as C_α–C–N–C_α, separated by three covalent bonds (Fig. 9.1a). X-ray diffraction studies of crystals of amino acids and simple dipeptides and tripeptides revealed that the amide (C–N) bond in a peptide is somewhat shorter than the same (C–N) bond in a simple amine. This result indicated the presence of resonance or partial sharing of two pairs of electrons between the carbonyl oxygen and the amide nitrogen (each pair of partial double-bond shown by a solid and a broken line, Fig. 9.1a). The partial negative charge of oxygen and the partial positive charge of nitrogen create a small electric dipole. The four atoms of the peptide group (C, O, N, H) lie in a single plane (coplanar) in such a way that the oxygen atom of the carbonyl group with a partial negative charge and the hydrogen atom of the amide nitrogen are trans to each other. From these findings, Pauling and Corey successfully concluded that the amide *C–N bonds are unable to rotate* freely because of their partial double-bond character. Thus, a series of rigid planes separated by substituted methylene groups, -CH(R)-, constitutes the backbone of a polypeptide chain, which can be described as the primary structure of protein (Fig. 9.1b). The rigid peptide bonds restrict the number of conformations that a polypeptide chain could otherwise assume.

Limited rotation can take place *around the N–C_α and C_α–C bonds*. By convention, the bond angle created by rotation of the N–C_α bond is called φ (phi) and of the C_α–C bond is called ψ (psi). Again conventionally, both φ and ψ angles are considered 0° when two peptide bonds connected to a single α-carbon are in the same plane. In principle, this configuration does not occur and

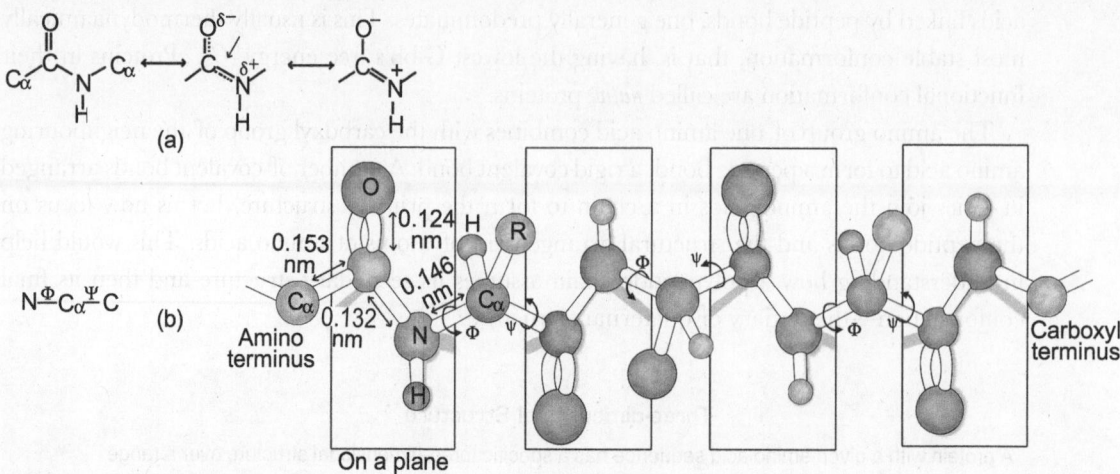

Fig. 9.1(a) The planar peptide group. Each peptide bond has partial double bond character (shown by one solid and one broken line). The carbonyl oxygen with a partial negative charge and the amide nitrogen with a partial positive charge set up a small electric dipole. The oxygen and hydrogen atoms in the plane are on the opposite sides of the C–N bond (trans configuration). Usually all peptide bonds in proteins exist in this configuration (except in β-turns). (b) Several amino acids are joined by rigid planar peptide bonds. The N–C$_\alpha$ and Cα–C bonds can rotate, with bond angles ϕ and ψ, respectively.

the angular value of ϕ and ψ may be anything between $-180°$ and $+180°$. But many values of ϕ and ψ are prohibited by steric hindrance between atoms in the polypeptide backbone and amino acid side chains (R). The value 0° for both the angles is meant as a reference point to calculate the angles of rotation in the actual situation.

In reality, the rotational bond angles ϕ and ψ, repeated at each amino acid residue explain the bending and turning of the primary structure of a polypeptide chain into every possible secondary structure (Fig. 9.2). The combinations of ϕ and ψ rotatory angles permitted for torsion of peptides have been shown by graphical plotting of ψ versus ϕ (Fig. 9.3). This graphical plotting is known as a *Ramachandran plot* after the name of the discoverer. Prof. G.N. Ramachandran established the Madras School of Structural Biology, at the Department of Crystallography and Biophysics, University of Madras, India in 1950. He worked with a team of gifted researchers to come up with the Ramachandran plot and the triple helix structure of collagen. When the Ramachandran plot (map) was originally proposed, only a few protein structures were known, and all the residues in each of these structures were found to fit very well into the allowed regions predicted in the map. Today, even with the hundredfold increase in the number of known structures of proteins, more than 98 per cent of the amino acid residues fit well within the conformation parameter limits proposed by the Ramachandran plot. The Ramachandran plot is now used as a touchstone for testing freshly solved protein structures before they are published.

9.2 STRUCTURAL ORGANIZATION OF A POLYPEPTIDE CHAIN

There are four levels in the structural organization of a polypeptide chain when it forms a protein molecule (Fig. 9.2).

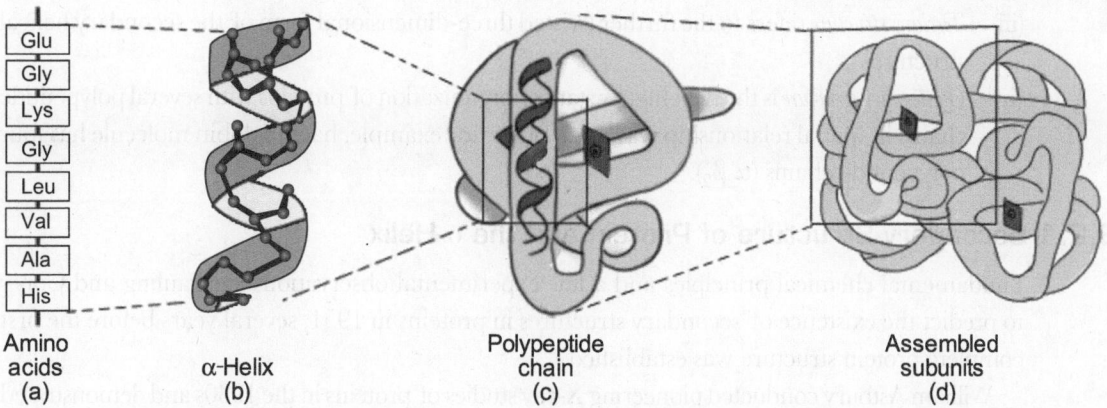

Fig. 9.2 Four levels of structural organization of polypeptide chain(s). (a) *Primary structure*—a sequence of amino acids linked together by covalent peptide bonds. (b) *Secondary structure*— e.g. α-helix, a coiled structure. (c) *Tertiary structure*—polypeptide helix folded further. (d) *Quaternary structure*—several polypeptide tertiary structures together.

    (i)   *Primary structure* is made up of peptide-bonded amino acids in a linear sequence (polypeptide chain).

    (ii) *Secondary structure* refers to the twisted primary structure with some recurring arrangements. Two most prominent secondary structures are the α-helix and the β-conformation.

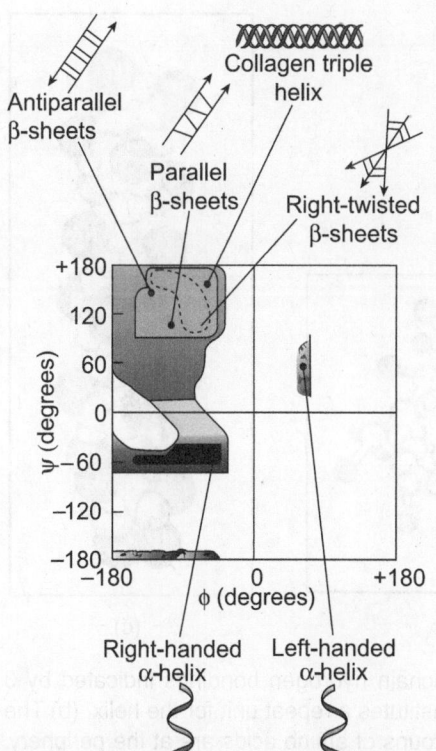

Fig. 9.3 A Ramachandran plot. The theoretical conformations of peptides allowed by the angles ϕ and ψ. The dark shaded areas indicate conformations that can be taken up by all amino acids in a peptide, or all except valine and isoleucine (medium shade); the lightest shading indicate somewhat unstable conformations found in some proteins.

(iii)   *Tertiary structure* refers to the further twisted three-dimensional form of the secondary helical structure.

(iv)   *Quaternary structure* is the next higher state of organization of proteins with several polypeptide chains in spatial relationship with each other; for example, haemoglobin molecule has four polypeptide chains ($\alpha_2\beta_2$).

9.2.1 Secondary Structure of Protein and the α-Helix

Fundamental chemical principles and a few experimental observations led Pauling and Corey to predict the existence of secondary structures in proteins in 1951, several years before the first complete protein structure was established.

William Astbury conducted pioneering X-ray studies of proteins in the 1930s and demonstrated that the protein that makes up hair and wool (the fibrous protein α-keratin) has regular repeats in its structure at every 0.54 nm. This information and data on the rigid peptide bond and other associated rotational bonds allowed Pauling and Corey to construct models for protein conformations. The simplest conformation that a polypeptide chain could assume was named by Pauling and Corey as *the α-helix* (Fig. 9.4). In this structure, the polypeptide backbone is tightly

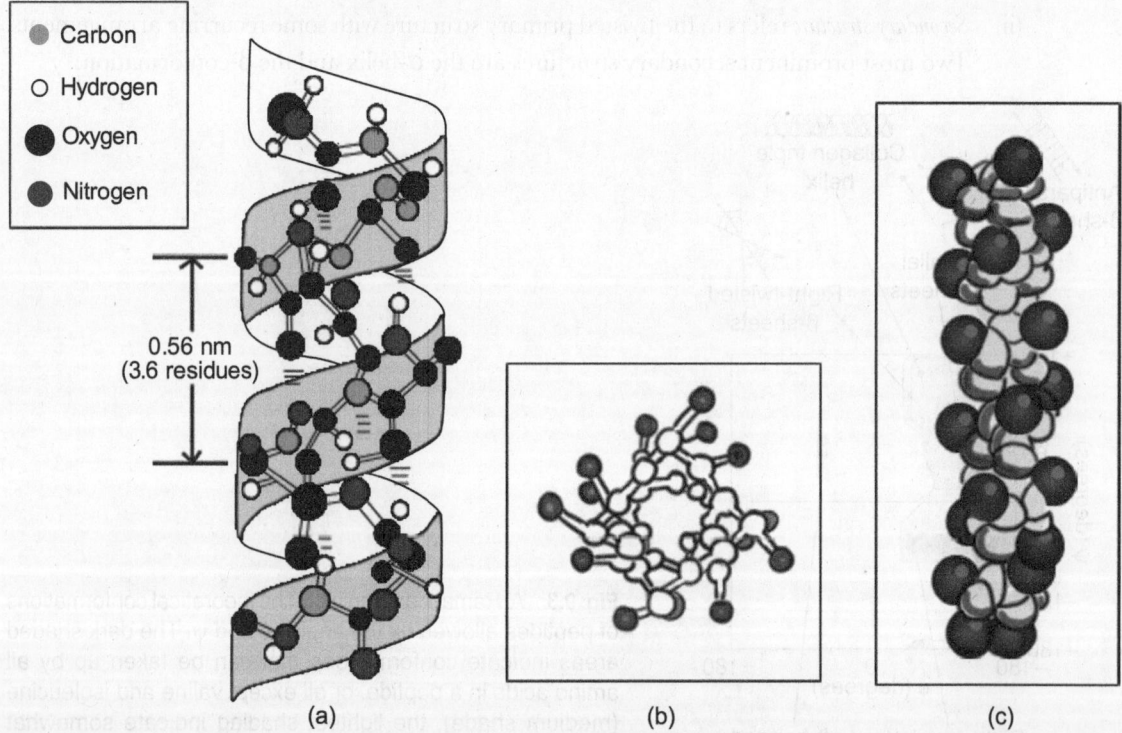

Carbon
Hydrogen
Oxygen
Nitrogen

0.56 nm
(3.6 residues)

(a)   (b)   (c)

Fig. 9.4(a) Ball and stick model of a right-handed α-helix, intrachain hydrogen bondings indicated by 3 dashes. A single turn of the helix consists of 3.6 residues, and constitutes a repeat unit for the helix. (b) The α-helix viewed from one end down the longitudinal axis. The R-groups of amino acids are at the periphery. (c) A space-filling model of the α-helix.

wound around the theoretical long axis of the molecule, and the R-groups of the amino acid residues protrude outward from the helical backbone. The repeating unit is a single turn of the helix extending 0.56 nm along the long axis, which corresponds closely with the periodicity Astbury observed in his X-ray analysis of hair keratin. Each helical turn (the repeating unit) includes 3.6 amino acids. The amino acid residues in an α-helix have ψ values in the range of $-45°$ to $-50°$ and ϕ values around $-60°$. The twist of the α-helix is right handed.

Proteins are classified into two major groups: (i) *fibrous proteins*—polypeptide chains in long strands or sheets, and (ii) *globular proteins*—polypeptide chains folded into a spherical or globular shape. Fibrous proteins are an integral part of the structural anatomy in higher vertebrates; they constitute 50 per cent or more of the total body protein. Fibrous proteins are usually made of a single type of secondary structure. This structural simplicity actually made the fibrous proteins interesting candidates for the analysis of protein structure. Globular proteins on the other hand are structurally complex, and often contain several types of secondary structures. They participate in the formation of enzymes, peptide hormones, antibody molecules, and secretory proteins. However, about one-fourth of all amino acid residues in globular proteins are in α-helical conformation.

Reasons for formation of helix in preference to other conformations

The α-helical structure is stabilized by a hydrogen bond between the hydrogen atom attached to the nitrogen atom of each peptide linkage and the electronegative carboxyl oxygen atom of the fourth amino acid on the amino-terminal side in the helix (Fig. 9.4a, the hydrogen bonds shown by horizontal dash marks). Every peptide bond of the chain participates in such hydrogen bonding. Several hydrogen bonds of a coiling part of the α-helix hold the successive coils of the helix in place. Thus, hydrogen bonds create the helical pattern, and provide considerable stability to the helix structure.

This also explains why polypeptide chains are not free to take up any three-dimensional structure at random. Steric constraints and many weak interactions make some arrangements more stable than others.

Further experiments with model building have shown that an α-helix can be formed with either L- or D-amino acids but not with a mixture of both. Naturally occurring L, D amino acids can form either right-or left-handed helices; but with rare exceptions, only right-handed helices are found in proteins.

Amino acid sequence affects α-helix stability

Till now we have discussed many weak interactions, especially the involvement of hydrogen bonds in stabilizing the α-helix structure. Other weak bonds are discussed later in the chapter. In the following text, the role of amino acid side chains (R) (see Fig. 4.13) in stabilizing the α-helix is discussed.

Interactions between amino acid side chains can stabilize or destabilize the α-helix structure; the charge and size of the side R-groups play a particularly important role. Thus, all polypeptides cannot form a stable α-helix. For example, negatively charged glutamate (Glu) residues in a long stretch do not allow the formation of α-helix at pH 7.0. The negatively charged carboxyl groups of adjacent Glu residues repel each other so strongly that they cancel the stabilizing influence of hydrogen bonds on the α-helix. Similarly, in a stretch many lysine (Lys) and/or Arginine (Arg) residues, with positively charged R-groups at pH 7.0, repel each other and prevent formation of the α-helix.

The critical interactions between side chains of amino acids that are three or sometimes four residues apart ensure twist formation in an α-helix. Positively charged amino acids often occur three residues away from negatively charged amino acids, and allow the formation of an ionic interaction.

Proline is rarely present in an α-helix because of the following two reasons: (i) nitrogen atom is part of a rigid ring and rotation about the N–C_α bond is not possible and (ii) nitrogen atom of a proline residue in a peptide linkage has no substituent hydrogen to form a hydrogen bond with other residues.

It has been indicated that a small electric dipole (opposite charges at two ends) exists in each peptide bond (see Fig. 9.1). These dipoles add across the hydrogen bonds in the helix so that the net dipole increases as helix length increases. The partial positive and negative charges of the helix dipole reside on the peptide amino and carbonyl groups at the amino-terminal and carboxyl-terminal ends of the helix, respectively, and contribute to the stability of the helix.

Constraints on Stability of α-helix

The summation of five different kinds of constraints imposed by amino acid sequence on the stability of an α-helix is as follows: (1) the electrostatic repulsion or attraction between amino acid residues that have charged R groups, (2) the size of adjacent R-groups, (3) the interaction between amino acid side chains located three (or four) residues away, (4) destabilization caused by proline residue in most cases, (5) the interaction between amino acids with inherent electric dipole at the ends of the helix.

Weak interactions stabilize a protein's conformation

The importance of hydrogen bonds in the stability of the α-helix has already been discussed. In fact, weak interactions, such as hydrogen bonds, and hydrophobic, ionic, and van der Waals interactions are important for folding of polypeptide chains into specific secondary, tertiary, and quaternary structures, and also for the stability of the structures.

Although covalent bonds including disulphide bonds are clearly much stronger, weak interactions predominate as a stabilizing force in protein structure because of their enormous number. In general, the most stable protein conformation with the lowest free energy is the one with the maximum number of weak interactions.

Of the different types of weak interactions, hydrophobic interactions are particularly important in stabilizing the protein structure. Hydrophobic amino acid side chains tend to cluster in a protein's interior, away from water. The interior of a protein is generally a densely packed core of hydrophobic amino acid side chains. Any polar or charged groups in the interior of the protein need to have suitable partners for hydrogen bonding or ionic interactions; this is important for the stability of the structure.

9.2.2 β-Conformation—Polypeptide Chains Organized into a Sheet

A second type of repetitive structure of polypeptide, β-*conformation* was also predicted by Pauling and Corey. Extended polypeptide chains, associated by hydrogen bonding form a sheet like structure (not helical). The polypeptide chains are planar and arranged side by side resembling a series of pleats, so the structure is often called a β-*pleated sheet.* Amino and carboxyl ends of adjacent polypeptide chains may be-oriented in the same direction in parallel β-pleated sheets or in opposite directions forming anti-parallel β-sheets (Fig. 9.5). The structure is stabilized by

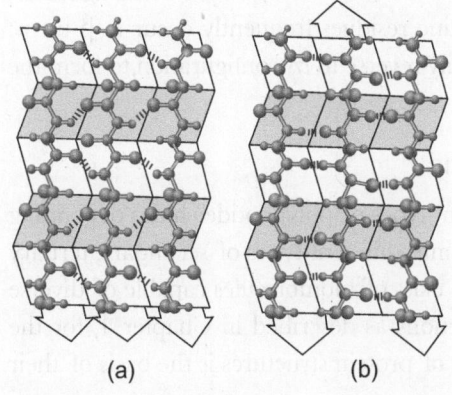

(a)   (b)

Fig. 9.5 β-pleated sheet (strand) of the polypeptide chain, in (a) parallel and (b) anti-parallel conformations.

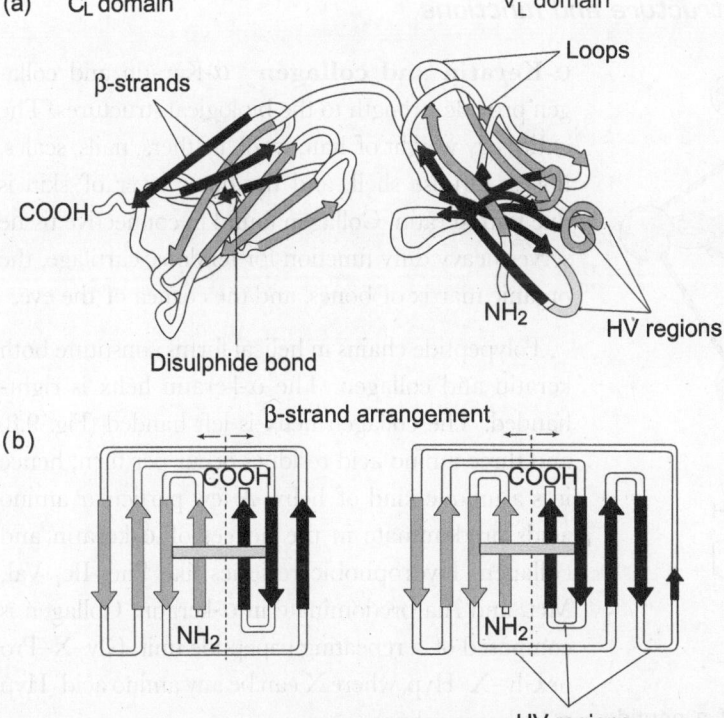

(a) C_L domain V_L domain

Loops

β-strands

COOH

Disulphide bond

NH₂

HV regions

β-strand arrangement

(b)

COOH   COOH

NH₂   NH₂

HV regions

Fig. 9.6 Characteristic organization of β-pleated sheets and loops in an immunoglobulin light chain.

hydrogen bonds between N–H and C–O groups of adjacent polypeptide chains. The distance between adjacent residues in a polypeptide chain along the direction of the sheet is 3.5 Å.

This laterally extended conformation of the polypeptide chains is found in many protein structures, sometimes in continuation with an α-helix as in an antibody molecule (Fig. 9.6). The silk protein fibroin (a member of a class of fibrous proteins called β-keratins) and the proteins of spider web are examples of β-pleated structure.

A β-bend or β-turn connects the ends of two adjacent segments of an anti-parallel β-pleated sheet, where a polypeptide chain takes an abrupt turn in the reverse direction. This is a tight turn (~180°) involving four amino acids. Glycine and proline residues frequently occur in β-turns; the former is small and flexible and the latter can readily reverse to *cis* configuration to form the turn (Fig. 9.7).

9.2.3 Structure–Function Relationship in Proteins

Structurally different proteins are produced by different types of polypeptide chains depending on amino acid sequences, and organization of chains into different types of secondary, tertiary and quaternary structures. Proteins represent a varied class of biomolecules capable of diverse functions—structural, catalytic, and many other functions as described in Chapter 4, for the existence of life and life processes. The diverse nature of protein structures is the basis of their diverse functions; examples are mentioned in the following text.

Secondary proteins—structure and functions

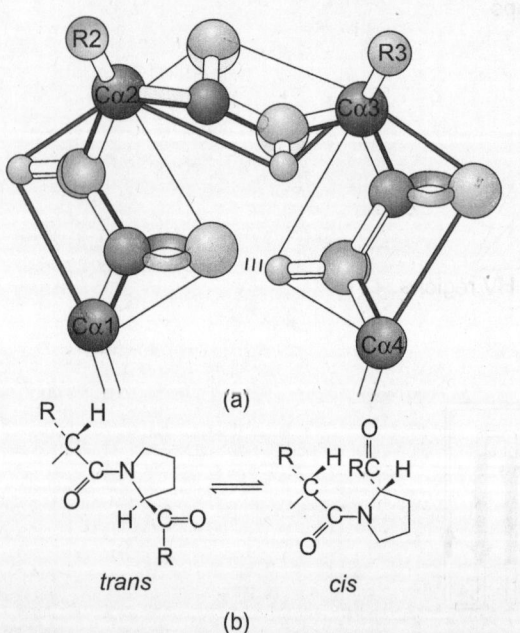

α-Keratin and collagen   α-Keratin and collagen provide strength to the biological structures. The entire dry weight of hair, wool, feathers, nails, scales, horns, tortoise shell, and the outer layer of skin is due to α-keratin. Collagen found in connective tissue serves heavy duty function for tendons, cartilage, the organic matrix of bones, and the cornea of the eye.

Polypeptide chains in helical forms constitute both keratin and collagen. The α-keratin helix is right-handed. The collagen helix is left handed (Fig. 9.8) and three amino acid residues occur per turn; hence it is a unique kind of helix. A few particular amino acids predominate in the helices of α-keratin and collagen. Hydrophobic residues like Phe, Ile, Val, Met, and Ala predominate in α-keratin. Collagen is composed of a repeating tripeptide unit, Gly–X–Pro or Gly–X–Hyp, where X can be any amino acid (Hyp = hydroxyproline).

Fig. 9.7 The trans and cis form of a peptide bond involving the imino nitrogen of proline at a β-turn or β-bend.

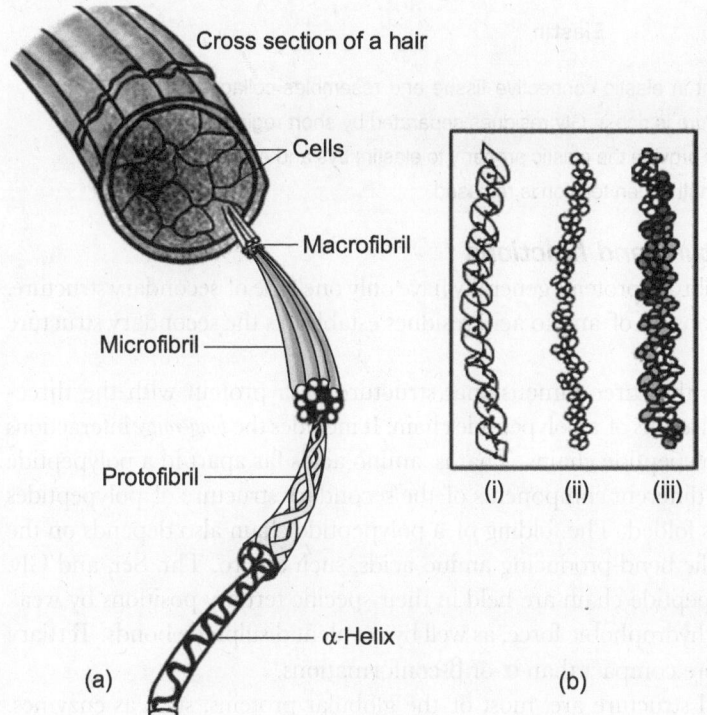

Cross section of a hair

Cells

Macrofibril

Microfibril

Protofibril

(i) (ii) (iii)

α-Helix

(a)

(b)

Fig. 9.8 (a) The structure of α-keratin and its organization in hair. Three α-helical chains form a supercoiled three stranded 'rope'-like protofibril, and 11 protofibrils constitute a hair microfibril. Bundles of microfibrils constitute macrofibrils which pass through and around the cells in the hair. The α-helices in a protofibril are linked by disulphide bonds and are present in enormous numbers in hard keratin-like tortoise shell. (b) The structure of collagen in 3 phases (i) left-handed helix, (ii) space filling model of the same, (iii) three of these helices wrap around one another in right-handed twist to form a three-stranded molecule of tropocollagen.

α-Keratin and collagen are insoluble in water, mainly due to the high concentration of hydrophobic amino acids on the surface, in addition to the interior. These proteins are an exception to the rule that hydrophobic groups must be buried within protein structures. In these cases, the hydrophobic core of the molecule therefore contributes comparatively less to the structural stability, and covalent bonds assume that function.

In both α-keratin and collagen, the strength of the fibril is increased by twisting three helical strands together in a superhelix, the way strings are twisted to make a strong rope (Fig. 9.8). The superhelical twisting is left-handed in α-keratin and right-handed in collagen. The collagen triple helices are tightly wrapped around each other due to great tensile strength with no capacity to stretch. This enables collagen fibres to support upto 10,000 times their own weight; they are considered to have greater tensile strength than a steel wire of equal cross section. Covalent cross-links, especially from disulphide bonds of cystine residues, between polypeptide chains within the triple helical 'ropes' and between adjacent ropes increase the strength of α-keratin and collagen to a significant level. The hardest and toughest α-keratin in tortoise shells and rhinoceros horns have as much as 18 per cent cystine residues involved in disulphide bonds.

Elastin

Elastin is a fibrous protein present in elastic connective tissue and resembles collagen to some extent. Its secondary helical structure is rich in Gly residues separated by short regions containing Lys and Ala residues. Gly residues provide the elastic property to elastin; Lys and Ala residues help the protein revert to the original length when tension is released.

Tertiary proteins—structure and functions

From our previous discussion, fibrous proteins generally have only one type of secondary structure. The *short range* structural relationship of amino acid residues establishes the secondary structure of polypeptide chains.

Tertiary structure presents the three-dimensional structure of a protein with the three-dimensional arrangement of all atoms of a polypeptide chain. It includes the *long-range* interactions of amino acid sequences in polypeptide chains. That is, amino acids far apart in a polypeptide sequence and even residing in different components of the secondary structure of polypeptides may interact when a protein is folded. The folding of a polypeptide chain also depends on the number and location of specific bend-producing amino acids, such as Pro, Thr, Ser, and Gly. Loops of a highly folded polypeptide chain are held in their specific tertiary positions by weak binding interactions including hydrophobic force, as well by covalent disulphide bonds. Tertiary folded structures are much more compact than α-or β-conformations.

Examples of tertiary folded structure are, most of the globular proteins, such as enzymes, transport proteins like myoglobin and haemoglobin, some peptide hormones, immunoglobulins, and serum albumin.

Myoglobin

Globular protein structure was first resolved by X-ray diffraction studies of myoglobin by John Kendrew and his colleagues in the 1950s. Myoglobin is composed of a single polypeptide chain of 153 amino acid residues (Mol. wt 16,700) and a single iron–porphyrin or haeme group. It carries oxygen to the muscle cells for cellular oxidation.

Enzymes   The highly specific catalytic activities of enzymes depend to a large extent on the three-dimensional configuration of the various reactive groups at the active site of the protein. The groups of catalytic amino acid residues are brought into the active site of an enzyme from different parts of the linear amino acid sequence in the polypeptide chain by the process of folding. Any change in the sequence and the consequent change in folding pattern of the protein may lead to large changes in the activity or even loss of function of the enzyme. Only precise three-dimensional arrangement of the active groups at the catalytic site allows an enzyme to lock to its substrate and carry out its function.

Protein quaternary structure

Two or more separate polypeptide chains (or subunits), identical or different, participate in the formation of quaternary structure of a protein. Haemoglobin, the oxygen-carrying protein of erythrocytes is one of the best examples of a multisubunit quaternary protein.

The size and number of subunits in quaternary proteins make the process of determination of structure with X-ray and other analytical methods more difficult and time-consuming.

Examples of larger, more complex multisubunit proteins are RNA polymerase enzyme of *E. coli*, for synthesis of RNA chains; the enzyme aspartate transcarbamoylase (12 chains), involved in the synthesis of nucleotides; and the enormous pyruvate dehydrogenase complex of mitochondria, as an extreme case, is a cluster of three enzymes containing a total of 102 polypeptide chains. Multiple non-covalent interactions, as in the case of tertiary structures, guide folding and stabilize the multiple subunits in a quaternary structure. The association of multiple polypeptide chains serves a variety of functions in many cases; ribosome is one such example.

Haemoglobin   Max Perutz, John Kendrew, and their colleagues at Cambridge resolved the quaternary structure of the first oligomeric protein—haemoglobin, in 1959. Haemoglobin contains four polypeptide chains, and four haeme prosthetic groups with iron atoms in the ferrous (Fe^{2+}) state, one attached to each chain. The protein chains are called globin and are four in number—two α-chains (141 residues each) and two β-chains (146 residues each). Here α and β are used as alphabets and do not refer to secondary protein structures. The chains bear similarity with the single chain myoglobin even in their amino acid composition; haemoglobin is four times bigger than myoglobin. The α- and β-chains have several segments of α-helix separated by bends. The haemoglobin molecule is roughly spherical with a diameter of about 5.5 nm.

The α-and β-chains have several contact points that consist mainly of the hydrophobic side chains of amino acid residues; ionic interactions involving the carboxyl terminal residues of the four subunits are also involved.

When a valine substitutes a glutamate residue at position 6 of the β-chain, a 'sticky' hydrophobic spot is formed on the surface of the molecule that causes abnormal quaternary association of haemoglobin resulting in sickle shaped erythrocytes. This is a natural example of the relationship between structure and function in proteins.

Denaturation of proteins—loss of structure and function

A classic demonstration of the dependence of function on the structure of a protein molecule can be revealed by the denaturation (loss of 3D

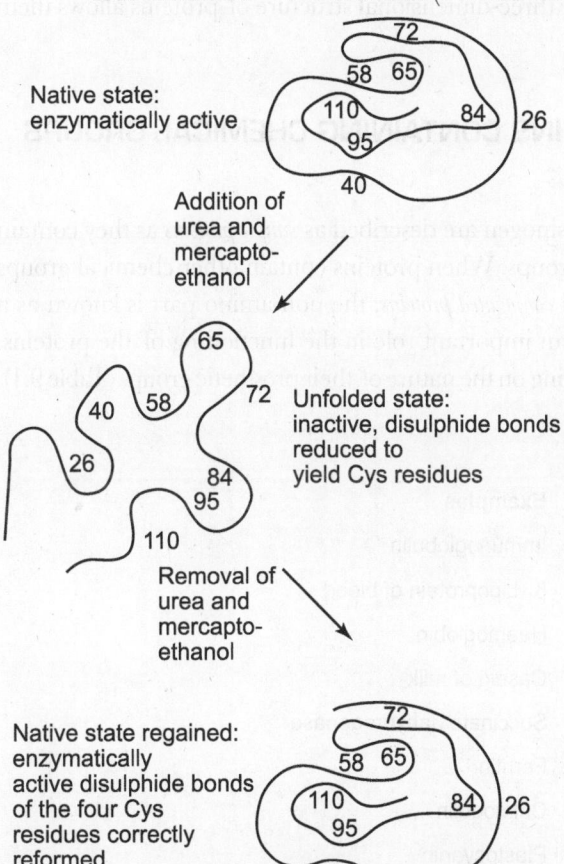

Fig. 9.9 Denaturation and renaturation of ribonuclease, demonstration of the dependence of the function of a protein on its structure.

structure) and renaturation of ribonuclease, a globular protein (Fig. 9.9). Purified ribonuclease gets completely denatured in a concentrated solution of urea in the presence of a reducing agent. Mercaptoethanol as reducing agent cleaves the four disulphide bonds to yield eight cysteine residues (marked with position number in Fig.9.9), and the urea disrupts the stabilizing hydrophobic interactions among amino acid residues, resulting in unfolding of the ribonuclease molecule. Although the completely unfolded peptide chain, may form a randomly coiled structure, the enzymatic activity is totally lost. After removal of the urea and the reducing agent, the randomly coiled, denatured ribonuclease spontaneously refolds into its correct tertiary structure, with full restoration of its catalytic activity. The refolding of ribonuclease during renaturation is so accurate that the four intrachain disulphide bonds (at cysteine residues) are formed once again in the same positions as in the original ribonuclease. Theoretically, the four disulfide bonds could be formed by the eight Cys residues at random in 105 different ways. This classic experiment was performed by Christian Anfinsen in the 1950s, and proves that the amino acid sequence in the polypeptide chain of proteins has all the necessary information for folding the chain into its native, three-dimensional structure. Proper three-dimensional structure of proteins allows them to function specifically.

9.3 CONJUGATED PROTEINS—PROTEINS CONTAINING CHEMICAL GROUPS OTHER THAN AMINO ACIDS

The enzymes ribonuclease and chymotrypsinogen are described as *simple proteins* as they contain only amino acids and no other chemical groups. When proteins contain other chemical groups in addition to amino acids, they are called *conjugated proteins;* the non-amino part is known as a *prosthetic group.* The prosthetic groups play an important role in the functioning of the proteins. Conjugated proteins are categorized depending on the nature of their prosthetic groups (Table 9.1).

Table 9.1 Classes of conjugated proteins

Class	Prosthetic group	Examples
Glycoproteins	Carbohydrates	Immunoglobulin
Lipoproteins	Lipids	β_1-Lipoprotein of blood
Haemoproteins	Haem (iron porphyrin)	Haemoglobin
Phoshoproteins	Phosphate groups	Casein of milk
Flavoproteins	Flavin nucleotides	Succinate dehydrogenase
Metalloproteins	Iron	Ferritin
	Calcium	Calmodulin
	Copper	Plastocyanin
	Molybdenum	Dinitrogenase
	Zinc	Alcohol dehydrogenase

9.4 PURIFICATION OF PROTEINS

Thousands of different types of proteins are present in cells. A pure preparation of a particular protein is needed to study its amino acid sequence, structure, and other characteristics.

Methods of separation of different proteins from a mixture have been devised on the basis of differential properties of proteins—size, charge, solubility, binding affinity, and so on. A *crude extract* from homogenized or broken cells is sequentially subjected to several different methods devised through many years of practice. The choice of method is somewhat empirical, and different protocols may be attempted to come up with the most effective one for a particular protein. It is prudent to start with purification by using procedures developed for similar proteins. Purification protocols for thousands of proteins are available in published literature.

The usual separation methods for purifying proteins based on their properties are listed in Table 9.2.

Table 9.2 Separation methods based on properties

Properties	Methods
Size and mass	Ultracentrifugation
	Dialysis and ultrafiltration
	Gel filtration
	Gel electrophoresis
Charge	Ion-exchange chromatography
	Electrophoresis
	Isoelectric focussing
Solubility	Change in pH
	Change in ionic strength
	Decrease in dielectric constant
Specific binding	Affinity chromatography

9.4.1 Protein Source

If a large amount of a particular protein to be purified is present in a particular source, its extraction becomes easier and more cost-effective. A tissue rich in a particular protein is collected and homogenized or, the organelle in which the protein is present is isolated by high-speed centrifugation on a specific gradient. If the protein is located in the cytosol of the cell, then the cells are lysed with hypotonic solution or by using a detergent.

For extraction, a suitable buffer solution in which the protein is stable is to be used. This is done to avoid denaturation of the protein by physical, chemical, or biological factors. The temperature of the buffer is usually maintained around 4°C or even less to avoid thermal denaturation and minimize the activity of proteases that are present in cell and tissue homogenates and extracts.

Modern technology like DNA cloning (see Chapter 6) can be used for expression of new or scarce proteins in large amounts in bacteria or eukaryotic cells.

9.4.2 Methods of Purification of Proteins

Different properties of proteins provide the basis for designing various methods of purification of proteins.

Ultracentrifugation

Centrifugation is a general procedure to precipitate the insoluble material in a liquid medium. This precipitable material can be cells, subcellular materials or large macromolecules like proteins and DNA. The smaller the material, the greater the centrifugation force that is required to cause it to sediment. An ultracentrifuge generates a higher rotatory speed and therefore a higher centrifugal force. Generally, sedimentation in an ultracentrifuge is carried out in a solution of an inert substance, sucrose or cesium chloride (CsCl), in which the density of the solution increases from the top to the bottom of the centrifuge tubes.

Ultracentrifugation may separate the particles on the basis of (i) *sedimentation coefficient* and (ii) *density* of the particles. Both the factors cause differential sedimentation velocity. Sedimentation velocity is indicated by *Svedberg* unit(s), named after the inventor of the ultracentrifuge, Theodor Svedberg. Sedimentation velocity is determined by both shape and size and hence is not an accurate measure of mass. The eukaryotic ribosome is composed of 60S and 40S subunits, which together form an 80S ribosome (Fig. 9.10).

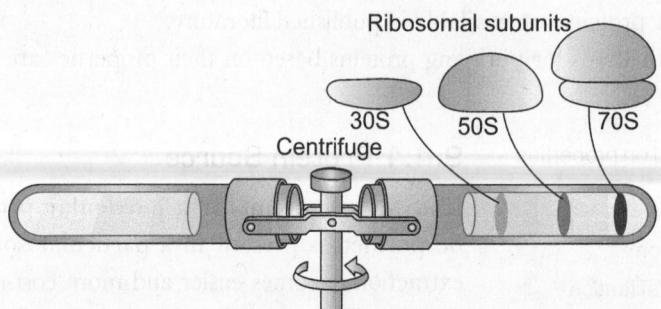

Fig. 9.10 Ultracentrifugation to separate ribosome subunits and the whole ribosome.

Equilibrium density gradient ultracentrifugation is used to purify or separate macromolecules on the basis of density of the substance as is often done for DNA purification (see Fig. 6.6). Solutions of varying density are layered on top of each other to form a density gradient, with the most dense solution at the bottom of the test tube; components of a sample form bands at positions where their densities are equal to that of the solution. The bands of different components are then removed separately.

Dialysis and ultracentrifugation

Dialysis is a process based on the principle of osmosis, that is, the movement of small molecules from an area of higher to an area of lower concentration. Dialysis membrane is a semipermeable membrane with small pores that act as a sieve to allow passage of substances weighing upto 10 kDa; it prevents the passage of larger molecules. This method only helps get rid of small molecules and salts from the protein solution. The pore size of the dialysis membrane may be changed by special mechanical and chemical treatment.

Ultracentrifugation can be used to generate pressure for faster dialysis.

Gel filtration

Column chromatography This the most common method of protein purification and is based on size and shape. The protein fraction is passed through a glass column filled with properly modified very small acrylamide or agarose (Sepharose) beads. Different categories of beads (supplied commercially marked with pore size of the beads) have different sized pores. Depending on the size of the pore, small proteins enter the beads and take longer to pass through the pore channels in the beads and elute (flow) out late from the column. This can also be described in another way—small proteins take more time to explore more space. Meanwhile, large proteins travel lesser space in the column (moving on the outside of the beads) and elute more rapidly (Fig. 9.11a).

During the run of a protein sample on the column, the buffer is continuously added at the top until small protein molecules elute out of the column. The eluted fractions are collected from the bottom of the column at different elution times and assayed for the protein of interest. The fractions with the highest concentration of protein are pooled for further purification to obtain the protein of interest.

If the shape of a protein molecule is markedly non-spherical, it may elute from the column at a different time than would a spherical molecule of equivalent size.

Ion-exchange chromatography   This is another modification of column chromatography, where proteins are separated on the basis of their surface ionic charge, using beads that are modified with either positively charged or negatively charged chemical groups. The electrostatic binding between beads and proteins of opposite charge is the basis of separation in ion-exchange chromatography (Fig. 9.11b). Thus, positively charged proteins bind to the negatively charged beads or matrix and are retained on the column, while negatively charged proteins pass through. Diethyl amino ethyl (DEAE) cellulose or Sephacel is negatively charged and retains positively charged proteins in the column. Carboxymethyl (CM) cellulose is positively charged and retains negatively charged proteins. The charge on a protein molecule comes from its constituent amino

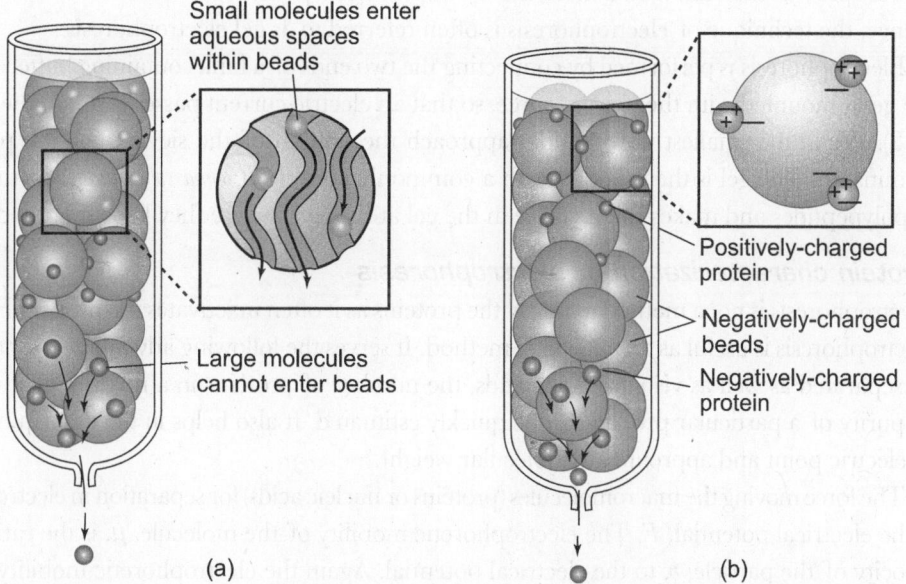

Fig. 9.11   Two commonly used gel filtration chromatography techniques. A glass tube is packed with beads in both the cases, and the protein mixture is passed through this matrix or column. The nature of the beads sets the basis of protein separation—on size or charge of the protein. (a) *Gel filtration:* The beads contain aqueous spaces into which small proteins enter and get slower in their progress through the column. Larger proteins cannot enter the beads and so pass through the column faster. (b) *Ion-exchange chromatography:* Negatively charged beads in the column bind positively charged proteins and retain them on the column, while negatively charged proteins pass through.

acids which may be neutral, positively or negatively charged; their overall charge is reflected on the protein.

Elution of the protein bound on the column can be brought about either by changing the pH of the input buffer and thus changing the charge or by increasing the strength of the salt solution so that the increased concentration of cations or anions will compete with the protein for the binding sites on the column exchanger. For example, proteins that interact weakly with the beads (being positively or negatively charged) are released from the beads (or eluted) in a low-salt buffer. Proteins that interact with the beads more strongly require a higher concentration of salts to be eluted. By gradually increasing the concentration of salt in the eluting buffer, even proteins with apparently similar charge characteristics get separated into different fractions as they elute from the column.

Electrophoresis

Electrophoresis is the migration of charged particles in an electric field. The technique is often used for the analytical separation of biological charged molecules like nucleic acids and proteins. Separation mainly depends on the magnitude of the charge carried by the ionizing groups on the molecules; the ionic strength and the pH of the buffer may influence the overall charge. Electrophoresis can be carried out using a variety of support media, such as paper, cellulose acetate, starch, polyacrylamide, and agarose. Of these media, gels made of polyacrylamide or agarose provide the best resolution for the electrophoretic separation of nucleic acids and proteins. Hence, the technique of electrophoresis is often referred to as gel electrophoresis.

Electrophoresis is performed by connecting the two ends of a tank containing buffer in which the gel is mounted with the power source, so that an electric current passes through the gel (Fig. 9.12). When the smallest polypeptides approach the bottom of the sieving gel, the process is terminated. The gel is then stained with a commonly used dye *Coomassie brilliant blue* that binds to polypeptides and makes them visible in the gel at the level where they have migrated.

Protein characterization by electrophoresis

Electrophoresis is not a method to purify the proteins as it often inactivates the protein. However, electrophoresis is useful as an analytical method. It serves the following advantages: proteins can be separated as well as visualized as bands, the number of proteins in a mixture or the degree of purity of a particular protein can be quickly estimated. It also helps in the determination of isoelectric point and approximate molecular weight.

The force moving the macromolecules (proteins or nucleic acids) for separation in electrophoresis is the electrical potential, E. The electrophoretic mobility of the molecule, μ, is the ratio of the velocity of the particle, v, to the electrical potential. Again the electrophoretic mobility may be expressed as the ratio of the net charge of the molecule, Z, to the frictional coefficient, f. These relationships can be expressed as follows:

$$\mu = \frac{v}{E} = \frac{Z}{f}$$

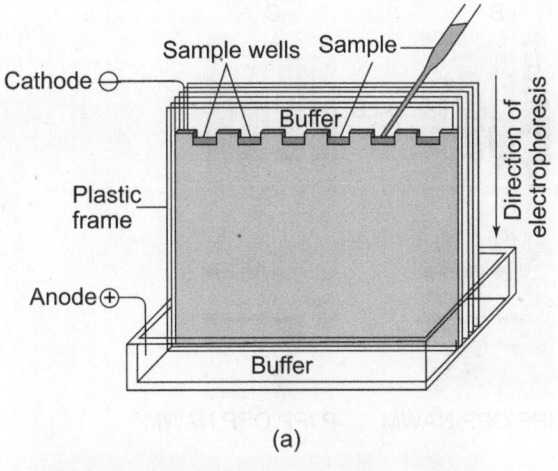

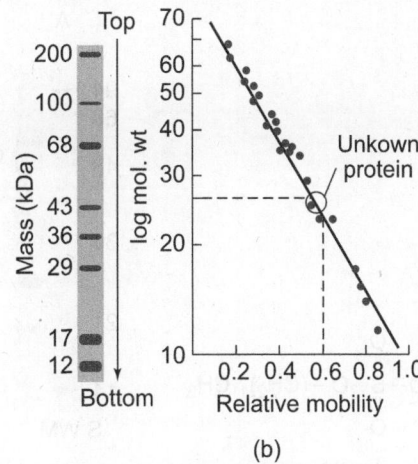

(a)

(b)

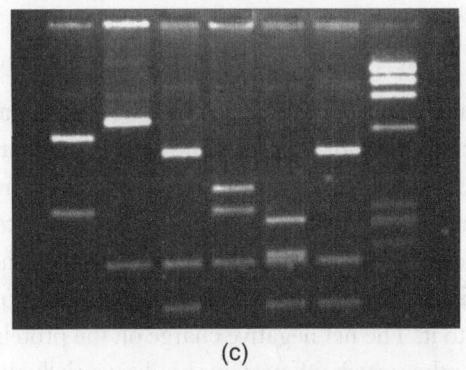

(c)

Fig. 9.12 (a) Apparatus for conducting gel electrophoresis to separate different proteins on the basis of their molecular weights in an electric field. (b) The molecular weight of a particular protein in a band is determined by the log of its migration distance against a standard plot. (c) Protein bands in different lanes of a gel after electrophoresis and Coomassie blue stain. Each band represents a different protein or protein subunit.

Polyacrylamide is a cross-linked polymer of acrylamide ($CH_2 = CHCONH_2$), usually used as a support matrix for electrophoresis. The microporosity in a polyacrylamide gel acts as a molecular sieve, slowing the migration of proteins in approximate proportion to their mass or molecular weight. Polyacrylamide gel electrophoresis (PAGE) is currently the most popular method for characterization of proteins. Summarily the separation in PAGE is on the basis of the charge on the molecules and sieving effect of the gel. Among molecules of equal size, the one with the higher net charge moves faster in an electrophoretic field. Small molecules move faster through the sieve of the gel than larger ones of the same net charge.

Electrophoresis can be (i) anionic and (ii) cationic. Basic proteins have an isoelectric point close to the basic pH region. So to increase the charge on the protein, the cationic system is preferable, while acidic proteins may be separated in an anionic system. Both the systems are used for an unknown protein. Three types of electrophoresis techniques are discussed below.

Sodium dodecyl sulphate polyacrylamide gel electrophoresis (SDS–PAGE)

Separation of proteins by SDS-PAGE is on the basis of molecular weights in an electric field. Hence determination of molecular weight of an unknown protein in reference to known markers

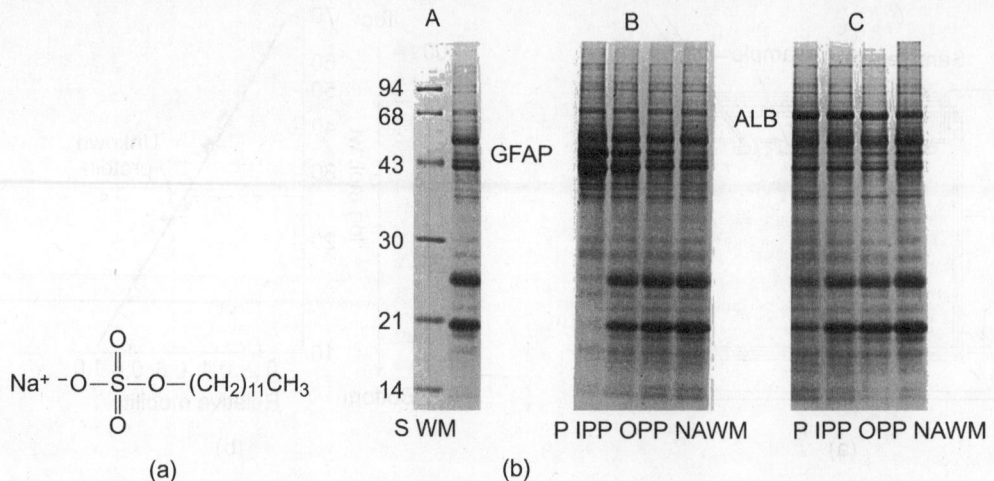

(a)   (b)

Fig. 9.13 (a) Sodium dodecyl sulphate (SDS) binds to most proteins probably by hydrophobic interactions. (b) SDS-PAGE of proteins.

is possible. The protein mixture in a sample is initially treated with a reducing agent such as β-mercaptoethanol or dithiothreitol to break the disulphide bonds. Then sodium dodecyl sulphate (SDS) (Fig. 9.13a), a negatively charged detergent, is added; SDS disrupts nearly all the non-covalent bonds (weak bonds) in the protein, unfolding the polypeptide chain into linear structure. SDS binds to a protein in amounts proportional to the length of the polypeptide chain. The intrinsic charge of the protein molecule becomes insignificant in the presence of the net charge of so many negatively charged SDS molecules bound to it. The net negative charge on the protein molecule is proportional to its molecular weight. In other words, all proteins bind essentially the same amount of SDS per gram of protein, and all SDS-protein complexes have essentially the same charge/mass ratio. On electrophoresis, the large molecules move slowly through the pores of the gel compared to the small molecules. The electrophoretic mobility or distance travelled by a protein molecule during SDS-PAGE is inversely proportional to the log of its molecular weight. Thus, the migration distance indicates the molecular weight of an unknown protein in reference to a set of proteins with standard weight (Fig. 9.13b).

Besides determination of molecular weights of the unknown proteins, SDS-PAGE is a rapid and sensitive method to determine the purity of a protein and the number of polypeptides within a protein.

Isoelectric focussing

Isoelectric focussing (IEF) is another electrophoretic technique which separates proteins solely by charge on a gel set-up with a pH gradient. An ampholyte solution containing polyamino acids with positive and negative charged properties [for example, alanine which in solution forms dipolar or zwitterions with properties of an acid (proton donor) and a base (proton acceptor)] is incorporated into a gel. The appropriate ampholyte establishes a stable pH gradient in an electric field (Fig. 9.14) in which the pH increases from anode to cathode. Proteins have either a net negative or a net positive charge and keep moving accordingly to the oppositely charged pole in the electric field. However, at a particular pH, each protein has an equal number of positive and negative charges, and this pH is called the *isoelectric point* (pI), at which the net charge on the protein becomes zero and it stops moving. If a protein molecule diffuses away from its isoelectric point, its net charge will change as it has moved into a region with a different pH; the electrophoretic force will then cause it to move back to its isoelectric point. Thus the protein is focused into a narrow band around its pI.

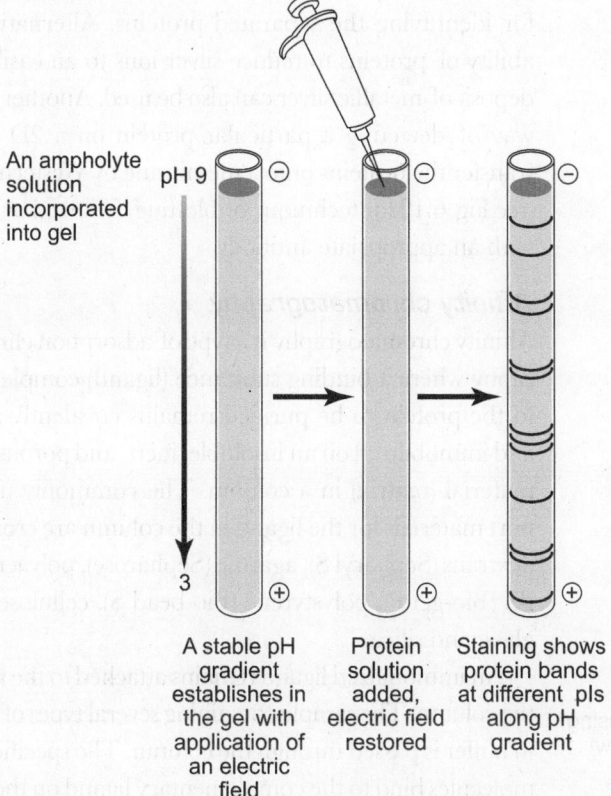

An ampholyte solution incorporated into gel

pH 9

3

| A stable pH gradient establishes in the gel with application of an electric field | Protein solution added, electric field restored | Staining shows protein bands at different pIs along pH gradient |

Fig. 9.14 Isoelectric focussing separates proteins according to their isoelectric point (pI) on a gel set up with a pH gradient. Electric field allows the proteins to enter the gel and migrate until each reaches a pH equivalent to its pI.

Isoelectric focussing is a sensitive and effective technique for the separation of different proteins.

Two-dimensional gel electrophoresis

This method combines isoelectric focussing and SDS-PAGE to obtain higher resolution in the separation of proteins, the procedure is known as *two-dimensional gel electrophoresis*.

A mixture of proteins is initially separated on an IEF gel, on the basis of the isoelectric points without regard to molecular weight. This is the first dimension of electrophoresis. Next, the IEF gel is placed horizontally on the top of a SDS-polyacrylamide slab gel and subjected to

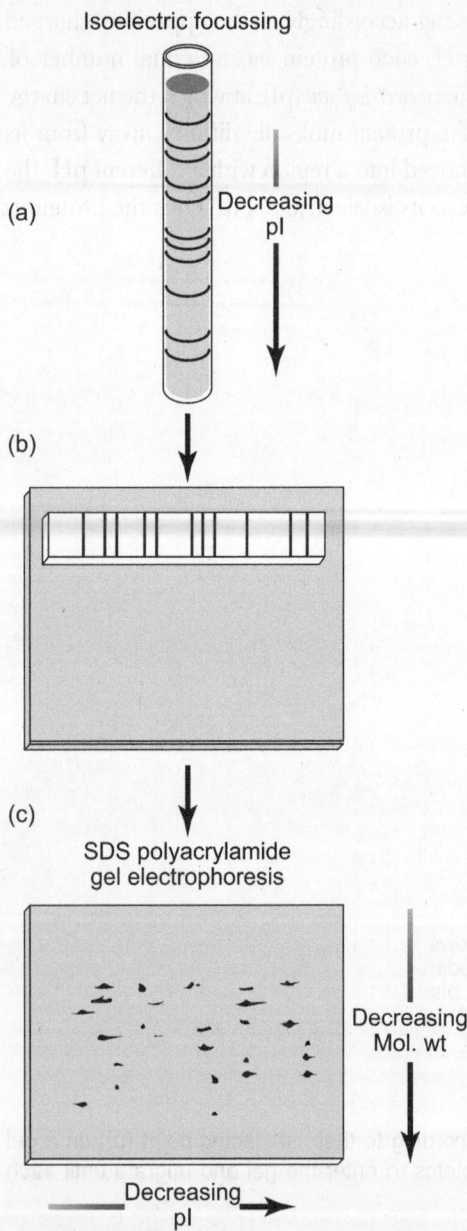

Isoelectric focussing

(a)

Decreasing
pI

(b)

(c)

SDS polyacrylamide
gel electrophoresis

Decreasing
Mol. wt

Decreasing
pI

Fig. 9.15 Two-dimensional gel electrophoresis. (a) A mixture of proteins is first separated by isoelectric focussing. (b) Next the IEF (pI) gel is placed horizontally on the second SDS polyacrylamide slab gel for electrophoretic separation of proteins. (c) Each band from pI gel gets further fractionated depending on the molecular weight of the components in a pI band.

SDS-PAGE separation (Fig. 9.15). At the beginning of this step, all proteins are allowed to react with SDS, so that the proteins migrating out of the IEF gel and running through the SDS polyacrylamide gel slab can segregate on the basis of their molecular weights. This is the second dimension. The proteins in the resulting 2D gel are usually viewed by staining with Coomassie blue (Fig. 9.15c). In case the proteins have been radiolabelled, autoradiography is done for identifying the separated proteins. Alternatively, the ability of proteins to reduce silver ions to an easily visible deposit of metallic silver can also be used. Another sensitive way of detecting a particular protein on a 2D gel is to transfer the proteins onto a membrane by Western blotting (see Fig. 6.19 for technique of blotting) followed by reaction with an appropriate antibody.

Affinity chromatography

Affinity chromatography is a type of adsorption chromatography where a binding substance (ligand) complementary to the protein to be purified remains covalently attached and immobilized on an insoluble, inert, and porous support material (matrix) in a column. The commonly used support materials for the ligand in the column are cross-linked dextrans (Sepharyl S), agarose (Sepharose), polyacrylamide gel (Bio-gel P), polystyrene (Bio-bead S), cellulose, porous glass, and silica.

An immobilized ligand remains attached to the matrix in the column. The sample containing several types of proteins in buffer is passed through the column. The specific protein molecules bind to the complementary ligand on the matrix. Then extensive washing of the column with buffer removes the non-bound proteins. The bound purified protein molecules are released by changing the pH of the buffer, or sometimes by adding a soluble ligand in the washing buffer to compete with the specific protein for binding to the immobilized ligand in the column.

Figure 9.16 exemplifies purification of a protein. Here the protein is an antigen, and is purified in an affinity chromatography column with beads coated with specific antibody molecules. Out of the mixture of antigens (proteins), a particular type of antigen molecule binds with the specific antibody molecules on the beads. The unbound antigens are washed away, and the bound antigen retained in the column is eluted by changing the pH or by adding a chemical that breaks the antigen–antibody bonds (weaker bonds, not covalent bonds). By this method, in a reverse way, desired antibodies can also be purified from cell culture supernatants, other body fluids, or serum, by coating the beads with the specific antigen.

Depending on the substance to be purified, different types of substances are used to coat the beads in affinity chromatography.

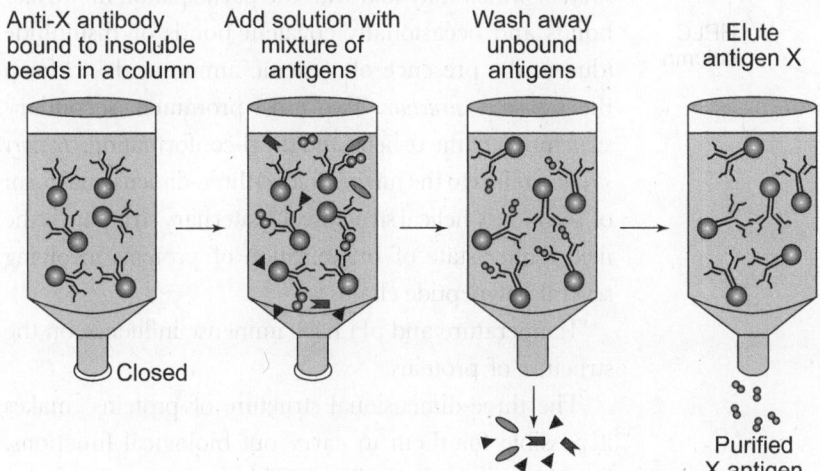

Fig. 9.16 Separation of an antigen from a mixture by affinity chromatography.

High performance liquid chromatography (HPLC)

The chromatography techniques described earlier are slow processes as they depend on gravity flow or sometimes need a very low pressure pump. These low-flow methods at times may take several hours to analyse a single sample. HPLC on the other hand uses greater pressure to force the sample through the column allowing better separation in shorter time. Thus it is possible to use non-compressible resin beads in the HPLC column (Fig. 9.17). HPLC systems have a notable limitation—less amount of protein is separated. The technique is more suitable for analytical purposes than for mass separation and purification.

The detector system can be *mass spectrometry* (*mass spec*) at the outlet of HPLC system. This method is highly sensitive in identifying trace elements and small peptide fragments in samples of large proteins that have been digested. The combined systems of HPLC and mass spec are used in the protein sequencer.

9.5 CHARACTERIZATION OF PROTEINS

Proteins are an important biochemical class of substances that have some generalized properties; they also differ among themselves structurally, in certain physical and chemical characteristics. Some of these differences among the proteins are the basis of their classification, methods of purification, and their functions. A summary of characteristic features of proteins is given below.

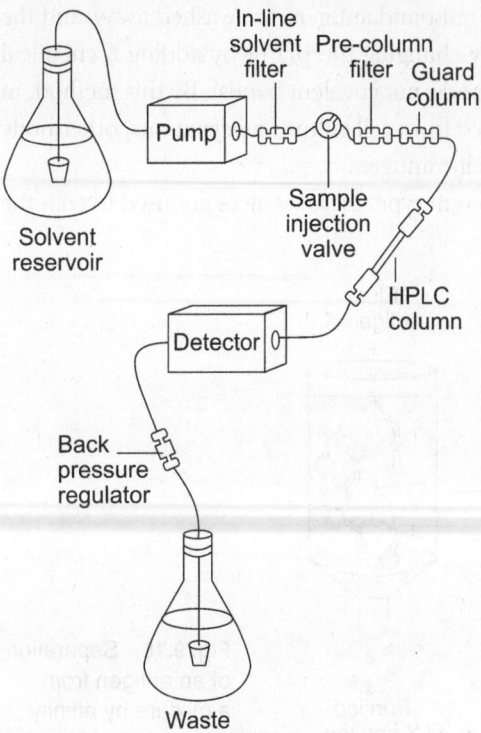

Fig. 9.17 High-performance liquid chromatography. This technique uses greater pressure to force the sample through the non-compressible resin beads for better separation of proteins in shorter time.

Biochemical nature The twenty different amino acids (L-form) joined in a specific sequence by peptide (covalent) bonds in a linear sequence form a polypeptide chain, the basic structure of a protein.

Structure Proteins may remain in four forms. Peptide-bonded amino acids in a linear sequence form the *primary structure* which may fold with the participation of weaker bonds and occasionally covalent bonds of disulphide (due to the presence of cysteine amino acids) to form the *secondary structure*. Two most prominent secondary structures are the α-helix and the β-conformation. *Tertiary structure* refers to the further folded three-dimensional form of secondary helical structure. Quaternary structure is the next higher state of organization of proteins involving several polypeptide chains.

Temperature and pH have immense influence on the structure of proteins.

The three-dimensional structure of proteins makes it possible for them to carry out biological functions. Proteins can be crystallized. This property of proteins helped in X-ray diffraction studies that revealed the atomic arrangement in protein structures.

Prosthetic group The non-amino acid part in many conjugated proteins is known as the prosthetic group. It plays an important role in the functioning of proteins.

Function Proteins are functionally more versatile than other macromolecules. Functions are diverse—as supporting structural elements, catalytic enzymes, transporting in blood plasma, nutrition and storage for developing embryos, contractile or motile function, defence as antibodies, regulatory function, and so on.

Solubility Proteins generally carry a large number of ionizable groups that have positive and negative charges. They are slightly soluble in pure water; the addition of salt promotes solubility, as the opposite charges of ions counter the charge on the proteins. The phenomenon is known as *salting in*. However, if the salt concentration is gradually increased to a point, the protein comes out of solution and precipitates. This reverse phenomenon is called *salting out*. Ammonium sulphate is used to precipitate proteins out of solution because of its high solubility. The concentration of this salt required for salting out varies from protein to protein.

Differential molecular sizes, density, and charge Depending on the differential molecular size, density, and charge, a mixture of heterogenous proteins can be separated by density gradient, ultracentrifugation, gel filtration, and ion-exchange chromatography. Electrophoresis separates different proteins on the basis of their charge.

Isoelectric point   Each protein has a characteristic isoelectric point (pI) which is a particular pH at which the protein has equal number of positive and negative charges. At the characteristic pI, the net charge on the protein is zero and it stops moving in an electrical field.

Binding affinity   Proteins have differential binding affinity to different substances including heavy metals, such as Zn^{2+}, Cu^{2+}, Cd^{2+}, Ni^{2+}, and Mn^{2+}.

9.6 PROTEIN-BASED PRODUCTS

Table 9.3 Essential and non-essential amino acids for humans and albino rat (mammal)

Essential	Non-essential
Arginine* (Arg)	Alanine (Ala)
Histidine (His)	Asparagine (Asn)
Isoleucine (Ile)	Aspartate (Asp)
Leucine (Leu)	Cysteine (Cys)
Lysine (Lys)	Glutamate (Glu)
Methionine (Met)	Glutamine (Gln)
Phenylalanine (Phe)	Glycine (Gly)
Threonine (Thr)	Proline (Pro)
Tryptophan (Trp)	Serine (Ser)
Valine (Val)	Tyrosine (Tyr)

*Essential for growth in young stage, but not in adults.

Mammals including human beings are incapable of synthesizing half of the 20 standard amino acids. These amino acids are called *essential amino acids* (Table 9.3) and have to be procurred from the diet or food. Thus proteins form a much needed food item for higher organisms. Meat and milk from animal sources, and cereals and vegetables as agricultural products meet the protein requirement of the billions of human beings all over the world. Plants supply the proteins necessary for herbivores in most cases.

Proteins are also the basis of multiple food items and pharmaceuticals produced in industries.

9.6.1 Milk Products

In homes and in the industry, boiled or pasteurized milk is inoculated with known cultures of microorganisms referred to as *starter cultures* for different types of fermented milk products. The important and mostly used fermented milk products are butter milk, curd (yogurt or *dahi*), and various kinds of cheese.

For production of curd, sour cream, and buttermilk, a particular species of *Lactobacillus* can be used as the starter culture; this bacterial species is a lactic acid fermenter. A mixed culture of *Lactobacillus bulgaricus* and *Streptococcus thermophilus* can also be used. Yogurt flavour is ascribed to the formation of lactic acid, acetaldehyde, and acetic acid due to fermentation. Fermented products develop a taste and texture very different from that of milk. Yogurt contains 3.25 per cent milk fat, a *low-fat yogurt* has between 0.5 and 2 per cent fat, and a *fat-free yogurt* has <0.5 per cent fat and has a good demand in the market.

For production of cheese, *Streptococcus lactis* or, *S. cremoris* or, and *S. thermophilus* in combination with *Lactobacillus lactis, and L. bulgaricus* are used as starter culture. The bacteria degrade lactose to lactic acid as in the case of yogurt making. Lactic acid lowers the pH to about 4.6. Then incubation with rennet (an extract from the stomach of calves containing chymosin enzyme) causes hydrolysis of κ-casein milk protein for removal of surface glycopeptides and easy coagulation of all the casein proteins—α-and β-caseins and hydrolysis products of κ-casein.

Nowadays, chymosin is produced by genetically engineered microorganisms.

9.6.2 Enzymes Used in the Food Industry

Different types of enzymes are in great demand in the food industry for processing of various foods. Some examples are given in Table 9.4.

Table 9.4 Enzymes used in food processing

Enzyme	Purpose/ Used for
Proteases	
Papain	Tenderization of meat
Endogenous proteases	Meat tenderization and development of flavour
Neutral or alkaline proteases	Production of meat slurry from mashed animal and fish bones that is used in canned meats and soups
Subtilisin	Partial hydrolysis of soy protein for increasing whipping expansion; Hydrolysis of RBC haemeolysate to yield haeme molecules to be spray-dried and used in cured meats, sausage, and meat-spread
Heat labile fungal protease	Hydrolysis of gluten in flour makes dough suitable for biscuit and pastry making
Glucose oxidase	Oxidizes D-glucose to gluconic acid; O_2 is utilized, H_2O_2 is produced
Amylases (Different types)	Production of glucose, maltose, and fructose syrup from starch, sucrose, and D-glucose.
Catalase	Degrades H_2O_2 into water and O_2; used with glucose oxidase to remove glucose and O_2 from foods and drinks

9.6.3 Recombinant Therapeutic Proteins

Recombinant DNA technology in the production of transgenic bacteria, plants, and animals has been discussed in detail in Chapter 6. The lists of biopharmaceuticals from recombinant microorganisms (Table 6.4), transgenic plants (Table 6.9), and transgenic animals (Table 6.10) have also been presented in that chapter. A list of human recombinant pharmaceutical proteins obtained from the milk of transgenic animals was produced in Table 6.10.

Protein-based Other Commercial Products

There are many other important protein-based products used for the well-being of man and animals. For example, gelatin obtained from the connective tissue of dead animals is used in the food industry and in photographic film. Monoclonal antibodies produced from cultures of hybridoma cells are used for various purposes—diagnosis of disease and pathogens, and treating cancer by conjugating toxin molecules to the antibodies. Endonucleases and other important enzymes used in research for molecular biology and for production of recombinant varieties of organisms are produced in large quantities, again by recombinant DNA technology.

Multiple uses of enzymes in the detergent industry, and forest products and paper industry can be found in Chapter 8.

Use of alkaline proteases to remove hair from hides is an improvement in the leather industry. Pancreatic enzymes are often used to increase suppleness and softness of dehaired hides, especially for the production of leather clothing, soft bags, and other items.

Protease activity of papain can provide wool with a soft silky appearance and add to its value. The high cost of papain restricts its use in the wool industry. Researchers are on the lookout for a cheaper enzyme for this purpose.

9.7 PROTEIN ENGINEERING

In the course of about three billion years of organic evolution, all the different kinds of proteins evolved to carryout complex functions in cells and in all the organisms. Random mutation of genes and natural selection of the gene products contributed to the size of today's large repertoire of proteins, glimpses of which you will get in this chapter.

Development of molecular genetics, particularly techniques of gene cloning and site-directed mutagenesis, have made it possible to design new gene products and research is no longer restricted to selecting useful genes that would only arise from rare events of natural mutation. In a way, the scientists or humans beings at large have entered a restricted domain by taking part in the course of organic evolution. This is the starting point of *protein engineering*—altering at will the structure of a protein in order to make it function 'better' for human benefit.

Protein engineering is the foremost front in the advances of biotechnology. It is an interactive multidisciplinary science, involving physics, biophysics, molecular genetics, organic chemistry, biochemistry, bioinformatics, genomics, proteomics, and computer science.

Protein engineering differs from protein design in the following ways:

(a) *Protein engineering* changes an existing protein, genetically or chemically to alter its function in predictable way.

(b) In protein design, a protein is designed with the guidance of databases of genomics and proteomics to perform a desired function.

The *concept of protein engineering* has already been introduced in Chapter 8. It is now about three decades, since biotechnologists have been trying to improve existing enzymes through gene manipulation to meet industrial requirements.

Objectives of protein engineering with enzymes are to improve enzymes on several counts:

- Faster kinetics for enzyme action
- Thermal stability
- Lesser energy consumption in reactions
- Stability under harsh conditions, such as extreme pH, temperature, and concentration of substrate
- To be able to act on substrate supplied in industry, that is slightly different from that in the cell
- Protease resistant, thus long life

Besides enzymes, protein engineering is also involved in improving the life of storage proteins and in drug designing; particularly pharmaceuticals of biological origin (see Chapter 6). Drug designing is involved in a big way in the multibillion dollar pharmaceuticals market. Protein engineering improves the target or receptor specificity of the drug.

9.7.1 Some Examples of Protein Engineering

Increasing catalytic function

Glucose isomerase converts glucose into other isomers like fructose and is used to make fructose corn syrup which is in much demand in the soft drink industry.

Glucose isomerase belongs to a *TIM* (triosephosphate isomerase) *barrel family* of enzymes which resemble each other in that they have a characteristic domain called TIM barrel, with an active catalytic site at one end. Enzymes of the family *Triosephosphate isomerase* are much more efficient than 'glucose isomerase'. Therefore, protein engineering redesigns glucose isomerase using the highly efficient domain of the TIM barrel family.

Engineering of lysozyme for more thermodynamic stability

(i) Lysozyme from bacteriophage T4 is 164 amino acid residues long and folds into two domains (Fig. 9.18). Two cysteine residues, Cys 54 and Cys 97, are not cross-linked by a disulphide bond.

Usually, reduction of the number of possible unfolded conformations increases thermal stability of a protein structure. These kind of stable enzymes are required by biotechnological industries. The stability of lysozyme T4 has been achieved by mutating the five amino acid residues—Ile3, Ile9, Thr21, Thr 142, and Leu164—each to cysteine in separate experiments. Cysteines at these residue locations lead to the formation of single, double, or triple disulphide bonds in lysozyme. In addition, the original Cys 54 is mutated to threonine to avoid formation of an incorrect disulphide bridge during folding of the engineered chain. The introduction of the disulphide bridges decreases the number of unfolded conformations and increases the thermal stability of the engineered proteins. It has been found that the effect is additive; mutant proteins with double disulphide bonds have a higher melting temperature (T_m) and hence higher thermal stability than those with a single engineered bridge. Similarly, molecules with engineered triple disulphide bonds have even higher stability.

(ii) Another method of engineering a protein to increase thermal stability is to replace glycine with any other residue or to increase the number of amino acids in a given sequence. Presence of glycine allows more conformational freedom than any other amino acid and therefore replacement of glycine by any other amino acid would decrease the number of conformational unfolded structures. Replacement with proline increases the stability. For example, the substitution of Gly 77 by alanine and Ala 82 by proline in T4 lysozyme increases the melting temperature (T_m).

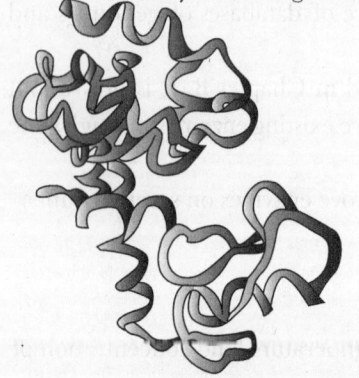

Fig. 9.18 Three-dimensional structure of T4 lysozyme folded into two domains.

Modification at the binding site of an enzyme with substrate

High specificity in complementary reactions between a protein and a ligand (or an enzyme and its substrate) is due to ionic interactions or formation of hydrogen bonds between the charged groups in the interaction area of the two components. By altering the amino acid constituents at the binding site, protein engineering may lead to positive (higher activity) or negative activity of the participating molecules.

Replacement of single amino acid, aspartic acid (Asp) by asparagine (Asn) causes thousand-fold decrease in the specific activity of *dihydrofolate reducatase* enzyme, indicating that aspartic acid is very important for the active site of the enzyme.

A single point mutation in bovine chymosin (digesting enzyme), replacing Val 111 by phenylalanine lowers the reaction efficiency of the enzyme by half. The X-ray diffraction studies showed the mutated phenylalanine side chain occupies a part of the binding pocket for the substrate.

9.7.2 Basic Consideration for Protein Engineering

Sometimes substitution of many amino acids, or deletions, or insertions lead to no change in enzyme activity, like silent mutations, and a single change at specific position(s) may lead to a drastic or desirable change. For example, replacement of glycine with aspartic acid in *E. coli aspartate transcarbamylase* leads to loss of activity of the enzyme.

9.7.3 Methods of Protein Engineering

Different techniques are employed for protein engineering to modify an existing protein—genetically or chemically—with the objective of altering its function in a predictable way.

Site-directed mutagenesis

Site-directed mutagenesis can induce a mutation at a nucleotide (point mutation) or at several nucleotides to alter the amino acid sequence in a protein product, almost at will. Alteration of even a single amino acid may lead to structural and functional changes in a protein. The details of site-directed mutagenesis have been presented in Chapter 6—at the end of the section (Section 6.12.6) on PCR.

Directed molecular evolution technology—adaptive mutation

Natural genetic engineering employs forced or directed evolution and adaptive mutation. Specific environmental stresses force microorganisms to mutate and adapt, thus creating microorganisms with new biological capabilities. The mechanisms of these adaptive mutational processes include DNA rearrangements in which transposable segments and various types of recombination play critical roles.

For example, scientists in biotech companies induce mutations randomly into genes and then select the microorganisms (bacteria) with the protein product, mostly enzyme, which has the highest activity. The mutations may be induced in the presence of environmental stress.

With this procedure, bacteria have been used to produce industrial enzymes that tolerate more than 1M cyanide concentration. This is a significant achievement as this would have never happened by natural slow mutation and molecular selection. The natural environment never contains cyanide at this high level. The cyanide-tolerant bacteria can be used to remediate cyanide contamination resulting from mining and other industrial waste accumulation.

If necessary, the gene with adaptive mutation can be introduced into other suitable hosts for the protein product.

Recombinant DNA techniques

Recombinant DNA technology (in Chapter 6) allows incorporation of a piece of new DNA into an existing DNA molecule leading to modified function of the original gene. The new recombinant DNA is introduced into suitable host cells for production of 'chimeric' gene products. Using this technology, chimera and hybrid antibodies can be produced.

Antibody engineering—chimeric and hybrid monoclonal antibodies

The mouse monoclonal antibodies (mAbs) cannot be injected in humans as such, say for fighting cancer cells, as they induce allergic and other harmful reactions. On the other hand, raising human monoclonal antibodies has numerous technical problems. To resolve the problem, recombinant DNA containing the promoters, leader sequence, and variable region sequences from a mouse antibody gene, and the human constant region sequence are cloned. Such a recombinant gene encodes a mouse–human chimeric antibody molecule, commonly described as a *humanized antibody*, because its constant region (major portion) is encoded by a human gene (Fig. 9.19). Thus, the chimeric antibody has fewer mouse antigenic determinants and is far less immunogenic than mouse monoclonal antibodies when administered to humans. Simultaneously, such a chimeric antibody retains the quality of a human antibody in triggering complement activation or Fc receptor binding.

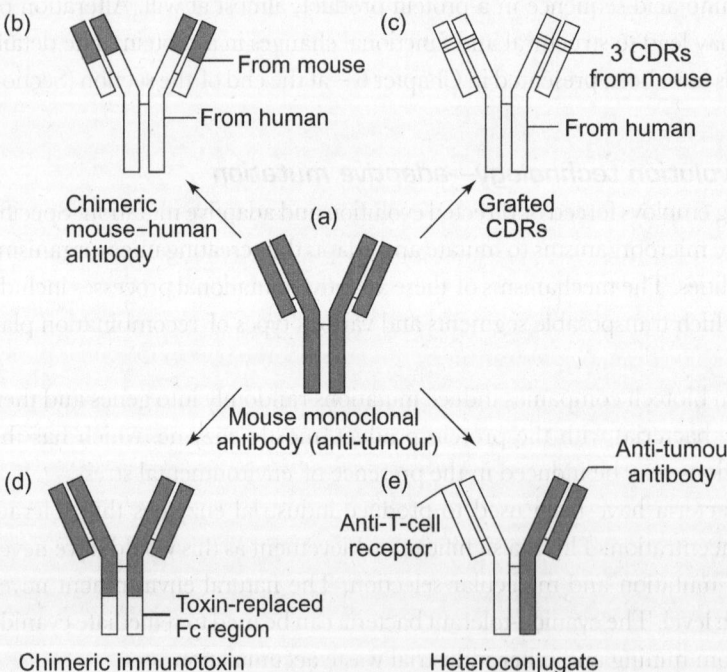

Fig. 9.19 Antibody engineering: final product obtained through r-DNA technology. (a) Mouse monoclonal anti-tumour antibody. (b) Chimeric mouse-human mAB– the variable part belongs to the mouse (black) and the constant part belongs to the human antibody (white). (c) Human Ab grafted with the CDRs (three black bands) of a mouse Ab. (d) Fc domain of a mouse mAb replaced by a toxin chain. (e) A heteroconjugate—one half of the mouse Ab specific for a tumour antigen and the other half specific for the T-cell receptor complex.

Figure labels:
(b) From mouse / From human — Chimeric mouse–human antibody
(c) 3 CDRs from mouse / From human — Grafted CDRs
(a) Mouse monoclonal antibody (anti-tumour)
(d) Toxin-replaced Fc region — Chimeric immunotoxin
(e) Anti-T-cell receptor / Anti-tumour antibody — Heteroconjugate

However, the mouse variable region in the chimeric antibody is capable of inducing an antibody response in humans. To improve the situation, only mouse CDRs (complementarity-determining regions of the variable part of the antibody molecule) are grafted in human framework regions in the DNA constructs (Fig. 9.19c). CDRs actually constitute the antigen-binding site. The encoded antibodies with lesser mouse proteins are less immunogenic in humans than humanized antibodies containing the entire mouse variable region. CDR-grafted mouse monoclonal antibodies are presently in clinical use. Such mAbs against CD20, a membrane-bound phosphoprotein, have been used for successful treatment of B cell lymphoma patients. Breast tumours over-expressing HER2 (human epidermal growth factor 2) are treated with engineered anti-HER2 mAbs.

Two other modifications have been made to build chimeric antibodies to carry out specific functions. In the first approach, the terminal constant region domain in a tumour-specific mAb is replaced with toxin chains (Fig. 9.19d). These rearranged mAbs attached to a toxin molecule are called *immunotoxins*. Since these immunotoxins lack the regular Fc domain, they do not bind to different cells bearing Fc receptors. Thus, being more tumour-specific, they are better therapeutic agents for cancer cells.

In the second approach, various heteroconjugates have been formed by combining one-half of the antibody specific for a tumour and the other half with a binding site for a surface molecule on an immune effector cell, such as an activated macrophage, NK cell, or a cytotxic T-lymphocyte (Fig. 9.19e). Then the heteroconjugate can cross-link the immune effector cell to the tumour cell leading to the destruction of the tumour cell.

Protein engineering through chemical modifications

It is a general knowledge that the proteins synthesized under the control of gene sequences undergo post-translational modifications within the cell. These modifications may affect one or multiple characteristics of the final protein molecules, such as stability, structural integrity, solubility, viscosity, and even chemical reactivity.

These kinds of alterations of proteins can also be made in vitro through chemical modifications of proteins. These alterations may also lead to creating new active sites or modifying the old ones; thus essentially creating a new enzyme.

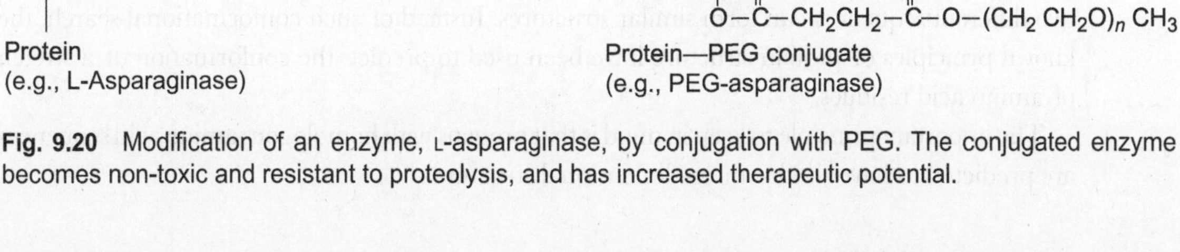

Fig. 9.20 Modification of an enzyme, L-asparaginase, by conjugation with PEG. The conjugated enzyme becomes non-toxic and resistant to proteolysis, and has increased therapeutic potential.

Cross-linking of lysine residues   Chemical linkers, such as simple aldehydes e.g., glycol, adipaldehyde, and glutaraldehyde, can cross-link lysine residues in proteins, and provide more stability to proteins. The chemical modification of lysine residues in proteins actually depends on the nature of amino group of the residue which reacts as a *nucleophile* to contribute an electron in a chemical reaction. Lysine is the second strongest nucleophile in a protein, next to cysteine. The modification reaction is performed at pH 8–9. In biotechnology, enzymes that have greater stability in a bioreactor are desirable.

Enzyme–PEG conjugates   *L-Asparginase*, an enzyme isolated from microorganisms, shows anti-tumour properties, but it is toxic and has a limited life time of about 18 hours. The last two limiting factors can be removed by conjugating the enzyme with polyethylene glycol (PEG) (Fig. 9.20), although its catalytic property is reduced to half. This PEG–asparaginase is used to treat malignant tumours in mice, cats, and humans. A large number of PEG-conjugated enzymes, such as uricase, catalase, adenosine deaminase, and so on are used in industry.

Other examples of chemical modifications of enzyme function   Phenylmethyl sulphonyl fluoride (PMSF) *modifies a protease (subtilisin) to a peptide ligase*. Protease 'subtilisin' is modified by converting a serine residue into cysteine or seleno-cysteine. A peptide ligase enzyme catalyses peptide ligation (addition) to a native enzyme that may lead to high specificity and selectivity in functions.

A protein capable of recognition and binding to specific DNA sequences may be modified by adding a chemical with a cleaving property to it. Then the *modified protein can perform like a restriction enzyme* (see Chapter 6) with three properties—recognition, binding, and cleaving at a particular sequence of DNA.

9.8 PROTEIN DESIGN

It has already been pointed out that protein engineering is concerned with modifying the structure of an existing protein, whereas protein design is about designing functional proteins on the basis of existing knowledge about proteins, mainly structure–function correlates and databases on genomics and proteomics. The function of a protein depends to a large extent on its 3D structure which has already been discussed. Thus, designing of a protein de novo from a sequence of amino acids should always be accompanied with 3D analysis. The prediction of protein structure from the amino acid sequence is extremely difficult as the possibilities of folding in different ways could be enormous in number. A search for the optimal conformation of a polypeptide out of all those possibilities is a problem of extreme computational complexity. No algorithm exists to resolve this problem in reasonable computer time, even with the most powerful computer. Sometimes two different sequences can form similar structures. Instead of such conformational search, the known principles of protein structure have been used to predict the conformation of a stretch of amino acid residues.

The important principle to keep in mind is that proteins with homologous amino acid sequences are predicted to have similar three-dimensional structures.

The method of P.Y. Chou and G.D. Fasman, USA, based on a statistical analysis of the protein structure database, is most frequently used for systematic prediction of the three-dimensional structure of an unknown sequence. Prediction of secondary and tertiary structure for a polypeptide chain is still in a hit-or-miss state; many attempts remain unsuccessful.

9.8.1 Computer Programs

Several software packages have been developed for prediction of protein structures; some of them may also be used in PCs.

Besides building, graphical display and manipulation with protein models, computer programing is extended to study dynamic motion in the conformation of proteins in the course of the protein's activity. This allows the study of the groups involved in the activity of a protein and the effects of chemical modifications.

9.8.2 Design of Peptide and Protein Mimics

Short synthetic oligopeptides assume many different conformations, out of which only one conformation can perform biological activity. The correct conformation of the synthetic peptides and protein mimics can be brought in by introducing stereo chemical constraints through the incorporation of residues such as α-amino isobutyric acid, α- and β-dehydro residues, proline, D-amino acids, and so on. These special residues can restrict the range of conformations that the peptide backbone can assume.

Correct placement of dehydro residues, such as Phe, Leu, and so on, may cause formation of β turns which are essential for particular conformations. Incorporation of α-amino isobutyric acid causes oligopeptides to adopt an α-helical conformation.

9.8.3 Benefits of Protein Design

Protein designing technology for drug design

Protein crystallographic studies revealed that structurally, the binding site of a substrate is complementary to the active site of an enzyme. Thus, computer modelling of the active site of an enzyme can suggest the structures of plausible molecules that interact with the site and inhibit the function of the enzyme. This is the way drugs like methotrexate and trimethoprim have been designed to inhibit the action of the enzyme dihydrofolate reductase.

Similarly drug molecules have been designed to inhibit the function of rennin for controlling hypertension. Normally, the enzyme rennin catalyses the first and rate-limiting step in the conversion of angiotensinogen to the hormone *angiotensin*–II which plays an important role in the regulation of blood pressure.

Peptide vaccines

Recombinant DNA technology is used to synthesize oligopeptides which can mimic antigenic epitopes for induction of immune responses. Such short peptides conjugated to carrier proteins can be used as vaccines.

9.9 PROTEIN ENGINEERING—FUTURE

Biochemical studies, X-ray crystallography, and NMR have established the structure and functions of many proteins, characterized new proteins and engineered or designed proteins. These studies have established protein databases which help in protein design. The design and de novo synthesis of novel proteins with enhanced functions are challenges that scientists are working on at present. These will find applications from ecology to health, from material science to bioelectronics. Protein engineering will remain at the forefront of the endeavours of biotechnologists in the coming years.

SUMMARY

♦ Proteins are more varied in structure and functions than the other macromolecules. The variations are based on different combinations of the twenty amino acid sequences in the primary structure and on the three-dimensional conformations of proteins.

♦ Peptide bonds and the rotational bond angles phi (φ) and psi (ψ) contribute towards linking of amino acids and towards bending and turning in the primary structure of a polypeptide chain.

♦ The structure of proteins is organized at four levels–primary, secondary, tertiary, and quaternary. The α-helix and the β-conformation are the two most prominent secondary structures. The three-dimensional form of secondary structure becomes the tertiary structure. Quaternary structure is the highest state of organization of proteins involving several polypeptide chains.

♦ Weak interactions, such as hydrogen bonds, and hydrophobic, ionic, and van der Waals interactions are important for the folding of polypeptide chains and stabilizing the protein's conformation.

♦ The diverse nature of protein structures are the basis of their diverse functions. Denaturation of protein structure leads to loss of protein function.

♦ Conjugated proteins have prosthetic groups which are chemical groups other than amino acids. The prosthetic groups play an important role in the functioning of the proteins.

♦ Sophisticated techniques separate and purify a particular kind of protein from a heterogenous mixture, and help in the characterization of a protein.

♦ The spectrum of protein based products ranges from dairy to pharmaceuticals.

♦ Protein engineering and protein designing have many promises for the present and future. Protein engineering changes an existing protein, genetically or chemically, to derive better function; whereas protein designing is concerned with the de novo synthesis of functional proteins.

William Henry Bragg and William Lawrence Bragg

Sir William Henry Bragg
(2 July 1862–10 March 1942)

William Henry Bragg and William Lawrence Bragg were a British father and son duo who were scientifically the most productive in history. They succeeded in constructing the first X-ray spectroscope in 1913, formulated Bragg's law, and established the science of X-ray crystallography. They were jointly awarded the Nobel Prize for Physics in 1915. The mineral Braggite was named after them as a token of honour. The father was knighted in 1920 and the son in 1941.

William Henry Bragg was born in Westward, Cumberland, England to a merchant marine officer and farmer, Robert John Bragg and his wife Mary. He lost his mother at the age of 7 and was raised in the family of his uncle also named William Bragg at Market Harborough, Leicestershire. His early education was at Market Harborough Old Grammar School, and he studied at King William's College, Isle of Man. He was always at the top of his classes, particularly talented in mathematics. He then joined Trinity College, Cambridge in 1881 and won a scholarship. Henry studied physics under John W. Strutt, Lord Rayleigh, and Sir Joseph J. Thomson.

Sir Thomson steered Bragg to an opening as Elder Professor of Mathematics and Experimental Physics at the University of Adelaide, Australia. He was 23 years old and undertook the long sea voyage, becoming a professor in 1886. Over the next 18 years he gained the reputation of a masterful lecturer; for a while he apprenticed himself with the firm of an instrument-maker to make apparatus for his under-equipped teaching laboratory. He did not conduct any original research or make any publication during this period. The turning point in Bragg's career came in 1904 when he delivered the presidential address to section A of the Australasian Association for the Advancement of Science at Dunedin, New Zealand, 'On some recent advances in the theory of the ionization of gases'.

In 1895 Bragg was visited by Earnest Rutherford, en-route from New Zealand to Cambridge; this visit was the commencement of a lifelong friendship. Bragg returned to England at the end of 1908 and accepted the Cavendish Chair of Physics at the University of Leeds in 1909. Around this time the British scientific community was excited with the discovery of X-ray diffraction by Max von Laue. Bragg and his son Lawrence became intrigued by this discovery and spent hours discussing its ramifications. Henry Bragg also got interested in radiation physics, particularly in the study of alpha particle emission. In 1912, Max von Laue showed that X-rays could be diffracted by crystals and established their wave nature. By 1913, senior Bragg had developed the X-ray spectrometer which allowed analysis of many crystals, and junior Bragg formulated Bragg's law for X-ray diffraction in 1912. They were awarded the Nobel Prize in 1915.

From 1914, Henry Bragg was occupied with war duty, connected with submarine detection. He was appointed as Quain Professor of Physics at University College London in 1915, and took up the job after World War I. From 1923 he was Fullerian Professor of Chemistry at the Royal Institution and director of the Davy Faraday Research Laboratory. This institution was rebuilt in 1929-30 under his supervision.

He was an FRS in 1907, and president of the Royal Society from 1935 to 1940. The lecture theatre of King William's College is renamed in memory of W.

Henry Bragg. In 1962, the Bragg Laboratories were constructed at the University of Adelaide, Australia to commemorate 100 years of Sir William Henry Bragg's birth. Since 1992, the Australian Institute of Physics has awarded the Bragg Gold Medal for Excellence in Physics for the best PhD thesis by a student at an Australian University. Images of Sir W.H.B. and his son Sir W.L.B. are on the two sides of the medal. Many other honours were bestowed upon him by different learned institutions.

Sir William Henry Bragg is often summarily described as a British mathematician, physicist, chemist, and sportsman.

Sir William Lawrence Bragg
(31 March 1890–1 July 1971)

William Lawrence Bragg was an X-ray crystallographer, and discoverer of Bragg's law of X-ray diffraction which makes it possible to calculate the positions of the atoms within a crystal. He received the Nobel Prize at the age of 25 jointly with his father (the photograph above is from around that time). To date he is the youngest Nobel Laureate. He was the director of the Cavendish Laboratory, Cambridge at the time of the epochal discovery of the structure of DNA made by James Watson and Francis Crick in February, 1953.

He was born in Adelaide, South Australia. Shortly after starting school, the five-year old Lawrence fell from his tricycle and broke his elbow. His father used the newly discovered X-ray equipment to examine the broken arm. This was the first recorded surgical use of X-rays in Australia. He was a bright student, studied at St Peter's Collegiate School, Adelaide. He entered the University of Adelaide at the age of 14 to study mathematics, physics, and chemistry in 1904. He graduated in 1908 and returned to England with his parents. Lawrence Bragg entered Trinity College, Cambridge in the autumn of 1909 and received a major scholarship in mathematics. After initially excelling in mathematics, he transferred to the physics course, and graduated with first class honours in 1911. He formulated Bragg's law in 1912, during his first year as a research student in Cambridge. He discussed his ideas with his father who developed the X-ray spectrometer in Leeds. J.J. Thomson and W. Henry Bragg were his doctoral advisors. In 1914 Lawrence Bragg was elected to a Fellowship at Trinity College after submission and defense of a thesis.

Bragg's research work was interrupted by both World War I and II. During both the Wars he worked on sound ranging methods for locating enemy guns. For his work during the World War I he was awarded the Military Cross and appointed Officer of the Order of the British Empire.

Between the wars, 1919–1937, Lawrence worked at the Victoria University of Manchester as Langworthy Professor of Physics. After World War II, he returned to Cambridge. In 1948 he became interested in the structure of proteins and was instrumental in assembling a group that used physics to solve biological problems. From this group, Max F. Perutz for his work on the structure of haemoglobin and John C. Kendrew for his work on myoglobin received the Nobel Prize for chemistry in 1962. Bragg's effort culminated in the discovery of DNA structure by Watson and Crick.

He had the very rare opportunity of celebrating a golden jubilee as a Nobel Laureate, a special function was arranged during the December ceremonies at Stockholm in 1965. He received many honorary doctorates and fellowships. He was an FRS in 1921, knighted by King George VI in 1941, and a recipient of Hughes Medal (1931), the Royal Society Medal (1946), and Copley Medal (1966).

Lawrence Bragg's hobbies included painting, literature, and a lifelong interest in gardening. He missed having a garden when he moved to London, and he started working as a part-time gardener, unrecognized by his employer, until a guest at the house expressed surprise at seeing him there.

EXERCISES

Objective Questions

1. Match the structures that occur in the course of organization of protein molecules in Column A with the correct definition in Column B.

Column A	Column B
(i) Primary structure	(a) Structural arrangement along a polypeptide chain caused by repeated patterns of atoms.
(ii) Secondary structure	(b) The sequence of amino acids in the polypeptide chain.
(iii) Tertiary structure	(c) Structural rotation around peptide bonds.
(iv) Quaternary structure	(d) Structural arrangement of polypeptides which are associated into a macromolecule.
	(e) The three-dimensional structure formed when a polypeptide folds upon itself

2. Which technique is employed to separate proteins on the basis of the following properties:
   (a) Size and shape
   (b) Magnitude of the charge due to the ionizing groups on the molecules
   (c) Molecular weight in an electric field
   (d) Surface ionic charge
   (e) Density
   (f) Surface net charge becomes zero at a particular pH

Review Questions

1. What is the status of Gibb's free energy (G) in a thermodynamically stable conformation of a protein?

2. Why are the amide C–N bonds unable to rotate freely?

3. If you discover a new macromolecule and subjected it to crystallographic study what criteria should it meet to be classified as a protein?

4. A long polypeptide chain contains several hydrophobic and hydrophilic domains. How would these domains affect the folding of the molecule in an aqueous environment?

5. Why is proline rarely found in an α-helix?

6. Name the interactions that are responsible for the folding and stability of a polypeptide chain.

7. Write a brief note on β-conformation of polypeptide chains.

8. What are the special features of a collagen helix in comparison with a α-keratin helix?

9. What are the advantages and disadvantages of an enzyme that has increased thermostability?

10. How would you prove that the function of protein is dependent on its 3D configuration?

11. What are the prosthetic groups in (a) Immunoglobulin, and (b) Casein of milk?

12. Is affinity chromatography more selective at separating proteins than ion exchange chromatography?

13. (a) Would you consider the modification of a part of protein for enhanced function, using chemical reaction as protein engineering? Elaborate such a modification with an example.
    (b) Can you differentiate between protein engineering and protein designing?
    (c) When protein engineering involves modifying a gene, what are the methods that have to be used?

14. How is a 'humanized antibody' produced?

10

Microbial Biotechnology

INTRODUCTION

Millennia ago, people of different cultures used the fermentation technology to produce wine, beer, vinegar, and *saki*, without much knowledge about the involvement of microorganisms (bacteria and yeast) in these processes and their mode of functioning. Then came important observations and contributions by many scientists towards the development of modern microbiology and its contribution to biotechnology.

The technology of microbial production of metabolites, such as ethanol, butanol, lactic acid, riboflavin, and so on, and enzymes, such as protease, amylase, and invertase, was developed in the early part of the twentieth century. Large-scale production of the antibiotic penicillin (a fungal product) began during World War II. Soon, production of other antibiotics, amino acids, enzymes, and nucleotides followed in the 1950s. Since the 1980s, genetically engineered microorganisms are being used for the commercial production of many useful non-microbial products such as insulin, human growth hormone, interferon, vaccines, and several other pharmaceutical items (see Chapter 6).

Microbes increase the productivity of crops by providing biofertilizers and bioinsecticides. They also contribute to the energy sector by producing biogas which can be used to generate electricity in rural areas. Microbes reduce environmental pollution to a large extent through bioremediation. There are particular types of bacteria to recover metals from polluted waters and mines. Mass-scale culture of microorganisms in bioreactors is being practised to obtain different types of commercial and pharmaceutical products (see Chapter 5). The 21st century will see further refinement of already discovered processes and techniques in microbial production, and many techniques and processes that are now only ideas will be put into practice.

Microbiology is an attractive, vast, and extremely promising field of study and research. Microbes or microorganisms are tiny organisms that are too small to be seen individually by the naked eye and are only visible through microscopes. Although the most abundant microorganisms are bacteria, microbes also include viruses (not organisms in the strict sense), algae, fungi such as single-celled yeasts and molds, and single-celled animals called protozoa. Bacteria were the first living organisms on our planet. They live virtually in any place where life is possible. They evolved into numerous species with diverse capabilities, and possibly constitute the largest component of the earth's biomass. The whole ecosystem depends on the activities of microorganisms. They influence the lives of other organisms as well as human society in countless ways. Modern microbiology has an immense impact on other fields, such as ecology, agriculture, food science, genetics, biochemistry, and molecular biology. *Microbial biotechnology is a discipline of biotechnology that involves the use of microorganisms*, primarily bacteria.

This chapter mainly deals with microbial culture, culture media, scale-up of microbial culture for industrial products, improvement of strains, and the bioethics of microbial technology.

10.1 MICROBIAL CULTURE

Microorganisms are grown in culture media in the laboratory to carry out (i) microbial identification, (ii) isolation of specific microorganisms, and (iii) scientific experiments. Many different media have been formulated for these and other purposes.

Microorganisms require about 10 elements known as *macronutrients* in large quantities, and several *micronutrients* or *trace elements* (Table 10.1). Nutrients are substances that are utilized in the process of biosynthesis and energy release by living organisms and are therefore required for microbial growth.

Table 10.1 Nutrient elements (and their functions) as required by microorganisms for growth in nature and culture

Nutrients	Functions
MACRONUTRIENTS Carbon (C) Oxygen (O) Hydrogen (H) Nitrogen (N) Sulphur (S) Phosphorus (P)	Contribute to components of carbohydrates, lipids, proteins, and nucleic acids
	Cations in the cell and play a variety of other roles:

(Contd)

Table 10.1   (*Contd*)

Nutrients	Functions
Calcium (Ca^{2+})	In cell activation, heat resistance of endospores For activity by a number of enzymes
Potassium (K$^+$) Magnesium (Mg^{2+})	Serves as a cofactor for many enzymes, stabilizes membranes and ribosomes
Iron (Fe^{2+} and Fe^{3+})	A part of cytochromes, a cofactor for enzymes, and electron-carrying proteins
MICRONUTRIENTS (Trace elements) Manganese (Mn^{2+}) Zinc (Zn^{2+}) Molybdenum (Mo^{2+}) Cobalt (Co^{2+}) Nickel (Ni) Copper (Cu^{2+})	A part of enzymes and cofactors, help in the catalysis of reactions and maintenance of protein structure Molybdenum for nitrogen fixation Cobalt a component of vitamin B$_{12}$

10.1.1 Requirements for Carbon, Hydrogen, and Oxygen

Carbon is the fundamental element present in the skeleton or backbone of all organic molecules.

Hydrogen and oxygen are required by the organisms as a source of electrons. Electron movements through electron transport chains and during other oxidation–reduction reactions provide energy for metabolism and work.

Carbon dioxide is one of the most important sources of carbon. On oxidation, it cannot provide hydrogen or energy. Thus, electrons are needed to reduce CO_2 to form organic molecules during biosynthesis.

Probably all microorganisms are capable of using CO_2, that is, they can reduce it and incorporate into organic molecules. By definition, *autotrophs* can use CO_2 as their sole or principal source of carbon; they carryout photosynthesis and use light as the energy source to reduce CO_2.

Organic nutrients that supply carbon, hydrogen, and oxygen at the same time are almost always reduced. They have electrons to donate to be used in energy release and biosynthesis. The more an organic molecule is reduced, the higher is its energy content; for example, lipids have a higher energy content than carbohydrates.

Electron transfer from reduced donors with more negative reduction potentials to oxidized electron acceptors with more positive potential releases energy.

As the reduction of CO_2 is a very energy-expensive process, many microorganisms cannot depend on CO_2 as their sole carbon source and thus rely on the presence of more reduced complex organic molecules, such as glucose. Organisms that use reduced, preformed organic molecules as carbon sources are called *heterotrophs*. The preformed organic molecules usually come from other organisms. Reduced organic compounds provide both carbon for biosynthesis of new molecules and energy as ATP and NADH.

Microorganisms are extremely flexible with respect to the carbon sources they can use to meet their nutritional requirement. Certain populations of microorganisms in nature would even metabolize relatively indigestible man-made substances such as pesticides.

10.1.2 Nutritional Types of Microorganisms

On the basis of the preferred *source of carbon*, microorganisms are classified into *autotrophs*—CO_2 users and *heterotrophs*—organic carbon users.

There are only two sources of energy available to organisms. Accordingly, microorganisms are classified into *phototrophs*—use light as the energy source and *chemotrophs*—obtain energy from the oxidation of chemical compounds (either inorganic or organic).

Further, microorganisms have only two sources electrons. With respect to this, there are two types of microorganisms: *lithotrophs* (rock eaters)—use reduced inorganic substances as their electron source and *organotrophs*—extract electrons from organic compounds.

The metabolic diversity of microorganisms may seem enormous. But most of them can be placed in one of the four nutritional classes on the basis of their primary sources of carbon, energy, and electrons (Table 10.2). The majority of the microorganisms studied so far are either *photolithotrophic autotrophs* or *chemoorganotrophic heterotrophs*.

Table 10.2 Classification of microorganisms based on nutrition types

		Nutritional types			
		Photolithoauto trophs	Photoorganohetero trophs	Chemolithoauto trophs	Chemoorganohetro trophs
S o u r c e	Energy	Light	Light	Inorganic chemical	Organic chemical
	Hydrogen (H)/ Electrons (e⁻)	Inorganic H/ e⁻	Organic H/e⁻	Inorganic H/e⁻	Organic H/e⁻
	Carbon	CO_2	Organic source (CO_2 may be used)	CO_2	Organic source
Examples		Cyanobacteria, purple and green sulphur bacteria, algae (eukaryotic)	Purple non-sulphur bacteria, green non-sulphur bacteria	Hydrogen bacteria, nitrifying bacteria, sulphur-oxidizing bacteria, Iron-oxidizing bacteria	Most non-photosynthetic bacteria (including most pathogens), protozoa, fungi

10.1.3 Culture Media

A culture medium is a liquid or solid (semi-solid, jelly-like) preparation used to grow, transport, and store microorganisms in the laboratory. The medium contains all the nutrients required by microorganisms for growth. Specialized media are needed for the isolation and identification of microorganisms, for the test of antibiotic sensitivities, water and food analyses, industrial microbiology for bioreactors, and other activities. Nutritional requirements may vary from species to species. The nutritional requirements of a particular type of microorganism is often related to its natural surroundings.

Defined or synthetic media

A medium in which all the components are known is a *defined* or *synthetic medium*. For example, the composition of defined media for the culture of cyanobacteria and *E. coli* has been presented in

Table 10.3. Photolithotrophic autotrophs such as cyanobacteria and eukaryotic algae can grow on relatively simple defined media containing CO_2 as a carbon source in the form of sodium carbonate or bicarbonate; nitrate or ammonia as a nitrogen source; and sulphate; phosphate, and other minerals in trace amounts. Not all defined media are so simple; some contain dozens of salts and other components.

Table 10.3 Examples of defined (synthetic) media

Ingredients	Amount (g/L)	Ingredients	Amount (g/L)
Cyanobacteria culture (BG 11)		**Escherichia coli culture**	
$Na\ NO_3$	1.5	Glucose	1.0
$K_2HPO_4 . 3H_2O$	0.04	$Na_2\ H\ PO_4$	16.4
$MgSO_4 . 7H_2O$	0.075	$KH_2\ PO_4$	1.5
$CaCl_2 . 2H_2O$	0.036	$(NH_4)_2SO_4$	2.0
Citric acid	0.006	$MgSO_4 . 7H_2O$	0.2
Ferric ammonium citrate	0.006	$CaCl_2$	0.01
EDTA (Na_2Mg salt)	0.001	$FeSO_4 . 7H_2O$	0.0005
Na_2CO_3	0.02		
Trace metal solution (H_3BO_3, $MnCl_2$, $ZnSO_4$, Na_2Mo_4, $CuSO_4$, $Co(NO_3)_2$	1.0 mL/L		
Final pH 7.4		Final pH 6.8 – 7.0	

Complex media

Complex media contain some ingredients of unknown chemical composition such as peptones, meat extract, and yeast extract (Table 10.4). Each of them contains many different chemicals that have not been completely characterized. Peptones are protein hydrolysates obtained by partial proteolytic digestion of meat, casein, soya meal, gelatin, and other proteins. Peptones serve as sources of carbon, energy, and nitrogen. Aqueous extracts of lean beef and brewer's yeast constitute beef extract and yeast extract, respectively. Beef extract supplies amino acids, peptides, nucleotides, organic acids, minerals, and vitamins. Yeast extract contains B vitamins, nitrogen, and carbon compounds.

A single complex medium may be sufficiently rich and complete to meet the nutritional requirements of different microorganisms. Such a medium is useful for culturing a particular microorganism whose nutritional requirements are yet to be determined.

The three commonly used complex media—(i) nutrient broth, (ii) tryptic soy broth, and (iii) MacConkey agar—have been mentioned in Table 10.4.

Agar media

Microorganisms are cultured on the surface of a solid (jelly-like) medium. Liquid media can be solidified with the addition of 1.0–2.0 per cent agar, most commonly 1.5 per cent. Agar is a sulphated polymer composed mainly of D-galactose, 3,6-anhydro-l-galactose, and D-glucuronic acid, supplied in powder form. It is extracted from red algae, and seaweeds belonging to the division *Rhodophyta*. Agar added to the media melts with rising temperature and solidifies when cooled down to about 40–42°C. Most microorganisms cannot degrade it.

Table 10.4 Commonly used complex media

Ingredients	Amount (g/L)	Ingredients	Amount (g/L)	Ingredients	Amount (g/L)
Nutrient broth		**Tryptic soy broth (General purpose)**		**MacConkey agar (Selective medium)**	
Peptone (gelatin hydrolysates)	5.0	Tryptone (pancreatic digest of casein)	17.0	Pancreatic digest of gelatin	17.0
Beef extract	3.0	Peptone (soybean digest)	3.0	Pancreatic digest of casein	1.5
		Glucose	2.5	Peptic digest of animal tissue	1.5
		$NaCl_2$	5.0	Lactose	10.0
		$K_2HPO_4 . H_2O$	2.5	Bile salts	1.5
				$NaCl_2$	5.0
				Neutral red	0.03
				Crystal violet	0.001
				Agar	13.5

Types of media

General purpose media   These support the growth of many types of microorganisms. When blood and other special nutrients are added to them to encourage growth of some fastidious heterotrophs, they are called *enriched media*.

Selective media   These support the growth of specific microorganisms. Bile salts and dyes, such as basic fuchsin and crystal violet, favour the growth of Gram-negative bacteria by inhibiting the growth of Gram-positive bacteria. MacConkey agar, endo agar, and eosin methylene blue agar are extensively used for the detection of *E. coli* and related bacteria in water supplies and other samples. Bacteria may also be selected by allowing them to grow in a medium containing nutrients that are specific for the particular bacteria. Cellulose-digesting bacteria can be selected or isolated in a medium containing only cellulose as the carbon and energy source. Dozens of special selective media are used for the isolation of specific bacteria.

Differential media   These distinguish between different groups of bacteria and permit primary identification of microorganisms based on their biological characteristics. Although, blood agar is an enriched medium, it can act as a differential medium as well. It distinguishes between haemolytic and non-haemolytic bacteria. Haemolytic bacteria, such as *Streptococci* and *Staphylococci* isolated from the throat, cause plaque-like clear zones around their colonies by destroying red blood cells in the agar medium. MacConkey agar is both selective and differential. As it contains lactose and neutral red dye, lactose-fermenting bacterial colonies grow and acquire a red or pink colour and thus can be easily differentiated from non-fermenting bacteria.

10.2 ESTABLISHING A PURE CULTURE

Several species of microorganisms in a mixed population are found in natural habitats. For experimental and industrial purposes a *pure culture*, which is a population of cells arising from a single cell, is needed. Such a culture actually represents an individual species. Robert Koch (1843–1910), a German bacteriologist, developed the pure culture technique and transformed microbiology. This technique allowed identification of most pathogens for the major human bacterial diseases within two decades. Three common ways of establishing pure cultures are described here.

Spread plate technique

A small volume of dilute microbial mixture containing about 30–300 cells is transferred with a pipette to the centre of an agar plate and spread evenly over the surface with a sterile bent glass rod (Fig. 10.1). Each of the dispersed cells develops into an isolated colony. The number of colonies is equal to the number of viable organisms in the sample cultured. Thus, spread plates are often used to count microbial populations. A *colony* is a macroscopically visible growth or cluster of microorganisms on a solid medium, derived from a single cell. A pure culture can then be obtained from each colony.

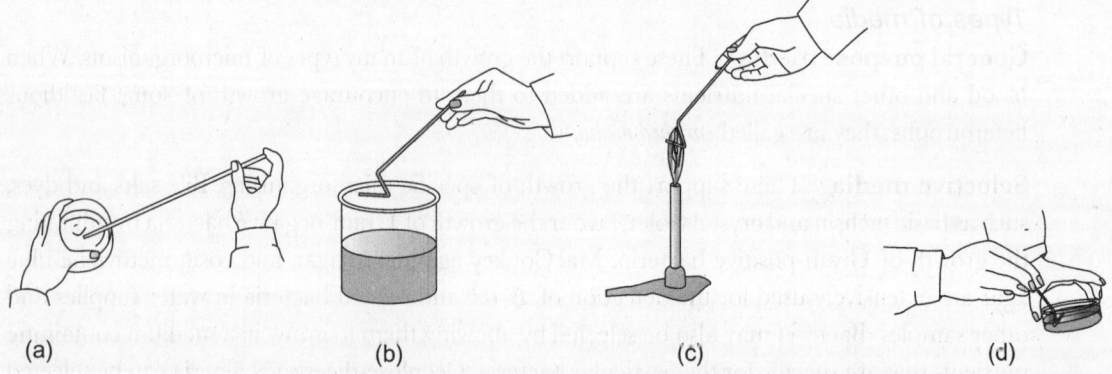

(a)   (b)   (c)   (d)

Fig. 10.1 Spread-plate technique. Steps: (a) A small volume of diluted microorganisms is transferred with a pipette onto the centre of an agar medium plate. (b) A bent glass spreader is dipped into alcohol in a beaker. (c) The ethanol-soaked spreader is briefly heated on a flame and then cooled. (d) The bacterial sample is spread evenly over the agar with the spreader. Then the Petri dish is incubated.

Streak plate technique

The diluted microbial mixture is transferred with a sterilized inoculating loop or swab to the edge of an agar plate and then streaked over the surface in one of several patterns (Fig. 10.2). Single cells along the streaking line develop into separate colonies. The proper spatial separation (that is, the gap between) of single cells in spread and streak plates is the basis of successful isolation of pure bacterial colonies.

Pour plate technique

The original sample of microbes is diluted several times to thin out the population sufficiently to obtain separate colonies on plating. The most diluted samples are then mixed with liquid warm agar at 45°C and poured into sterile Petri dishes (Fig. 10.3). After the agar has solidified, isolated cells are fixed in the plate and grow into colonies, and each of the colonies can be used to establish a pure culture. In the agar plate, the colonies on the surface are circular and those at the subsurface assume a lenticular or lens shape.

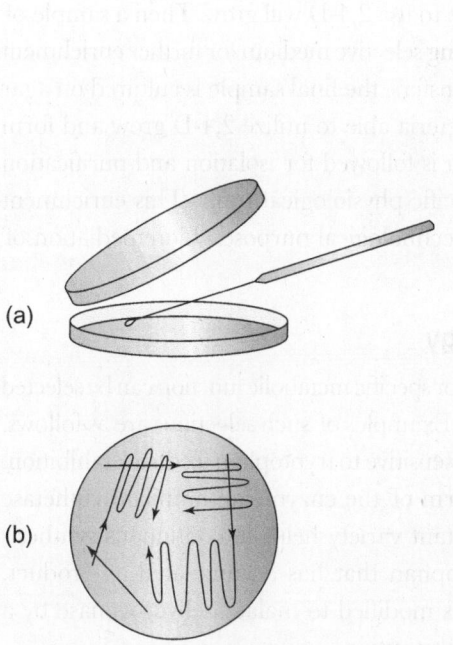

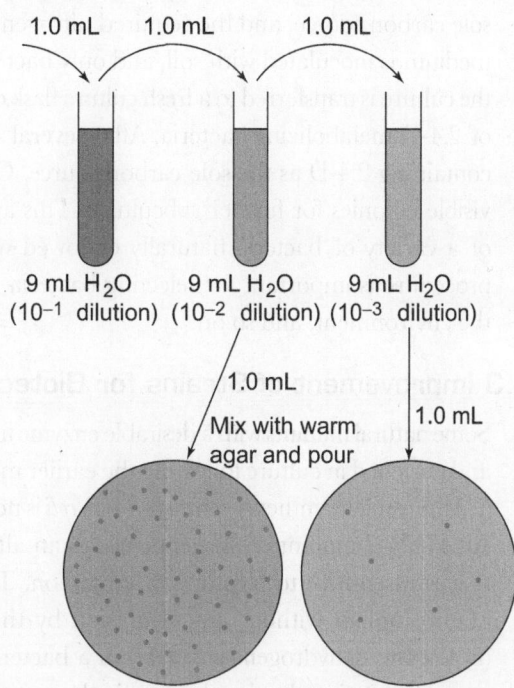

Fig. 10.2 Streak-plate technique. (a) Agar in a Petri dish being streaked with an inoculating loop. (b) A streaking pattern on the agar.

Fig. 10.3 Pour-plate technique. Original sample is diluted to thin out the bacterial population. The extremely diluted samples are then mixed with warm agar and poured into Petri dishes. The colonies grow and can be used to establish pure cultures.

All three techniques require the use of circular culture dishes, named Petri dishes after the name of their inventor Julins Richard Petri, a member in Robert Koch's laboratory. Petri invented the dishes around 1887. Petri dishes are very easy to work with, can be stacked over each other during storage to save space, and are an indispensable item in a microbiology laboratory.

10.2.1 Cell Growth in Colonies

The most rapid cell growth occurs at the edge of a colony. Cell growth is much slower in the centre, and cell death can commence in the centre of older colony. Gradients of oxygen, nutrients, and toxic metabolic products are responsible for the differential growth of cells in a colony. Oxygen and nutrients are plentiful at the edge of a colony. Towards the centre, the cell density is higher, oxygen and nutrients cannot diffuse readily, and toxic accumulation cannot be eliminated rapidly. Thus, growth in the centre of a colony slows down.

10.2.2 Enrichment and Isolation of Pure Cultures

When a particular type of microorganism is present in very low numbers in a sample, the plating methods use selective or differential media to enrich and isolate the rare microorganism. For example, isolation of the bacteria that metabolize and degrade the herbicide 2,4-dichlorophenoxyacetic acid (2,4-D) can be achieved by culturing the bacteria in a liquid medium containing 2,4-D as the sole carbon source, and the required nitrogen, phosphorus, sulphur, and minerals. Initially, this medium is inoculated with soil, and only bacteria able to use 2,4-D will grow. Then a sample of the culture is transferred to a fresh culture flask containing selective medium for further enrichment of 2,4-D metabolizing bacteria. After several such transfers, the final sample is cultured on agar containing 2,4-D as the sole carbon source. Only bacteria able to utilize 2,4-D grow and form visible colonies for further subculture. This approach is followed for isolation and purification of a variety of bacteria naturally endowed with specific physiological traits. This enrichment procedure is important for selecting bacteria for biotechnological purposes, bioremediation of the environment, and so on.

10.2.3 Improvement of Strains for Biotechnology

Some natural mutants with a desirable enzyme function or specific metabolic function can be selected and enriched in culture following the earlier methods. Examples of such selections are as follows. (i) Anthranilatesynthetase enzyme of *E. coli* is normally sensitive to tryptophan feedback inhibition. An MTR 2 mutant of *E. coli* possesses an altered form of the enzyme anthranilatesynthetase that is insensitive to tryptophan inhibition. This mutant variety helps in continuous synthesis of tryptophan without any inhibition by the tryptophan that has accumulated as product. (ii) Lactate dehydrogenase (LDH) in a bacterium was modified to malate dehydrogenase by a natural mutation that involved a single amino acid substitution.

In these types of naturally occurring mutant enzymes, single amino acid substitution or addition/ deletion has been found. If improvement in protein function requires changes in several amino acids, the corresponding mutant will be rare or non-existent; such modification is possible only through gene modification.

Different types of in-vitro mutagenesis techniques can be employed to improve the physical and chemical properties of proteins. Combinations of gene targeting, and site-specific recombination contribute to site-directed mutagenesis (see Chapters 6 and 9).

10.3 MICROBIAL GROWTH

Growth may be defined as an increase in cellular constituents that may result in an increase in a microorganism's size, population, or both.

When microorganisms are grown in a liquid medium in a closed system—they are incubated in a closed culture vessel with a single batch of medium—the growth of the population remains exponential only for a few generations, and then enters a stationary phase due to limited availability of nutrients and waste accumulation. The resulting growth curve with four distinct phases is depicted in Fig. 10.4. In an open system with a continuous supply of nutrition and removal of waste, the exponential phase can be maintained for a longer period.

10.3.1 Four Phases of Microbial Growth

Microbial growth in a batch culture shows four phases as a function of time and is presented in Fig. 10.4.

Lag phase

Immediately after introduction of microorganisms into fresh culture medium, no cell division takes place, there is no net increase in mass, and the cells only synthesize new cellular components. This is the lag phase, prior to the start of cell division.

The duration of the lag phase differs considerably depending on the condition of the microorganisms and the nature of the medium. This phase gets longer when the inoculum is from an old culture or from a refrigerated stock. If the fresh medium is different from the one in which the microorganisms were growing previously, the lag phase becomes longer. When inoculum from a young, vigorously growing exponential phase culture is added to a fresh medium of the same composition, the lag phase is short or almost non-existent.

Exponential or log phase

Microorganisms grow and divide at the maximal rate during the exponential phase. The microorganisms divide and double in number at regular intervals and their rate of growth is constant during this phase. The increase in number is plotted on a log scale. The population remains most uniform in chemical and physiological properties during this phase. This is why microorganisms in the exponential phase of growth are used for biochemical and physiological analyses.

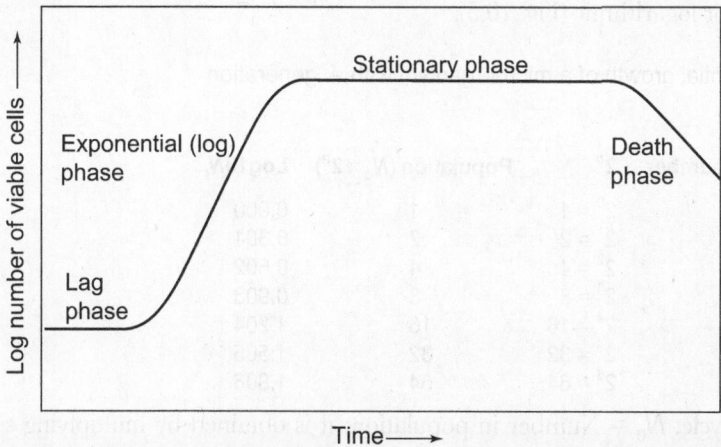

Fig. 10.4 The four phases of the microbial growth curve.

Stationary phase

The exponential phase eventually ends up in a stationary phase when population growth ceases; the total number of viable microorganisms remains constant. The growth curve becomes horizontal. This may be due to a balance between cell division and cell death or because the population does not divide but remains metabolically active.

In the stationary phase, a population of bacteria may reach a density of 10^9 cells per mL. Protozoan and algal cultures in this phase may have maximum concentrations of about 10^6 cells per mL.

Population growth enters the stationary phase due to several factors operating in concert—(i) depletion of essential nutrients, (ii) low availability of oxygen, or (iii) accumulation of toxic waste products. It is observed that growth may cease when a critical population density is reached.

Death phase

Nutrient deprivation and accumulation of toxic metabolic wastes lead to a decline in the number of viable cells, a typical feature of the death phase. Usually, a constant proportion of cells die every hour and the curve in the death phase assumes a logarithmic nature similar to the growth curve during the exponential phase.

10.3.2 Measurement and Kinetics of Microbial Growth

Measurement of the growth rates of microorganisms during the exponential phase is necessary for understanding physiological aspects, and for identifying solutions to applied problems in the biotechnological industry.

Each microorganism divides at constant intervals during the exponential phase. Thus, the population doubles in number in a specific period of time known as the *generation time* or *doubling time*. For example, if a culture tube is inoculated with one cell that divides every 20 minutes; the population becomes two cells after 20 minutes, four cells after 40 minutes, and so forth (Table 10.5). As the population doubles in every generation, the increase in population is always $2n$ where n is the number of generations. The increase in the population can be plotted, and it is either exponential or logarithmic (Fig. 10.5).

Table 10.5 Exponential growth of a microorganism with a generation time of 20 minutes

Time[a]	Division number	2^n	Population ($N_o \times 2^n$)	$\text{Log10}N_t$
0	0	$2^0 = 1$	1	0.000
20	1	$2^1 = 2$	2	0.301
40	2	$2^2 = 4$	4	0.602
60	3	$2^3 = 8$	8	0.903
80	4	$2^4 = 16$	16	1.204
100	5	$2^5 = 32$	32	1.505
120	6	$2^6 = 64$	64	1.806

[a]20 min for each cycle; N_0 = Number in population; it is obtained by multiplying seeding in number with the duplication value of the cell cycle (2^n); N_t = Number in population ($N_0 \times 2^n$)

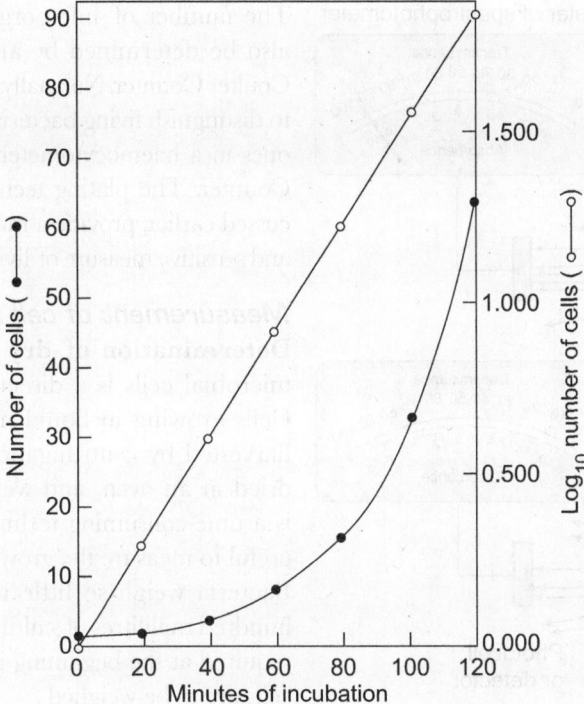

Fig. 10.5 The data from Table 10.5 for six generations of growth plotted directly (•) and in the logarithmic form (o). The linearity of the log plot indicates that the growth curve is exponential.

The rate of growth during the exponential phase in a batch of culture can be expressed in terms of *mean growth rate constant (k)*. This is equal to the number of generations per unit time, usually expressed as the generations per hour.

$$k = \frac{n}{t}$$

where *n* is the number of generations in time *t*.

The *mean generation time* or mean doubling time (*g*) is the reciprocal of the mean growth rate constant (*k*).

$$g = \frac{1}{k}$$

10.3.3 Quantitation of Microbial Growth

Microbial growth can be measured in several ways to determine the growth rate and generation time. The number of microorganisms or their mass is taken into account as growth by definition leads to an increase in both.

Measurement of cell number

The number of growing bacteria in a diluted sample can be directly counted under a microscope at 400× to 500× magnification after charging them in a haemocytometer.

The number of bacteria observed in a specified small area of the slide should be multiplied with the dilution factor (the number of times the original culture sample is diluted for easy counting), and the factor used to raise the volume of the fluid in the specified counting area to 1 mL.

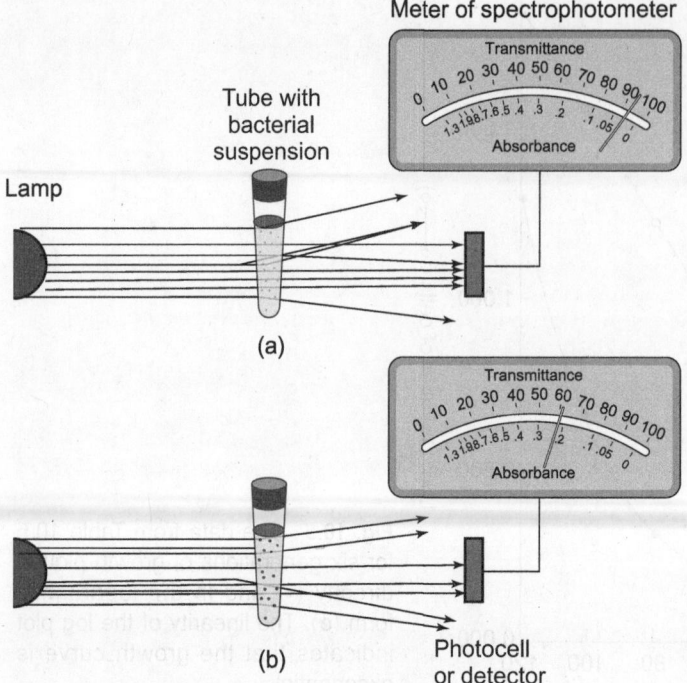

The number of microorganisms can also be determined by an electronic Coulter Counter. Normally, it is difficult to distinguish living bacteria from dead ones in a haemocytometer or Coulter Counter. The plating techniques, discussed earlier, provide a more accurate and sensitive measure of living bacteria.

Measurement of cell mass

Determination of dry weight of microbial cells is a direct approach. Cells growing in liquid medium are harvested by centrifugation, washed, dried in an oven, and weighed. This is a time-consuming technique and is useful to measure the growth of fungi. Bacteria weigh so little that several hundred millilitres of culture would be required at the beginning if harvested cells are to be weighed.

Fig. 10.6   Determination of microbial mass by spectrophotometer. As the population and turbidity increase (tube 2), more light is scattered and the absorbance reading given by the spectrophotometer increases. The meter of the spectrophotometer has two scales, the bottom scale is for absorbance and the top scale for percentage transmittance of light. Absorbance increases as percentage transmittance decreases.

Spectrophotometric determination of cell mass is a better option due to its sensitivity and speed. As the population of microorganisms and turbidity of the medium increase, more light is scattered and the reading for absorbance of light increases in the spectrophotometer. Thus, the absorbance is directly proportional to the cell mass present and is obviously related to the cell number. There are two scales in the meter of the spectrophotometer. The bottom scale is to read absorbance and the top scale reads percentage transmittance (light received by the photocell or detector). Absorbance increases as percentage transmittance decreases (Fig. 10.6).

The total amount of a cell constituent is directly related to the total microbial cell mass. Thus, total protein or nitrogen of a sample of washed cells collected from a known volume of medium will give an idea about the increase or decrease of cell mass. Similarly, chlorophyll determination can be used as a measure of algal and cyanobacterial populations. The quantity of ATP is another index to measure living microbial mass.

Exhibit 10.A   Environmental Factors and Eco-distribution of Bacteria

The manner in which gradients of nutrients, oxygen, and accumulation of metabolic toxic waste influence the growth of bacteria in a colony has already been discussed. The growth of microorganisms is greatly influenced by the chemical and physical nature of the surroundings. This knowledge is helpful in controlling microbial growth and in understanding the ecological distribution of microorganisms, especially in extreme conditions.

Prokaryotes are present almost anywhere, in all types of habitats. *Bacillus infernus* live more than 1.5 miles below the earth's surface, without oxygen and at temperatures above 60°C. Microorganisms capable of growing in such extreme conditions are called extremophiles. Thermophile prokaryotes grow in sulphide chimneys or 'black smokers', located along rifts and ridges on the ocean floor, that spew sulphide-rich superheated vent water with temperatures above 350°C. Interestingly, the sea water in this habitat does not boil below 460°C as the pressure is as high as 265 atm.

There is evidence that the microbes of this region can grow and reproduce at or above 113°C. The stability of proteins, membranes, and nucleic acids of these prokaryotes is intriguing and raises the possibility of designing enzymes operative at very high temperature in the near future. Already Taq polymerase from the thermophile *Thermusaquaticus* is used extensively in the thermocycler for polymerase chain reaction (PCR).

There are certain bacteria which inhabit deep layers of snow in the freezing temperatures of Antarctica. Some of the most important environmental factors that affect microbial growth and their ecological distribution in nature are: solutes and osmoregulation, pH, temperature, oxygen concentration, pressure, and radiation including light. Microorganisms can be categorized in terms of their response to these factors. For example,

Factor	Category	Response
Solute and osmoregulation	Osmotolerant	Grows over a wide range of osmotic concentration
	Halophile	Requires high level of NaCl (above 0.2M)
pH	Acidophile	Grows in acidic pH, between 0 and 5.5
	Neutrophile	Growth optimum between pH 5.5 and 8.0
	Alkalophile	Growth optimum between pH 8.5 and 11.5
Temperature	Psychrophile	Grows at 0°C, optimum upto 15°C
	Mesophile	Growth optimum 20–45°C
	Thermophile	Growth optimum 55–65°C
	Hyperthermophile	Growth optimum 80–113°C
O_2 concentration	Obligate aerobe	Growth dependent on atmospheric O_2
	Obligate anaerobe	Cannot tolerate O_2 and dies in its presence
Pressure	Barophilic	Grows better at high hydrostatic pressures

Exhibit 10.B   Quorum Sensing

A bacterial population was considered as a collection of individuals growing and performing activities independently. Recently, bacteria in a population have been found to communicate and cooperate through chemical signalling. The process of chemical mediated communication between individual bacteria to sense the density of the population is known as *quorum sensing*.

Bacteria monitor their own population density by sensing the levels of signal molecules. The signal molecules are also called autoinducers as they stimulate the cell that releases them to continue the synthesis of these molecules and other substances. The concentration of these signal molecules increases with increase in the number of bacteria in the population till a specific threshold is reached which acts as a signal to the bacteria that the population density has reached a necessary critical level or quorum. The bacteria then begin to express sets of quorum-dependent genes. Thus, quorum sensing allows the bacteria to reach a high population density before synthesis and release of the gene encoded products. As a consequence the levels of the products are high enough to be effective. It is suggested that when there is too much diffusion of autoinducer molecules in a thin population of bacteria, it would not be sensible to release molecules like proteases, antibiotics, and siderophores which are effective in interacting with the surroundings and the host, for better survival of bacteria. Perhaps, the signal molecules (autoinducers) serve both functions—to sense population density and diffusion rates.

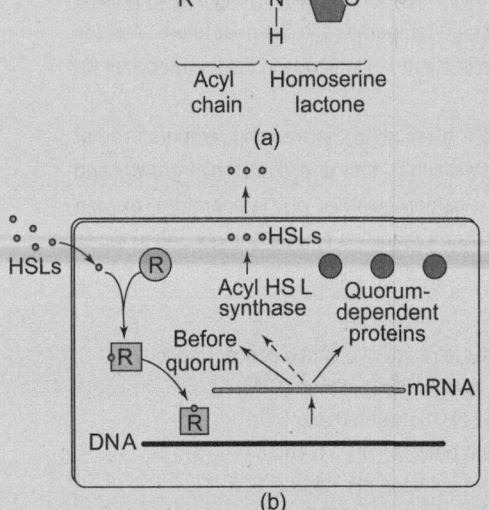

Fig. 10.B   Quorum sensing in Gram-negative bacteria. (a) Structure of acyl homoserine lactone (HSL), the best known quorum sensing signal molecule or autoinducer. (b) Schematic presentation of quorum sensing process. HSLs diffuse into the target cell; when in sufficiently high level, HSLs bind to the special receptor (R) proteins and trigger a structural change in R. The activated complex of HSL and R binds to a target site on DNA and stimulate transcription of the quorum sensing gene for further synthesis of HSLs.

Quorum sensing was first discovered in Gram-negative bacteria and was later found in Gram-positive bacteria. The schematic presentation of quorum sensing functions in Gram-negative bacteria is shown in Fig. 10.B. Acyl homoserine lactones (HSLs) are the common signal molecules (autoinducers) in gram-negative bacteria. These are small molecules composed of a 4- to 14-carbon acyl chain joined by an amide bond to homoserine lactone (Fig. 10.B.a). The acyl chain may have a keto or hydroxyl group on its third carbon. Acyl HSLs diffuse into the target bacterial cell (Fig. 10.B.b). When the level of acyl HSLs within the cell is sufficiently high, they bind to specific receptor proteins and cause a conformational change in the proteins. These activated proteins plus acyl HSL complexes act as inducers; they bind to the target site on the DNA and initiate transcription of quorum-sensitive genes. The gene encoding acyl HSL may also be activated for production and release of more autoinducer molecules.

Some examples of the processes initiated by acyl HSL signals and activation of quorum sensing genes in Gram-negative bacteria are (i) bioluminescence production by *Vibrio fischeri,* (ii) antibiotic production by *Erwinia carotovora* and *Pseudomonas aureofaciens,* (iii) event of conjugation for transfer of genetic material by *Agrobacterium tumefaciens,* and (iv) production of virulence factor by *Pseudomonas aeruginosa.* The quorum sensing process is likely to promote biofilms by the pathogen *Pseudomonas aeruginosa,* and this may play a role in cystic fibrosis.

An oligopeptide pheromone has been identified which acts as a quorum sensing signal in Gram-positive bacteria. It induces many processes in different bacteria, competence in *Streptococcus pneumonia,* conjugation in *Enterococcus faecalis,* sporulation in *Bacillus subtilis,* and production of toxins and virulence factors by *Staphylococcus aureus.* The quorum sensing event activates the development of aerial mycelia and production of streptomycin by *Streptomyces griseus*; the signalling molecule, here, is γ-butyrolactone.

Quorum sensing is analogous to the behaviour of cells in a multicellular organism, in that many individual cells communicate and coordinate among themselves to act as a unit. Quorum sensing is likely to take part in the development of pattern formation in bacterial colonies.

10.4 SCALE-UP OF MICROBIAL CULTURE FOR INDUSTRIAL PURPOSE

Industrial processes require culture of microorganisms in bulk quantity in specifically designed media under sterile and precisely controlled conditions of temperature, aeration, pH, and nutrient feeding.

Depending on the requirement, microorganisms can be grown in culture tubes, shake flasks, and stirrer fermenters or other mass culture systems. A stirrer fermenter with its control system for temperature, aeration, pH, and nutrient feeding has been described in Chapter 5. The word fermentation has many connotations. It may tell about a process of mass culture of aerobic or anaerobic microorganisms, or it may indicate production of alcoholic beverages, and so on. Here, fermentation mainly indicates mass culture of microorganisms in a specially designed vessel (see Fig. 5.15), which is often called a *bioreactor*. Plant and animal cells can also be cultured in bulk in these reactors. Bulk culture of microorganisms in a bioreactor produces *primary* and *secondary* metabolites. Primary metabolites are formed during the active growth phase of microorganisms, and secondary metabolites are formed after growth is completed.

Stirred fermenters may vary in size from 3 or 4 litres to 100,000 litres or more, depending on the production requirements. During scale-up, stirring is necessary to ensure that critical physical conditions, such as temperature, aeration, pH, and so on, are uniform throughout the large volume of medium in a bioreactor.

10.4.1 Culture Medium for Mass Culture

The economic competitiveness often guides the composition of the culture medium that is used for a particular process. Lower-cost crude materials are preferable as sources of carbon, nitrogen, vitamins, trace elements, and so on as indicated in Table 10.6. Crude plant hydrolysates are often used as a complex source of carbon, nitrogen, and other growth factors.

Table 10.6   Crude components of growth media used in industrial bioreactors

Source	Carbon and energy	Nitrogen	Vitamins	Iron and trace salts	Buffering agents	Antifoam agents
Material used	Molasses, whey, agricultural wastes (corncobs), and grains	Soybean meal, corn-steep liquor, stick liquor (slaughter house products), ammonia and ammonium salts, nitrates, distiller's soluble from brewery	Crude plant hydrolysates and animal products	Crude inorganic chemicals	Chalk or crude carbonates and fertilizer grade phosphates	Higher alcohols, natural esters, lard and vegetable oils, and silicons

Often, a critical component in the medium, even the carbon source needs to be added continuously in the culture. The *continuous feed* is necessary so that excess substrate is not provided to the microorganisms at the beginning of the fermentation. For example, excess glucose and other carbohydrates can easily be catabolized into ethanol, which is lost as a volatile product and results in a reduction in the final yield.

10.4.2 Alternate Methods of Mass Culture

Besides the frequently used stirred bioreactor, there are other methods of mass culture of microorganisms in industrial processes (Fig. 10.7). Often these alternate methods incur lower operating costs and are handy when specialized growth conditions need to be created for the microorganism that is required for the synthesis of a particular product. In addition to the four processes presented in Fig. 10.7, there are certain other methods. Lift-tube fermenter has been discussed in detail in Chapters 11 and 12, where gas bubbles entering from the bottom of the culture vessel cause circulation of the culture fluid.

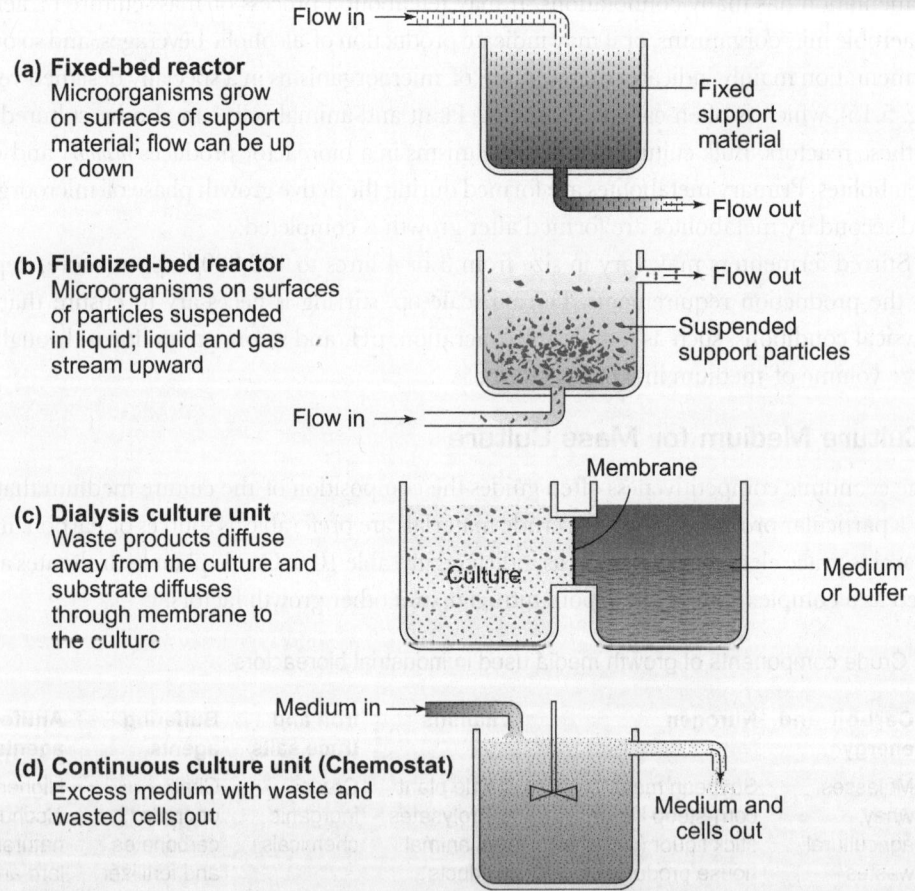

Flow in ⟶

(a) Fixed-bed reactor
Microorganisms grow
on surfaces of support
material; flow can be up
or down

Fixed
support
material

Flow out ⟶

(b) Fluidized-bed reactor
Microorganisms on surfaces
of particles suspended
in liquid; liquid and gas
stream upward

⟶ Flow out

Suspended
support particles

Flow in ⟶

Membrane

(c) Dialysis culture unit
Waste products diffuse
away from the culture and
substrate diffuses
through membrane to
the culture

Culture

Medium
or buffer

Medium in ⟶

(d) Continuous culture unit (Chemostat)
Excess medium with waste and
wasted cells out

Medium and
cells out

Fig. 10.7 Alternate methods of mass culture of microorganisms in industrial processes. These alternate methods usually operate at lower costs.

10.4.3 Industrial Microbial Products

The list of products from industrial microbiology is now quite long. It includes different kinds of beverages, industrial chemicals, enzymes, food additives, pharmaceuticals for human and animal health, plant hormones, and other chemicals for agricultural use, biofuels, and so on (Table 10.7). These products have had an immense impact on the well-being of humans and animals. Medical products, particularly antibiotics, have increased the life span of the human population. Some examples of biopharmaceuticals and commercial products produced by recombinant microorganisms have been discussed in Chapter 6.

Table 10.7 Some biotechnological microbial products

Products	Microorganisms
Industrial chemicals	
Ethanol (from glucose)	*Saccharomyces cerevisiae*(yeast)
Ethanol (from lactose)	*Kluyveromyces fagilis*
Acetone and butanol	*Clostridium acetobutylicum*
Enzymes	*Asperillus, Bacillus, Mucor, Tricoderma*
Food additives	
Amino acids	
Lysine	*Corynbacterium glutamicum*
Glutamic acid	*C. glutamicum* (mutant)
Organic acids	
Citric acid	*Aspergillus niger*
Acetic acid	*Acetobacter*
Polysaccharides	*Xanthomonas*
Vitamins	*Ashbya, Eremothecium, Blakeslea*
Pharmaceuticals	
Antibiotics	
Penicillin	*Penicillium* (Fungi)
Bacitracin, Polymixins	*Bacillus*
Cephalosporins	*Cephalosporium* (Fungi)
Amphotericin B, Chloramphenicol, Erythromycin, Kanamycin, Neomycin, Nystatin, Rifampin, Streptomycin, Tetracyclines, Vancomycin	*Streptomyces*
Gentamycin	*Micromonospora*
Alkaloids	*Claviceps purpurea*
Insulin, human growth hormone, somatostatin, interferons	*Escherichia coli, Saccharomyces cerevisiae,* etc. (involving rDNA technology)
Avermectins (anti-helminths)	*Streptomyces avermitilis*
Statins (cholesterol lowering)	*Pencilillium citrinum*

(*Contd*)

Table 10.7   (*Contd*)

Cyclosporin A (immunosuppressant)	*Tolypocladium inflatum*
Doxorubicin (anti-cancer)	*Streptomyces peucetius caesius*
Agriculture supporting agents	
Bialaphos (bioherbicide)	*Streptomyces hygroscopicus*
Gibberellins	*Gibberella fujikuroi*
Biofuels	
Hydrogen	Photosynthetic microorganisms
Methane	*Methanobacterium*
Ethanol	*Zymomonas*
Biopolymer (Polyesters)	*Pseudomonas oleovorans*

Bioremediation of polluted land, water bodies, and air by using microorganisms as such or genetically modified microorganisms may also be included under the applications of microbial biotechnology. This is dealt with in detail in the chapter on *Environmental biotechnology* (Chapter 13).

Isolation of Microbial Products

The technology to recover microbial products from mass culture in fermenters may differ depending on the types of microorganisms cultured, their life cycle, types of culture media, and the design and working principles of the bioreactors (fermenters). The cell cycle, and the growth phase of the organisms, guide the timing of collection of primary or secondary metabolites; the primary ones are produced during the active exponential growth phase, whereas the secondary metabolites are formed after growth is completed.

Here, for example, the *production and isolation* of a prime antibiotic, *penicillin* has been discussed. Regulation of the medium composition can achieve maximum yields and variations in products from the culture of *Penicillium chrysogenum* in a bioreactor.

The slowly hydrolysable disaccharide lactose, in combination with limited nitrogen supply, stimulates a greater accumulation of penicillin after growth has stopped. A slow continuous feed of glucose produces a similar result. Addition of phenylacetic acid maximizes production of penicillin G, which has a benzyl side chain. Maximum stability of the newly synthesized penicillin is achieved by maintaining the pH of the culture medium around neutral.

The fermentation of penicillin usually continues for about a week. Then the fungal mycelia are separated from the culture broth. The broth is processed by absorption, precipitation, and crystallization to purify the final product, penicillin. This again can be modified by chemical procedures to produce different variety of penicillins.

10.5 STRAIN ISOLATION AND IMPROVEMENT

Usually, the strains of microorganisms with desirable (metabolic) characteristics for microbial biotechnology are isolated from natural sources such as soil samples, water, spoiled bread, and fruits. The search for new microorganisms with different capabilities is continuing throughout the world and is often called *bioprospecting*. Microbiologists are relentlessly trying to culture different microbes, so that they can be well characterized, classified, and probed for their new products. Diverse microorganisms from different niches including extreme environments are being collected for the purpose.

Once a microorganism is found to flourish in culture and synthesizes promising product(s), multiple techniques based on modern genetics and molecular biology are applied to improve the performance of the microorganism, mostly in terms of the type and amount of the product. The techniques can be serialized as follows:

Mutation   Mutagenic agents like X-rays, UV light, and chemical mutagens are employed to mutate the gene for a desired trait to improve the strain of microorganism.

For example, the first cultures of *Penicillium notatum* in static conditions produced low amounts of penicillin. A strain, NRRL 1951, of *Penicillium chrysogenum* was isolated in 1943 and improved through mutations induced by X-ray, UV light, and mustard gas. The yield of penicillin increased from 120 International Units (IU) to 2,580 IU per mL, a 20-fold increase. This increased yield was in turn 55 fold higher than the penicillin produced by *Penicillium notatum* under original static culture conditions. Nowadays, most of the penicillin is produced with a genetically improved version of *Pencillium chrysogenum*, grown in aerobic stirred fermenters.

Protoplast fusion   The technique of protoplast fusion is described in Chapter 11. This technique makes it possible to combine the genetic material of two different species in a single cell, even when the two species are not closely linked taxonomically. This technique is often used for yeasts and fungi (molds) in microbial biotechnology.

Site-directed mutagenesis   Insertion of a short sequence of chemically synthesized DNA can produce genetic alterations of the recipient microorganism leading to a change of one or a few amino acids in the target protein. Such minor amino acid changes sometimes cause remarkable changes in protein characteristics, and can sometimes give rise to new functions. Site-directed mutagenesis comes under the purview of protein engineering.

Transgenic technique   This technique transfers genes from one organism into another allowing the recipient to transcribe the transferred genes. The technique and its utility in microbial technology have been discussed in great detail in Chapter 6. Creation of 'Superbug for Clearing Oil Spills' (see Exhibit 6.B) is an interesting topic in this regard.

Modification in gene regulation   In transgenic technology, a specific class of modifications for gene expression is possible, in addition to the modification of structural genes which encode proteins. Modification of gene regulation involves changes in gene transcription, fusion of protein products, creation of hybrid promoters, and removal of feedback regulation controls. This often allows overproduction of certain desirable products.

Inactivation or deregulation of specific genes in metabolic pathways may be altered by this technique. Understanding of *pathway architecture* for a metabolic process helps in designing the most efficient metabolic pathway and avoiding low productivity or energetically more costly routes. This is described as *metabolic pathway engineering* (MPE). Metabolic engineering can be used for synthesis of modified antibiotics.

Directed or adaptive mutation   It has been observed that specific environmental stresses may cause microorganisms to mutate and adapt, thus creating microorganisms with new biological capabilities. These new capabilities can be harnessed in industrial microbiology. The mechanisms of these adaptive mutational processes include DNA rearrangements, activity of mutator genes for *hypermutation*. Induction of adaptive mutation is described as *natural genetic engineering*.

Strain Preservation

When a strain of microorganism is isolated and selected, or improved to produce a specific substance, it must be preserved for further use and improvement.

Transfer from an older culture to a fresh new culture from time to time is the age old method for propagation of microorganisms. In this method of preservation through serial passages, the microorganisms get a chance to be mutated and undergo changes in phenotypic or metabolic characteristics.

At present, *lyophilization* or freezing and storage in liquid nitrogen is the usual practice for preservation of microbial strains. In this way, microbial cultures can be stored for years without loss of viability or an accumulation of mutations.

10.6 BIOETHICS IN MICROBIAL TECHNOLOGY

It has already been mentioned in the introduction of this chapter that the whole ecosystem depends on the activities of microorganisms and they influence the life of other organisms including human beings in countless ways. Some of the microbes are dreaded pathogens causing diseases in human beings and other organisms, and others are agents of spoilage. In this chapter, we have learnt the immense contributions of the microbial world through beneficial microbial products. One of the products, antibiotics, combat harmful microorganisms and help in increasing the life span of human beings and animals.

However, biotechnologists need to take into consideration not only the profitable part of the technology but also the long-term potential ecological impacts of *genetically engineered organisms* (GEOs). Invasiveness of the newly modified microorganisms, variations in the competitive environment, and timing of introductions can confound predictions. Restricted release of microbes outside the laboratory or production unit is of utmost importance for the microbial industry. Possible ethical and ecological impacts of a particular product or process should also be a matter of concern.

Exhibit 10.C   Synthetic Genomics: Synthesis of a Bacterial Genome—Craig Venter

The human genome project worked out the nucleotide sequence of the entire human genome consisting of about 3.2 billion base pairs, in 2003. The project involved many scientists all over the world (see Exhibit 3.A, Chapter 3). The National Centre for Genome Research of the National Institute of Health, USA under the leadership of James D. Watson guided and coordinated the efforts of the scientists. The project was completed two years ahead of schedule. The speeding up of the project was possible because of *Craig Venter* who worked at a private biotech company *Celera Genomics* and used his 'Whole Genome Shotgun' (WGS) sequencing technique. The dialogues between the leaders of the public funded genome project and the private company led to simultaneous announcement of the release of first drafts of the human genome on 26 June 2000. The two drafts were published simultaneously on 15/16 February, 2001 in the two most prestigious science journals, *Nature* and *Science*. The draft sequences had a number of gaps which were subsequently worked out, and the final complete sequence of the human genome was published in April, 2003.

In 2003, Craig Venter and his colleagues reported another breakthrough. They had synthesized the entire genome of the smallest bacterium, *Mycoplasma genitalium*, consisting of 582,970 bp (~583 kb) which was twenty times longer than any DNA molecule synthesized earlier in vitro.

This was also the first report of synthesizing the full set of genes needed to make a living organism in the laboratory. This suggested initiation of creating the first artificial organism.

The complete genome of *M. genitalium* was constructed from 10,000 synthetic oligonucleotides, each made up of about 50 nucleotides. The synthesized oligonucleotides were assembled into as many as 101 custom-made fragments or 'cassettes', each in the range of 5,000 to 7,000 bases. All this assembling was outsourced and was done commercially. The cassettes with overlapping sequences were placed next to one another and joined by enzymes.

This discovery by Craig Venter has initiated a new era of *Synthetic Genomics* in biotechnology. This will enable design and creation of naturally occurring or completely novel genomes of microorganisms with special attributes. Bioinformatics will soon come up with a 'Registry of Standard Biological Parts' called *BioBricks*, which are essentially DNA sequences capable of encoding specific products. These BioBricks can be put together to build more complex structures to realize some specific functions; in other words, to obtain products as per order. In all these ventures, the computer, its programming, and the latest information on genomics will be the supporting essentials.

The synthetic genomics involving a dozen microbial genes, genes from worm-wood (*Artemisia annua*) plant, and several expression control genomic elements, are used for introduction of the synthesized genome in bacteria and yeast to obtain artemisinic acid. This acid can easily be converted into *artemisinin,* a potent and modern antimalarial drug.

Craig Venter's recent company *Synthetic Genomics Inc* intends designing microbes to convert sunlight and water into hydrogen for use as fuel, and other microbes to capture carbon emissions from coal plants. Other companies are trying to use synthetic genomics to develop microbes capable of producing gasoline, diesel, and other biofuels.

Venter and his colleagues are searching around for useful genes for synthetic genomics.

SUMMARY

- Microbiology is an attractive and vast field of study and research with many promises for biotechnology.
- Microbial culture, culture media, scale-up of microbial culture for industrial products, improvement of microbial strains, and bioethics applicable to microbial technology are the main topics discussed in this chapter.
- Different basic aspects of microbial culture and growth including kinetics of growth have been dealt with.

- Industrial mass-scale products from microbial sources meet various needs of human beings including important pharmaceuticals.
- The quality and quantity of the products can be improved through improvement of the strains of microorganisms.
- It is mainly the immense ecological impacts of manipulated strains of microorganisms that come under ethical consideration.

Robert Koch

**Heinrich Hermann Robert Koch
(11 December 1843–27 May 1910)**

Robert Koch was a German physician, microbiologist, and founder of bacteriology. He is famous for the isolation and culture of anthrax, tuberculosis, and cholera germs, and for the postulates of the Germ Theory that showed the relation between microorganisms and diseases. He received the Nobel Prize in Physiology or Medicine in 1905 for his findings on tuberculosis which helped in tackling the disease which caused numerous deaths in the mid-19th century.

Robert Koch was born in Clausthal, Germany, then in the Kingdom of Hanover (part of Prussia) to a family of a mining officials. He studied medicine at the University of Göttingen, and graduated in 1866; his doctoral advisor was Friedrich Gustav Jakob Henle. He started his practice in 1869 in the Province of Posen after serving as an assistant in the General Hospital at Hamburg. Then he served in the Franco–Prussian War and later became the district medical officer in Wollstein, Prussian Poland from 1872 to 1880.

While he was posted in Wollestein, he studied anthrax closely and developed methods for purifying the bacillus from blood and grew pure cultures. When he published his findings that endospores persisting in the soil are responsible for the spontaneous outbreak of anthrax in 1876, he was offered a job at the Imperial Health Office in Berlin. It was in Berlin that he discovered the causative organism of tuberculosis, *Mycobacterium tuberculosis*, in 1882.

He urged sterilization of surgical instruments using heat. He introduced staining and purification techniques for bacteria, growth media, agar plates, and the use of the Petri dish named after its inventor, his assistant Julius Richard Petri. These devices are still in use for microbiological work.

While studying cholera with a French research team in Alexandria, Egypt in 1883, Robert Koch identified *Vibrio cholerae*, the bacterium that causes cholera. Previously, the cholera bacterium had been isolated by the Italian anatomist Filippo Pacini in 1854. Koch was unaware of Pacini's work which was an independent discovery. However, in 1965, the bacterium was formally renamed *Vibrio cholerae Pacini* 1854.

In 1885, he became professor of hygiene at the University of Berlin, and in 1891 he was made honorary professor of the medical faculty and Director of the new Prussian Institute for Infectious Diseases (eventually renamed as the Robert Koch Institute). He resigned the position in 1904 and began travelling around the world, studying diseases in South Africa, India, and Java. On the request of the British India Government, he visited the institute at Mukteshwar to investigate cattle plague. The institute is now called Indian Veterinary Research Institute (IVRI); the museum of IVRI has preserved the microscope used by Robert Koch.

Koch's pupils identified the organisms responsible for diphtheria, typhoid, pneumonia, gonorrhoea, cerebrospinal meningitis, leprosy, bubonic plague, tetanus, and syphilis using Koch's methods.

The messiah who won over microbiological diseases received many honours across the globe.

The crater Koch on the moon is named after him. The Robert Koch Prize and Medal to honour microbiologists who have made groundbreaking discoveries or contribution to global health are a unique way to honour and remember Koch. He was a member of the Royal Swedish Academy of Sciences. In many countries, his statues were erected and commemorative coins and postal stamps were brought out.

EXERCISES

Objective Questions

1. Mention at least two contributions from microorganisms in each of the following categories:
   (a) Commercial
   (b) Pharmaceutical
   (c) Agricultural

2. Tick the substances that are micronutrients in the culture medium of microorganisms:
   (a) Sulphur (S)
   (b) Magnesium (Mg^{2+})
   (c) Manganese (Mn^{2+})
   (d) Phosphorus (P)
   (e) Copper (Cu^{2+})
   (f) Iron (Fe^{2+})

3. Mark the following statements as True or False.
   (a) The reduced form of carbon is incorporated into organic molecules.
   (b) Organisms that use reduced, preformed organic molecules as carbon sources are heterotrophs.
   (c) Chemotroph microorganisms obtain energy from light.
   (d) Synthetic medium contains peptone.
   (e) Agar is obtained from algae.
   (f) The most rapid cell growth takes place at the centre of a colony of microorganisms.
   (g) In pour-plate culture, the colonies on the surface of the agar are circular, and the colonies at the subsurface are lens shaped.

Review Questions

1. Define nutrient.

2. What do you understand by a 'pure culture' of microorganisms?

3. What is a 'differential medium' used for?

4. What is a 'colony' of microorganisms?

5. Discuss the procedure of enrichment and isolation of pure cultures of a rare variety of microorganisms.

6. In microbial growth in a closed system—
   (a) What are the factors that cause a longer lag phase?
   (b) What are the limiting factors for the exponential growth phase?
   (c) Why does the growth curve become horizontal in the stationary phase?

7. During the exponential phase of growth in a batch of microbial culture—
   (a) What is the 'mean growth rate constant'?
   (b) Show the relation between 'mean generation time' and 'mean growth rate constant'.

8. Name the methods employed for determination of cell number in a culture of microorganisms.

9. Distinguish between primary and secondary metabolites in a culture of microorganisms.

10. What are the benefits of continuous feed for industrial mass culture of microorganisms?

11. Name five industrial microbial products.

12. Write a note on the techniques that are used to improve microbial strains for types and amount of the product.

Plant Biotechnology

LEARNING OBJECTIVES

♦ The aseptic measures taken for plant cell and tissue culture
♦ Tissue culture laboratory and media
♦ Plant tissue culture for in-vitro micropropagation
♦ The process of regeneration of plants using various plant materials through tissue culture
♦ Protoplast culture and somatic hybridization
♦ Ex-situ conservation through long-term storage

INTRODUCTION

The animal world is dependent on plants primarily for its food, oxygen, and dwelling. Human beings depend on plants in many other ways—for sources of medicine, building materials, fibres for clothes, and as a source of fuels and bulk chemicals like cellulose, amylose, and rubber. The list of usage of plants, their products and fine chemicals for various purposes including pesticides, dyes, and so on is getting longer day by day.

Plant biotechnology utilizes the *totipotency* of plant cells for micropropagation of plants by tissue culture, clonal expansion, and production of transgenic plants. These contribute to better agrihorticultural and forest products and serve as better sources of medicinal and other chemical products. *Totipotency* is the potentiality of a cell to grow and develop into a complete organism. Thus, in-vitro plant tissue and cell culture is at the heart of plant biotechnology. *Plant tissue culture is the culture of cells and tissues from plant organs, such as shoot tips, leaves, roots, embryos, axial buds, anthers, and so on.* The cells of the meristem and vascular cambial tissues divide faster and in a coordinated manner. Embryonic cells at the early stage of development remain in an undifferentiated and proliferative state. The rapidly dividing cells from these sources can produce a *callus* which is a cell mass that can form shoot, roots, and eventually a whole plant, when cultured on semi-solid nutrient agar medium.

11.1 DEVELOPMENT OF PLANT TISSUE CULTURE

In the early 1930s, White in USA and Gautheret and Nobercourt in France, each independently, initiated the culture of tissues collected from several plants on defined nutrient media for a long period. Use of growth hormones for rapid multiplication of totipotent cells, and certain factors that came to light through research in plant physiology and molecular biology improved plant tissue culture by the 1950s. By that time, the single cell culture technique was also introduced. In the 1960s, the use of cellulase and pectinase enzymes allowed improvement of certain aspects of plant tissue culture, especially protoplast culture. In 1966 Guha and Maheshwari, developed the technique of production of vast numbers of embryos from cultures of pollen and sporogenous tissue of anthers.

11.2 OBJECTIVES OF PLANT CELL AND TISSUE CULTURE

Plant tissue culture techniques serve multiple purposes. The most important are mentioned below; others are discussed later in the text, under applications of the technology.

Micropropagation This is 'in-vitro propagation of the selected *true-to-type* genotypes from organs, tissues, cells, or protoplasts'. It helps in

(i) mass multiplication of plants with specific traits which are commercially very profitable
(ii) production of pathogen (virus)-free plants
(iii) clonal propagation of parental stock for production of hybrid seeds, if necessary
(iv) year-round production of nursery plants
(v) long-term preservation of germplasm, through cryopreservation (freezing); needed more for threatened wild varieties of plants.

Protoplast fusion This involves cells of two different plant species and is known as *somatic hybridization*. The somatic hybridization can overcome the inter-species incompatibility barriers that arise in conventional breeding. It offers opportunity for

(i) creation of diverse new plants through novel hybrids
(ii) somatic cell genetics and crop improvement
(iii) genetic manipulation of vegetatively propagated crops including correction, for sterile or subfertile varieties
(iv) mitochondria-based cytoplasmic sterility may also be transferred from one species to another.

Genetic modifications of clonal cultures are initiated in in-vitro cell culture.

Mass cell cultures for secondary metabolites and fine chemicals These are performed with suspension cultures in bioreactors for commercial purposes. Genetic engineering is also used to improve yields of required products.

11.3 TISSUE CULTURE

Plant tissue culture refers to the culture of cells, protoplasm, and plant organs such as roots, shoot tips, leaves, and anthers in defined media in vitro (outside the organism). Plant tissue culture has become a major biotechnological technique for agriculture, horticulture, forestry, and industry.

Plant cell culture requires highly aseptic sterile conditions in each step. In addition to the plant material, the nutrient media and other chemical substances used in the cultures are a rich resource of nutrients for microorganisms. Controlling microbial contamination in the tissue culture laboratory is one of the prime tasks. Three steps in plant tissue culture are crucial for maintaining sterility of the culture: (1) collection of explants for tissue culture that are sterile or axenic (free of other organisms), (2) preparation of sterile culture media and culture vessels, and (3) maintenance of aseptic conditions of the cultures and avoidance of contamination during transfer of cultures. Steps 2 and 3 are performed in a laminar flow hood in a 'sterile transfer room' in the tissue culture laboratory.

11.3.1 General Aseptic Measures to be Followed

Tissue culture is done in absolutley sterile conditions. Contamination of microbes or other organisms in culture vessels or in the culture room is not allowed. Persons carrying out tissue culture should be very careful and follow certain aseptic measures for the purpose, considering themselves as a source of contamination.

(i)   A person working in a tissue culture laboratory should have a fair knowledge about culture and aseptic techniques.

(ii)   Only clean and essential items should be brought into the tissue culture laboratory.

(iii) Clean and uncontaminated lab coats and gloves should be worn.

(iv)   Hair should be clean and may be covered with a cap; a facemask may also be worn.

(v)   The cultures should regularly be checked and contaminated ones should be removed in a disposal container for further sterilization and disposal.

(vi)   The number of persons in a tissue culture laboratory should always be minimum.

(vii) Prior to start, the hands and arms up to the elbow should be thoroughly washed with regular soap and water; if necessary they must also be swabbed with 70 per cent alcohol.

(viii) Any plant and other material that is accidentally dropped on the working surface of the laminar hood or on the floor should be considered contaminated and discarded.

11.3.2 Tissue Culture Laboratory

Plan of the laboratory

A tissue culture laboratory should be an extremely clean facility, free from dust and microbial contamination and unnecessary chemicals, for carrying out various aseptic operations and for incubating growing tissues. A moderate-size tissue culture laboratory to conduct research activities is depicted in Fig. 11.1. The size of the laboratory mainly depends on the scale and kind of tissue culture. A plant tissue culture laboratory needs to have three distinct areas within the confines of three interconnected rooms for specific purposes—(1) media preparation room, (2) sterile transfer room, and (3) culture room for growth.

An explant preparation room may be adjacent to the media preparation room. Two small rooms, one for shaker cultures and incubators and another for photography and computer(s), may be assigned within the set-up of the tissue culture laboratory.

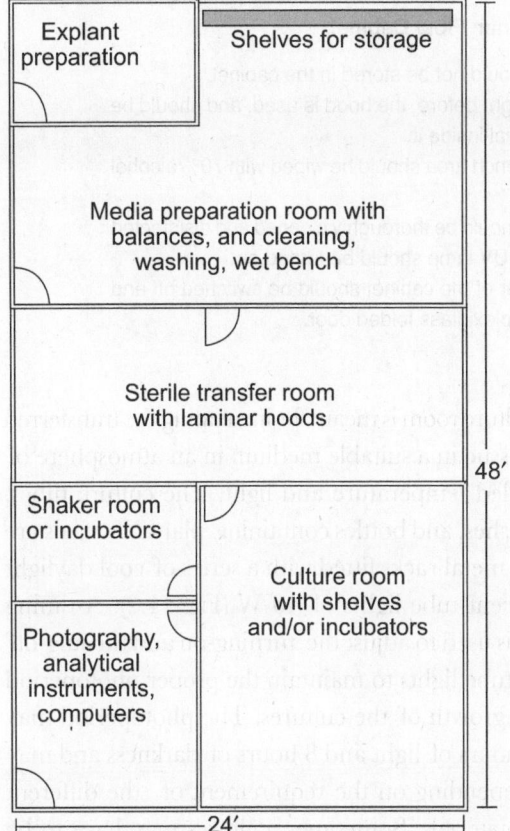

Fig. 11.1   Plan of a moderate sized plant tissue culture laboratory.

The floor, walls, and ceiling of the tissue culture facility need to be light coloured (to spot dirt easily) and easily washable with detergent and disinfectants from time to time. Standard synthetic materials for the walls and ceiling, and tiles for the floor are used nowadays. Air from the outside coming through air conditioners or windows, into the three major rooms needs to be filtered, preferably by HEPA (high-efficiency particulate air) filter to cut down the particulate matter.

Media preparation room

The room should be equipped with sophisticated balances, a centrifuge, pH meters, magnetic stirrer, deionized distilled water supply, vacuum, and compressed air lines to generate suction or pressure for filter sterilization and transfer of solutions. A pressure cooker or an autoclave for sterilization of media, a refrigerator, and a deep freezer for storing chemicals and media need to be in the room. There should be a clean sink with running hot and cold water and a stove or a burner.

The working bench for balances, pH meters, and other purposes may be of usual height with stool facilities; a table of suitable height to work while standing is also necessary. Media that have been prepared and ready to use can be kept on clean storing shelves.

Sterile transfer room

The transfer chambers or horizontal laminar airflow cabinets are placed along the wall, away from the doors, and foot-traffic. The cabinet is enclosed on all sides except the side occupied by a worker on a stool. A continuous horizontal filtered airflow over the working surface towards the worker prevents contamination from outside (see Chapter 12 for laminar hood). When the hood is not in operation, the front part is kept closed by a horizontally folding plexiglass door. The hood is fitted with a germicidal UV-light facility, besides the regular fluorescent tube light. The UV-light should be turned on before and after working, to keep the hood in sterile condition.

A stainless steel working surface in the hood is the most durable and easiest to keep clean. A usual laminar flow hood has 4 ft × 2 ft working surface area and is fitted with a gas line for a Bunsen burner. The processing and transfer of plant materials to culture bottles and transfer of media are performed in the laminar flow hood. Laminar flow hoods of different specifications are supplied by many manufacturers. The laminar flow cabinet should be serviced by factory personnel twice a year.

Aseptic Measures for Laminar Flow Cabinet

- Instruments, chemicals, or any other materials should not be stored in the cabinet.
- The UV tube light should be on, preferably overnight before the hood is used, and should be turned-off before working and placing plant material inside it.
- The interior of the cabinet including the working bench area should be wiped with 70% alcohol before the blower fan is switched on.
- At the completion of the work, the working bench should be thoroughly cleaned and disinfected with 70% alcohol or other suitable chemicals; the UV lamp should be turned on.
- When the room is not in use, the lights and blower of the cabinet should be switched off and the front of the cabinet should be closed with the plexiglass folded door.

Culture or incubation room

Fig. 11.2 Cell culture rack fitted with fluorescent tube lights. (Also see colour plate 3.)

The culture room is meant for incubating the transferred plant tissue in a suitable medium in an atmosphere of controlled temperature and light. The culture tubes, Petri dishes, and bottles containing plant materials are kept in metal racks fitted with a series of cool daylight fluorescent tube lights of 40 W (Fig. 11.2). A timing device is used to adjust the turning-on and turning-off of the tube lights to maintain the proper photoperiod for the growth of the cultures. The photoperiod may be 16 hours of light and 8 hours of darkness and may vary depending on the requirement of the different plant materials. Sometimes, cultures may have to be incubated in continuous darkness. The most desirable temperature for the culture room is around $25\pm2°C$, though it varies according to the requirement of plant species.

The temperature is maintained by air-coolers or air conditioners. Nowadays, an elaborate thermostat control maintains the exact temperature of the culture room by controlling heating and cooling equipment. The heat given off by the tube lights in the culture racks helps in warming up the culture room. A humidifier maintains the humidity of the room.

Interestingly, the culture or incubation room for plant materials occupies a large area in comparison to the animal cell cultures which are mostly done within the confines of average sized incubators with an inlet pipe for CO_2 and air, in a humidified atmosphere, usually at 37°C for mammalian tissue and cell culture (see Chapter 12).

Other necessary facilities

Certain other facilities are required to support the operations in a plant tissue culture laboratory. They have to be located outside the tissue culture laboratory.

Washing and drying room   The washing room must have provision for water, electricity, lines for gas and compressed air, and proper ventilation. There should be a huge washing sink in one corner of the room with various facilities (including a washing machine if possible), different sizes of brushes to be used for different culture wares, bottles, pipettes, and so on, and running hot and cold water.

An autoclave machine is to be used to liquefy the agar and kill any microorganisms in irregular as well as contaminated cultures that have been discarded.

The culture vessels should be emptied, rinsed, and soaked in a detergent bath overnight. The glassware should be scrubbed with a proper brush and rinsed thrice with tap water followed by three rinses in distilled water. New culture vessels also need to be washed thoroughly before they are used.

To drain and dry the different types of clean glassware, drying racks could be placed adjacent to the sink. When dry, the glassware should be stored in clean and closed cabinets.

The pipettes should be plugged at the drawing-in end with small non-absorbent cotton and placed in aluminum canisters for sterilization. Other glassware should be wrapped in aluminum foil for sterilization.

Sterilization room   Preferably two autoclaves, one big and one moderate, are required for sterilizing the glassware and media. The oven for drying or heat sterilization of the glassware is also maintained in this room. Since a large amount of heat is generated in the room, it should be well ventilated with a provision for exhaust fans.

Shaker room   The shaker room is a cubicle in the main tissue culture laboratory, that has rotary shakers with variable speed controls ranging from 80 to 220 rpm for cell culture. Fluorescent tube lights are fitted about 60 cm above the shaker to provide light of 2,000 lux intensity to the cultures. The platform of the shaker is fitted with rubber discs to hold flasks with capacities of 50 to 1,000 mL. The platform with the flasks rotates in an orbital circular motion.

Acclimatization room   The cultured plants are removed from the rooting media and transferred to pots in the acclimatization room. This room should have the provision for high-intensity light around 3,000 lux and a humidity of about 90 per cent.

11.3.3 Plant Tissue Culture Media

A defined nutrient medium for tissue culture usually contains inorganic salts, a carbon source, some vitamins, growth regulators (hormones), and some organic supplements.

A medium containing only 'chemically defined' compounds is referred to as a *synthetic medium*. The following are the various ingredients of media.

Inorganic salts

Relatively small numbers of inorganic salts are used for plant tissue culture media. The inorganic salts are categorized as *macronutrients* and *micronutrients*, depending on the amount required by plant cells.

Macronutrients   These provide the major ions needed by plants. Macronutrients generally include nitrate, ammonium (2–20 mM), 1–3 mM of calcium, sulphate, potassium, phosphorous, and magnesium. Sodium and chloride ions are sometimes included in macronutrient compositions, but they are not essential for most plants.

A given set of balanced macronutrient ion concentrations can be prepared from several different combinations of salts (shown as alternatives in Table 11.1). There is a large variety of plant cell culture media published in the literature, depending on the specific needs of cell types from different plant species.

As macronutrients are present in relatively large amounts, they have an appreciable effect on the osmotic potential of a medium, and in turn, can influence the growth and morphogenesis of the cells and tissues. The other components present in large amounts are sugars (for example, sucrose) that also affect the osmotic potential of a medium.

Micronutrients   These are compounds required by plants and their cells in very small quantities, but are essential for the biochemical synthesis of cellular elements and physiological processes. Essential micronutrients are manganese (Mn), zinc (Zn), boron (B), copper (Cu), cobalt (Co), and molybdenum (Mo). The amounts of all these metal salts in plant cell culture *basal media* are mentioned in Table 11.1, along with macronutrients and other supplements. Iron is often required in small quantities.

Physiological role of micronutrients   At least five of the metallic elements are necessary for chlorophyll synthesis and chloroplast function. Manganese is needed for the maintenance of the ultrastructure of chloroplasts and the photosynthetic process. Molybdenum and zinc deficiency lead to poorly developed chloroplasts and a reduction in their number.

Iron is required for the formation of amino laevulinic acid, protoporphyrinogen, and ferredoxin proteins, which play the role of electron carriers in photosynthesis. Iron is also a component of many proteins that participate in the regulation of oxidation or reduction reactions. Copper atoms occur in plastocyanin, a pigment participating in electron transfer. Micronutrients are required for the activity of DNA and RNA polymerase and other enzymes.

Sucrose or glucose as carbon source

Cultured plant tissues and cells are usually unable to synthesize their own carbohydrates through photosynthesis. Therefore, sucrose or glucose is added to the media as carbon source. Fructose or other forms of carbohydrates are not used readily except by some variant cell lines. As sucrose is added in large amounts, it affects the osmotic state of the media, which might influence the rate of cell division.

Vitamins

Although plants synthesize all vitamins, plant cells in culture require thiamine. Vitamins affect the growth and development of cells in culture.

Amino acids

Amino acids can serve as a readily available source of reduced nitrogen. For example, L-Glutamine can act as the sole source of nitrogen for cell growth and differentiation in vitro.

Table 11.1   Composition of media for plant tissue culture
(three alternative formulations)

Concentration (mg L^{-1}) in medium*			
Components	MS	B5	N6
Macronutrients:			
KNO$_3$	1900	2500	28030
NH$_4$NO$_3$	1650	–	–
MgSO$_4$.7H$_2$O	370	250	185
KH$_2$PO$_4$	170	–	400
NaH$_2$PO$_4$.H$_2$O	–	150	–
CaCl$_2$,2H$_2$O	440	150	166
Ca(NO$_3$)$_2$.4H$_2$O	–	–	–
(NH$_4$)$_2$.SO$_4$	–	134	463
KCl	–	–	–
Na$_2$SO$_4$	–	–	–
Micronutrients:			
H$_3$BO$_3$	6.2	3	1.6
MnSO$_4$.H$_2$O	16.9	10	3.3
ZnSO$_4$.7H$_2$O	8.6	2	1.5
Na$_2$MoO$_4$.2H$_2$O	0.25	0.25	0.25
CuSO$_4$.5H$_2$O	0.025	0.025	0.025
CoCl$_2$.6H$_2$O	0.025	0.025	–
KI	0.83	0.75	0.8
FeSO$_4$.7H$_2$O	27.8	–	27.8
Na$_2$.EDTA	37.3	–	37.3
NaFe.EDTA salt	–	40	–
Vitamins and organics:			
Thiamine HCl	0.5	10	1
Pyridoxine HCl	0.5	1	0.5
Nicotinic acid	0.5	1	0.5
Myo-inositol	100	100	–
Glycine	–	–	40
Ca Panthothenate			
Cysteine HCl	–	–	–
Sucrose	30×10^3	20×10^3	50×10^3
pH	5.8	5.5	5.8

Growth regulators and other additives are added to the basal media as required.
*Sources: MS = Murashige and Skoog
B5 = Gamborg *et al.*
N6 = Chu

Growth regulators

Growth regulators are endogenous compounds which regulate the growth and development of plants and their cells at extremely low concentrations. There are five classes of growth regulators—auxins, cytokinins, gibberellins, ethylene, and abscisic acid (Fig. 11.3). They are selectively added in low quantities to the culture media. The type and concentration of growth regulators to be added to the media depend on the cell type and purpose of the culture.

Auxins and cytokinins are the most important regulators for growth and morphogenesis (steps for differentiation of organs) in plant tissue and organ cultures. Auxins are necessary for cell division in cultured tissues and often for the induction of root formation. Cytokinins are adenine derivatives and are essential for the differentiation of plant organs in most species. Auxins and cytokinins are used in different combinations with varying quantities. Gibberellins are effective in plant regeneration from meristem and after shoot primordia formation starts. Although specific functions of abscisic acid in tissue culture have yet to be defined, it is needed in in-vitro somatic embryogenesis. Chemical alternatives to the natural gibberellins and abscisic acid are not yet available. Gibberellins extracted from cultured fungi are sometimes used in plant tissue culture.

Certain organic components

For the purpose of buffering, when pH stability of the medium is necessary, organic acids and salts of organic acids are added to the medium. Acids of tricarboxylic acid cycle intermediates, such as citrate or malate or fumarate are used in media for protoplast cultures.

Undefined Supplements

Sometimes raw extracts from plant sources are used in media. For example, juices of various fruits, coconut milk, plant sap, corn endosperm (corn milk), cornstarch, potato extract, yeast extract, and so on are used as additives in different cell cultures. Raw banana homogenate is also used as a supplement in media for orchid culture.

Activated charcoal is used in rooting media to adsorb root-inhibiting factors. Usually, it is used at 0.6 g L^{-1} concentration.

11.3.4 Media Preparation

Readers can consult Section 11.1 for early development of plant tissue culture. Earlier researchers used to weigh the ingredients for culture media and then dissolve them in a specified quantity of sterile distilled water.

Nowadays, commercially available dry powdered media, containing inorganic salts, vitamins, and amino acids are dissolved in a specified volume of deionized double distilled water. The powder is first dissolved in distilled water in a volume 10 per cent less than the final volume; then sugar, agar, and other desired supplements are added, and the final volume is made up with distilled water. Powder media are used for routine purposes of micropropagation and tissue culture. For experimental works and specific requirements of some cultures, suitable powder media are not always available and the media need to be prepared by mixing the requisite ingredients.

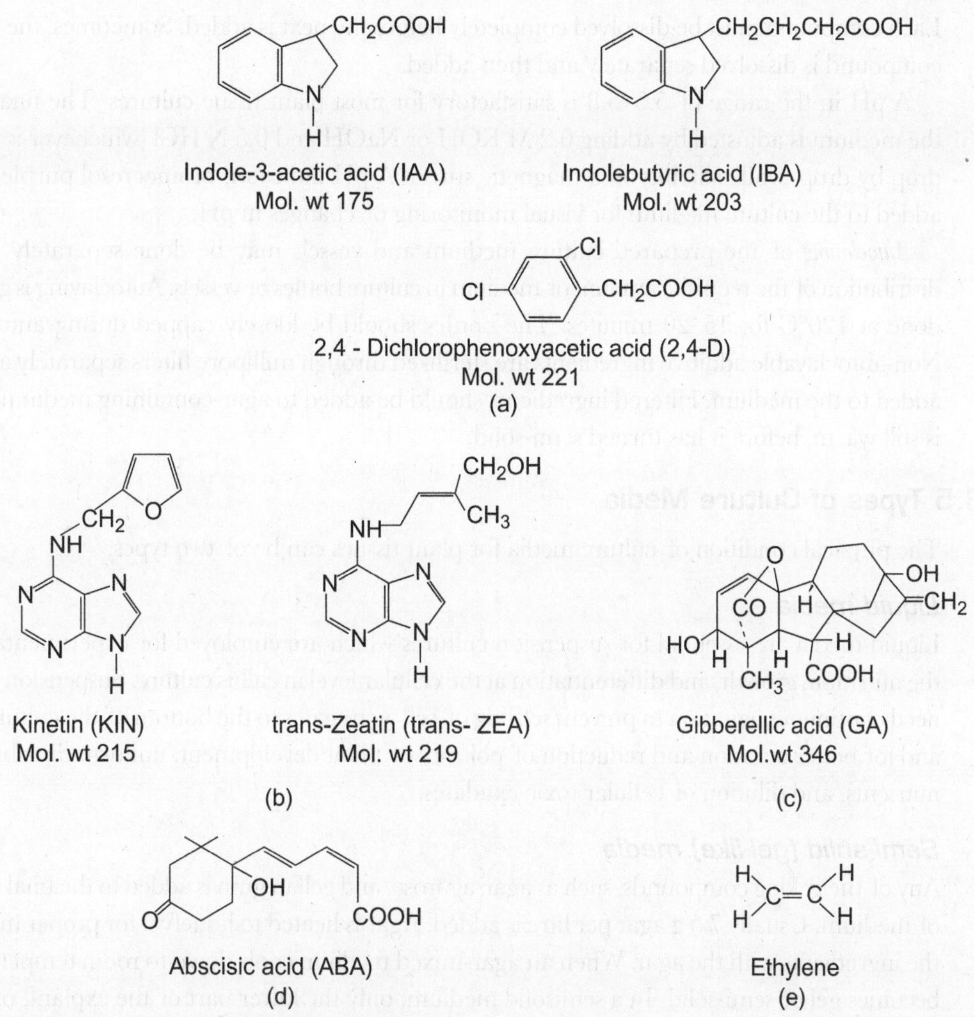

Fig. 11.3 Five classes of plant growth regulators (hormones) and their chemical structure. (a) Auxins, (b) Cytokinins, (c) Gibberellic acid, (d) Abscisic acid, and (e) Ethylene.

This can be done in two ways: (1) the required quantities of the ingredients are weighed separately and mixed to make the medium. (2) 10 or 100 times concentrated stock solutions of ingredients are made separately and required amounts are mixed to prepare the final culture solution. This second procedure is more convenient and popular as it eliminates the need to weigh so many different chemicals for each batch of medium, and the inaccuracies that could arise while weighing small quantities. The stock solutions can be stored for a while in the refrigerator.

The salts must be dissolved one at a time. The problem of precipitation in the medium (due to complexes formed between calcium, phosphate, and magnesium) can be avoided if the compounds are added in the following order—inorganic nitrogen compounds first, followed by the magnesium compound and the calcium compound, and finally the phosphate compound.

Each compound must be dissolved completely before the next is added. Sometimes, the calcium compound is dissolved separately and then added.

A pH in the range of 5.5–5.8 is satisfactory for most plant tissue cultures. The final pH of the medium is adjusted by adding 0.2 M KOH or NaOH and 0.5 N HCl (whichever is needed) drop by drop, while stirring on a magnetic stirrer. A pH indicator, bromocresol purple, can be added to the culture medium for visual monitoring of changes in pH.

Autoclaving of the prepared culture medium and vessels may be done separately or after distribution of the requisite amount of medium in culture bottles or vessels. Autoclaving is generally done at 120°C for 15–20 minutes. The bottles should be loosely capped during autoclaving. Non-autoclavable additive ingredients are sterilized through millipore filters separately and then added to the medium. Filtered ingredients should be added to agar-containing medium while it is still warm, before it has turned semi-solid.

11.3.5 Types of Culture Media

The physical condition of culture media for plant tissues can be of two types:

Liquid media

Liquid media are essential for suspension cultures which are employed for experimentation on the nutrition, growth, and differentiation at the cellular level in callus culture. Suspension cultures need continuous agitation to prevent settling of cell aggregates to the bottom of the culture flasks, and for better aeration and reduction of polarity of plant development, uniform distribution of nutrients, and dilution of cellular toxic exudates.

Semi-solid (gel-like) media

Any of the gelling compounds, such as agar, agarose, and gellan gum is added to the final solution of medium. Usually 7.5 g agar per litre is added. Agar is heated to liquefy it for proper mixing of the ingredients with the agar. When an agar-mixed medium cools down to room temperature, it becomes gel or semi-solid. In a semisolid medium, only the lower part of the explant, organ, or tissue remains in contact with the medium.

The date of preparation of the media should be indicated on the bottles or Petri dishes containing them.

11.4 EXPLANT PREPARATION AND TISSUE CULTURE INITIATION

An *explant is a piece of a living plant that is used to initiate tissue culture.* Explants can be from a wide range of plant organs and tissues (Fig. 11.4). *They can be almost any part of a plant*—shoot tips, meristems, stem pieces (may be macerated or teased), nodes, buds, different parts of a flower (peduncle or flower stalk, petals, anthers, nucellus, or the central part of an ovule), seeds, embryos, seedlings, leaf or petiole, bulblets, bulb scales, radicle, stolons, rhizome tips, and root pieces.

Explants may range in size from one-tenth of a millimetre to several centimetres.

11.4.1 Collection of Explants

It is desirable to use explants that are not contaminated with microorganisms. Plants grown in natural environments are usually contaminated with microorganisms and pests, which are normally confined to the outer surface of the plants; sometimes microbial contamination may exist within the tissue system. Meristems are usually virus free as they are devoid of vascular tissue that transports viruses and are in the active state of mitosis.

Well in advance, before collecting explants, stock plants in pots need to be moved to a clean greenhouse for special care. The plants are initially washed with clean water, the foliages are then allowed to dry, and the plants are watered regularly at the base. The new juvenile shoots are ideal as a clean source of explants. Juvenility of the explant tissue increases the success rate of the cultures.

Usually, explants are cut bigger or longer than the final size required to facilitate processing. After proper cleaning, parts of the leaves can be used for tissue culture; discs from the leaves may be drilled with the help of a cork borer. Leaf tissue is a usual source of protoplasts.

11.4.2 Disinfection of Explants

The commonly used bleach or disinfectants for explants are 5 per cent sodium hypochlorite, 70 per cent ethyl alcohol or isopropyl alcohol, 5–10 per cent calcium hypochlorite, 3 per cent hydrogen peroxide, and 0.1–0.2 per cent mercuric chloride.
The general procedure of disinfection is as follows:
1. 100 mL of 1 per cent bleach (sodium hypochlorite) and two drops of Tween 20 are poured into a sterile 150 mL beaker or a conical flask with a magnetic stir bar (Fig. 11.5).
2. Explants are placed in a beaker or conical flask and stirred on a magnetic stirrer for 10 minutes. Alternatively, the explants and cleaning solution can be placed in a tightly closed flask and shaken by hand.
3. After pouring off the bleaching solution, the explants are placed in a clean beaker and rinsed with 70 per cent ethyl or isopropyl alcohol (or 3 per cent H_2O_2) for 5 seconds. The rinsing solution is then poured out and the explants are again placed in a clean sterile beaker.
4. The explants are rinsed in distilled water and then placed in a clean 150 mL beaker with a magnetic stir bar.
5. 100 mL of 10 per cent bleach and two drops of Tween 20 are added to the beaker containing the explant and stirred for 15 minutes.
6. The beaker with the bleach solution containing the explants is moved to the transfer hood.

11.4.3 Transfer of Explants into the Growth Medium

The last step of processing and transfer of explants into the growth medium are performed in a laminar flow hood under extremely sterile conditions, following the procedure outlined below.

Procedure of explant transfer
1. Ten minutes before transfer, the blower fan of the laminar hood should be turned on.

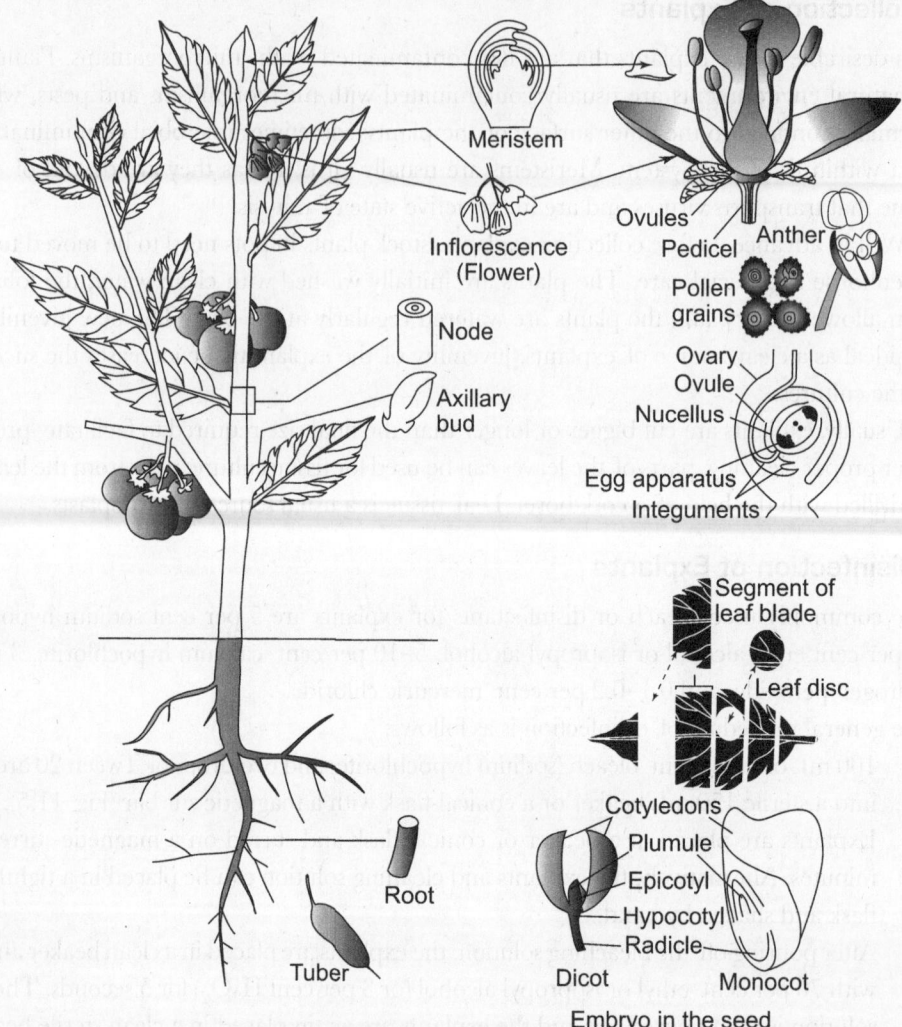

Fig. 11.4 Explant source from different parts of a plant.

2. The inside of the hood is wiped with cotton dipped in 10 per cent bleach (sodium hypochlorite) or 70 per cent alcohol.

3. After gloves are worn they are rinsed in 10 per cent bleach. The tools should be sterilized in 10 per cent bleach or 70 per cent alcohol.

4. The explants in the beaker should be taken out with sterile forceps and rinsed in 1 per cent bleach, followed by two rinses in sterile distilled water. The explants should be kept in sterile water during the next two steps.

5. A sterile paper towel is spread using forceps and a knife on the working area of the hood as far back as possible.

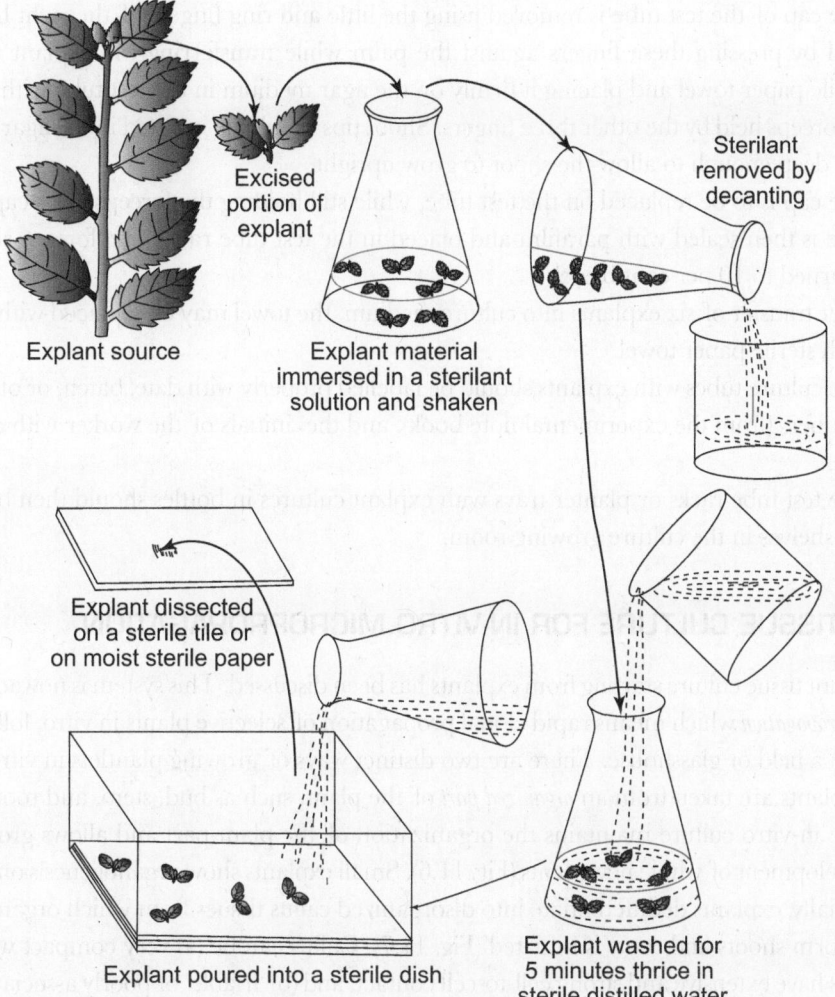

Fig. 11.5   The explants in disinfectant solution on a magnetic stirrer and washed.

6.   The forceps and the knife should be returned to 10 per cent bleaching solution for 1 minute, and then rinsed in 1per cent bleach.

7.   An explant is taken out of sterile water and placed onto the paper towel with forceps.

8.   The explant is trimmed to the appropriate size—1 mm to a few cm for shoot tips, 2–3 cm for stems, and just a few mm for petioles, using forceps and knife.

9.   The forceps need to be rinsed in 1 per cent bleach for 1 minute (or a fresh pair of sterilized forceps can be used) for final transfer of the explants into the test tubes or bottles containing the culture medium. The forceps should be held by the thumb and index fingers of the right hand.

10.   The test tube should be held at the lower half in the left hand at an angle of about 50°, and parallel with the frontline of the hood (neither facing the air blower at the back, nor the worker), to avoid contamination.

11. The cap of the test tube is removed using the little and ring fingers of the right hand and held by pressing these fingers against the palm while transferring the explant from the sterile paper towel and placing it firmly on the agar medium in the test tube with the help of forceps held by the other three fingers. Shoot tips should be inserted in the agar medium just deep enough to allow the shoot to grow upright.

12. The cap is to be replaced on the test tube, while still holding the forceps. The capped test tube is then sealed with parafilm and placed in the test tube rack. The forceps should be returned to 10 per cent bleach.

13. After transfer of six explants into culture medium, the towel may be replaced with another fresh sterile paper towel.

14. The culture tubes with explants should be labelled properly with date, batch, or other code numbers (from the experimental note book), and the initials of the worker with a marker pen.

15. The test tube racks or planter trays with explant cultures in bottles should then be put on the shelves in the culture growing room.

11.5 PLANT TISSUE CULTURE FOR IN-VITRO MICROPROPAGATION

So far, plant tissue culture starting from explants has been discussed. This system is nowadays used for *micropropagation* which means rapid clonal propagation of selective plants in vitro, followed by growth in a field or glasshouse. There are two distinct ways of growing plantlets in vitro:

(i) Explants are taken from an *organized part* of the plant, such as bud, stem, and root cutting. The in-vitro culture maintains the organization of the plant part and allows growth and development of whole new plants (Fig. 11.6). Small explants show organogenesis only rarely.

(ii) Initially, explants dedifferentiate into disorganized callus tissues from which organogenesis to form shoots and roots is initiated (Fig. 11.7). Callus can be (a) very compact where the cells have extensive and strong cell-to-cell contact, and (b) 'friable' or poorly associated small disintegrating aggregates with fast growing potential. The latter is best suited for initiation of cell suspension culture (dissociated cells in liquid culture medium).

All these are possible due to the totipotency of plant cells.

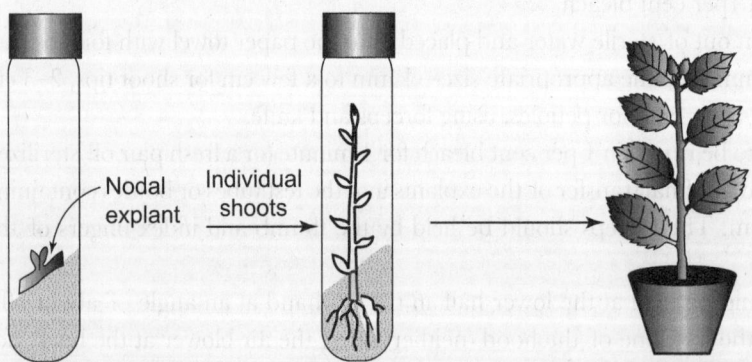

Fig. 11.6 Micropropagation from axillary bud, an organized part of a plant.

Nodal explant   Individual shoots

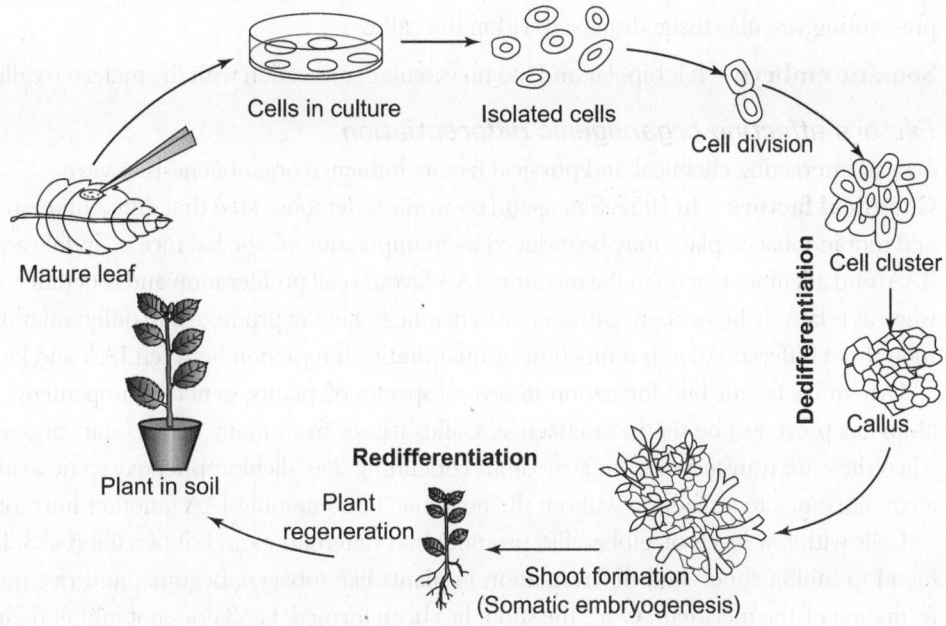

Fig. 11.7(a) Micropropagation through callus formation.

Fig. 11.7(b) Callus culture (Also see colour plate 3.)

Totipotency

Totipotency is the inherent capacity of a plant cell to give rise to all the different types of cells and tissue and ultimately a whole plant.

Totipotency is usually retained by the cells even after their differentiation to a particular tissue in a whole plant. However, sieve tube and xylem elements, whose nuclei have disintegrated or become fibres, lose their totipotency.

11.5.1 Morphogenesis and Organogenic Differentiation

Morphogenesis literally means creation of new forms. Callus tissue can generate new organs like shoot (caulogenesis) and root (rhizogenesis; organogenic differentiation) through the process of morphogenesis. Newly formed organs are called *adventive* or adventitious.

Shoot bud and somatic embryogenesis

Whole plant regeneration from cultured cells or callus usually occurs through shoot bud differentiation or somatic embryogenesis. A shoot bud and a somatic embryo can be morphologically distinguished (see Fig. 11.8).

Shoot bud It is monopolar and develops procambial strands that establish connection with pre-existing vascular tissue dispersed within the callus.

Somatic embryo It is bipolar and has no vascular connection with the maternal callus tissue.

Factors affecting organogenic differentiation

Several interacting chemical and physical factors influence organogenesis in vitro.

Chemical factors In 1957, Skoog and co-workers demonstrated that differentiation of shoot and root in tobacco plant may be induced by manipulation of the balance of indole acetic acid (IAA) and adenine/kinetin in the medium. IAA favours cell proliferation and root differentiation, whereas relatively higher concentrations of adenine or kinetin promote bud differentiation. Thus, root–shoot differentiation is a function of quantitative interaction between IAA and kinetin.

Cytokinins favour bud formation in several species of plants; generally isopentenyl adenine (2-ip) has proved to be the most effective. Callus tissues from many cereals start organogenesis when they are transferred from a medium containing 2,4- dichlorophenoxyacetic acid (2,4-D) auxin hormone to a medium without the hormone or containing IAA (another hormone).

Cells with low levels of gibberellin promote bud differentiation. Gibberellin (GA3) has been found to inhibit shoot–bud differentiation in plants like tobacco, begonia, and rice during the formation of the meristem. Once the shoot has been formed, GA3 does not inhibit their further development.

The variability in chemical control of organ formation across the species does not allow proposing a universal formula.

Physical factors Consistency of the media has striking effects on organogenesis in culture. In some cases, liquid medium allows formation of leafy shoot buds when compared to solid medium

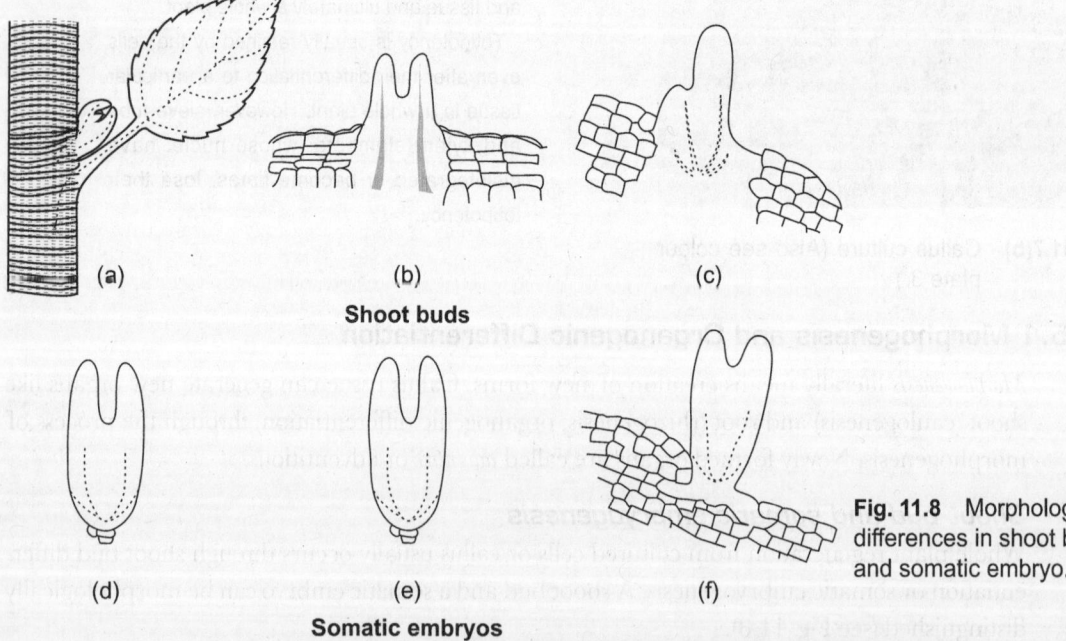

(a)   (b)   (c)

Shoot buds

(d)   (e)   (f)

Somatic embryos

Fig. 11.8 Morphological differences in shoot buds and somatic embryo.

of identical composition. Agar concentrations influence organogenesis in tissues of tobacco. Flowers are formed only in 1 per cent agar; lower concentration of agar favours vegetative bud formation.

Osmolarity of the medium and light intensity are some important factors that influence bud formation. High-intensity light inhibits formation of shoot buds. The quality of light also affects organogenesis. Blue light promotes shoot formation, and red light stimulates root formation in tobacco. The growth and differentiation of the callus increase with rise in temperature upto 33°C.

11.5.2 Subculture

A subculture is made when the density of cells, tissues, or organs become excessive for a limited amount of medium in a culture and it is necessary to enlarge the culture volume or increase the number of organs for micropropagation. Subculturing is applicable for callus culture, suspension cell cultures, and cultures of indeterminate organs. The time period from the initiation of a culture to transfer for subculture is often called *passage*. Suspension cell cultures are regularly subcultured at the end of the period of exponential growth for many passages. Another important reason for subculture is to reduce the accumulation of toxic metabolites arising from the growing tissue, in the medium.

11.5.3 Stages of Micropropagation

In literature, stages of micropropagation are often marked as O–IV stages.

Stage O: Selection and preparation of explant for in-vitro culture

Stage I: Establishment of aseptic culture

Stage II: Production of new plant outgrowths (adventitious shoots)

Stage III: Shoot elongation, root formation, and preparation for growth in the natural environment

Stage IV: Transfer to the glasshouse and field

Details of the processes up to Stage II have already been discussed. Certain discussions about Stage III and IV are to follow.

11.5.4 Hardening Off for Greenhouse or Field Transfer

The tissue culture atmosphere contains (a) high humidity, (b) comparatively lesser light than the atmosphere outside, and (c) the plantlets draw sucrose in the medium for energy instead of using CO_2 in the process of photosynthesis. Leaves that develop in tissue culture usually have abnormally high number of stomata per unit surface area leading to more water loss. Epicuticular wax on leaves and stems of the cultured plantlets is thin and cannot provide transpiration protection, as is the case in plants grown naturally.

Thus, acclimatization of the plantlets grown in vitro for transfer to the greenhouse on the way to the field is a must; otherwise the micropropagation project may suffer high losses. Usually, acclimatization does not modify existing organs of the plantlets, rather it induces changes in new growths that are produced. Acclimatization is gradual exposure to normal conditions of nature that leads to progressive morphological and physiological adaptation, that is, hardening off.

Steps that are taken for hardening off are as follows:

Lowering of humidity   Covers or caps of tissue culture containers are loosened for escape of water vapour (which causes high humidity).

The same result can be achieved in some cases by hanging small bags of desiccant (silica gel) in the containers.

Another effective measure is to use a lid with a filter, which allows escape of water vapour, some desirable air exchange, and minimizes entry of contaminants in the culture vessel.

Induction of rooting   Decrease in salt levels, especially nitrates, often helps to induce rooting. Growth regulators, such as hormones and cytokinins, are effective in the induction of rooting. Isopentenyl adenine (2-iP) or kinetin instead of 6-benzyladenine (BA) are better inducers of rooting in stage II; all these are cytokinins. In stage III, cytokinins are usually eliminated and the auxin level is raised. The level of auxin is critical; too much of it is worse than none at all. Phloroglucinol has been found to be beneficial for rooting in some fruit plants.

Increase in light intensity   The acclimatization room has a provision for higher intensity of light than the culture room. Increased light sometimes favours rooting; however, some bulbous and rosaceous plantlets form root initials, more readily in darkness. Higher intensity of light induces formation of new chlorophylls, and better photosynthesis by the newly formed leaves.

Lowering the sucrose level in the media   Lowering the level of sucrose in culture media encourages the utilization of CO_2 in plantlets and thus, in photosynthesis.

Final procedure of plantlet transfer from the culture medium to soil in the greenhouse

The uncapped culture vessels taken out of the acclimatization room are placed in the greenhouse in a highly humid and shaded condition for a few days. Then the agar from the roots is washed away gently as the agar is likely to serve as a substrate for growth of disease-causing organisms. Usually, the plantlets with rootings from stage II or stage III are used for planting.

The tiny plantlets are planted in special soil bed on planter trays. The trays are placed in highly humid condition in tents on benches in the shaded greenhouse. Humidifiers are less expensive and do well in smaller areas. For bigger areas, an automatic fog system is ideal. When a sophisticated method is not available, glass beakers may be used to cover the pots to maintain conditions of high humidity initially.

Conditions of high humidity and low-intensity light are maintained over a 2–4 weeks period. A progressive reduction in humidity and an increase in light intensity is achieved over the following few weeks. The sides of the tent are gradually opened and the amount of mist is gradually reduced. The gradual processes allow the existing leaves to adjust and assist the new leaves to grow and adjust to the natural environment.

The whole procedure of micropropagation (in-vitro propagation), so far discussed, has been schematically summarized in Fig. 11.9.

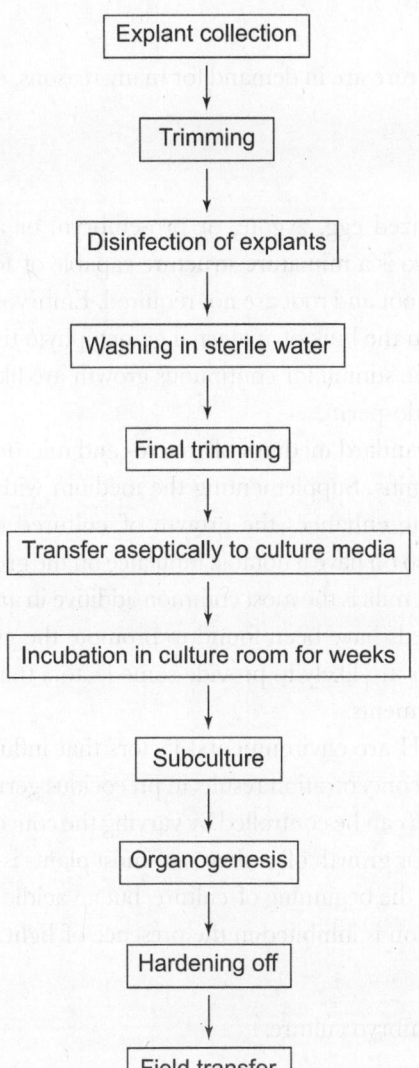

Explant collection

↓

Trimming

↓

Disinfection of explants

↓

Washing in sterile water

↓

Final trimming

↓

Transfer aseptically to culture media

↓

Incubation in culture room for weeks

↓

Subculture

↓

Organogenesis

↓

Hardening off

↓

Field transfer

Fig. 11.9(b) Plant differentiation in in-vitro culture. (Also see colour plate 3.)

Fig. 11.9(a) Schematic summary of the procedure of micropropagation.

11.6 REGENERATION OF PLANTS USING VARIOUS PLANT MATERIALS THROUGH TISSUE CULTURE

Using in-vitro methodologies, numerous types of plant materials can be used as explants for micropropagation, starting with seeds, embryos from seeds, the usual meristems, any vegetative tissues, callus, and cell suspension somatic embryos. An account of this aspect is given below and a discussion on protoplast culture will follow.

11.6.1 Seed Culture

Sometimes seedlings produced from seeds in culture are in demand for many reasons, especially for exotic plants like orchids.

11.6.2 Embryo Culture

In-vitro culture of either an immature (polarized egg, zygote, or pro-embryo) or a mature zygotic embryo is relatively simple as the embryo is a miniature structure capable of forming a complete plant and de novo differentiation of shoot and root are not required. Embryos nearing maturation are separated with relative ease from the bulk of maternal gametophyte tissue, and they are completely autotrophic. In embryos, the stimuli for continuous growth are likely to be present within their cells or in the associated endosperm.

Usually, mature embryos are cultured on a standard medium of macro- and micronutrients, containing sugar (sucrose, is the best) and vitamins. Supplementing the medium with organic nitrogen like amino acids or casein hydrolysate enhances the growth of cultured embryos. Phytohormones such as IAA, NAA, 2,4-D, and so on have a notable influence on the growth and development of the embryos in culture. Coconut milk is the most common additive in an embryo culture medium. Alcohol-diffusates of some seeds have been found to promote the growth of embryo. Coconut milk and diffusates from seeds are likely to provide some factors that are not available from standard media and usual supplements.

Osmotic pressure, light, temperature, and pH are environmental factors that influence the growth of embryos in vitro. Often, low osmotic concentration results in precocious germination of embryos. The osmotic pressure in the medium can be controlled by varying the concentration of salts and sugars. The optimum temperature for growth of embryos of most plants is between 27°C and 30°C. A pH near neutral is optimal at the beginning of culture, but an acidic medium is suitable as the embryos mature. Root elongation is inhibited in the presence of light.

Purposes of embryo culture

There are numerous practical applications of embryo culture.

Rescue of embryo from premature death to obtain rare hybrids   Fertilization in many interspecies and intergenic crosses proceeds normally, but abnormal development of endosperm (due to incompatibility) leads to premature death (abortion) of hybrid embryos. Such embryos when rescued before abortion and cultured in vitro provide the rare opportunity of growing adult plants from intergenic crosses. This would not be possible by raising seeds. This method allows crosses of various agronomic plants with their wild relatives, for example, wheat with barley, oats with wild oats, pearl millet with elephant grass, peanuts with wild *Arachis*, and so on.

Speeding up development of rare plants with slow-developing embryo   Seedlings can be easily obtained by culturing the excised embryos, much faster than from slowly developing seeds.

To overcome seed dormancy   Normal germination in many species is delayed due to the long dormancy periods of their seeds. Seed dormancy is usually caused by chemical inhibitors or mechanical resistance in the structures covering the embryo. Excision of embryos from the testa and then culturing them on media help overcome seed dormancy.

Study of nutritional requirement and morphogenesis of the developing embryos
The in-vitro study of embryonic development provides knowledge about the developmental requirements of the embryo, which in turn will help in agronomical management.

Embryo culture also helps in certain other aspects, including maintenance of some plants from which mature seeds are not available for various reasons and in which diseases set in before their maturation.

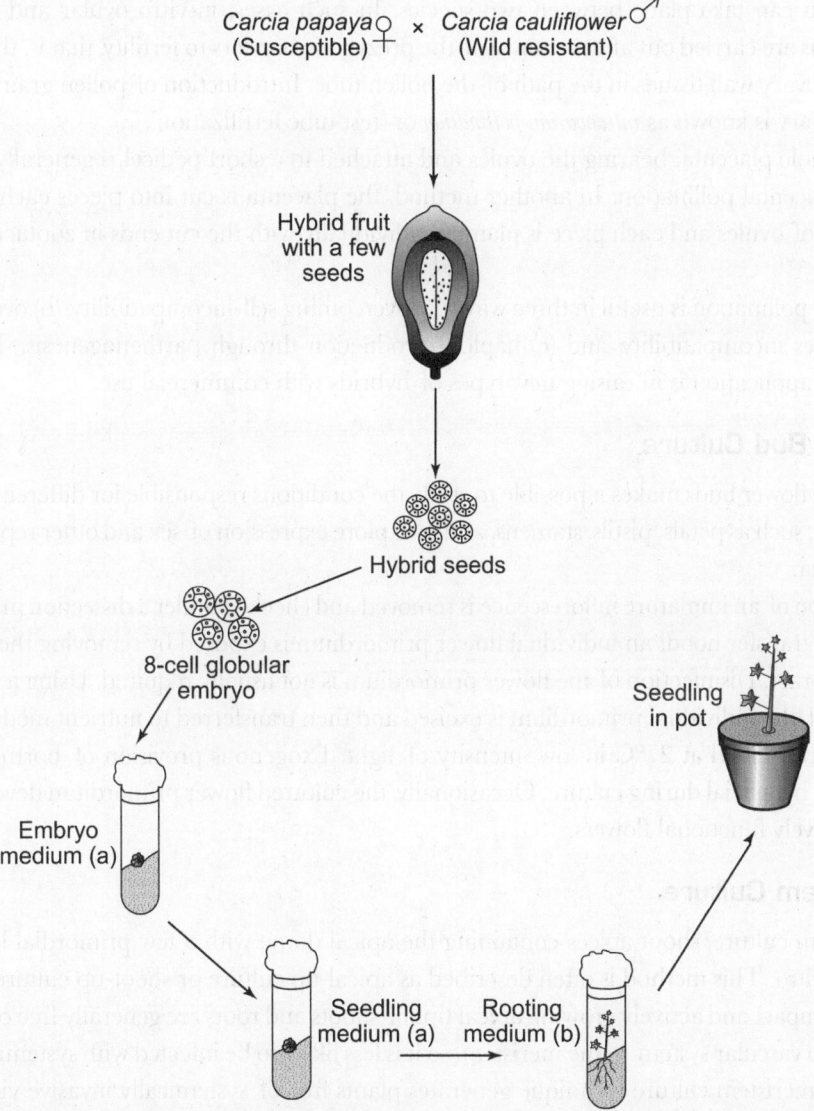

Fig. 11.10 Embryo culture of hybrid papaya.

The cultivable papaya (*Carcia papaya*) is susceptible to the papaya leaf-mosaic virus and the ring-spot virus diseases. Using the embryo culture technique (Fig. 11.10), interspecies hybrids are obtained from immature hybrid embryos of *C. papaya* (susceptible) crossed with *C. cauliflower* (a wild resistant species).

11.6.3 Ovary and Ovule Culture

The main objective of embryo culture as discussed is to rescue the embryo from premature death due to incompatibility of the endosperm in interspecies crosses. Similarly, in hybridization programmes, when the incompatibility lies in the stigma, style, or ovarian tissue, no successful fertilization can take place between two species. In such cases, in-vitro ovular and placental pollinations are carried out after removal of the prezygotic barriers to fertility, that is, the stigma, style, and ovary wall tissues in the path of the pollen tube. Introduction of pollen grains directly into the ovary is known as *intraovarian pollination* or 'test tube fertilization'.

The whole placenta, bearing the ovules and attached to a short pedicel is generally used for in-vitro placental pollination. In another method, the placenta is cut into pieces each carrying a number of ovules and each piece is planted individually with the cut ends in contact with the medium.

In-vitro pollination is useful in three ways: (a) overcoming self-incompatibility, (b) overcoming inter-species incompatibility, and (c) haploid production through parthenogenesis. The most important application is in raising new types of hybrids with commercial use.

11.6.4 Flower Bud Culture

Culture of flower buds makes it possible to study the conditions responsible for differentiation of floral parts, such as petals, pistils, stamens, and to explore expression of sex and other reproductive phenomena.

A portion of an immature inflorescence is removed and checked under a dissection microscope in a sterile transfer hood; an individual flower primordium is exposed by removing the cover of the floral bract. Disinfection of the flower primordium is not usually required. Using a surgeon's fine scalpel, the individual primordium is excised and then transferred to nutrient medium. The culture is continued at 27°C in low intensity of light. Exogenous provision of hormones and kinetin are beneficial during culture. Occasionally, the cultured flower primordium develops into reproductively functional flowers.

11.6.5 Meristem Culture

In meristem culture, shoot apices containing the apical dome with a few primordial leaves are grown in vitro. This method is often described as apical-tip culture or shoot-tip culture.

The compact and actively growing apical tips of shoots and roots are generally free of viruses. There is no vascular system in the meristem, so it is less likely to be infected with systemic viruses. Thus, the meristem culture technique generates plants free of systemically invasive viruses, for various crops, especially vegetatively propagated crops such as potatoes, pineapple, strawberry, sugarcane, dahlias, chrysanthemum, carnation, and so on.

Furthermore, the size of the initial explant is a critical factor for success of meristem culture. The smaller the size (below 0.1 mm × 0.25 mm), the greater the chance of virus elimination, but the lower the chance of its survival.

For complete elimination of viruses, the parent plants (from which meristem explants are collected) are grown in a controlled temperature cabinet at 30–40°C for 6–12 weeks till the new shoots develop. Alternatively, meristems can also be cultured at 30–40°C. Sometimes lower temperature treatment at 5°C can block the synthesis of viral proteins leading to degradation of virus particles. The treatment of the explant with antiviral chemicals prior to culture has also been found to be effective.

11.7 SUSPENSION CULTURE

Suspension culture is initiated by transferring the *callus*, a mass of undifferentiated cells, to a liquid medium in a conical flask and gently agitating the culture on an orbital shaker. A loosely organized *friable callus* is a better option for starting suspension cultures (Fig. 11.11). The swirling motion of the shaker keeps the cells in suspension and oxygenates the medium; the movement eliminates any polarity of the cells due to gravity, and nutrient gradients in the medium or at the surface of the cells. The orbital shaker rotates in a horizontal plane at a selected speed from 50 to 500 rpm in a room or chamber with a stable environment. There should be provision for its

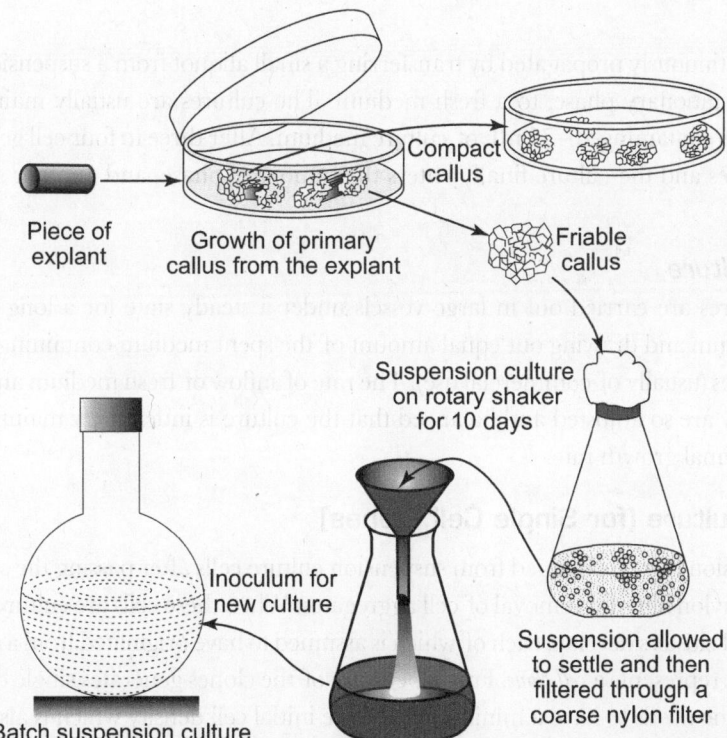

Piece of explant

Growth of primary callus from the explant

Compact callus

Friable callus

Suspension culture on rotary shaker for 10 days

Inoculum for new culture

Batch suspension culture

Suspension allowed to settle and then filtered through a coarse nylon filter

Fig. 11.11 Steps for initiation of suspension culture.

functioning 24 hours a day and around the year, if necessary. Suspension cultures are useful in harvesting *secondary metabolites* as plant cell products in plant biotechnology.

Ideally, the plant cell suspension culture should consist of single cells; however, some amount of cell aggregation takes place or cell clumps are formed and that has been found to be beneficial for the enhanced yield of secondary metabolites, and for retaining the totipotency of the cells. Cell suspension culture provides an excellent starting material for isolation of protoplasts for fusion, embryogenesis, genetic manipulation, and so on.

The growth curve of a cell suspension culture has a characteristic shape consisting of a lag phase, an exponential phase with rapid increase in fresh weight, dry weight and DNA content, followed by a stationary phase, and then a decline in growth rate (refer Fig. 12.5). During the stationary phase, small aliquots of cell suspensions are subcultured in new flasks containing fresh medium, and cell growth is revived.

To make synchronously growing cell suspension cultures, generally cell cycle inhibitors are used to block cell division at G1/S or G2/M phase. On removing the inhibitors, cells enter the next phase of the division cycle synchronously. Tobacco BY2 cells are the most commonly used synchronous cell cultures for various cell biology experiments.

11.7.1 Types of Suspension Cultures

There are two types of suspension cultures—batch culture and continuous culture.

Batch culture

The culture is continuously propagated by transferring a small aliquot from a suspension culture that has reached stationary phase, to a fresh medium. The cultures are usually maintained in 100–250 mL flasks containing 20–75 mL of culture medium. After three to four cell generations, the growth declines and the culture finally enters the stationary phase, and another subculture is initiated.

Continuous culture

Continuous cultures are carried out in large vessels under a steady state for a long period by adding fresh medium and drawing out equal amount of the spent medium containing cells and cellular metabolites (usually of commercial use). The rate of inflow of fresh medium and culture-harvest by outflow are so adjusted and balanced that the culture is indefinitely maintained at a constant, submaximal growth rate.

11.7.2 Single Cell Culture (for Single Cell Clones)

Single cell suspension can be obtained from suspension culture cells after passing the suspension through a coarse nylon filter for removal of cell aggregates. When these cells are cultured on agar medium, small cell clusters develop, each of which is assumed to have originated from a single cell. Thus, each cluster represents a *cell clone*. Further culture of the clones generates single cell clones.

Each initial clone needs to have a minimum effective initial cell density which is also referred to as the minimum inoculation density (10–15 cells/mL), below which successful onward culture is not possible. The minimum inoculation density can be lowered when a filtered medium from a previous cell culture (conditioned medium) is added to a standard medium, or special organic supplements and growth regulators are used.

Provision of a special environment

The plated single cells often do not enter into spontaneous cell division due to the suboptimal cell density around them. Cell division in these cells can be initiated by allowing them to be *nursed* by tissue growing nearby. There are several variations of the single cell culture technique as presented in Fig. 11.12.

Filter paper raft-nurse tissue culture   Inoculum of single cells on a sterile filter paper raft (8 × 8 mm) is placed in contact with an established callus culture of the same species in a Petri dish (set up in advance to act as nurse cells or a feeder layer; Fig. 11.12a). After several days of incubation, macroscopic colonies form from single cells on the filter paper, and each clone or colony is isolated individually to start a fresh culture in culture medium.

Culture in Petri dish segments   Single-cell cultures are maintained with provision for diffusates from nurse cells in nearby segments (Fig. 11.12b).

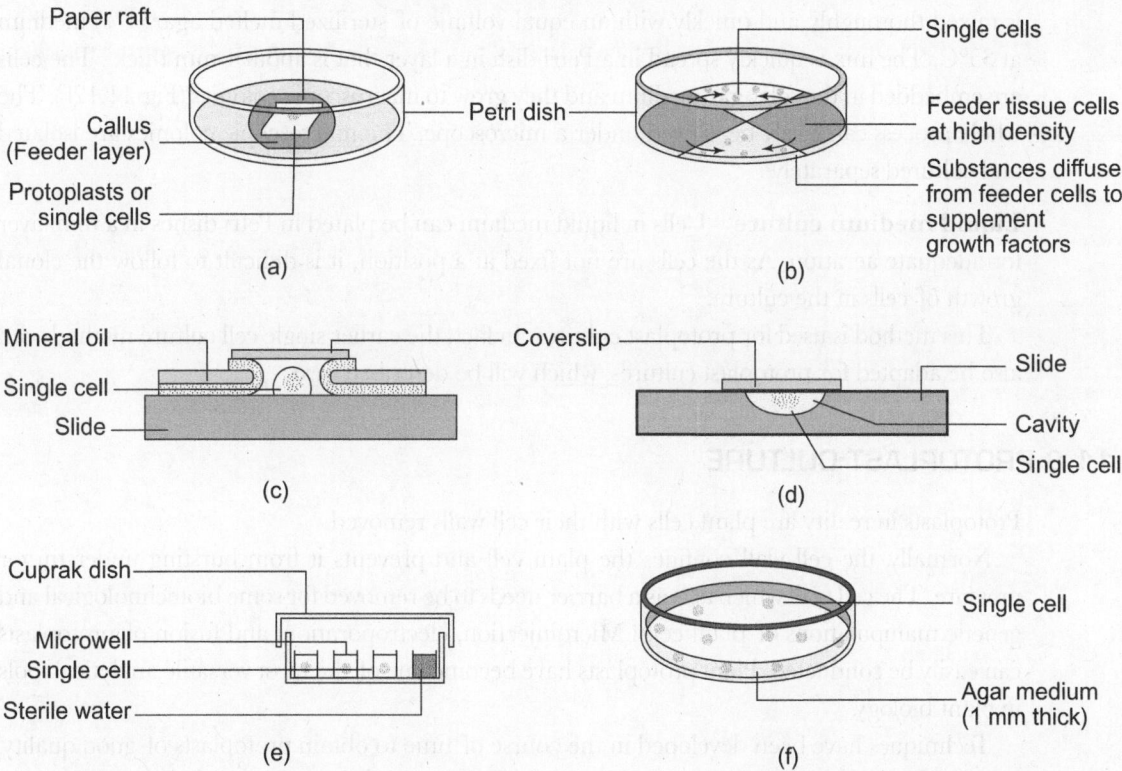

Fig. 11.12 Variations in techniques for culture of single cells. (a) Filter Paper Raft-nurse culture (b) Cells and feeder layer in different segments (c) and (d) Micro chamber cultures (e) Microdrop culture (f) Bergmann's agar plating

Microchamber culture (a) A microchamber is created between a microscope slide and a coverslip, held together by mineral oil (Fig. 11.12c). The coverslip is placed after a single cell in a drop of enriched conditioned medium is put on the slide.

(b) Another way of microchamber culture is to use a cavity slide. A drop of medium containing a single cell is placed on a coverslip which is then inverted and placed over the cavity of the slide (Fig. 11.12d).

The advantage of the microchamber technique is that microscopic observations can be made while the single cell divides and forms the clone. In this regard, a point of caution is to culture in the dark, and make only a limited number of microscopic observations.

Microdrop culture A specially designed dish, known as 'Cuprak dish', is used that contains a number of microwells, and a smaller peripheral chamber filled with distilled water to prevent desiccation of the cell culture in the microwells (Fig. 11.12e). The cells are diluted with medium in such a way that a microdrop of 0.25–0.5 mL contains a single cell. The cells in microdrops are distributed in the microwells and the dish is sealed with parafilm.

Bergmann's plating technique This is the most widely used method for single cell culture. A liquid medium containing single-cell suspension at a density twice the final density in culture is mixed thoroughly and quickly with an equal volume of sterilized melted agar (1%) medium at 35°C. The mix is quickly spread in a Petri dish in a layer that is about 1 mm thick . The cells are embedded in the soft agar medium and they grow to macroscopic colonies (Fig. 11.12f). The whole process can easily be viewed under a microscope. The macroscopic colonies are isolated and cultured separately.

Liquid medium culture Cells in liquid medium can be plated in Petri dishes in a thin layer for adequate aeration. As the cells are not fixed at a position, it is difficult to follow the clonal growth of cells in the culture.

This method is used for protoplast cultures. In fact, the earlier single cell culture methods can also be adapted for protoplast cultures, which will be described next.

11.8 PROTOPLAST CULTURE

Protoplasts in reality are plant cells with their cell walls removed.

Normally, the cell wall confines the plant cell and prevents it from bursting under turgor pressure. The cell wall which acts as a barrier needs to be removed for some biotechnological and genetic manipulations of plant cells. Microinjection, electroporation, and fusion of protoplasts can easily be conducted. Plant protoplasts have become one of the most versatile analytical tools in plant biology.

Techniques have been developed in the course of time to obtain protoplasts of good quality. Here, two major techniques are considered.

Mechanical abrasion or cutting Earlier, tissues were kept in a solution of higher osmotic potential for a while to cause shrinkage of the cells within the cell walls. The cell walls were broken mechanically by cutting or abrasion of the tissue. The osmotic pressure of the medium was then lowered to allow swelling of the cells so that they could pop out of the broken cell walls. This method gave a low yield of viable protoplasts. So there was a need for a better technique.

Enzymatic breakdown of cell walls   The main idea of this technique is to hydrolyse or digest the structural elements in the plant cell wall by enzymes to release the cell or so called protoplasm. The digestion mixtures usually include several enzymes, mainly cellulase, macerozyme, and pectinase for digestion of cell walls, and sorbitol and mannitol as osmoticum to protect the naked and fragile plant cells in solutions of low osmotic potential. In recent years, sugar alcohols are also included as they are metabolically inert and infuse into protoplasts rather slowly. Inclusion of Ca^{2+} and Mg^{2+} ions improve the stability of the protoplasts. Sometimes, the use of potassium dextran sulphate helps in the adsorption of phenols.

Enzymatic digestion in a dilute solution in a Petri dish is often carried out overnight at 20°C. The next morning, gently shaking the dish helps release the protoplasts. The content of the dish (protoplasts, cell debris, and undigested tissue) is placed on a sterile 64 μm pore-size filter inside a sterile funnel. The filtrate is transferred to a 15 mL round bottom glass centrifuge tube.

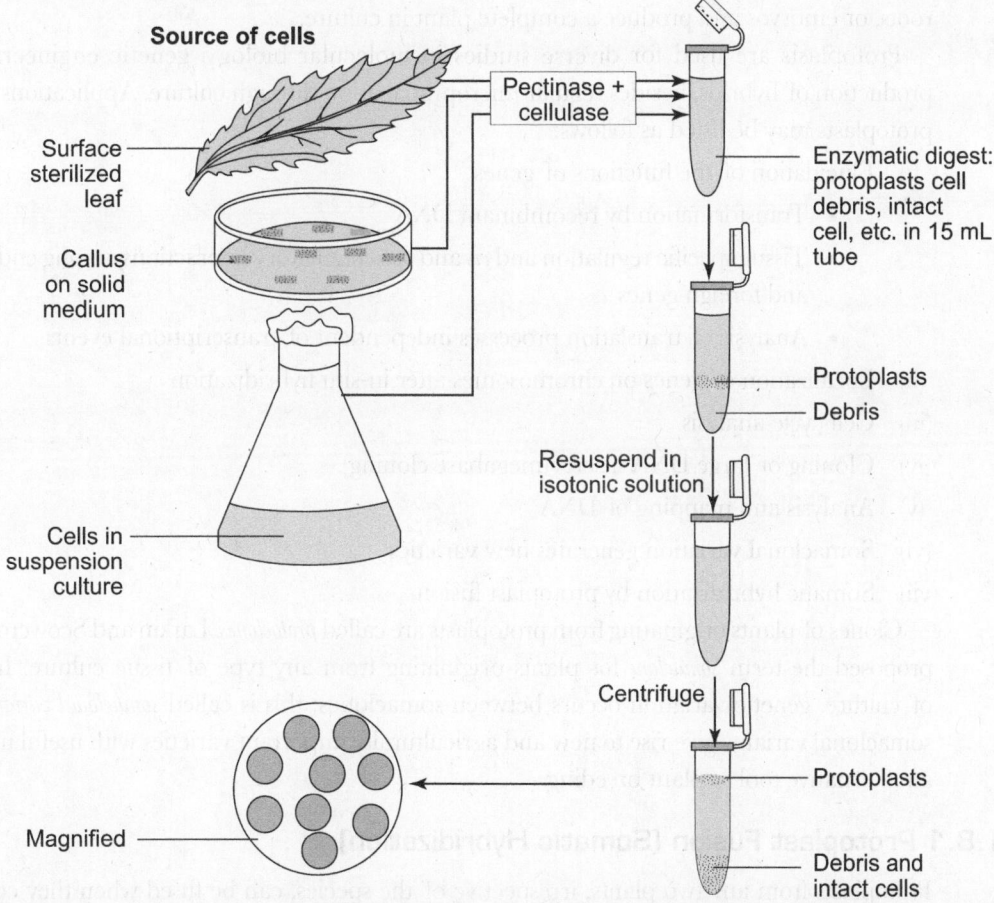

Fig. 11.13   Outline of the enzymatic digestion procedure for protoplast isolation.

After centrifugation at 700 rpm for 5–10 minutes, the enzymatic solution is removed carefully with a sterile Pasteur pipette. 5 mL of 0.4 M sucrose is added to the pellet for gentle resuspension and dispersion of the protoplasts. The resuspended solution is carefully layered over 0.6 M sucrose solution in MES (2-N-morpholinoethanesulfonic acid) buffer at pH 5.8 with a Pasteur pipette and centrifugation is done at 700 rpm for 5–6 minutes. A band of protoplasts will be formed in the denser layer of the sucrose (Fig. 11.13). It is isolated, washed, and suspended in isotonic solution or culture medium and counted with a haemocytometer (see Fig. 12.3) to adjust the final density of protoplasts at 105 mL^{-1} for culture.

Source materials for isolation of protoplasts are often the cells from callus or suspension culture, fresh leaf (7–8 days old) and a variety of other tissues. In the case of fresh leaf, it is sterilized with chemical solution first, then the lower epidermis is peeled off for better penetration of digesting enzymes.

The fate of the isolated protoplasts may be different, depending upon the species and culture conditions: (i) regenerate a cell wall in culture within 2–4 days (entire cell regenerated from naked protoplast is called *plastocyte*); (ii) divide and dedifferentiate into callus; (iii) differentiate into shoots, roots, or embryos and produce a complete plant in culture.

Protoplasts are used for diverse studies in molecular biology, genetic engineering, and production of hybrids, besides regular micropropagation through culture. Applications of plant protoplasts may be listed as follows:

(i) Elucidation of the functions of genes
- Transformation by recombinant DNA
- Tissue specific regulation and *cis* and *trans* regulatory interactions among endogenous and foreign genes
- Analysis of translation processes independent of transcriptional events

(ii) Localization of genes on chromosomes after in-situ hybridization

(iii) Cell cycle analysis

(iv) Cloning of large DNA inserts (megabase cloning)

(v) Analysis and mapping of DNA

(vi) Somaclonal variation generates new varieties

(vii) Somatic hybridization by protoplast fusion

Clones of plants originating from protoplasts are called *protoclones*. Larkin and Scowcroft (1981) proposed the term *somaclone* for plants originating from any type of tissue culture. In course of culture, genetic variation occurs between somaclones; this is called *somaclonal variation*. The somaclonal variants give rise to new and agriculturally important varieties with useful traits. It is an alternative tool to plant breeding.

11.8.1 Protoplast Fusion (Somatic Hybridization)

Protoplasts from any two plants, irrespective of the species, can be fused when they come into contact with each other. Production of hybrid plants through the fusion of protoplasts of two different species is called *somatic hybridization*. Somatic hybridization permits reproducibly isolated plant genomes (two species) to be combined at the protoplast level for the generation of novel hybrids.

Objectives of Somatic Hybridization

Protoplast fusion and somatic hybridization provide new approaches for crop improvement.

Overcome barriers of sexual incompatibility   Incompatibility barriers do not normally allow sexual crossing between two species. This problem can be overcome by somatic cell fusion. For instance, fusion of the protoplasts of potato and tomato has produced a hybrid plant, *pomato*.

Genetic recombination in sexual or sterile plants   Fusion of protoplasts is the only option to cause recombination between two different plants which cannot reproduce sexually or are sexually sterile.

Direct transfer of cytoplasmic genetic factors   The genotype of cytoplasm (present in mitochondria and chloroplasts) codes for a number of practically important traits, such as the rate of photosynthesis, male sterility, low or high temperature tolerance, and resistance to disease or herbicides. These traits can easily be transferred to a hybrid through somatic hybridization.

11.8.2 Protoplast Fusion Techniques

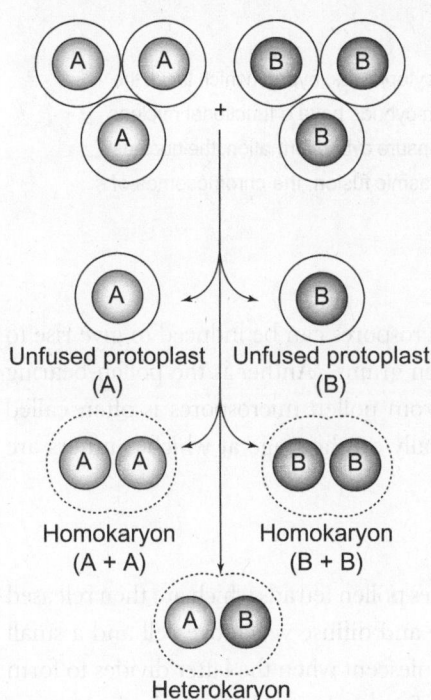

Fig. 11.14 Different types of products after fusion process of protoplasts of two different species (A and B) mixed at 1:1 ratio.

Normally, protoplasts do not fuse in vitro due to the presence of negative charge on the outside surface of the plasma membrane which causes repulsion between protoplasts. Therefore, a fusion-inducing agent (*fusogen*, usually charge neutralizing) needs to be used to induce fusion of protoplasts. Various agents (polycations) can act as fusogen; of them, $NaNO_3$, high pH and high Ca^{2+}, and polyethylene glycol (PEG) have been found to be the most effective.

PEG is the most effective but cytotoxic; use of carbonyl-free PEG (Mol.wt 1,540) improves the survival of fusion products. PEG is useful for inducing non-specific cell fusion, especially between unrelated taxa, such as soybean–tobacco, animal cells–yeast protoplasts, and animal cells–plant protoplasts.

Equal aliquots of protoplasts from two sources, in 1:1 ratio at a density of 2×10^5 mL^{-1} are taken in a 15 mL capacity screw-capped centrifuge tube for the fusion process.

Selection of somatic hybrids

In spite of the best efforts to increase protoplast fusion frequencies, the number of viable, binucleated, heterokaryons (hybrid protoplasts) remains less than 5 per cent of the protoplast population. At the end of the fusion process, mainly three cell (protoplast) types are available—parental (non-fused), *homokaryones* (fusion between similar genotypes), and *heterokaryones* (fusion between two genotypes; Fig. 11.14). Therefore, heterokaryons (hybrid) have to be selected from all the different products at the end of fusion. Several selection methods have been developed, mostly on the basis of genetic markers. One easy visual method to locate the heterokaryones has been described below; this method is also useful for demonstrative purposes.

One of the fusion protoplast partners is from an established callus culture; the culture conditions induce a chlorophyll-deficient condition in the cells, hence the white colour. The other protoplast partner is isolated from fresh mesophyll leaf tissues that are green. White protoplasts from the callus and green protoplasts from leaf mesophyll make it easy to identify heterokaryones, homokaryons, and parental protoplasts.

Somatic protoplast fusion which occurs between normal and almost intact protoplasts of both the parents is known as *symmetrical hybridization*. When somatic protoplast fusion takes place between one parent that partially donates only a few chromosomes, or subchromosome fragment(s) to the other intact partner, it is referred to as *asymmetrical hybridization*.

In contrast to sexual hybrid cells (which are heterozygotes for nuclear genes and contain organellar genes only from female plants), symmetrical hybridization of protoplasts produces hybrid cells containing nuclear and cytoplasmic complements from both the parental plants.

Cybrids

The fusion of the cytoplasm of a cell with an intact cell creates a cytoplasmic hybrid (which lacks the nucleus from one source) known as a cybrid. Plants formed from cybrids have a functional nuclear genome of one species and are hybrids for cytoplasmic traits. To ensure cybrid formation, the nucleus of the cells donating cytoplasm is irradiated so that after protoplasmic fusion, the chromosomes of the donor cells are lost during division.

11.9 Anther Culture

Anther or pollen culture is possible when the pollen microspores can be induced to give rise to vegetative (somatic) cells, instead of gametophytic pollen grains. (Anther is the pollen-bearing male gametophytic organ.) The formation of plants from pollen microspores is often called *androgenesis*. The success of androgenesis depends critically on the stage at which anthers are obtained from flower buds.

11.9.1 Androgenesis

Meiotic cell division in pollen mother cells in vivo produces pollen tetrads which are then released in the form of microspores. By the first mitosis, a large and diffuse vegetative cell and a small dense generative cell are formed. The former remains quiescent when the latter divides to form sperms. Microspores obtained from the flower bud, just before, or during the first mitosis, are most suitable for the induction of androgenesis. The microspores (pollen grains) become multicellular through the division of the vegetative and generative cell; ultimately, multicellular pollen grains burst open to release the cell mass. The cell mass sometimes assumes the shape of a globular embryo and passes through the developmental stages of an embryo to produce a plant. In other cases, it may develop into a callus. The anthers are cultured on an agar medium in a glass tube or small Petri dishes, in an 18-hour photoperiod.

The change from a normal gametophytic development to a vegetative sporophytic pattern seems possible in an early phase of the cell cycle when transcription of genes for gametophytic development is blocked and sporophytic genes are activated. The plants, thus, produced are haploid (with gametic *n* number of chromosomes). Guha and Maheswari, in 1964, were able to produce such haploid plants from anthers of *Datura innoxia*. More success for production of haploid plants from anthers has come from *Solanceous* species. The presence of the anther wall provides a stimulus for sporophytic development; this is why anther culture is more successful than pollen culture in the production of haploid plants.

11.9.2 Diploidization

Haploid plants raised through culture of anthers are sterile and cannot produce seeds. Spontaneous duplication of chromosomes occurs within anther culture-derived callus cells. This leads to the production of fertile *doubled haploid plants*. The two sets of chromosomes within such plants are identical; the plants are fully homozygous and breed true. To induce diploidization, 0.5 per cent colchicines can be used to treat the young haploid plantlets, while they are still attached to the anther in culture. The doubled haploid plants harbour no hidden trait, and are used in breeding programmes to improve crop production. Usually, haploids are cultured in large numbers and screened for desired traits, such as disease resistance, form, chemical content, and so on, and then diploidized to produce homozygous plants.

11.9.3 Applications of Anther Culture

Anther culture mainly produces haploid plants, which are very useful in several ways for genetic and plant breeding research.

(i)   It helps in establishing almost isogenic plant lines (breed true).

(ii) Homozygous diploids from anther culture cut short the time period required (several generations for conventional breeding) to come up with such homozygosity.

(iii) Already doubled diploids helped in raising strains of different crops with improvements in yield, and/or in disease resistance, cold resistance, adaptability, maturity duration, and so on.

11.10 POLLEN CULTURE

Pollen grains isolated from the anthers can also be cultured in vitro to give rise to haploid embryos or callus. Anthers from a flower bud are collected and squeezed with a glass rod in a medium. The pollen suspension is passed through a nylon filter of 25–100 μm porosity to get rid of debris, and the filtrate is centrifuged at 500–800 rpm for 5 minutes. The pollen pellet is washed twice and suspended in medium at a density of 10^3–10^4 pollen mL^{-1}. In float culture technique, anthers are allowed to float on a shallow liquid medium in Petri dishes and dehisce in a few days, gradually releasing their pollen into the medium.

Anther cultures need to be supplemented with glutamine, L-serine, asparagine, and inositol.

Anther culture usually produces higher yields, that is, more plantlets than pollen culture. However, there are a few advantages of pollen culture over anther culture: no callus formation takes place from anther wall tissue, and there is no mix up or chimera formation from products of different pollen grains.

11.11 STORAGE OF PLANT MATERIALS

Plants propagate naturally through two developmental cycles—*sexually by* producing seeds and *asexually* through vegetative propagation. Propagation by seeds is the most economical as large numbers are produced, and they are easily distributed, usually pest and disease-free, and can be stored for long periods. Plant materials can be stored for both consumption and as a resource for further consumption by two ways. *In-situ conservation* maintains the plants in their natural habitat. The limitation of this method is the risk of loss due to environmental hazards and the cost of maintenance over the years. The other method is *ex-situ conservation*, which maintains seeds or in vitro raised plant cells, tissues, and organs under appropriate conditions for long-term storage as a gene bank or germplasm bank.

Tissue culture techniques provide a new approach for storing the plant material for short or long term so that propagation can be carried out at the desired time. Short-term storage of the materials is done by continuing to maintain them in in-vitro culture system with some modifications in certain aspects, such as temperature (colder), lower intensity of light, reduced oxygen tension, modifications in amount and composition of nutrients, and so on. Freezing or *cryopreservation* at ultra-low temperature is the alternative method of storing important genotypes and materials from tissue culture, for a long time.

11.11.1 Cryopreservation

In the cryopreservation technique, the temperature is slowly lowered to freezing point and then decreased further, converting water into the solid or vitrified phase. At this phase, water is unavailable for metabolic processes of the cells; absence of energy in cells does not allow molecular movement, biochemical reactions, and diffusion. Thus at ultra-low temperatures, living cells remain indefinitely in suspended animation, without losing their viability.

Cryopreservation of plant cell cultures and eventual regeneration involves the following critical processes.

(i)   Raising of the cells and tissue culture aseptically.

(ii) Addition of cryoprotectants to avoid freeze damage; cryoprotectants significantly increase the chances of plant cells surviving freezing and thawing. Various chemicals serve as cryoprotectants. They are usually highly soluble in water, reduce the concentration of salts within the cell, and restrict the growth of ice crystals which cause a lot of damage to protein structure. They are usually of low molecular weight, non-toxic, can be easily washed away from the cells, and also permeate the cells rapidly. DMSO, glycerol, mannitol, sorbitol, and polyethylene glycol are usually used as cryoprotectants. Sometimes, it is better to use more than one cryoprotectant in a protocol. The total concentration of the protectants needs to be 1–2 M. DMSO (5–15 per cent) is the most commonly used cryoprotectant. Sugars, amino acids (particularly proline), and methanol are also used as cryoprotectants.

(iii) Step-wise cooling, followed by freezing in liquid nitrogen at –196°C for long-term storage.

(iv) Revival by thawing or rapid rewarming of cells and removal of cryoprotectant by repeated washing: Rapid thawing seems to counter the formation of ice crystals and favours viability of the cells. Ampoules containing the cell samples or plant tissues are placed in water at 35–40°C until thawed and then transferred to room temperature.

(v) Viability test by microscopic observation.

(vi) Reculture of the retrieved cells.

(vii) Induction for growth and regeneration of plants.

Prospects of cryopreservation are diverse. Cryopreservation is for long-term storage, retaining morphogenic potential and genetic stability. It reduces the space required for maintenance, saves culture medium, avoids subculturing, and prevents ageing. This technology helps in the preservation of rare genomes.

It may be mentioned that some genetic variations may arise by the accumulated mutations caused by background ionizing radiation during the period of storage.

SUMMARY

♦ Plant biotechnology utilizes the totipotency of plant cells, that is, the potential of a cell to grow and develop into a total plant.

♦ Plant tissue culture refers to the culture of cells and tissues from plant organs, such as shoot tips, leaves, roots, axial buds, embryos, anthers, and so on.

♦ Plant tissue culture is used in micropropagation of plants, clonal expansion, production of transgenic and pathogen (virus)-free plants, and mass cell cultures for secondary metabolites and chemicals. All these contribute to better agrihorticultural and forest products, and serve as a better source of medicinal and other chemical products.

♦ In-vitro plant cell tissue cultures require highly aseptic, sterile conditions in each step. Control of microbial contamination in the specially designed tissue culture laboratory is one of the prime tasks. A laminar flow cabinet is essential for final transfer of aseptic tissue pieces in cultures or pouring the culture medium in the sterile vessels, and any other aseptic operations.

♦ A defined nutrient medium for tissue culture usually contains inorganic salts, a carbon source, some vitamins, growth regulators (hormones), and some organic supplements. Sometimes special supplements are required for some types of cells, depending on the plant species.

♦ Culture media can be of two types—liquid media and semi-solid (gel like) media containing some gelling compounds such as agar or agarose.

♦ For micropropagation, plantlets can be grown in two ways: (i) culturing explants from an organized part of the plants, such as bud, meristem, root cuttings and (ii) *callus,* dedifferentiated rapidly dividing cell mass, derived initially from part of organized tissue. Different types of physical and chemical factors influence the differentiation and growth of plantlets in vitro.

♦ After in-vitro growth of plantlets under controlled conditions in a culture room, they are conditioned step by step for final transfer from the culture media to the soil in a greenhouse.

♦ Multiple types of plant materials, such as ovule, seeds, embryos, anthers, and pollen, can be cultured in-vitro to circumvent some practical difficulties or in some cases to facilitate germination.

♦ Suspension cultures are useful for mass culture of cells and for harvesting secondary metabolites (chemical) as plant cell products in plant biotechnology.

♦ Protoplasts in reality are plant cells after removal of their cell walls. The plant cell walls can be removed by mechanical abrasion or in a more sophisticated manner by enzymatic digestion. Microinjection, electroporation, genetic manipulation, and fusion of protoplasts can be easily conducted. Plant protoplasts have become one of the most versatile analytical tools in plant biology. Protoplast fusion is a novel way to generate somatic hybrids across the species.

♦ Cryopreservation is for long-term storage of plant cell cultures; it retains morphogenic potential and genetic stability. It also helps in the preservation of rare genomes.

Barbara McClintock

**Barbara McClintock
(16 June, 1902–2 September 1992)**

Barbara McClintock, an American cytogeneticist, was awarded the Nobel Prize in 1983 'for her discovery of mobile genetic elements' (transposons) and study of genetic regulation in maize. She is the only woman to receive an unshared Nobel Prize in Physiology or Medicine.

Barbara McClintock was born in Hartford, Connecticut, as the third of the four children to physician father Thomas Henry and mother Sara Handy McClintock. From an early age, she stayed with an aunt and uncle in Brooklyn, New York to ease the burden on her father while he established his medical practice. She had an independent nature from a very young age, a trait in her word 'capacity to be alone'; she was described by many people as a tomboy. She went to Erasmus Hall High School in Brooklyn and was impressed with science. She entered Cornell University College of Agriculture, Ithaca, New York in 1919. Her interest in genetics had been sparked when she took her first course in that field in 1921 taught by C.B. Hutchison, a plant breeder and geneticist. She received a BSc degree in botany in 1923. During her graduate studies and postgraduate appointment as a botany instructor, Barbara was instrumental in assembling a group of plant breeders, and cytologists who worked in the new field of cytogenetics in maize. Some of them became famous later; one of them George Beadle became a

Nobel Laureate in 1958 together with Tatum formulating the 'one gene one enzyme hypothesis'. Barbara received her MS and Ph D from Cornell itself in 1925 and 1927, where she was a leader in the development of maize cytogenetics. The field remained the focus of her research for the rest of her career.

McClintock was the first to demonstrate many fundamental cytogenetic ideas in the coming years. She was the first person to show chromosomal crossing over during meiosis using a microscope (1930), and with Harriet Creighton she linked crossover with genetic recombination (1931). She built the first genetic map for maize showing the order of three genes on maize chromosome 9 (1931).

She first realized the functional role of the telomere (1932) and centromere (1938) of chromosomes. She described the nucleolar organizer on maize chromosome 6.

Several post-doctoral fellowships from the National Research Council allowed her to continue to study genetics at Cornell, University of Missouri and the California Institute of Technology. In 1931–1932, Lewis Stadler, a geneticist at Missouri introduced her to the use of X-ray as a mutagen, which helped her make some marvelous discoveries. As a Guggenheim Fellow for six months (1933–1934), McClintock intended to spend time in Germany with Curt Stern who showed crossing over in *Drosophila* weeks after she and Creighton demonstrated it in maize. In the meantime, Stern emigrated to the US; she joined the laboratory of geneticist Richard B. Goldschmidt in Germany. She came back to Cornell and stayed there until 1936. Then an assistant professorship was offered to her by Lewis Stadler in the Department of Botany at the University of Missouri. There she expanded her research on the effect of X-rays on maize cytogenetics and discovered extensive mutation and variegation in the endosperm.

In early 1941, she was invited by the Director of the Department of Genetics at Cold Spring Harbour Laboratory (CSHL) to spend her summer there. She also accepted a visiting professorship at Columbia University. After a year-long temporary appointment, McClintock accepted a full time research position at CSHL as Staff Member, Carnegie Institution of Washington for 1942-1967. In 1944, in recognition of her prominence in the field of genetics, McClintock was elected as member of the National Academy of Sciences; at that time she was the third woman to be elected. In the same year, she undertook a cytogenetic analysis of *Neurospora crassa,* widely used as experimental material for genetic analysis at the suggestion of George Beadle and successfully described the number of chromosomes of this fungus. In 1945, she became the first woman president of the Genetics Society of America.

In the summer of 1944 at CSHL, McClintock began systematic studies on the mechanisms of the mosaic colour patterns of maize seeds and unstable inheritance of this mosaicism. This led her to discover 'transposons'(wandering genes) and their role in turning physical characteristics on or off from one generation of maize plants to the next. She published this in the Proceedings of the National Academy of Sciences in 1953. This discovery challenged the concept of genome as a static set of instructions passed between generations. Encountering skepticism of contemporaries about her findings, she stopped further publication on the topic. She remained busy in extensive study of cytogenetics and ethnobotany of maize races from South America. With advancement in genetics her research on transposons became well understood in the 1960s and 1970s. In 1973, after 20 years of silence McClintock again talked about controlling elements. Appreciation of her work by the scientific community led her to become a recipient of the Nobel Prize at the age of 81.

McClintock received many honours and recognitions, two of them may be specified here. She was recipient of the National Medal of Science from President Richard Nixon in 1971. The Cold Spring Harbor named a building in her honor in 1973. She received several honorary doctorates from Universities and academic institutions. The US Postal Service commemorated her on a 37 cent postage stamp on 4 May 2005. A street was named after her in Germany.

EXERCISES

Objective Questions

1. Match the statements in Column A with the items in Column B.

Column A	Column B
(i) To obtain secondary metabolites, plant cells need to be cultured in	(a) gel media.
(ii) Plant embryos are cultured in	(b) suspension media.
(iii) In modern tissue culture laboratory, the wall and ceiling are to be	(c) white washed regularly.
(iv) HEPA filter is used to	(d) made up of standard synthetic materials.
(v) Horizontal laminar flow hood is more to protect	(e) sterilize the air.
(vi) Vertical laminar flow hood is more to protect	(f) cut down the particulate matter in the air.
(vii) Tissue culture room needs to be illuminated by	(g) the worker
	(h) the cultures.
	(i) natural light.
	(j) fluorescent tubes.

2. Indicate whether the following statements are correct. If not correct then write the correct version.
   (a) The inorganic salts constitute the macronutrients only, not the micronutrients in the plant tissue culture media.
   (b) Iron molecules as micronutrients are components of many plant proteins participating in the regulation of oxidation or reduction reactions.
   (c) Amino acids in tissue culture media can serve as readily available source of reduced nitrogen.

(d) Cytokinins are amino acid derivatives essential for differentiation of plant organs in most species.

(e) For media preparation, it is better to prepare a solution of calcium compounds separately, and then add it to the media.

(f) Meristems are usually virus free.

(g) 50 % sodium hypochlorite or 70 % ethyl alcohol is commonly used as disinfectant for explants.

(h) In in-vitro culture, the plantlets draw sucrose in the medium for energy instead of using CO_2 in the process of photosynthesis.

(i) The smaller the size of the meristem explants in culture, the greater are the chances of virus elimination and the chance of explants' survival.

Review Questions

1. 'Biotechnology is utilizing the *totipotency* of plant cells'—Justify this statement.

2. What is a callus?

3. What was the contribution of Guha and Maheswari in 1966, towards the development of plant tissue culture?

4. What are the objectives of plant tissue culture?

5. Why is continuous agitation needed for suspension cell cultures?

6. Write down a few important steps to be followed for transfer of explants into the growth medium.

7. Provide a brief note on 'somatic embryo'.

8. What are the purposes of 'subculture'?

9. What are the steps to be taken for hardening off plantlets in-vitro culture for greenhouse or field transfer?

10. What are the purposes of culture of zygotic plant embryos?

11. What are the basic differences in 'batch' and 'continuous' suspension cell cultures?

12. Name the different procedures employed in single cell cultures.

13. Write down the protocol to obtain plant protoplasts with enzymatic treatment.

14. Elucidate the utilities of the protoplast fusion technique.

15. What is a cybrid?

16. What are the applications of anther culture?

17. What is cryopreservation?

18. What are the prospects of cryopreservation?

12

Animal Biotechnology

LEARNING OBJECTIVES

♦ The basics of animal cell and tissue culture
♦ Concept of primary culture and cell lines
♦ Different cell culture techniques
♦ The process of cell synchronization
♦ The scale-up of cell cultures for biotechnology industries
♦ Methods of cryopreservation of cell lines
♦ In-vitro tissue engineering
♦ Clinical uses of stem cell culture

INTRODUCTION

Animal biotechnology is the application of biotechnological principles for production of materials from animals. It has several components, such as animal cell and tissue culture, cell lines, application of cell culture for stem cell technology for production of monoclonal antibodies, cytokines, vaccines, and so on. Fundamentally, the productivity of animal cells in culture is the main focus.

The animal cell culture technique was initiated with undisaggregated fragments of tissues to allow migration of cells in culture. The technique of 'tissue and cell culture', thus, refers to this past attempt.

In 1907, Ross Harrison published his observations about the culture of animal cells. He first cultured embryonic nerve cells of frog by the hanging drop method. Then he used connective tissue cells for extended periods. In 1885, Wilhelm Roux attempted culture of medullary plate of chick embryo on warm saline, and in 1903 Jolly studied in-vitro (outside the organism) cell survival and cell division in leucocytes obtained from salamander, an amphibian.

In 1912, Alexis Carrel showed contractility of heart muscle cells in culture for over two to three months. He introduced aseptic techniques, use of trypsin to dissociate the tissue, and use of embryo extract and animal serum in culture. Gradually, the tissue culture technique was developed over the years by using tissue from warm blooded animals, particularly mammals. The implications for medical science was mainly behind this development.

The development of cell culture in the recent past has been primarily to meet the needs of two major branches of medical research: the production of antiviral vaccines and the understanding of neoplasia or cancer. The standardization of conditions and cell lines for the production of large numbers of cells necessary for, biochemical analysis and assays for cell products at the micro level provided much impetus for the development of modern tissue culture technology. The commercial supply of reliable culture media, sera, and use of antibiotics, air-filters, and laminar flow hoods contributed towards improvement of the technology.

12.1 Cell Culture

The in-vitro cell culture technique is used to grow cells in isolation, devoid of the effects of multiple systems, and to judge the cell's functions under the influence of various factors in a controlled environment. The advantages of the system are many.

Advantages of cell culture

(i)   A substance/factor/drug can be tested in replicate cultures of a cell population obtained from an animal instead of sacrificing several animals in an experiment. A sample of cells may be obtained by biopsy without causing harm to an animal or an individual.

(ii) A minor subpopulation of cells can be used for its functional evaluation. For example, cell culture has made it possible to define the functional differences between $CD4^+$ T helper cells and $CD8^+$ T cytotoxic cells, and cell–cell interactions between different categories of lymphocytes and phagocytic cells.

(iii) A large homogeneous population is raised from a cell type present in vivo in low numbers. Large homogeneous populations of cloned B and T cells in vitro provide the opportunity to decipher the molecular mechanism of cell activation in great detail.

(iv) The in-vitro system categorically assesses the functions of cytokines of different cell types. Some of the cytokines, such as IL-2, IL-4, and IL-5, have even been discovered in the course of the functional assessment of the factors in cultures containing different cell types.

(v)   The genetic expression at different maturational stages of a cell type can be clearly studied in the cell culture system.

(vi) Genetic manipulation and transformation studies in cells can be easily done in cell culture.

(vii) The cell lines provide large numbers of homogeneous cells for testing the efficacy of drugs or biological products.

(viii) The cell lines produce large quantities of specific cytokines, vaccines, and so on in cultures and the products can be isolated and purified for research and commercial purposes. Bioreactors have been designed for commercial use of cell products in bulk, in biotechnology industries.

(ix) Tissue culture technology has been adopted for routine analysis of chromosomal or genetic disorders in the unborn child in the womb from the cells in amniotic fluid obtained by amniocentesis. If the disorder is of a serious nature and likely to affect the child with an inborn defect or a life-threatening condition, then termination of pregnancy is suggested in early months.

(x) Tissue culture technology offers possibilities in homografting and reconstructive surgery using an individual's own cells, particularly for severe burns.

A notable limitation of the cell culture technique is that variant cells arise spontaneously in the course of the prolonged culture of a cell type. That is why frequent subcloning needs to be done to limit cellular heterogeneity.

12.2 Cell Culture and Cell Lines

Starting from a collection of a particular cell type or a cell line, in-vitro cell culture involves many technicalities, such as aseptic measures, preparation of culture media, choice of cell culture vessels, gas phase, cell density, scaling up of cell cultures for industrial purposes, and so on.

12.2.1 Primary Cultures

Freshly isolated cells in a nutrient medium constitute a primary culture and represent the characteristics of the original tissue. They grow slowly and soon stop proliferating.

12.2.2 Cell Lines

As soon as the cells from a primary culture are subcultured in new Petri dishes or cell culture bottles, the subcultured cells technically establish cell lines. Finite cell lines die after a few subcultures. Some genetically altered cell types may continue indefinitely in the subcultures, and thus, establish continuous cell lines. *An immortalized clone of cells grown in culture is called a cell line.* Certain mutations, induced by chemical carcinogens or viruses, alter the growth pattern of the cells, causing continuous cell lines to originate from primary cultures.

First Cell Line

The first cell line, mouse fibroblast L-929, was derived in the 1940s from a primary culture exposed to methylcholanthrene, a chemical carcinogen, over a period of a few months. This cell line is used even now in DNA transfection studies and to assay tumour necrosis factor (TNF).

In the early 1950s, the HeLa cell line was derived by culturing human cervical cancer cells.

Scientists have now established hundreds of cell lines, each consisting of a population of genetically identical cells capable of indefinite growth in culture. These are used for various purposes in research and biotechnology. For immunological research, cell lines have been derived from spontaneously occurring tumours of lymphocytes, macrophages, and other accessory cells. Normal lymphoid cells can also be transformed by viruses such as Abelson's murine leukaemia virus (A-MLV), simian virus 40 (SV40), Epstein-Barr virus (EBV), and human T-cell leukaemia virus type 1 (HTLV-1) to produce cell lines.

12.2.3 Types of Tissue Culture

Tissue culture can be initiated usually in three ways (Fig. 12.1):

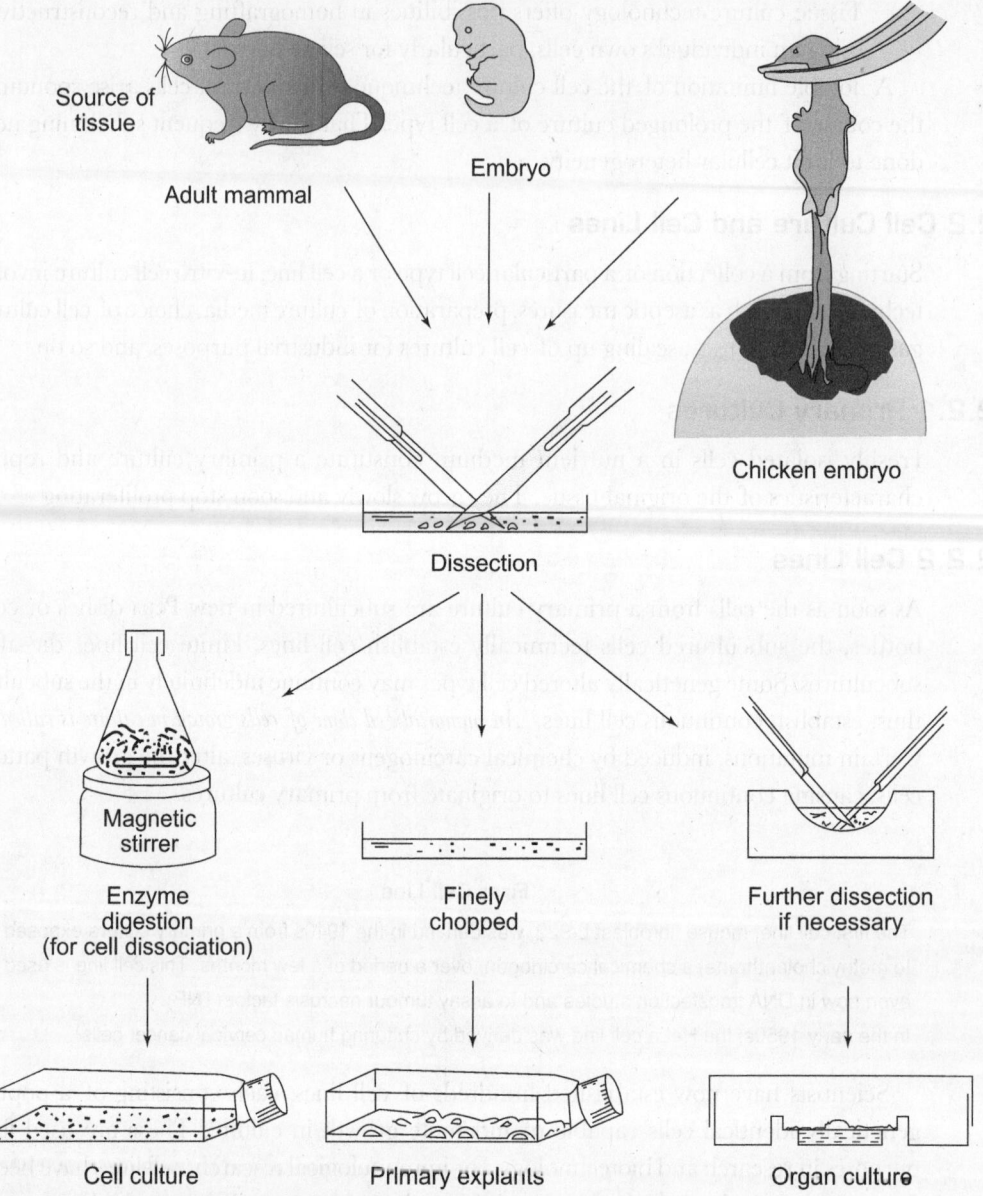

Source of tissue

Adult mammal

Embryo

Chicken embryo

Dissection

Magnetic stirrer

Enzyme digestion (for cell dissociation)

Finely chopped

Further dissection if necessary

Cell culture

Primary explants

Organ culture

Fig. 12.1   Three types of tissue culture.

Organ (bit) culture

A small part of an organ is cultured at the liquid-gas (atmosphere) interface on a raft, grid or gel. The raft can be made up of a thin strip of plastic with a central hole of about 3 mm diameter sealed on one side by microfilter paper of 0.45 μm porosity. A grid may be made up of stainless wire mesh of small porosity, and the gel may be prepared by mixing serum and chicken embryo extract in a 1:1 ratio. The culture medium must touch the organ bit. The culture at the gas-liquid interface helps the explant to retain a spherical geometry and tissue architecture.

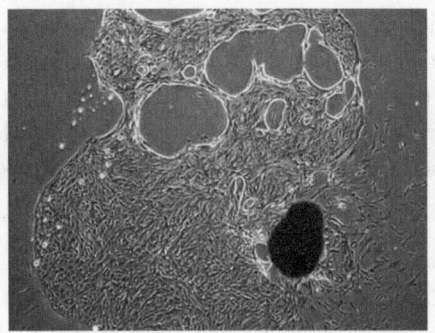

Fig. 12.2 Cells migrate out of an explant and attach to the substratum of the culture vessels.

Primary explant culture

A fragment of tissue (~1 mm^3) is placed in a glass or plastic culture Petri dish or a cell culture bottle with the required amount of culture medium. The cells soon migrate out of the explant and attach to the substratum of the culture vessel (Fig. 12.2). Several explants of a tissue can be used at a time in a culture to obtain a higher number of cells.

Cell culture

The cells are obtained by mechanical dispersion, or enzymatic digestion of tissue fragments, or even by scraping, or mild enzyme digestion of the cells growing at the substratum of the culture vessels with tissue explants as discussed in the previous paragraph. After proper wash with balanced salt solution, the cells are suspended in culture medium and the desired density of cells is obtained by counting the cell number in a haemocytometer and adding further culture medium if necessary. A drop of 1 per cent trypan blue is mixed with a drop of cell suspension and the mixture is charged into the counting chamber of a haemocytometer to count the number of live cells (Fig. 12.3); the dead cells appear blue.

12.2.4 Cell Dissociation

The cells in a tissue must be dispersed first to obtain the cells to set up cultures.

Mechanical disaggregation of a tissue The cells can be collected from spill out when the tissue is sliced into smaller pieces in buffered saline solution or culture medium, or by pressing (sometimes scraping) the dissected tissue against sieves of stainless steel with a sterilized forceps. Then the cells and very fine fragments of tissue are passed through a series of sieves with gradually reducing mesh size, to obtain a uniform cell suspension. Alternatively, the coarse suspension of cells and small tissue fragments may be passed several times through a syringe fitted with a needle of 27 or higher gauge.

The mechanical dispersion of tissue into cell suspension is less time-consuming than enzymatic digestion and also avoids the risk of damaging cells during the relatively longer period required for enzymatic digestion. However, the mechanical procedure may cause mechanical damage such as leakages, to the cells.

Enzymatic disaggregation of a tissue Enzyme treatment disrupts the intercellular matrix involved in cell–cell adhesion in a tissue. The intercellular matrix (in a way extracellular matrix) contains interacting glycopeptides, often described as cell adhesion molecules or CAMs, which are Ca^{2+} dependent. Different enzymes for digesting specific substances in the extracellular matrix, alone or sometime in combinations are used for dissociation of tissue cells. They are usually proteases, glycanases (hyaluronidase, heparinase), collagenase, and so on. The chelating agent EDTA is used for removal of Ca^{2+}.

Usually, 0.25 per cent crude trypsin solution in phosphate buffered saline (PBS) is used. For different types of tissue, the percentage of trypsin may vary between 0.01 per cent and 0.5 per cent. The optimum amount for maximum yield of viable cells may be decided by trial and error. The purer the trypsin, the less toxic it is to the tissue. But crude trypsin is more effective due to the presence of other proteases. Enzymatic disaggregation yields a higher number of cells from tissues. After enzymatic treatment for a definite period, the cells are centrifuged, the trypsin solution is discarded, and the cells are washed two to three times with PBS. Crude trypsin is mostly used in tissue disaggregation, as it is effective for many tissues and if any of its residual activity is retained after washes, it is neutralized by the serum of the culture medium. In serum-free culture, soya bean trypsin inhibitor is used.

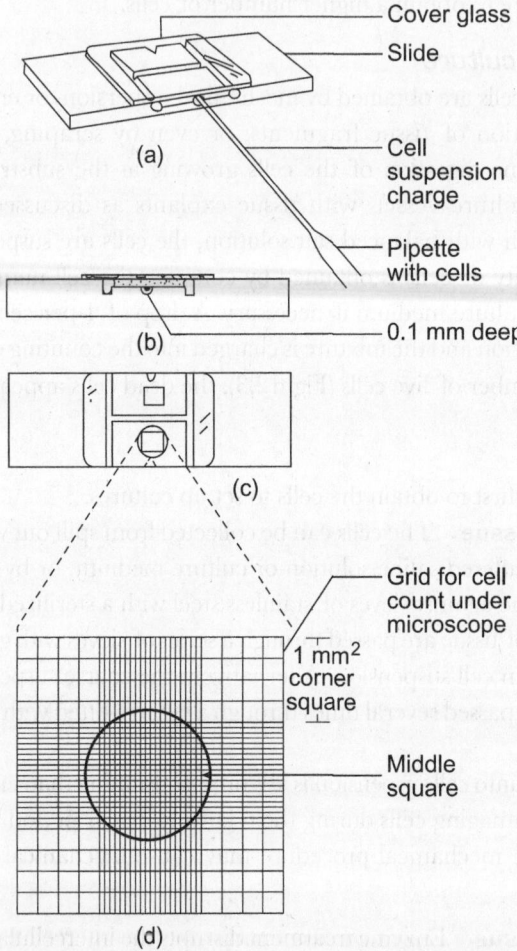

Fig. 12.3   Counting cells in a haemocytometer. (a) Charging a haemocytometer with cell suspension mixed with viability stain (trypan blue). (b) Side view of haemocytometer slide showing the cell sample in a 0.1 mm deep counting chamber. (c) Top view of the slide. (d) Grid of a Neubauer haemocytometer under a microscope. The number of cells in a corner square or middle square are counted. The volume of fluid in a square is 1 x 1 x 0.1 (depth) mm. Thus, number of cells in a square is in 0.1 mm^3 volume of fluid; this volume needs to be multiplied by 10^4 to provide the number of cells in 1 mL suspension.

12.2.5 Aseptic Measures for Cell Culture

- The dissecting tools are sterilized in boiling water. Just before use, the tools should be kept immersed more than half way in 70 per cent alcohol in a beaker (usually 125 mL size), keeping the part of the tools used to handle them, outside the alcohol.

- When tissue is collected from a sacrificed animal, the external surface of the animal needs to be swabbed thoroughly with 70 per cent alcohol. (Embryos from the womb or egg are normally in aseptic condition.)

- Starting from the point when the tissue is collected, all the tools, wares, and fluids to be used must be pre-sterilized; and the cap of the bottle, tip of the pipette, or any other tools should not touch the surface of the culture hood or any unsterilized spot.

- The tools, pipettes, syringe, or any other reusable items are to be discarded after one use in a deep tray half filled with detergents.

- At the time of opening a bottle, the cap is to be gripped by the little finger and partially by the ring finger of the right hand.

- The rim and neck of the bottles containing culture fluid or any other additives are to be passed briefly through the hottest part of the flame of a Bunsen burner each time a bottle is opened or closed. The tip of the pipettes should also be passed through the flame before use and pipettes are to be discarded once they are used. The plastic pipettes are supposed to be pre-sterilized and individually packed by the company. The plastic tip of a micropipette should also be discarded after it is used once.

- Culture fluid or any other liquid must not spill on the work area. In case of a spill, it should be cleaned with cotton and 70 per cent alcohol.

12.2.6 Laminar Flow Hood

Laminar flow hoods or cabinets were first developed in the 1960s. Now they are an essential item for cell culture laboratories across the world. A laminar flow cabinet reduces the level of particulates, microbes, and contaminating agents in the air to a minimum by constant air filtration with industrial-grade filters, and then blows this pre-filtered air into the working space of the cabinet under positive pressure. The positive pressure prevents infiltration of the air in the room into the cabinet. The laminar flow cabinet is usually enclosed on all sides, the front screen can be raised and opened when the hood is in use and otherwise kept closed (Fig. 12.4a).

A hood with a standard working surface of 4 ft width and 2 ft depth is appropriate for one person to use at a time comfortably. The hood is fitted with tube lights and a UV light, and a gas inlet for a Bunsen burner. The UV light needs to be turned on for some time before and after use to keep the hood sterile.

There are two main types of laminar flow hoods depending on the direction of airflow entering into the work space: (i) horizontal and (ii) vertical. So far, the horizontal laminar flow has been discussed, where the airflow blows from the side facing the user, parallel to the work surface, and is not recirculated. This provides the most stable airflow and the best protection in terms of sterile conditions to the culture and reagents.

In vertical laminar flow (Fig. 12.4b), the air blows down from the top of the hood onto the work surface and is drawn back through the work surface and to some extent from the front opening where the user faces the laminar flow. The air that is drawn back is either recirculated or vented. In most hoods, 20 per cent is vented and made up by drawing in air at the front from the user's side. Vertical flow hoods give more protection to the operator. Vertical hoods are classified on the

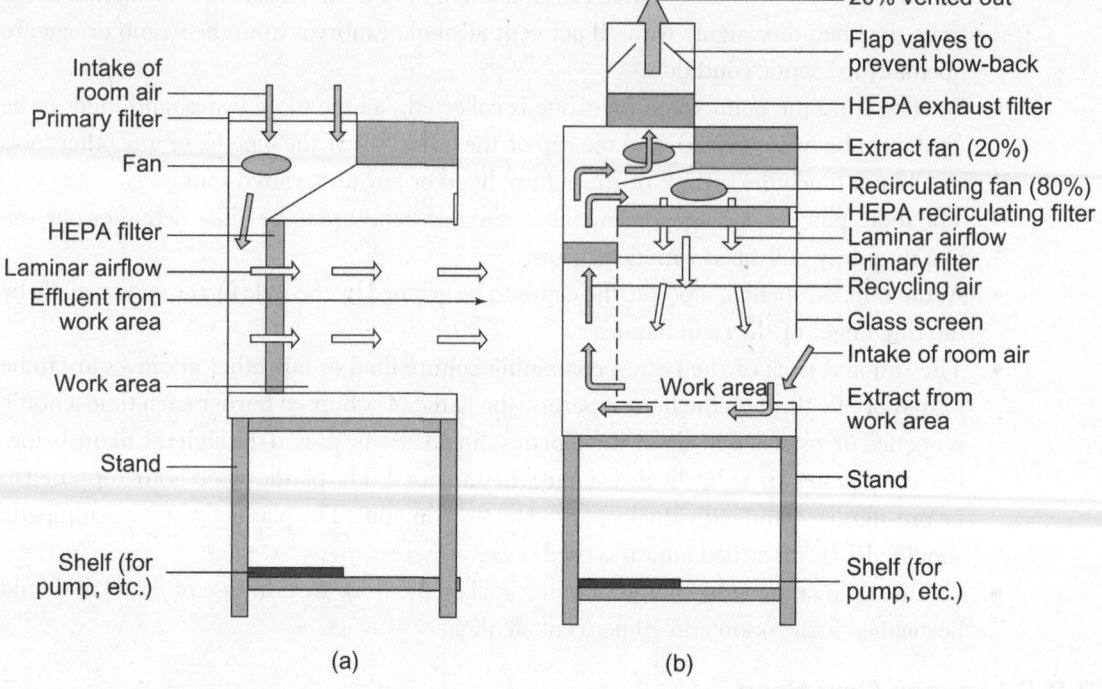

Fig. 12.4(A)   Laminar flow hood. (a) Horizontal; (b) Vertical. Arrows indicate direction of airflow.

Fig. 12.4(B)   Photograph of a laminar hood. (Also see colour plate 4.)

basis of the degree of safety provided to the user. A class II vertical-flow biohazard hood needs to be used to handle potentially hazardous materials, such as human or primate derived cultures, virally infected cultures, carcinogenic (cancer producing) reagents, and so on.

Regular cleaning of the primary filter is required, every three to six months on an average. After shutting off the horizontal laminar flow, the filter is removed and may be washed in soap and water or discarded. Airflow through the main high-efficiency particulate air (HEPA) filter located at top of the hood should be monitored every six months, and it should be replaced if necessary. It is best to use the services of professional personnel or an engineer for this purpose.

12.3 CELL CULTURE TECHNIQUES

Cells in cultures can be maintained for growth in two ways: (i) monolayers (for primary cell culture) and (ii) suspension culture (mostly for transformed cell lines, to be stirred).

Monolayer culture is anchorage dependent, that is, the cells attach to the substratum and this is a prerequisite for cell proliferation. On the other hand, suspension cultures are anchorage independent; haematopoietic cells, transformed cell lines, and cells from malignant tumours grow in this type of culture.

12.3.1 Culture Vessels

Cells are cultured in different types of vessels, such as culture tubes, Petri dishes, microtitre plates, flasks, and bottles (for mass culture).

The vessels are usually made up of glass, plastic, or metal. The vessels need to be chemically inert and negatively charged. Since the animal cell surface bears a net negative charge, cross-linking with glycoproteins and divalent cations such as Ca^{2+} and Mg^{2+} is a must for cell adhesion.

Cultures of fibroblasts or endothetial cells produce an extracellular matrix that coats the surface of culture vessels. The matrix coating helps attachment, growth, and differentiation of some cells in monolayer culture, particularly when they are at low cell densities. Coating with synthetic matrices and collagen favours epithelial proliferation.

Glass

It is easily washable, readily sterilized, and optically clear for microscopic observations of cultures. Alum borosilicate glass is preferred over soda lime glass, which releases alkali into the culture medium and needs boiling in weak acid prior to use. Repeated use reduces cell adherence, following which treatment with 1 mM magnesium acetate for several hours is necessary. Prior to each use, the glassware must be washed in a solution of non-toxic detergent and then with water followed by double-distilled water, and then dried and sterilized.

Plastic

Plastic wares made up of polystyrene, polyethylene, polycarbonate, Perspex, polyvinylchloride (PVC), cellulose acetate, and so on are used. These are non-autoclavable, and supplied sterile—they are sterilized by UV-irradiation or a high-voltage discharge—and treated to make them negatively charged and wettable, for a single use.

Metal

Stainless steel and titanium are used. Prior to use, metalwares are washed with acid to clean the surface and remove the impurities. In the special cases of cultures of neurons and muscle cells, the surface of the vessels is coated with gelatin, polylysine, or collagen to make them positively charged.

12.3.2 Culture Media

The cell culture medium usually provides a liquid medium with osmotically balanced salts and nutrients for proper growth and propagation of cells in vitro.

A balanced salt solution (BSS) with specified pH (7.2–7.4) and osmotic pressure provides the minimum essentials for cell culture, which is composed of inorganic salts and may include sodium bicarbonate and, in some cases, glucose. The salts are mainly $CaCl_2$ (anhydrous), KCl, $KHPO_4$, $MgCl_2$, $MgSO_4$, NaCl, $NaHCO_3$, $NaHPO_4$, and Na_2HPO_4. BSS forms the basis of many complete media. Complete media range in complexity from the relatively simple Eagle's minimum essential medium (MEM). MEM represents the basic balanced salt solution plus essential amino acids and water soluble vitamins. It has been modified and enriched by various scientists, and different specified media, such as Dulbecco's minimum essential medium (DME), RPMI 1640, McCoy's 5A, and CMRL 1066, have been developed. All these media are commercially available.

Phenol red (0.01 g/L) is added to the medium as an indicator for pH. It is red at pH 7.4 and becomes orange at pH 7.0, yellow at pH 6.5, lemon yellow below pH 6.5, pink at pH 7.6, and purple at pH 7.8.

The nutrient media contain nutritional, hormonal, and stromal factors for the survival and growth of cells in culture. They are supplemented with serum (5–20%), sometimes chick embryo extract, and penicillin/streptomycin (anti-bacterial), and anti-fungal agents for inhibiting contamination. Serum is the source of plasma proteins, peptides, lipids, carbohydrates, minerals, some enzymes, and cell growth factors. Serum increases the viscosity of the medium and, thus, protects the cells from mechanical damage, especially from the shear force generated due to the agitation of suspension cultures. Proteases in serum protect the cells, especially cells that have been dissociated with trypsin, from proteolysis. Chick embryo or other tissue extracts can be substituted by a mixture of amino acids and certain other organic compounds.

12.3.3 Sterilization

All apparatus, glass ware, and liquids that come in contact with cultures or other reagents must be sterilized by proper methods to kill microorganisms and spores.

Glassware is kept in containers (metallic canisters) or wrapped in aluminium foil and usually sterilized at 160°C for one hour in an oven. The plastic screw caps of cell culture tubes must be autoclaved separately.

The media minus serum and organic additives are sterilized by autoclaving at about 120°C for 20 minutes under 15 lb pressure. The caps on the bottles and tubes should be kept slack to prevent them from breaking in the course of autoclaving. Heat-labile serum, proteins, trypsin, and other additives are sterilized by filtration through a 0.2 μm porosity membrane filter. Nowadays, most of the media are supplied in a pre-sterilized condition. The cell culture preparations are made in laminar air flow cabinets to avoid contamination. The mouths of bottles containing the culture solution and the sterilized pipettes used to withdraw the culture medium are heated for a second on the open flame of a Bunsen burner or spirit lamp before each use, for maintaining sterility.

12.3.4 Atmosphere and Gas Phase

Animal (mammalian) cell cultures are maintained at 37°C in a humidified atmosphere of air and 5 per cent CO_2. O_2 and CO_2 are critical gases and their percentages may be varied according to the requirement of the cell type. Since culture media are usually buffered with bicarbonate, the concentration of bicarbonate in the medium and CO_2 tension in the gas phase must be at equilibrium. About 23 mM HCO_3^- is required when 5 per cent CO_2 is in the gas phase. This achieves the correct pH and osmolarity for a growing cell culture.

HEPES is a much stronger buffer in the pH 7.2–7.6 range and is now used in many cases at 10 or 20 mM, CO_2 is not necessary to stabilize the pH. When HEPES is used with exogenous CO_2, the HEPES concentration must be more than double that of the bicarbonate for adequate buffering. In spite of comparatively poor buffering capacity at physiological pH (7.2–7.4), bicarbonate buffer is preferred over other buffers, because of its low toxicity, low cost, and nutritional benefit to the culture.

12.3.5 Serum-free Media

It has already been mentioned that the culture media need to be supplemented with serum as a source of nutrients, growth factors, certain elements, and for higher viscosity for protection of cells from mechanical damage. However, in many instances, cells are propagated in serum-free media. The serum contains many undefined substances; thus, for certain in-vitro studies serum is not added to the standard cell culture medium; besides, a natural product like serum may sometimes be a source of contamination (mainly virus). Culturing the cells in serum-free medium and testing for the more specific requirements of primary culture make it possible to formulate a standard culture medium more suited for a particular cell type.

Appropriate nutritional and hormonal modifications are made for serum-free culture media. Thus, serum-free media become more complex, in terms of their constituents. Often they are more expensive than conventional media. For adhesion cell cultures in serum-free media, the plastic ware needs to be evenly coated with fibronectin present in the extracellular matrix and serum.

Transformed cell lines produce their own growth factors and may be cultured in serum-free media, supplemented with a mixture of amino acids, vitamins, and other organic compounds as per the requirement of a cell line.

Serum-free media also have some disadvantages. Each cell type requires a different recipe, the requirements may vary for malignant cell lines, even within one class of tumours. Standardization of a culture solution is a time-consuming and labour-intensive affair. This is the most serious limiting factor for serum-free culture media. Quite a few serum-free media with defined formulations for certain cell types are commercially available now.

12.3.6 Cell Density in Culture

An optimal density of the cells, which varies with cell type, is to be maintained in cultures for the proper growth and differentiation of a particular type of cells. The pH of the culture solution should be regularly checked by visual examination of the colour of phenol red. If there is a fall in pH, the old medium in the culture needs to be removed aseptically by suction and replaced with fresh medium warmed up at 37°C in an incubator prior to use (to avoid temperature shock to the cultured cells). The culture should be immediately returned to the incubator. Changing of the medium is termed as feeding. There are three phases (lag, log, and stationary) of growth for a cell type (Fig. 12.5). The cell density is highest in the stationary phase, and that is the time to relieve the density pressure by harvesting the cells to introduce them in a lower number in fresh cultures; this process is known as subculturing.

As the cell density reaches the maximum, for a monolayer culture, all of the available substrate is occupied, and for a suspension culture the cell number exceeds the capacity of the medium. This marks the beginning of the stationary phase when cell growth ceases or is greatly reduced and the cells need to be subcultured. Subculture involves removal of the medium along with the cells from the suspension culture and dissociation of the cells in the monolayer culture with trypsin,

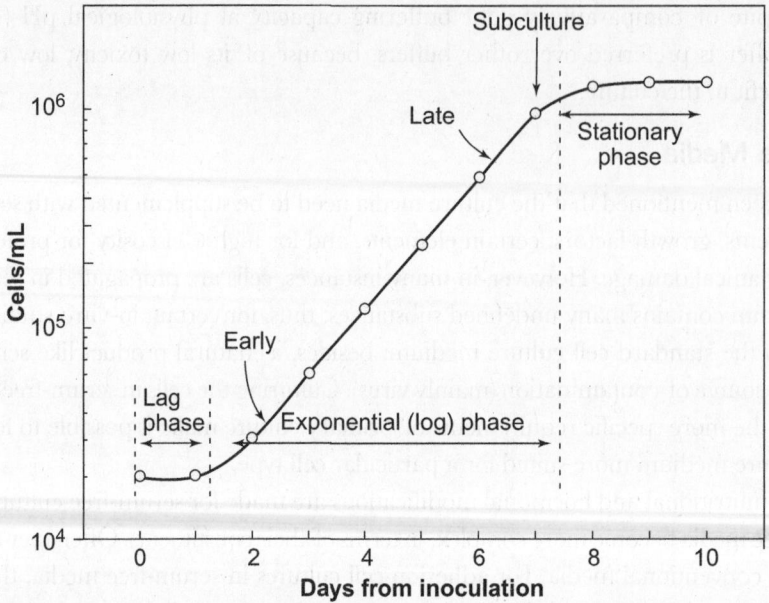

Fig. 12.5 The pattern of growth curve of a cell culture, showing three phases—lag, log, and plateau or stationary phases, and indicating the time of initiation of subculture. It has assumed a seeding density of 10^4 cells/mL, cell doubling or generation time of 24 hours, and a lag phase of 24 hours; all these may change with different cell lines.

dispase, and collagenase, or sometimes pronase, depending on the nature of the cell type. After treatment with proteases, the cells are washed with BSS and suspended in the appropriate volume of fresh medium so that the cell number is suitable for seeding. At low cell densities and under frequent subculture, normal cells are retained in culture. Normal cells gradually tend to decline at higher cell densities and under prolonged subculture, as they are overrun by transformed cells, whose growth is not sensitive to high cell densities. Normally, subcultures are done after every seven days; this could be done after every three to four days for fast-growing cell lines.

12.3.7 Contamination

In spite of all the precautions and addition of antibiotics, contamination with bacteria, fungi, mycoplasma, and viruses may occur. Cell lines are often maintained without antibiotics to avoid the persistence of some undetected contaminants. The cell cultures should be regularly checked for any drastic change in the pH (colour change) of the medium, cloudiness in the medium, granular structures on the outside of the cells (under microscope), and non-regular floating objects in the medium as signs of contamination. For a continuous culture, mycoplasma contamination is checked after every three months by a fluorescent DNA technique.

12.3.8 Eradication of Contamination

Contamination with other microorganisms and cells foils the very purpose of culture of a particular cell type. People experienced in tissue culture have developed several measures to fight the problem of contamination.

Bacteria, Fungi, and Yeasts

The general rule is that contaminated cultures are discarded. Decontamination can only be attempted when it is absolutely necessary to retain the cell strain or to save a cell line. This should be done under quarantine and with expert supervision.

The cultures should be washed several times with a high concentration of antibiotics by rinsing the monolayer or by centrifugation of cells in suspension. Then three subcultures with, and three without antibiotics are to be grown. Every two days, the culture medium is to be replaced with high-antibiotic medium. Then subcultures should be done without antibiotic and checked by phase-contrast microscopy for elimination of all contamination.

Sometimes, complete decontamination is not possible, particularly with yeast, and in due course hardier, antibiotic-resistant strains may appear.

Viruses

There are no reliable methods for elimination of viral contamination. Disposal or tolerating them, if possible, is the only means.

Mycoplasma

For mycoplasma contamination, the general rule of eradication is as with other forms of contamination. The cultures should be discarded for autoclaving or incineration. In extreme necessity, decontamination may be attempted with agents like kanamycin, gentamycin, tylosin, polyanetholsulfonate, and 5-bromouracil in combination with UV light. Tylosin is quite effective. Ciprofloxacin and BM-Cycline are also used.

Sterilization prior to disposal of contaminated cultures

The contaminating agents grown in cultures supplemented with antibiotics as a rule are considered potentially biohazardous. These cultures must be sterilized by autoclaving or immersing in a sterilizing agent such as hypochlorite before disposal. Hypochlorite is effective and can be easily washed off the items that need to be reused; at the same time, it is highly corrosive. It can even corrode stainless steel. So gloves and a lab coat should be worn while using hypochlorite, and soaking baths and cylinders are to be made of polypropylene.

Cross contamination by other cell lines

Cross infection of a cell line by other fast growing cell lines maintained in the laboratory may occur. HeLa and other rapidly growing cell lines have been found to cause cross contamination of other cell lines. For prevention of such contamination, one important precaution is not to simultaneously open media bottles and culture flasks that contain different cell lines. The same pipette should never be used more than once, and for different cell lines.

12.4 CELL SYNCHRONIZATION

All the cells in a synchronous culture divide at the same time. A synchronously growing population is essential for studying several facets of cell growth and metabolism, cellular proteins regulating cell cycle, expression of particular genes or enzymes related to a particular cellular event, and so on.

Two principles are adopted to initiate a synchronized culture: (i) a cell population at a particular stage of the cell cycle is fractionated or (ii) cells are allowed to accumulate at a particular point of the cell cycle by using metabolic blockers, so that on return to regular culture, the cells will be at the same phase.

12.4.1 Fractionation

Mitotic shake-off

It is the simplest technique to collect synchronized cells. Monolayer cells tend to round up at metaphase and are loosely attached to their substrate; they can be easily shaken off by tapping; 90–97 per cent of cells isolated in this fashion are in mitotic phase. They can be grown synchronously into the next G_1 phase.

Yield of cells obtained by this method is low as only a small fraction of the entire population of the cells undergoes mitosis at a given point of time. However, the cells can be collected repeatedly with some time gap. The batches of collected cells should be kept in ice, for a maximum of 2 hours, before they are transferred to cultures at 37°C. Another way of increasing the yield is to use the drug nocodazole which disrupts non-kinetochore microtubules (necessary for spindle formation to separate the chromosomes) to arrest the cells at metaphase.

Sedimentation

Cells increase in size linearly as they proceed through the cell cycle. Based on this fact, the cells can be separated. Layering cells over a serum gradient in medium, at unit gravity (1g, without centrifugation), allows the cells to settle through the medium according to their size and sedimentation velocity. Usually, small and dense cells sediment at the bottom. However, unless the cell sizes are very different at different phases of cell division, the separation of cells by this method is not very effective. Rather, nowadays, centrifugal elutriation (giving moderate resolution, but a high yield) or fluorescence-activated cell sorter (high resolution with a low yield) is used for cell separation based on cell size.

Centrifugal elutriation

The centrifugal elutriator is a device for increasing the sedimentation rate in a specially designed centrifuge and rotor. The major advantage of this technique is that no chemicals are used for separation.

Fluorescence-activated cell sorting

A fluorescence-activated cell sorter (FACS) separates the cells rapidly on the basis of size and fluorescence or light emission from individual cells. All cells with similar properties are collected into the same tube. The DNA content of a cell varies depending on the stage of cell division. A non-toxic, reversible DNA stain, such as Hoechst 33342, can be used for differential emission of light from the stained cells that differ in their DNA content.

12.4.2 Metabolic Blockade

Two types of metabolic blockades can be used.

G_1 synchronization by isoleucine deprivation

Cells can be arrested in G_1 phase when isoleucine is removed from the medium for 24 hours and then replenished. After restoration of isoleucine, the cells enter S phase within a few hours. This kind of nutritional deprivation for synchronization may also be performed by withdrawal of serum from the medium.

S phase synchronization by inhibition of DNA synthesis

Synthesis of deoxyribonucleotide triphosphates can be inhibited by using excess thymidine and hydroxyurea. The inhibition efficiently synchronizes the cell population in the S phase. The plant protein mimosine, other agents like cytosine arabinoside and aminopterin are also used for synchronization of cells at G_1/S and S phase.

12.5 SCALE-UP OF CELL CULTURE FOR BIOTECHNOLOGY INDUSTRY

The laboratory scale for usage of cell culture has been discussed so far. The basic principles remaining the same, the cell culture technique can be scaled up for production of cells in bulk for industrial purpose.

12.5.1 Suspension Cell Culture

Suspension cell cultures of 1×10^9–1×10^{10} cells can be produced in a stirrer culture flask of 1–10 L capacity (Fig. 12.6). Larger scale cultures of 1×10^{11} to 1×10^{12} cells need apparatus ranging from 100 L to industrial pilot plants with a capacity of 1,000 L. Industrial production uses 5,000 to 20,000 L bioreactors. The scaled up version of a culture flask is referred to as a bioreactor. Often the bioreactor is termed as a fermenter, although the term fermenter is derived from microbiological culture systems, designed originally for growth of bacteria and yeast. The large-scale culture of mammalian cells is far more expensive than microbial culture.

12.5.2 Large Stirrer Culture Flask

When the depth of the culture medium exceeds 5 mm (as the ratio of surface area to volume decreases), agitation of the medium is necessary. Stirring of such cultures is done by slow movement of a magnet encased in a glass pendulum hanging from the top and induction of a magnetic stirrer. A rotating paddle-like structure can also be used. The stirring speed is kept between 30 and 100 rpm, to prevent cell sedimentation; it should never produce a shear force as this can damage the cells. Bubbling in of CO_2 and air is required to maintain adequate gas exchange. An inlet gas tube with a 2–3 mm internal diameter reaches almost to the bottom of the vessel, but remains clear of the pendulum when it is stirring. An in-line sterile micropore filter, 25 mm diameter, 0.2 μm porosity, is fitted to the gas entry line.

An antifoam agent needs to be added to the culture when serum concentration exceeds 2 per cent. In serum-free culture, 1–2 per cent carboxymethyl cellulose is added to increase the viscosity of the medium for protection of cells.

The cells can be harvested by removing the in-line filter from the gas line or from another non-engaged port and attaching a peristaltic pump to take the cells and medium out. Then the

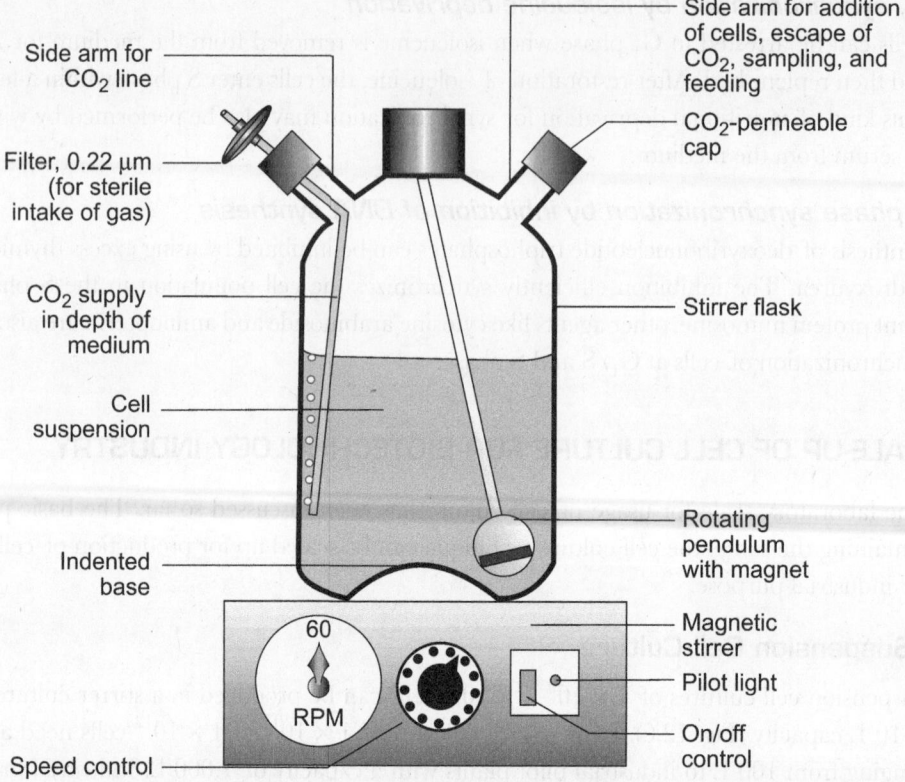

Side arm for addition of cells, escape of CO_2, sampling, and feeding

CO_2-permeable cap

Side arm for CO_2 line

Filter, 0.22 µm (for sterile intake of gas)

CO_2 supply in depth of medium

Stirrer flask

Cell suspension

Indented base

Rotating pendulum with magnet

60

RPM

Magnetic stirrer

Pilot light

Speed control

On/off control

Fig. 12.6 Suspension cell culture in a large stirrer culture flask.

cells can be separated by centrifugation. The supernatant medium can be used for isolating of any specific cell product, if necessary.

12.5.3 Biostat

A large stirrer culture flask can be modified into a chemostat or biostat for continuous suspension culture (Fig. 12.7) with continuous quantity matched input of fresh medium and output of cell suspension. Initially the cells are grown to the mid-log phase (Fig. 12.5) (monitored by daily cell counts), then the continuous, or intermittent input and output systems are turned on. The flow rate is maintained by variable peristaltic pumps on the inlet and outlet line. There are certain other modifications of the suspension cell culture technique for industrial purposes.

12.5.4 Air-lift Fermenter

The bioreactor has two concentric cylinders, with the inner cylinder shorter at both ends than the outer one creating two interconnected chambers (Fig. 12.8). Five per cent CO_2 in air is pumped into a porous steel ring at the bottom of the central cylinder. The bubbles make an upward stream

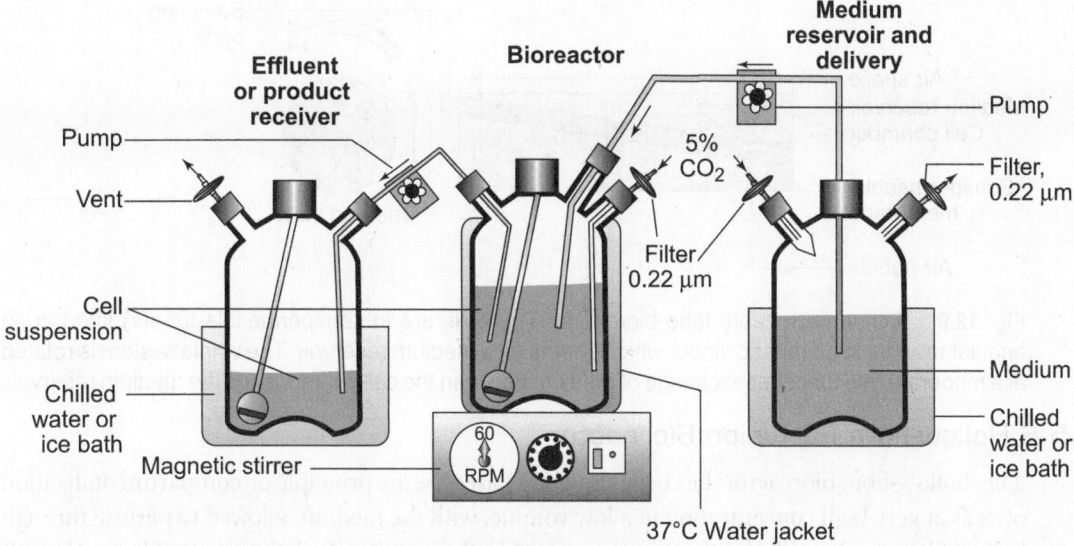

Fig. 12.7 Biostat for continuous suspension culture.

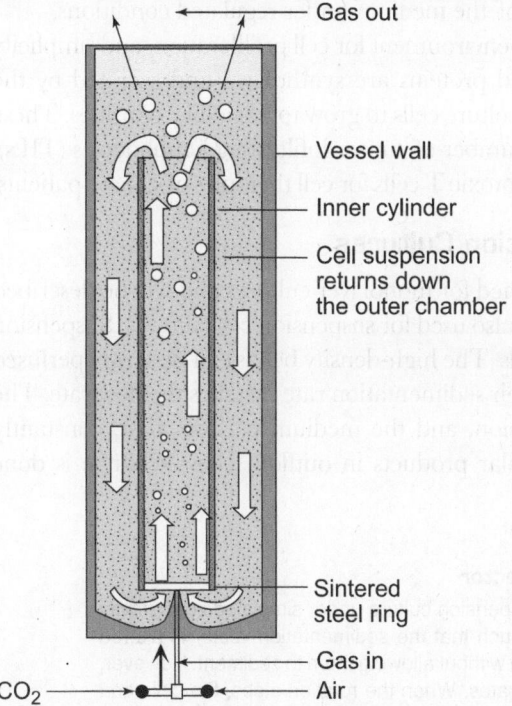

Fig. 12.8 Air-lift fermenter or bioreactor for large-scale suspension cell culture.

of medium along with the cells. The cell suspension recycles back to the bottom along the outer chamber. CO_2 is vented from the top. Biotechnology industries use this apparently simple type of fermenter quite extensively.

12.5.5 Rotating Chamber

The conventional roller bottle culture usually used for monolayer cell culture can also be employed for suspension cell culture with the purpose of proper mixing and aeration (see monolayer culture). The rotating chambers can have different designs. One such design is the Techne permeable tube bioreactor.

The cells in high concentration are enclosed in semipermeable tubes fitted in an angular manner in an outer cylinder of the Techne system, containing the culture medium. An air bubble is maintained in each tube (Fig. 12.9); the bubble moves up and down the tubes during rotation, ensuring mixing and movement in the cell mass of the tube. The whole Techne system is mechanically rotated by a motor for easy percolation of nutrients in and diffusion of waste products out of the cell culture chamber to the outer culture fluid. The cellular macromolecular products, such as antibodies, accumulate in a semipermeable cellular compartment till they are collected. From time to time, the medium may be replaced without disturbing the cells or product in the semipermeable tubes.

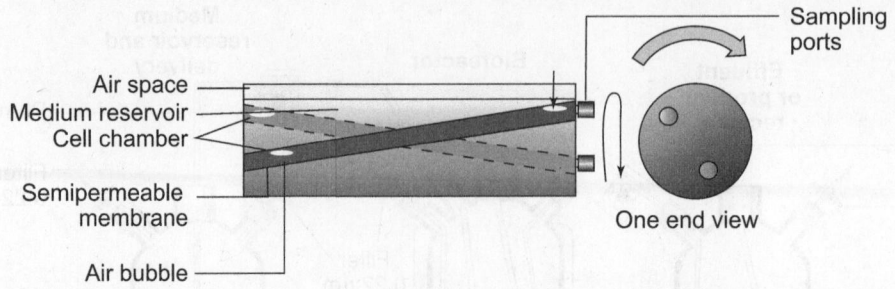

Fig. 12.9 Techne permeable tube bioreactor. The cells are in semipermeable tube(s) fitted in an angular manner in an outer cylinder which serves as a medium reservoir. The whole system is rotated at a moderate rate for better exchange of medium between the cell chamber and the medium reservoir.

12.5.6 Hollow-fibre Perfusion Bioreactor

The 'hollow-fibre bioreactor' has been designed based on the principle of compartmentalization of cells at very high concentration in a low volume, with the medium allowed to perfuse through hollow fibres to the cell compartment, as stated in the rotating Techne system. The medium is circulated from a large reservoir to the culture chamber by a peristaltic pump (Fig. 12.10). Aeration and CO_2 exchange take place in the reservoir of the medium under regulated conditions.

This type of bioreactor provides a shear-free environment for cell proliferation, and simplicity of operation. Large quantities of concentrated proteins are synthesized and secreted by the cells. Hollow-fibre perfusion bioreactors allow culture cells to grow to tissue-like densities. These bioreactors have been used for enriching the number of tumor infiltrating lymphocytes (TILs), lymphokine activated killer (LAK) cells, and cytotoxic T cells for cell therapies of cancer patients.

12.5.7 Fluidized Bed Reactors for Suspension Cultures

Originally the 'fluidized bed reactor' was designed for monolayer cultures and will be described later in that connection. This type of reactor is also used for suspension cell cultures. Suspension cells occupy the interstices of microcarrier beads. The high-density beads can be slowly perfused with medium from below, at such a rate that their sedimentation rate matches the flow rate. The beads, therefore, remain in stationary suspension, and the medium perfuses past, constantly replenishing nutrients and collecting the cellular products in outflow. Gas exchange is done outside the reactor.

NASA Bioreactor

NASA fabricated a rotating chamber for growth of suspension culture under simulated zero gravity by slowly rotating the chamber; the rotation rate is such that the sedimentation vector is altered continuously, to keep the cells stationary in suspension without allowing them to sediment. However, the cells tend to form three-dimensional small aggregates. When the rotation stops, the cells and aggregates sediment, and the medium can be replaced.

12.5.8 Bioreactor Process Control

Computer aided process control is a regular feature for a modern bioreactor of large capacity. The process control regulates input and output of medium and gas on the basis of conditions

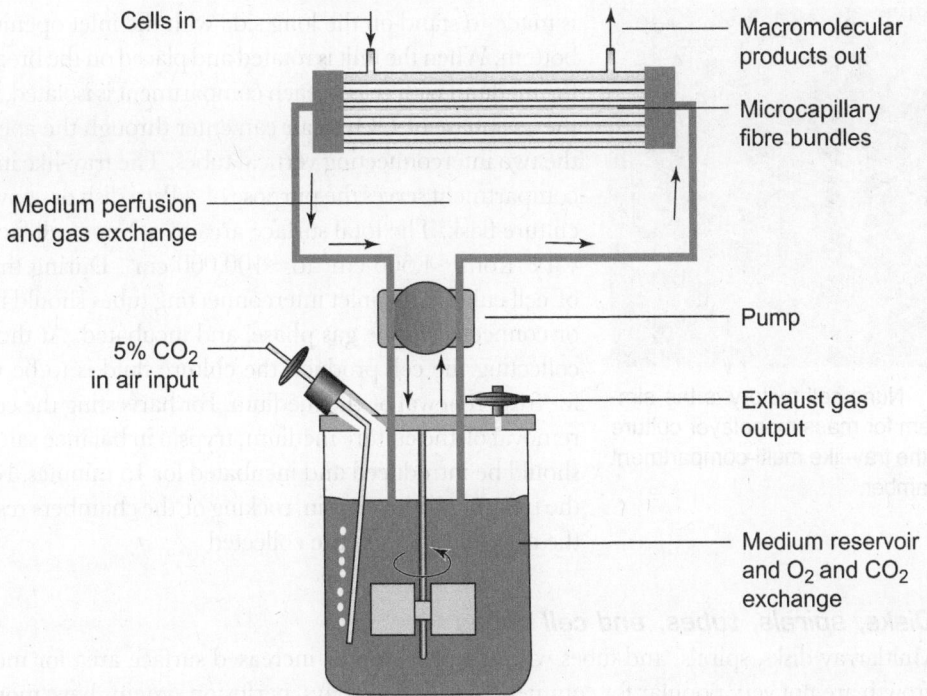

Fig. 12.10 Hollow-fibre perfusion bioreactor. Cells in high concentration are cultured in the chamber of the microcapillary fibre bundles, medium is circulated from a reservoir to the culture chamber by a peristaltic pump. Aeration and CO_2 exchange are regulated in the reservoir of medium.

within the bioreactor, as revealed by readings from oxygen, CO_2, glucose, pH, viscosity, and turbidity electrodes. The stirring speed and temperature of the medium can also be monitored and regulated through the process control.

12.6 MONOLAYER CELL CULTURE (SCALE-UP)

Scale-up of suspension cell culture for industrial purpose is comparatively simpler than that of monolayer cell culture. The main objective for monolayer culture is provision of extensive surface for adherence and proper growth of the cells. This has been achieved by designing culture vessels differently with the optimal provision of medium and gas.

12.6.1 Multisurface Propagator

Nunc cell factory

It provides the simplest system for mass monolayer culture of cells. Nunc cell factory consists of a rectangular tray-like multi-compartment culture chamber, interconnected at two adjacent corners by vertical tubes (Fig. 12.11). The positions of the apertures for each tray from the vertical tubes are such that the medium can flow between the tray-like compartments only when the whole unit

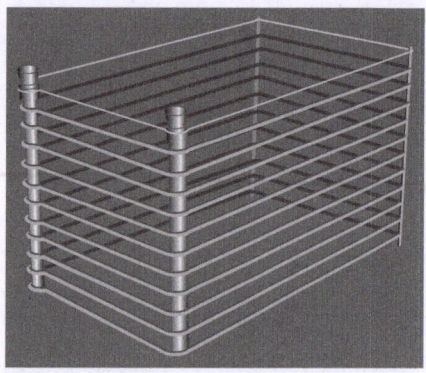

Fig. 12.11   Nunc cell factory is the simplest system for mass monolayer culture of cells in the tray-like multi-compartment culture chamber.

is made to stand on the long side with the inlet opening at the bottom. When the unit is rotated and placed on the broader end, the medium with cells in each compartment is isolated, and only the gas phase of CO_2 in air can enter through the apertures in the two interconnecting vertical tubes. The tray-like individual compartment serves the purpose of a Petri dish or conventional culture flask. The total surface area of a Nunc cell factory can vary from ~1,500 cm^2 to ~100,000 cm^2. During the period of cell culture, the inlet interconnecting tubes should be sealed or connected to the gas phase, and incubated. At the time of collecting the cell product, the culture fluid is to be removed for fresh renewal of the medium. For harvesting the cells, after removal of the culture medium, trypsin in balance salt solution should be introduced and incubated for 15 minutes. Following the treatment with trypsin, rocking of the chambers resuspends the cells and the cells are collected.

Disks, spirals, tubes, and cell cubes

Multiarray disks, spirals, and tubes with the provision of increased surface area for monolayer growth are not very popular for commercial use. Nowadays, perfusion systems have more or less replaced these multisurface propagators.

Cell cube, marketed by Corning Costar is a multisurface perfusion system. It is a hollow polystyrene cube with multiple inner lamellae. The lamellae support monolayer cell growth on both surfaces when perfused with oxygenated, warm medium.

Roller bottle culture

The cells in medium are placed in a roller culture bottle and the cap is secured in place, then the bottle is placed on a roller rack for slow rotation at 20 revolutions per hour (Fig. 12.12) until the cells attach to the inner surface of the bottle (24–48 hr). Then the rotational speed is increased to 60–80 revolutions per hour with the increase in cell density. If the cells are non-adherent, they will be agitated along with the medium by the rolling action.

Advantages of the roller culture over static monolayer culture

- Increase in the surface area.
- The constant, but gentle, agitation of the medium effectively increases the ratio of the medium's surface area to its volume.
- Gas exchange takes place at an increased rate through the thin film of medium over cells in the course of rotations.

Variations   There could be many variations for roller cultures, such as the size of the roller bottles, appropriate roller racks, rotational speed, precoating the inner surface of the roller bottles with fibronectin or polylysine for better attachment of the cells.

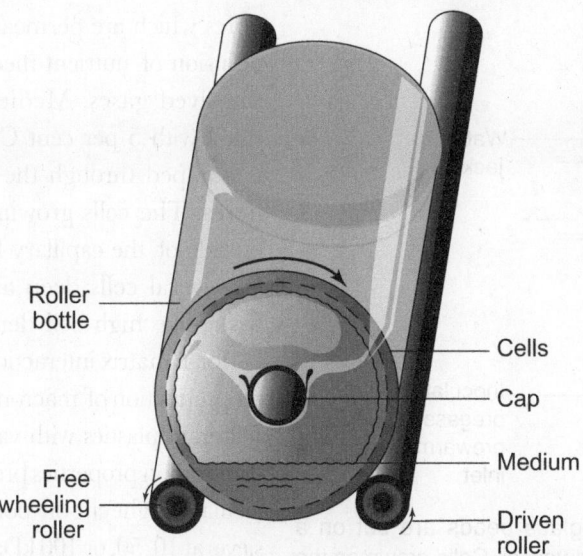

Roller bottle

Cells

Cap

Medium

Free wheeling roller

Driven roller

Fig. 12.12 Roller bottle culture. Monolayer cell culture on the inner wall by bathing with constant and gentle agitation of the medium. The roller is driven by a motor.

12.7 MICROCARRIERS

Microbeads ranging in diameter from 90–300 micrometres, and made of glass, plastic, gelatin or collagen can be used in culture vessels for monolayer cells, to effectively increase the surface area around the circumference of the beads. Thus, microbeads maximize the ratio of the surface area of the culture to the volume of the medium; it could be upto 90,000 cm^2/L, depending on the size and density of the beads. Furthermore, the cells adhered to the beads provide a situation analogous to suspension cultures, especially when a pendulum or paddle is in operation to achieve efficient stirring without damaging the beads. Stirring speed needs to be increased with higher density beads.

Growth of some cells is influenced by the radius of curvature and they may prefer larger bead diameters. Cells are usually seeded in higher concentrations to begin with.

12.8 PERFUSED MONOLAYER CULTURE

Perfusion of the static cell culture with medium facilitates continuous replacement of aerated medium and product for better cell growth and output. There are several variations of the method for culturing monolayer cells. Here, three principal methodologies that are used in the biotechnology industry, are discussed.

12.8.1 Hollow-fibre Perfusion

This technique has already been stated in connection with the suspension culture bioreactor (Fig. 12.10). Adherent cells grow on the outer surface of bundles of plastic microcapillary

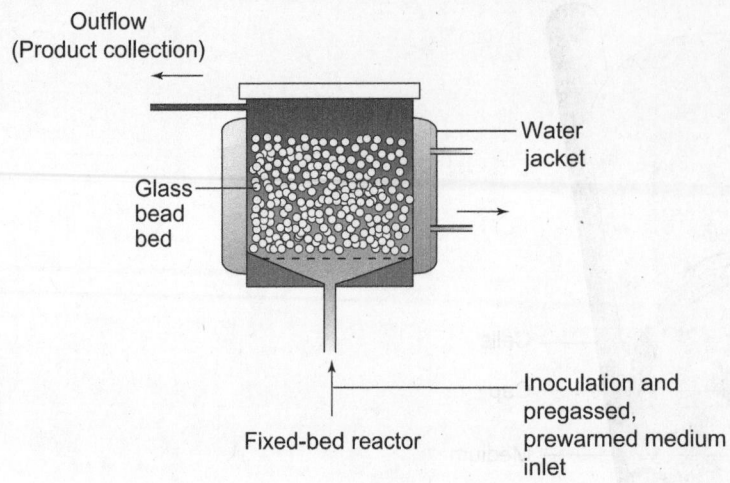

Fig. 12.13 Fixed-bed reactor. Usually glass beads are set on a perforated base at the bottom of the culture vessel. Cells grown on the surface of the beads are perfused with the medium.

fibres which are permeable to the diffusion of nutrient medium and dissolved gases. Medium, saturated with 5 per cent CO_2 in air is pumped through the capillary fibres. The cells growing on the outside of the capillary fibres can be several cells deep and reach tissue-like high cell density, and perform matrix interaction for better production of macromolecules. Different plastics with variable ultrafiltration properties provide molecular weight cut-off points (like a sieve) at 10, 50, or 100 kDa, regulating the diffusion of macromolecules in the outflow of medium.

High-density, tissue-type disposition of the cells in this type of culture, allows them to behave as they would in vivo. For example, in such cultures, choriocarcinoma cells release more human chorionic gonadotrophin than they do in conventional monolayer culture and colonic carcinoma cells produce more of carcino embryonic antigen (CEA). However, several technicalities are involved in setting up the chambers, and the system is expensive.

12.8.2 Fixed-bed Reactor

A dense bed of glass beads is set on a perforated base at the bottom of the culture vessel (Fig. 12.13). The aerated medium is allowed to perfuse upward through the bed or percolate downward by gravity to the cells on the surface of the beads. Once the culture is set, the beads are not moved, and the product is collected with the spent medium in a reservoir.

12.8.3 Fluidized-bed Reactor

Porous beads of lower density, made of ceramics or ceramics and collagen are used so that they remain suspended in an upward stream of medium when the flow rate of the medium matches the sedimentation rate of the beads. In monolayer culture, cells adhere to the outer surfaces as well as the interstices of the porous beads. (If the system is used for suspension culture, cells lodge in the beads).

Problem of monitoring cell growth in fixed-bed and hollow-fibre bioreactors

In scaled-up versions of monolayer culture in a fixed-bed or hollow-fibre bioreactor, the cells cannot be directly observed, and thus, neither progress of cultures, nor the number of cells in culture can be ascertained. Today, a nuclear magnetic resonance (NMR) assay is being used for the purpose.

12.9 APPLICATIONS OF CELL CULTURE

Cell culture provides both cellular products and number of cells. Depending on the type of cell, there are product variations. Cell culture technology provides the opportunity to expand a particular cell type or combination of cell types into an organ for tissue engineering. Multipotent embryonic stem cells can be propagated in vitro for differentiating into different types of tissues under the influence of various differentiating factors. Thus, in-vitro cell culture products have opened a new horizon for treating different kinds of human ailments.

Some cell culture products (Table 12.1) and commonly used cell lines are listed in Table 12.2.

Table 12.1 Cell culture products with applications

Products	Applications
Cytokines	
Interleukin-2 (IL-2)	Stimulate lymphocytes for better immunity
Interleukin-3 (IL-3)	Promotes haematopoietic progenitor cells
Interferon-γ (IFN-γ)	Inhibits viral replication, and used for cancer treatment
Erythropoietins	
Erythropoietin -α	Treating anaemia resulting from cancer and chemotherapy
Erythropoietin -β	Anaemia arising out of kidney diseases (kidney is the site of production)
Blood clotting factor	
Factor VIII	Treats haemophilia
Human growth hormones	
hGH	Treatment of growth deficiency in children, and renal cell carcinoma
Somatotropin	To cure chronic renal insufficiency, Turner's syndrome
Monoclonal antibodies therapeutic	
Anti-lipopolysaccharide	Treatment of sepsis
Murine anti-idiotype/human B-cell lymphoma	Treatment for B-cell lymphoma
Monoclonal antibodies diagnostics	
Anti-fibrin 99	To detect blood clot
PR-356 CYT-356-in-111	Diagnosis of prostate adenocarcinoma
Plasminogen activator	
Urokinase-type plasminogen activator	To treat acute myocardial infarction
Vaccines	
HIV vaccines (gp120)	For AIDS prophylaxis and treatment (under trial)
Polio vaccines	For poliomyelitis prophylaxis
Rubella vaccine	For measles prophylaxis

Table 12.2   The most commonly used cell lines in biotechnology

Cell line	Mammalian source
CHO	Chinese hamster ovary
MDCK	Dog cocker spaniel kidney
HeLa	Human cervix carcinoma
NS0	Myeloma
BHK21	Syrian hamster kidney fibroblast
HEK293	Human embryonic kidney
Vero	Monkey kidney cells
GH3	Rat pituitary tumour
WI-38	Human foetal lung cells
J558L	Mouse myeloma
HepZ	Rat liver cells
CPAE	Cow endothelium
STO	Mouse embryonic fibroblast

12.10 CRYOPRESERVATION OF CELL LINES

Cell lines are a valuable resource for research and industrial production of many materials. Raising a cell line is an expensive and time-consuming proposition. If a cell line is unique, it might be impossible to replace. Thus, maintenance and preservation of cell lines is essential. Instead of continuing with culturing and sub-culturing a cell line, which is expensive both in terms of time and money, nowadays, a stock of cells is frozen following a protocol and preserved in liquid nitrogen. This is known as cryopreservation.

Cells are usually collected from late log phase at a high cell density. It is better to have the cells of suspension culture from a clonal culture (derived from a single cell). The cells should be healthy, free of contamination and with specific characteristics (microscopically observed). The freezing medium is made by diluting the preservative dimethyl sulfoxide (DMSO) to 5–10 per cent or glycerol to 10–15 per cent, with culture medium. DMSO solution should be stored in a glass or polypropylene tube; it is a powerful solvent and leaches impurities out of rubber and some plastics. It should be handled carefully. Addition of serum upto 50 per cent or more, even with cells grown in a serum-free medium, improves survival of the cells after freezing. Serum should be washed off the cells obtained initially from serum-free cultures, immediately after thawing (at the end of freezing, to bring the cells back to the active phase).

The cells from a line are to be aliquoted in small polypropylene ampules, properly marked and cooled step by step and ultimately frozen following the standard protocol. Then the ampoules are put in a suitable basket to transfer to the tank of liquid nitrogen. Prior experience of handling the canister with liquid nitrogen is essential, otherwise there can be an accident. An entry with all the details of the item that has been frozen should be made in a record book.

12.11 RAISING MONOCLONAL ANTIBODIES FROM A CULTURE OF HYBRIDOMA CELLS

The fusion of a malignant cancer cell (immortal by nature) with a B lymphocyte immunized with an antigen immortalizes the hybrid cell for propagation in culture and continuous production of monoclonal antibodies. The fused hybrid cell is called a hybridoma and the monospecific antibodies produced by the clone of a hybridoma cell is called a monoclonal antibody. A clone is a population of cells derived as the progeny of a single cell. Köhler and Milstein's (1975) hybridoma technology revolutionized the use of mammalian cells in biotechnology with profound medical and commercial potential. Monoclonal antibodies can be generated on different scales for commercial purposes.

12.12 VACCINES RAISED FROM CELL CULTURE

Large-scale production of vaccines from cell cultures under controlled conditions is nowadays an industrial practice. Cell culture technology is mainly employed to raise safe virus vaccines. For virus vaccine production, cell cultures made from monkey kidney and chick embryos are used. Human diploid cells are also extensively used. They are carefully checked for pre-infection with any kind of viruses, if so, the cultures are rejected. The cells are grown in a monolayer, and then infected with the desired variety of virus.

Interestingly, the tissue culture process often attenuates the virus (turns into less virulent). For example, the polio viruses used in the orally administered Sabin vaccine are attenuated by growing the viruses in monkey kidney epithelial cells. A strain of rubella virus, grown in duck embryo cells and later in human cell lines, constitutes the attenuated measles vaccine.

12.13 IN-VITRO TISSUE ENGINEERING

A characterized cell line propagated at high density in the presence of appropriate extracellular matrix and soluble growth factors is called a histotypic culture. Basic histological architecture of a tissue is maintained in this type of culture. For example, vascular endothelial cells can form capillary tubules when grown in a collagen matrix in the presence of appropriate soluble factors. High density of homologous cells allows interactions among the cells. A three-dimensional matrix (architecture) helps in maintenance of the differentiated phenotype of cells in vitro. This research information leads to the concept of 'in-vitro tissue engineering'. Tissue Engineering is an emerging field and soon will help in the manufacture of implants for doctors to routinely repair or replace failing or ageing body parts.

Using this technology, in the near future, laboratory grown skin, cartilage, bone, blood vessels, liver and other organ parts will be available.

12.13.1 Skin

Skin is an important organ for mechanical and immunological protection, and maintenance of aseptic conditions of the internal structures of an organism and other necessary physiological functions. Burn patients usually die due to injury and loss of the skin. It has already been pointed out that in-vitro propagated skin can effectively treat serious burn injuries.

In the early 1970s, skin was the first organ to be grown in vitro. An irradiated (to stop cell division) 3T3 fibroblast cell line is grown with keratinocytes which constitute 90 per cent of the cells in the skin. 3T3 cells stimulate the growth of skin cells and their differentiation into epidermis.

The extracellular supporting matrix could be natural or synthetic, such as collagen, polygalactic acid (PGA), and synthetic polymers. Neonatal foreskin can be used as source of skin cells.

Several companies are involved in growing skin in culture. Living skin equivalent (LSE) comprising dermis epithelial layer and collagen matrix has been developed by a company. An LSE graft has a plus point; it does not elicit an immunological rejection response in the host and is gradually replaced by host cells.

12.13.2 Cartilage and Bone

The chondrocytes, precursor cells for cartilage, are grown in combination with PGA or collagen in culture. When all the factors and a suitable microenvironment is provided, the cartilage can be grown in vitro.

By shaping the extracellular matrix (collagen) or biodegradable polymer according to the need, seeding it with living cells, and bathing the whole matrix in culture medium with growth factors, tissue engineers are expecting to end up even with tissues such as bones for implantation. Blood vessels from the host will penetrate the graft and it will blend with the original tissue (under repair) and the artificial scaffold will dissolve with time (Fig. 12.14).

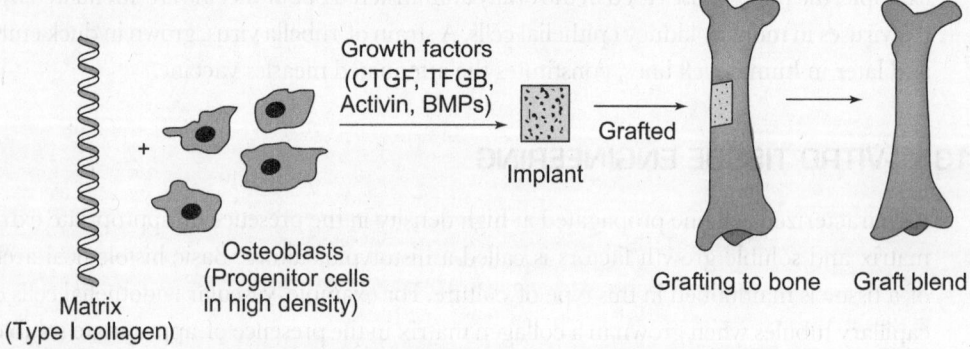

Fig. 12.14   Bone growth in vitro for repair by tissue engineering.

12.13.3 Liver

Experimentally a working liver (organ relica, 'organoid') was made from a combination of liver cells, collagen, and Gore-tex fibres, and growth factors, and was surgically implanted into the peritoneal cavity of a rat. These organoids have the potential to replace diseased organs or deliver genetically altered cells into a patient's body.

Implanting stem cells (totipotent cells) in the appropriate location can generate any kind of tissue, such as bone, cartilage, muscle and neural tissue, and so on.

The success of tissue engineering will depend on the understanding of complex cellular interactions, choice of the right kind of scaffold materials and proper growth factors.

12.14 CLINICAL USES OF STEM CELL CULTURE

Stem cells are self-renewing (proliferative) undifferentiated pluripotent cells, capable of differentiating into different types of cells under the influence of various factors and cytokines and in-situ microenvironments. Stem cells can be divided into two categories depending on the source of the cells.

(i) Embryonic Stem Cells, and (ii) Adult stem cells. Embryonic stem cells are derived from a blastocyst, which is in an early stage of development. In the blastocyst stage an embryo is a hollow sphere of cells with an inside cavity (blastocoel); there are a limited number of stem cells at the

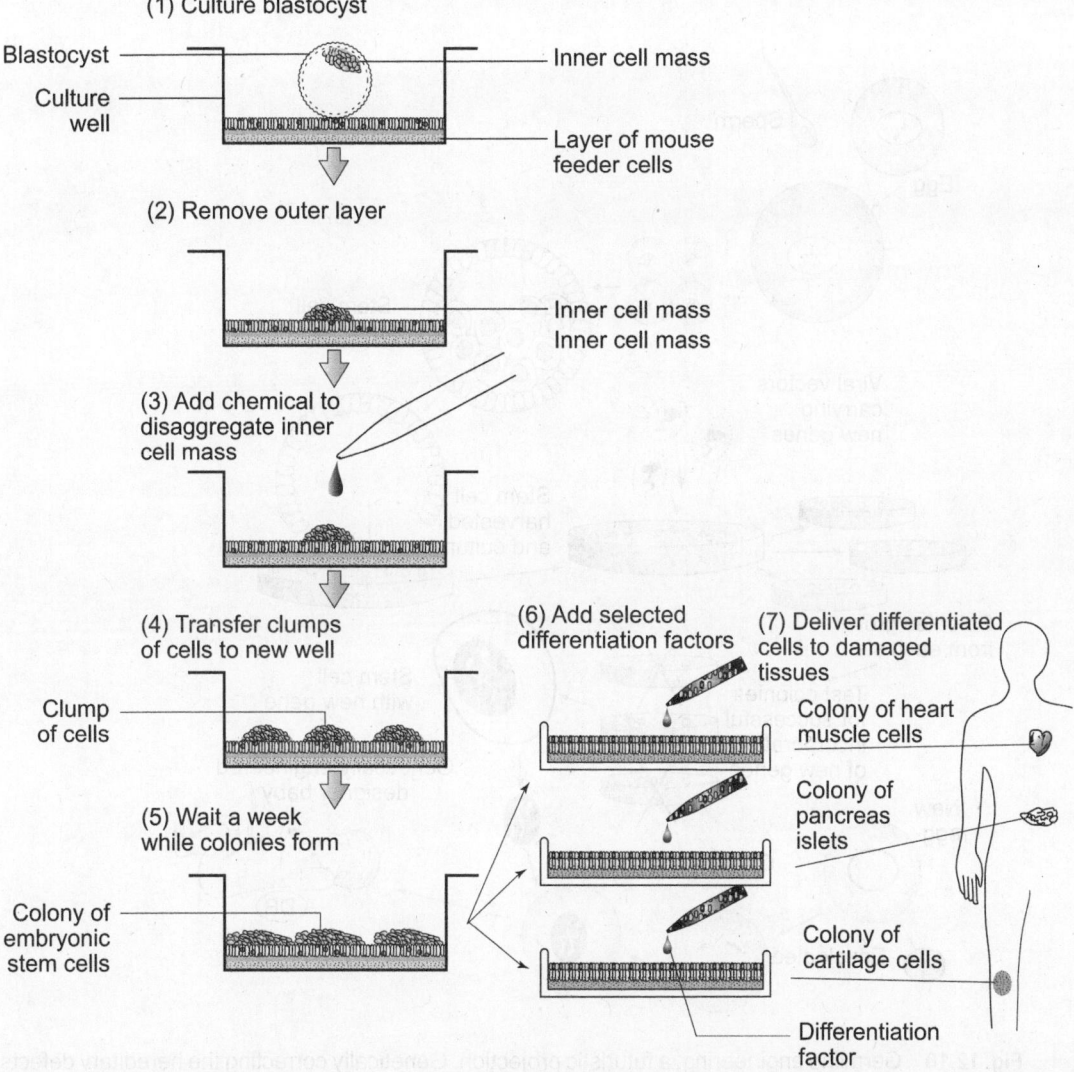

Fig. 12.15 Scheme of propagation and differentiation of embryonic stem cells in vitro for clinical use in man to repair damaged tissue in the future.

bottom of the blastocoel, which form the future embryo. Any of these cells is potentially able to form the total embryo or any tissue cell type under specific induction as already indicated. Human embryonic stem cells (hESC) offer the scope of expansion of the cells in culture to a desired cell type so that they can be used to treat various disease conditions. Stem cell technology can provide skin replacement for burn injuries, myocardial cells for repairing heart or neural tissue for treating Alzheimer's disease and Parkinson's disease, and hepatocytes for handling actue liver failure (Fig. 12.15). The strategy is successful in the mouse model, but there is a ban on the usage of human stem cells for clinical purposes, for ethical reasons. The in-vitro system can also be used for genetically correcting hereditary defects in human stem cells and implanting the corrected nucleus in an enucleated ovum to obtain genetically engineered designer babies (DB) (Fig. 12.16).

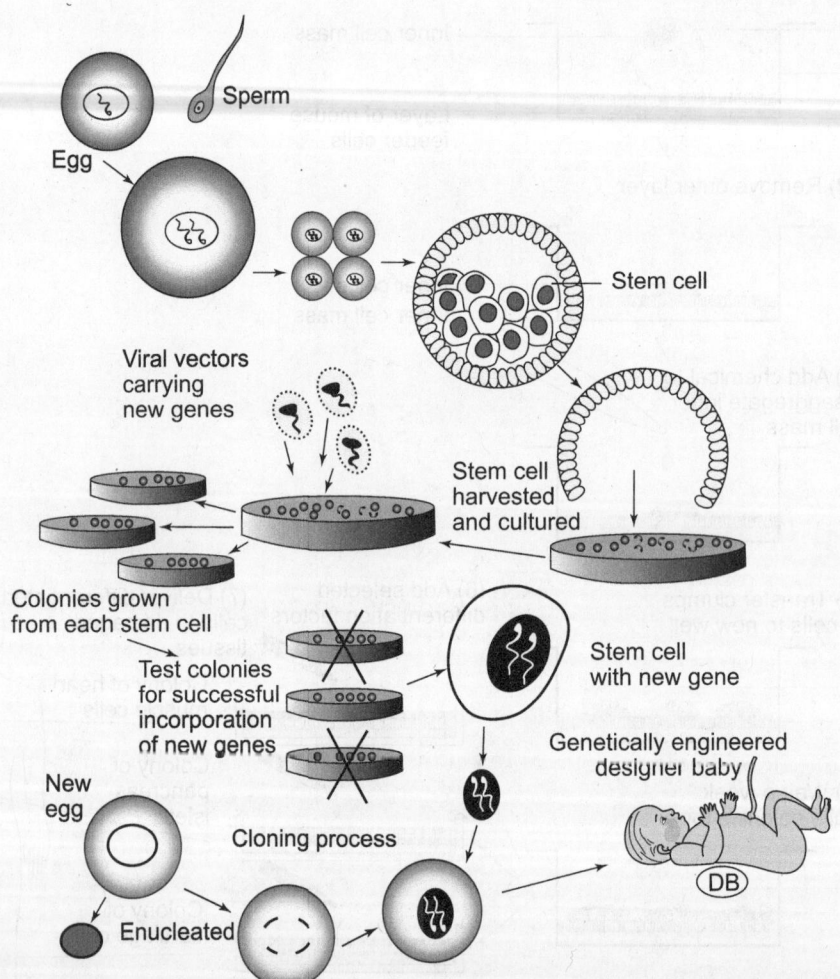

Fig. 12.16 Germline engineering, a futuristic projection. Genetically correcting the hereditary defects in human stem cells and then implanting the corrected nucleus into an enucleated ovum to produce a genetically engineered baby.

Adult Stem Cells

Haematopoietic stem cells (HSCs) in the bone marrow of adult persons were the first adult stem cells to be identified and categorized in accordance to the source. HSCs have the remarkable capability of regeneration and can completely restore the erythroid (RBCs) and leukocyte populations. Merely 10 per cent of a donor's total volume of bone marrow can restore the haematopoietic system. Bone marrow is simply transferred by intravenous injection to the histocompatible host. Bone marrow transfer can cure different disease conditions, including leukaemias. No cell culture is involved in the process. HSCs from umbilical cord blood can be transplanted for rejuvenation of damaged or defective bone marrow. The blood samples may also be cryopreserved for transplantation to the same individual at adulthood in the case of medical emergency.

Nowadays, different research groups have identified adult stem cells in other organs and a group has even claimed to induce stem cell status in skin cells.

SUMMARY

- The productivity of animal cells in culture comes under the purview of animal biotechnology.
- Freshly isolated cells from a tissue in a nutrient medium constitutes primary culture. They are slow to grow and soon stop proliferating.
- Some genetically altered cell types may continue indefinitely in the subcultures and establish continuous cell lines.
- The tissue culture may be initiated in three ways: (i) organ (bit) culture, (ii) primary explant culture, and (iii) cell culture.
- Sterilization of cell culture vessels and other wares, aseptic measures and use of laminar flow hood are a must for any successful tissue or cell culture.
- Cells in cultures can be maintained for growth in two ways: (i) monolayers (anchorage dependent, for primary cell culture), and (ii) suspension culture (anchorage independent, mostly for transformed cell lines).
- Culture or nutrient media contain balanced salt solution, and nutritional, hormonal, and stromal factors for the survival and growth of cells in culture.

- Animal (mammalian) cell cultures are maintained at 37°C in a humidified atmosphere of air and 5 per cent CO_2.
- All the cells in a synchronous culture divide at the same time. Synchronously grown cells are essential for studying cell growth, metabolism, protein molecules that regulate the cell cycle, expression of genes, and so on. A synchronized cell culture is initiated with a fractionated cell population at a particular stage of the cell cycle. This can also be done by cells at a particular stage by depriving a particular nutrient or by inhibiting DNA synthesis.
- The technique for cell culture can be scaled up for production of cells or their products in bulk for industrial purpose. There are many methods and bioreactors for the mass production of cells for biotechnological purposes. The designing of culture vessels differs for adherent and non-adherent cell types.
- Biotechnological and clinical usage of different types of cells from established cell lines or primary cultures including stem cells are diverse and are raising hopes for the success of in-vitro tissue engineering.

Georges Köhler and Cesar Milstein

Georges J. F. Köhler was born in Munich Germany. He began his studies in biology at the University of Freiburg in April 1965. He received a diploma in biology in 1971. After receiving his Ph D in 1974 from the University of Freiburg, Köhler did postdoctoral work in cell biology in Dr Milstein's laboratory at the Medical Research Council, Laboratory of Molecular Biology.

César Milstein was born in Bahía Blanca, Argentina, into a Jewish family. César obtained his undergraduate

degree from the University of Buenos Aires. On a British Council fellowship, he worked on the metal activation of phosphoglucomutase—a liver enzyme critical for glycogen breakdown. During this period, he collaborated with Fred Sanger and after graduating, joined his group at Medical Research Council (MRC), Cambridge. In his own laboratory at MRC, he focussed on antibody structure and the mechanisms of generation of antibody diversity. Along with Georges J. F. Köhler,

Georges Jean Franz Köhler
(17 March 1946 – 1 March 1995)

César Milstein
(8 October 1927 – 24 March 2002)

Source: http://www.nobelprize.org/nobel_prizes/
medicine/laureates/1984/kohler.html

then a postdoctoral fellow in Milstein laboratory, César developed the hybridoma technique for production of monoclonal antibodies. This technology revolutionized the use of antibodies in science and medicine, in the recombinant DNA era. Consequently, the duo of Köhler and Milstein along with Niels Kaj Jerne received the Nobel Prize for Physiology and Medicine in 1984. César continued to make significant progress in the field of

antibodies and was bestowed with several honours including Fellow of the Royal society, the Copley Medal (1989), and the Companion of Honour (1995).

Prior to joining the Milstein laboratory, Köhler had obtained his PhD in 1974 from the University of Freiburg after completion of his undergraduate degree in the same university. In 1984 he became the director of the Max Planck Institute for Immunobiology where he worked until his death.

EXERCISES

Objective Questions

1. Multiple Choice Questions
   (i) Fluidized-bed reactor allows culture of
       (a) adherent cells
       (b) non-adherent cells
       (c) both types of cells

   (ii) For production of monoclonal antibodies of a particular specificity, hybridoma cells are produced by fusion of malignant cells with
       (a) Virgin B lymphocytes
       (b) B lymphocytes immunized with an antigen
       (c) T cells immunized with an antigen

2. Match the statements in Column 2 with the items in Column 1.

Column 1	Column 2
(i) Tissue cultures include only	(a) histotypic tissue culture (b) organ culture (c) cell suspension culture (d) tissue and cell culture
(ii) The first published account on animal cell culture used	(a) muscle cells of mouse (b) embryonic kidney cells from chicken embryo (c) embryonic nerve cells of frog (d) human leukocytes

3. Mark the following statements as True or False.

(a) Frequent subculturing of a cell line limits the chances of arising of spontaneous variant cells.

(b) Presence of divalent cations, Ca^{2+} and Mg^{2+} is essential for cell adhesion.

(c) There is simultaneous need of EDTA, Ca^{2+}, and Mg^{2+} for adherent cell culture.

(d) Soda lime glass is preferred over alum borosilicate glass for glassware used for cell culture.

(e) Residual activity of trypsin is neutralized by the serum of the culture medium.

(f) A syringe fitted with a fine gauge needle may help in preparing a finer cell suspension.

(g) Phenol red indicator in culture solution becomes lemon yellow when pH is more than 7.6.

(h) Usually heat sterilization is done at 360°C for an hour.

(i) Subculture from a cell culture is made at the early log phase.

(j) Sterilization of contaminated cultures before disposal is essential.

(k) Conventional roller bottle culture is used only for suspension cell culture.

(l) The bubble maintained in 'Techne permeable tube bioreactor' helps in aeration.

(m) 'Hollow-fibre perfusion bioreactor' provides a shear-free environment for cell proliferation.

(n) Nunc cell factory can be used for both adherent and non-adherent cell cultures.

(o) In roller bottle culture, there is an increase in the ratio of the medium's surface area to its volume.

(p) Microbeads used for monolayer cell culture, have a diameter in the range of 500 µm to 5 mm.

(q) In hollow-fibre perfusion technique, cells grow with very high cell density.

(r) The tissue culture process often attenuates the virus to be used as vaccine.

Review Questions

1. Indicate the factors responsible for the development of cell culture techniques in the recent past.

2. Mention five important points to show the advantages of the cell culture system.

3. (i) What is the purpose of a laminar flow hood in cell culture technology? (ii) List the basic differences in the make and use of horizontal and vertical laminar flow hoods.

4. (i) What is the purpose of adding serum in certain types of cell culture. (ii) For what kind of cell cultures, are serum-free media used?

5. What do you understand by cell synchronization in culture? Mention briefly the techniques used for synchronization of cells in cultures.

6. Describe in short the 'Air-Lift Fermenter' for mass cell culture. Indicate why this type of fermenter is preferred over the 'Large Stirrer Culture Flask' in biotechnology industries.

7. What are the functions of bioreactor process control?

8. What are the advantages of perfusion of a static cell culture with medium?

9. Provide a list of five cell culture products with their applications.

10. What is meant by cryopreservation of cell lines? What are the precautions one should take while carrying out cryopreservation?

11. Write short notes on the following:

(a) In-vitro tissue engineering

(b) Clinical uses of stem cells

13

Environmental Biotechnology

LEARNING OBJECTIVES

♦ Components of environment
♦ Environmental pollution
♦ World conventions on environment
♦ Involvement of biotechnology as a remedy for pollution
♦ Treatment of solid waste and waste water
♦ Bioremediation of contaminated land and water

INTRODUCTION

The earth which originated about 4.6 billion years ago, is so far the only planet that supports life, that too in many forms. The origin of life was possible because of the presence of a proper environment through two billion years of physical and chemical evolution. The abiotic environment allowed the transformation of inorganic molecules to organic compounds through chemical evolution and made the origin of life possible. Following the origin of life, evolution took more than two billion years to add the diverse variety of biotic components to the environment. Thus, today's environment on the earth is the outcome of the natural processes of more than four billion years.

Human beings evolved approximately 50,000 years ago and are the latest arrival on the earth. With a rapidly growing population that has already exceeded six billion, they are causing much harm to the global environment by their industries which cater to their day-to-day needs and their greed. The Industrial Revolution took place about 150 years ago and consumerism has flourished since the end of World War II, around the middle of the twentieth century. In the last century, human beings claimed that they had 'conquered' nature and have only recently become aware about the extent to which they have harmed nature in the course of their activities. The industries of 'developed nations' and the enormous populations of developing countries are being held responsible for this extensive damage to the environment. In the Earth Summit at Rio de Janeiro in 1992, participating countries acknowleged the alarming deterioration of the world's environment which had already started spelling doom for many species including human beings. The countries became aware of the need to adopt measures to stop any further abuse to nature as enough pollution has already taken place.

Environmental biotechnology has been developed in the course of the past three decades by scientists and engineers to tackle the problem of environment pollution. Some of the measures are now mandatory for industries, for example, treating polluting effluents and gaseous and particulate emission, recovery of polluted soil, and so on; it is also mandatory for municipalities to treat sewage. This chapter discusses the involvement of living organisms from bacteria to higher forms of life in all these processes related to curing the earth's environment.

13.1 ENVIRONMENT

Environment may be defined as all the external conditions, both abiotic and biotic, that influence organisms and their perpetuation. In short, environment is the surroundings on the earth that sustain life.

13.1.1 Components of Environment

The components of environment constitute two major categories.

Abiotic

Inorganic substances such as carbon, nitrogen, oxygen, hydrogen, carbon dioxide, and water, involved in building organic compounds, present in the crust and the atmosphere of the earth.
Organic compounds such as proteins, carbohydrates, lipids, and humic substances.
Climate regime with a source of energy, temperature range, and other physical factors.

Biotic

Producers, largely autotrophic green plants that manufacture food from inorganic substances and fix solar energy in the process.
Macroconsumers, heterotrophic (eating on other organisms), chiefly animals obtaining food and therefore energy from plants or other organisms.
Microconsumers, also heterotrophic, are mostly bacteria and fungi which break down complex organic compounds and dead organisms through the process of decomposition to release inorganic materials to be recycled again through the living world.

13.1.2 Interacting Subsystems

The environment of the earth is often categorized into the following interacting subsystems:
Atmosphere   (composition in Table 13.1)
Hydrosphere   This includes all the water available on the earth in the three physical states (liquid, solid, and vapour) found in different reservoirs. The largest amount, in the form of salt water, is confined to the world's oceans (97.2 per cent) covering three-fourths of the surface of the earth. Fresh water constitutes just 2.8 per cent of the total water volume on the earth.

Lithosphere   This is the outer earth shell made of rigid brittle rocks, it is topped with the crust and the cooler upper part of the mantle. It ranges in thickness from 60 to150 km; it is thickest under the continents and thinnest under the ocean basins.

Biosphere   This includes all forms of life—microorganisms, plants, and animals including human beings.

Atmosphere

The atmosphere enwraps the earth and extends up to many kilometres upwards from the earth's surface. The atmosphere is held by the gravitational force of the earth and is densest near the earth's surface where gravity is supposed to be maximum and thins out gradually at higher levels. The maximum mass of the atmosphere, about 99 per cent of the total lies within a height of about 30 kilometres. The outer limit of the atmosphere cannot be fixed precisely as traces of the atmosphere have been found even at a height of 10,000 kilometres. The atmosphere has been subdivided into six distinct layers—*troposphere, stratosphere, mesosphere, thermosphere (ionosphere) magnetosphere,* and *exosphere.* The lowest layer, nearest to the earth, is called the troposphere. It extends up to a height of about 18 kilometres at the equator and 8 kilometres over the poles. Most of the weather phenomena take place in this layer and affect life on earth. The ozone layer found in the stratosphere absorbs the harmful ultraviolet radiation from the sun.

Temperature   The earth is neither too close nor too far from the sun; thus, it receives the right quantum of solar radiation conducive for life activities. As the earth rotates on its axis in the course of a day, all the sides get exposed to the solar radiation. The chemical reactions necessary for living organisms can occur within the range of temperatures found on the earth. At very low temperatures, the chemical reactions cease and at high temperatures, compounds are too unstable for the survival of life.

Table 13.1   Composition of the earth's atmosphere

Type of gas	Percentage by volume
Nitrogen (N_2)	78.09
Oxygen (O_2)	20.95
Water (H_2O)	Up to 1.00 (variable)
Carbon dioxide (CO_2)	0.03
Argon (Ar)	0.93
Neon (Ne)	0.002
Other trace gases: Methane (CH_4), carbon mono-oxide (CO), nitrous oxide (N_2O), ozone (O_3) hydrogen (H), helium (He) krypton (Kr), xenon (Xe)	Less than 0.001

Note: The atmosphere also contains suspended solid particles of dust, smoke, salt grains, pollen grains, and emission from factories and cars.

13.2 ENVIRONMENTAL POLLUTION

The ecologist Odum held the view that 'Pollution is an undesirable change in the physical, chemical, or biological characteristics of our air, land and water that may or will harmfully affect human life or that of desirable species …'

Environmental pollution, by definition, is the condition in which any gas, chemical, or material or a form of energy accumulates in the environment (comprising atmosphere, hydrosphere, lithosphere, and biosphere) beyond a defined threshold concentration. Scientifically, threshold concentrations of different substances and physical conditions have been determined, beyond which a pollutant will adversely affect human beings, directly or indirectly. Toxicity, chemical and physical reactivity of pollutants manifest as adverse effects. Presence of one or more pollutants in the environment is referred to as environmental pollution.

13.2.1 Nature and Source of Pollutants

Although sometimes pollutants are categorized as land polluters, water polluters, and air polluters, all these polluters are interconnected and pollutants for one sub-system can very well pollute others. For example, fertilizers and pesticides primarily affect land but they are gradually washed away in streams and lakes, and into deep subsoil water reserves. Oxides of carbon, sulphur, and nitrogen, and suspended particles are emitted from industries, notably from petroleum refineries and thermal power plants as air polluters. The oxides of nitrogen and sulphur are converted to acidic gases. When it rains, the acidic gases form nitric and sulphuric acid, come down in the rain water and acidify both the soil and the water bodies. A high level of acidity in the soil directly affects agriculture, growth of plants, and herbivore animals and indirectly affects animals and even human beings. Acid rains corrode heritage buildings, monuments, and statues.

13.2.2 Major Pollutants and Sources

Human activities produce most of the wastes and pollutants. Wastes are the by-products of housekeeping and industrial activities and usually add to the pollutants.

Industrial wastes   These include industrial effluent containing solid wastes, hazardous heavy metals depending on the type of industry, various inorganic and organic wastes from dye industries and tanneries, toxic compounds; emission of gases and flying ashes due to burning of fossil fuel during energy production; thermal pollutants associated with water used for cooling of coal and nuclear power plants and discharged in nearby rivers or lakes; radioactive wastes; oil spills.

Urban wastes   These include organic and inorganic solid waste from construction, households, markets, and commercial establishments; special category of wastes from hospitals; plastics and sewage; smoke and emissions from households, cars and airplanes using fossil fuels (petroleum, diesel) producing large amounts of CO, various oxides of sulphur and nitrogen; chloro fluoro carbons (CFC) in aerosol and refrigeration (these deplete ozone in the ozone layer).

Agricultural waste   These include chemical fertilizers and pesticides.

Biological waste   These include dead animals lying on land and in water bodies, excretory products of animals including man.

Arsenic pollution   This occurs in underground water due to overuse of ground water for agricultural purpose, inert arsenopyrites in submerged ground water get oxidized when water is pumped out and the vacuum that results is filled with oxygen; oxidized arsenopyrite is soluble in water.

Noise pollution   Intensity of sound up to 65 decibel (dB) is considered safe for our hearing system. Intensity of sound beyond this limit is unlawful and causes noise pollution which is considered an air polluter. Certain industries, motor vehicles, particularly horns produce monotonous, sharp and disturbing sounds that are classified as noise pollution. Legislation and the development of machines to reduce noise are the means to curb noise pollution. Usually this does not come under biotechnological skills, except the knowledge that a tree line absorbs a considerable level of noise. Thus, modern factories and cities adopted the technique of tree plantation as a policy to keep the environment free from many kinds of pollution including noise and recharge it with fresh oxygen.

13.3 WORLD CONVENTIONS ON ENVIRONMENT

Three important world conferences have been held to resolve the constant threat to the environment and the depleting natural resources.

- The 'First Conference on the Human Environment' was convened at Stockholm, Sweden in 1972.
- Twenty years later, the 'United Nations Conference on Environment and Development' (UNCED), popularly referred to as the 'Rio Earth Summit', was held in Rio de Janeiro, Brazil from June 3 to 14, 1992 with the heads of the states and representatives from 166 countries to examine the issues involved in maintaining a clean environment in the world.
- The 'World Summit on Sustainable Development' (WSSD) was arranged in Johannesburg, South Africa between August 26 and September 4, 2002 to assess the change in the global environment since the Rio Earth Summit. There were concerns for the increasing problem of regulating environmental pollution, and the conservation of nature and natural resources.

Both these problems now get constant attention from environmentalists and governments of different countries. An additional area of concern is the release of genetically engineered organisms in the environment, particularly in effluents from biotechnology companies. Thus, the developments in biotechnology come under the purview of environmental pollution. There is a continuing debate on the safety of the use of the products of biotechnology. This is discussed under bio-safety in Chapter 15. In this chapter, the applications of biotechnology to protect the environment from pollution and to conserve natural resources are taken up.

In the last two world summits, the disparity of socio-economic conditions and poverty in different countries and the lifestyle of the economically developed countries came up in the discussion as a contributing factor to pollution and depletion of resources in the world. It is high time we realized that the countries on the both the sides of the economic divide should jointly find a path of development which 'meets the needs of the present without compromising the ability of future generations to meet their needs' (World Commission on Environment and Development).

A variety of measures adopted by all the countries of the world can make this laudable goal a reality, and biotechnology will be one of these measures.

13.4 GREENHOUSE GASES (MAINLY CO_2) AND GLOBAL WARMING

Atmospheric air contains 0.03 per cent CO_2 (Table 13.1). CO_2 occurs at much higher concentrations in oceans and aquatic water bodies as dissolved CO_2 or in combination with calcium as bicarbonates and carbonates, at much higher concentrations. The carbon cycle which involves the atmosphere, water, soil, producers, consumers, and decomposers normally remains in a finetuned balance and maintains the CO_2 content of the air at a steady state.

Consumption of fossil fuels (coal, petroleum oil, and gas) by industries and human activities is continuously releasing tremendous amount of CO_2 in the atmosphere. The concentration of CO_2 in the atmosphere is estimated to have increased from 280 ppm in 1800 to 360 ppm in 2000, and is likely to increase further to ~ 550 ppm by 2050 and ~ 800 ppm by the end of this century. The higher concentration of CO_2 in the air readily absorbs long wave or thermal radiation radiating from the surface of the earth and results in the greenhouse effect—a warm condition is produced on the earth similar to that in a greenhouse, and CO_2 is called a greenhouse gas. In addition to CO_2, methane (CH_4), nitrous oxide (N_2O), sulphur dioxide (SO_2), chlorofluorocarbon (CFC), and water vapour contribute to the greenhouse effect. The increased level of CO_2 is mainly responsible for the increase in average global temperature, often described as *global warming*.

Global warming has already resulted in an average rise in temperature of 1°C from AD 1800 to 2000. The Intergovernmental Panel on Climatic Change (IPCC) has predicted an increase in average global temperature of 1°C to 3.2°C by 2100. Global warming has serious implications for the survival of the human population and the biodiversity of the planet. The IPCC has already estimated that global warming would cause melting of polar ice caps and glaciers and a 0.3 to 1.1 m rise in sea level by the end of the 21st century. Obviously this would seriously affect coastal areas, erode beaches, and cause extinction of coral reefs and extensive inundation of low lying areas including some of the world's most populated areas and megalopolises like New York and Kolkata. Global warming would also affect the physiological and behavioural responses of many organisms ranging from plants to human beings.

13.5 KYOTO PROTOCOL (1997)

Keeping all the points discussed above in view, a treaty popularly described as *Kyoto Protocol* was negotiated among participating countries in Japan in 1997. An effort was made to make it mandatory for the developed countries to reduce greenhouse gas emission to 5 per cent below the 1990 level by 2012. There were also suggestions to the developing countries to put a limit on the emission of greenhouse gases. This protocol and other related issues of energy use were also the focus of discussion at the world summit at Johannesburg in 2002. The US and a few other developed countries, unfortunately, were not ready to accept the provisions of the Kyoto Protocol.

The hegemony of the western world is such that it may not be willing to adopt the precautions necessary to control global warming which is a serious threat to the global ecosystem in terms of climate change and health hazards. However, the western countries are very concerned about the rapid depletion of energy resources coupled with the increasing demand for energy. That is why they are searching for and developing alternate, renewable energy sources from solar radiation, wind power, thermal difference in the different levels of the ocean, and so on. The alternate renewable sources are not as polluting as the fossil fuels.

The conflict between economic development and environmental issues, particularly protection of the environment, biodiversity, and other resources will probably continue for quite a long time.

All these issues are very important and solutions are being investigated. These topics are more relevant for ecology and other fields. This chapter will discuss the involvement of biotechnology in remedial measures for pollution and rejuvenating the environment. Other eco-friendly aspects of biotechnology, such as bio-fertilizers, bio-pesticides, biomass production, single cell protein (SCP) technology, bio-fuels and biogas as alternatives to fossil fuels will be discussed in the next chapter.

13.6 REMEDY FOR POLLUTION AND ROLE OF BIOTECHNOLOGY

The wastes from industries and urban settlements are of enormous proportions, expensive to manage and infective and hazardous for human beings and other species. Awareness about pollution will definitely reduce the volume of unwanted wastes but will not bring it to a zero-waste level, especially in modern society which depends on a huge range of gadgets and is used to a life of comfort. In this context, biotechnology can provide a means of treating waste products or chemically contaminated soil and ground water reservoirs by introducing the right kind of microorganisms or other types of organisms. Biotechnological procedures to manage pollution, at times may seem as a boon when different world bodies are striving to control and limit pollution and ban polluting industries. The objectives of such biotechnological measures are to reduce the toxicity or hazardous aspects of wastes and at the same time to come up with useful by-products.

13.6.1 Waste Treatment and Biotechnology

The huge volumes of wastes generated from industrial and urban sources have already been mentioned. The wastes may occur in a concentrated and localized fashion as in industrial and urban areas or they may be present in low concentrations or dispersed, in other places. The wastes may be biodegradable, non-biodegradable, or a mixture of both. Biotechnology can be introduced for biodegradable wastes; non-biodegradable heavy metals, some man-made chemicals including organic ones are difficult to treat and they also inhibit the growth of microorganisms.

The physical nature of the wastes may be gas, liquid, and solid. Accordingly, the waste treatment procedures vary as stated below.

(a)   Gases—biofilters
(b)   Solids—biodegradation, incineration or burning, landfill
(c)   Liquids—aerobic and anaerobic digestion

13.6.2 Biofilters for Pollutant Gases

Volatile organic compounds, such as acetone and hydrocarbons, produced from food and chemical factories are absorbed efficiently by microorganisms in a large filter mesh. The filter area needs to be kept moist and supplied with nitrogen and trace elements. The filter is otherwise self-sustaining, as the organisms get carbon from the gases and oxygen from the air. Usually this filter is cheaper than the conventional chemical process of cleaning.

Biofilters convert gaseous wastes into non-hazardous compounds and can be of two types—solid-support or two-phase (gas/liquid) biofilter.

Solid-support system   It consists of peat or some other solid material to support the growth of microorganisms. The gaseous waste is passed through the solid support for the pollutants to be broken down by microbes.

Results of bioconversion of gases are variable as the type and amount of various microorganisms are not always controllable.

Two phase (gas/liquid) biofilter   This is an improved version of the solid-support system. The liquid phase is separated from the incoming gas phase by a membrane, and the biological agents remain immobilized on the membrane on the side of the liquid phase. When the pollutant gases enter the liquid phase, they get consumed or broken down by the biological agents on the membrane.

13.6.3 Solid Waste Treatment

The treatment techniques for solid wastes differ on the basis of the nature and bulk of the wastes, and availability of certain facilities.

Biodegradation   Biodegradable and non-biodegradable components of solid waste are sorted out first. Biodegradable organic wastes are dumped in a place and covered with earth for rapid anaerobic biodegradation into compost manure. In households, the organic matter, especially cow dung and excreta can be decomposed in a small gas plant (cow dung or 'gobar' gas plant) to generate methane gas as alternate source of energy for cooking. The organic waste materials may also be subjected to *vermicomposting* using suitable earthworm species and conditions for their proper growth (see Exhibit 13.A).

Incineration   The organic solid waste may also be burnt down (incineration). In general, non-degradable and non-recyclable wastes including biomedical wastes should be incinerated properly. Recyclable materials, such as plastics, paper, glass, rubber, and ferrous and non-ferrous metals can be separated from the solid waste for recycling.

Landfill   The collected solid wastes are subjected to pretreatment—sorting of recyclable material, mechanical pulverization (crushing into dust), and incineration. Often without all these treatments, the solid wastes are transferred to a landfill. In a landfill, the solid waste is dumped in a natural or manmade pit and then covered with soil. Usually landfill sites are located far away from human dwellings and water bodies, in derelict land to avoid subsequent problems.

Two types of pits may be made for landfills, and accordingly landfill practice is classified into two types.

Cell emplacement method   Cells about 2.5 m deep of suitable size depending on the amount of the waste to be dumped each day are excavated. If the cell is filled with the waste of a single day, the cell is compacted and covered with a layer of soil about 20 cm deep. In the case of multi-layered cells, the waste accumulated in one day is compacted and covered with 30–40 cm soil, so that the waste of the next day can be compacted on the first layer; when the cell gets full, it is covered with a layer of soil 60–90 cm deep.

Trench method   Long trenches are made, filled with waste and covered with soil.

In landfill treatment of solid wastes, several hundred species of microorganisms are involved in anaerobic digestion and biogas production. The microorganisms mainly belong to four trophic groups: (a) hydrolytic and fermentative bacteria, (b) syntrophic H_2 producing bacteria (c) methanogenic bacteria, and (d) acetogenic bacteria. These bacteria sequentially digest cellulose, starch, proteins, lipids, and nucleic acids present in the biodegradable component of solid wastes. The anaerobic digestion produces methane and carbon dioxide as major gaseous by-products.

Hazards of landfills and remedial measures   The landfill method for disposal of wastes presents several hazards, such as (i) leakage of methane, and dry materials posing fire threats; (ii) foul smell; (iii) leaching of toxic and corrosive materials such as heavy metals, mercury, cyanides, and arsenic into soil and underground water; and (iv) breeding ground for disease vectors like flies. The risks of fire, offensive odours, and vector populations can be done away with by covering the wastes with soil on a daily basis. Some planning needs to be done to collect methane or the wastes need to be burned systematically. The problem of polluting chemicals leaching into ground water may be avoided by lining the pit at the beginning with an impervious material such as clay, soil–cement mixture, concrete, polymeric material, and asphalts. The clay lining is not considered as a long-term solution. It is best to remove toxic items, if possible, and treat them specifically before they are dumped.

Caution should be maintained for a suitably long period before the dumping grounds are used by the public. The landfill sites are sometimes developed into gardens.

13.6.4 Waste Water Treatment

Waste water from urban and industrial sewerages carry solid materials, plastics, organic material, inorganic nutrients, inert sedimenting particulate matter, and sometimes hot water from industrial cooling plants. None of these are eco-friendly substances. Often, sewerages pour the materials into a nearby river or a large water body. The solid and sediment materials gradually choke the river. Large amount of organic and inorganic materials can cause *eutrophication*, which promotes extensive microbial and plant growth leading to depletion of oxygen from water resulting in the death of fish and other aquatic faunae. The water in sewage is contaminated with all kinds of bacteria and microbes including some that cause deadly diseases such as cholera, typhoid, and so on. Since sewage water contains many kinds of contaminants, it is necessary to treat it before it is allowed to flow into water bodies.

Parameters determining the level of pollution in waste water   The important parameters that determine pollution in the waste water are (i) biochemical oxygen demand (BOD), (ii) chemical oxygen demand (COD), (iii) suspended solids, (iv) ammoniacal nitrogen

Exhibit 13.A   Vermicomposting

Vermicomposting is a method of producing compost using earthworms. The prefix *vermin* is Latin for worm. Long back earthworms are considered friendly to the farmers. Earthworms are known to recycle food and backyard wastes in an eco-friendly manner and the compost is useful as fertilizer and maintains the soil structure which is often degraded by the large-scale use of chemical fertilizers, especially in the era of the green revolution.

Earthworms consume organic matter with a relatively wide C:N ratio and convert it into a lower C:N ratio, and thereby increase the nitrogen content in the soil. Vermicomposting quickly reduces the volume of household garbage without producing foul smell, or pathogenic microbes. It spawns worms for fish and poultry feed, free of cost. It contains immobilized enzymes, such as amylase, protease, lipase, lichenase, and chitinase which help in the biodegradation of macromolecules in the soil.

Outline of the vermicomposting technique

Worm pits are prepared in a shady place. The pit may be made of concrete rings or dug in the ground, or bins of wood or plastic can be used. Pits or bins need to be 20 to 30 cm deep for *epigeic* (surface dweller) earthworms for efficient composting. Drilling air/drainage holes (6 to 12 mm diameter) in the bottom and sides of the bin allows drainage of excess water and proper air circulation. The excess water that drains from the bin (vermiwash or compost tea) can be used as liquid fertilizer. Each bin or pit should be covered with straw, leaves, or moist burlap to ensure darkness while allowing good air ventilation for proper respiration of the worms. Worms prefer darkness. Outdoor bins should be insulated from the cold during the winter to protect the worms.

Bedding materials and introduction of worms   Bedding materials for vermicomposting include a diverse range of organic materials of high cellulose content, such as shredded newspaper (non-glossy), cardboards, shredded leaves, vegetables and fruit wastes (citrus products in excess may turn the compost too acidic), straw, forest products, sawdust, and so on. Grass clippings are to be allowed to age before they are used, otherwise they may decompose fast causing the compost to heat up. For the same reason the bedding materials are also allowed to set for several days. Some soil or sand can be added to provide grit for the worms' digestive system. (Night soil, bones, meat, and dairy products are not used in vermicomposting to avoid certain problems, flies, and pests.)

The bedding material is thoroughly moistened till it has the consistency of a damp sponge, before worms are added to it. The composting worms need to be collected from the soil or purchased from suppliers.

The pH of the compost should be maintained at 6.5 in the range of 6 to 7. If the compost is too acidic crushed egg shells can be added to decrease the acidity. The optimum range of temperature for earthworm is between 13° C and 15° C. A small light over the pit prevents the worms from leaving the pit at night.

Metals, plastics, glass, soaps, poisonous plants or plants sprayed freshly with insecticides should never be put in the compost as many of these are detrimental to worms.

Fig. 13. A

and phosphate content, and (v) relative abundance/absence of certain organisms (bio-indicators) in the river water.

Dissolved oxygen is an essential requirement for aquatic life. The optimum dissolved oxygen in natural water is 4–6 parts per million (ppm).

Biochemical oxygen demand (BOD) and chemical oxygen demand (COD)   These are measures to estimate biologically oxidizable organic matter (in other words, carbon content) in a water sample. BOD and COD are determined differently. A higher index in both scales indicates the presence of more organic matter as pollutant.

BOD assays the amount of O_2 used up in a sample of water over a definite period of time. The BOD is defined as the amount of dissolved oxygen consumed when a (waste) water sample is seeded with active bacteria and incubated at 20°C for five days in the dark. This is sometimes known as BOD_5.

Usually serial dilutions of the water sample are incubated to ensure that O_2 is not completely depleted before all the bioxidisable organic material is oxidized.

Oxidation of NH_4^+ by the joint action of *nitrosomonas* $(NH_4^+ \rightarrow NO_2^-)$ and *nitrobacter* $(NO_2^- \rightarrow NO_3^-)$ also uses O_2; this process is called nitrification. When NH_4^+ concentration in the waste water is high, nitrification should be inhibited by using allyl thiourea, before BOD determination.

The measure of BOD is extremely important to evaluate the self-purification capacity of a water body.

COD measures the amount of chemically oxidizable organic matter present in a water sample. Instead of an index for utilization of oxygen as in BOD, COD estimates the total organic matter in a sample that can be oxidized chemically irrespective of whether it is biodegradable or not.

An excess of acidic potassium dichromate is added to the test sample, and after two hours the amount of unused dichromate is measured by titration with ferrous ammonium sulphate to determine the COD value.

The COD value of a sample is always higher than its BOD value.

13.6.5 Processes of Waste Water Treatment

Once the type and the level of pollutants in waste water are determined, the water is processed or treated in different steps. The type of treatment required is decided on the basis of the purpose for which the treated water is to be used. The steps of treatment are described in Fig. 13.1.

Primary treatment   Primary treatment is for the mechanical removal of large waste materials and involves a few steps:

Screening of waste water is for the removal of large floating objects, such as plastic bags, rags, floating wood, floating shoes or other household or industrial objects, which may damage pumps and clog pipes carrying waste water to the next stages. For this purpose, screens of varying sizes are used; they consist of steel bars spaced 2–8 cm apart, followed by wire meshes or screens with reduced spacing.

An additional step of **comminution** grinds the materials to a size that is small enough to allow easy flow of the waste water.

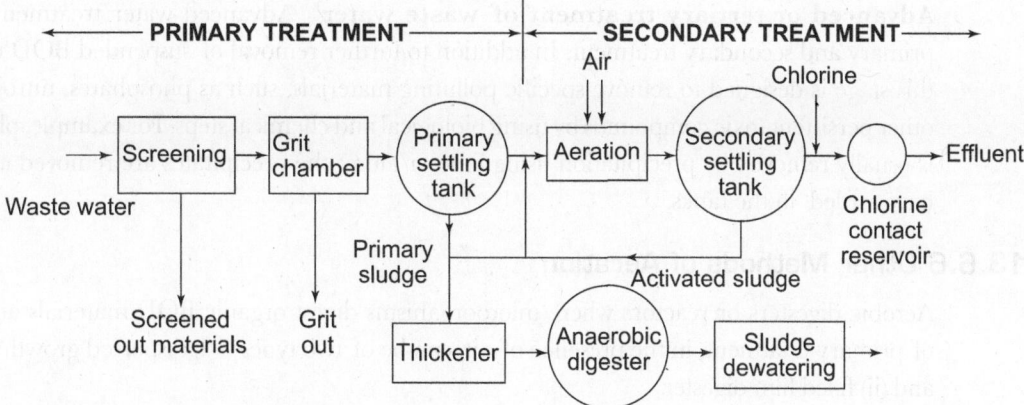

Fig. 13.1   Process of waste water treatment using activated sludge process.

Grit chamber is the next step in the treatment of waste water; the water is held for a few minutes to allow the sedimentation of grit and coarse particles under gravity. The material that settles as sediment in this chamber can be disposed of in landfills.

Primary settling tanks receive the waste water and it is allowed to stand for 2–3 hours so that most of the suspended material settles down. The suspended material consists of 50–60 per cent solids and 20–40 per cent BOD material. A chemical coagulant may be added to promote settling. The settled material is called primary sludge and is removed for further processing. The effluent at this stage is chlorinated to control odours before release.

Secondary or biological treatment   Secondary treatment reduces the suspended or dissolved organic matter (BOD material) in the waste water by aerobic and anaerobic digestion by bacteria. The ability of microorganisms to convert organic waste into low-energy compounds is exploited by the activated sludge process in the aeration tank or trickling filter. The biodegradable organic material is oxidized to CO_2 and H_2O, and biomass and nitrogenous compounds are also produced.

There are certain other methods of aeration at the secondary stage of waste water treatment. These are discussed under other methods of aeration.

Activated Sludge Process

The effluent from primary treatment enters the aeration tank unit. A mass of biological organisms, called activated sludge, is also added to the aeration tank from the secondary settlement tank (next unit). Aerobic conditions are maintained in the tank by adding air and the mixture is constantly agitated for 6–8 hours for digestion by aerobic microorganisms. The microorganisms are the same as those found in nature—bacteria, fungi, cyanobacteria, protozoa, and algae.

Then waste water flows from the aeration tank into the settling tank where solids, mainly bacterial mass, are separated from the liquid part by subsidence. It has already been mentioned that a portion of the bacterial mass is recycled as *activated sludge* into the aeration tank. The rest of the sludge is sent to an anaerobic digester tank, where the sludge is removed for drying and disposal in landfills or agricultural land.

From the secondary settling tank, the effluent is chlorinated before release.

Advanced or tertiary treatment of waste water   Advanced water treatment follows primary and secondary treatment. In addition to further removal of suspended BOD material, this stage is designed to remove specific polluting materials, such as phosphates, nitrogen, and other persistent toxic compounds by using biological and chemical steps. For example, phosphate is usually removed by precipitation using lime or alum; the precipitates are removed after they have settled in the tanks.

13.6.6 Other Methods of Aeration

Aerobic digesters or reactors where microorganisms digest organic BOD materials at the end of primary treatment, in the presence of air, can be of two types—(i) dispersed growth digester, and (ii) fixed film digester.

Dispersed growth digester   The *activated sludge process* described earlier is a typical example of a *dispersed growth digester*. It has a large vessel for oxidation of waste water in the presence of a recycled mass of microorganisms and air or oxygen that is bubbled in. Waste water continuously flows into the vessel and treated water flows out at a predetermined rate that facilitates optimal digestion of the suspended organic matter. The microbial flocks also flow out with the treated water.

Fixed film digesters   The biological components or microorganisms remain in the form of a fixed film on filter particles or large discs in the digesters. These are again basically of two types—(a) trickling filter digester and (b) rotating disc (contactor) digester.

Trickling filter   This is a bed of a graded mixture of crushed stones, gravel, clinkers, blast furnace slag, and so on. It can be circular or rectangular and about 3 m deep. The concept of the trickling filter is schematically presented in Fig. 13.2. The waste water influent is evenly sprayed over the bed matrix. Air flow is maintained to ensure aerobic conditions. The microorganisms (fungi, bacteria, protozoa, and algae) are present as a layer of *biofilm* on the surface of the filter bed material. The waste water percolates through the bed and the microbes metabolize the organic waste. With microbial growth, the biofilm grows thick and peels off the filter particles. A new biofilm grows continuously. Thus, continuous growth of the biofilm leading to death and detachment of the thickened biofilm indicates excess growth of microorganisms and clogging of the filter. The peeled-off biofilm flows out with the sewage. The percolated water from the filter bed collects at the bottom and is taken to the sedimentation tank to allow the fragments of film and other solids from the effluent to settle. The settled solids are called *humus sludge*.

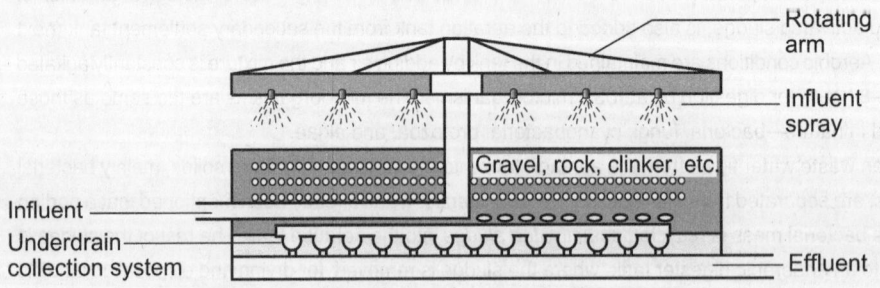

Fig. 13.2   Diagram of a trickling filter.

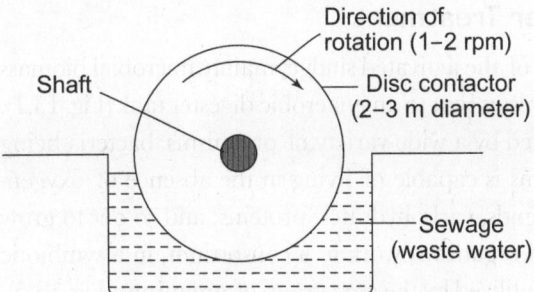

Shaft

Direction of rotation (1–2 rpm)

Disc contactor (2–3 m diameter)

Sewage (waste water)

Fig. 13.3 Diagram of a rotating disc contactor (digester).

Well-maintained trickle filters can last for many years. In recent times, plastic support materials for the filter bed are used in industrial waste water management, as they have some advantages over the conventional supports.

Rotating disc A set of discs made of wood, metal or plastics, about 2–3 m in diameter is oriented in the direction of waste water flow (Fig. 13.3). The biofilm grows on the disc surface. The discs keep rotating at a low speed of 1–2 rpm and thus allows extended contact between the microorganisms and the waste water. The biofilm on the surface of the discs moves into and out of the tank containing waste water. When microorganisms are submerged in the waste water, they absorb organic material; when they rotate out of the waste water, they obtain the oxygen required for metabolism and growth from the air.

The rotating disc (contactor) digester has two main advantages—small space requirement and low maintenance cost.

Oxidation pond An oxidation pond is a large shallow, 1–2 m deep pond, where raw and partially treated sewage is decomposed by all sorts of microorganisms under aerobic conditions by surface air exchange. The oxidation ponds may be designed to be aerobic and partially anaerobic (Fig. 13.4). The deeper part of the pond serves as the anaerobic zone for the action of anaerobic microorganisms.

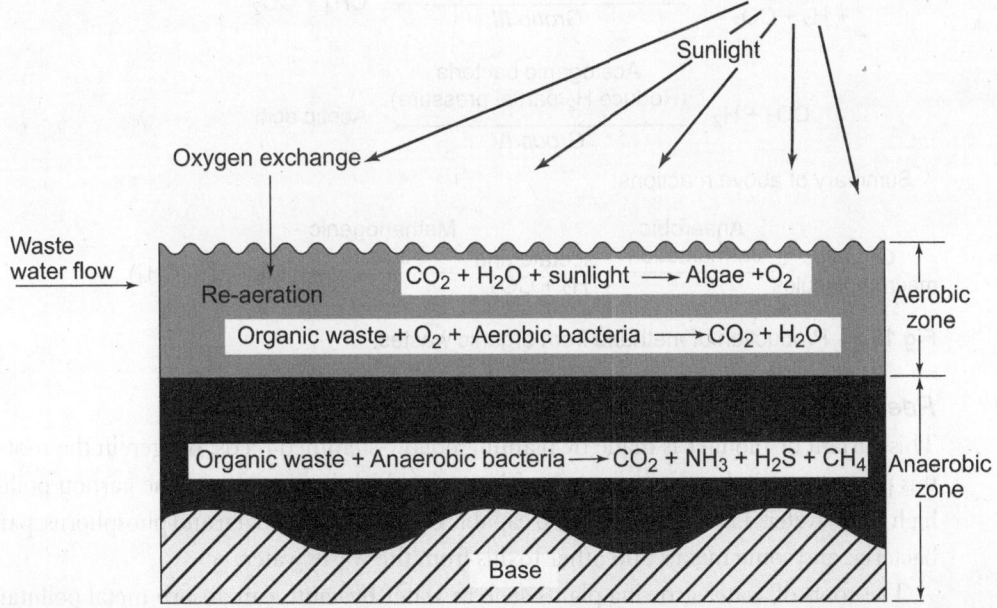

Sunlight

Oxygen exchange

Waste water flow

Re-aeration

$CO_2 + H_2O + sunlight \longrightarrow Algae + O_2$

$Organic\ waste + O_2 + Aerobic\ bacteria \longrightarrow CO_2 + H_2O$

Aerobic zone

$Organic\ waste + Anaerobic\ bacteria \longrightarrow CO_2 + NH_3 + H_2S + CH_4$

Anaerobic zone

Base

Fig.13.4 Oxidation pond for decomposing organic materials by microorganisms in aerobic and anaerobic conditions.

13.6.7 Anaerobic Process in Waste Water Treatment

In the secondary stage of treatment, a portion of the activated sludge, mainly microbial biomass is reintroduced into the aeration tank and the rest is sent to an anaerobic digester tank (Fig. 13.1).

The anaerobic digestion process is conducted by a wide variety of organisms, bacteria being the most predominant. This class of organisms is capable of living in the absence of oxygen. They digest complex organic molecules like lipids, carbohydrates, proteins, and so on to grow and reproduce. Different species of anaerobic organisms work in a consortium, in a symbiotic fashion. The digested product of one group is utilized by the next group of organisms (Fig. 13.5).

Sulphate (SO_4^{2-}) and nitrate (NO_3^-) present in the waste water are digested by the activities of bacteria like *Desulphovibrio* and denitrifying bacteria.

Anaerobic biological processes have several advantages over aerobic processes. They use less energy, produce less biological sludge, and produce energy through the production of methane (CH_4).

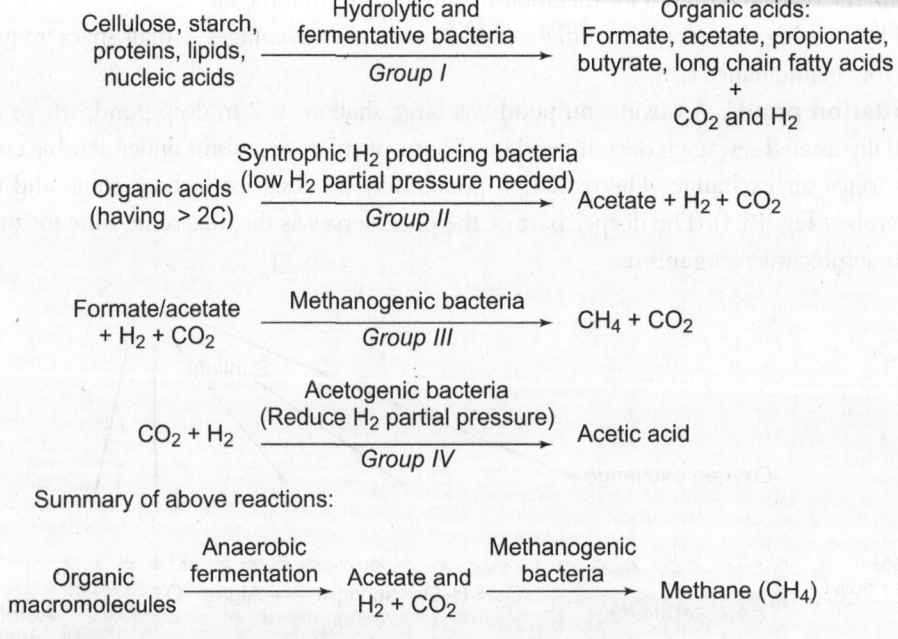

Fig.13.5 Production of methane from organic wastes.

Reed bed biofilter

This variant of biofilter is made by planting hollow-stemmed reeds; oxygen in the root hairs of this plant supports luxurious growth of bacteria which in turn use up the carbon pollutants in high BOD water. The bacteria are also capable of removing nitrogen and phosphorus, pathogenic bacteria, and some metals and other toxins from the waste water.

The roots of water-growing plants such as water hyacinth can absorb metal pollutants from water bodies. This is known as **rhizofiltration** and is used in reed bed biofilters.

Exhibit 13.B Drinking Water Quality

The World Health Organization (WHO) produces international norms on water quality and human health in the form of guidelines that are used as the basis for regulation and setting standards, in developing and developed countries world-wide. This is done to safeguard the health and life of living beings. The standards used by different countries differ in amounts of different substances in small range, sometimes depending on the available water source and geographical distribution of the elements in the earth's crust. There are European standards, and a USA standard. The Indian standard based on WHO norms has been presented in Table 13.B. The standard setting tables are updated from time to time.

Table 13.B Indian Standard Specifications for Drinking Water

(IS: 10500 - 1993)			
Parameters	**Units**	**In drinking water**	
		Desirable	**Maximum**
Colour	Hazen units	5	25
Odour	-	Unobjectionable (No offensive odour)	-
Taste	-	Agreeable	-
Turbidity	NTU	5	10
pH value	-	6.5 to 8.5	No relaxation
Dissolved solids	mg/L	500	2000
Total hardness (as $CaCO_3$)	mg/L	300	600
Iron (Fe)	mg/L	0.3	1.0
Chlorides (Cl)	mg/L	250	1000
Residual free chlorine	mg/L	0.2	-
Calcium (Ca)	mg/L	75	200
Copper (Cu)	mg/L	0.05	1.5
Manganese (Mn)	mg/L	0.1	0.3
Magnesium (Mg)	mg/L	30	100
Sulphate (SO4)	mg/L	200	400
Nitrate (NO_3)	mg/L	50	No Relaxation
Fluoride	mg/L	1.0	1.5
Phenolic compound (C_6H_5OH)	mg/L	0.001	0.002
Mercury (Hg)	mg/L	0.001	No relaxation
Cadmium (Cd)	mg/L	0.01	No relaxation
Selenium (Se)	mg/L	0.01	No relaxation
Arsenic (As)	mg/L	0.05	No relaxation
Cyanide (CN)	mg/L	0.05	No relaxation
Lead (Pb)	mg/L	0.05	No relaxation
Zinc (Zn)	mg/L	5	15

(Contd)

Table 13.B   *(Contd)*

Anionic detergents	mg/L	0.2	1
Chromium (Cr)	mg/L	0.05	No relaxation
Polynuclear aromatic hydrocarbons	mg/L	-	-
Mineral oil	mg/L	0.01	0.03
Pesticides	mg/L	0	0.001
Radioactive materials			
(a) Alpha (α) emitters	Bq/L	-	0.1
(b) Beta (β) emitters	Pci/L	-	0.037
Alkalinity	mg/L	200	600
Aluminium (Al)	mg/L	0.03	0.2
Boron	mg/L	1	5

Water in rivers and open water bodies is often contaminated with several kinds of microorganisms, such as bacteria, viruses, protozoa, fungi, causing various diseases in man and animals. The water from these open water bodies is pretreated with sedimentation and various germicidal agents in the water treatment plants in urban areas to provide safe drinking water and to prevent water-borne diseases. Guidelines to keep the level of the microorganisms in drinking water as low as possible are also prescribed in different countries under the public health system.

13.6.8 Bioremediation of Contaminated Land and Water

Bioremediation is the process of reducing or eliminating chemical pollutants in the environment by using biological systems, usually microorganisms. Biodegradation of wastes by microorganisms is an extremely old technology, used for sewage treatment. Microorganisms are capable of degrading all naturally occurring compounds; this was proposed as the *principle of microbial infallibility* by Alexander in 1965. In recent decades, serious and large-scale efforts have been made to harness this ability of bacteria to degrade various compounds, for effective restoration of chemically polluted environments, especially environments polluted by xenobiotic compounds (Fig. 13.6).

Xenobiotic compounds

These are man-made chemicals, structurally resistant to microbial degradation and hence persist in the environment for many years to cause severe chemical pollution; they are described as 'recalcitrant'. They are highly stable because they are chemically and biologically inert due to the presence of substitution groups like halogens, nitro-, sulphonate, amino-, methoxy, and carbamyl groups. They are insoluble in water, but are adsorbed by the soil. Xenobiotic compounds are highly toxic and produce toxic compounds on limited microbial activity. Usually their large molecular size and lack of the permease needed for their transport into the microbial cells, prevent their entry into microbial cells. When they are not degraded they cause the problem of bioconcentration and biomagnification in the food chain.

Types of xenobiotic compounds   Usual examples of xenobiotic chemicals are petroleum products, pesticides, detergents, solvents, polychlorinated biphenyls (PCBs), plastics, and

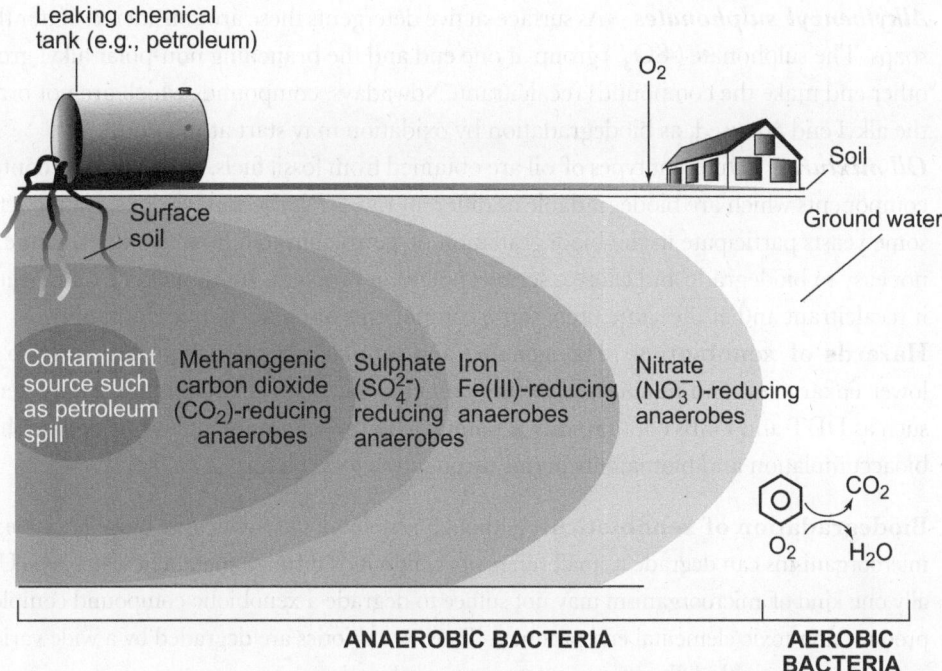

Fig. 13.6 Contamination of soil and ground water, say by petroleum spill from a leaking tank, can be degraded by anaerobic and aerobic bacteria.

nitrogenous compounds. They can be grouped into five major types depending mainly on the structural properties of the chemicals.

Halocarbons   Halogens like Cl, Br, F (fluorine), and I atoms are present in the place of H atoms. They are used as solvents like chloroform ($CHCl_3$), as propellants in perfumes and body spray cans, paints, in condenser units of refrigeration systems (Freons, CCl_3F, CCl_2F_2, $CClF_3$, CF_4), as insecticides like DDT (dichloro-diphenyl-trichloroethane), BHC (benzene hexachloride), lindane, and as herbicides like dalapon. Chloroform and freons being volatile, escape into the atmosphere and destroy the ozone (O_3) layer, which protects the living world from the hazards of UV radiation. Pesticides and herbicides leach into water bodies causing accumulation and biomagnification in the food chain.

Polychlorinated biphenyls (*PCBs*) Two covalently linked benzene rings with halogens for H make these compounds biologically and chemically inert, and the presence of a number of chlorine atoms adds to this inert property. The compounds are mostly used as insulator coolants in transformers and as heat exchange fluids.

Synthetic polymers   Polyethylene, polystyrene, polyvinyl chloride, and so on as plastics are used for many items of domestic and industrial purposes; nylons are used as garment and wrapping materials. These materials are insoluble in water and add to the pollution problems.

Alkylbenzyl sulphonates   As surface-active detergents these are considered better than usual soaps. The sulphonate ($-SO_3^-$) group at one end and the branching non-polar alkyl group at the other end make the compounds recalcitrant. Nowadays, compounds which are not branched at the alkyl end are used, as biodegradation by oxidation may start at this end.

Oil mixtures   Different types of oil are obtained from fossil fuels, and oil wastes contain many components which are biodegradable at different rates. *Pseudomonas*, various cyanobacteria, and some yeasts participate in the biodegradation of petroleum to a limited extent. Large spills are not easy to biodegrade and cause a serious pollution problem. Insolubility of oil in water makes it recalcitrant and at the same time, some components are toxic to microorganisms.

Hazards of xenobiotics   Halogenated and aromatic hydrocarbons are toxic to bacteria, lower eukaryotes, and humans. They are carcinogenic (cancer producing). Many xenobiotics, such as DDT and PCBs continuously accumulate in the environment and being lipophilic cause bioaccumulation and biomagnifications through the food chain.

Biodegradation of xenobiotics   Although xenobiotics are not easily biodegradable, some microorganisms can degrade a small range of compounds through metabolic pathways. Usually one kind of microorganism may not suffice to degrade a xenobiotic compound completely to produce non-toxic elemental end products. Thus, xenobiotics are degraded by a wide variety of microorganisms with different enzyme systems, working in sequence.

Oxic (in presence of O_2) degradation of aliphatic and aromatic hydrocarbons   Some microbes, such as *Pseudomonas, Mycobacterium, Nocardia*, yeasts, and molds have enzymes for sequential oxidation of the terminal carbon of long chain aliphatic hydrocarbons to transform them into fatty acids. The fatty acids are then degraded by the β-oxidation pathway to yield acetyl CoA which is consumed in the citric acid cycle. Additional groups like CH_3 and Cl atoms at the positions which participate in the β-oxidation pathway have been found to resist degradation.

In the degradation of alkanes and aromatic hydrocarbons, an oxygenase first introduces a hydroxyl (OH) group to make the compound reactive. The OH group is then oxidized by another oxygenase to a carboxyl group, and the ring structure opens up for further degradation by β-oxidation to yield acetyl CoA which is metabolized in the usual manner as indicated earlier. A dioxygenase may participate in the oxidation of benzene ring; this enzyme adds oxygen at two positions in a single step.

It may be mentioned here that the halogens or other substituent groups in hydrocarbons are either modified or removed in the initial steps of metabolism.

Anoxic degradation of xenobiotics   In the absence of oxygen as an electron acceptor, the microbes adopt alternate forms like nitrates, sulphates, and ferric ions for the role of O_2. Xenobiotics can be used by microbes as electron acceptors and this prepares the xenobiotics for degradation by the metabolizing pathway.

Glycol (petrochemical) biotreatment process   Propylene oxide (PO), propylene glycol (PG), and polyols are produced in petrochemical industries and are released in the effluents. These chemical effluents have high biological oxygen demand (BOD) and chemical oxygen demand (COD) and are hazardous to organisms living in water bodies.

Biotreatment of Hazardous Chemicals

Southern Petrochemical Industries Corporation (SPIC) in Chennai, India devised a two-step biotreatment process for these obnoxious chemicals by using a mixed culture of *Pseudomonas* and *Aerobacter*. (i) *Pseudomonas* metabolizes PO/PG to volatile acids like lactic acid, pyruvic acid, and formic acid. (ii) *Aerobacter* degrades the volatile acids into CO_2 and H_2O. This treatment allows more than 95 per cent biodegradation in terms of reduction in initial BOD/COD values. SPIC's effort can be described as a pioneering attempt in the world. SPIC also developed a phenol biotreatment process with the same logistic of using a mixed population of bacteria to get rid of toxic and foul smelling phenolic wastes from oil refineries.

Most bioremediation projects involve three crucial steps

Selection of the microorganisms   The right kind of microorganism can be obtained from soil that has been contaminated with the target chemical for some time. Such soil exists near pipeline junctions or tank overflow valves in plants that manufacture the chemical. Variants of an organism, which grow faster or degrade the target chemical more efficiently, are then created in the laboratory, by combinations of traditional microbial genetics, recombinant DNA technology, and selection. Usually, a consortium of microorganisms, rather than a single variety is used for bioremediation. Such mixtures of organisms can catalyse the breakdown of different components of a pollutant or different sequential breakdown products of a complex molecule.

Bioremediation genome projects under 'Microbial Genome Programme' (MGP) have been undertaken by the Department of Energy, USA to identify microbes and their genes responsible for bioremediation and to improve them.

Biostimulation of physiology of the organism   Sometimes the right organism collected from the site does not grow and work fast enough. A cocktail of nutrients and other growth limiting co-substrates are added to stimulate the growth and catalytic ability of the microorganisms. Designs of the cocktail are complex; the aim should be to condition the bacteria towards digesting the target chemicals, and not just to feed on the nutrient cocktail.

Inoculation of the polluted environment   The selected microorganism is introduced into the polluted site, usually with a nutrient cocktail to support its growth and to encourage it to break down the targeted (polluting) compound. For bioremediation, mostly complex hydrocarbon-based compounds which are metabolized by oxidation are targets. Thus, oxygen is a requirement for the process. Nitrogen and phosphorus are also commonly provided so that the bacterium is under continued selection pressure to utilize all the carbon available at the polluted site for growth, including that present in the target (polluting) compound. This phase of bioremediation is as critical as identifying a suitable microorganism, and requires good understanding of microbial physiology and ecology.

Bioremediation projects fail when the selected organism cannot break down the polluting compound at a useful rate at the site, despite performing well under laboratory conditions. Densely packed clays are not a good site for bioremediation as air is hardly present and penetration by water is not easy; besides, these conditions do not allow bacteria to flourish. Similarly, very highly contaminated soil restricts the growth of bacteria in too cold or dry conditions. In such cases, the soil can be placed in a bioreactor tank where bioremediation can take place. The bioreactor is essentially a large, insulated tank into which soil or waste is placed with a bacterial inoculum. Air is blown in to keep the mass oxygenated and a nutrient cocktail is supplied. Such a process has been successfully employed to remove aromatic hydrocarbons, such as benzene, toluene, and xylene from contaminated soil. This bioreactor process is also known as *ex-situ bioremediation*.

The cost and hazards of excavation of potentially large quantities of contaminated soil needs to be balanced against the benefits of this approach.

Bioremediation that has hit the headlines on several occasions has involved degradation of hydrocarbons, in particular petroleum products, which enter the environment from offshore oil drilling, loading of tankers or accidental spilling from tankers. These oil-eating bacteria break down the long chain hydrocarbon molecules of oil into soluble molecules that other bacteria can digest. This type of bioremediation works more effectively in estuarine and coastal areas, rather than in open-ocean oil spills (see Superbug in Chapter 6).

13.6.9 Heavy Metal Bioremediation

Toxic metals are a special class of environmental pollutants and cause much harm to living organisms including human beings. Metals cannot be degraded; they can only be converted to different oxidation states. The objective of bioremediation for metal-contaminated environments is mainly to seclude the metals to make them unavailable to biological components of the ecosystem or to mobilize them so that they can be flushed out from the system for collection and disposal. This immobilization of heavy metals is often done by sulphate–reducing bacteria that produce hydrogen sulphide. By the action of these bacteria contaminating metals are transformed into metal sulphides which are extremely insoluble and remain out of reach of the organisms that depend on water.

Phytoremediation of heavy metal contaminated soils and waters   Various plants can take up and concentrate heavy metals, including radionuclides and incorporate them in the plant biomass. Thus, the concentrations of hazardous metals in soils can be prevented by some specific plants. This process is often referred to as *phytoremediation*. Once the plants accumulate the heavy metals, they are harvested and the plant biomass on being incinerated leaves behind the metals as residue. These recovered metals can be reused. However, incineration of radioactive plants poses many technical difficulties.

Indian mustard (Brassica juncea) and ragweed (various *Ambrosia* species) among other plants, are used for phytoremediation. Indian mustard can accumulate heavy metals, including lead to levels that correspond to as much as 40 per cent of their biomass.

Demonstration projects have been carried out at various sites, including Chernobyl in Russia where heavy contamination with radionuclides occurred from a nuclear reactor accident.

13.6.10 Biohydrometallurgy

Biohydrometallurgy includes two broad areas of bacterial activity: (a) *Biosorption*—selective absorption of metal ions and (b) *Redox reactions*—use of metal ions or minerals for oxidation and release of energy for bacterial metabolism.

Biosorption   Many organisms have components that bind metal ions; for example, human bone matrix material binds strontium very well. There are sulphur-binding proteins in many organisms. Some bacteria can actively accumulate (using energy) a particular metal ion (*biosorption*). Extraction of the biosorptive material from the organism can be done so that the particular material can be used. This is a part of bioremediation. But this process is yet to be industrially viable. For this, removal of metals from the waste stream needs to be at least 90 per cent efficient and metal must account for at least 15 per cent of the organism's weight.

Redox reactions   Some bacteria use a metal ion, or a mineral in which the metal is immobilized, for their metabolism. The oxidation of sulphide to sulphate releases substantial chemical energy which can be used by the bacteria. Consequently, transformation of insoluble sulphide ore to the soluble sulphate state allows easy extraction of the ore. The same reactions can also be used to oxidize sulphides in one compound, causing formation of sulphuric acid, which then dissolves another compound, or to preoxidize a metal ore to make it more amenable to procure and process further. The whole process is often known as *leaching* and *microbial mining*.

Microbial mining is used commercially to recover copper and uranium from low-grade ores, especially chalcopyrite ($CuFeS_2$), covellite (CuS), chalcocite (Cu_2S), and uraninite (UO_2). A number of other metals, such as antimony, arsenic, cadmium, cobalt, molybdenum, and nickel, can be extracted by bacteria, but are yet to be used to a reasonable extent.

13.6.11 Biomineralization

Biomineralization is an extension of biohydrometallurgy, when the metallic inclusions of bacteria are laid down as extremely small crystals or granules. Magnetic bacteria take up the magnetic material, swim preferentially along the lines of a magnetic field, and then deposit the magnetite that they produce along these lines.

> The manganese nodules on the sea floor and the banded iron formation rock strata, laid down some 1000 million years ago are likely to be the result of bacterial reduction of manganese and oxidation of iron, respectively.

SUMMARY

- Environmental biotechnology involves living organisms from bacteria to higher forms of life that are involved in reducing or eliminating the pollution in the environment.
- The environment comprises all the composite external conditions, both abiotic and biotic, that influence organisms and their perpetuation.
- The environment of the earth is often classified into atmosphere, hydrosphere, lithosphere, and biosphere.
- The condition in which any gas, chemical, or material or a form of energy accumulates in the environment is called environmental pollution.

- The major sources of pollutants are industrial wastes, urban wastes, agricultural wastes, biological waste, and arsenic pollution.
- There were several world conventions to resolve the constant threat to the clean environment and to the depleting natural resources.
- Consumption of fossil fuels (coal, petroleum oil, and gas) by industries and human activities is steadily increasing the CO_2 in the atmosphere leading to the greenhouse effect and global warming.
- Biotechnology can be used to break down biodegradable wastes; non-biodegradable heavy metals and

some man-made chemicals including organic ones that, are difficult to treat and they also inhibit the growth of microorganisms. Non-biodegradable components of solid waste can be sorted out first.

◆ The collected solid wastes are subjected to incineration, and transferred to landfills for biodegradation.

◆ Once the type and the level of pollutants, BOD or COD in waste water are determined, the water is treated through different steps in a treatment plant.

◆ Both aerobic and anaerobic processes are used for the digestion of the organic wastes in treatment plants.

◆ The sludge, which is the end product of the treatment plant is taken out for drying and disposed of in a landfill or on an agricultural land.

◆ In recent decades, bioremediation has assumed the role of a serious and large-scale effort to harness the ability of bacteria to degrade a variety of compounds for effective restoration of a chemically polluted environment.

◆ In heavy metal bioremediation, sulphate-reducing bacteria transform metals to metal sulphides which are extremely insoluble and remain out of reach of the organisms dependent on water.

◆ Various plants can take up and concentrate heavy metals by the process of phytoremediation of heavy metal contaminated soils and waters.

◆ Microbial mining is used commercially to recover copper and uranium from low-grade ores.

Al Gore and Rajender Pachauri

Albert Arnold 'Al' Gore, Jr
(Born 31 March 1948)

Rajendra Kumar Pachauri
(Born 20 August 1940)

Albert Gore Junior was born in Washington, DC, USA. After attending St. Albans School, he joined Harvard in 1965 and graduated with majors in government in 1969. In August 1969, Gore enlisted in the army and was stationed at 20th Engineer Brigade in Bien Hoa during the Vietnam War. In May 1971, he received an honourable discharge from the army. At the age of 28, Gore joined the US Congress and served for sixteen years. He attained the office of Vice-president of the United States in 1993 and served till 2001 under the President Bill Clinton. Al Gore's concern for the world's environment culminated in a documentary—'An Inconvenient Truth', which brought to the world's notice, the problems of global warming and climate change.

Rajendra K. Pachauri was born in Nainital, India. Educated from La Martiniere College, Lucknow, and The Indian Railways Institute of Mechanical and Electrical Engineers, Bihar, Pachauri went on to receive his master's degree in Industrial Engineering from North Carolina State University in 1972. After obtaining PhD in 1974 in industrial engineering and economics, he served as assistant professor and then as visiting faculty in the Department of Economics and Business at North Carolina State University. From 1975 to 1979, Pachauri served in the Administrative Staff College of India. In 1981, he became the director of TERI (The Energy and Research Institute), Delhi and currently heads the organization. Since 2000, Pachauri has served as the chairman of the Intergovernmental Panel on Climate Change (IPCC). He has also headed the climate-related institutes at various Ivy League schools like Columbia and Yale.

Al Gore and IPCC (Chair held by Rajendra Pachauri) jointly received the Nobel Peace Prize in 2007 'for their efforts to build up and disseminate greater knowledge about man-made climate change, and to lay the foundations for the measures that are needed to counteract such change'.

EXERCISES

Objective Questions

1. Multiple Choice Questions (in some questions more than one answer may be right).

(i) Chemical evolution started
   (a) 2 billion years ago
   (b) 2×10^6 years ago
   (c) 4 billion years ago
   (d) 4.6 billion years ago

(ii) Microconsumers are
   (a) autotrophic bacteria
   (b) heterotrophic bacteria and fungi
   (c) organisms that incorporate inorganic substances and fix solar energy in the process.

(iii) Percentage of carbon dioxide in atmosphere
   (a) 0.93
   (b) 0.3
   (c) 0.03
   (d) 0.003
   (e) none of these

(iv) Xenobiotic compounds are resistant to microbial degradation due to
   (a) their chemical structure.
   (b) the presence of substitution groups like halogens.
   (c) the presence of OH group at reactive sites.
   (d) their insolubility in water.
   (e) their toxicity to microorganisms.
   (f) the absence of specific oxidizing enzymes in the microorganisms.
   (g) the lack of permease needed for their transport into microbial cells.

Review Questions

1. What are the interacting sub-systems of environment?
2. Define environmental pollution.
3. What measures would you take to reduce arsenic pollution of ground water?
4. What were the objectives of 'Rio Earth Summit' and 'World Summit on Sustainable Development'?
5. Provide a note on greenhouse gases and global warming.
6. What types of solid wastes are subjected to biodegradation?
7. What are the hazards of using landfills for the management of solid wastes?
8. What is eutrophication?
9. Define and differentiate BOD and COD.
10. Mention the types of organisms involved in the aerobic process of waste water treatment.
11. Describe the Trickling Filter Bed and its working principle.
12. What happens in the anaerobic process of waste water treatment?
13. What are the types of situations that can be controlled by bioremediation?
14. What is bioremediation genome project?
15. What do you understand by microbial mining?
16. Give three examples of synthetic polymers.
17. In oxic degradation of aliphatic hydrocarbons, what is the substrate produced for β oxidation?
18. What pioneering attempt was made by Southern Petrochemical Industries Corporation (SPIC), Chennai, for biotreatment of some effluents of the petrochemical industries?
19. What conditions are not suitable for bioremediation of polluted soil?

Horizon of Biotechnology

LEARNING OBJECTIVES

♦ Biofertilizers and biopesticides and their applications
♦ The use of single cell protein (SCP) as a food additive
♦ The fundamentals of aquatic biotechnology, biotransformation, and biofuels (bio-diesel)
♦ Biosensors such as ISFET, thermal sensors, and immunosensors and their working
♦ Application of DNA fingerprinting
♦ Importance of monoclonal antibodies and vaccines
♦ Importance and utilization of stem cells for human welfare

INTRODUCTION

Going through the earlier chapters, the readers must have definitely realized the span of biotechnology. It covers basic principles in cell biology, genetics, biochemistry, molecular biology, biochemical process engineering, and then reaches aspects of manipulation of genes or recombinant DNA technology to achieve almost any kind of desirable combination of genes, normally not achievable through conventional breeding. Transgenic prokaryotes, plants, and animals can deliver many products, such as food, beverages, enzymes, pharmaceuticals, and industrial items. The transformed cells of these organisms in mass cultures or in different types of bioreactors can also provide the substances in bulk. Bioinformatics has made it possible to fish out the right genes from a genome to transcribe for specific purposes. Furthermore, the manipulated microbes offer solutions to the indiscriminate pollution of the environment by human beings. Biotechnology lays the path for a future without chemical fertilizers and pesticides that have undesirable side effects and pollute land, water bodies, and organisms.

The principles of biotechnology have allowed imaginative researchers to innovate in the areas of agriculture, biomimetics, biofuels, biosensors, monoclonal antibodies, stem cell technology, and so on. A thorough knowledge and understanding of these topics is essential as they are going to expand the horizon of biotechnology in the coming years. Biotechnology is a promising field that will offer solutions to the various problems faced by mankind from food to health and meet the increasing demands for many items by the growing population all over the world. The ever increasing horizon of biotechnology will justifiably mark the 21st century as the century of biotechnology.

14.1 AGRICULTURAL BIOTECHNOLOGY

The world population has doubled in the past 40 years, while the amount of agricultural land has increased by a meagre 10 per cent and in many cases by undesirable deforestation. In fact, world food production has increased reasonably over the past 40 years. This has been possible because of high-yield crop strains raised by conventional cross breeding methods. In recent times, the productivity of crops has been accelerated by the direct transfer of genes. The use of rDNA technology for improvement of cultivable varieties and their food value, raising disease resistant crops, and production of some important items through transgenic plants has been discussed in detail in Chapter 6. Biotechnologists have also taken steps to improve biofertilizers and reduce the need for chemical fertilizers and pesticides.

14.1.1 Biofertilizer

Biofertilizer is a natural organic fertilizer, often derived from biological wastes, that helps provide all the nutrients required by plants and improve the quality of the soil using a natural population of microorganisms.

Biofertilizers support organic agriculture, sustainable agriculture, green agriculture, and non-pollution agriculture. For centuries, soil bacteria, fungi, earthworms and bio-organisms, in the course of degrading organic waste, have enriched the soil by producing safe, nutritious fertilizer that favours the abundant growth of crops.

Overuse of petrochemical-based fertilizers and toxic pesticides has had a detrimental effect on our soils, river and other water supplies, food, animals, and on humans. Chemical fertilizers cause loss of organic matter and microbial activity, soil fertility, and irreparable damage to the overall system. Nowadays, farmers are more sensible, use a sustainable approach and employ the resources of both science and nature to allow better production.

Soils with lots of organic matter remains loose and airy, hold more moisture and nutrients, favour faster growth of soil organisms, and promote healthier plant root development.

Some free-living and symbiotic prokaryotic microorganisms (bacteria) are capable of reducing atmospheric nitrogen to ammonia. This ammonia can be used for synthesis of amino acids and other nitrogenous compounds such as purines and pyrimidines (for synthesis of DNA and RNA), essential for plant growth. The whole process is known as *nitrogen fixation by microorganisms.*

It has been estimated that the bacteria belonging to the genus *Rhizobium* fix up to 200 million tonnes of nitrogen globally per year, much more than the 40 million tonnes of nitrogen fertilizer synthesized annually, mainly by the energy-intensive Haber-Bosch process. This illustrates the efficiency with which microorganisms fix nitrogen for a sustainable environment. The major focus of biotechnological research for increasing biological nitrogen fixation has been on *Rhizobium*. This bacterium is associated as a symbiont with roots (nodule) of leguminous plants such as soybean, peas, clover, and so on. Soon biotechnological research will transfer nitrogen-fixing (*nif*) genes to non-leguminous plants and other soil organisms.

Research is also being carried out on a number of nitrogen-fixing bacteria, such as *Frankia* sp. and *Azospirillum* sp. that have an asymbiotic relationship with non-leguminous trees or crop plants. The blue-green alga *Anabaena azolla* can also fix nitrogen and remain in association with the water fern, *Azolla caroliniana*, which is traditionally used as a source of nitrogen for rice crops, particularly in South-East Asia.

Summary of the types of biofertilizers

For nitrogen
- *Rhizobium* for legume crops
- *Azotobacter/Azospirillum* for non-legume crops
- *Acetobacter* for sugarcane only

For phosphorus
- *Phosphotika* for all crops along with nitrogen-fixing varieties
- *Pseudomonas fluorescens/P. putida*, phosphate-solubilizing bacteria—convert non-available inorganic phosphates into soluble form for utilization by crop plants

For enriched compost
- Cellulolytic fungal culture
- *Phosphotika* and *Azotobacter* culture
- Plant growth promoting rhizobacteria produce *siderophores* (iron chelating substances, for example, pseudobactin) and make iron unavailable to harmful fungi like *Erwinia*, resulting in an increase in plant yield

14.1.2 Biopesticides

Pests cause economic destruction, mainly by attacking crops and grain. Thus, synthetic pesticides, fungicides, and herbicides are required for enhanced production. The presence of residues of these chemicals has risen at an alarming level in many organisms, including human beings, and in rivers, lakes, and ground water, and also in the atmosphere. Many of these chemicals have been shown to be carcinogenic (cancer producing).

Biopesticides are compounds derived from natural sources, such as animals, plants, bacteria, and living organisms and are capable of limiting the growth of pests. Biopesticides can decrease the harmful effects of synthetic peptides and provide a more sustainable environment.

Biopesticides are different from *biocontrol agents*. Biocontrol is the control of one species by another which has been deliberately introduced for that purpose. Pharaoh ants that have been used to combat destructive insects in grain stores of China since ancient times are one such example. Biopesticides are non-living, whereas biocontrol agents are active, living organisms that reproduce and seek out the target pest to be destroyed.

There are a wide range of biochemicals of plant origin that foil pest attack—the caffeine in coffee beans is one such example. Some biochemical pesticides of plant origin may interfere with the mating cycles of pests, for example, pheromones and scented plant extracts.

Bacillus thuringiensis

Bacillus thuringiensis is a naturally occurring bacterium which on sporulation produces a protein that has an insecticidal effect. Breakdown of this spore protein in the gut of lepidopteran larvae (caterpillars) produces a highly specific (Bt) toxin which interferes with ion transport and absorption of food from the gut of the larvae and kills them within 4–7 days of ingestion. Bt toxin seems to have no effect on vertebrates or invertebrates other than some *Lepidoptera*, *Coleoptera*, and *Diptera* insects.

B. thuringiensis has been used as a biocontrol agent for many years, but recently biotechnology has isolated the Bt toxin to be used as biopesticides. The isolated toxin has been applied directly over insect-infested fields by spraying, but its effect does not persist for long. Biotechnologists have introduced the gene encoding Bt toxin protein into plant cells making these plants more resistant to pest attack. The insect pests die when they nibble the leaf. The gene has been successfully introduced and expressed in petunia, cotton, tobacco, and tomato plants. There is detailed information about Bt based biopesticides in Chapter 6.

Advantages of biopesticides over chemical pesticides

- Biopesticides are derived from natural sources or living organisms; in general they are less toxic than chemical pesticides, and also eco-friendly due to certain other properties.
- Biopesticides are highly specific for target pests, and are unlikely to directly affect non-target organisms.
- They are biodegradable and consequently do not pose the problems of bioaccumulation and biomagnification and accumulation in the soil as seen in the case of broad spectrum chemical pesticides like DDT.
- After more than two decades of use, it is claimed that there is no pest resistance to Bt toxin.
- Since the toxin is specific it is only required in small quantities to control pests. There are some counter arguments regarding the use of Bt pesticides.

Herbicide-resistant Plants

A variety of agronomic crops have been developed by transferring the resistance gene from soil bacteria to the crop plants, so that herbicides may be sprayed in the fields to kill the weeds without affecting the crop plants. This has been discussed in Chapter 6. This is to save on the labour and wages required for deweeding the cultivated land.

14.2 BIOMASS

Biomass is the mass of organic material in any bulk biological material. Fermentation generates biomass (more of fermenting organisms) as well as the fermentation product (alcohol). Usually, in a fermentation system, the biomass is a side product, more of a waste as the alcohol is the desired product and not the yeast (fermenting organism). But some processes grow bacteria, fungi, and plants specifically for their biomass. Biomass production employs several organisms and systems and serves many purposes for human beings.

14.2.1 Single Cell Protein—As a Food Additive

Single cell protein (SCP) technology uses bacteria to grow on a cheap carbon substrate and with a cheap nitrogen source such as ammonia, to make protein fit for human or animal consumption. Gas oil, that is, the oil that remains after petrol and paraffin have been distilled out of crude oil, has been tried as a carbon source for the organism *Candida lipolytica*. Methanol made from natural gas is a good potential substrate since bacteria find it easier to use and need less oxygen. The Imperial Chemical Industries (ICI) Limited, Britain, developed a large-scale biomass process based on the bacterium, *Methylophilus methylotrophus* that utilizes methanol, to produce a partially purified protein product 'Pruteen'. ICI also inducted genetic engineering to improve the effectiveness of the bacterium's metabolism in using ammonia to improve protein production. All these efforts were at best only marginally economical. Furthermore, the nucleic acid (DNA and RNA) content of microorganisms is much higher than that of animals or plants; this may pose a health problem if they are eaten raw. The cells of microorganisms can be extremely hard to digest. Thus, most effort on SCP has gone into using it as an animal feed supplement.

14.2.2 Algal Biomass

Single cell plants such as *Spirulina* and *Chlorella* are grown commercially in ponds to produce food materials for consumption by man and fish respectively. Use of *Spirulina* as a health food was much hyped a few years ago, because of an unfounded belief that it was extraordinarily nutritious.

Scientists are examining the ways in which algae can produce alternate fuels. Then the marine biomass can be converted into ethanol. *Bioprocessing* is an engineering approach to produce a biological product such as a recombinant protein, related to applications of biomass. Scientists have succeeded in producing large amounts of antibodies in marine algae which can be grown on a large scale.

14.2.3 Plant Biomass

Growing crop plants for food is called farming. Sometimes, the production of crop plants provides the base for processing certain chemicals which can be included under biotechnological activities. For example, Brazil engages considerable amount of effort and money for growing sugarcane to make ethanol by fermentation. The countries with little or no hydrocarbon reserve (petroleum) may use the biomass of agricultural and forestry wastes and convert it into alcohol.

Biotechnological research is continuing in different parts of the world to increase plant biomass production by rapid propagation of fast growing species; this can be multifunctional in relation to energy production (including burning) and in achieving a sustainable environment.

14.3 AQUATIC BIOTECHNOLOGY

Three-fourths of the earth's surface is covered with water which is the natural abode for an enormous number of species representing a majority of the phyla of the plant and animal kingdoms. Some that are exotic, some the largest of the living world; and many that have adapted to different depths, temperatures, and other conditions add up to a long list for biodiversity; a good number of aquatic species satisfies the hunger and other needs of land dwellers. There is a lot to explore about aquatic species. More than 80 per cent of the earth's organisms live in aquatic ecosystems. Although we have reached the 21st century and scientists have explored a great deal about oceanography, the vast majority of marine organisms, particularly microorganisms, is yet to be identified.

14.3.1 Aquaculture

Raising finfish (a vertebrate class that have fins with fin rays) and shellfish (invertebrate) in controlled conditions for use as a food source is the oldest practice of aquatic biotechnology. (Culture of pearl oysters is also considered as aquaculture.) Aquaculture has been popular in the third world for a long time and is getting popular in the developed countries. More and more people are finding a cheaper and healthier source of protein in fish and a good number of people feel that soon this source will replace a good share of beef and other meats in the food market.

Aquaculture is the cultivation of aquatic animals, such as finfish and shellfish, and aquatic plants for food, other commercial purposes, recreation, and sports.

Specifically, marine aquaculture can be called *mariculture*. Aquaculture conventionally may be considered under agriculture.

14.3.2 New Developments in Aquatic Biotechnology

Genetic engineering and new discoveries in biology introduced some fascinating developments in modern aquatic biotechnology and have inspired new ideas to identify more products from the aquatic world. Some of the examples are as follows:

- Disease-resistant strains of oysters (shellfish)
- Vaccines against viruses that infect large finfish, such as carp, salmon, and so on
- Transgenic salmon capable of overproducing growth hormone leading to extraordinary growth rates in a short growing period, thus reducing the time and cost involved in growing them on a commercial scale
- Antifreeze promoter of ocean pout, a fish native to cold habitats, cloned with the growth hormone cDNA of salmon allows cultivation and fast growth of salmon in cold climates
- Certain species of marine plankton and snails which have been found to be a rich source of anti-tumour and anti-cancer molecules

Marine microbes, algae, shellfish, finfish, and countless other organisms live in some of the most extreme and harsh environmental conditions. Extreme cold, excessive pressure at great depths, abysmal darkness, high salinity, extreme temperature, and volcanic conditions in parts of the ocean floor have influenced the genome of marine organisms to come up with different types of adaptations through the millions of years of evolution. As a result, all these organisms are likely to be rich and valuable sources of new genes, proteins, and metabolic processes. Scientists are striving to identify certain fish genes with important applications that would benefit human beings.

Aquaculture is an emerging frontier in biotechnology research which can tap the potential wealth to meet the food requirements of a growing population, human medical needs, and restore habitats and denuded environments. Several *research priorities* identified for aquaculture biotechnology are as follows:

- Improving strains for quality and quantity of food supply for the world
- Improvement of safety and quality of seafood, particularly shellfish so that it is without pathogens and contaminating agents
- New approaches to monitor and treat diseases in aquatic creatures
- Discovering novel substances or compounds with medicinal properties
- Developing new products with applications in industries
- Restoration of habitat loss due to overharvesting and pollution
- Further identification of the species and geochemical processes influencing aquaculture

Among Asian countries, Japan is at the forefront of research and development in aquatic biotechnology, spending to the tune of $ 1 billion annually. The US spends almost half that amount for the same purpose.

14.3.3 Improving Strains for Aquaculture

Improvement of certain qualities, such as growth rate, fat content, taste, texture, colour, and disease resistance of finfish or shell fish are the desired goals of fish farming.

Selective breeding   This is an age-old technique and may be used in combination with some modern techniques to raise strains with desirable characteristics. For example, ultrasound machines can be used to measure muscle mass as a means to estimate fillet yield. Fish showing the highest yields are then bred to produce offspring with increased muscle mass. *Bioimpedance* tracks movement of low-voltage electrical current through the tissue of a living fish to measure lean muscle mass and fat content. Dense tissue with a high muscle content and low fat content interferes with or impedes movement of electric current through the tissue. This technique makes it possible to develop strains of the best tasting, low-fat catfish and other fish species. This technique has also been used for the selective breeding of cattle, swine, and many other farm animals.

Food additive   *Astaxanthin* is the pigment that gives shrimps their pink colour. Astaxanthin in fish food imparts multicoloured or reddish hues to the flesh of salmon and trout. Most people prefer this colour with a belief that these salmon are of better quality. Igene Biotech of Columbia, Maryland uses gene-cloning techniques for mass production of astaxanthin. The Swiss drug company Roche Holding AG is a leading producer of the pigment which gives a choice of shades ranging from light pink to dark crimson. Aquaculturists purchase fish food containing astaxanthin at concentrations that will produce fish with flesh colour of their choice. Interestingly, astaxanthin is also a potential antioxidant.

Identifying and cloning novel genes   Scientists are constantly on the look out for novel genes and the environmental factors that control gene expression at extreme temperature and water pressure, at great depths in the ocean. They are also trying to identify mutations associated with diseases in fish. This information will help them control or eradicate diseases from breeder stocks of fish.

Scientists can incorporate a novel gene in a fish stock by genetic manipulation, and modify or remove deleterious genes affecting quality, growth, health, and longevity of fish.

Cloning of the salmon *growth hormone gene* can be cited as an excellent example of developing a transgenic strain of salmon showing accelerated growth rate in comparison to the native strains. A genetically engineered salmon breed grows to the adult size suitable for market sale in half the time that a normal salmon does. The short growing period lowers the cost of raising and feeding the fish.

Another example involves research to block the release of *molt-inhibiting hormone* (MIH) in crabs. Decrease in MIH triggers molting (shedding of exoskeleton), leading to soft-shelled crabs that can be eaten whole and are much in demand in the seafood industry. Researchers at the University of Alabama Birmingham are involved in this project.

14.3.4 Antifreeze Protein Gene

Antifreeze proteins (AFPs) protect fishes from freezing at cold temperatures in a variety of ways. Seawater freezes at approximately −1.8°C. AFPs typically lower the freezing point of fish body fluid (blood) by approximately 2–3°C. AFPs can bind to the surface of ice crystals to modify or block ice crystal formation in blood and cells of fishes adapted to the cold and protect the cells from cold damage.

Several types of AFPs have been discovered. Structurally, they are mostly dominated by extensive alpha helices, held together by large numbers of disulphide bridges.

In the first round, AFPs were isolated from a bottom-dwelling fish species, northern cod that lives off the coast of eastern Canada and Antarctica. Subsequently, AFPs have been isolated from other cold water species, such as ocean pout (*Macrozoarces americanus*), herring (*Clupea harengus*), winter flounder (*Pleuronectes americanus*), and smelt (*Osmerus mordax*). Interestingly, mealworm beetles have been found to possess similar proteins.

AFP genes have been used to produce transgenic fish and plants. Transgenic salmon that can withstand the cold sea in the Antarctic region has been produced with success. Normal salmon cannot produce this cryoprotective protein. Furthermore, the AFP gene promoter sequences have been ligated with the growth hormone gene of salmon in recombinant DNA experiments. As a result, AFP promoter sequences which are activated under cold water conditions lead to transcription of 'downstream' ligated genes including growth hormone gene (Fig. 14.1). This will allow aquaculture of salmon beyond its traditional territory.

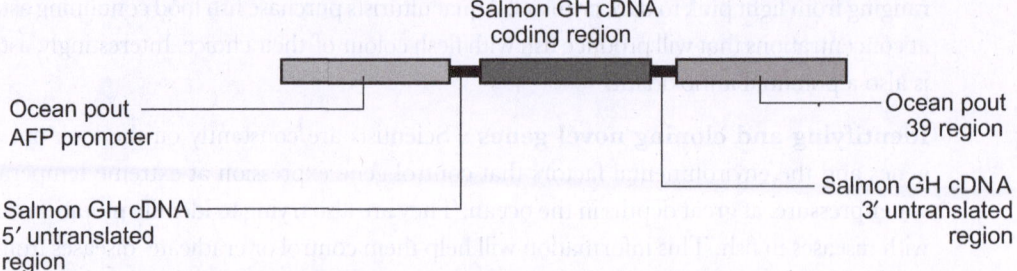

Fig.14.1   A recombinant DNA plasmid construct containing an antifreeze protein (AFP) gene promoter from ocean pout fish ligated to the cDNA for salmon growth hormone (GH) gene to produce transgenic salmon for rapid growth in cold conditions.

AFP genes have been experimentally introduced into plants to produce cold-resistant transgenic strains, for example, in tomato plants. Loss of many popular crops, such as wheat, coffee, corn, soybean, potatoes, and fruits like apples, pears, cherries, and peaches at cold temperatures is a worldwide problem. Scientists hope to solve this problem with transgenic varieties containing AFP genes.

Cryoprotective AFPs will have multifarious applications. Biotechnologists are already working on recombinant AFP production in bacterial and mammalian cells to meet the anticipated future needs of AFPs.

Cryoprotectants like AFP which prevent formation of ice crystals will be extremely useful in medical science. These could be used for cryogenic storage of human organs such as heart and liver before transplantation surgery, to store oocytes and sperms before in-vitro fertilization, or stem cells for future use.

AFPs will also find applications in the improvement of the shelf life and quality of frozen foods, including ice cream. AFPs may alter the ice crystallization in frozen foods to improve their overall quality. AFPs might also be used to control ice formation on aircraft and roadways.

14.3.5 Reporter Gene—Green Fluorescent Protein (GFP)

A novel gene from the bioluminescent jellyfish (marine invertebrate medusa) *Aequorea victoria*, encoding *green fluorescent protein* (GFP), is ligated to a gene of interest in a plasmid construct, and the construct is introduced into cells in culture (Fig. 14.2). These cells will then express mRNA molecules, which will be translated to produce a fusion protein of GFP and the cloned protein of interest. This fusion protein will fluoresce when exposed to ultraviolet light, '*reporting*' the location of the expressed protein of interest. This reporter gene protein is used as a marker to study the basic processes of gene expression and regulation.

GFP-expressing embryonic stem cells make tracking of their differentiation into different cell types during development of the embryo easy. The GFP reporter gene constructs are being used for medical diagnostics and disease treatment. Tumour formation in mice, bacterial infections in the gastro-intestinal tract, or the death of bacteria following antibiotic treatment can be pin pointed by using GFP gene constructs.

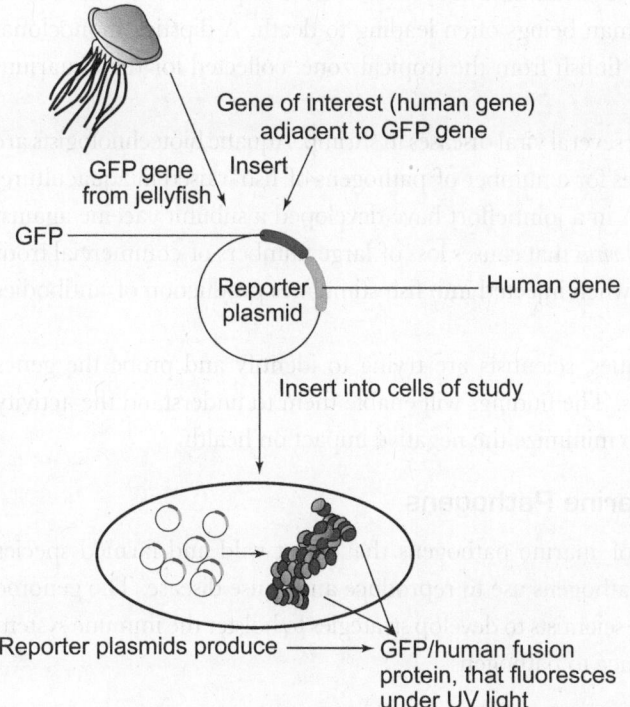

Fig. 14.2   Green fluorescent protein (GFP) gene from bioluminescent jelly fish acts as a novel reporter gene, when ligated to a gene of interest (GI) in a plasmid and inserted into test cells. The fusion proteins (GF + GI) fluoresce under UV light in the cells reporting the expression of the gene of interest.

GloFish with Red Fluorescent Protein Gene

A transgenic strain of zebra fish containing the red fluorescent protein gene from sea anemone was raised by Yorktown Industries of Austin Texas. These fish fluoresce (glow) bright pink under ultraviolet light. This was the first genetically modified pet sold in the US. This became news headlines in 2004; anti-genetic engineering groups voiced their protests against this novel feat of genetic engineering.

14.3.6 Enhancing Safety for Aquaculture and its Products

Enhancing the safety of the organisms in aquaculture as well their products is as important as increasing the quality of aqua-products. Thus aquatic biotechnologists are concerned about infection to the organisms, and contamination and toxic content of the organisms from the perspective of their consumption by humans.

Many molecular probes and PCR-based assays are being developed to detect bacteria, viruses, and other parasites that infect finfish and shellfish. For example, PCR-based approaches have been very effective for sensitive and rapid detection of different protozoan parasite species causing substantial damage to the oyster population in the eastern US. Proper identification enables scientists to control the infections.

An antibody test kit can easily detect *Vibrio cholera* in oysters, which is responsible for cholera—a severe disease of dehydration in human beings often leading to death. A dipstick monoclonal antibody test can detect *ciguatoxin* in finfish from the tropical zone, collected for the aquarium industry.

Gene probes are used for detecting several viral diseases in shrimp. Aquatic biotechnologists are developing detection kits and vaccines for a number of pathogens of fish raised in aquaculture. Scientists in several institutes in USA in a joint effort have developed a subunit vaccine against the *infectious hematopoietic necrosis (IHN) virus* that causes loss of large numbers of commercial trout and salmon each year. This vaccine when injected into fish stimulates production of antibodies against IHN virus.

Using molecular biology techniques, scientists are trying to identify and probe the genes encoding toxins in marine organisms. The findings will enable them to understand the activity of the toxins in human beings and to minimize the negative impact on health.

14.3.7 Cloning the Genomes of Marine Pathogens

Cloning the genomes of a variety of marine pathogens that affect wild and farmed species will help us learn about genes that pathogens use to reproduce and cause disease. The genome information of pathogens will enable scientists to develop strategies to bolster the immune system of farmed species to improve resistance to pathogens.

14.3.8 Polyploidy for Rapid Growth in Fish

Prior to the transgenic technique, induction of the polyploidy condition (additional complete sets of chromosomes in cell) in fish was used to achieve a higher rate of growth in the organisms.

Diploid sets ($2n$) of chromosomes are present in somatic or body cells of many animals and plants, whereas gametes of these organisms have a single set (n) or haploid number of chromosomes. Polyploidy refers to the presence of three ($3n$) or more ($4n$, and so on) sets of chromosomes.

The majority of polyploid fish created to date are triploid ($3n$), with three sets of chromosomes. Different techniques are adapted to create triploids. Usually, eggs are subjected to temperature or chemical treatment to interfere with egg cell (meiotic) division and the egg cell ends up with two sets of chromosome ($2n$) instead of becoming haploid (Fig. 14.3a).

Colchicine is a chemical derived from the crocus flower (*Calchicum autumale*); it blocks cell division by interfering with the formation of microtubules of spindle fibres that are necessary for cell division. As a consequence, the chromosomes replicate into the $2n$ state in treated egg cells, but the cells do not divide into two. The egg cells end up with a diploid number of chromosomes. Fertilization of such an egg cell by a normal haploid sperm cell results in a triploid organism. A triploid embryo can also be produced by the fertilization of a normal haploid egg by two normal haploid sperms (Fig. 14.3b).

Triploids grow more rapidly. In most species, triploids grow 30–50 per cent faster and larger than the normal diploids. But triploids in most cases are sterile as they produce gametes with

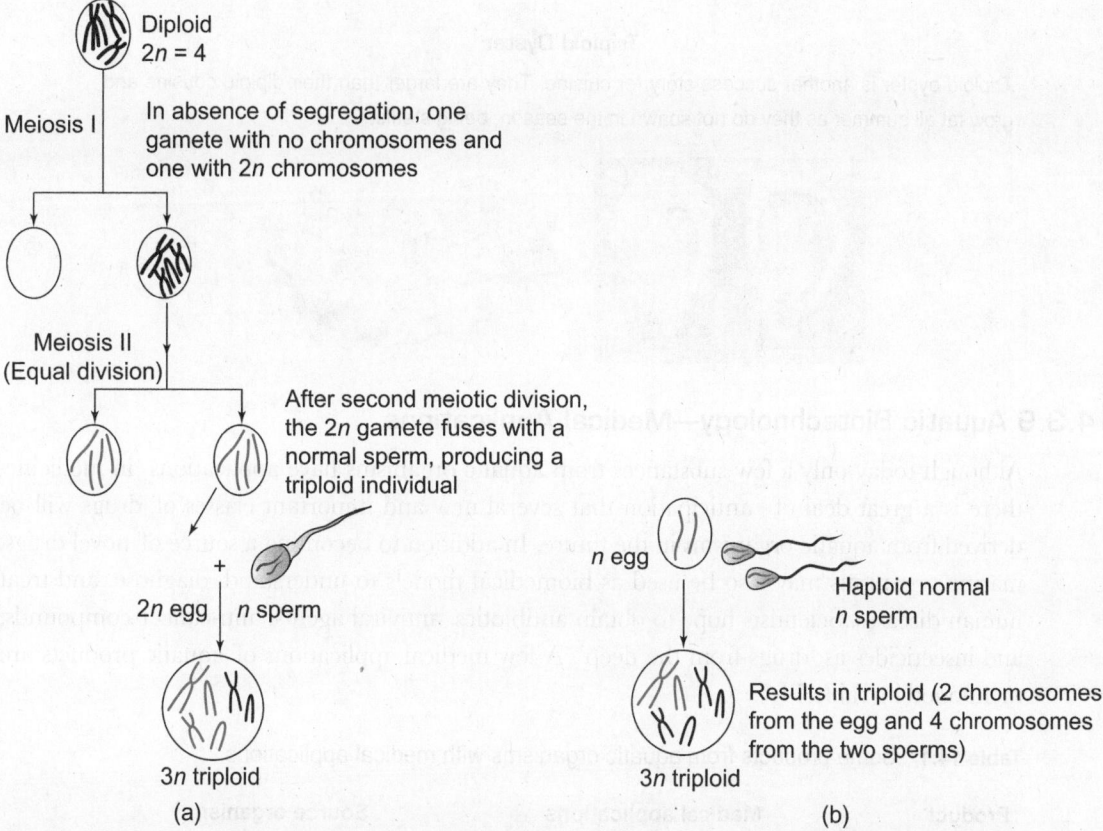

Fig. 14.3 Two techniques to produce triploid organisms (fish): (a) Chemical or heat treatment blocks chromosome separation during meiotic I cell division to produce diploid eggs; when these eggs are fertilized by a normal haploid sperm cell, triploid offsprings are produced. (b) A triploid created by fertilization of a normal haploid egg by two normal haploid sperms.

an abnormal number of chromosomes or in some cases no gametes at all. In a way, the sterility safeguards against undesirable breeding with wild counterparts in a natural habitat. The triploid grass carp (*Ctenopharyngodon idella*) developed by giving eggs a temperature shock is the first widely used triploid strain of fish. It has a voracious appetite for many types of aquatic vegetation, and thus have been very useful in controlling the enormous growth of aquatic weeds in freshwater bodies and lakes and have brought down the use of chemical herbicides. Sometimes, dramatic decreases in weeds may affect the habitat of other types of fishes, leading to a decline in their population.

Transgenic finfish and shellfish are mostly modified with growth hormone for enhancing the growth rate, anti-freeze gene to confer cold adaptation, disease resistance genes to raise disease-free fishes. No transgenic fish has been approved by the Food and Drug Administration (FDA) of the US. Cuba is liberal about it and has already allowed sale of genetically modified *Tilapia*. Moreover, many products from fishes are used for multiple purposes.

Triploid Oyster

Triploid oyster is another success story for cuisine. They are larger than their diploid cousins and grow fat all summer as they do not spawn in the season, being sterile.

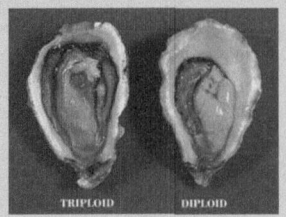 

14.3.9 Aquatic Biotechnology—Medical Applications

Although today only a few substances from aquatic organisms have applications in medicine, there is a great deal of anticipation that several new and important classes of drugs will be derived from aquatic organisms in the future. In addition to becoming a source of novel drugs, marine organisms may also be used as biomedical models to understand, diagnose, and treat human diseases. Scientists hope to obtain antibiotics, antiviral agents, anti-cancer compounds, and insecticides as 'drugs from the deep'. A few medical applications of aquatic products are mentioned in Table 14.1.

Table 14.1 Some products from aquatic organisms with medical applications

Product	Medical applications	Source organism
Calcitonin	Thyroid hormone, stimulates calcium uptake by osteoblast bone cells and calcification. To treat osteoporosis.	Coho salmon produces calcitonin with 20 times higher activity than that of human calcitonin.
Hydroxyapatite (HA)	HA implants act as bone grafts for repair of fractured bones.	Skeleton of corals.
Prialt (a peptide conotoxin)	Neurotoxin, used as painkiller to treat chronic pain and arthritis.	Marine cone snail (*Conus magus*).
Manoalide (non-steroid)	Anti-inflammatory and analgesic (under clinical trials).	A Pacific sponge.
Neovastat (anti-angiogenic compound)	Block blood vessel formation. To treat cancer.	Hammerhead shark, cartilage extract.

14.3.10 Other Utilities

Aquatic biotechnologists are constantly on the lookout for many other utility items from marine organisms. Already potential applications of quite a few substances have been identified and some are under experimental investigation. A few such promising products are mentioned in Table 14.2. At present some of these items are prohibitively expensive to produce. For example, about 10,000 mussels would be required for 1 gm of adhesive from byssal fibres. The scientists have used rDNA technology to express the byssal fibre gene in bacteria and yeast for mass production of the adhesive protein.

Scientists are studying marine organisms that developed tolerance to ultraviolet light as a potential source of natural sunscreens to be used in cream.

Seaweeds (*Rhodophyta*) have been used for a long time as a source of *carrageenan*, an agar used in many preserved foods, toothpastes, cosmetics and for laboratory use in bacterial culture plates, and agarose gels for electrophoresis. Carrageenan is a sulphate-rich polysaccharide. Carrageenan is also used as a stabilizing and bulking agent in chewing gum, chocolate milk, syrups, sauces, salad dressing, adhesives, polishes, and many other products of daily use.

Researchers have found that algae may be potentially very valuable for expressing recombinant proteins, such as antibodies and other proteins. Algae and different types of bacteria are being extensively studied by aquatic biotechnologists to identify novel products or genes.

Table 14.2 Promising applications of aquatic biotechnology in medicine and industry

Product	Source	Properties	Applications
Super-adhesive protein-rich glue	Adhesive *byssal fibres* of mussels *Mytilus edulis*	Several times tougher and more extensible than human tendons (tougher than steel)	Automobile tyres to shoes, soft body armour for soldiers, artificial tendon and ligament grafts, bone and teeth repair, surgical sutures
Anti-cancer compounds (a dozen)	Marine invertebrates (sponge, tunicates, mollusks, dinoflagellate-plankton)	Anti-cancer	Some under clinical trials
Venom	Venomous marine creatures	Neurotoxin	Analgesic
Tetrodotoxin (TTX)	*Fugu rubripes* (Japanese puffer fish or blowfish)	Deadly toxin (10,000 times more lethal than cyanide), blocks sodium channels in nerve cells	Anaesthetic, treatment of chronic pain, also as anti-cancer agent
Anti-leukaemic	A marine invertebrate *Bugula neritina*	Active against leukaemia	Anti-leukaemic drug
Squalamine	*Squalus acanthias* (Dogfish shark)	Steroid	Anti-fungal agent to treat life-threatening fungal infection
Chitin and chitosan (in ground form)	Exoskeleton of Arthropoda—crabs, shrimps, lobster, etc.	Complex carbohydrates, similar to cellulose	Dietary fibres in skin cream and contact lens, for non-allergenic dissolvable stitches that stimulate healing in human beings
Limulus ameobocyte lysate (LAL)	Limulus (king or horseshoe crab–'Living Fossil' since 300 million years ago) Ameobocyte–blood cell	Detect endotoxin from bacterial cell wall	LAL test—rapid test for endotoxins in human blood or body fluid samples, raw milk and beef; to ensure absence of cytotoxins in recombinant therapeutic proteins and drugs, surgical instruments

14.3.11 Aquatic Biotechnology for Certain Environmental Applications

Attachment of organisms to surfaces is called *biofilming* or *biofouling*. This occurs all the time everywhere. Bacteria that coat the teeth all day long and bacteria that adhere to the contact lenses are some examples. Also, implanted surgical devices and prostheses are examples of biofilming. Similar biofilming takes place in the plumbing and walls with many different types of organisms.

Algae, barnacles, mussels, clams, and bacteria produce biofilms in marine environments including the outer surface and equipment of ships. Attachment of marine organisms to the hull of a ship technically increases the resistance of the ship in the course of its movement through the water, and thereby lowers the fuel efficiency and increases the cost of operation. Biofilming also corrodes metal surfaces and increases the maintenance cost. Copper and mercury-rich paints are used to minimize the growth of fouling organisms. The toxic metal ingredients in paints leach and accumulate, and pollute the environment. Scientists are trying to understand how marine organisms such as mussels, clams, corals, and turtles prevent biofilming on their own surface. Many stationary marine organisms have been found to release defensive chemicals to create a protective zone around them against the predators. These chemicals may also deter fouling micro and macro organisms.

Marine sea grasses, such as the eelgrass *Zostra marina*, and algae produce compounds capable of deterring bacteria, algae, and fungi from adherence. Marine biotechnologists are trying to producing biofilm deterring chemicals from biological sources to provide protective coatings for ship hulls, aquaculture equipment, and other surfaces to prevent biofilming as well as pollution from heavy metal-based paints that are currently in use.

14.4 BIOMIMETIC

Biomimetic means mimicking or imitating life. An area of chemistry is trying to develop reagents that would perform some of the functions of biological molecules. This is to overcome the difficulty of producing and handling large biological molecules with useful functions and applications and to cut down the cost for commercial purposes.

Mimetics

Arris Pharmaceuticals is developing a small molecule 'mimetic' of erythropoietin and DuPont Merck is working on a neurotensin mimetic using small, stable, cheap organic molecules. Both the mimetics are being developed to mimic the molecular effect of large natural protein molecules so that they can be used for treating some medical conditions.

Some of the biomimetics that are currently under production are
- *Cofactor substitutes* for enzymes
- *Peptide biomimetics* (peptomimetics) to stimulate the shape and functions of some regular peptides, not broken down easily by proteases (enzyme), potential for drugs
- *Synzymes*, low molecular weight molecules that act as artificial enzymes with highly specific catalytic functions

- *Glycomimetics* with properties of sugars and affecting interactions normally mediated by sugars
- *DNA substitutes* with a polyamide chain instead of the standard sugar–phosphate backbone, binding tightly to single-strand DNA to act as antisense and resistant to breakdown by nucleases or proteases enzymes

14.5 BIOTRANSFORMATION

The use of biological catalysts to convert a chemical (substrate) into a product in a limited number of enzymatic steps is the domain of biotransformation. Usually, the catalyst is an enzyme, or an intact, dead microorganism that contains an enzyme or several enzymes. In recent times, the catalytic role of antibodies and ribozymes has broadened the definition and scope of biotransformation.

(*Bioconversion* is the conversion of one chemical into another by living organisms; this is different from the conversion by enzymes as in biotransformation. Conversion of sugar to alcohol by using microorganisms is a typical example of bioconversion. Bioconversion involves several enzymatic steps and may involve some unstable enzymes within the bacteria.)

Biotransformations are performed by a variety of biological catalysts, such as isolated enzymes, immobilized enzymes and cells. Recombinant DNA technology helps improve enzyme production by catalytic organisms. Biological catalysts are energy effective, that is, they work at moderate temperatures, pressure, and pH values. They produce higher quality products with lower amounts of toxic wastes, and decrease the quantity of waste water that is generated. Thus, biotransformation is a cost-effective and environment-friendly process.

A key use of biotransformation is the important advantage of biocatalysts over chemical catalysts. Biocatalysts function in a regiospecific and stereospecific manner leading to higher purity of the products. *Regiospecificity* means that biocatalysts change only one particular part of a large molecule. A *steriospecific* reaction produces only one optical isomer (enantiomer) of a chiral compound. A chiral compound has two isomers–distinct left and right hand forms, containing the same atoms and are chemically similar but physically and optically different due to different orientation of the atoms. The two isomers are denoted L and D, or + and – forms.

Although there is no chemical difference between the isomers, their form has substantial implications for pharmaceuticals and agrochemicals. Different enantiomers of the same drug can affect biological systems in quite different ways. Thalidomide is an example; it is an effective and safe anti-nausea agent, but its other enantiomer (mirror opposite form) has teratogenic side effects. The drug was given as a racemic mixture of thalidomide and its enantiomer mixture, so the patients showed both therapeutic effects and side effects. In 1960s, there was an uproar due to the birth of malformed babies (due to the teratogenic effect) whose mothers had used the anti-nauseating drug thalidomide.

Legislative pressure grew for the use of one enantiomer and not a racemic mixture for medicinal food and agricultural purposes. Thus, biotransformation is being used to meet the industrial requirement for purity by converting a racemic mix of a chiral compound into one of the optical isomers.

14.6 BIOFUELS

The concept of biofuel, particularly *biodiesel* dates back to 1885 when Dr Rudolf Diesel built the first diesel engine with the intention of running it on a vegetative source. He first displayed his engine at the Paris show of 1900 and astounded everyone by running it on any hydrocarbon fuel available, which included gasoline and peanut oil. He prophesied in 1912 that the use of vegetable oils for engine fuels would be as important as petroleum in the course of time. Recent environmental (Kyoto Protocol) and economic concern due to the hike in the price of crude oil have prompted resurgence in the use of biodiesel throughout the world.

Viscosity (thickness) of vegetable oils can be reduced by *transesterification*, which removes glycerin and creates alcohol ester. This is accomplished by mixing methanol with sodium hydroxide to make sodium methoxide. This liquid is then mixed into the vegetable oil. When the mixture settles, glycerin is left on the bottom and methyl esters or biodiesel move to the top and need to be washed and filtered.

Biodiesel is a substitute for, or an additive to, regular petro-based diesel fuel, and can be derived from the oils and fats of plants, like sunflower, canola or *Jatropha*.

As India is deficient in edible oils, non-edible oil is the main choice as defined by the government policy for producing biodiesel. Vegetable oil from *Jatropha* has been chosen as a source for biodiesel. Several research institutes including Indian oil corporation have taken up research and development work in this regard. Species of *Jatropha* can be grown on degraded land, mostly in areas with adverse agro-climatic conditions. Genetic technology can improve the oil-yielding capacity of the plant species and research institutes are searching for other species, like *Mahua* to use for the purpose.

Ethanol made from cane sugar by fermentation and distillation is another form of biofuel. Brazil is producing this in commercial quantities. The US took the initiative to promote *gasohol*, various gasoline–ethanol mixes. But this could not be promoted owing to changing political support and general disapproval from the petroleum industry. Most fuel alcohol made in the US is produced by fermentation of corn (maize) starch. Sweet sorghum (a tall, grass-like plant) has been targeted for ethanol production in some other countries. Sugar cane, maize, sorghum, and so on are often called energy crops and their products are known as 'agricultural ethanol'.

Methanol has also been suggested for use as a biofuel, but it is harder to make and is more corrosive. Other methodologies for production of biofuel are under trial but without much competitive success. For example, if any dry biological matter is heated up slowly, it undergoes 'pyrolysis' and produces a complex mixture of oily materials and charred polymers. Then the oils can be distilled in the same way as conventional mineral oil to produce fractions with properties similar to gasoline, diesel, lubricating oil, and so on. The charred remains can be burned further to heat up the pyrolysis reactors and stills. The process of production in this way is so far not competitive with mineral oil production.

Hydrogen made primarily by the photolysis of water can be used as gaseous biofuel. Photosynthesis also involves the photolysis of water, but does not release hydrogen as gas; rather it uses it to make sugars. One aim of biofuel research is to get organisms, probably single celled algae, to release hydrogen when exposed to sunlight. In 1995, scientists showed that the enzymes hydrogenase and glucose dehydrogenase could use glucose to generate hydrogen. They speculate that wood pulp could be a source of glucose. In theory, an enzymatic process could use the recycled newspaper generated in the US to power a city of one million people by hydrogen alone. Hydrogen produced in this way will inevitably be called 'biohydrogen', although in terms of its properties it would be exactly the same as normal hydrogen. Hydrogen is difficult to store safely in a tank as it is potentially explosive. Scientists are trying to resolve this problem.

14.6.1 Biogas

Biogas as a source of energy needs to be discussed along with biofuels. Methane ('natural gas') produced by fermenting waste by the activity of microorganisms is primarily named as biogas. The organic matter in the waste is converted by anaerobic fermentation mainly into methane and carbon dioxide, and the methane can be burned to provide power and heat. *Methanogenic bacteria*, an unusal group, generate methane. Other bacteria break wastes down into substances that the methanogens can use.

14.7 BIOSENSORS

A biosensor is a sensor with a biological element as an intimate part and is used to detect something in an extremely sensitive and specific way. The practical use of biosensors has been limited by the biological element being very prone to destruction.

There are different classes of biosensors:

- Ion-sensitive field-effect transistor (ISFET)-based sensors—using immobilized enzyme
- Thermal sensors (physical sensors)—using enzyme
- Enzyme electrodes
- Immobilized cell biosensor—different types of bacteria are used depending on what has to be detected
- Immunosensors—antibody coated
- Optical biosensors—also antibody coated

14.7.1 Ion-sensitive Field-effect Transistor (ISFET)

Ion-sensitive field-effect transistors (ISFET) monitor ion concentrations in a range of biotechnological processes. They can be turned into biosensors by replacing the ion-selective layer with an enzyme that generates ions by its action (Fig. 14.4). For example, urease acts on uncharged urea molecules and splits them into ammonia and carbon dioxide; the ammonia picks up a proton to become a charged ammonium ion. The build-up of ions is detected by the electrode. As an enzyme is attached to it, this type of sensor is also called an enzyme FET (or ENFET).

In principle, ENFET may be manufactured on a large scale by the technique of silicon chip manufacture; several sensors can be put on one chip together with control and reference electrodes, miniaturization would make it more sensitive to measure extremely small charge changes. However all these ideas are yet to become reality.

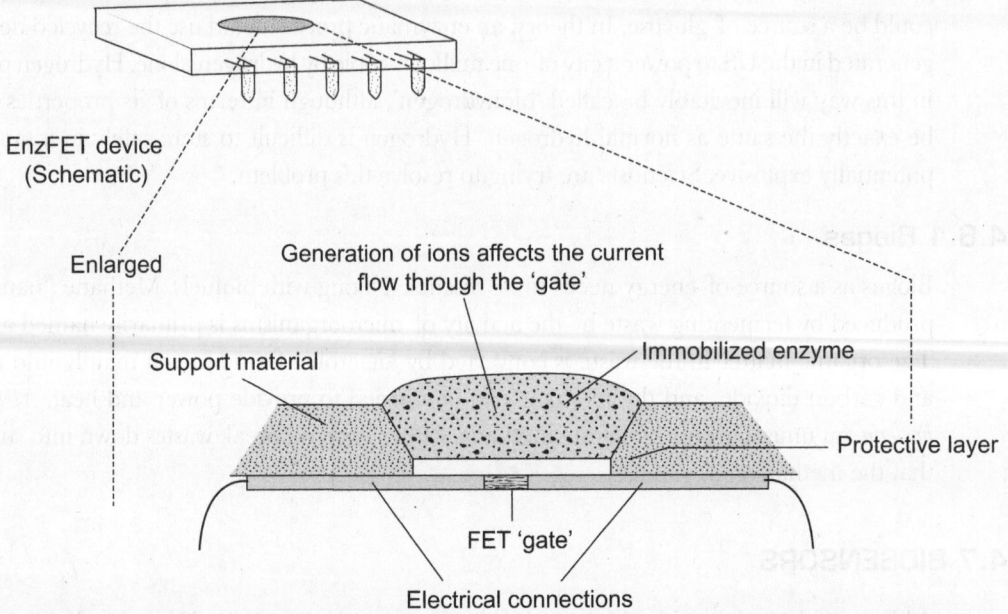

Fig. 14.4   Diagram of enzyme-coated biosensor (EnzFET or ENFET, enzyme field-effect transistor). (Based on Bains, W. 2001. *Biotechnology From A to Z.* Oxford Univ. Press, Oxford, 2nd ed. p. 226.)

14.7.2 Thermal Sensors (Physical)

Thermal sensors detect tiny changes in heat or temperature and are already employed in many situations. Attempts have been made to harness thermal sensors as biosensors to detect heat given out when an enzymatic reaction occurs. This could be more efficient than enzyme electrodes since relatively few enzyme reactions involve the transfer of electrons that could be picked up by an electrode, whereas nearly all result in the release of heat. The amount of heat released in samples of dilute material is meagre and difficult to measure; very sensitive heat sensors are required.

14.7.3 Enzyme Electrode

In a biosensor, an electrode could have an enzyme immobilized on its surface. When the enzyme catalyses its reaction, electrons are transferred from the reactant to the electrode, and a current or voltage is generated. *Amperometric sensors* measure the current generated by reactions and are operated at as low a voltage as possible. *Potentiometric sensors* measure the voltage generated by the reaction and are operated with as little current flowing through them as possible.

Usually enzymes transfer electrons inefficiently to the electrode, so a mediator compound is coated on the electrode to facilitate the transfer. Mediators like ferrocenes can easily carry a single electron at the electrode potential suitable for enzyme oxidations and reductions. The 'organic metals', that is, organic compounds that conduct electricity hold promise as electrode materials.

Immobilization of enzymes on the electrode is done in several ways—physical adsorption, chemical cross-linking, immobilization in a gel, or capture behind a selective membrane. Although there was a surge in developing various enzyme electrode systems since the early 1980s, the practical design for commercialization has not met with much success. The most prominent exception is the *glucose biosensor* for diabetic monitoring. It is an enzyme electrode based on glucose oxidase and has been commercialized by several companies.

14.7.4 Immobilized Cell Biosensor

Living cells, usually bacteria, are used for detector devices; that is why these devices are often called microbial biosensors. The device has two parts: detection or sensing by immobilized cell sensors, and amplification of the generated signals which are very weak. Specific types of cells are chosen depending on what is being detected; typical examples are given below.

- Amino acids—bacteria that metabolize them
- Glucose—almost any cell
- Toxic chemicals—bacteria sensitive to the chemical
- Carcinogens—bacteria with a defect in DNA repair genes
- BOD (biological oxygen demand)—measuring how much organic matter is present in waste waters
- Heavy metals—metal resistant bacteria
- Herbicides—plant cells or blue-green algae
- Toxicity—cultured animal cells

Only a few of these could be translated into realistic sensor devices. The readout methods are diverse:

- Depletion of the amount of oxygen or generation of CO_2 by the bacteria
- For toxicity measure, degree of well-being of naturally luminescent bacteria or bacteria genetically engineered with luciferase (the light generating enzyme) genes
- Effort to hook the electrode directly into the bacterium's own electron transport system, a more sophisticated version of measuring oxygen uptake

Bacterial biosensors are usually much less specific than other biosensors. Two luminescence-based bacterial biosensors for toxicity and BOD measurement are in use in the water industry.

14.7.5 Immunosensors

In immunosensors, the biological part is an antibody, which can bind specifically to an antigen. Measure of binding of an antibody to its antigen is based on a *physical mass detection system* or an optical device (Fig. 14.5).

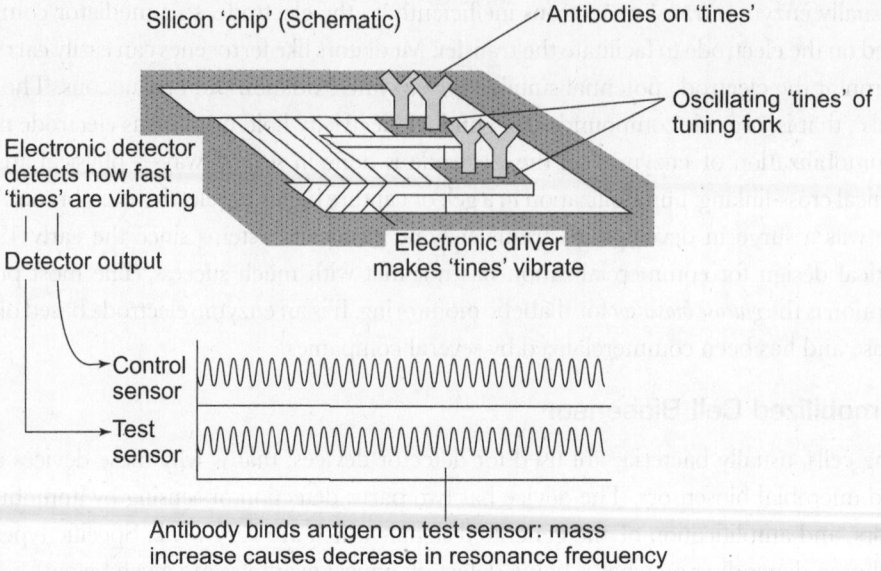

Fig. 14.5   Immunosensor and its working principle.

Tiny changes in mass occur when an antibody binds to an antigen. The tiny changes are measured through resonance devices in a mass detection system. The resonance device is based on the principle of the tuning fork. The sound note produced by a tuning fork depends on the mass of the tines (slender projections of a fork). When the mass increases, the note goes down. The sensors have the equivalent of a microscopic tuning fork with the antibody coated on the tines. The silicon surface from which the tines are made detects the frequency of vibration of the tines. Electronic driver makes the tines vibrate (Fig. 14.5). When a substance (antigen) binds to the antibody on the tines, the change in mass causes the note to fall, and this is picked up by the circuit. Surface acoustic wave (SAW) devices are a modification of this theme.

Usually, these detectors are extremely small manufactured on a silicon 'chip', and hence called 'microchip biosensors'. Since tuning forks are made of piezoelectric material, they are sometimes called 'piezoelectric sensors'.

Despite the use of very specific monoclonal antibody as a biological element, these sensors are very prone to interference from anything landing on the tines. Thus, they are yet to be developed as reliable biosensors.

14.7.6 Optical Biosensors

Optical biosensors use changes in the emission pattern of light to detect biological substances. Two types of sensors are discussed here.

Evanescent wave

When light is trapped inside an optical fibre or a prism, a little bit of it temporarily leaks to the outside. The light outside the trap is called an *evanescent wave*. The wave only occurs right next to the optical fibre or prism and it can be absorbed by certain chemicals stuck to the fibre or prism. If the optical fibre is coated with an antibody, then when that antibody binds its antigen, the absorbance of the evanescent wave will change, which can be detected. Thus, the antigen in a test sample can be detected by fibre optic sensors.

A lot of research has gone into making enzyme sensors work on optical fibres. *Fibre optical chemical sensors* (FOCSs) have been in use to measure pH, oxygen, and carbon dioxide for process monitoring and medical purposes. These are much stronger than ion-sensitive electrodes and small enough to be inserted into veins for diagnosis.

Surface plasmon resonance (SPR)

The amount of light scattered at different angles from a surface depends on the exact nature of the surface and its absorbance. If specific antibody molecules are stuck on the surface, the scattering of light will vary depending on whether or not the antibody has bound to its antigen (Fig. 14.6). Although these biosensors are far from the 'dipstick' type easy to use device, Pharmacia and Affinity Sensors Ltd. launched a commercial sensor project based on this SPR principle. They are particularly useful in research to gain insights on the kinetics and thermodynamics of binding of different molecules. Much of biochemistry and molecular biology deals with molecules binding to each other.

The major problem with optical sensor systems is the false signals from anything that absorbs light and non-specifically sticks to them.

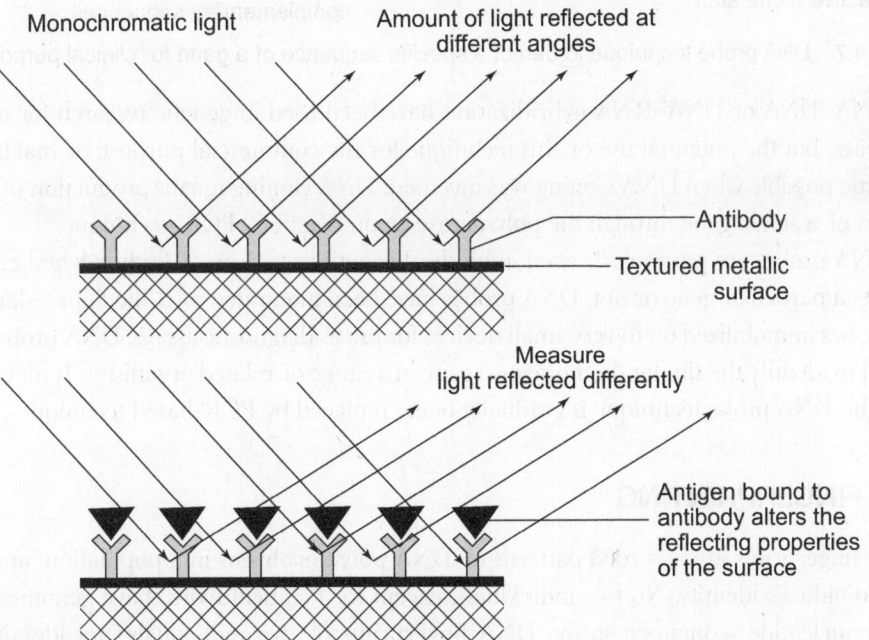

Fig. 14.6   Optical biosensor (antibody coated).

14.7.7 DNA Probes

DNA probes as the biological element can identify the complementary nucleotide segments (where A matches with the T nucleotide of the DNA probe and G with C) in genomic materials, and thus may also be considered as biosensors. The probe contains one strand of a specific length of DNA double helix, which will bind to a target single strand of DNA to form a double helix (Fig.14.7). This process is called hybridization. If the DNA strand on the probe is not complementary (matching) to the target DNA, then no helix will be formed. Thus the DNA probe is used as a reagent to detect the presence of a specific sequence in a mixture of sequences. The DNA probe can also hybridize with mRNA encoded by the specific sequence.

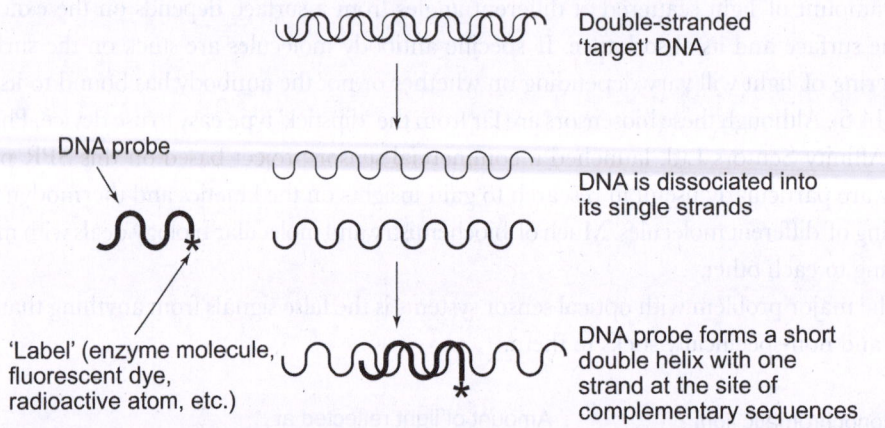

Double-stranded 'target' DNA

DNA probe

DNA is dissociated into its single strands

'Label' (enzyme molecule, fluorescent dye, radioactive atom, etc.)

DNA probe forms a short double helix with one strand at the site of complementary sequences

Fig. 14.7 DNA probe technique to detect a specific sequence of a gene for clinical purpose.

DNA–DNA or DNA–RNA hybridization have been used in genetic research for more than 35 years, but the potential use of this technique for the commercial purpose of making probes became possible when DNA cloning was invented. DNA cloning means production of a million copies of a single gene through the polymerase chain reaction (PCR) technique.

DNA probes are particularly used in medical genetics as a way of finding whether a person carries a particular gene or not. DNA probes have been integrated into 'biochips'—large arrays of probes immobilized on to very small devices for use as diagnostic assays. DNA probes are also useful to identify the similar (homologous) gene in a range of related organisms. It may be added that the DNA probe technique is gradually being replaced by PCR-based techniques.

14.8 DNA FINGERPRINTING

DNA fingerprints are recorded patterns of DNA polymorphisms in a population, and provide an individual's identity. No two individuals, except for identical twins, have genomes with the same nucleotide sequences; so the DNA fingerprints of two individuals (not identical twins) always differ. In other words, DNA fingerprints are as unique as conventional fingerprints from where the name is derived. DNA fingerprints establish the identity of an individual; they provide evidence for a criminal's presence in the crime scene from a specimen of blood, semen, or hair of the criminal, left on a victim or at the site, or from the victim's blood found on the criminal. DNA fingerprint establishes the identity of a missing or dead/murdered person even from DNA samples of hair bulb or a piece of bone several years after the incident and relates it with the living members of the family of the victim. Now the fingerprint technique is routinely used for forensic cases and to resolve paternity disputes. The DNA fingerprinting technique is often commonly referred to as 'DNA test'.

The technique involves digestion of DNA from a cell with an endonuclease. The different lengths of DNA fragments that are produced are then separated by gel electrophoresis into separate bands. The number, size, and position of bands are specific for an individual and treated as a DNA fingerprint (Fig. 14.8).

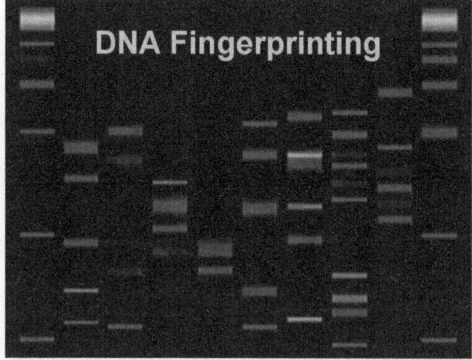

Fig. 14.8   DNA fingerprints.

Although the differences in DNA sequence between any two persons are quite small, they alter the lengths of some of the fragments produced on digestion by restriction endonucleases. These differences in fragment length are called *restriction fragment length polymorphisms* (RFLPs).

VNTRs

There is another variation of the fingerprinting technique. Minisatellites consisting of variable numbers of tandem repeats (VNTRs) are dispersed throughout the genome. A repeat unit may be about 10–100 base-pairs long. The number of times one repeat unit is repeated sequentially in a row varies from person to person. The differential repeat number causes differential restriction fragment patterns following gel electrophoresis of DNAs from different persons. This technique differs from the traditional RFLP technique which judges only the absence or presence of endonuclease cleaving sites.

14.9 MONOCLONAL ANTIBODY

Normal mammalian cells, with a few exceptions like fibroblast cells, do not propagate in in-vitro cell culture. Georges Köhler and Cesar Milstein succeeded in the continuous culture of antibody secreting B lymphocytes of mice after they fused them with mouse myeloma cells, that is, bone marrow cancer cells, in 1975. Each B cell is unique in synthesizing antibodies of a particular specificity against an antigenic epitope (antigenicity determining part of the antigen). The fusion of a malignant cancer cell (immortal by nature) with a B cell immunized with an antigen immortalizes the hybrid cell for propagation and continuous production of monoclonal antibodies. The fused hybrid cell is called *hybridoma* and the monospecific antibodies produced by the clone of a hybridoma cell are called *monoclonal antibodies* (Fig. 14.9). (A clone is a population

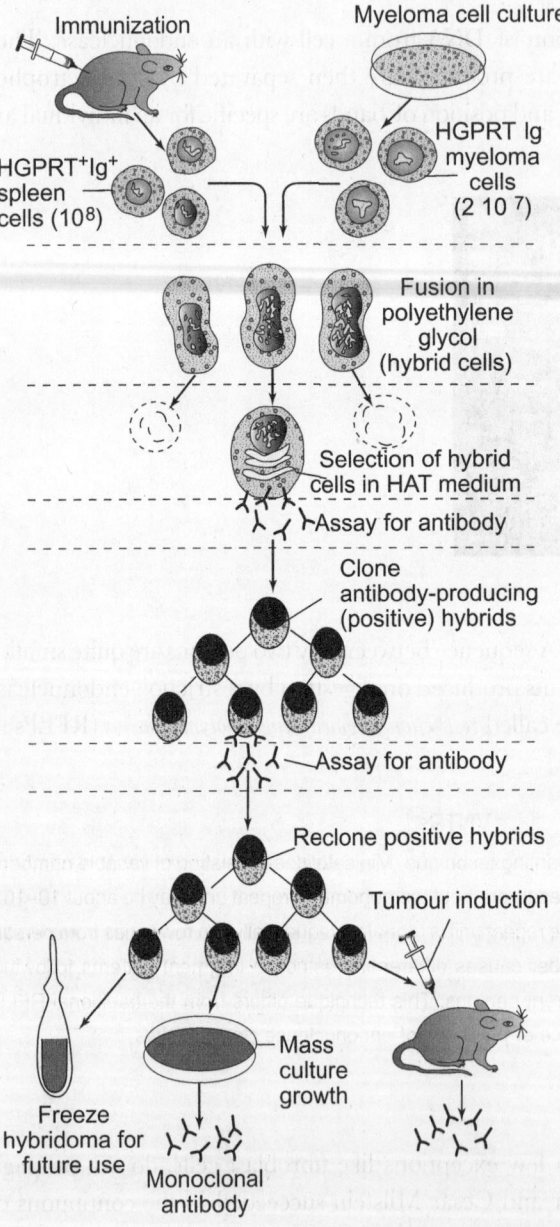

Fig. 14.9 Outline of hybridoma technology to produce monoclonal antibody (mAb). (Based on Chakravarty, A.K. 2006. *Immunology and Immunotechnology*. Oxford University Press, New Delhi. p. 537.)

of cells derived as the progeny of a single cell.) The *hybridoma technology* fulfilled one major goal that immunologists were trying to achieve for years—the routine production of large amounts of homogeneous antibodies of a single specificity against any antigen of choice.

Köhler and Milstein's hybridoma technology revolutionized the use of mammalian cells in biotechnology with profound medical and commercial potential.

Monoclonal antibodies can be generated on different scales for commercial purposes.

The hybridoma cell line producing specific antibodies, once established, can be grown in the peritoneal cavity of histocompatible mice (which will not reject hybridoma cells) in cell flasks or in a suspended cell fermenter. (The hybridoma cell line may be cryopreserved in liquid nitrogen for future use.)

The hybridoma cells are injected in the peritoneal cavity of mice where they grow and secrete monoclonal antibodies in the ascitic fluid of the abdomen; the antibody is then purified from the fluid. On an average, 50 mg of monoclonal antibody is obtained from each mouse. The supernatant of the tissue culture flask in which the hybridoma cells are being cultured is the source of antibody, which needs to be concentrated. When hybridoma cells are grown in a traditional biotechnology bioreactor with the right kind of adaptation, a large amount of monoclonal antibody is produced. Only a very small amount of monoclonal antibody is needed for diagnostic or clinical purposes.

14.9.1 Uses of Monoclonal Antibodies

The usage of monoclonal antibodies is varied and the total potential of their applicability is yet to be realized. Some of their current uses are as follows:

- Identification of microorganisms and their products for clinical diagnoses. (Also see colour plate 4.)
- Assay of hormone levels, tumour antigens, and blood levels of various drugs.
- Cell and tissue typing depending on characteristic cell-surface antigens; necessary for tissue grafting, detection, and targeted killing of cancer cells.
- Radiobelled (I^{131}) monoclonal antibody for early detection of primary or metastatic breast cancer including tumours spread to regional lymph nodes in imaging analysis (*immunoscintigraphy*).
- Therapeutic use as targeted immunotoxins, *Ricin*, *Shigella*, and *Diphtheria* toxin coupled to tumour-specific monoclonal antibodies reach the tumour cells and are endocytosed by them; then the toxin molecules inhibit protein synthesis and kill the tumour cells
- Identification of the major histocompatibility complex (MHC) products on cells, thus allowing more accurate tissue typing for successful transplantation
- Monoclonal antibodies acting as enzymes (abzymes). One main goal of current research is to come up with a series of abzymes, each capable of hydrolysing specific amino acid residues, just like restriction endonucleases cut DNA at specific sites. These will be immensely helpful tools for the structural and functional analysis of proteins, to dissolve blood clots, or to cleave viral glycoproteins to block their infectivity.
- Monoclonal antibody and *antibody engineering*; recombinant DNA technology can produce hybrid monoclonal antibodies that contain just the variable region or CDR of the mouse

monoclonal antibody, the rest of the antibody corresponding to the human antibody molecule. In 1986, G. Winter and colleagues pioneered the technique to 'humanize' mouse antibodies, in which except for the antigen binding portion of the antibody which is of mouse origin, the rest of the antibody molecule is of human origin. Such engineered antibody molecules when injected are well tolerated by human hosts, do not induce hypersensitivity (allergic) reactions, and can be used for immunotherapy.

14.10 VACCINES

The total eradication of smallpox and the dramatic success of active vaccines against polio, measles, mumps, rubella, diphtheria, tetanus, and most recently hepatitis B virus testify to the value and utility of vaccination in the prevention of diseases. The induction of immunity by way of vaccination has become a cornerstone of modern medicine and an essential part of primary health care in most countries.

A *vaccine* consists of the organism that causes the disease, but suitably attenuated (living but incapable of producing disease) or killed, or some subcellular part of the organism. Administration of a small quantity of a vaccine through various routes (intra-dermal, subcutaneous, intra-muscular, intravenous, and intra-peritoneal) in a host elicits a subdued immune response and generation of memory cells which can react faster and in much higher numbers on subsequent entry of the pathogens.

Ancient India and China used to practice some forms of vaccination to provide protection against smallpox. Around 1798, Edward Jenner demonstrated that cowpox viruses as vaccine would protect man against smallpox. Then credit of using live, attenuated, and heat-killed vaccines went to Louis Pasteur. Since then the discovery of various vaccines progressed steadily. More recently, techniques of genetic manipulation and recombinant DNA technology have contributed in a big way to designing effective vaccines for different diseases and have generated a multi-billion dollar market for biotech companies, throughout the world. Information on different type of vaccines developed and those under research is given in the following text.

14.10.1 Live, Attenuated Bacterial and Viral Vaccines

Live but attenuated viruses and bacteria become avirulent but retain the ability to grow transiently within inoculated hosts. Thus, attenuated vaccines mimic original microbes without causing the disease. They often present antigens to immunocompetent cells in a much better fashion than dead organisms and their protection sometimes lasts for many years.

Among the attenuated bacterial vaccines in use today are, *Mycobacterium bovis, M. tuberculosis*, bacilli Calmette-Guérin (BCG), avirulent mutants of *Salmonella typhi*, inactivated *Vibrio cholerae*, and so on. The Sabin polio vaccine consists of three attenuated strains of poliovirus and is administered orally to children in sugar solution. The attenuated strains colonize the intestine and induce protective immunity to the virulent forms of all the three strains of poliovirus.

Distinct methods have been developed over time for attenuation of different types of pathogens.

Since Jenner's time cowpox is known to be avirulent in man, but it would protect man against smallpox. That is, microbes virulent for another species, but avirulent in man, were used as a vaccine for human beings. By varying the culture conditions attenuated strains of pathogens are raised to produce vaccines. Genetic recombination between two different strains can often generate an attenuated strain.

The major risk of attenuated vaccines is the possibility of their reversion to a virulent form. However, the frequency of reversion, even if it does occur, is pretty low.

14.10.2 Killed Microorganisms as Vaccines

Killing microbes in an appropriate manner destroys their ability to multiply and cause disease, but maintains the structure of epitopes of antigens on their surfaces, which is sufficient to stimulate an individual's immunity. The use of formaldehyde and some other chemicals have been found to be better than heat inactivation. Vaccines for polio (Salk), measles, pertussis, typhoid, cholera, and so on are produced in this manner.

14.10.3 Toxoid Vaccines

In some bacterial diseases such as tetanus, diphtheria, and gas gangrene, exotoxins or enterotoxins from the bacteria are responsible for the pathogenicity and disease symptoms. Chemically modified toxins that are inactivated but retain the intact antigenic determinant can be used as vaccines. This modified form of toxin known as *toxoid*, can stimulate the immune system and does not cause any harm to the recepient. Toxoids are often adsorbed onto aluminium hydroxide, which is known to act as an *adjuvant* to produce stronger immune responses. The toxoids made out of exotoxins from diphtheria and tetanus bacilli are two of the most successful vaccines. The limitation in obtaining the diphtheria and tetanus toxoids has been overcome by cloning the exotoxin genes and expressing them in easily grown host cell culture.

14.10.4 Purified Antigen Vaccines

An isolated and purified antigen as a vaccine is sometimes more desirable than vaccination with a whole microorganism or parasite. The entire organism contains many more irrelevant antigens which might suppress the response to protective antigens or elicit a detrimental hypersensitivity reaction. Once identified and isolated, protective antigens can even be produced synthetically, especially when bulk growth of the organisms or isolation of a large amount of the antigen is technically impractical or is not cost-effective. Purified capsular polysaccharides of different bacteria are being used as vaccines.

Pneumonia Vaccines

The vaccine for *Streptococcus pneumonia* (causing pneumonia) consists of 23 antigenically different capsular polysaccharides. It is marketed by the name of *Pneumovax* 23 by Merck and Pnulmmune 23 by Lederle Laboratories..

Purified protein vaccines for many diseases are being produced with rDNA technology. The gene encoding the target antigen on pathogens can be isolated and cloned in bacterial, yeast, or mammalian cells using rDNA technology; the cloned cells produce enormous amounts of that target protein antigen, which can be used as vaccines. The first such recombinant antigen vaccine approved for human use is the hepatitis-B virus surface antigen (HB_sAg) produced by cloned yeast harbouring the specific viral gene. The recombined yeast cells are grown in large fermenters, harvested, and disrupted under high pressure to release the synthesized recombinant HB_sAg, which is then purified following conventional biochemical techniques. Such protein subunit vaccines have also been used for influenza viruses and are under trial for *Bordetella pertusis* and cholera. Subunit vaccines for other serious diseases including HIV are under consideration.

14.10.5 Synthetic Peptides

Instead of manipulating the genes, small peptide sequences bearing similarity to antigenic epitopes on a microbial antigen can easily be synthesized and often effectively used as vaccines by coupling them to a suitable carrier. Such vaccines have been developed for influenza viruses, several picornaviruses, and for certain other pathogens infecting humans and animals.

Designing a vaccine against the malarial parasite which consists of a synthetic peptide mimicking a 12 amino acid long sequence of the circumsporozoite antigen specific to the *sporozoite* stage of the parasite's life cycle is underway. The long slender sporozoites enter the human bloodstream when an infected mosquito takes a blood meal. They remain in the circulation for about 30 minutes and then infect hepatocytes of the liver and disappear from the blood. The immune responses of the host can target sporozoites during this short period of 30 minutes.

14.10.6 Recombinant Vector Vaccines

The *Vaccinia* virus has a comparatively large genome of about 200 genes, and additionally it can carry several dozens of transferred foreign genes and retain its ability to infect host cells and replicate. In the early 1980s, Bernard Moss and Enzo Paoletti independently introduced genes for target antigens of other pathogens into *Vaccinia* virus and opened up a major chapter in vaccinology. Genes for viral envelope proteins from hepatitis-B virus, influenza virus, herpes simplex virus, and HIV-1 virus have been inserted into the genome of *Vaccinia* virus to raise recombinant vector vaccines. These vector vaccines can be administered by dermal scratching, just like the smallpox vaccine. The foreign genes are expressed in host cells infected with genetically engineered *Vaccinia* virus. Specific viral coat proteins are expressed on the surface of the host cells and induce immune responses in the host.

There are certain reservations regarding the use of the recombinant vaccinia vector vaccine. One serious concern is the revival of virulence in the *Vaccinia* virus. Alternate vectors are being searched; canary pox virus has been found to be reasonable. The bacterial species of *Salmonella*, *Shigella*, and *Mycobacteria* are also currently under experimental considerations for recombinant vector vaccines.

14.10.7 Anti-idiotype Antibodies as Vaccines

The paratope (antigen-binding site) of a specific antibody is a three-dimensional reverse image of the antigenic *epitope*. An epitope is a small part of an antigen which induces the antibody and cell-mediated immune response. When an anti-antibody is raised against an antibody, the antigen binding site of the anti-antibody is a reverse image of the paratope of the original antibody, and mimics the original antigenic epitope that induced the formation of the first antibody (Fig. 14.10). Here, the first antibody is called idiotype and the anti-antibody is called the anti-idiotype. By earlier arguments anti-idiotype being equivalent to the antigenic epitope, can be used as vaccines when the original pathogen cannot be used to stimulate the host due to safety and other considerations.

The specificity of anti-idiotype vaccines is appealing. An anti-idiotype antibody is specific only for one antigenic determinant, whereas an attenuated infectious agent may carry a number of antigenic determinants besides the one that induces the protective immunity. Some may even resemble the determinants on a tissue of the recipient and thus trigger an autoimmune disorder.

Anti-idiotype vaccines might be of value in newborn infants, particularly when their immature immune system is not capable of making antibodies to the carbohydrate moieties of the envelopes of certain pathogenic bacteria. On the other hand, several considerations could limit the use of anti-idiotype vaccines in human beings.

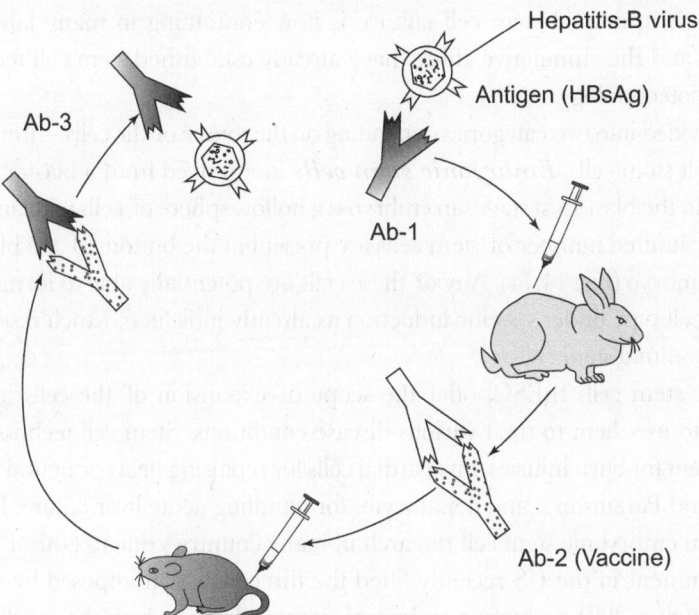

Fig. 14.10 Anti-idiotype vaccine against hepatitis-B virus raised by injecting human antibody (Ab-1) into a rabbit, and then injecting antibody from rabbit (Ab-2) (anti-idiotype antibody to human Ab-1 antibody) into a mouse (or a chimpanzee) which produces antibody (Ab-3) to the anti-idiotype (Ab-2) antibody. Ab-3 can bind to the hepatitis-B virus and neutralize the virus. Thus, Ab-2 when injected as a vaccine in humans can also produce antibody equivalent to Ab-3.

14.10.8 Edible Vaccines

C.J. Arntzen and coworkers introduced the concept of edible vaccines, resulting from the expression of antigenic proteins in transgenic plants. The first experiment involved the insertion of the gene for HB_sAg into cells of tobacco plants; the cells produced an antigen which was found to be immunogenic on injection to animal models. Similarly, the gene for the heat-labile enterotoxin of *E. coli* has been engineered into potatoes, and 1 mg enterotoxin can be obtained per raw potato. Much effort is going into producing genetically engineered bananas with desired genes for various immunizing antigens. Biotechnologists hope that edible vaccines will lead to a much easier, cheaper, and more widely acceptable procedure of vaccination.

14.11 STEM CELL

Stem cells are self-renewing (proliferative) undifferentiated pluripotent cells, capable of differentiating into various types of cells under the influence of various factors and cytokines in an in-situ microenvironment.

In case of plants, almost any cell can dedifferentiate into a callus in in-vitro culture and differentiate into a whole plant. In case of animals, the paraphernalia for stem cell culture for tissue growth is more complicated, and the requirement of growth factors and cytokines is also elaborate. The research in animal stem cell culture is now continuing in many laboratories throughout the world and the cumulative efforts have already established stem cell technology as a new branch of biotechnology.

Stem cells can be divided into two categories depending on the source of the cells – (i) embryonic stem cells, and (ii) adult stem cells. *Embryonic stem cells* are derived from a blastocyst (early developmental state). In the blastocyst stage, an embryo is a hollow sphere of cells with an internal cavity (blastocoel); the limited number of stem cells are present at the bottom of the blastocoel, and form the future embryo (Fig. 14.11). Any of these cells are potentially able to form the total embryo or any tissue cell type under specific induction as already indicated. Much research has been carried out with animal stem cells.

Human embryonic stem cells (hESC) offer the scope of expansion of the cells in culture to a desired cell type to use them to treat various disease conditions. Stem cell technology can provide skin replacement for burn injuries, myocardial cells for repairing heart or neural tissue for treating Alzheimer's and Parkinson's, and hepatocytes for handling acute liver failure. But there is restriction on human embryonic stem cell research in many countries due to ethical reasons.

The Obama government in the US recently lifted the three-year ban imposed by the Bush administration on 9 August, 2001 on human embryonic stem cell research on the condition that pre-existing embryonic stem cell cultures could be used. Adult tissues also harbour stem cells; the pluripotent stem cells in the hematopoietic system in bone marrow (1 stem cell in 100,000 bone marrow cells) make the precursors of all types of cells in the blood. Human hematopoietic stem cells (HSC) isolated from umbilical cord blood obtained at the time of child birth can reconstitute the haematopoietic system of cancer patients following high-dose radiotherapy. HSC transfusion is most effective in children and has limited application in adults due to the low yield of HSC from cord blood. Expansion of the cell number in culture allows therapy of adults to minimize the short-term effects of neutropaenia and thrombocytopaenia (lowering in number of neutrophils and platelets respectively). Cord blood banks collect and preserve the cord blood of an individual at the time of birth to provide HSC for treatment at any point of life. Using one's own cord blood stem cells eliminates the problem of histoincompatible rejection.

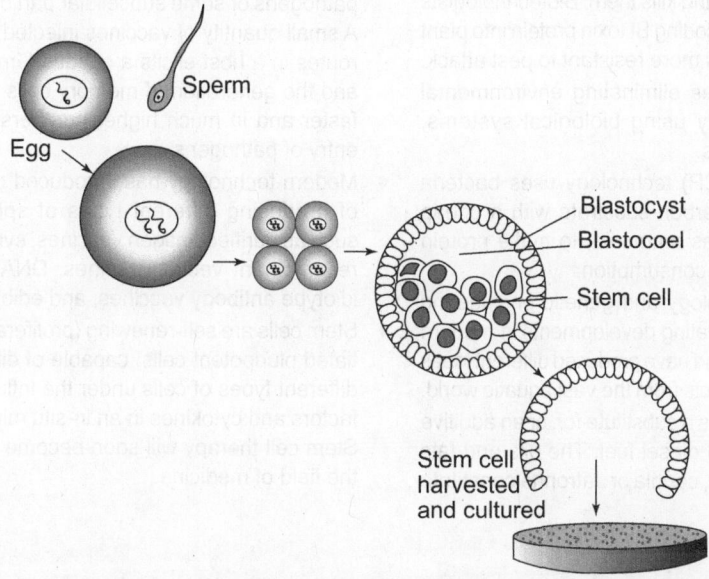

Fig. 14.11 Location of stem cells in blastocoel (internal cavity of a blastocyst) (Stem cells drawn larger for emphasis).

Some companies have developed a serum-free, animal protein-free medium for hematopoietic stem cell expansion. This medium supports high viable cell densities and eliminates the safety risks associated with possible adventitious agents present in the serum normally used for cell culture. Autologous chondrocytes are expanded in-vitro before being reintroduced for treating severe cartilage defects.

In the area of stem cell research, the world is yet to see many novel discoveries in the near future.

Stem Cells from Skin Cells

Very recently, in 2009, different groups of scientists claimed that they have produced stem cells from mouse and human skin cells. These embryonic like stem cells have been called *induced pluripotent stem cells (iPS cells)*. One group from China has even claimed of creating mice from the iPS cells.

SUMMARY

♦ A biofertilizer is a natural organic fertilizer, often from biological wastes, that helps provide all the nutrients required by the plants and helps increase the quality of the soil using a natural environment containing microorganisms. This organic fertilizer supports sustainable agriculture.

♦ Bt toxin obtained from *Bacillus thuringiensis* interferes with ion transport and the absorption of food from the gut of the insect larvae and kills them. Biotechnologists introduced the gene encoding Bt toxin protein into plant cells to make the plants more resistant to pest attack.

♦ Bioremediation involves eliminating environmental chemical pollutants by using biological systems, usually microorganisms.

♦ Single cell protein (SCP) technology uses bacteria to grow on a cheap carbon substrate with a cheap nitrogen source such as ammonia, to make protein fit for human or animal consumption.

♦ New discoveries in biology and genetic engineering have led to some fascinating developments in modern aquatic biotechnology and have produced different ideas for exploring new products from the vast aquatic world.

♦ Biodiesel can be used as a substitute for, or an additive to, regular petro-based diesel fuel. The oils and fats of plants, like sunflower, canola or Jatropha constitute sources of biodiesel.

♦ A biosensor is a sensor with a biological element as an intimate part to detect something in an extremely sensitive and specific manner.

♦ *Hybridoma technology* routinely produces large amounts of homogeneous (monoclonal) antibodies of single specificity against any antigen of choice. Monoclonal antibodies have multiple uses.

♦ Vaccines consist of killed or suitably attenuated pathogens or some subcellular part of the pathogens. A small quantity of vaccines injected through various routes in a host elicits a subdued immune response and the generation of memory cells which can react faster and in much higher numbers on subsequent entry of pathogens.

♦ Modern technology has introduced many processes of producing different types of specific vaccines, such as purified antigen vaccines, synthetic peptides, recombinant vector vaccines, DNA vaccines, anti-idiotype antibody vaccines, and edible vaccines.

♦ Stem cells are self-renewing (proliferative) undifferentiated pluripotent cells, capable of differentiating into different types of cells under the influence of various factors and cytokines in an in-situ microenvironment. Stem cell therapy will soon become a big chapter in the field of medicine.

Kary Mullis

Kary Banks Mullis
(Born 28 December 1944)

Kary B. Mullis was born in Lenoir, North Carolina, USA, in a family with a farming background. He went to Dreher High School in Columbia, South Carolina. Mullis obtained his Bachelor of Science majoring in chemistry in 1966 from Georgia Institute of Technology, Atlanta and subsequently received PhD in biochemistry from University of California, Berkeley in 1972. After PhD, he joined Kansas City medical school as a biochemist. Taking an off-beat career, he worked as a bakery manager for the next two years. With the help of an old friend, Mullis got a job in Cetus Corporation, California—an upcoming biotechnology company. He worked at Cetus for seven years as a DNA chemist, where he conceived the idea of the polymerase chain reaction (PCR). In 1986, he left Cetus and became the director of molecular biology for Xytronyx Inc., San Diego.

For conceptualizing and improvising the polymerase chain reaction using a thermostable DNA polymerase, Kary B. Mullis shared the Nobel Prize in chemistry in 1993 with Michael Smith. The concept of amplifying a gene was first described in a paper by Kjell Kleppe and Har Gobind Khorana (Nobel Laureate, 1968) in 1971.

Mullis' improvization of the PCR technique made it a key tool for biochemistry and molecular biology.

EXERCISES

Objective Questions

1. Multiple Choice Questions
   (a) *Rhizobium* helps in
       (i) development of rhizome system
       (ii) production of siderophores only
       (iii) solubilizing phosphate
       (iv) providing *nif* gene
   (b) A reporter gene like green fluorescent protein (GFP) gene from the bioluminescent jellyfish
       (i) is picked up by the cells directly
       (ii) helps in reporting the presence of a specified gene in a particular chromosome
       (iii) allows detection of the distribution of the expressed protein of interest
       (iv) allows the study of different phases of the replication cycle of a cell
   (c) Which of the following statements about DNA fingerprints is incorrect?
       (i) DNA fingerprints of two allogeneic individuals always differ.
       (ii) DNA fingerprints help establish the identity of a criminal.
       (iii) DNA fingerprints help in determining the size of the genome of an individual.
       (iv) DNA fingerprints can resolve paternity disputes.
   (d) Which of the following statements is not applicable for Sabin polio vaccines?
       (i) Three strains of poliovirus are used.
       (ii) The viruses are heat killed.
       (iii) The vaccine is administered to the children orally in sugar solution.
       (iv) The strains of poliovirus colonize the intestine and induce protective immunity to the virulent forms for a long time.
   (e) *Rhizobium* bacteria fix 40/200/400 million tonnes of atmospheric nitrogen globally per year. Circle the right answer; if none of these is correct then mention the correct one.

2. Mark the following statement as True or False.
   (a) ISFET sensor is made by coating the electrode with monoclonal antibody

3. Match the statements in Column 2 with the items in Column 1.

Column 1	Column 2
(a) Pneumovax 23	(i) Antibody
(b) HBsAg	(ii) Endotoxin
(c) Toxoid	(iii) Recombined *Vaccinia* virus
(d) Recombinant vector vaccine by cloned yeast	(iv) Hepatitis-B virus surface antigen
(e) Anti-idiotype vaccine	(v) Capsular polysaccharides of *Streptococcus pneumonia*

Review Questions

1. What is the important component for sustainable or green agriculture?
2. Differentiate between *biopesticides* and *biocontrol* agents.
3. Write a brief note on the activity of Bt toxin on lepidopteran caterpillars.
4. What are the topics one should include under aquaculture?
5. Suggest a technique by which salmon can be cultured in the cold temperatures of the Arctic Ocean.
6. What is 'bioimpedance'? How does it help in the development of aquaculture?
7. What are the uses of 'astaxanthin' and 'carrageenan'.
8. Elaborate the differences between transgenic fish and triploids.
9. How are triploid fishes produced?
10. Indicate at least four fascinating developments in aquatic biotechnology that occurred with the introduction of genetic engineering.
11. Mention some applications of the anti-freeze protein gene from a biotechnological perspective.

12. What are biofilms?

13. Give a few examples and mention the steps taken by aquatic biotechnologists to fight the menace of biofilms.

14. Differentiate between bioconversion and biotransformation.

15. What is a biosensor?

16. What are monoclonal antibodies? Mention some of their uses.

17. What is 'transesterification' of vegetable oils and why is this done?

18. What is the function of a vaccine?

19. Explain a toxoid vaccine.

20. What kind of virus will you choose to produce a recombinant vector vaccine? Give reasons for your choice.

21. What is the working principle of a DNA vaccine?

22. Explain the use of stem cells as therapeutic agents.

Bioethics and Biosafety

LEARNING OBJECTIVES

♦ Bioethics and biosafety concerns related to biotechnology
♦ Genetically manipulated organisms (GMOs)
♦ The regulatory aspects of GMOs and other biotechnological products
♦ The GMOs in India

INTRODUCTION

Ethics signifies a code of values for our actions in a societal system. It helps distinguish between right and wrong, good and evil.

People from different cultures might have some differences in their approach to ethical issues. Religion, philosophers' thinking, collective or personal experience, and inner feelings of individuals, all contribute to objectivism and defining moral imperatives to resolve ethical issues or courses of action.

In a philosophical way, ethics is an inner guidance to live without inflicting harm to others. Indian philosophy transpires into prayer—*Sarbe bhabantu sukhinah, sarbe shantu niramaya….*, that is 'Let everyone be happy, let all be well'. Ethically the well-being of all is desired in the Indian tradition. In the deeper sense, 'all' is not restricted to humanity alone, often it includes other living organisms; moreover unnecessary harm is not to be committed to the earth (*Dharitree*).

Ethics dealing with the implications of biological research and biotechnological applications constitutes *bioethics*.

15.1 BIOTECHNOLOGY AND ETHICS

In certain situations, it is difficult to decide in favour of life or death, such as choosing between suffering without a solution and euthanasia, life versus capital punishment in criminal offense, serving someone or many at the cost of leading life as a saviour. Modern biotechnology with a history of a little over three decades has put researchers and the society at large in an almost similar situation—to choose between starvation and new technology, immediate benefit versus far-reaching consequences beyond the existing generation.

Classical use of biotechnology in fermentation, breeding, agriculture, and waste management did not throw researchers/users out of the scope of basic ethics. But genetic manipulations, transferring genes between distant species which nature normally does not allow, has definitely raised strong ethical questions—what to do and what not to do.

Here, we may take into account that nature has been carrying out biotechnological experimentation since life appeared on the earth. Bacteria routinely exchange plasmids among related groups to allow recombination, and also mutate, creating new combinations of genes that did not exist before. However, as soon as human beings interfere in the name of rDNA technology to create new genetic combinations, the bioethical considerations loom large. The human attempt might alter the genomic integrity and coexistence of species achieved through million years of evolution.

Bioethics deals with preventing the creation of some dangerous bugs that could adversely affect other living organisms or the ecosystem, and to ensure that newly created organisms do not get out of control. Again, often, biotechnologists need to do something which has not been done before and sometimes with little idea of the consequences of their act. Bioethics may need to put a limit on the curiosity of scientists and prevent them from letting engineered living organisms escape beyond the laboratory door. For example, scientists are now capable of implanting mouse cells into chicken embryos to induce the development of teeth—something that does not normally take place in chickens. In such situations, bioethics comes into the picture, sometimes in a strong way. Just because the technology is at hand, should such a manipulation be done? Although so far the experimental chicken embryos were not allowed to develop beyond the embryonic stage, should the embryos be allowed to grow into adult chickens with a series of teeth, just to satisfy the curiosity of some scientists?

More information, rational thinking, awareness of the consequences, other's views, objections, objective of experiments, and a basic sense of ethics are likely to help in taking bioethical decisions at critical junctures.

Initially, a general sense of fear of biotechnological products prevailed in the society. Gradually, people became aware of the benefits of biopharmaceuticals, and the microbial contributions in many other aspects to satisfy the everyday needs of human beings. In spite of the contributions of biotechnology to society, the question of safety of transgenic organisms and their products remains.

15.2 BIOTECHNOLOGY AND SOCIETY

So far we have realized the potentiality of the multifaceted field of biotechnology to deliver products on a scale which was beyond imagination fifty years ago. Bioremediation technologies are used to clean our environment by removing toxic substances from contaminated soils and ground water. Agricultural biotechnology reduces our dependence on pesticides and on physical labour by introducing herbicide-resistant varieties of crops. Biotechnology-based manufacturing processes produce paper and chemicals with less energy input, less pollution output, and less waste. Biofuels will provide an alternative for dwindling stocks and reduce the pollution caused due to the use of fossil fuels. Forensic technologies based on the growing knowledge about DNA help exonerate the innocent and bring criminals to justice.

Modern biotechnology has tinkered with the genetic material of different organisms by introducing foreign genes. Organisms can be made to perform new functions or produce new products which they are otherwise incapable of producing. Genetic engineering manipulates organisms from viruses to mammals and has ushered in the genomics revolution, often described as a major part of the third technological revolution following the industrial revolution and computer revolution. Its impact on our daily life has already been felt in several ways:

(i) Production of Golden rice by incorporating three genes required for production of Vitamin A in Taipei rice (Fig. 15.1).

(ii) Production of recombinant proteins such as insulin from bacteria, known as *biopharming*, has saved millions of pigs from being slaughtered for the purpose.

(iii) The number of human proteins that can be made by genetically modified organisms (GMOs) will be increased to cater to the therapy of many more diseases.

(iv) Recombinant DNA technology has come up with newer and effective methods for raising vaccines against dreaded diseases.

(v) Insect-resistant plants engineered by introducing genes for the insecticidal toxins of *Bacillus thuringiensis* (Bt) have improved the yield of cotton and reduced the cost of insecticides.

(vi) Monoclonal antibodies obtained by fusing cancerous cells with B cells of mice immunized with specific antigen, introduced refined techniques for clinical diagnosis, and therapies including immunotherapy of cancer.

(vii) Creation of transgenic plants for production of vaccines including edible vaccines will go a long way.

(viii) Stem cell-based therapy made us hopeful about leukaemia and thalassemia patients, and will be useful for treating persons inflicted with heart attack and Alzheimer's diseases.

(ix) Similarly, tissue and organ culture will soon come up with replicas for damaged tissue or organs.

(x) The human genome project's results have allowed sequencing of growing numbers of disease-causing and disease-susceptible genes and characterization of mutations.

Fig. 15.1   Golden rice. (Also see colour plate 4.)

(xi)   It will be possible to develop DNA probes to diagnose most of the genetic disorders and for somatic cell gene therapy.

Thus, biotechnology has promising solutions to many problems, mainly related to food (agriculture), medicines and therapies, and for improving the quality of the environment to meet the growing demands of the world's increasing population. Modern biotechnology is expected to provide sustainable development and a sustainable lifestyle as the population goes up from the current six billion to a predicted ten billion by 2050.

Countries with a reasonable science base and a sustainable economy are indulging in the practice of and research in biotechnology. USA currently employs more than 150,000 people and invests to the tune of $10 billion a year on research and development, and has already created a market of more than 35 billion dollars for biotechnology products.

Biotechnology is, by definition, primarily based on applied research rather than basic research. However, for continued development of biotechnology, much further basic research is required.

15.3 BIOETHICS AND BIOSAFETY

Ethics is a code of values based on human conscience for the actions of human beings, particularly towards others and their surroundings. Bioethics refers to our ethical considerations in relation to biological inventions, applications, biotechnology, and so on. New technology brings in both benefits and risks. The risks involved in the introduction of a new technology (say biotechnology) need to be well assessed before its introduction in the society; in a way, this is a part of ethics.

Issues in adoption of biotechnology: international trade and developing countries

Biotechnology will affect every area of a country's economy and society at large. Plant and animal breeding by humans is associated with commerce. Biological products have been at the base of international trade for many countries for a long time. International trade pressure and competition have changed the internal economic and social structure of many countries, and they will continue to do so even in the arena of biotechnological products.

Use of biotechnology could have both positive and negative effects for different countries. Some examples of these effects on developing countries have been presented in Table 15.1. International competition will enable some countries to improve their international trade based on biotechnology; however not all countries would be able to improve their trade in this manner. Today, the developing countries may not be in a state to gain much out of biotechnology related trades. Even then they should strive to adopt biotechnology to the benefit of their own people and not be swamped by trade and the discoveries of developed nations. Developing countries should adopt biotechnology more wisely to maintain a reasonable lifestyle and quality of the environment which is consistent with a sustainable way of life in the international community; only then can the question of distributive justice in the world get attention. The proper practice of biotechnology in different countries may ultimately provide rich dividend in commerce and in the protection of the resources and environment of the planet. This approach is likely to earn a steady state in international trade and technologies in the near future.

Table 15.1 Probable effects of biotechnology and biotechnology based international trade on developing countries

Positive effects
1. Availability of more food by increased crop production for the growing population
2. Introduction of drought-and salt-tolerant crops to increase food production and employment
3. Increase in nutritional qualities of agricultural products
4. Higher standard for livestock health and production
5. Increased storage life of foodstuffs so they can be sent to distant places in a country or be used for international trade, without sophisticated storing facilities
6. Reduced dependence on imported fertilizers and chemical pesticides
7. Faster growing trees, for generation of biomass and energy
8. Better public health through provision of cheaper vaccines and diagnosis kits
9. Production of health care products (antibiotics, antibodies, cytokines, hormones, and so on) from genetically engineered bacteria, plants and animal cell culture, and transgenic plants and animals
10. Use of biofertilizers and bio-pesticides would help reduce pollution
11. Biofuel will save exchequers by reducing oil import
12. Bio-remediation will conserve soil and water resources
13. Biodegrading materials for packing would cut down on the use of plastics
Negative effects
1. Loss of export markets for agro-products due to substitution by alternative products from industrialized countries.
2. Thus, loss of foreign income and employment (reduced labour for cultivation).
3. Production of more cash crops may push out traditional agro-products.
4. In the process of strengthening large agricultural estates for economic management small-scale farmers and land owners would be displaced and marginalized.
5. Farmers may lose control of raising seeds for cultivation, if they are compelled to purchase high yielding varieties of seeds all the time.
6. Privatization may increase both legal and financial barriers to the use of varieties.
7. Introduction of herbicide-tolerant plants may reduce the employment of labour for weeding and would increase dependence on foreign imports of chemicals (herbicides).
8. Gradually bio-diversity may be replaced by more efficient mono-crop systems.
9. Loss of natural eco-systems in marginal lands where new biotech crops are introduced.
10. The ecological characteristics of new organisms are unknown; there may be problems associated with introducing them in new environments.
11. International companies are interested in exploitation rights in return for conservation of the biodiversity of developing countries. This may be serious as it entails mortgaging one country's biodiversity to other countries.

15.3.1 Biosafety in the Application of Biotechnology

Several ethical issues have come forth regarding the applications of biotechnology, particularly in agriculture and health care. They need to be addressed properly for the benefit of the planet and mankind in the long run.

Risk assessment and containment

1. To carry out genetic engineering experiments in microorganisms, certain guidelines have been laid down, initially formulated at the Asilomar Conference (1977), USA, to prevent escape of engineered organisms. As per these guidelines, experiments are to be conducted in a laminar flow cell culture hood in an isolated clean room and the experimental organisms are to be conditioned such that they would die outside in case they escape. Even then there is an apprehension that the use of recombinant microorganisms for various commercial purposes can accidentally create new infectious agents.

2. The use of genetically engineered plants and animals for agricultural and other purposes may disturb the existing ecological balance in an unpredictable way. A relevant point in this regard is the introduction of high-yielding (breed) varieties of wheat and rice which require large amounts of water. Shallow pumps are used to meet this excessive requirement, sometimes leading to leaching of arsenic in the underground water.

3. Introduction of genetically engineered herbicide-tolerant plants and use of herbicides may result in abundant growth of selected weeds that are more fit.

4. Modified animals, such as transgenic fish could raise problems. This is anticipated as the introduction of certain fish species as biological control agents occasionally had disastrous consequences. Introduction of voracious giant catfish proved to be disastrous for other species of fish.

5. Cloning of plants and animals is likely to lead to a decline in biodiversity; cloned materials will get preference over natural species. This may be disastrous if the monoclonal varieties lack the genetic make-up to counter unforeseen viruses or other pathogens, and changes in the environment in the future.

The question of cloning human beings may be at a distant horizon of thinking. It will have to cross many hurdles or ethical questions before being seriously concerned. Currently it is banned in many countries.

15.3.2 Introduction of Genetically Manipulated Organisms (GMOs)

Genetically manipulated (GM) crops have raised many questions about propriety regarding their use. Public opinion was sought in different European countries for application of GM technology through questionnaires during the last decade. Euro barometer surveys revealed a plurality of beliefs and viewpoints, mostly guided by cultural and political traditions (nationality), religion, and knowledge of the subject. Many people are concerned about the ever-growing influence of technology, in general, in their lives, and at times mistrust the technology. They like to restrain people from playing God in the arena of genetic engineering, out of unknown fear. The public is aware from past experience that in the long run some technology might bring adverse effects such as those being felt due to pollution or global warming, and atomic explosions, and the occurrence of higher percentage of cancers.

The majority of Europeans tend to believe in continuing traditional breeding methods rather than changing the hereditary characteristics of plants and animals through modern biotechnology. People of developing countries including India with a shortage in food production, are confused about GM crops and in the core of their hearts have reservations against them; they, particularly scientists and knowledgeable persons became vociferous against the introduction of Bt brinjal in January 2010 and convinced the Government of India to keep it in abeyance.

Pro-GM groups including scientists and business personnel feel that public perception of genetic engineering is low and naive. For centuries now, human beings have manipulated the genomes of plants and animals by guided mating primarily to enhance desired characteristics or minimize unwanted traits. In this way, food plants and domestic animals now bear little resemblance to their predecessors; importantly food became progressively less expensive. Traditional breeding is made at the level of the whole organism; desired phenotypes are selected for without much knowledge about precise genetic changes and at times undesired genetic changes are perpetuated. Pro-GM groups maintain that biotechnology, in contrast, enables genetic materials to be modified at the cellular and molecular level with more precision, producing desired characteristics in a predictable manner, in accordance with the aims of classical breeders.

There is now a worldwide debate on the *safety aspects of genetically engineered crop plants* and derived food products. All the groups agree that there should be reasonable certainty that no harm will result from the consumption of GM foods. Foods or food ingredients derived from GM plants must be as safe as, or safer than, their traditional counterparts. This concept of 'substantial equivalence' will be judged at the molecular and protein characterization level along with toxicological and nutritional factors.

Consumer rights are now recognized by all member states in the European Union (EU). How will these operate in developing countries, and will these be violated by the multinational companies by the introduction of GM food? When will GM food be the order of the day, and will consumers have the right to choose non-GM foodstuff? To resolve all these issues, proper regulatory bodies have been constituted with experts in some countries and at the international level. Use of GMOs is regulated in Europe under the Health and Safety at Work Act through the Genetically Modified Organisms (Contained Use) Regulations which are administered by the Health and Safety Executive (HSE) in the UK. The HSE receives advice from the Advisory Committee on Genetic Modification.

15.3.3 Regulation of Introduction of GMOs and Biotechnological Products

Every country maintains different agencies under the control of the government to monitor and permit any products for agricultural and pharmaceutical use. The US has several organized agencies with well laid-out norms, rules, and facilities to oversee the introduction of any new product as food, pharmaceuticals for consumers or any new item for agriculture. Some of these agencies existed before the era of biotechnology and they have taken charge of biotechnological products in recent times.

Regulatory agencies

A voluntary conference was organized at Asilomar, California, in 1974 by researchers involved in gene transfer experiments to frame guidelines to introduce safeguards against the spread of genetically manipulated organisms and the consequences of such spread which are unknown. The National Institute of Health (NIH), USA was the first federal agency to take up regulatory responsibility over biotechnology. NIH published research guidelines for recombinant DNA techniques keeping the safeguards in focus. NIH was the sole monitor and reviewer of all DNA research until 1984. By 1986, the US government established the formal policy as the 'Coordinated Framework for Regulation of Biotechnology' for evaluating products developed using modern biotechnology, and placed it under the joint responsibility of NIH, USDA, and EPA. Today, three agencies are responsible for most of the biotechnology products (Table 15.2) — The US Department of Agriculture (USDA) since 1862 and the Food and Drug administration (FDA) are the two old agencies in the US, and The Environmental Protection Agency (EPA) is a recent one, established in 1970.

Table 15.2 Regulatory agencies for introduction of biotechnology products in USA

Agency	Products regulated
US Department of Agriculture (USDA)	Plants, plant pests including microorganisms, animal vaccines
Environmental Protection Agency (EPA)	Microbial/plant pesticides, microorganisms and animals producing toxic substances
Food and Drug Administration	Food, food additives, animal feeds, human (FDA) and animal drugs, human vaccines, medical devices, transgenic animals, cosmetics

The *Animal and Plant Health Inspection Service* (APHIS) as a branch of the USDA is responsible for protecting US agriculture from pests and diseases. Considering that genetically engineered insects and plants are potentially invasive, they treated as plant pests and regulated by APHIS. Approval of both APHIS (USDA) and EPA is needed for the environmental release of GMOs.

APHIS issues permits for developing and field testing genetically engineered plants under rigorous conditions. If any experimental plant poses any kind of potential threat to pre-existing agriculture, the agency terminates its further trials and ensures that the safeguards are maintained. APHIS sometimes requires several years of field trials to investigate many relevant features of the plants, such as disease resistance, drought tolerance, reproductive rates, and any adverse effects on other plants. Field trials must be carried out with precautions, mainly to prevent accidental cross-pollination. Trial fields must be at least one mile away from normal cultivation fields, and a wide fallow (unplanted and tilled) land must surround the entire field to prevent 'plant pest' risks. The flowers need to be bagged to confine the pollen. The experimental plants at the end should be destroyed and the trial field should be kept fallow and vacant for the next season following the harvest of the trial crop.

Food and Drug Administration (FDA)   Both food and drug products are important categories of biotechnology products and are under the vigil of (FDA). FDA provides consultancy to biotechnology developers on new food products or new food additives. It advices them on testing practices. New products should not have unexpected and undesirable effects. Biotechnology products need to be at par with their normal counterparts, but they should not contain any allergy causing compounds.

The Flavr Savr tomato (see Chapter 6) is the first commercial plant product approved by FDA. FDA has the responsibility and power to remove from the market any item previously approved but proved to be unsafe.

The regulatory agencies in the US for biotechnological products are pioneers and often serve as role models for creation of such agencies in other countries.

Other important regulatory agencies   An agency of the United Nations (UN), the *Food and Agricultural Organization* (FAO) provides guidance on safety and risk assessment on biotechnology in developing nations.

The European Union (EU) was formed in 1993. It streamlined the process of introducing new biotechnology products into the market. A key part of the new system is the European Agency for the Evaluation of Medicinal Products (EMEA), an equivalent of the FDA. The EMEA certifies medicinal products for human and veterinary use. It provides scientific information about drugs in all 11 official EU languages and helps an approved drug to be marketed in the 15 countries of the EU.

Most countries have now established agencies for regulation of biotechnological products and introduction of GMOs.

15.4 INTRODUCTION OF GMOs IN INDIA

In India, the *Genetic Engineering Approval Committee* (GEAC) under the Union Ministry of Agriculture provides the clearing signal to GM food crops for field trials.

The procedure for regulation of GM crops in India, was laid down in 1989 under the Environment Protection Act.

Outline of the stepwise procedure

1. **Review Committee on Genetic Modification (RCGM)**   Assesses and decides on applications to test a genetically modified crop.
2. **Genetic Engineering Approval Committee (GEAC)**   After RCGM's clearance, GEAC considers the application for field trials. If approved, the GEAC assesses the field trials and decides whether or not to commercialize the GM crop variety.
3. **Monitoring and Evaluation Committee (MEC)**   Monitors small-scale trials under RCGM and those under GEAC. It looks at bio-safety and crop yield among other responsibilities and reports to RCGM.
4. **GEAC**   Based on the MEC report, GEAC clears the crop for commercial release.

5. **Union Ministry of Agriculture**   Receives the application with GEAC clearance, checks compliance with the Seed Act and releases the crop for cultivation.

Bt Brinjal

India's first GM food crop Bt brinjal came for consideration by the GEAC in a meeting on 14 January 2009. Mahyco developed Bt brinjal under a license from the US agri-biotech giant Monsanto. As in the case of Bt cotton, the Bt gene (*Cry1AC*) from *Bacillus thuringiensis* was incorporated in the genome of brinjal. The Bt gene product has insecticidal properties. The Bt toxin disrupts the digestive system of insects that feed on the crop containing the Bt gene. Thus, Bt gene provides the crop with in-built pesticides and cuts down on the cost of pesticides. GEAC was forced to delay a decision on the approval of Bt brinjal. Anti-GM groups, mostly non-governmental and consumers' rights organizations have alleged that GEAC works under the influence of the GM seed industry and hence its decisions were not independent. In 2006, it permitted Mahyco, the seed company itself, to carry out field trials of Bt brinjal, ignoring protests. Non-profit independent organizations and scientists found that Mahyco had left out statistically significant differences between the GM brinjal and control groups in its report to GEAC, and found errors in Mahyco's research methodology, including inadequate sample size. Human safety trials for Bt brinjal have not yet been conducted.

There are 238 varieties of 56 GM plants at different stages of research and trials in India. This includes 169 varieties of 41 different types of food crops — Bt-cauliflower, cabbage, rice, chickpea, groundnut, maize, mustard, okra, pigeon pea, potato, tomato, watermelon, papaya, apple, chilli, coffee, and so on.

India at present is at the cross road to the introduction of GM crops. Agri-biotech industry personnel are extremely eager to promote GM crops as a means of food security in Asia and Africa where farmers are less suspecting and less critical of the GM technology and the governments tend to be more pliable. Quite a few well-informed persons and researchers are of the opinion that GMOs have nothing to do with food security and there is not enough research on the safety aspect for human consumption. A strong vigilance from researchers, non-profit organizations, and public forums is necessary for the introduction of GMOs, particularly GM crops.

Pollen transfer from GM plants to compatible wild relatives is possible, in theory. Could pest or herbicide-resistance genes incorporated into transgenic plants be transferred into other closely related non-desired plants and increase their resistance and 'weediness'? Under normal conditions gene transfer by way of pollen between close relatives is exceptionally rare and this is expected to be true for transgenic plants. However, all commercially released and growing transgenic plants need to be regularly monitored to confirm this assumption.

15.5 CONTROL OVER HUMAN DNA

The sequencing of the human genome ranks in importance with the Manhattan Project that produced atomic weapons during World War II, and the space programme that sent men to the moon. The joint announcement of the complete draft sequence of the human genome by the then Prime Minister of the UK, Tony Blair, and the President of the United States, Bill Clinton, on 26 June 2000 stimulated new concepts in medicine and bioethics. For instance, it would enable genetic screening of individuals for disease traits and susceptibility to diseases. Automatically, the question of confidentiality of the screening data would arise. Confidentiality would protect normal individuals found to be carriers of alleles for a genetic disease, so that they are not discriminated against. The question of fairness in the use of genetic information for insurance, employment, criminal law, adoptions, entrance to the education system, and other areas must be addressed. Education and laws should ensure respect for equality; stigmatization or ostracism and labelling in general on the basis of genetic information, should not be allowed in society under any circumstances. Social attitudes need to be attuned to take care of possible individual psychological responses.

The society should be cautious of unregulated eugenics. The ability to genetically engineer all organisms including ourselves might need to be restrained with our sense of morality or ethics of responsibility to future generations. We need to ensure that future generations retain the same power over their destiny that we have enjoyed so far, while reaping the benefits of the culture, knowledge, and technology that the human race has developed.

SUMMARY

♦ Bioethics is concerned with preventing the creation of dangerous organisms that would adversely affect other living organisms or the ecosystem.
♦ Genetic engineering manipulates organisms from viruses to mammals and has ushered in the age of modern biotechnology which has an impact on our daily life in several ways including trade in biopesticides, agricultural and 'biopharming' products.
♦ The time has come to be aware of the probable effects of biotechnology and biotechnology-based international trade on developing countries.
♦ To carryout genetic engineering experiments in microorganisms, certain guidelines have been laid down; this includes guidelines to prevent the escape of these experimental organisms.
♦ There is now a worldwide debate on the safety aspects of genetically manipulated crop plants and derived food products.
♦ Every country maintains different agencies to regulate the introduction of genetically modified organisms (GMOs) and biotechnological products.
♦ In India, Genetic Engineering Approval Committee (GEAC) under the Union Ministry of Agriculture provides the clearance to GM food crops for field trials.
♦ India's first GM food crop Bt brinjal came up for consideration by GEAC. Mahyco at Mysore developed Bt brinjal under a licence from the US agri-biotech giant Monsanto.

Paul D. Boyer

Paul D. Boyer an American biochemist and analytical chemist shared the 1997 Nobel Prize in chemistry with John E. Walker and Jens C. Skou for elucidating the enzymatic mechanism responsible for the synthesis of adenosine triphosphate (ATP), (enzyme ATP synthase).

Boyer was born in Provo, Utah and attended Provo High School. In 1939, he received a bachelor degree in chemistry from Brigham Young University, located just a few blocks away from his home. His enthusiastic teachers and interest in organic chemistry played a big role in his appreciation of chemistry. He went on to pursue graduate studies at the University of Wisconsin (UW), Madison on a Wisconsin Alumni Research Foundation Scholarship. While working under Paul Philips at UW, Madison, he was exposed to cutting-edge research in enzymology and metabolism

Paul D. Boyer
(born 31 July 1918)

through discussions with young faculty members and by attending seminars from the pioneers of respiratory enzymes like Fritz Lipmann, Otto Meyerhof, and Carl Cori. His experience in graduate school at Wisconsin primed him for a career in science, and he obtained Ph D in biochemistry from UW, Madison in 1943. He spent 2 years at Stanford University on a war-related research project to stabilize serum albumin for transfusions. From this study he 'gained experience with proteins and a growing respect for the beauty of their structure'.

Boyer started his own lab at the University of Minnesota, Minneapolis in the spring of 1946. In 1955, he received a Guggenheim Fellowship and worked with Professor Hugo Theorell at the Nobel Medical Institute in Sweden on the functional mechanisms of alcohol dehydrogenase. Incidentally, Theorell was honoured with the Nobel Prize in that year. In the same year Boyer received the Award in Enzyme Chemistry of the American Chemical Society. He accepted the Hill Foundation Professorship and moved to the medical campus of the University of Minnesota.

Since 1963, Boyer has been a professor in the Department of Chemistry and Biochemistry at the University of California, Los Angeles (UCLA), where he became the Founding Director of the Molecular Biology Institute in 1965 and spearheaded the construction of a new building for the Institute and initiated an interdepartmental PhD programme. Continued efforts in his own lab and Skou and Walker labs consolidated the mechanism of ATP synthase. Rotary catalysis and binding change mechanisms were postulated for the function of ATP synthase and tested experimentally by 1971. In a true tribute to his alma mater, Paul mentioned an interesting anecdote about the term 'rotational catalysis' in his Nobel lecture in 1997—'Conferences at the University of Wisconsin provided opportunity to publish thoughts about rotational catalysis that had not been enthusiastically endorsed at Gordon Conferences, ...'

Paul D. Boyer was associated with many learned bodies throughout his active career. He was the Chairman of the Biochemical Section of the American Chemical Society during the period 1959-60; he acted as the President of the American Society of Biological Chemists in 1969–70. He was the Editor or Associated Editor of the Annual Review of Biochemistry from 1963 to 1989, and Editor of the classic series 'The Enzymes'.

[In the course of working at UCLA in the mid 1970s, the author of this book had the privilege to experience Boyer's enthusiasm and meticulousness during inauguration of the new building of the Molecular Biology Institute 'in a molecular way'. The gentleman who had donated money for the building was requested to add a fluid drop by drop from a test tube on a gluey material in a hole of a plastic slab holding a chord. Within two minutes, to everybody's surprise, the chord got loose and released the restrained bottle of champagne hung from a lofty branch of a tree in front of the building; the bottle swung and smashed on the wall of the new building to the claps of many gathered for the occasion that afternoon. Interestingly, the gluey material was discarded DNA, and the fluid in the test tube was DNase, that had been kept secret by Boyer and his research associates.]

EXERCISES

Objective Question

1. Cross the incorrect statement:
   (i) Golden rice is produced
      (a) by breeding between strains of Taipei rice
      (b) by incorporating a single gene
      (c) by incorporating three genes
      (d) to make it Vitamin A rich

Review Questions

1. Why should ethics have a role in the run of biotechnology?
2. Assess critically the impact of biotechnology in the sphere of international trade of developing countries.
3. Discuss the safety aspects of genetically engineered crop plants.
4. Elaborate the new concepts in medicine and bioethics with introduction of genome analysis of an individual after completion of the human genome project.
5. What is 'biopharming'?
6. 'Clonal propagation of plants may bring disaster'. Explain.
7. Name the regulatory agencies for introduction of biotechnology products in USA and India.
8. What is FAO and what are its functions?
9. State briefly the utilities of:
   (a) Monoclonal antibodies
   (b) Stem cells
   (c) Results of human genome project

Intellectual Property Rights and Biopatents

LEARNING OBJECTIVES

♦ The main forms of protection provided by Intellectual Property Rights (IPRs)—Patents, trade secrets, copyright, trademarks, and plant breeder's rights
♦ International organizations and conventions for patents
♦ The patenting strategy
♦ GATT, Uruguay Round, TRIPs, WTO, and Doha Round
♦ The process of patenting in India
♦ Biodiversity in the light of IPR

INTRODUCTION

Usually the term 'property' refers to physical objects such as land, house, household goods, and money for which ownership and associated rights are guaranteed and protected by the laws in a country. Such property is often referred to as *tangible property* or *physical property*. *Intellectual property*, on the other hand, is an idea, a design, an invention, a manuscript, an art material, and so on, which can ultimately come up with a useful product or application. Intellectual inputs, ingenuity, innovativeness, and sometimes substantial sums of money are involved in the development of such property. Societies and governments have long devised various ways to reward the inventors of intellectual property so that they and in general others, are encouraged to work with greater zeal and devotion towards newer inventions. This attitude of the society gave rise to the concept of passing laws for the protection of intellectual property rights. The intellectual property right (IPR) provides protection against copying, imitation, or reproduction of intellectual properties, the major problems associated with such properties. IPR ensures the right of an inventor to derive economic benefits from his/her invention (i.e., intellectual property). However, IPR is maintained by the governments as long as it is not detrimental to the society. Intellectual property is classified into two categories: (i) industrial property and (ii) literary and artistic work.

It should be clear from the preceeding chapters of this book that biotechnological processes and products involve a great deal of ingenuity on the part of researchers and capital investments of private and government sectors. Obviously private companies and government institutes look forward to realizing the money involved in the research for an item and to gain profit for further investment. Thus, the ideas and techniques involved in genetic engineering or broadly biotechnology come under the purview of IPR.

16.1 HISTORY OF PATENT FOR BIOTECHNOLOGY DISCOVERIES

Earlier, living things were not patentable in any country, but biological materials obtained by non-biological means, in which 'the hand of human' had a part, were patentable. The first patent for an organism altered by genetic manipulation was granted to Ananda Chakraborty by the US Supreme Court in 1980, for a hydrocarbon degrading *Pseudomonas* strain to be used to clean up oil spillage. The patent claim of Chakraborty working in the research laboratory of General Electric company (GE) was

'A bacterium from the genus *Pseudomonas* containing therein at least two stable energy-generating plasmids, each of the said plasmids providing a separate hydrocarbon degradative pathway' (US pat. 4,259, 444).

In approving this claim, the Court said 'the distinction is not between living and inanimate things but between products of nature, whether living or not, and human made inventions'.

This judgement of the US Supreme Court opened the scope of patenting biotechnology techniques and products.

A patent for 'oncomouse' invented by the scientists of the Harvard University was granted in 1992 and became another milestone in patenting of life forms.

16.2 PROTECTION OF INTELLECTUAL PROPERTY RIGHTS

The protection of IPR may be enacted in several forms depending mainly on the type of intellectual property and the type of protection sought. Each form of protection has its own advantages and pitfalls. The main forms of IPR protection are provided under (1) Patents, (2) Trade secrets, (3) Copyright, (4) Trademarks, and (5) Plant breeder's rights (PBR).

16.2.1 Patent

In principle, *patents are based on claims of novelty, utility, and enablement*; that is a patent must describe something new—information which was not in possession of the public, something useful, and enable someone else to reproduce it.

Patents provide protection to inventions, processes, or products produced. Patents compensate for the time and money spent on inventions. A patent gives its owner the exclusive right to make and sell or use the patented product or process of an invention for a limited period of a specified number of years. The period of patent protection varies widely between countries, but in Europe, USA, and Japan it has recently been standardized to about fifteen years after product approval.

The Indian Patent Act of 1970 did not allow product patents in pharmaceuticals, foods, and agrochemicals. This Act has been amended as Indian Patents (amended) Act, 1999, which allows product patents except for some specified medicines/drugs. With the growing number of different types of discoveries in biotechnology, the concept of patentable items is bound to change with time.

Interestingly, a patent entitled 'Basmati Rice Line and Grain' for a novel rice variety was awarded by the US court to Rice Tech, Texas on 2 September 1997. In the patent application, this novel rice line is claimed to have been developed by using a novel criterion and its description covers the complete range of features present in the entire germplasm of *basmati*, the world famous brand of quality rice from India and Pakistan. For this patent protection, India is likely to face difficulty in the export of *Indian basmati*. Azadirachtin, purified from 'neem', *Azadirachta indica*, a native plant of India has also been patented in the US as a purified natural product. After a prolonged battle in court, India regained the patent right on neem and *haldi* (curcumin) from the US (see Exhibit 16.A).

All these instances imply that the government needs to be cautious and introduce the right kinds of patent laws for the produce of a country, and at the same time prevent other countries from freely using the bio-materials of the country.

Specialist patent attorneys are now coming to the fore, those who can file patent applications properly, contest/defend such applications and enforce the rights accruing from patents to the inventors.

The principle of 'enablement' as stated at the beginning also means disclosure. Disclosure of the invention in sufficient detail enables other skillful persons to reproduce the invention. In addition, a sample of the microorganisms, cell line, and so on being genetically modified and patented may be required to be deposited in the designated culture collection. The deposited material serves as a reference in cases of disputes with novelty or unauthorized use, and as a source material for the authorized users.

Processing of patenting

The patent application for an invention is filed in a prescribed pro forma with the patent office of the country along with some payment. The patent officials scrutinize and assess the application. If for some reason the application is found unsuitable for patenting, it is returned to the inventor with reasons for the rejection. The application may be withdrawn, or modified according to the objections raised by the patent office, and resubmitted.

If an application qualifies for patenting, the invention with adequate details is published in the patent office's publication for the information of all concerned. In India, this is usually done 18 months after filing the application. Anyone can challenge the award of a patent within a stipulated time, which is four months in India. The publication also helps the other competitors improve the patented invention, and thus improve the technology.

In case a patent application is not challenged, the patent is granted immediately after the expiry of the stipulated period. In case of a challenge, the arguments of both the applicant and the person challenging the application are heard by the competent authority of the patent office and the final decision of granting the patent is taken. Following the publication of the application, the invention stands disclosed, and if objection to the application is sustained the invention will no longer be patentable and will not even be considered for the option of trade secret.

16.2.2 Trade Secrets

Trade secrets allow secrecy about private proprietary information on physical material. Trade secrets provide certain advantages to the owner. Unlike patents, trade secrets have an unlimited duration, and need not satisfy the stringent conditions laid down by the law for patent applications. Disclosure of a trade secret and its unauthorized use are punishable offence, and are compensable. The best and popular example of a trade secret is the formula of the Coca-Cola brand syrup.

Coca-Cola Trade Secret

John S. Pemberton in Atlanta, USA invented Coca-Cola in 1886, and managed the secret formula for the syrup as a trade secret. The close secret is only shared with a small group of executives and not written down. In 1891, Asa Candler purchased the proprietorship of Coca-Cola and the rights to the business. Then, in 1919, Earnest Woodruff and a group of investors purchased the company from Candler and his family. Woodruff financed the purchase by arranging a loan and as collateral he submitted documentation of the formula committed to paper by Candler's son on request. This was placed in a vault in the Guaranty Bank in New York until the loan was repaid in 1925. Woodruff reclaimed the secret formula and took it to Atlanta to place in the Trust Company Bank, now Sun Trust, where it remained until its recent move to the World of Coca-Cola, where it is zealously protected. This is the essence of the classic case of the Coca-Cola trade secret.

Selling an enterprise built around a totally protected trade secret has been found to increase the value of the firm substantially. Coca-Cola, Betty Crocker, Duncan Hines, Oil of Olay, Ben and Jerry's, and Schlitz are only a few examples of renowned brands built around a trade secret.

Trade secrets for biotechnology may include materials such as hybridization conditions, cell lines, corporate merchandising plans, customer lists, and so on. It may be difficult to maintain trade secrets in biotechnology as a large number of scientists and young researchers are involved in the field and they often exchange their ideas and methodologies in conferences and in publications. In view of the long time it takes for a patent to be granted (~2 years in the US, and at least 5 years in India), trade secrets are sometimes more practical.

If a trade secret comes under public knowledge by independent discovery or in some other way, it would no longer be protectable.

16.2.3 Copyright

Certain intellectual properties like authored and edited books, audio and video cassettes, and so on are not patentable, but can be protected under the copyright act. The copyright of a book may be held by the author, editor or the publisher, and the right excludes others from reproducing the book in any form for the specified number of years enacted in the law. More recently, computer software has been included in the list of copyright properties (the Information Technology Act, 2000).

In biotechnology, copyright protection may be extended for DNA sequence data which may be published. One may get around this protection by designing an alternate sequence of DNA to encode the same protein; this is possible because of the degeneracy of the genetic code, that is, more than one code (triplet codons) can specify a particular amino acid.

16.2.4 Trademarks

All of us are aware of the trademarks on items like dresses and household goods purchased in our daily life. A trademark is a symbol or word 'adopted and used by a manufacturer or merchant to identify his goods and distinguish them from those manufactured or sold by others'. In the field of biotechnology, laboratory equipment and reagents carry trademarks that develop familiarity with the product among researchers. Certain vectors that are used in recombinant research can also have trademarks. Although a trademark can protect the intellectual property right of an industry, it is not strong enough to provide legal protection.

16.2.5 Plant Breeder's Rights (PBR)

The development of crop varieties is another form of intellectual property and is protected through 'plant breeder's rights' or PBR. This empowers a plant breeder, an originator or owner of a variety, to exclude others from producing or commercializing the propagating material of that variety for a minimum period of 15–20 years. At the same time, the person holding the PBR title to a variety can authorize other interested parties to produce and sell materials produced by the variety, on his terms. The government may get involved in issuing licenses of the titles to others in public interest. The *object of protection in PBR is the variety*, and not the genetic components and the breeding procedure.

PBRs have recently been allowed in India since 2001 under the new 'Protection of Plant Varieties and Farmers Right Act' (PPV & FRA).

16.3 OTHER VIEWS ABOUT IPR

Recently in other countries, *utility patents* for genetic material, both plant and animal, have been allowed, barring the use of patented material for further breeding or to save and use as seeds for cultivation. These IPRS may be good for the original breeder but adversely affect the availability and use of plant and animal genetic resources. These patents, if allowed for transgenic animals and plants, may work as impediments in the free exchange of genetic material for improvement of crops and livestock.

Some times, it seems that IPRs may prevent improvement of cultivable materials due to patent restrains, and adversely affect food security, biological diversity by monoculture practice, and in consequence the ecological balance. Above all, IPRs may severely affect the livelihood of the poor in developing countries.

The current IPR systems do not promote the protection of diversity of whole ecosystems or natural unmodified plants.

16.4 BROAD PATENT COVERAGE IN BIOTECHNOLOGY

Usually, the patent applications in agricultural biotechnology are prepared with the widest possible coverage available. Such protections, if allowed and continued in future, will only favour financially powerful corporations allowing them exclusive control over biotechnological modifications even of such crops that feed and sustain mankind. The business houses, in the near future, may acquire legal rights to determine the course of high-tech research for agriculture and plant breeding. Such a situation of monopoly would be undesirable for scientific and technological progress and undermine the existing system of agriculture, farmer's rights to the seeds, independent planning of a country, and above all global food security.

16.4.1 Farmer's Rights

Since the dawn of civilization, some 10,000 or more years ago, the genetic resources in plants and animals have been selected, developed, used, and conserved by farmers and farming communities especially in the countries rich in biodiversity—mostly gene-rich developing nations. The same materials are now being used as raw materials to evolve the modern high-yielding varieties of crops and resources for biotechnology. Seed sales of the improved varieties earn huge profits for seed corporations.

> In 1983, at a UN Food and Agricultural Organization Conference, representatives from 156 countries recognized that 'plant resources were a part of the common heritage of mankind and should be respected without any restriction'.

Protection of Plant Varieties and Farmers' Right Act, 2001   This Act gives concurrent attention to the rights of farmers, breeders, and researchers to breed new varieties of crops, and the protection of public interest. This includes compulsory licensing of rights and prevention of import of varieties incorporating the Genetic Use Restriction Technology (GURT). An example is the terminator technology of the US company, Monsanto, which does not allow the crop to bear viable seeds. Farmers and researchers have exemptions from the payment of royalties on seeds that they save from their harvest. However, so far no reward is given to the farmers who for millennia have established crop varieties which plant breeders use as starting materials.

The key questions of quantum of reward, whom to pay, to what extent, and in what manner have not yet been answered. The farmer's rights are yet to be legalized. The governments of different countries and corporations maintain silence for their own reasons.

16.5 INTERNATIONAL ORGANIZATIONS AND CONVENTIONS FOR PATENTS

Worldwide the key organizations in patenting are European Patent Office (EPO), US Patent and Trademark Office (PTO), the various European National Patent Offices, and the Japan Patent Office.

Apart from the differences and ambiguities in patent law in different countries, the time between filing a patent and getting it granted is longer for biotechnology companies than for companies in most other fields. This delay means that biotechnology companies cannot defend a patent in court till it is sanctioned, even after it has been made public.

There are quite a few high-profile patent cases in biotechnology. Some examples are mentioned here.

Patent was not granted for the PCR process, which is of enormous utility for modern molecular biology on the ground that is it an invention (clause of novelty). Although Amgen and Genetics Institute worked roughly at the same time to develop *genetically engineered erythropoietin* (EPO), complete patent right was granted to Amgen, because according to the Court the technical information in the patent application of Genetics Institute did not enable someone else to reproduce it (enablement clause). Genentech, Scripps Clinic and Chiron developed methods for purifying *Factor VIII* to treat haemophilia. The US Appeals Court did not allow the right to patent the product but opined that the specific ways for purifying were patentable. There are many demands to patent DNA sequences that have been discovered with funding to biotechnology companies and institutes. Gene sequencing is a leading edge of modern biology. Whenever a human gene is sequenced, the questions of ethics are raised. In one tactical move, the US pharmaceutical company Merck is placing all the DNA sequences it discovers into the public domain as a 'spoiling action' to stop others from patenting them (once in the public domain, they are no more novel). The fierce debate on the issues of patenting different types of biotechnology products will continue for some time as the whole gamut of patenting was based on products of physical and chemical sciences; biotechnology is a recent entrant in the arena and the jury needs to be well informed about some extremely complex technical issues so that it does not base its judgement solely on instinct.

The US patent to Harvard for the *oncomouse*, in 1988, was the first patent granted for a transgenic animal. The *myc* gene is an oncogene that contributes to the development of cancer. The gene is spliced (joined) together with the promoter from MMTV virus, which causes the gene to express its protein specifically in the mammary gland leading to cancer. Thus, the transgenic oncomouse develops mammary cancer at a very high rate. This in turn makes it a useful animal model both for detecting the events that lead to cancer and for developing treatment strategies. The US Patent and Trademark Office (PTO) has granted patents on approximately a dozen transgenic non-human animals. Initially, the EPO in Europe was reluctant to allow animals to be patented.

16.6 PATENTING STRATEGIES

The party applying for a patent should not have any ambiguity in its approach. The following steps are very important for seeking a patent on a product or discovery.
- The application must have a clear title.
- Prior novelty search should be conducted.
- The feasibility or reproducibility of the finding should be checked.
- A patent attorney should be appointed.

16.7 INTERNATIONAL PATENTING

In the field of patent law, ever since the Paris Convention was signed in 1883, there has been a strong tradition of international cooperation by means of international conventions to regulate formal and substantive patent matters between member states.

The *Paris Convention* is also known as the 'The International Convention for the Protection of Industrial Property'. There are 151 member states now, the majority from the industrial world, and they enjoy equal terms with respect to the protection of industrial property. One of the most important articles of this convention is that the first filing date of a patent application in any member state which establishes a right of priority will also be accorded to a corresponding patent application in any other member state if the latter is filed within 12 months of the first filing. The text of the Paris Convention has been modified several times, the latest being after the Stockholm Convention in 1979.

The other important international conventions for regulation of patent matters are the following:

Strasbourg Convention is the 'Convention on the Unification of Certain Points of Substantive Law on Patents for Invention' and dates back to November 1963. Many features of this convention have been incorporated into the European Patent Convention signed ten years later; the notable is the definition of the 'state of the art' against which the degree of novelty and inventiveness of the subject matter of a patent application must be judged. The exclusion of plant and animal varieties from patent protection also stems from this convention.

The *Patent Co-operation Treaty* (PCT) was signed in 1970 and came into force in June 1978 along with the European Patent Convention. The PCT has broad international scope and more than 100 member states. Patent applications filed under PCT are considered as 'international', because an international body, the World Intellectual Property Organization (WIPO) based in Geneva administers and processes the applications in an international phase before they are formally introduced into national systems. The international phase is primarily concerned with preliminaries, a prior art search and publication of the application.

The *European Patent Convention* (EPC) was signed in October 1973 and established the European Patent Organization as a legal entity comprising the European Patent Office (EPO) and an Administrative Council as its two organs. The convention came into force in June 1978. An application may be made in any of the regional offices of the EPO (National Patent Offices) which in due course is examined by the EPO in Munich. Upon grant, the patent right enters the national phase in each designated state and emerges as a 'bundle' of national patents, such as the European patent (UK), European Patent (France), and so on , and is thus divided into independent objects of property.

Under the community patent convention, a single application filed through the EPO matures into a single unitary indivisible object of property covering the whole of the European Economic Community.

The *Budapest Treaty* or 'The Budapest Treaty on the International Recognition of the Deposit of Microorganisms for the Purpose of Patent Procedure', was signed in April 1977 and came into force at the end of 1980, and now has 45 member states. The Treaty is for the recognition of culture collections at International Depository Authorities (IDAs), into any of which a new strain of microorganism can be deposited for the purpose of a patent application in any member state. Deposited strains are allotted accession dates and numbers by which they are identified and referred to in the patent specification.

16.8 INTERNATIONAL HARMONY FOR PATENT LAWS

Usually, patents are valid in a country or in a territory that has a union of several nations as in case of Europe (EPC). It is a costly and time-consuming affair to get patents from different countries. This is critical for all biotechnology processes and products which have an element of intellectual property, and at the same time have been traded across national boundaries.

To provide protection for IPRs under an international umbrella, 'Trade-Related IPRs (TRIPs)' came into existence on 1 January 1995. TRIP has its roots in the 'General Agreement on Tariffs and Trade' (GATT), framed in 1948, and was meant to be a temporary arrangement to settle disputes among countries participating in trade. It was supposed to determine both tariff rates and quantitative restrictions on imports and exports globally. Thus, GATT was a provisional treaty for contractual agreement till it took the shape of the *World Trade Organization* (WTO) in 1995. The following brief discussion on GATT, WTO, and TRIPs is for the awareness of students of biotechnology.

16.8.1 GATT, 'Uruguay Round' of Negotiations, and TRIPs

There have been eight rounds of negotiations during 46 years of GATT's existence, from 1948 to 1994. The final and eighth round was the longest round of trade negotiations, starting at Punta del Este (a city in Uruguay) in 1986. It continued till 1994, mainly because of disagreement between the US and the EU. The Uruguay negotiations (1986–1994), popularly known as the 'Uruguay Round', was finally signed by 123 member countries in 1994. This led to the establishment of WTO on 1 January 1995 with its office in Geneva. (WTO replaced GATT.) TRIPs form a part of the Uruguay Round of GATT and became operative with the WTO, and is the most comprehensive multilateral agreement on IPR.

The member countries of WTO are supposed to meet all the articles of TRIPs. They were allowed to suitably amend their IPR laws within five years; the least developed countries were given an additional five years. A variety of intellectual properties, usual patents, and new varieties of plants came under the provision of TRIPs.

The developing countries were expected to provide a 'mail box' protection to member nations by accepting patent applications for pharmaceutical and agricultural chemicals from 1 January 1995 till they modified their patent laws as per TRIPs provision. Furthermore, they were also to provide an exclusive marketing right for five years for each invention applied for under the 'mail box' provision provided the same had been granted marketing approval by another member state after 1 January 1995.

16.8.2 WTO and 'Doha Round' of Negotiations

WTO replaced GATT as an international body to promote free trade by persuading its member countries to abolish import tariffs and other barriers like subsidies, thus promoting globalization in world trade. GATT originally had 23 member countries in 1948. The number of members swelled to 123 with WTO in 1995 and to 152 by 2008.

The *Ministerial Conference* (MC) is the highest decision-making body of the WTO, which meets every two years. Six Ministerial Conferences took place during 1996–2005. MC4 was the most important for many reasons and was held in Doha, a city in Qatar, in the Middle East, in November 2001, and is often referred to as the *Doha Round*. The objective of the Doha Round was to reduce trade barriers across a wide range of sectors and to address the needs of developing or economically poor countries, with the aim of better and effective integration of these countries into the global market. The developed nations were to abolish barriers in import and subsidies in export of farm items. However, the recommendations at the Doha Round could not be agreed upon and finalized by the member countries in MC5 at Cancun, Mexico in 2003 and MC6 in Hong Kong in 2005 or even later. Developing countries including India have put forward a demand to implement the decisions taken in the Doha Round.

The Omnibus Trade and Competitiveness Act, enacted by the US Congress in 1988 may be mentioned here. This trade act empowers the US government to 'investigate' trade and investment-related laws of the other countries to ascertain if they have any element that maybe detrimental to the interests of the US. If the investigated country refrains from changing its laws during a 'warning period', it would face US retaliatory action.

EXHIBIT 16.A   Neem Patent Case

Indian farmers staged one mass demonstration after another in 1994, against the proposed GAAT Uruguay Round agreement. About 200,000 farmers gathered at Delhi in March, with several demands, a major one was that the draft treaty—known colloquially as 'the Dunkeldraft' after the chief negotiator Arthur Dunkel—should be translated into all Indian languages. On 2 October, about half a million Indian farmers gathered at Bangalore to voice their fears about the impending legislation and the threat that GAAT poses to their livelihoods, by allowing multinational organizations to enter Third World markets, which would earn at their expense. The protesters carried twigs or branches of the neem tree, as they were aware of recent patents granted for several extracts of neem to US companies; they considered it as intellectual piracy. They were incensed as the village neem trees are a symbol of Indian indigenous knowledge, and the US companies would expropriate this knowledge for their own profit and would deprive local farmers of their ability to produce and use neem-based pesticides. Neem, *Azadirachta indica* is a plant of the Mahogany family *Meliaceae*. It is an evergreen, grows upto 20 metres and is found all over India, particularly in semi-arid areas. It is useful to humanity as it serves many purposes due to its astonishing versatility. It has been considered as 'Sarva Roga Nivarini', that is, the curer of all ailments, by Indians for millennia. The chemical azadirachtin from neem seeds is used for many purposes.

In 1971, Robert Larson, a US timber importer observed the usefulness of the neem tree in India and began importing neem seeds to his company headquarters in Wisconsin. A pesticidal neem extract called Margosan-O was tested for safety and efficacy over the years by Larson. In 1985, he got clearance for the product from the US Environmental Protection Agency (EPA). Three years later he sold the patent for the product to the multinational chemical corporation W.R. Grace and Co. Interestingly, since 1985, more than a dozen US patents have been acquired by US and Japanese firms on formulae for stable neem-based solutions and emulsions and also for a neem-based toothpaste.

Armed with US patents and with the prospect of a patent from the European Patent Office (EPO), Grace Co. set about manufacturing and commercializing their product by establishing a base in India. The Company made offers to several Indian manufacturers to buy their technology.

The patent to neem oil as fungicide was granted by the EPO to the United States Department of Agriculture and the chemical multinational W.R. Grace, in 1995. Since then Vandana Shiva, Director, Research Foundation for Science, Technology and Ecology in India along with the International Federation of Organic Agriculture Movement and the Green Party in European Parliament opposed the patent on the ground that it was pure and simple piracy. The oil from neem seeds has been used traditionally by Indian farmers and village folk to prevent fungus. It was neither a novel idea nor was it invented. After several years of court battles, in 2000, the EPO revoked the patent. But the victory was short-lived as the revocation was followed by an appeal of the Grace Co.

In the next round multiple evidences were presented. Shiva and others presented a series of evidences to show that farmers were using the knowledge about neem for a long time. Neem's many virtues were mentioned in many ancient texts, and in Indian traditional medicine and Ayurveda. The bark, leaves, flowers, seeds, and fruit pulp are used to treat a wide range of diseases and complaints ranging from leprosy and diabetes to ulcers, skin disorders, and constipation. Millions of Indians use neem twigs every morning as an antiseptic toothbrush; the oil is used in the preparation of toothpaste and soap. Neem oil is also used for contraception purpose, as lamp oil, a repellant to insects, and as fungicide. Neem fruit pulp is used for the manufacture of methane. Neem timber is hard and termite resistant and used as construction material.

They also provided information about the research with ethanolic extract of neem seed kernel and repellant action of neem oil by Indian scientists in the 1960s and 1970s, much before the patent had been filed.

On 8 March 2005, the European Patent Office in Munich dismissed the appeal against revoking the patent granted by it in 1995 for the preparation of a fungicide derived from the seeds of the neem tree.

The Hindu, a national daily reported on 9 March 2005, that 'India wins the neem patent case'. Vandana Shiva commented from Germany on the judgement that this will go a long way in giving confidence to traditional users. Shiva considered it a historic moment, and added 'Patenting is one of the ways through which traditional users can be threatened. But now, such patents will no longer be a threat for traditional users'. It may be mentioned here that more than 200 organizations from 35 nations supported the Indian cause.

In May 1995, the US patent Office granted to the University of Mississipi Medical Center a patent (No. 5,401,504) for 'Use of Turmeric in Wound Healing'. Imagine the implications—an Indian using turmeric in the usual customary manner would infringe US patent laws and would become open to prosecution.

The US patent was promptly challenged by R.A. Mashelkar, Director, Council of Science and Industrial Research (CSIR), New Delhi, who has done much to awaken India to IPR issues. After four months of submissions it was established that the use of turmeric as a healing agent was well known in India for centuries. The patent was annulled.

What next?

Two battles of neem and turmeric were won indeed, but it seems there are many such battles ahead for India. London's *Observer* reported that there are more than 100 Indian plants awaiting patent grants at the US patent office. In fact, patents have already been granted to uses of Amla, Jar Amla, Anar, Salai, Dudhi, Gulmendhi, Karela and so on; all these are Indian household names.

Vandana Shiva is of the opinion that the West has a clever structure in place. Using convenient patent laws as a system, the TRIP instrument as a stick and the WTO as the enforcing authority, the developed countries are seeking to rob the Third World Countries.

Mashelkar as the Director of CSIR took a very meaningful step to create a massive database for plants that will record all practical ideas proposed in Indian knowledge systems including earlier literatures. Once created the database will deny bio-pirates—on the basis of prior knowledge of their prior use and lack of novelty—the patents that they seek surreptitiously.

Indian campaigners against bio-piracy are fighting a nationalistic battle. They believe, the profit if any must go to the communities which have striven to experiment, refine, and preserve over the millennia, the practical uses of nature's gifts. Two Indian scientists in Kerala discovered the commercial potential of a herb called *arogyapachcha*. Kani's Tribal Welfare Trust was made co-holder of this patent, as the Kani tribe has preserved this herb through the ages.

For centuries British, French, and Portuguese colonists ignored neem and other Indian plants as a gesture of racist dismissal of indigenous knowledge. With increasing consciousness for environmental protection, the West has realized that the days of the commercial potential of chemical pesticides are numbered. They are now looking for biological sources like neem that have been identified in the traditional knowledge of the Third World Countries. Besides India, much of Africa and Latin America are now victims of biopiracy. Interestingly, there are reports that some Third World Countries have also started cultivating neem trees in millions in their land.

The conclusion to this Exhibit may be drawn with a story, probably apocryphal. The Portuguese anchored off the Malabar coast in Kerala in the 17 th century. They were received warmly by the Zamorin. After a few days the palace guard rushed with panic into the court to report 'Your Majesty the foreigners are on the hill slopes, uprooting pepper vines and carrying them away to the ships. If they begin to grow these in their lands we will lose our trade'. The Zamorin unperturbed, said 'Ah, don't worry too much. They may take the vines but how can they take our monsoons'. The West must be aware of various 'Zamorins' present in all ages in benevolent India!

Abduction of turmeric provoked India

Turmeric (*Curcuma longa*) is a condiment in Indian food and helps digestion; the powder heals open wounds, when drunk with warm milk it stems coughs, cures colds, and comforts the throat. The prescription of *haldi* (turmeric) can be found in the very many pages of the voluminous Ayurvedic literature that is several millennia old. Turmeric is used in India as paint, insecticide, in depilatory skin cream, and *as kumkum* for *bindi*. Turmeric is used symbolically for religious ceremonies and in many other chores including pasting it on the couple to be married before the ritual bath. For Indians turmeric is like a benevolent goddess. Recently, it has been described as the gold in Indian kitchens.

16.9 SCENARIO OF PATENTING IN INDIA

There are some recent developments regarding the patent system in India in view of globalization of the economy—greater foreign investment in research, joint ventures, and facilitation of technology transfer.

India acceded to the Paris Convention and Patent Cooperation Treaty (PCT) on 7 December 1998. Before that the Government of India notified the Convention countries to claim priority with respect to patents and trademarks. The reciprocity with the member countries has been established. The advantages of joining PCT include (i) filing of a single international application for protection in designated member countries and (ii) 75 per cent reduction in fees for individual inventors of developing countries.

There are some common misconceptions in India. A patent granted on an application in India is not automatically protected in the Paris Convention and PCT member countries. It is also not true that applicants in India cannot file PCT applications with respect to biotechnological inventions, particularly for living organisms and software, as India does not grant patents in these fields.

16.9.1 India in the Context of TRIPs

A complaint was lodged by the US to WTO in 1997 that India has failed in basic commitments to TRIPs. The Dispute Settlement Body of WTO found that India did not provide the 'mail box' provision (see discussion earlier) for exclusive marketing rights to the concerned products (patents). India was asked to accommodate the 'mail box' provision by April 1999, failing which the US might go for appropriate sanctions.

> The Indian Patent Act, 1970 has been amended in the context of TRIPs in 1999 and came into effect from 1 January 1995 the time of enactment of TRIPS. Now India allows product patents.

16.10 BIODIVERSITY IN THE LIGHT OF IPR

Biodiversity is the source which provides genes from wild species for biotechnology exercises. The developed countries of the globe have been utilizing the biodiversity of the developing nations without paying any compensation. Since 1992, several biodiversity conventions have been held to discuss measures to preserve the biodiversity, but not much progress has been made regarding the issue of compensation from the developed nations to the biodiversity rich developing countries for using their gene resources.

The 'Convention of Biological Diversity' (CBD) assigned the responsibility to individual nations to conserve and use their own biodiversity. This convention essentially replaced the 'principle of *common heritage* of genetic resources' to humanity. In article 15 of the CBD, it is provided that the access of one state to genetic resources of another state will be allowed for 'environmentally sound use' with 'prior informed consent' and will be based on 'mutually agreed terms'.

The 'International Undertaking on Plant Genetic Resources' (IUPGR) was established by FAO in 1983, to be named later as 'International Undertaking on Plant Genetic Resources for Food and Agriculture' (IUPGRFA). It was adopted by the member countries of FAO as a non-binding agreement to promote conservation, exchange, and use of 'plant genetic resources for food and agriculture' (PGRFA). The Undertaking was modified to accommodate 'breeders' rights' and 'farmers' rights'. The IUPGRFA is now renamed the 'Interntaional Treaty on PGRFA' (ITPGRFA). This is also known as the 'International Seed Treaty' (IST). It was approved by FAO on 3 November 2001 and came into effect from 29 June 2004. The central theme of the Treaty is 'Multilateral Systems (MLS) for Access and Benefit-Sharing' which for certain categories of PGRFA guarantees facilitated access in return for sharing commercial benefits.

The concept of sovereign rights of a nation over its genetic resources also led to the concept of biodiversity prospecting, thus conceptually giving a price tag to the genetic diversity.

16.10.1 Biodiversity Act, 2002

The Ministry of Environment and Forestry (MoEF), Government of India drafted a 'Biodiversity Bill' in 1998, with the objective to regulate collection, documentation, conservation, and sustainable use of the country's biodiversity. The Bill was passed by both houses of the Parliament in December 2002. India took the first lead among the 12 mega-diversity countries, to come up with such a legislation. This will allow India to deal properly with different international instruments including CBD, UPOV (International Union for Protection of New Plant Varieties), TRIPS, and so on to which India is a signatory.

The Bill is likely to provide measures to curtail free access to foreigners including multinational companies, institutions, and individuals to India's rich and diverse biological resources, and also prevent biopiracy of the resource materials. The Bill includes protection measures for local people and farmers, who contributed to the generation of biological diversity and researchers in the country.

The Bill has instructions to create a three-level structure for management.

National Biodiversity Authority (NBA)   Set up at Chennai, NBA is meant for approval of the applications or requests from foreign nations and individuals for obtaining IPRs on inventions based on biological resources or traditional knowledge of Indian origin. NBA is also entrusted with setting guidelines for granting approval and sharing benefits from an invention.

State Biodiversity Board (SBB)   It needs to be informed by local firms or individuals about any commercial utilization of biological resources. They need to register with the SBB; no approval of NBA is needed normally. *Hakims, vaids,* practitioners with *Ayurveda* or other Indian systems of traditional medicine are exempted from registering with SBBs. Agriculture, animal husbandry, poultry, and bee-keeping are also exempted.

Biodiversity Management Committee (BMC)   This is a local body at the *Gram Panchayat* or *Mandal Panchayat* to maintain registers of local resources and knowledge. The register is called a 'Peoples' Diversity Register'. Besides documentation, this body is supposed to promote conservation and sustainable use of biological resources at the level of the farmer or household. There would be consultation between NBA and BMCs before granting permission to any foreign firm or individuals for use of a biological material in the jurisdiction of the BMCs.

The Bill also has provisions for setting up 'Biodiversity Funds' at the central, state, and local levels, that define the sources and purpose of utilization of these funds.

There are also prescribed punishments to the violators in the Bill; offences are cognizable and non-bailable.

For some reason agro-biodiversity has been kept out of the purview of the Bill and the Bill is often criticized on this count by different NGOs.

SUMMARY

♦ Intellectual property is an idea, a design, an invention, a manuscript, an art material, and so on, which can ultimately come up with a useful product or commercial application. Intellectual property may be of two types–industrial, based on scientific innovations, and literary and artistic works.

♦ The society protects the intellectual property rights (IPRs) to the inventors by enacting acts and laws, so that the inventors can enjoy monetary or other benefits from their innovations for a certain specified number of years.

♦ The ideas and techniques involved in genetic engineering or broadly biotechnology come under the purview of IPR.

♦ The main forms of IPR protection are provided under: (1) Patents, (2) Trade Secrets, (3) Copyright, (4) Trademarks, and (5) Plant Breeder's Right (PBR).

♦ Patents are based on claims of novelty, utility and enablement, that is, a patent must describe something new, something useful, and enable someone else to reproduce it.

♦ Breeder's Rights and Farmer's Rights are comparatively new clauses for providing benefits to the respective groups.

♦ The key organizations in patenting are the European Patent Office (EPO), the US patent and Trademark Office (PTO), the various European national and other national patent offices, and the patent office of Japan.

♦ In the field of patent law, since the Paris Convention was signed in 1883, there is a strong tradition of international cooperation by means of international conventions to regulate formal and substantive patent matters between member states.

♦ Ever since living material became patentable after the landmark judgement of the US Supreme Court in 1980, many conventions were held among different countries, and new laws, institutions (like WTO), and methods for IPR protection (TRIP) have been established.

♦ India is a signatory to major world organizations involved with patenting and IPRs.

♦ In view of TRIPs for biological materials, India amended its patent laws, and enacted the Biodiversity Act, 2002.

Ananda Mohan Chakrabarty

Ananda Mohan Chakrabarty
(born 4 April 1938)

Ananda Mohan Chakrabarty is an Indian–American microbiologist. In 1979, while working in the Research and Development Centre at the General Electric

Company, New York, he developed a bacterium strain from four different oil-metabolizing strains of *Pseudomonas putida* through plasmid transfer; the new strain was faster at digesting (or degrading) the hydrocarbons present in oil spills. According to Chakrabarty the new strain was 'multi-plasmid hydrocarbon-degrading *Pseudomonas*', popularly known as 'superbug'. The bacteria drew international attention when he applied for a patent—the first ever patent for a living organism.

Ananda Mohan Chakrabarty was born in Sainthia, Birbhum, West Bengal, India. He attended Sainthia High School, Belur Bidyamandir and St. Xavier's College, Calcutta. He received his MSc and PhD degrees (1965) from the University of Calcutta.

His application for patent on the bacteria was initially denied by the patent office as it was thought that the patent code precluded patents on living organisms. But then the United States Court of Customs and Patent Appeals took a decision in Chakrabarty's favour. This decision was further challenged but finally the patent was granted to Ananda Mohan by the US Supreme Court and it was determined that

'A live, human-made micro-organism is patentable subject matter under [Title 35 U.S.C.] 101. Respondent's micro-organism constitutes a "manufacture" or "composition of matter" within that statute.'

This judgement paved the way for many patents on genetically modified microorganisms and other life forms.

Ananda Mohan is currently a Distinguished University Professor in the Department of Microbiology and Immunology at the University of Illinois at Chicago College of Medicine. He is pursuing research with bacterial products in cancer regression and arresting

cell cycle progression. Five patents have been generated by his work at the University of Illinois. Professor Chakrabarty founded CDG Therapeutics Company, Incorporated at Delaware. In 2008 he co-founded a second bio-pharmaceutical discovery company Amrita Therapeutics Ltd., registered in Ahmedabad, Gujarat.

Apart from being an eminent scientist, he has been an advisor to judges, governments, and the UN. He is one of the founding members of a UNIDO committee that proposed the establishment of the International Centre for Genetic Engineering and Biotechnology (ICGEB) with three components at Trieste, Italy; New Delhi, India; and Cape Town, South Africa. He is associated with many distinguished scientific bodies, and the scientific advisory boards of many academic institutions. He served the Stockholm Environmental Institute at Sweden. Besides many awards from the US for his work in genetic engineering technology, he is a recipient of the Padma Shri from the Government of India in 2007.

EXERCISES

Objective Questions

1. Multiple Choice Questions
   (i) The first patent for an organism was granted for:
       (a) *Streptomyces* spp.
       (b) *Pseudomonas* spp.
       (c) Oncomouse
       (d) *Escherichia coli*
       (e) *Drosophila melanogaster*
   (ii) Stipulated period for objection to the award of a patent in India is
       (a) 4 months
       (b) 1 year
       (c) 18 months
   (iii) The object of protection in Plant Breeder's Rights is
       (a) Breeding procedure
       (b) Genetic components
       (c) Variety

   (iv) Tick the right statement(s) in reference to the area of biotechnology
       (a) US Patent and Trademark Office (PTO) grants patents on bacteria and plants only.
       (b) European Patent Office (EPO) prefers to grant patents on plants and plant products only.
       (c) EPO is reluctant to allow patent on animals.
       (d) US-PTO grants patents on prokaryotes as well as organisms of plant and animal kingdoms.
       (e) US-PTO does not allow patents on animals.

2. Complete the sentence:
   Patents are based on claims of _____, _____, and _____.

3. Copyright protection may be extended to DNA sequence data. --- Yes / No.

Review Questions

1. What are intellectual property rights? Who are the beneficiaries?

2. Are there demerits of IPRs?

3. In which way are GATT and TRIP related?

4. If someone is not sure about the novelty of his/her discovery but is sure about its commercial potential, what is the best option of IPR for the person to seek?

5. Elaborate on WTO and the 'Doha Round' of negotiations.

6. Assess the benefits of IPRs in the developed versus the developing countries.

7. How do patents differ from other IPRs?

8. Discuss the Biodiversity Act, 2002 legislated in India.

9. Write down your views about patenting a terminator gene.

10. IPRs may adversely affect food security. Explain.

11. Why does the US pharmaceutical company Merck shares all the DNA sequences it discovers in the public domain?

12. Provide a brief note on establishing the European Patent Organization and its functions.

13. Who was Arthur Dunkel?

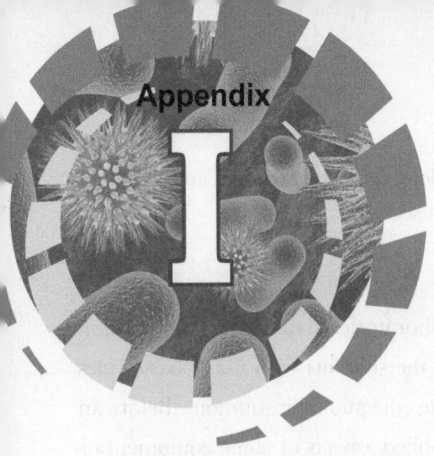

I

Lab Work

INTRODUCTION

This appendix is to familiarize the students with the methods and materials used in biotechnology research and industry.

Starting with an experiment explaining the bacterial culture procedure, we move on to experiments such as the study of bacterial growth and various staining techniques for microscopic studies. Prokaryotic (bacterial) culture is a necessary step in many molecular genetics experiments as well as for producing biopharmaceuticals.

Next an experiment to judge the sensitivity of bacteria to different antibiotics is presented. The antibiotic sensitivity of bacteria is primarily used for selecting clones of recombinant bacteria in genetic engineering processes.

In plant and animal biotechnology, in-vitro tissue culture techniques are primarily important. The 3methodologies for plant tissue culture from explants, production of callus in culture, and suspension cell culture are presented. Protoplast culture is important for genetic manipulation in plant biotechnology, and this technique is also explained here.

The preparation of cell suspension from spleen and study of different aspects of lymphocytes and phagocytic cells are contents of the next experiment. These are included to introduce in-vitro techniques for animal cells used in animal biotechnology. After this, lymphocyte culture to test the effects of some substances (here Con A) is presented in a separate experiment. The objective is to inculcate the idea of planning in-vitro experiments with animal cells to find out the effects of some drugs or chemicals on cells.

Antibody synthesis against an injected antigen in a mouse as test animal, collection of antiserum, and testing it in haemagglutination assay are objectives of Experiment 11. Experiment 12 helps the readers to understand the logical steps of SDS-PAGE technique to separate proteins of different molecular sizes. To carryout this experiment, a good biochemistry or molecular biology laboratory set-up is required.

Experiment 13 explains the process to visualize mitotic chromosomes in plant tissue—here the tissue used is from root tips.

Extraction procedure for DNA from animal tissues in described in the last experiment. Extraction of DNA is a necessary technique to proceed with many other advanced techniques, such as PCR, hybridization, endonuclease digestion, fingerprinting, DNA synthesis, and so on.

Sometimes the students may get glimpses of the micro world of biology very easily without any elaborate experiment. Some such **Fun Experiments** are presented in the last section.

Visit to Institutes and Industry

Visits to certain well-equipped biochemistry and molecular biology laboratories in universities or research institutes, brewery industry, and biotechnology industry will enrich the students with many basic rules and their applications. Such visits should always be encouraged by the educational institutions. Before an intended visit, the students need to study the basics, theories, and applied aspects of some equipment or methods they will be exposed to. They should carry a note book and a camera during such visits to take necessary notes and photos, wherever required.

EXPERIMENT 1   BACTERIAL CULTURE

Objective

Successful propagation of prokaryotes or bacterial cells in culture is an essential step in molecular biology and recombinant DNA technology experiments. The most commonly used bacterium for such experiments is *Escherichia coli* (*E. coli*).

The survival and continued growth of microorganisms in a culture depend on adequate supply of nutrients and favourable environment conditions.

Maintenance of Sterile or Aseptic Conditions

Maintaining sterile conditions while performing any experiment is very important to get the desired results. To maintain the same, sterilized equipment, media, and solutions should be used throughout the culture, so that it does not get contaminated by other microorganisms or some other contaminants. Petri plates used for cultures are made of either plastic or glass. Plastic petri plates may be provided in sterilized condition by the supplier. The glass petri plates can be sterilized along with the other glassware in an autoclave at high temperature.

The tungsten (or some other metal) loop or the needle used to transfer bacteria from one petri plate to another is sterilized instantly over the flame of a Bunsen burner just before use.

Autoclave   Autoclaving is the process of sterilizing media and equipment used for growth and culture of microorganisms. In this process, the material to be sterilized is subjected to high-pressure saturated steam at 121°C for about 15–20 minutes.. This can easily be done in a regular big pressure cooker. At the time of autoclaving, the caps on bottles or tubes should be in loosened condition so as to avoid accidental burst due to pressure build-up.

During the autoclaving process, the solutions, media, and other materials (not the temperature-sensitive materials, such as vitamins and enzymes.) are kept on a platform with some water at the bottom. The heater is turned on to reach 15 lb pressure (psi), which is maintained for 10 minutes. It should be made sure that the safety valve and pressure valves in the autoclave machine or the pressure cooker are in place and operative before the heater is turned on.

Another precaution taken to maintain sterility of the culture is that whenever a cotton plug from a flask mouth or a screw cap from a culture tube is opened for dispensing the medium into sterile Petri plates, the mouth of the flask or the tube should to be held briefly above the flame for sterilization.

Sterilization by filter   For sterilization of a liquid medium or a liquid substance, not capable of withstanding heat sterilization, porous filters with a pore size of about 0.2 μm diameter are used. These filters retain undesired microorganisms and let the solution pass through. A small amount of liquid taken up in a syringe may be sterilized quickly by using Millipore filter assembly fitted on the syringe.

Bacteria culture hood or laminar air flow hood   The laminar air flow hood is used for preparation of medium, distribution of media to culture plates, and inoculation of culture plates with bacteria so as to maintain aseptic conditions.

Culture Media

The culture medium used can be liquid (or broth), semisolid, and solid.
Broth (liquid) medium   The components are dissolved in purified water.
Semisolid medium <1% agar is added to the broth medium as a solidifying agent.
Solid medium 1.5–1.8% agar is added to the broth medium as a solidifying agent.

Preparation of defined (M9) medium for E. coli culture
1. Take 450 mL distilled water in a 1 L beaker. Dissolve 0.5 g NH_4Cl, 3 g Na_2HPO_4, 1.5 g KH_2PO_4, and 0.25 g NaCl in it.
2. Adjust the pH of the solution to 7.4 with 1 N NaOH. Bring the final volume to 494 mL by adding distilled water. After this, add 7.5 g agar to this solution and transfer the final solution to a 1 L Erlenmeyer flask and keep for autoclaving.
3. Prepare 50 mL solutions of each of the following in 100 mL screw cap bottles: 1 M $MgSO_4 \cdot 7H_2O$, 20% (w/v) dextrose (glucose), and 1 M $CaCl_2$.
4. Autoclave solutions prepared in Steps 2 and 3.
5. Before use, the contents of the flask prepared in Step 2 are to be kept in an oven or a water bath maintained at 50°C.
6. Following are to be aseptically added to the culture medium just before using it: 1 mL of $MgSO_4$, 5 mL of glucose, and 50 mL of $CaCl_2$ (prepared in Step 3).

Establishing a Bacterial Culture

There are several methods to transfer bacteria for establishing new cultures. Three such important techniques have been elaborated here.

Sterile Plating and Streaking
1. Remove the flask containing the molten agar from water bath and wipe off the water on the outside wall with a paper towel.
2. The molten agar should be immediately distributed in the 3″ to 5″ diameter Petri dishes with a 10 mL pipette, pre-warmed over the flame, under a culture (laminar) hood; 5 mL medium for a 3″ dish and 10 mL for a 5″ Petri dish. The Petri dishes are kept either at room temperature in the culture hood (or at 37°C in incubator overnight) for cooling and solidifying of agar.
3. The bacterial stock used for culture is stored in agar or broth in a test tube and kept in a refrigerator in aseptic conditions (keep the test tube properly capped in a large sterile beaker with proper covering).

. The stock should be obtained from a laboratory regularly using it for experimental purpose or from a reliable supplier.

4. The bacteria from the stock culture tube is transferred to the fresh agar Petri plates prepared in Step 1 with the help of the inoculating loop heated on the flame and then cooled (Fig. E.1).

5. The inoculating loop with the inoculum is moved on the surface of the agar in Petri dishes in different streaks as shown in Fig. E.2. The cover of the Petri dish should be immediately replaced to avoid contamination. The inoculating loop or wire should be kept in a sterilized test tube while not in use; the part of the handle may be outside the test tube. A straight wire mounted on a handle (transfer needle, Fig. E.1) or a sterilized cotton swab may also be used for transfer of bacteria instead of the inoculating loop.

6. The Petri dishes inoculated with bacteria are kept in the incubator at 37°C to let *E. coli* grow (some other bacteria need to be incubated at 30°C). The incubation period can be 24–48 hours. Then observation is made, and the details recorded.

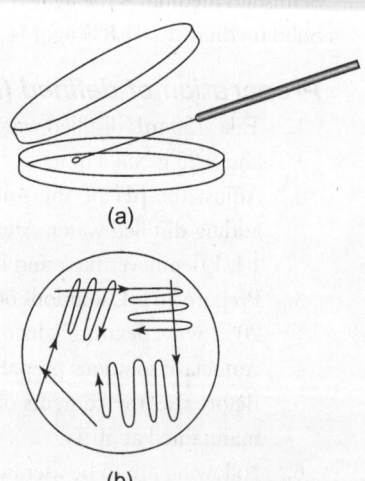

Fig. E.2 Streak-plate technique: (a) A Petri dish of agar culture streaked with an inoculating loop. (b) A common pattern of streaking. The Petri dish may be turned around with the left hand to streak at different regions of the plate.

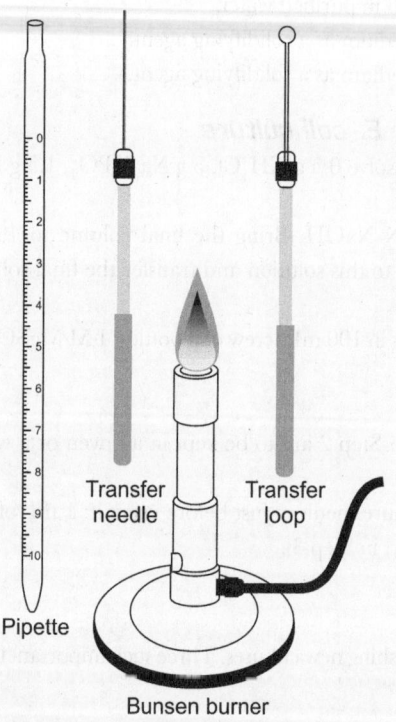

Fig. E.1 Instrument for bacterial transfer.

Spread Plate Technique

The sterile agar Petri plates are prepared following the protocol indicated earlier (Steps 1 and 2). A small sample of original broth culture is transferred with a sterile pipette to the centre of a Petri plate containing solidified agar medium. A bent spreader, made up of a glass rod, is dipped into 95% ethyl alcohol in a beaker and the bent portion in passed through the Bunsen burner flame and cooled for 10–15 seconds for sterilization. The transferred bacterial sample at the centre of the agar plate is then evenly spread over the agar surface by back and forth movement of the bent portion of the glass spreader (Fig. E.3).

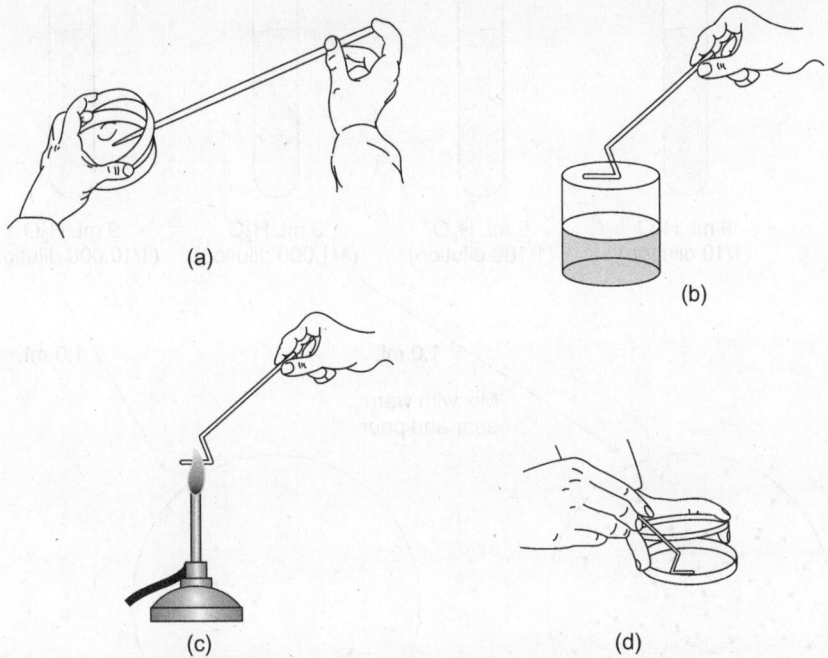

(a)

(b)

(c)

(d)

Fig. E.3 Spread-plate technique. (a) A small sample of bacterial culture pipetted onto the centre of an agar medium plate. (b) A glass spreader dipped into a beaker of 95% ethanol for sterilization. (c) The ethanol soaked spreader is briefly held on flame and cooled. (d) The transferred sample of bacteria is evenly spread over the surface agar medium with the sterilized spread. The Petri dish is then incubated.

Pour Plate Technique

In this technique, the original sample of bacteria in liquid culture is diluted several times. This is done to reduce the population density sufficiently so that when the final solution is spread on a plate, there would be enough space for growth of individual colonies. The process of dilution is shown in Fig. E.4. With each step of dilution, the cell number reduces by one-tenth of that in the previous tube (serial dilutions).

The most diluted samples are then mixed with molten agar at 45°C in different test tubes. Usually, 0.1 mL of a dilution (contains approximately 2000–3000 cells per mL) is good for inoculation of each Petri dish. Amount of 0.1 mL is supposed to contain 200–300 cells for a dish. The number of bacteria in a solution can be determined with a haemocytometer slide (discussed later) and accordingly the dilution can be adjusted.

The agar tubes with bacterial inoculation are rolled between palms keeping the tube upright for proper mixing and then poured into individual sterile Petri dishes. The dishes are incubated for 24–48 hours.

Keep one or two agar Petri dishes without any inoculation of bacteria as blank control. No bacterial colony should grow in the blank control Petri dish(es). If it does then this would indicate that the culture conditions are not sterile.

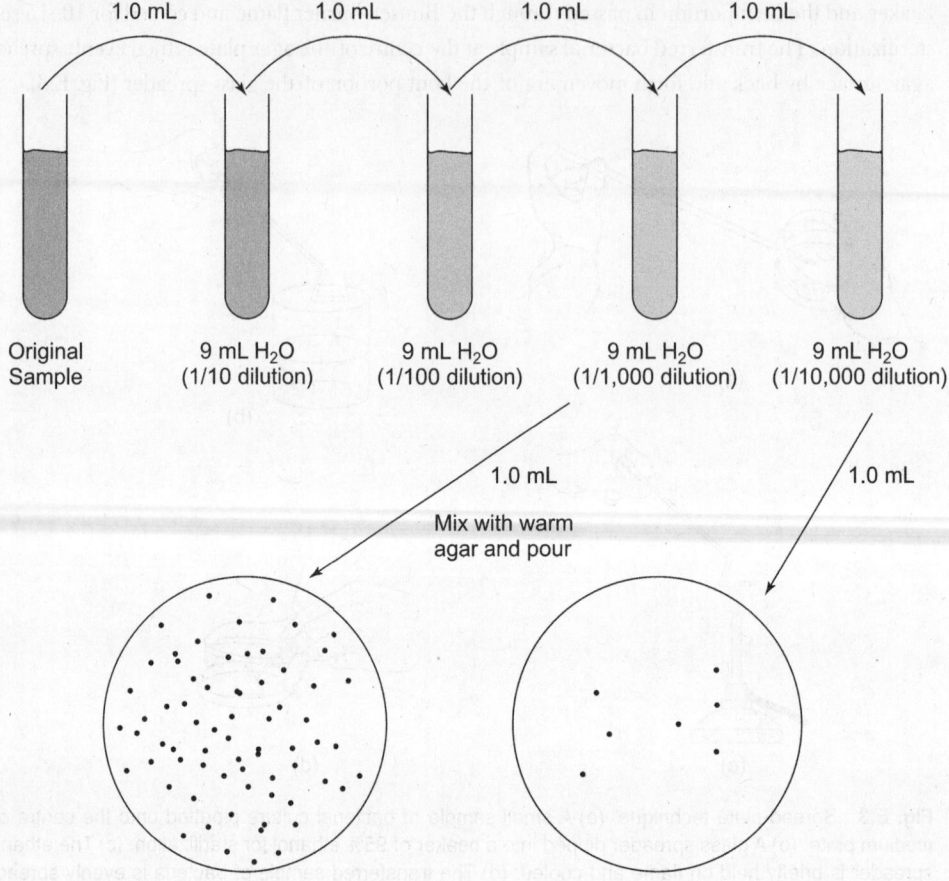

Fig. E.4 Pour-plate technique. The original bacterial sample is diluted several times to thin out the population density. The more diluted samples are then mixed with warm agar culture medium at 45°C and poured into Petri dishes. Isolated cells grow into colonies as indicated by black dots and can be subcultured to establish pure cultures. A colony count helps in the finding the number of bacteria in the original sample.

Observations

Let us now learn how observations are made for a culture experiment.

1. After 24 hours and 48 hours of incubation, respectively, the Petri dishes containing bacterial culture may be observed under binocular dissecting microscope or magnifying glass of a colony counter.

2. Each individual bacterial cell will divide by binary fission and form a colony during the incubation period.

    The following observations should be made:

    Size of the colony   Pinpoint, small, moderate, or large.

    Shape of the colony   Circular, irregular periphery, schizoid or root-like spreading.

    Elevation of the colony   Flat, raised, convex dome-shaped.

    Colour of the colony   If any.

[The above-mentioned characteristics may vary from species to species or sometimes due to certain culture conditions.]

Each colony is the clonal cell population derived from a bacterium at the time of inoculation.

The number of colonies are counted using the bacteria colony counter with square grid marks. The number of colonies in a way indicates the initial number of bacteria in the inoculum.

At the end of culture, the culture dishes are immersed in 70% alcohol and after half an hour, the contents of the dishes are discarded with a scraper.

EXPERIMENT 2   STUDY OF BACTERIAL GROWTH

Dynamics of the microbial growth can be charted by means of a population growth curve. The growth curve depicts the different stages of the growth cycle of microorganisms (cells in general)—lag phase, log phase, stationary phase, and decline or death phase (Fig. E.5). The growth curve provides the cell number at different phases and the rate of growth expressed in *generation time* (or doubling time, the time required for a microbial population to double).

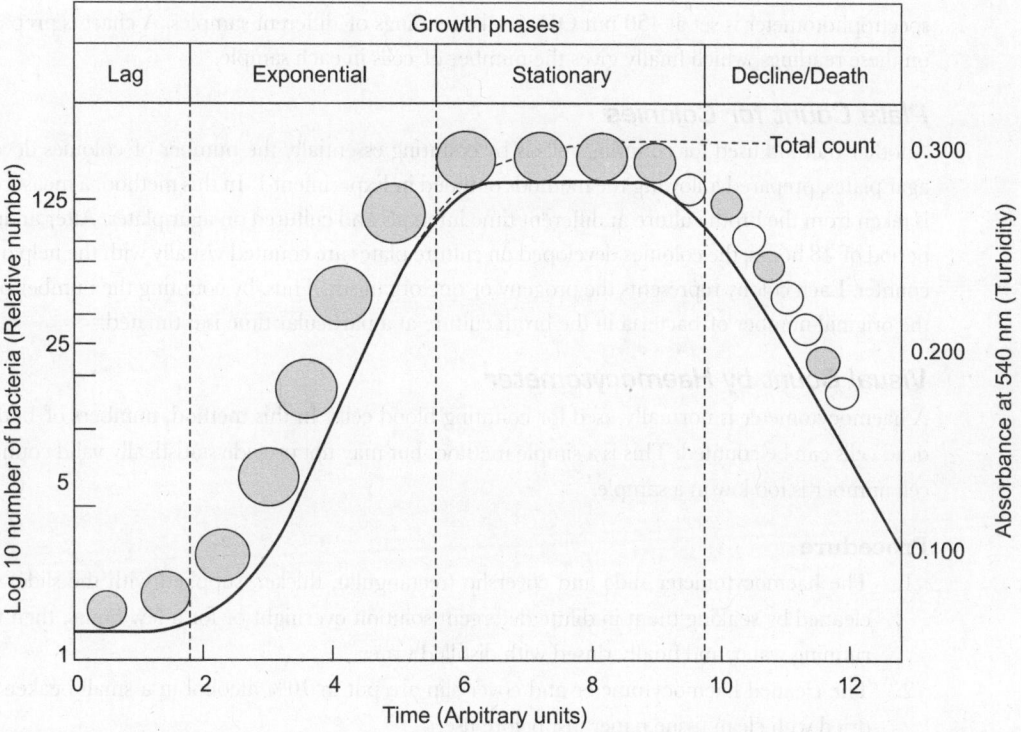

Fig. E.5 Growth curve of bacteria with four phases after inoculation of a fresh culture. The changes in cell size in different phases are indicated by the oval bodies on the curve; the empty bodies in the declining phase refer to the dead cells.

The cell sizes change during the growth cycle. Initially, the cells are smaller, grow during the lag phase, and become largest during the log phase. In stationary phase, the cells become smaller (the round bodies along the outside of the growth curve in Fig. E.5 schematically represent cell sizes in different phases).

The growth curve can be obtained by counting the bacteria in the culture at different time intervals, such as 30, 60, 90, 120, 150 minutes. Usually, the liquid or broth cultures that are used for such studies are maintained at a specific temperature and pH.

Methods of Cell Counting

Cell count can be of two types: viable count and total count. In *viable count*, the total number of only living cells is taken into account, whereas in *total count* the total number of both living and dead cells is counted. The latter is easier to measure. There are several methods to count bacterial cells.

Spectrophotometric Total Cell Count

This is a modern, sensitive, and rapid technique. It is based on the property of microbial cells to scatter incident light. The amount of scattering is proportional to the concentration of cells in an aqueous medium.

The higher the number of cells present in a suspension, the more is the turbidity or 'cloudiness'. In other words, the cell mass is directly proportional to the optical density (OD) of the suspension. To measure the cell density, a spectrophotometer is used, which correlates the number of bacterial cells with the change in OD values. The cell suspension is taken in a sample cuvette in the spectrophotometer for reading. The spectrophotometer is set at 450 nm OD to take readings of different samples. A chart is prepared based on these readings, which finally gives the number of cells in each sample.

Plate Count for Colonies

Another method used for counting cells is by counting essentially the number of colonies developed on agar plates, prepared following the methods outlined in Experiment 1. In this method, a measured sample is taken from the broth culture at different time intervals and cultured on agar plates. After an incubation period of 48 hours, the colonies developed on culture plates are counted visually with the help of a colony counter. Each colony represents the progeny of one organism. Thus, by counting the number of colonies the original number of bacteria in the broth culture at a particular time is estimated.

Visual Count by Haemocytometer

A haemocytometer is normally used for counting blood cells. In this method, numbers of both live and dead cells can be counted. This is a simple method, but may not provide statistically valid counting when cell number is too low in a sample.

Procedure

1. The haemocytometer slide and coverslip (rectangular, thicker, supplied with the slide) should be cleaned by soaking them in dilute detergent solution overnight or for a few hours, then washed in running water, and finally rinsed with distilled water.
2. The cleaned haemocytometer and cover slip are put in 70% alcohol in a small beaker. They are dried with clean tissue paper just before use.
3. The cell suspension to be counted is taken up in a sterile Pasteur pipette and a small drop is transferred to the loading groove on the edge of haemocytometer counting chamber (Fig. E.6). The coverslip is placed over the drop, taking care that no air bubble in introduced in between.
4. The drop spreads evenly under the cover slip by capillary action.
5. There are two counting chambers on a haemocytometer, opposite to each other. The second counting chamber may also be filled with the same cell suspension for more counting of the same sample to get better statistical count or it may be filled with another sample with another fresh and sterilized Pasteur pipette.

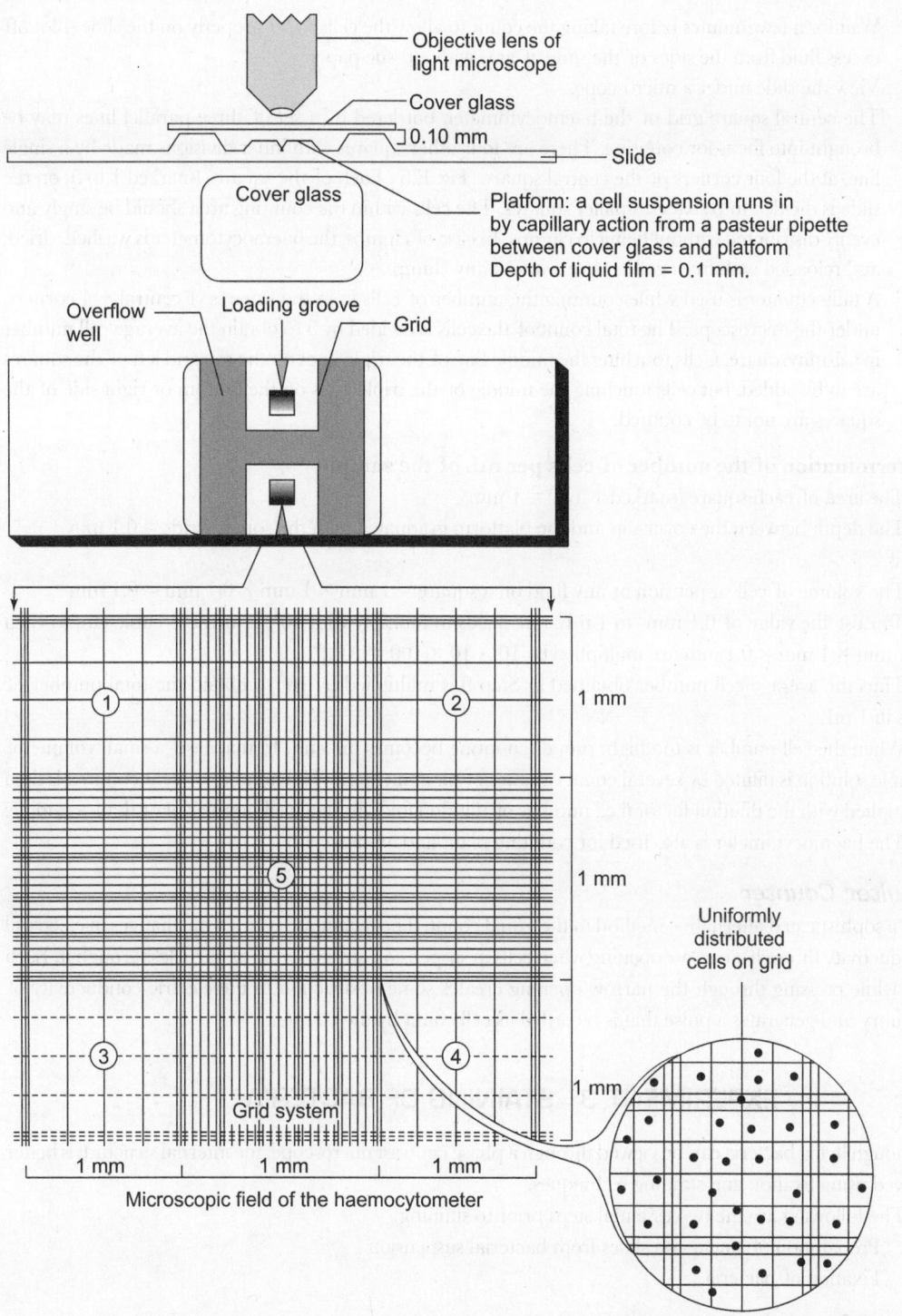

Fig. E.6 Haemocytometer counting chambers.

6. Wait for a few minutes before taking the count to allow the cells to set properly on the slide. Blot off excess fluid from the sides of the slide, if any, with a tissue paper.

7. View the slide under a microscope.

8. The central square grid of the haemocytometer, bordered by a set of three parallel lines may be brought into focus for counting. There are four other squares with inner divisions made by a single line, at the four corners of the central square (Fig. E.6). Each of the squares (marked 1 to 5) on the slide is divided into sixteen smaller squares. The cells within the counting area should be singly and evenly distributed without being in clumps. In case of clumps, the haemocytometer is washed, dried, and reloaded with fresh cell sample without any clump.

9. A tally counter is used while counting the number of cells in in the squares (1 central + 4 corners) under the microscope. The total count of the cells is divided by 5 to obtain the average cell number in a 1 mm square. Cells touching the middle line of the triple lines on the top and left of the squares are to be added, but cells touching the middle of the triple lines on the bottom or right side of the squares are not to be counted.

Determination of the number of cells per mL of the sample

The area of each square (marked 1 to 5) = 1 mm^2

The depth between the cover slip and the platform graduated with the square grids = 0.1 mm

Thus,

The volume of cell suspension or any fluid on a square = 1 mm × 1 mm × 0.1 mm = 0.1 mm^3

To raise the value of 0.1 mm^3 to 1 mL, one needs to multiply with 10^4. [1 mL is 1 cubic cm, so each of 1 mm × 1 mm × 0.1 mm are multiplied by 10 × 10 × 100 = 10^4.]

Thus the average cell number obtained in Step 9 is multiplied by 10^4 to obtain the total number of cells in 1 mL.

When the cell number is too high, proper counting becomes difficult. In that case, a small volume of sample solution is diluted by several equal volumes of plain medium. The number of cells counted is then multiplied with the dilution factor (i.e., number of dilution fold) for final cell number per mL of a sample.

The haemocytometer is also used for counting plant and animal cells.

Coulter Counter

It is a sophisticated alternative method to the visual count. The technique is based on changes in electrical conductivity through a narrow opening when cells in suspension are sucked in as particles through it. Each cell while crossing through the narrow opening creates some obstruction to the electric conductivity of circuitry and generates a pulse that is recorded as cell count by the counter.

EXPERIMENT 3   STAINING OF BACTERIA

Although living bacteria can be viewed through a phase contrast microscope, the internal structure is better viewed using fixation and staining techniques.

The following are the two essential steps prior to staining:

(i) Preparation of smears on slides from bacterial suspension

(ii) Fixation of bacteria

Materials

(i) Bacteria   :   From 24 hours nutrient agar slant cultures in test tube or broth culture

(ii) Staining reagents   :   Methylene blue, crystal violet, and carbol fuchsin

(iii) Equipment   :   Microscope, glass slides, inoculating wire loop, staining tray, tissue paper, lens paper to clean the objectives lens of microscope, Bunsen burner, forceps.

Procedure

1. The glass slides are cleaned with detergent, followed by cleaning in water. They are then soaked in 95% alcohol, dried, and placed on laboratory towel.

2. Preparation of thin smear of bacteria from broth or agar culture on glass slide:

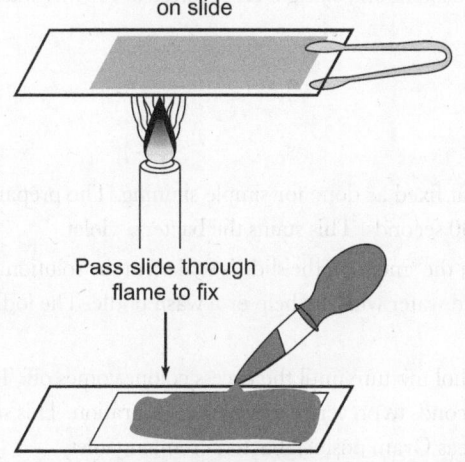

From *broth culture*, one or two loopfuls of suspended bacterial cells are taken on a slide and spread evenly with the inoculating loop over an area of about 1.5 cm diameter.

From *agar culture*, a loopful of bacteria is diluted on the slide with a loopful of water kept on the slide. By circular motion of the loop the diluted suspension of cells is spread over an area of about 2 cm diameter.

A good smear preparation should appear as a semi-transparent, confluent, whitish film. The smear is completely air dried.

Fig. E.7 Fixation of air-dried smear of bacteria over flame, followed by staining.

3. The air-dried smear is heat fixed by holding the slide with forceps and passing the slide horizontally just over the yellow flame of a Bunsen burner two or three times in quick successions (Fig. E.7).

By heat fixation, the bacterial proteins are coagulated and fixed to the surface of the glass slide, and does not allow the bacteria to be washed away during the staining procedure.

4. Staining can be of two types: (a) simple staining—involving a single stain and (b) differential staining—using two contrasting stains.

Simple Staining

Basic stains with a positively charged chromogen are usually used for staining bacterial cells as they bind better to the negatively charged nucleic acids and cell wall components of bacteria. Simple staining reveals the morphology and arrangement of bacterial cells.

Procedure

1. The slide with heat-fixed bacterial smear is placed in the staining tray and flooded with any one of the following simple stains:

   Methylene blue—for 1 to 2 minutes

   Crystal violet—20 to 60 seconds

   Carbol fuchsin—15 to 30 seconds

2. The excess stain is then removed by washing with tap water. The slide should be held parallel to the stream of water.

3. The slide is then blotted with tissue paper without rubbing the smear.

4. To visualize the stained bacteria, the slide is examined under microscope using oil immersion magnification.

5. The outline and any other structural details of bacterial cells are drawn in the laboratory note book.

Differential Staining

Materials

(i) Gram stain:0.5% aqueous crystal violet

(ii) Mordant: (Lugol's iodine)/Gram's iodine—0.33 gm iodine and 0.66 gm KI dissolved in 100 mL distilled water

(iii) Decolouriser:acetone:alcohol 50:50

(iv) Counterstain:Safranin 1% aqueous

Procedure

1. A bacterial smear is prepared on a slide and heat fixed as done for simple staining. The prepared slide is then flooded with crystal violet stain for 30 seconds. This stains the bacteria violet.

2. The slide is washed with iodine solution such that the smear on the slide is soaked in the solution for 30 seconds. Then the slide is washed with distilled water with the help of a wash bottle. The iodine fixes the stain more permanently.

3. The slide is then flooded with the acetone : alcohol mixture until the excess colour comes off. The slide should be washed with water only for few seconds to prevent excessive decolouration. This step decolourizes only Gram negative bacteria, whereas Gram positive bacteria remain violet.

4. Safranin solution is then added on the slide and kept for 1 minute. Safranin counterstains Gram-negative bacteria to red. The stain is washed off with water. The slide is dried gently in between blotting paper sheets and finally air dried.

5. A drop of immersion oil is added on the slide to examine under oil immersion 100 objective lens.

Observation

Gram positive bacteria appears violet.

Gram negative bacteria appears red.

NEGATIVE STAINING (WITH NIGROSIN OR INDIA INK)

Some bacteria are difficult to stain. For such bacteria, acidic stain, such as India ink or nigrosin, is used that imparts usually black colour to the background against which the unstained bacteria are easily visible.

The acidic stain with its negatively charged chromogen does not enter into the cells as the bacterial surface itself bears negative charge.

The negative staining not only helps in identifying non-stainable bacteria, but also it does not affect the size and shape of the bacteria as these may change in course of heat fixation and other chemical treatments.

Procedure

1. A drop of nigrosin is placed close to one end of a clean slide (Fig. E.8).

2. A loopful of bacteria is placed in the drop of nigrosin and mixed well.

3. Another clean slide is taken and its smaller edge is placed in front of the bacteria and nigrosin mixture at a 45° angle; the mixture spreads along the edge of the second slide. The second slide is then moved away from the mixture towards the other end of the first slide making a thin smear of the mixture (as it is done at the time of preparing blood film).
4. The smear on the slide is air dried (should not be heat fixed).

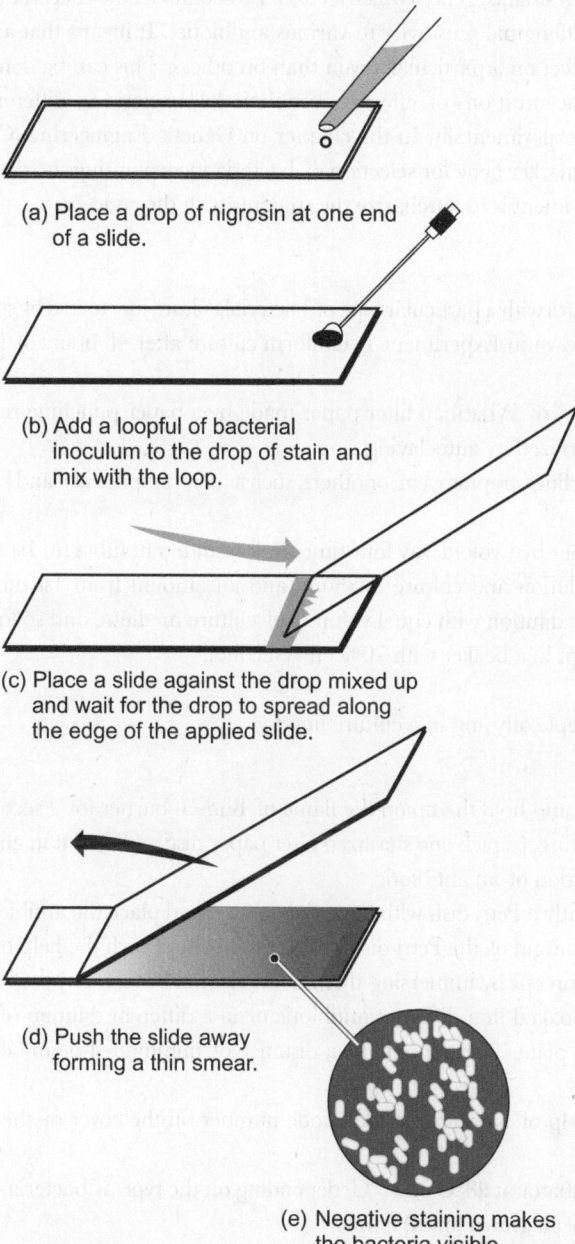

(a) Place a drop of nigrosin at one end of a slide.

(b) Add a loopful of bacterial inoculum to the drop of stain and mix with the loop.

(c) Place a slide against the drop mixed up and wait for the drop to spread along the edge of the applied slide.

(d) Push the slide away forming a thin smear.

(e) Negative staining makes the bacteria visible.

Fig. E.8   Negative staining procedure.

5. The slide is then examined under oil immersion.

6. Draw bacteria as observed under the microscope in the laboratory note book.

EXPERIMENT 4   STUDY OF SENSITIVITY OF BACTERIA TO ANTIBIOTICS

Bacterial genome harbours antibiotic resistance gene(s) which let bacteria counteract the effects of antibiotics. Different strains of bacteria have differential sensitivity to various antibiotics. It means that a particular antibiotic can have a more lethal effect on a particular strain than on others. This can be demonstrated experimentally. Again, different concentrations of effective antibiotic kill bacteria to different degrees and this can also be demonstrated experimentally. In the chapter on Genetic Engineering (Chapter 6), the use of antibiotic sensitivity as a marker gene for selection of bacteria incorporating the desired DNA was elaborated. The following experiment is to familiarize the students with the concept.

Materials

(i) 9 cm-diameter Petri culture plates with a particular type of bacterial culture inoculated by spread plate or pour plate technique, as shown in Experiment 1. Uniform culture after 48 hours of incubation is suitable.

(ii) Small disc (4 – 5 mm diameter) of Whatman filter paper made by a paper punching machine (or from commercial supplier) sterilized by autoclaving.

(iii) Antibiotics: ampicillin, tetracycline, streptomycin (or others, such as chloramphenicol and kanamycin) in solution.

Each antibiotic is diluted further by twofold, say four times in four different tubes (in 1st tube equal volume of antibiotic stock solution and culture medium, and an amount from 1st tube is to be transferred to the 2nd tube for dilution with equal volume of culture medium, and so forth).

(iv) Forceps cleaned, sterilized, kept in a beaker with 70% ethyl alcohol.

(v) Bunsen burner.

(vi) Everything should be done aseptically and in a culture hood.

Procedure

1. Take the forceps from alcohol and hold the tip on the flame of Bunsen burner for 2 seconds. Then with the help of the sterilized forceps pick one sterilized filter paper disc and soak it in an antibiotic solution or in a particular dilution of an antibiotic.

2. In the culture hood, open slightly a Petri dish with bacterial culture, and place the antibiotic soaked disc at least 1 cm away from margin of the Petri dish on the agar culture with the help of forceps.

3. Close the dish and clean the forceps by immersing them in an alcohol beaker. Repeat Steps 1 and 2 to place another filter disc soaked in a different antibiotic or in a different dilution of the same antibiotic on the same culture plate. There should be a distance of minimum 1.5 cm between two discs.

4. Seal the Petri dish with the help of parafilm. Mark a code number on the cover of the Petri dish with a marker pen.

5. Place the Petri dish in the incubator at 28°C or 37°C, depending on the type of bacteria, i.e., at the temperature the culture had been grown for 48 hours.

Observation

After 24 hours, observe the bacterial culture plate with the antibiotic soaked discs. Usually, the normal bacterial culture plate looks opaque. A clear zone surrounding a disc indicates lysis of bacteria due to presence of an antibiotic. The diameter of the circular lysis zone increases with the effectiveness of the antibiotic (or its dilution) towards the bacteria (Fig. E.9).

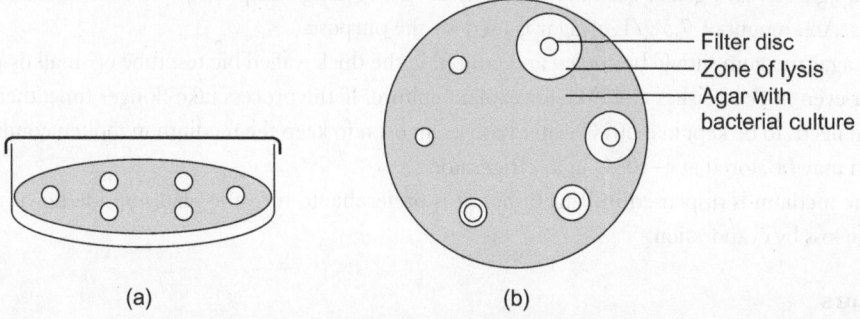

(a)   (b)

Fig. E.9   Test for antibiotic sensitivity of bacteria. (a) Small filter paper discs soaked with different concentrations of an antibiotic or different types of antibiotics placed at certain distance on a bacterial culture (apparently opaque in look). (b) After a day a clear circular zone around a disc develops due to lysis or death of bacteria sensitive to the antibiotic. The different sizes (diameter) of clear zones are proportional to the concentration of an antibiotic (e.g. first 4 clear zones surrounding the discs clockwise) or due to effectiveness of different antibiotics.

EXPERIMENT 5   PLANT TISSUE CULTURE WITH EXPLANTS

Plant tissues from different parts and organs excised from plants are called **explants**. They can be grown in-vitro on a suitably prepared artificial nutrient medium or culture medium (see Ch. 11) in a sterile and controlled environment.

Maintenance of Sterile Conditions

Sterility of the explants, culture media, culture room is an absolute necessity for any kind of in-vitro culture of living tissue. Otherwise, contamination with all kinds of undesired microorganisms spoils the effort. Basic steps for maintenance of sterility in course of plant and animal tissue cultures have already been discussed in the respective chapters (see Chapters 11 & 12).

The equipment—forceps, stainless steel knife with disposable blade, Petri dishes, culture tubes, flasks, paper towels and anything likely to be in contact with the explants should be pre-sterilized by cleaning, heating, and autoclaving.

The medium should be sterilized by autoclaving at 120° C at 15 psi for 15 to 20 minutes. Allow a vacant air space at the top of the medium in the flask to prevent burst due to pressure and spilling.

The culture hood should be wiped with a disinfectant such as 10% bleach, 70% alcohol, or lysol or some other household disinfectant, on daily basis, before and after each use.

Preparation of Medium

A requisite medium as dry powder containing inorganic salts, vitamins, and amino acids are nowadays available from commercial laboratory suppliers. Otherwise the medium needs to be prepared following appropriate composition list (see Chapter 11); the components are dissolved in double distilled water, the pH of the medium is adjusted to be at neutral, if necessary, by adding dropwise either 0.2 M KOH/0.2 N NaOH or 0.5 N HCl.

Agar, agarose, and gellan gum are normally used as gelling compounds for the culture medium of explants. An amount of 7.5 g/L of agar is used for the purpose.

The agar medium should be cooled and poured in the thick walled big test tube or small tissue culture flasks or even in Petri dishes at 45° C, for explant culture. If the process takes longer time, then the agar medium needs to be kept in a 50° C water bath or an oven to keep the medium in molten condition. The medium may be stored at 4–10° C in a refrigerator.

If the medium is dispensed in Petri dishes, it is preferable to turn the plates upside down to prevent moisture loss by evaporation.

Explants

Small explants from different kinds of plant tissue are the starting material. Explants can be obtained from any part of plant meristems (at growing points), shoot tips, macerated stem segments, nodes, buds, flowers, peduncle (flower stalk), petals, anthers, portions of leaf or petiole, seeds, central part of an ovule (nucellus tissue), embryos, seedlings, hypocotyls, stolons, rhizome tips, root pieces, etc. Meristems are usually free of virus infection as they do not have vascular tissue that transports viruses.

Size of the explants varies from microscopic 0.1 mm to several centimetres of stem pieces.

The stock plant for explants should be maintained in a clean area or greenhouse. The plant is to be washed with clean water to begin with, then watered at the base. The new shoots are supposed to provide clean and juvenile explants. The more juvenile is the explant material, the greater the chance of better tissue culture.

Procedure

Disinfecting the explants

1. Several pieces of explants taken out of a plant are to be put into a flask or a jar with disinfectant or bleach (5% sodium hypochlorite/70% ethyl alcohol/70% isopropyl alcohol/ 5–10% calcium hypochlorite / 3% hydrogen peroxide / 0.1–0.2% mercuric chloride). The materials in the flask may be stirred for 10 minutes with a magnetic stirrer, or after closing the jar tightly the materials with bleach may be shaken by hand (Fig. E.10).
2. The bleach solution is then poured off and the explants are transferred to a fresh beaker and washed with 70% isopropyl alcohol or 3% hydrogen peroxide for 5 seconds. After pouring off the alcohol, the explants are placed in a clean beaker.
3. The explants are rinsed with distilled water and transferred into a 150 mL beaker containing bleach and 2 drops of Tween 20.
4. The beaker with the explants is kept in sterile culture hood.
5. A sterile paper towel is spread with sterile forceps on the counter of the hood. (The towel is to be replaced with a new one after preparing a number of explants.)

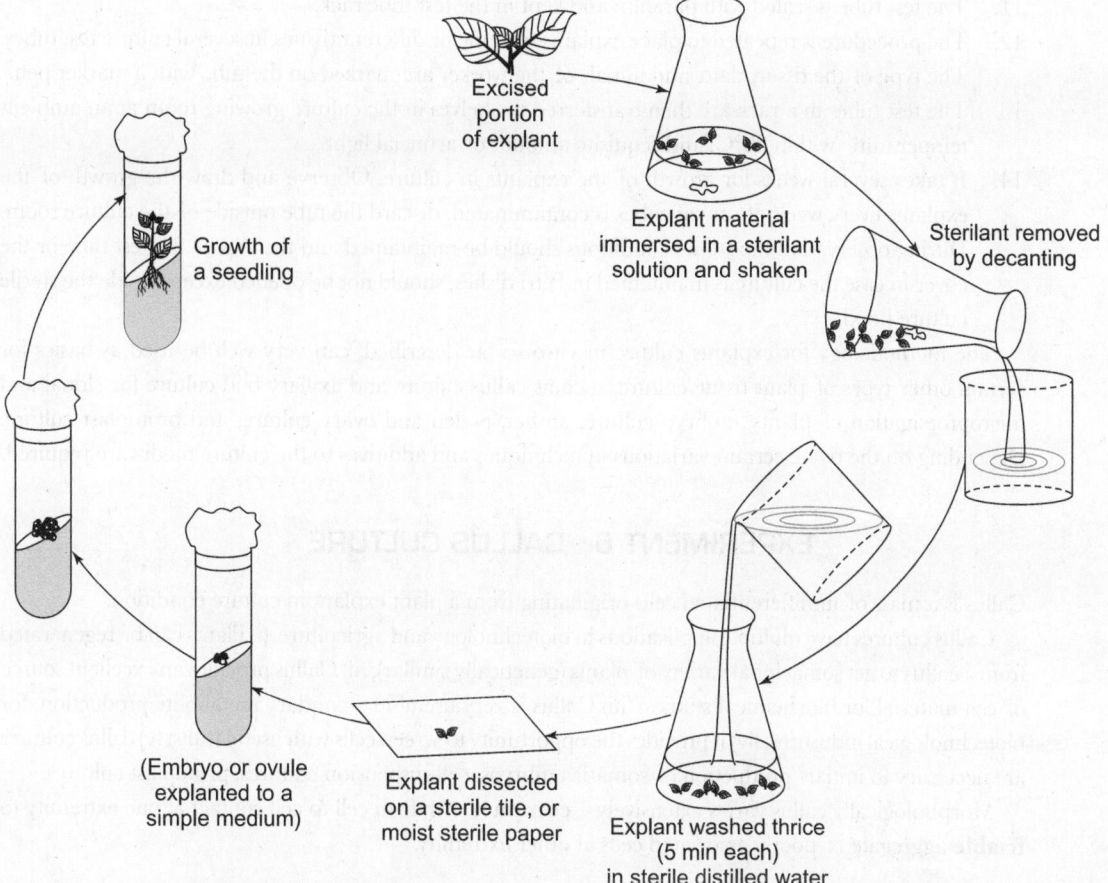

Fig. E.10 Technique of disinfecting an explant from terminal part of a plant and starting tissue culture.

Labels in figure:
- Excised portion of explant
- Explant material immersed in a sterilant solution and shaken
- Sterilant removed by decanting
- Explant washed thrice (5 min each) in sterile distilled water
- Explant dissected on a sterile tile, or moist sterile paper
- (Embryo or ovule explanted to a simple medium)
- Growth of a seedling

6. An explant from the beaker is placed on the paper towel with sterile forceps. Holding the forceps by left hand and knife in right hand, the explant is cut into requisite sizes and trimmed. For shoot tips, from 1 to several mm, for stem 2 to 3 cm, a few mm for petiole pieces are usually right sizes for explants.

7. A test tube containing the culture medium for the explants is picked up and held at its lower half parallel with the frontline of the hood at an angle of 50°. This is done so that the mouth of the test tube remains away from incoming flow of air from the back of the hood and expiration of the worker. Moreover, the 50° angle lessens the chance of settling of falling microbes (whichever minimum may be) in the test tube.

8. Take the forceps with right hand and employ the little finger of the right hand to unscrew the cap of the test tube held by the left hand.

9. An explant from the sterile paper towel is picked up using forceps and placed firmly on the agar medium in the test tube.

10. Still holding the forceps, the cap is replaced on the test tube. The forceps is then returned to the bleach.

11. The test tube is sealed with parafilm and kept in the test tube rack.

12. The procedure is repeated to place explants of same or different tissues in several culture test tubes. The type of the tissue, date, and initials of the worker are marked on the tube with a marker pen.

13. The test tubes in a rack are then transferred to shelves in the culture growing room at an ambient temperature (within 28°C) and requisite amount of artificial light.

14. It takes several weeks for growth of the explants in culture. Observe and draw the growth of the explants every week. If any of tubes is contaminated, discard the tube outside of the culture room. During observation the sterile conditions should be maintained and the cap of the test tube or the cover in case the culture is maintained in Petri dishes, should not be opened except inside the sterile culture hood.

The methodology for explants culture in vitro so far described, can very well be used as basics for certain other types of plant tissue culture, such as callus culture and axillary bud culture for clonal and micropropagation of plants, embryo culture, anther, pollen and ovary culture, and protoplast culture. Depending on the tissue certain variations in techniques and additives to the culture media are required.

EXPERIMENT 6   CALLUS CULTURE

Callus is a mass of undifferentiated cells originating from a plant explant in culture condition.

Callus cultures have multiple applications in biotechnology and agriculture. (i) Plants can be regenerated from a callus to get somaclonal variety of plants (genetically similar); (ii) Callus provides an excellent source of cell material for biochemical studies; (iii) Callus is very useful in secondary metabolite production, for biotechnological industries; (iv) it provides the opportunity to screen cells with useful traits; (v) callus cultures are necessary to initiate production of somatic embryos, cell suspension cultures, protoplast cultures.

Morphologically callus varies extensively—**compact** with firm cell to cell contact at one extremity to **friable** aggregate of poorly associated cells at other extremity.

Equipment and Materials for Callus Culture

Media preparation, glassware sterilization, and maintenance of sterile conditions, use of culture hood for callus culture are similar to that used for plant tissue (explants) culture as outlined in the earlier experiment (Experiment 5).

Essentially all organs and tissues of plants can be used as explant sources for callus cultures. In this experiment carrot taproot is used as the source material.

Procedure

1. A healthy undamaged carrot of 3 cm or more in diameter at the widest part of the taproot is selected, thoroughly cleaned under running tap water by scrubbing with a small brush (nylon nail brush).

2. The carrot is placed in a 1,000 mL beaker to submerge with the solution of sodium hypochlorite for 30 minutes.

3. The sterilized carrot in sodium hypochlorite solution is then transferred to the sterile room. After decanting off the solution, the carrot is washed thrice with sterilized distilled water, agitating the beaker with hands to completely remove the hypochlorite. The excess water on the carrot is dried by sterile tissue paper.

4. The uppermost 1 cm of the root (including the shoot portion) is removed by a sharp sterilized knife. The portion of the tap root at the distal end with less than 2 cm diameter is also removed.

5. The remaining portion of the carrot is removed aseptically with a large forcep on to a sheet of sterile filter paper or a sterile tile in the hood.

6. 1 cm-thick slices from both the ends is then removed with a sterile sharp knife.

7. The remainder is cut into as many 1 mm-thick transverse slices as possible (Fig. E.11). The slices are kept in a Petri dish with a filter paper moistened with sterile distilled water (not shown in Fig. E.11).

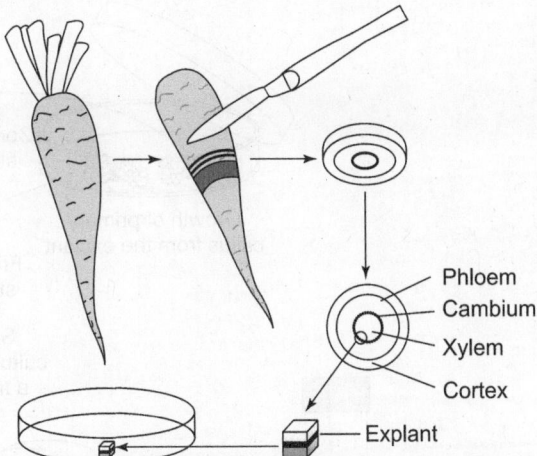

Fig. E.11   Collection of explants from taproot of carrot to start a callus culture.

8. A single slice is placed on a sterile tile and 8 × 8 mm explants are then cut out as shown in the diagram. Each explant should contain a segment of the cambial ring and some tissues on both sides with xylem and phloem. Two to three such explants may be taken from a slice.

9. About 5 such explants are placed aseptically in a 9 cm diameter Petri dish containing 20 mL of agar culture medium. It is preferable to maintain the polarity of the explants by placing the root pole downward in contact with the culture medium, at the time of the transfer. The dishes are then sealed with parafilm and incubated at 25° C under low intensity light.

10. The Petri dishes with explant culture are observed next day and at a few days intervals. If any culture shows signs of contamination, it should be discarded promptly. Usually the callus formation is visible within seven days. Draw the growth pattern of the calluses.

11. After three weeks, properly grown calluses are taken to cut out 5 × 5 × 2 mm pieces, five or six of such pieces are placed in a new petri dish with fresh medium. Petri dishes are sealed along the circumference with parafilm for further incubation and subculture (Fig. E.12).

12. Friable callus develops in three weeks in 4–5 weeks it develops better. The most friable callus may also be taken separately for sub-culturing or as source material for suspension cell culture (Fig. E.12).

13. The repeated subcultures maintain the callus for longer period.

Regeneration of Somaclonal Plants from a Callus

Parts of callus can be grown to a complete plant in the culture, due to inherent totipotency of plant cells. The regenerated plants from a callus are used for micropropagation as somaclonal variety (genetically similar). After regeneration of these plants in culture medium, the plantlets are transferred from the in-vitro to the ex-vitro external environment of a glasshouse. Once the plantlets acclimatize to the environment of the glasshouse and show signs of growth, the shoots as micro-cuttings or rooted plants as such can be transferred to soil. Precautions to avoid stresses to the plants during the transfers from one environment to other are the basic of successful micropropagation.

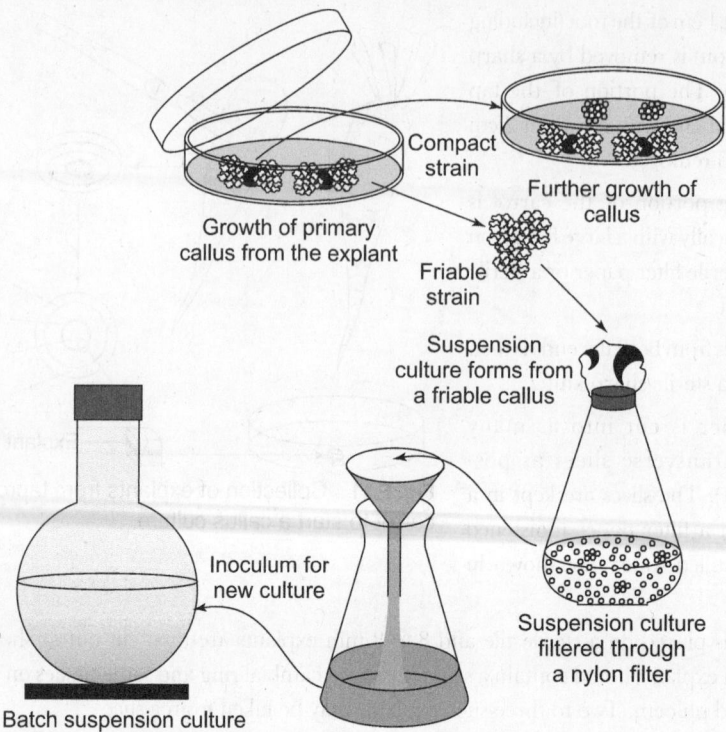

Fig. E.12 Initiation of callus culture from plant explants and suspension culture from friable callus.

EXPERIMENT 7 SUSPENSION CULTURE FROM CALLUS

The generation of suspension cell culture from 'friable' callus has already been mentioned in Experiment 6.

Suspension culture of plant cells in liquid medium in a flask is continuously agitated on a rotary shaker. This agitation of cells eliminates certain disadvantages of the tissue culture on agar plates. The movement of the cells facilitates gaseous exchange, eliminates gravity-based polarity effect to the cells, and also removes nutrient gradients within the medium and at the surface of the cells. Although suspension of single cells is theoretically ideal, some degree of aggregation of cells is found in the suspension culture. About 5–200 cells can be found in the aggregates. Interestingly, cell aggregation is often beneficial for maintaining totipotency and enhanced production of secondary metabolites for biotechnological purposes.

Cells grown in a suspension culture usually assume the characteristics of parenchymatous cells with large vacuoles, thin layer of cytoplasm, and thin and round cell wall. These features may differ depending on the plant type and culture conditions. The cells in suspension culture manifest the four phases of growth —lag, log, stationary and decline or death phases.

Procedure

The procedure is summarily presented in Fig. E.12.

1. Friable callus is the starting material. About 2 gm of the peripheral material from friable callus is aseptically transferred with a forceps or a long spatula into the conical flask with liquid culture medium.

2. The transfer from Petri dish to the culture flasks and other steps are performed in the sterile environment of laminar flow hood. The aluminum foil cap from the culture flask is partially opened and the neck of the flask is passed over the flame before the transfer for maintenance of sterility. Then the flask is closed with aluminum foil or sterilized cotton plug.

3. The flask is firmly secured on a rotary shaker under low intensity light.

4. After 10 days, the suspension culture is filtered through a sterile 250 μm nylon/stainless steel mesh. The filtrate is collected in a sterile flask.

5. Wait 10 minutes to allow the cells to settle at the bottom of the flask. The medium from the top is pipetted off as much as possible. The cells at the bottom are taken by a pipette to inoculate fresh culture in a round bottomed large flask.

6. The process of reinoculation should be performed after every 10 days.

Observation

At the time of inoculation take a few drops of cell suspension on slides and a haemocytometer for microscopic observation of cells, cell aggregates, and cell count in per mL. Draw the cells in a note book.

EXPERIMENT 8 PROTOPLAST CULTURE

Protoplast is a plant cell from which cell wall has been removed. Normally the cell wall confines the plant cell protoplast, protects against turgor pressure that may burst the cell. The cell wall becomes a formidable barrier to many biotechnological manipulations. Freed from the confines of cellulosic cell wall, the plasma membrane becomes accessible for investigation or purification, and the plant cells can be manipulated by microinjection, electroporation, and fusion. Furthermore, the protoplasts can be easily transformed by recombinant DNA technology for biotechnological products.

The first step in protoplast culture technique is to isolate protoplasts in good state, free from contaminating debris.

The plant tissue is placed in a high osmotic solution that causes shrinkage of cells within the cell wall. Then, the cell walls are broken mechanically by abrasion. In next step, the osmotic strength of the medium is reduced, the cells swell up and pop out of the broken cell walls. The yield of viable protoplasts following this protocol is low. Modern technology uses enzymatic digestion of cell walls to obtain protoplasts from plant cells.

Materials

(i) Enzyme solution: Mixture of 1% (w/v) macerozyme R-10 and 1% cellulase R-10 in 0.4 M mannitol (72.8 g/L), sterilized by passing through 0.22 μm porosity filter.

Macerozyme helps in loosening the cells from tissue organization. Cellulase breaks the cell walls. Mannitol (or sorbitol) as sugar alcohol is metabolically inert and infuses slowly into protoplasts and protects them from osmotic shock (maintains osmotic balance of protoplasts as osmoticum).

Salts for Ca^{2+} and Mg^{2+} ions may also be included in the solution to improve the stability of the protoplasts. Similarly, potassium dextran sulphate inclusion may help in the adsorption of phenols.

Some other enzymes, such as hemicellulase H-2125, Meicelase–P, pectinase, and zymolase, may also be included in enzyme solution for digesting different components in plant tissue and cell walls for obtaining better-quality protoplasts.

All the enzymes and other ingredients are commercially available.

(ii) Washing solution: W5 salt solution: 154 mM NaCl; 125 mM $CaCl_2$. $2H_2O$; 5 mM glucose; pH 5.85-6.0

(iii) Equipment: Sterilized forceps, scalpels, blades, 100 mL beaker, capped plastic centrifuge tubes, pipettes (1 mL, 5 mL, 10 mL); 6 cm and 9 cm Petri dishes; funnel fitted with 64 μm pore size filter and Pasteur pipettes. Haemocytometer for cell and protoplast counting; parafilm for sealing; Incubator.

(iv) Plant materials as source for protoplasts: Plant cells from three sources can be used as starting material:
  (i)   Plant tissues; here leaf is to be considered
  (ii)   Compact callus grown on solid agar medium
  (iii) Cell suspension culture which may derive from friable callus as shown in Experiment 7.

Procedure
Preparation of leaves as explants (Fig. E.13)

1. Fully expanded healthy leaves that are seven to eight days old are isolated from a plant and surface sterilization is done in 7% w/v of saturated calcium hypochlorite solution for 10 minutes. The leaves are then washed with sterile distilled water thrice for 5 minutes each.

2. A big sterile Petri dish may be used for preparing the sterile leaves for enzymatic digestion. First the lower epidermis of the leaves is removed with a pair of fine forceps and then cut into small pieces. Removal of the lower epidermis allows easy penetration of the enzymes in cell layers of the leaves.

3. 10 mL enzyme solution is added to the leaves in a 9 cm Petri dish and incubated at 20°C for overnight. The hours of treatment with enzyme solution may need to be adjusted with plant materials from different species and for cells obtained from culture, so that the protoplasts are not damaged. The addition of MES (2-N-morpholinoethanesulfonic acid) buffer at 1–5 mM, pH 5.8 helps in stabilizing protoplasts during isolation, washing, and fusion steps.

    All these steps should be carried out in extreme sterile conditions so that no contamination of bacteria and fungal spores take place.

Isolation of viable protoplasts

4. Next day the Petri dish is gently shaken to free the protoplasts from the bottom and the content of the dish (the protoplast, enzyme solution, and undigested tissues) is transferred with a sterile Pasteur pipette onto the sterile funnel fitted inside with a sterile 64 μm filter. The protoplasts are supposed to be in the filtrate. Care should be taken not to damage the fragile protoplasts at this stage during pipetting, washing, and centrifugation. Nothing should be done vigorously.

5. The filtrate is to be transferred to 15 mL round bottomed centrifuge tubes with cap and spun at 500*g* (or 700 rpm) for 5 minutes. The supernatant, mainly the enzyme solution, is removed with a sterile Pasteur pipette.

6. Add 5 mL of 0.4 M sucrose solution to the pellet and gently shake the test tube for dispersing and suspending of the protoplasts.

7. Completely dispersed protoplasts are layered with a sterile Pasteur pipette over the 0.6 M sucrose solution in a centrifuge tube and centrifuged at 500*g* for 6 minutes.

8. The band of floating protoplasts is removed with a sterile Pasteur pipette to fresh sterile tube. It is easier to suck mixture of air and protoplast suspension at the time of transfer by the pipette.

9. W5 salt solution or liquid culture medium is added to the protoplasts to four to five times volume.

10. Centrifuge at 500*g* (700 rpm) for 5 minutes. The supernatant is then removed with a pipette. (It is better to attach the sterile pipette with a tubing to a vacuum pump working at medium suction to remove the supernatant). Shake the pellet gently and resuspend it in 5–10 mL liquid culture medium.

11. The number of the protoplasts per mL is counted by haemocytometer and the final volume of the suspension should be adjusted such that the density of protoplasts becomes 10^5 per mL in culture solution.

    Observe the nature and size of the protoplasts under the microscope and note them in laboratory note book.

Protoplast culture

12. The protoplasts may be cultured by the following three protocols:

    (i) In microdroplets of 10–25 μL each, spotted in a Petri dish.

    (ii) 10 mL of liquid culture medium containing 5×10^4 protoplasts per mL is transferred into a 9 cm Petri dish. Incubate for 72 hours in the dark, then in low light/dark alternate cycles. Alternatively 1.5 mL protoplast suspension as thin layer may be cultured in a 6 cm Petri dish.

    (iii) A thin liquid layer of 1.5 mL protoplast suspension is layered on an agarose plate. Usually 0.6g agarose powder, autoclaved earlier is dissolved in sterile warm medium to prepare the agarose to lay in the Petri dishes. The temperature of the agar at the time of embedding with the protoplasts should be below 37°C.

Observation

Under microscope

(i)   Cell walls regenerate surrounding the protoplasts within two to four days in culture.

(ii) Note the increment of the size of the protoplasts.

(iii) After 72 hours of incubation find the dividing cells. The seven to eight days, the cells undergo four or five division cycle. So the cell density will increase.

(iv) Experiments may be designed at this stage for testing or selecting protoplast cultures resistant to antibiotics, herbicides, etc., especially when genetic transformations have been induced.

Macroscopic

(v) If the protoplast culture is continued upto three weeks, small colonies of cells will be visible in culture to the naked eyes.

(vi) Microcalluses of 2–3 mm formed out of the protoplast culture may be transferred with a wide-mouth Pasteur pipette on top of a fresh agar plate. Soon the calluses will increase in size and regenerate multiple shoot structures (Fig. E.13).

EXPERIMENT 9   PREPARATION OF SPLEEN CELL SUSPENSION

Procedure

1. Kill the mice by cervical dislocation or by chloroform soaked in a cotton ball in a 250 mL beaker covered with a lid of a Petri dish.

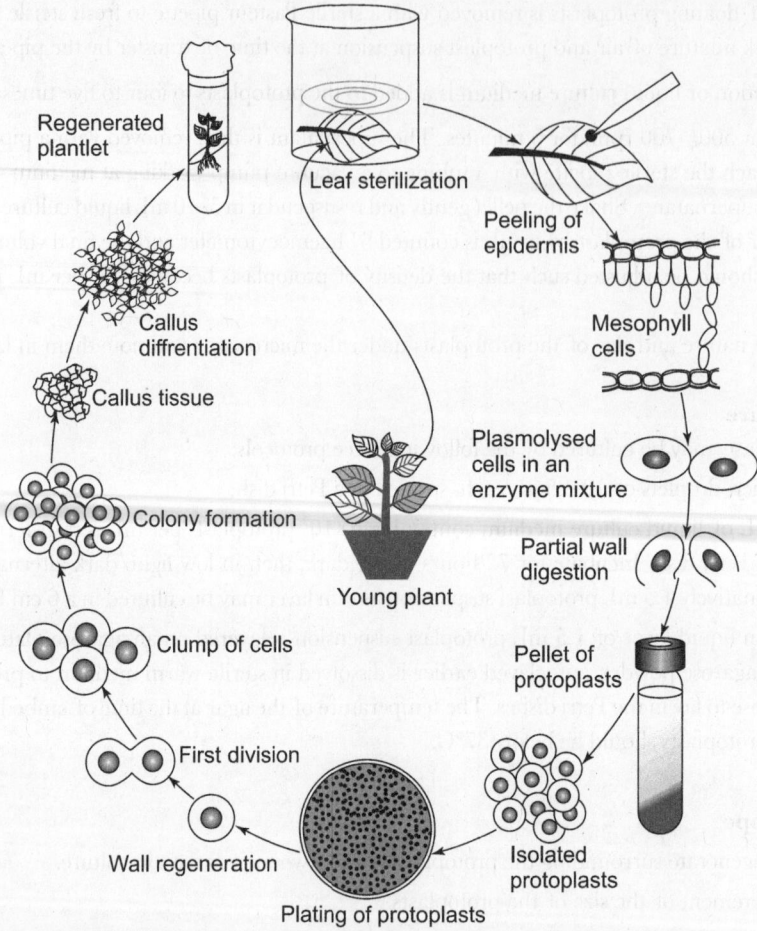

Fig. E.13   Procedure for isolation and culture of protoplasts from leaf cells.

2. Open the abdomen.

3. Lift the spleen with a pair of forceps and separate it from the vessels and the membrane with the fine scissors (Fig. E.14).

4. Put the spleen in a Petri dish containing 5 mL of PBS. Inject a small volume of PBS into the spleen with a 27 G needle on a 1 mL syringe for perfusion and loosening the cells within.

5. Dissociate the cells mechanically by scraping the spleen against a stainless steel fine wire mesh on a coarse net, fitted together between the stainless rings of a tissue dissociator (Fig. E.15), cells may also be dissociated chemically.

6. Repeatedly take the cell suspension in the Petri dish in and out of the small syringe with a 27-gauge needle for dissociation of the small cellular clumps.

7. Wash the splenocytes in cold PBS. Keep the suspension in cold, sterilized Tris- NH_4Cl (0.84% pH 7.4) solution for 5 minutes to lyse the red blood cells completely.

8. Wash twice with PBS.

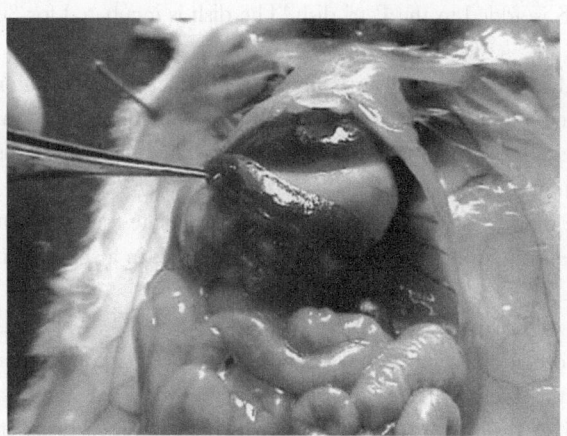

Fig. E.14 Spleen of a mouse being removed from the anatomical site underneath the left side of its stomach.

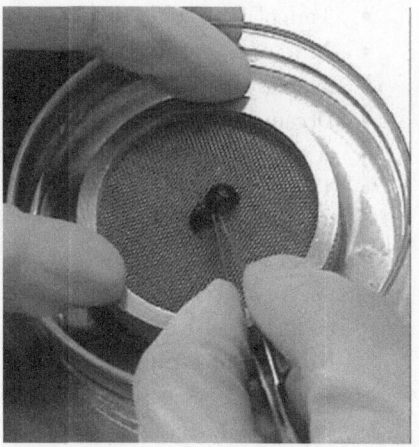

Fig. E.15 Cells being dissociated mechanically by scraping the spleen against the stainless steel wire mesh of a tissue dissociator.

9. Whenever necessary to remove the debris, layer the cell suspension on Ficoll and Hypaque solution and spin down at 3000 rpm for 5–8 minutes.

10. Collect the lymphocytes from the interface and wash twice with PBS.

11. Count the viable cells with the help of a haemocytometer in the presence of 1% trypan blue (dead cells become blue).

12. Adjust the cell number to 10^6 cells/mL.

[An appropriate culture solution like MEM (minimum essential medium) with 10% foetal calf or goat serum may be used instead of PBS for better survival of the lymphocytes over the hours.]

Composition of Tris-NH₄Cl solution   Tris-base 206 mg, NH_4Cl 770 mg. Make the total volume of the solution 100 mL with double distilled H_2O.

Chemical Dissociation of Cells

Incubate several small pieces of spleen in 5 mL of 0.25–1% Trypsin solution in culture medium in a small Petri dish at 37°C for 30 minutes. Dislodge the dissociated cells from the Petri dish by gently forcing the medium and suction using a Pasteur pipette. Collect the cells in a test tube, spin at 500 rpm for 5 minutes. Then wash twice with culture medium with serum. Then follow from Step 9 stated earlier.

Living cells   The cells are mostly WBC; they may be observed by phase contrast microscope.

Staining the cells   A thin smear of the cell suspension is made on a glass slide and air dried. Then stain with commercially supplied Giemsa. Extra stain is washed off, slide is dried, and viewed under the microscope. If necessary, oil immersion lens may be used for higher magnification

Study of phagocytic cells   5 mL cell suspension is plated in a 9 cm Petri dish and incubated in a incubator at 37°C for 45 minutes.

- The phagocytic cells, such as macrophages, neutrophils adhere to the base of the Petri dish.
- Lift the Petri dish with right hand and gently generate a swirl motion in the medium by twisting of the wrist; non-adherent cells get loosened from the substratum. The fluid is removed by a 10 mL pipette into a test tube.

- 3 mL of fresh culture fluid is added to the Petri dish and the previous step is repeated.
- 2 mL of 1% neutral red solution in PBS is added to the Petri dish. The dish is incubated for 30 minutes at 37°C.
- Decant off the neutral red solution from the Petri dish. Add 5 mL of chilled PBS (kept in refrigerator at around 4°C) to the Petri dish. The chilled solution sets the adherent cells free from the substratum.
- Take a drop of cell suspension to charge a haemocytometer and observe that the phagocytic cells have engulfed the neutral red in their cytoplasm.
- One can take a count of the cells in the marked field of haemocytometer, calculate the adherent phagocytic cells in the original cell suspension, and even the total number of cells in a spleen can also be calculated.

Number of lymphocytes and ratio of lymphocytes to the phagocytic cells   The non-adherent cells set aside in a test tube before adding neutral red solution to the Petri dish in earlier steps, mainly constitute lymphocytes (B and T cells). The test tube is spun at 500 rpm for 8 minutes, the pellet is diluted with 5 mL of PBS. A drop from this suspension is mixed with a drop of 1% Trypan Blue and charge into a chamber of haemocytometer for cell count in per mL.

This count will allow to determine the ratio of lymphocytes to phagocytes (adherent cells) in spleen. Phagocytic cells like macrophages, and neutrophils provide innate or natural immunity against foreign pathogens forms the first line of defence of the body. Lymphocytes (B and T cells) provide specific and adaptive immunity that is more effective and generates memory cells.

Separation of B and T cells   B and T cells can also be separated from lymphocyte population by nylon wool column. Nylon wool is commercially available and used to fill in a column made by a one or two mL syringe or even a straw used for cold drinks. The mixed population of lymphocytes in culture solution is loaded in the column suitably closing the down outlet and incubated for 45–60 minutes at 37°C. At the end of incubation further volume of culture solution is added and the lower outlet is opened on a test tube to collect the run out cells from the column. These cells are primarily T cells, i.e., nylon wool column non-adherent cell population. After running certain volume of culture solution, the column is flushed with chilled medium at 4° C. The nylon wool adhering B cell population will now loosen out from the nylon wool and is collected in a separate test tube.

Note:

1. Fluorescence-activated cell sorter (FACS) is used in recent years as sophisticated rapid technology for the separation of different sub-populations of lymphocytes. Separation is done on the basis of the size and fluorescence of individual cells, as they flow in a single file in a stream past a laser illuminator system coupled with a set of highly sensitive detectors.

2. Spleen cell preparation when performed under sterile condition aseptically with sterilized equipment, glasswares, and medium, can be used for cell culture and different studies including in-vitro sensitization with an antigen to raise antibody producing cells and testing different chemicals and drugs on the immunocompetent cells.

EXPERIMENT 10   BLASTOGENIC TRANSFORMATION OF LYMPHOCYTES IN CULTURE WITH CON A

When lymphocytes are stimulated or activated by an antigen or a mitogen (causing mitosis) the lymphocytes carry on with synthesis of multiple cellular substances including DNA and undergo transformation to blasts—enlarged cells, as preparation towards mitotic cycles. The whole process can be studied in in-vitro cell culture. In fact in-vitro studies established the processes of blastogenesis on the basis of cell size measurement; DNA, RNA, and protein synthesis by incorporation of radioactive precursors for these macromolecules; and many other functions of different cellular components.

The in-vitro study of blastogenesis will provide the students a firsthand knowledge of aseptic cell culture technique and its use in understanding cellular functions. Here, only the physical measurements of the blast cells and their percentage count have been taken into consideration. Study of DNA or other macromolecular syntheses can be performed in a sophisticated laboratory with better supply of chemicals, equipment, and facility for working with radioactive materials.

Concanavalin A (Con A) is a carbohydrate moiety derive from *Phaseolus* bean seeds and a mitogenic substance for lymphocytes, particularly mouse T lymphocytes. In culture all white blood cells, not just T lymphocytes, are used.

Equipment

(i)   Animal cell culture room with laminar flow hood

(ii) Autoclave

(iii) 5% CO_2 incubator at 37°C

(iv) Microscope with occulometer fitted with an eyepiece and a stage micrometer for measuring the diameter of the cells.

(v)   Centrifuge machine

(vi) Mono-pan or a good chemical balance

(vii) Eppendorf pipette of 50 μL, and other volumes

Materials

1″ diameter sterilized plastic Petri dishes (as such from supplier); sterilized plastic tray to hold the Petri dishes; 100 and 250 mL flasks for culture medium; sterilized 5 and 10 mL pipettes in a canister or individually packed; haemocytometer.

Procedure

(i)   Cell suspension from a mouse spleen aseptically* prepared in Eagles or MEM or other suitable animal cell culture medium supplemented with 10% faetal calf or goat serum and antibiotics[†], following the method outlined in Experiment 9, is the starting material. After counting the cells in haemocytometer, the final cell density is adjusted to 10^6 cells /mL in a 15 mL screw capped test tube. [*The dissection of the mouse is to be performed outside of the culture room. The skin on abdomen of the mouse is to be washed with 70% alcohol. The spleen should be dissected out with sterilized scissors and forceps. [†]**Antibiotics:** Penicillin 100 U/mL streptomycin 100 μg/mL, Nystatin (anti-fungal) 50 μg/mL.]

(ii) Take a 5 or 10 mL pipette out of the canister (the cap of the canister to be closed immediately) and the tip is passed over the flame of a burner. Next, with oral suction take the cell suspension out of the screw capped tube up to the top mark and distribute 2 mL of cell suspension in each Petri dish by opening the cover from one side for a short time. A pipette should be used only once and then discarded in a bucket with water placed underneath the hood.

(iii) Usually Con A comes as a powder, 20 or 50 mg is weighted in a mono-pan balance and diluted with culture medium or PBS into a stock solution. The calculation needs to be done to find out how much of the stock solution is to be added to each Petri dish containing 2 mL of cell suspension, so that final concentration of Con A becomes 5 μg/mL. It is desirable the final volume of Con A for each Petri dish remains restricted to 50 μL.

(iv)   The Petri dishes to which Con A has been added are marked as Expt (experimental) with a marker pen. The Petri dishes without any addition of Con A are treated as control.

(v)   The Petri dishes placed on the tray are put in an incubator with 5% CO_2 and rest air at 37°C in a humidified atmosphere.

(vi)   The culture medium, Con A stock solution is refrigerated. The other items should be in proper racks. The base of the hood is wiped with a cotton ball soaked in 70% alcohol.

(vii)   The UV and fluorescent lamps of the hood and Bunsen burner are turned off.

Observation

After 24 hours of incubation one of the experimental and one control Petri dish are taken out of the incubator. (In case of any contamination, the pink colour of culture medium usually turns to yellow colour.) Add 70% alcohol from a wash bottle to the contaminated culture and discard the Petri dish in a basin outside of the culture room.

A haemocytometer is charged with cells from experimental dishes and another with cells from control dishes.

Examine under microscope [with 10×40 (objective) magnification] and measure the diameter of the living cells. The marks of occulometer at this magnification should previously been equated with the scale inscribed on the stage micrometer. Try to identify four categories of cells size wise – upto 5 μm normal cells, 5–7 μm diameter cells as small blast cells, 8–10 μm as medium sized blasts, and beyond 10 μm as large blast cells.

Count these four categories of cells in the different fields of haemocytometer and record with a multiple key tally counter. This count helps one to find out the percentage of blasts of different sizes, and total blast cell percentage in a population. The percentage of blast cells count at different hours provides the rate of blastogenesis with activation of Con A. The count can be performed at 24, 48, and 72 hours taking out a Petri dish from the incubator.

At every hour of observation the count from Petri dishes with normal cells, without any treatment should also be maintained to understand the background count of blast cells. Usually the normal lymphocytes without any activation in culture dies much faster and the total count is always lower than the stimulated cells in the experimental Petri dishes,

The proportion of dead cells in normal and experimental categories can also be determined with Trypan Blue exclusion test (dead cells become blue).

EXPERIMENT 11   RAISING ANTIBODY IN MOUSE AND HAEMAGGLUTINATION TEST

Antibody synthesis can be induced in a mouse by injecting a particulate antigen like RBCs from another animal or a soluble protein antigen.

Preparation of antigen

- Antigen (Ag): sheep red blood cell (SRBC), host: mouse.
- SRBC collected from the jugular vein in a sterilized conical flask containing Alsever's solution; can be stored at 4°C in the refrigerator for several days.

- For use, centrifuge SRBC in Alsever's solution at 1500 rpm, wash twice with sterile phosphate buffer saline (PBS) in a conical graduated centrifuge tube.
- Re-suspend to different percentage (v/v) of SRBC assuming the packed volume to be 100%.

[Note: Sheeps' erythrocytes are particulate antigen. Soluble Ag, like BSA, can be prepared as a % solution.]

Composition of Alsever's solution (for preserving SRBC) Dextrose 20.5 g, NaCl 4.2 g, sodium citrate 8.0 g. Make a solution of volume 1000 mL by adding double distilled water. Sterilize by membrane filtration.

Composition of phosphate buffer saline (0.85%, pH 7.2; for living cell suspension.) NaCl 8.0 g, KCl 0.2 g, disodium hydrogen phosphate 1.15 g, potassium hydrogen phosphate 0.2 g. Make a solution of volume 1000 mL by adding double distilled water to the salts. (10% serum may be added for keeping the cells in suspension for long; it is preferable to keep the test tubes of cells in an ice bucket.)

Immunization

0.1 mL of 20–25% SRBC is injected intravenously into the lateral tail vein of a mouse (Fig. E.16). The mice may be pre-warmed in a 37°C incubator for 3–4 minutes or xylene-soaked cotton may be rubbed on the tail for vasodilatation. Other routes of immunization are intraperitoneal (i.p.) (Fig. E.17) intramuscular (i.m.), and subcutaneous (s.c.).

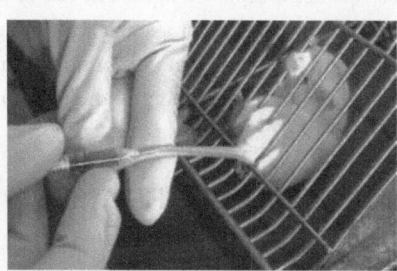

Fig. E.16 Antigen (SRBC) being injected intravenously into the lateral tail vein of a mouse.

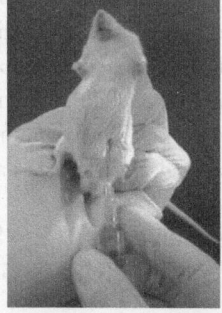

Fig. E. 17 Antigen being injected intraperitoneally into a mouse.

Collection of antiserum

- After 5 days of immunization (injection or SRBC) blood is collected from either of the following:
  - Tail vein of the mouse (by cutting the tip of the tail with a pair of scissors)
  - Retro-orbital plexus (behind the eyeball) by suction with a Pasteur pipette
  - Cardiac puncture by a 27-gauge needle fitted on a 1 mL syringe after anaesthetizing the animal
  - Ear vein of rabbit or cardiac puncture
- The blood collected is stored in a sterilized glass tube.
- It is kept for 45 minutes at room temperature for clotting.
- The clot is separated from the wall of the glass tube by a fine mounted needle.
- The tube is centrifuged at 1000 rpm for 5 minutes.
- The serum from the top is collected by a Pasteur pipette carefully.

- Antisera collected from immunized mice are incubated for 45 minutes at 56°C in a hot water bath to inactivate the complement and freeze in small aliquots (–20°C) for future use.

[*Serum for complement* raised from the blood of normal mice (usually Guinea pig or rabbit) following the same procedure, except heat inactivation.]

Haemagglutination test

This titre is used to determine the presence of antibody in immunized serum. This is a qualitative test to measure the level of antibody produced by an animal. The procedure is as follows.

- Ten small test (Khan) tubes are set in a series in a test tube rack.
- In the first test tube, 0.1 mL of antiserum and 0.9 mL PBS are kept (one-tenth dilution of antiserum).
- In the second and other tubes, 0.5 mL PBS (to begin with) is kept.
- Double-fold serial dilution from second tube onwards: 0.5 mL solution from the first test tube is transferred to 0.5 mL PBS of the second tube with a pipette, mixe, and again 0.5 mL of the mixture is transferred to the third tube. This procedure is carried on down the tubes. 0.5 mL mixture withdrawn from the last tube can be kept in a fresh tube in the refrigerator; in case the titre continues up to the last tube in the series, this diluted serum may be used.
- 0.1 mL of 1% SRBC is added to each tube.
- The entire set of Khan tubes is incubated overnight in a 37°C water bath.

Observation

SRBC (Ag)–Ab complexes precipitate at the bottom of the tube as agglutinated lattices, visible to the naked eye, better seen under a dissecting binocular microscope in transmission light. The size of the lattices represents the degree of reaction and is indicated qualitatively by multiple or a single + or ± sign. There will be the least agglutination in one of the tubes down the row, beyond which no agglutination is visible. The dilution (x) of the serum in the tube with the least agglutination is considered the end point of the serum titre and represented as a reciprocal of serum dilution (1/x).

EXPERIMENT 12   SDS-PAGE FOR SEPARATION OF PROTEIN MOLECULES OF DIFFERENT MOLECULAR WEIGHTS

The students are requested to read the procedure to appreciate the logic of the steps involved in a sophisticated biochemical experiment. It may not be possible for them to carry out the experiment without a biochemistry or molecular biology laboratory.

Principle

A charged molecule in an electrophoretic field moves towards the oppositely charged electrode. Protein molecules are charged variously depending on the constituent amino acids. The rate of electrophoretic mobility of a molecule depends on (i) the sign (+ or –) and net electrical charge on it and (ii) its size and shape. Keeping all the other factors equal among molecules of equal size, the one with the higher net charge moves faster in an electrophoretic field. Small molecules move faster through the sieve of the gel than larger ones of the same net charge. The shape of a molecule may increase or decrease its mobility due to frictional drag and end up in atypical bands.

Sodium dodecyl sulphate

Sodium dodecyl sulphate (SDS) is a negatively charged detergent and binds to protein in amounts proportional to the length of the polypeptide chain. This binding destroys the characteristic tertiary and secondary foldings of the protein, transforming it into a negatively charged linear structure. The intrinsic charge of the protein molecule becomes insignificant in the presence of the net charge of so many negatively charged SDS molecules bound to it. The net negative charge on the protein molecule is proportional to its molecular weight, and thus a mixture of heterogeneous sized proteins can be separated in electrophoresis. The electrophoretic mobility or distance travelled by a protein molecule during SDS-PAGE is inversely proportional to the log of its molecular weight. Thus, the distance indicates the molecular weight of the protein and can easily be compared with a set of proteins with standard weight.

Preparation of denaturing SDS-PAGE: Vertical gel electrophoresis

Preparation of resolving gel (10%) Tris-HCl (1.5 M, pH 8.8) 2.5 mL, acrylamide: bis-acrylamide 3.3 mL, SDS (10%) 0.1 mL, ammonium persulphate (10%) 25 μL, TEMED 5 μL, distilled water 4 mL.

Preparation of stacking gel (4%) Tris-HCl (1.5 M, pH 6.8) 1.25 mL, acrylamide: bis-acrylamide 0.65 mL, SDS (10%) 0.05 mL, ammonium persulphate (10%) 25 μL, TEMED 5 μL, distilled water 4 mL.

Preparation of acrylamide stock solution (Acrylamide: bis-acrylamide) Dissolve 29.2 g acrylamide and 0.8 g bis-acrylamide in 100 mL distilled H_2O.

- First seal the glass plate with spacer and then 1% agarose.
- Cast (pour) the resolving gel with a Pasteur pipette by one side of the 1-mm spacer to form a thin gel slab between the glass plates up to the level where the stacking gel is to be poured.
- Now pour the stacking gel above the resolving gel.
- Insert a comb carefully (before the gel polymerizes) between the two glass plates at the stacking gel region, avoiding air bubbles being trapped under the teeth.
- Allow the gel slabs to polymerize for 60 minutes at room temperature.
- Remove the comb after gel has polymerized.
- Thus, wells are formed, ready for the protein samples to be pipetted in.

Composition of sample preparation buffer Tris buffer 1.0 mL (0.5 M Tris-HCl, pH 6.8), 10% SDS 1.6 mL, glycerol 0.8 mL, mercaptoethanol 0.4 mL, de-ionized water 4 mL, bromophenol blue 5 μL (1%).

Composition of running buffer (pH 8.3) (Tris-glycine buffer) 4X, Tris (0.025 M) 12 g, glycine (0.192 M) 57.6 g, SDS (0.1%) 4 g. Dissolve all the chemicals in de-ionized water to make the final volume equal to 1000 mL.

Denaturation of protein sample

- Adjust the protein samples to approximately 0.5 mg/mL.
- Add one part of the sample preparation buffer to three parts of protein sample.
- Heat the mixed sample at 95°C in a water bath for 5 minutes.

Loading the sample

- Pour running buffer into the chamber of the vertical unit.
- Carefully pipette the denatured protein samples into separate precast notches (well) on top of the gel (Fig. E.18).
- Run the gel at 150 V till the bromophenol dye reaches up to the end of the gel.
- Then remove the gel from between the glass plates and stain it with dye that binds to proteins (Coomassie blue, Silver stain).

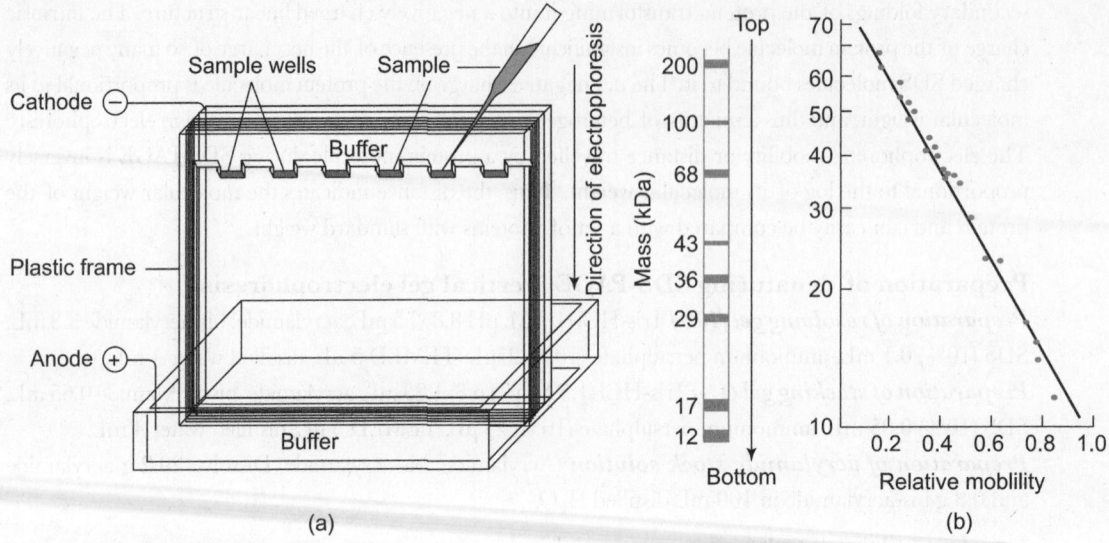

Fig. E.18 (a) Loading of protein sample on gel to run SDS-PAGE to separate different proteins on the basis of their molecular weights. (b) The molecular weight of a particular protein in a band is determined by the log of its migration distance against a standard plot. (Based on Weber, K. and M. Osborn 1975, *The Proteins*, 3rd edn, Academic Press, New York, vol. 1, p. 179; Goldsby, R.A., T.J. Kindt, B.A. Osborne, and J. Kuby 2003, *Immunology*, 5th edn, W.H. Freeman & Co., New York, p.530.)

Coomassie blue staining

Composition of Coomassie blue stain Coomassie brilliant blue 1 g, methanol 200 mL, acetic acid 100 mL. Dissolve all the chemicals in 1000 mL distilled water.

Composition of destaining solution Methanol 100 mL, acetic acid 70 mL, water 530 mL.

- After electrophoresis, the gels are stained overnight in Coomassie blue stain, followed by destaining solution.
- Gels are visible with a clear background revealing the bands to be photographed.

Silver staining

This method gives better resolution and detectability of proteins due to the diffusion of silver salt throughout the gel matrix. Silver ions tend to associate with the trapped protein; thus the concentration of silver around the protein bands is higher than that in the gel matrix.

Composition of gel fixing solution Methanol (CH_3OH) 50%, acetic acid 10%, water 40%.

Composition of silver nitrate stock Silver nitrate 10% in water (store in amber bottle).

Composition of stop solution Citric acid (2.3 M) 4.376 g in 10 mL distilled water.

Composition of dithiothreitol (DTT) stock Dithiothreitol 5 mg/mL of water (store at–20°C).

Composition of developer Formaldehyde (37%) 50 µL, sodium bicarbonate (3%) 100 mL.

Staining procedure Perform all the staining steps in glass trays on a rocker or shaker at a very gentle speed at room temperature.

- Place gels into a tray confining fixation solution for 15 minutes in a plastic bag.
- Rinse with de-ionized water.

- Dilute 10 mL DTT stock in 100 mL water, pour onto gel; shake for 15–30 minutes.\
- Pour out the DTT.
- Add 0.1% $AgNO_3$ to gel and shake for 15–30 minutes.
- Rinse in distilled water.
- Add 75 mL developer, shake until developed for 15–30 minutes.
- Add 5 mL citric acid to stop. After 5 minutes wash several times with water.
- Dry gel after soaking in 2% glycerol.

EXPERIMENT 13   MITOTIC CHROMOSOMES IN PLANT TISSUE—ROOT TIP

Cytological techniques for demonstration of chromosomes involve four essential steps—collection of right material, fixation, staining, and squash preparation for microscopic observation. For studying mitotic chromosomes, a fifth step of pre-treatment to arrest mitosis at metaphase for revealing chromosomes at most condensed form is necessary.

Plant material
Usually roots tips from germinating cereal seeds or regular plantlets or an onion bulb kept in moistened sand or sandy soil for a few days are collected for chromosome preparation.

Cereal seeds may be allowed to germinate on two layers of filter paper in a covered Petri dish at room temperature around 22°C for 2 days.

Procedure
1. Take 3–4 roots from germinating seeds in cold water in a small glass vial and keep in a refrigerator at 2°C for 24 hours. This cold pre-treatment is to arrest cell division and accumulation of cells at metaphase. Different chemicals may also be used as mitotic arrestor as mentioned later for root tips from plantlets.
2. Wash the root tips briefly with distilled water.
3. Transfer the root tips to a small watch glass or a vial with 1% aceto-orcein, fixative cum stain. The watch glass needs to be covered with the cover of a small Petri dish. (1% aceto-carmine solution can also be used in place of aceto-orcein). Aceto-orcein may be obtained from a supplier. Otherwise, dissolve 2 g of orcein in 100 mL of 45% acetic acid by boiling gently. Shake and filter. Before use this should be diluted 1:1 with 45% acetic acid, bring the final concentration to 1% orcein.
4. Stain of the root tips in aceto-orcein should continue for 30 minutes.
5. Then place 1 or 2 root tips on a clean glass slide in a small drop of 45% acetic acid. Tap the root tips with a flat-ended glass rod. The root tips need to be completely macerated; any remaining hard tissues should be removed by a needle.
6. A coverslip should be placed gently on the macerated root tip materials. Place a piece of filter paper over the coverslip and gentle pressure with a finger tip would remove any excess stain and acetic acid.
7. Hold the slide with the fingers of one hand, keeping the filter paper on the coverslip, use the thumb of the other hand to squash the materials creating vertically downward pressure on the coverslip with the overlaying filter paper.
8. The pressure of the thumb causes spreading of the chromosomes.
9. Examine under microscope to find out whether the preparation is good enough to observe chromosomes in cells. If necessary make another fresh squash preparation with a separate stained root tip. A good preparation on a slide should be sealed around the coverslip with nail polish or

rubber solution, so that the chromosome preparation does not get dried quickly during observation.

10.   Observe under microscope the mitotic chromosomes at different stages of cell division in different cells. Draw the chromosomes in different stages of cell cycle. Count the number of chromosomes in a cell.

Root tips from an onion bulb or plantlets

The method of squash preparation of regular root tips is more or less the same as in case of root tips from germinating seeds, with a few modifications.

1.   Healthy roots are collected in a small vial containing distilled water and within 15 minutes healthy roots are selected and transferred to pre-treatment solution for arresting mitosis at metaphase.

2.   0.05% Colchicine solution in distilled water is often used for pre-treatment for 4–6 hours at 18°–20°C. (8-hydroxyquinoline, α-bromonaphthalene solutions are also used for roots of certain plants.)

3.   Then transfer the roots into 3:1 absolute alcohol : glacial acetic acid for fixation for 24 hours at 4°C. The root tips can be stored as such in the fixative for months.

4.   Remove the roots from fixative, wash thoroughly in water.

5.   Put the roots in a test tube to hydrolyse in 1N HCl at 60°C for 8–10 minutes.

6.   Then wash the roots thoroughly in distilled water.

7.   Put the roots in Feulgen stain for 1 hour. Feulgen stains specifically DNA.

Composition of Feulgen Stain

0.9 g basic fuchsin, 4.8 g sodium metabisulphite, 250 mL 0.15 M HCl—mixed in a conical flask, covered with foil, and shaken for 24 hrs. 5 g activated charcoal added and mixed well. Filter in a fume hood. Filtrate will be colourless; if not, the charcoal step should be repeated. The final filtrate is stored at 4°C with a dark cover on the bottle in a dark corner of the refrigerator.

8.   Take a stained root on a clean glass slide. The translucent root cap at the extreme tip and the unstained portion of the root are removed.

9.   Place the stained root in a small drop of 45% acetic acid and mix with a drop of aceto-carmine. Aceto-carmine colours cytoplasm and improve the staining.

Then follow the protocols indicated earlier in procedure from Step 5 onwards.

EXPERIMENT 14   DNA EXTRACTION FROM ANIMAL TISSUE

Nowadays, a minimal quantity of DNA is sufficient to carry out experiments for modern biotechnological procedures including genomic studies. If necessary a single specific DNA sequence can be multiplied to millions by using PCR.

Here, a procedure of DNA extraction following a high salt method is discussed. The whole procedure can be carried out in small microcentrifuge tube. There are many modifications of DNA extraction method. The students are asked to go through this topic in the text.

For this technique a centrifuge machine that can hold microcentrifuge tubes is essential.

Materials   (i) TNES buffer (Tris, NaCl, EDTA, SDS): 10 mM Tris, pH 7.5; 400 mM NaCl, 100 mM EDTA, 0.6% SDS.

(ii) 6 M NaCl saturated solution stored at 37°C: 6 M NaCl is weighed and dissolved in sterile distilled water and the solution is heated until all the salt dissolves. Allow to cool at room temperature. Some crystals may form as it is a saturated solution.

(iii)   Proteinase – K
(iv)   100% ethanol
(v)   70% ethanol
(vi)   Sterile distilled water
(vii)   Tris-EDTA

Procedure

1.   Fresh tissue or tissue that has been stored in absolute ethanol at room temperature is used. In case of storage in ethanol, soak out excess ethanol by using sterile tissue paper.

2.   Take a small portion of tissue ($\sim$ 0.5 cm^2) and chop it as small as possible with a sterile scalpel. Transfer the sample to a 1.5 mL microcentrifuge tube and mark the tube. [Too much tissue can give a worse yield of DNA].

3.   Add 600 µL of TNES buffer and 35 µL of proteinase-K (20 mg/mL). Close the cap of the microcentrifuge tube and mix the sample by inverting the tube several times.

4.   Incubate the samples overnight (or 5–24 hours) in a water bath at 50°C. If possible occasionally mix the samples by inverting the tube. Make sure the cap is closed, as it may get loose when warm. [More proteinase-K may be added to speed up tissue digestion and reduce the incubation time to 2–4 hours.]

5.   Add 166.7 µL of 6 M NaCl. Shake the samples vigorously for 20 seconds. [Shake harder than simply inverting the tubes, but not too roughly to damage DNA.]

6.   Microfuge the samples at full speed (12–14,000 rpm) for 5–10 minutes at room temperature. [Label corresponding new tubes whilst centrifuging samples.]

7.   Remove supernatant to the new, labelled 1.5 mL microcentrifuge tube. [Be careful not to transfer any of the tissue debris to the new tube, it is better to leave a little supernatant rather than transfer the cell debris.]

8.   Add an equal volume ($\sim$ 800 µL) of cold 100% ethanol and gently mix by inverting the tube couple of times – white DNA precipitates out of solution. [Better yield is obtained if the ethanol is kept at –20°C before use and then in an ice bucket when on the bench. If the samples are left at –20°C for a few hours or overnight, DNA yield becomes more.]

9.   Centrifuge the sample at full speed (12–14,000 rpm) for 10–20 minutes at 4°C.

Note: It is very helpful to place the hinge of the microcentrifuge tube towards the outside of the rotor while pelleting the DNA. The subsequent washing steps are made much easier since the DNA pellet can be readily seen at the bottom of the centrifuge tube below the hinge.

10.   Pour (or pipette) off the supernatant, without dislodging the pellet of DNA.

11.   Wash the DNA pellet adding 200–700 µL of 100% ethanol (Close cap of the tube and invert gently, or roll the tube between your palms. Pour (or pipette) off the ethanol and briefly spin the samples to keep the pellet at the bottom of the tube.

12.   Wash DNA pellet with 70% ethanol as above. After removing the supernatant of 70% ethanol, briefly centrifuge the samples to pipette off the remaining small amount of ethanol.

13.   Leave the sample to air dry—usually 10–30 min depending upon the temperature. [Do not dry out too much or the DNA will be hard to re-suspend, remember any ethanol left in the tube will inhibit subsequent PCR].

14.   As soon as the sample is *just* dry add 100–200 µL (or more if the pellet is large) of sterile distilled water or Tris-EDTA to resuspend the DNA.

Appendix

II

Fun Experiments

EXAMINATION OF PLANT CELLS

Procedure

1. Take an onion bulb and remove the dry covering scales. Make cuts at both the ends and take off an overlapping scale of onion.
2. Hold the onion piece by the left hand and peel off a thin membrane from the inner surface with a bent forcep.
3. Place the thin membrane on a slide and observe under microscope to see the rectangular onion cells with the nucleus at one side, large vacuoles, and streaming movement of the cytoplasm. If you want to observe for a long time put a drop of water and a drop of glycerine on the membrane and cover it with a coverslip without forming bubbles.
4. Draw and label the cells.

Observe the Tissue Organization in Stem and Root Cells

1. Take a soft bodied small plant or herb. Clean it with running tap water and keep it in a clean beaker or a ceramic bowl.
2. Cut out with a sharp razor or a blade 2 or 3 stem pieces, each of about 1.5″ long. Place them in a watch glass with tap water. Similar 1 or 2 pieces from the root is cut out and kept in a separate watch glass.
3. Hold a stem piece (or root piece) with the finger tips of left hand and try to make thin transverse (cross) sections with a sharp razor or a blade holding close to the plant piece and moving it on your left direction quickly.
4. After cutting a few thin slices, float them on water in a fresh watch glass. Repeat the process in Step 3 to produce a good number of thin cross sections.
5. Select the thinnest possible intact section by visual estimate and place it on a clean slide with the help of a camel brush in your biological dissecting box.
6. Put a drop of water on the thin section, place a cover glass on it from a side without producing bubbles.
7. Observe under the low magnification of a microscope. Draw the diagram in the note book. Consult a botany book and label the epidermis with hairs, other cell layers beneath, xylem and phloem, and so on. You will get a fair idea of the organization of plant cells in its body.

8. For studying the tissues in root, follow the Steps from Steps 3 and observe in a similar fashion. Coloured and permanent preparation of sections from stem and root can also be done following some elaborate procedures.

MOVEMENT OF BLOOD CORPUSCLES IN CAPILLARIES

Materials

During rainy season, collect some early tadpoles of toad or frog from water bodies. Keep then in a beaker with clean pond or tap water.

Transfer one or two tadpoles in a watch glass with a wide mouth pipette.

Procedure

1. Take one tadpole out of the watch glass with the same pipette on a clean slide.
2. There should be only a drop of water surrounding the tadpole. Excess water is to be removed by a fine-tip Pasteur pipette.
3. The tadpole may be turned upside down with a needle without injuring it, so that it becomes comparatively immobilized, the tail should be laterally placed.
4. Examine under the lowest power of a compound microscope. Try to focus on the laterally placed tail of the tadpole.

Observation

Blood corpuscles, especially the RBCs will be seen moving in the small blood vessels and capillaries. They move in a file in the capillaries. They may be seen to move with a jerk due to their transmission by systolic pressure from heart.

At the end of the experiment, the tadpoles can be released into the same water body wherefrom they have been collected so as not to disturb the biotic component of the environment as far as possible.

VISUALIZATION OF CELLS IN DIVISION

During rainy season toads and frogs lay their eggs in water and they are externally fertilized by the sperms. The fertilized eggs remain in jelly-like coat, the eggs in series in the jelly-like coat look like a transparent thread with dark beads of the fertilized eggs that remains entangled in aquatic plants or even at the bottom in shallow water ponds and lakes.

Collect about a foot length of string of eggs in jelly coat in a beaker and bring it to the laboratory. With a scissors shorten the length of the string of eggs to 1 to 2 inches.

Take a small piece in a watch glass with tap water and try to visualize the eggs individually. They are usually black or grey.

Observation

1. The egg is round and a few mm in diameter. One side is comparatively lighter indicating the presence of yolk for larval development. This side is often called vegetal pole that develops into the ventral or abdominal side of the animal. The darker side, animal pole, is destined to form the dorsal side of the animal.
2. With close observation one egg may be found intact with round shape or divided into two halves by a furrow or divided into four parts with two furrows at 180° apart. The furrows indicate the cell divisional planes dividing the original intact zygote into cells.

The divisional planes (furrows) and divided cells can be seen by naked eyes. If necessary help of a magnifying lens or low-power dissecting binoculars can also be used to see better.

COLLECTION OF CELLS FROM MOUSE PERITONEAL CAVITY AND STUDY OF MACROPHAGES

The stomach and intestine reside in the peritoneal cavity in the abdomen beneath the diaphragm. Spleen and the largest lymph node—mesenteric lymph node are, two important lymphoid organs of the immune system that are located in the peritoneal cavity. Some wandering macrophages and lymphocytes can also be found in the fluid in this cavity. The mobilization of these cells in large number can be induced by injecting liquid paraffin in the abdominal cavity.

Procedure

1. One day (24 hours) prior to the experiment, inject 2 mL of liquid paraffin into the intra-peritoneal cavity of a normal mouse with a syringe in the same manner as shown in Fig. E.17. (The peritoneal cavity lies beneath the skin and abdominal muscle. The tip of the syringe should be slightly angular and not so deep, so that it does not cause any damage to the intestine or other organs lying within the peritoneal cavity.)

2. Next day take 5 mL of normal saline or PBS in a 5 mL syringe fitted with 20-gauge (large bore) needle and inject in the mouse already injected with liquid paraffin, in the same manner as indicated in the earlier step.

   Do not take out the syringe after injection of normal saline. Try to aspirate the fluid in the abdominal cavity as much as possible with the same syringe. Then take out the needle.

3. Collect the abdominal aspirate in a 15 mL centrifuge tube. Add normal saline to the tube approximately up to 10 mL mark.

4. Spin the test tube at 500 rpm in a clinical centrifuge for 8 minutes. Aspirate out the supernatant from the top with a Pasteur pipette without disturbing the pellet of the cells.

5. Add fresh saline or PBS up to the 10 mL mark of the centrifuge tube. Holding the tube by the left hand strike it with the index finger of the right to dissociate the cell pellet. This can also be done with brief vortexing, if the equipment is in the laboratory. Centrifuge the tube at 500 rpm for 5 minutes.

6. Repeat Step 5 once more.

7. To the final pellet add 5 mL of PBS. Count the cells in a haemocytometer and observe their nature lowering the condenser of the microscope to cut the light intensity for better observation.

8. Transfer 2 mL aliquot of cell suspension obtained in Step 7 to a fresh test tube. Spin the tube for 5 minutes at 500 rpm and aspirate out the supernatant. Then add 1% solution of neutral red in culture solution supplemented with 10% serum to the tube and incubate for 45 minutes at 37°C in an incubator.

9. At the end of incubation suck in and out the culture medium with a Pasteur pipette to dislodge the cells from the surface of the test tube. Add 1 mL of PBS. Take a drop to charge a haemocytometer to observe the cells and to find out the percentage of cells engulfing neutral red indicating their phagocytic nature. Usually macrophages belong to the phagocytic group of cells.

10. Take a drop of cells on a clean slide and mix with a drop of 1% Trypan Blue and take the mixture in counting chambers of a haemocytometer. Calculate the percentage of living and dead cells. Trypan blue imparts blue colour to the dead cells.

SEE YOUR OWN CELL AND BARR BODY (IN FEMALES)

Procedure

1. Take a fresh pre-cleaned small (~ 15 mm diameter) round cover glass by the tip of the thumb and index finger of the right hand and scrape lightly 2 to 3 times (not more than that) the inside of your lower lip holding the portion of the lip extended outside with the fingers of your left hand.

2. A clean glass slide with a small drop of saline at the centre should be kept ready to transfer the scraped material from the cover glass. The materials can be immediately seen after putting the cover glass on it, preferably under phase contrast microscope. Or spread the material with the help of the cover glass at the centre of the slide.

3. The film of cells should be air dried and the slide is to be put on a staining tray or in a Petri dish and covered with Giemsa stain for 10–15 minutes.

4. The excess stain should be drained off and tap water added to the slide for a few minutes.

5. Dip the slide in tap water in a beaker and keep it standing at one end against a support for drying.

6. Put a drop of DPX as mounting medium and a cover glass on the stained smear.

7. Examine under the high power of a compound microscope.

Observation

Your epidermal cells with nucleus will be visible.

If you are a female, a stained small bar like structure will be seen at the periphery of the nucleus. This is actually one of the two X chromosomes of a female, which is non-functional, heterochromatinized, and known as Barr body after the name of the discoverer.

In case of a male (XY), there will be only one X chromosome which is functional and not condensed.

This is one of the easiest way to determine the sex of a person at the cellular level, and used for establishing sex of the athletes.

DETERMINE YOUR BLOOD GROUP

Materials

(i) The kit for blood group determination, usually three small bottles with three different coloured fluid containing anti-A, anti-B, and anti-Rh antibodies (diluted)

(ii) Clean slides

(iii) Sterilized hypodermic needle

(iv) 70% ethyl alcohol and sterilized cotton

Procedure

1. Clean the tip of the index or middle finger of your left hand with cotton soaked in 70% alcohol. Allow the tip of the finger to dry.

2. Take out the needle from a pack and hold it at the base with the fingers of your right hand. Then prick the clean fingertip of your left hand with the needle. Blood appears at the tip of the finger as a drop.

3. Apply one drop of blood at a place and put three such drops in three places on the slide with some gap.

4. Add three different anti-sera on the three drops of blood; this may be in the order of anti-A, anti-B, and anti-Rh.

5. Then mix the blood drop and anti-serum with a clean needle or a clean toothpick, mixing should be done separately with three different clean needles or toothpicks.

Observation

Within a minute of mixing, agglutination (small aggregates form) of red blood cells will start on the slide.

Agglutination with	Blood group of the person
Anti-A	A
Anti-B	B
Both Anti-A & Anti-B	AB
No agglutination	O
Anti-Rh	Rh$^+$
No agglutination with Anti-Rh	Rh$^-$

Caution It is better to carry out this experiment under the supervision of a teacher. Do not use the finding of your experiment for clinical purposes; for such purposes blood group typing should be done by a qualified person or a doctor.

Glossary

Abzymes   Antibodies capable of catalysing specific chemical reactions like enzyme. A catalytic abzyme is produced against the expected transition state off a specific substrate.

Accession number   Unique identifying code number of each cloned DNA sequence is catalogued in databases such as GenBank. This number is used to retrieve database information on a particular sequence.

Acridine   A chemical, with dye like properties, that intercalates in the double-stranded (ds) DNA and causes mutation involving usually a single base pair.

Active site   The part of an enzyme that binds to its substrate for catalytic action.

Adenosine triphosphate (ATP)   Triphosphates attached to the nitrogenous base adenine. It stores and transfers energy in cell system.

Adenovirus   Virus that causes the common cold.

Adult stem cells (ASCs)   Stem cells with totipotency that derive from adult tissues, including bone marrow.

Aerobes   Organisms that grow in the presence of molecular oxygen, which acts as terminal electron acceptor in aerobic respiration.

Agar   Extracted from marine red algae of family *Rhodophyceae*. It is used in powder form as a solidifying agent for microbiological culture medium.

Agarose   Isolated from marine red algae (*Gelidium* and *Gracilaria*) and is highly purified. Basically, it is a linear polysaccharide comprising repeating units of D-galactose and 3,6-anhydro-L-galactose joined by β(1→4) linkage. It is used for separation of different types of biological macromolecules (see gel electrophoresis).

Agricultural biotechnology   A sub-discipline of biotechnology that involves plants for improvement of agriculture and their application for utility products. It takes the assistance of genetic engineering.

Agrobacterium tumefaciens   A rod-shaped spore-forming Gram-negative pathogenic soil bacterium, capable of genetically transforming a plant cell and inducing gall tumours in plants.

Alkaline phosphatase   An enzyme that removes phosphate groups from the 5′ ends of DNA molecules.

Allele   One of the alternative forms of a gene present at a particular locus on a chromosome (DNA).

Allele-specific oligonucleotide (ASO)   Oligonucleotide specific to a disease gene for analysing a person's DNA using PCR.

Alternative splicing   Splicing joins certain exons with *cut out* other exons being treated as introns. This way multiple mRNAs of different sizes from the same gene originate and produce different proteins with different functions. The net result is that alternative splicing produces several different proteins from the same gene sequence.

Amino acids   Organic acids bearing both amino and carboxyl groups, and act as the monomeric units (building block) of proteins. Twenty different amino acids join together by covalent bonds to make a polypeptide chain or protein.

Aminoacyl transfer RNA (tRNA)   Transfer RNA (tRNA) molecule with an activated amino acid at one end and anticodon at the other end to fit to codon of mRNA.

Amniocentesis   A technique to obtain foetal cells in amniotic fluid from a pregnant woman for analysing foetal chromosomes and abnormalities if any, sex and biochemical constituents of the amniotic fluid, metabolic status of the cells.

Amino terminus (N-terminus)   The end of a polypeptide or protein chain with amino group, the other end is with carboxyl group.

Ampicillin   A semisynthetic penicillin that inhibits cell wall synthesis in bacteria, which is used in molecular biology experiments for selection of ampicillin-resistant bacteria.

Anaerobes   The organisms that carry out metabolic activities as well as growth in the absence of molecular oxygen.

Angiogenesis   The process of growth of new blood vessels.

Annealling   Denaturing of ds DNA by heating and allowing the formation of hybrid DNA of two different DNA strands or DNA:RNA molecules in course of denaturing by lowering the temperature.

Anode   The positive electrode, usually red coloured, in gel electrophoresis apparatus.

Antibiotic   A substance produced microbially or synthetically, having antimicrobial properties.

Antibody (Immunoglobulin)   A class of serum protein, produced by B lymphocyte cells, has specific binding site for 'foreign antigen', and provides humoral (serum mediated) immunity in vertebrate classes.

Anticodon   A sequence of three nucleotides present at one end of a tRNA molecule that binds to the complementary bases in an mRNA molecule bound to the ribosome during the translation process for protein synthesis.

Antifreeze proteins (AFPs)   A category of proteins that have the unique property of lowering the freezing temperature of body fluids and tissues in aquatic organisms, including fishes in cold environment.

Antigen   Any foreign substance, after entry into an organism triggers an immune response for production of lytic substances in invertebrates, and antibodies and cell-mediated response in vertebrates.

Antigenic determinant (epitope)   The small portion of an antigen molecule that is recognized by the antigen receptor on lymphocytes or immunocompetent cells.

Antisense RNA   An RNA molecule that is complementary to all or a part of a pre-mRNA or mRNA encoded by a gene. Presence of antisense RNA in a cell leads to annealling with normal mRNA and formation of duplex molecules, which either get degraded or suppress translational ability of normal RNA. This introduced the first reverse genetic procedure for selective shutting off the expression of specific genes.

Antitermination   Antitermination proteins allow RNA polymerase to transcribe through certain terminator sites to read into the adjacent genes.

Apoptosis (Programmed cell death)   A cell responds to certain stimuli by initiating a cascade of biochemical reactions that lead to the cell death through a characteristic set of cellular events. Controlled apoptosis occurs during the development of an organism to remodel the tissues, whereas uncontrolled apoptosis is involved in degenerative disease conditions such as arthritis.

Arabidopsis thaliana   A plant (common name - mouse ear cress) with a very small genome that is used as a model organism for the study of eukaryotic genes and their control. This plant is also used for the study of plant growth and development.

Aquaculture   Farming procedures for finfish, shellfish, and aquatic plants for commercial or recreational uses.

Aquatic biotechnology   Biotechnological applications involving finfish, shellfish, marine bacteria, and aquatic plants.

Archaea   A group of prokaryotes that diverged from all others at an early stage in evolution, and they have a number of significant differences from others. They can live in extreme environments, like boiling hot springs.

Aseptic techniques   Practical measures to prevent the contamination and growth of unwanted microbes from the surroundings in course of microbial or cell cultures and surgical operations.

Aspiration   Removal of liquid supernatant from a sample by suction using a pipette attached to a vacuum source with a trap for effluent.

Astaxanthin   A pigment that causes pink colour of shrimps. Recombinant astaxanthin gene is used in aquatic biotechnology to impart colour in fish like salmon, for commercial appeal.

att sites   Loci on a phage and the bacterial chromosome for site-specific integration or excision of the phage genome to or from the bacterial chromosome.

Attenuated vaccine   Modified, weakened (for virulence) but alive microorganisms used as vaccine, with intact antigenic epitope to elicit immune response for virulent form of the organisms.

Autoclave   A device for sterilization with steam under pressure.

Autograft   A graft from one region to another of the same individual, for example, artery for coronary bypass surgery, skin or hair grafts.

Autoradiography   The technique to record the spatial distribution or location of radiolabelled compound within a specimen or an object by exposing photographic film to the emission from radioactive material.

Autosomes   Chromosomes not harbouring sex determining genes; chromosomes 1–22 in humans.

Auxotroph   A mutant that lacks the ability of synthesizing an important nutrient, such as an amino acid or a vitamin, and needs supplementation of that nutrient for normal growth like the wild type.

Avidin   A glycoprotein present in egg white that binds strongly to biotin and inhibits bacterial growth.

Bacillus (pl. bacilli)   A rod-shaped, Gram-positive, non-pathogenic, sporulating, obligate bacterium. Its nucleic acid has low GC content.

Bacmid   A shuttle vector used for genetic engineering, based on the *Autographa californica* multiple nuclear polyhedrosis virus (AcMNPV) genome, which can be propagated in both *E. coli* and insect cells.

Bacteria   Primitive, relatively simple, unicellular prokaryotes, without any nuclear membrane, and diverse in structure and adaptability.

Bacterial artificial chromosome (BAC)   A vector system designed from *E. coli* F-factor having the sequences needed for replication and segregation in bacteria. It is used for cloning large DNA inserts; extensively used in the human genome project.

Bacteriocin   A toxic protein produced by one bacterium that kills closely related bacteria.

Bacteriophage (phage)   Virus that infects bacteria.

Baculovirus   A virus that infects and destroys insect cells.

Basic Local Alignment Search Tool (BLAST)   Internet-based program designed by the US National Center for Biotechnology Information (NCBI) for searching sequences similarity from protein or DNA databases (GenBank).

Batch process (large-scale or scale-up process)   Scale-up of the culture of organisms, such as bacteria or yeast or mammalian cells, in large quantity in a bioreactor for isolation of useful products in a batch.

β-sheet   One of the two secondary structures of proteins; not helical in shape.

Bed volume   The volume occupied by the packed matrix in a column; it includes the volume of the beads (matrix) plus the void volume.

Bidirectional replication   The origin point of DNA replication produces two replication forks that proceed away from the origin in opposite directions.

Bioaccumulation   Progressive increase in concentration (accumulation) of a substance, such as a chemical pollutant (DDT), as it moves up the 'food chain' from organism to organism.

Binary vector system   The engineered T-DNA region is on one plasmid, the virulent genes for entry in host cells are on the other plasmid. A two-plasmid system is used in *Agrobacterium* sp. for transferring a T-DNA region with cloned genes into plant cells.

Biocapsules (microcapsules)   Tiny spheres or small tubes made up of permeable material, filled with desired drug or therapeutic cells, for slow release of the drug or cell products, are implanted in the patients for release of the therapeutic agents in course of long duration.

Biodegradation   The use of microorganisms to break down chemicals including pollutants.

Biodiversity   The range of different species present in an area. Biodiversity is sometimes described as genetic resources, as the diverse species can be used as the starting point for new products or processes.

Bioethics   In the social context, code of values for implications of biologic, biomedical research, and particularly biotechnological applications are focusing points for bioethics. The industry and regulators of biotechnology have a substantial concern in bioethics. The serious and practical side of bioethics is sometimes called ELSI—ethical, legal, and social issues.

Biofilm   A layer of organisms growing on a surface, for example, growth of bacteria on the teeth, growth of organisms including algae on domestic plumbing line, and shellfishes on the hull of a ship.

Biofuels   Fuels made from bulk biological materials such as cane sugar, corn, wood pulp (for 'agricultural ehtanol'), and sewage (for biogas methane).

Biohydrometallurgy   Microorganisms like bacteria are used commercially to recover metals, such as sulphur, copper, antimony, molybdenum, zinc, cadmium, and uranium from low-grade ores.

Bioimpedance   A technique that involves low-voltage electricity to measure the density of muscle and fat content of farm animals including fishes for commercial purpose.

Bioinformatics   Interdisciplinary science that involves molecular biology, mathematics, statistics, and computer, to develop and apply information technology of enormous data on DNA and protein sequences. It creates DNA and protein databases for storing and sharing through the Internet allowing search for similar sequences and structures across the species at different levels of phylogeny and predicting the function to a particular sequence.

Biolistics (microprojectile bombardment by gene gun)   A technique for introducing DNA into plant and animal cells, and organelles by means of DNA-coated gold or tungsten microparticles (~1 μm diameter) that are fired under pressure at high speed by the gene gun. This method was develop at Cornell University and commercialized by DuPont.

Biological containment   Prevention of the self-catalysed transfer of recombinant DNA molecule (gene) from one host to another, and thus, preventing the possibility of catastrophe from an engineered gene.

Bioluminescence   Emission of light by living organisms such as bacteria and insects, usually catalysed by the enzyme luciferase.

Biomass   Large mass of biological matter in a bulk. Fermentation process generates biomass of microorganisms as well as the fermentation product like alcohol. It is often measured as the dry weight of living material in a defined mass.

Bioprocessing   The use of biologic systems to manufacture a product.

Bioprospecting   Efforts to find out natural resources for useful materials including medicines on the basis of indigenous or traditional knowledge.

Bioreactor   A specially designed vessel in which cells, cell extracts, or enzymes carry out a biological reaction in bulk, including fermenter.

Bioremediation   A process of using living organisms (often bacteria) to degrade and clean up contaminant, man-made pollutants, or harmful substances from soil or water.

Biosensors   Living organisms or products used for detection of some chemicals for clinical purposes or for detection of pollutants on the basis of biological, physical, and chemical principles.

Bioterrorism   The use of obnoxious and harmful live organisms or their toxins as weapons against enemy or other party.

Biotin   Widely used vitamin to label nucleic acids; it also binds very tightly with avidin.

bla gene (amp^r gene)   Gene encoding β-lactamase, which causes resistance to ampicillin.

Blastocyst   Ball-like cluster of dividing cells from a zygote, containing a hollow space blastocoel inside.

Blotting   Transfer of macromolecules (DNA, RNA, and protein) from a gel to a overlying membrane under capillary action generated by a stack of blotting papers.

Blunt end   Double stranded ends of a DNA molecule created by the cut of certain endonuclease restriction enzymes.

BRCA 1 gene   The gene associated with the majority of human breast tumors; often its detection is part of diagnosis.

Broad host range plasmid   A plasmid capable of replicating in a number of different species of bacteria.

Bromophenol blue   A dye commonly used for tracking the bands of nucleic acids in electrophoresis gels.

Buffer   A conjugate acid–base pair that stabilizes the pH during biological studies in vitro.

Buoyant density   The density of a particle or a molecule when suspended in an aqueous salt or sugar solution.

Byssal fibres   Protein-rich ultra-strong fibers secreted by mussels like *Mytilus* and other shellfish for adherence to rocks and withstand great stress and shearing force of sea waves. Biotechnologists are planning to produce this material with unique strength by genetic engineering to manufacture lots of utility items.

CaMV 35S promoter   A constitutive promoter obtained from cauliflower mosaic virus and often used for expression of cloned genes in plants.

CAAT Box   Short repetitive nucleotide sequence (CAAT), located 80–90 base pairs upstream (towards 5′end) of the start site of many eukaryotic protein-coding genes; part of promoter sequence in CAAT box bound by transcription factors help in the initiation of the function of RNA polymerase.

Calcitonin   A thyroid hormone that stimulates calcium uptake by digestive organs for promoting calcification or hardening of bone.

Callus   When a plant tissue is aseptically cultured in-vitro, on a defined medium, an aggregate of dividing cell mass, poorly differentiated is formed, termed as callus. Plant regeneration from callus is done to obtain 'somaclonal variants'.

Cancer Genome Atlas Project   The project is to map relevant genes and genetic changes responsible for cancer, by the National Institute of Health (NIH), USA.

Capsid   The unit of external protein coat of a virus.

Carcinogen   A chemical inducing higher frequency of mutations in cells leading to cancerous condition.

Carrageenan   A polysaccharide derived from seaweeds and used as a bulking or thickening agent in syrups, sauces, and adhesives.

Cassette   Various DNA elements are genetically manipulated to bring into a clonable unit for performing certain specific functions.

Cathode   The negative electrode, usually black coloured in a gel electrophoresis apparatus.

cDNA   Complementary DNA, synthesized on the template of mRNA with the help of reverse transcriptase enzyme (RNA based DNA polymerase).

Cell culture   The technique to grow cells taken out of an organism in defined culture media under specified laboratory conditions (in-vitro conditions).

Cell line   A cell lineage maintained in culture represents a particular cell type. The original cell acquires the ability to proliferate indefinitely, either through random mutation or deliberate modification. There are many established cell lines throughout the world to carry out experimental works.

Cell lysis   The breakdown and dissolution of a cell.

Cellulase   Bacterial enzyme that degrades the polysaccharide component of cellulose in plant cell wall.

Centrifugation   Rotational centrifuge force generated in a centrifuge equipment to separate the components in a liquid medium on the basis of their weight (separation of different cell types, proteins from DNA). Centrifugal force is measured in revolutions per minute (rpm) or times gravity (g).

Centromere   Constricted region of a chromosome that holds together two sister chromatids after replication of a chromosome and binds to the spindle fibres for separation of chromatids to the opposite poles of a dividing cell. Thus, the correct distribution of chromosomes between daughter cells is ensured by centromere during cell division.

Cephalosporin   Antibiotics of β-lactum type that inhibit cross-linking of the peptidoglycan of the bacterial cell wall and inhibits growth of bacteria.

Cesium chloride (CsCl)   A heavy inorganic salt used for making density gradients for purification (separation) of nucleic acid in a high-speed ultracentrifuge.

Cetyl trimethyl ammonium bromide (CTAB)   Detergent that participates in forming insoluble complex with nucleic acids and allows separation of polysaccharide contamination.

Chain termination sequencing Method of sequencing DNA that involves enzymatic synthesis of polynucleotide chains terminating at specific nucleotide positions by using *dideoxynucleotides* (nucleotides lacking a 3' hydroxyl group, that interfere with the normal enzymatic synthesis of DNA). The method was introduced by Frederick Sanger.

Chemical degradation method of DNA sequencing The chemical method used by Allan Maxam and Walter Gilbert to cleave DNA preferentially at specific bases with the help of (non-protein) chemicals, for sequencing purpose.

Chemiluminescence The emission of light due to a chemical reaction.

Chemotherapy The treatment of cancer and other diseased cells by generating toxicity to them by specific chemical agents or drugs.

Chimera Combination of more than one type of cells (differing in genotypes) in an organism.

Chimeric plasmid (hybrid plasmid) A plasmid that includes DNA from other source.

Chitin A complex polysaccharide comprising repeated units of a sugar, N-acetylglucosamine, is present in exoskeleton of crustaceans like shrimps, lobsters, and crabs. It is also found in insects and fungi.

Chitosan Polysaccharide polymer obtained from chitin, used from agriculture to health care purposes—to dye fabrics, dietary supplements for weight loss, etc.

Chloramphenicol A broad spectrum antibiotic obtained from *Streptomyces venezuelae* that binds to 23S rRNA of prokaryotic ribosome and interferes with peptide bond formation in protein synthesis. This is used as a selecting agent for bacterial colony.

Chloroplast DNA (cpDNA) Genome of circular DNA molecule present in the chloroplasts of plants.

Chromatography A column filled with micro beads with different properties for separation of macromolecules.

Chromosome Organization of a DNA molecule with associative specific proteins to carry the genes and for their proper segregation after cell division.

Chromosome walking A technique for genome analysis to construct a clone *contig* by identifying overlapping fragments of cloned DNA.

Clone A genetically identical population of cells derived from a single cell.

Genetically identical cell populations or organisms result from asexual reproduction, isogenic pure breeding, and somatic nuclear transplantation to enucleated ovum. Making copies of a gene in multiplying prokaryotes or cells is also known as *gene cloning*.

Clone library A collection of clones in bacteria or cells, representing an entire genome of an organism, from which individual clone (gene) of interest may be obtained for biotechnological purposes.

Coding strand The non-template strand of a DNA molecule having equivalent sequences as in the mRNA encoded by the DNA template strand. (By convention it is called coding strand although it is not directly involved in transcription).

Codon (coding triplet) Three nucleotides that specify for an amino acid or a terminal signal.

Cohesive (sticky) ends Overhanging single stranded ends of a DNA molecule created by cut from certain restriction endonucleases. The sticky ends help in insertion of a DNA fragment (containing gene) in a DNA molecule for rDNA technology.

Colchicine Substance obtained from crocus flowers and blocks the formation of microtubules as spindle during cell division and thus prevents cell division.

Colicin Toxic protein from *E. coli*, acts as bacteriocin (killing) for related bacteria.

Collagenase Protease for digesting collagen and used in a number of biotechnological procedures.

Colony Usually refers to a visible growth of microorganisms or cells in specks on a semi-solid growth medium.

Colony hybridization A procedure for binding of a single-stranded DNA probe to DNA molecules of bacterial colonies containing cloned gene(s).

Competent cells Chemically treated or physiologically able bacterial cells that can accept DNA (transforming) from surrounding environment.

Complementary DNA (cDNA) DNA synthesized on mRNA template by reverse transcriptase enzyme.

Complex medium Nutritional medium for culture of microorganisms that contains such ingredients as yeast extract and Bacto Peptone, the exact chemical composition of which is not known.

Concatemer (catenanes) A tandem array of repeating unit lengths of DNA.

Conjugation (mating) A process of DNA (gene) transfer in bacteria through conjugating tube arising of a sex pilus.

Conserved positions Particular locations in nucleic acid or protein where the same individual bases or amino acids are always found across the individuals or species.

Contig (contiguous sequence) Restriction enzyme generated DNA pieces of entire choromosome are sequenced separately, and computer programs align the DNA fragments sequentially on the basis of overlapping sequence pieces called contigs.

Coomassie blue   A blue dye for staining protein.

Cosmid   A large hybrid cloning vector (ds DNA) that contains *cos* end sequences of bacteriophage λ, and is used for gene cloning and construction of genomic libraries.

Covalent bond   A bond formed between atoms by sharing a pair of electrons (strong bond).

Cre-lox system   A genetic recombination system and attached to a particular DNA (gene) sequence, and splicing with Cre recombinase enzyme the function of the particular DNA sequence can be regulated in cell types or tissues.

Crown gall   A tumour induced in many plants by infection with the bacterium *Agrobacterium tumefaciens*. Originally the bacterium was found to induce tumour at crown region, junction of shoot, and root of a plant.

Cryopreservation   Storage of cells or tissue samples after stepwise processing, at ultra low temperature of liquid nitrogen, for reviving and use in future.

C-terminus   The end of a polypeptide chain with a free carboxyl group.

Culture medium   A semi-solid or liquid medium of specified contents that is used for growth of microorganisms, cells, tissues, or organisms.

Cuvette   A small quartz or plastic tube (usually square in shape) with specific light-absorbing quality holds a sample for spectrophotometric analysis.

Cytokines   Low-molecular-weight (usually within 30 kDa) soluble regulatory proteins or glycoproteins secreted by leukocytes and a variety of other cell types in response to inducing stimuli, that play a significant role in cell cooperation during the development of immuncompetent cells.

Cytokinin   A plant hormone for stimulating cell division.

Defined medium (synthetic medium)   A nutrient medium for microorganisms, the chemical composition of which is known.

Degenerate primer   Primer having several alternate bases at certain positions.

Denaturation of nucleic acids   A process that use high heat (about 90°C) or alkali treatment to break the hydrogen bonds in DNA or RNA molecules to separate complementary base pairs and create single-stranded nucleic acid for hybridization with labelled probe or primer in PCR technique.

Denaturation of protein   Changing of the physiological conformation of protein to inactive conformation through heat or chemical treatment.

Density gradient centrifugation   A centrifugation technique that separates macromolecules on the basis of difference in their density; the solution of heavy compound such as CsCl is used for the purpose.

Deoxyribonucleic acid (DNA)   A high-molecular weight double-stranded biopolymer of nucleic acid consisting of nitrogenous bases (adenines, guanines, cytosines, thymidines), sugar (deoxyribose), and phosphate groups, that serves as the carrier of 'genes' which direct the production of proteins and RNAs in a cell and maintains molecular mechanism of inheritance.

Diabetes mellitus   A physiological condition of elevated blood sugar is called diabetes. In type I or insulin-dependent diabetes mellitus, little or no insulin hormone is synthesiszed by pancreatic β-Langerhan's cells.

Dialysis   A technique to remove impurities like salts from a solution of macromolecules by allowing diffusion of smaller molecules across a semipermeable membrane into water or a proper buffer.

Dicotyledon   A seed with two embryonic seed leaves, as in gram, groundnut, etc.; they supply nutrients to the germinating seedling.

Dideoxyribonucleotide (ddNTP)   Modified nucleotide in the form of nucleoside triphosphate that lacks hydroxyl groups (OH) on both 2′ and 3′ carbons of the pentose sugar. This is used for chain termination technique of Sanger for sequencing DNA in genome projects.

Differentiation   Structural and functional maturation of different cells in a multicellular organism following changes in gene expression with differential inductions from microenvironment.

Diploid   A cell or an organism having two homologous sets of each chromosome, representing one set of chromosomes from each parent (mother and father). This is often referred as *2n* condition.

DNA fingerprinting   Cleavage of genomic DNA by restriction endonuclease produces fragments and band pattern in gel electrophoresis which is unique for an individual. This unique pattern is used as a characteristic marker or DNA fingerprint for identification of an individual in case of forensic analysis and disputed paternity. The presence of a particular subset of bands in any two individuals speaks for their common lineage (inheritance); similarly nearness of two species can also be established.

DNA helicase   Enzyme that is responsible for separation of two strands of a DNA double helix structure during DNA replication.

DNA library   Collection of cloned DNA fragments of all DNA sequences of an organism's genome, which can be 'screened' to isolate genes of interest.

DNA ligase   An enzyme that makes a covalent bond between an adjacent 3′-OH and 5′-phosphate end of nucleotides in a DNA strand at a point of nick or for Okazaki fragments during DNA replication. This is routinely used for joining DNA fragments in recombinant DNA experiments.

DNA microarry (gene chip) Thousand pieces of single-stranded DNA of unique sequence are attached to a spot on a glass slide for hybridization with test samples of nucleic acid sequence to be identified.

DNA polymerase Key enzyme that copies complementary DNA strand on a template DNA strand during DNA replication. It is used for sunthesizing DNA in molecular biology experiments.

DNA sequencing Determination of sequential arrangement of nucleotides (A, G, T and C) in a DNA molecule.

DNA size markers DNA fragments of known sizes that are used to quantitate approximate sizes of unknown DNA fragments in bands, run on agarose gels.

DNA synthesizer (gene synthesizing machine) Automated machine to perform the steps of chemical reactions for DNA synthesis, to produce single stranded olionucleotides.

DNA topisomerase When two DNA molecules get interlocked during replication, this enzyme cuts one DNA molecule to pass out another DNA molecule through the cut, and reseals the break.

DNA (nucleic acid) vaccine Construct of plasmid DNA and DNA encoding specific antigen of a pathogen including its promoter in saline solution constitutes DNA vaccine. The expressed pathogenic antigen on the cells of immunee stimulates both antibody and cell mediated immune responses.

DNase footprinting Analysis of binding of a protein to DNA on the basis of protection of DNA from DNase 1 (non-specific) degradation, provided by the binding.

Domain A segment of a protein with discrete conformation and function.

Domain also represents the highest level of taxonomic groupings, above the level of kingdom; examples of domains are *Archaea*, *Bacteria*, and *Eukarya* (including both plant and animal kingdom).

Domain Archaea Prokaryotic cells that live in extreme environments, for example, boiling hot water springs, ice of Antarctica.

Dot/slot blot A detection strip coated with DNA sequences for detection of specific biomolecules.

Double-blind trial In a clinical trial of a drug the patients or the researchers do not know which patients received the drug and which patients received a placebo; the researchers learn of the treatment at the end of the study.

Doubling time (generation time) The time taken by single cell organisms to double in number.

Downstream sequences The nucleotide sequences of DNA that lie in the 3′ direction of the site of initiation of transcription which is marked as +1 nucleotide; downstream nucleotides are marked sequentially as +2, +3, etc.

Down syndrome Human genetic disorder resulting from the presence of three copies of chromosome 21 (trisomy 21) in an individual. This was first described by John Langdon Down. The characteristics of the affected persons are a flat face, shortness; short, broad hands and fingers; impaired physical and mental development; and protruding tongue.

DPT vaccine (triple antigen) A vaccine for immunization of children to protect against three types of bacterial diseases. It includes *diphtheria* toxin (disease affect upper respiratory tract), killed *Bordetella pertusis* (for whooping cough), and *tetanus* toxoid (to prevent tetanus – muscular spasm, and paralysis).

Early genes Genes transcribed before the replication of phage DNA, which provide regulators and other proteins for infection in later stages.

Edible vaccine Recombinant DNA technology allows expression of antigenic proteins from dreaded pathogens in transgenic plants and their fruits; eating of such raw fruits and vegetables (as vaccine) can induce immunity.

EDTA (Ethylene diamine tetraacetic acid, disodium salt) A chelating agent that removes divalent metal cations such as Mg^{2+} and Ca^{2+}, and disrupts the structure of cell envelope; removal of Mg^{2+} inactivates DNases in cellular exudate and prevents degradation of DNA during extraction.

Effluent Treated water as it leaves a treatment plant to remove chemical and biological pollutants, as from a sewage-treatment plant.

Electrophoresis A technique of separation of molecules on the basis of the electrical charge on them in an electric field.

Electroporation High-voltage short-time electrical treatment of cells including bacteria that cause transient pores, through which extraneous DNA is taken into the cells and genetic transformation of the cells takes place.

Elution volume (Ve) The volume of solvent used to elute out a solute from the time of its entry into the gel to the time it starts to emerge at the bottom of the column.

Embryonic stem cells (ES cells) Cells derived from the inner cell mass of a blastocyst (early embryonic stage) that can undergo differentiation to form all cell types in the body.

Encode Through the process of transcription and translation gene specifies the sequence of amino acids in a protein.

End-labelling A radioactively labelled group is added at one end (5′ or 3′) of a DNA strand.

Endonucleases Enzymes that cleave (cut) the phosphodiester bonds at specific site of an intact nucleic acid molecule.

Endoprotease Enzyme that cleaves the peptide bonds between amino acids at one or more specific sites within a protein.

Endotoxins   Heat-stable lipopolysaccharides (LPS) that are derived from the cell wall of Gram-negative bacteria; their toxicity resides in the lipid moiety (lipid A). They are toxic to the cells.

Enhancer   A *cis* sequence of nucleotides in DNA that binds transcription factors and enhances the function of eukaryotic promoters, and it can influence the promoter located upstream or downstream.

Enucleation   Removal of the nucleus from a cell.

Environmental genome project   The project to study the impact of environmental chemicals on human genetics and disease, under the supervision of NIH, USA.

Environmental genomics (metagenomics)   Genome sequencing for entire microbial communities in all kinds of environment—water, soil, and air throughout the world, oceans, glaciers, hot springs, mines.

enviroPig   A transgenic eco-friendly pig expressing phytase enzyme for digestion of phosphorus and excreting lesser amount of it through urine and faeces.

Enzyme   A cellular catalyst, usually protein that acts on specific substrate.

Episome   A plasmid that can integrate into bacterial chromosomal DNA.

Epitope (antigenic determinant)   The restricted portion of a complex antigen that binds to the paratope of the specific antibody and determines the specificity of Ag–Ab reactions.

Escherichia coli   Rod-shaped, Gram-negative, unicellular, facultative anaerobic chemo- organotroph, capable of respiratory and fermentative metabolism, mesophylic, non-photosynthetic, and non-sporulating eubacterium. These bacteria are natural flora in the intestine of animals including humans, and serve as experimental model for studies in genetics and molecular biology.

Ethidium bromide (EtBr)   An orange dye that intercalates (penetrates) between the base pairs of nucleic acids (efficient with double stranded) and fluoresces under UV light, and thus allows visualization of nucleic acid fragments in bands in gels and caesium chloride density gradients. The intensity of fluorescence also helps in quantization of nucleic acid in bands in reference to known standards.

Ethylene   A gaseous compound that acts as a plant hormone and has a role in fruit ripening, seed germination, senescence of flower, root elongation, and for withstanding environmental stress.

Euchromatin   In the interphase state of cell division, euchromatin remains in less tightly coiled than heterochromatin, and contains the active genes.

Eukaryotes   Animals, plants, protists, and fungi, that have chromosomes enclosed in a membrane bound (true) nucleus, and they also have organelles like mitochondria and chloroplasts (in plants and protists).

European Molecular Biology Laboratory (EMBL)   Maintains a comprehensive nucleotide (DNA and RNA) sequence database.

Eutrophication   Nutrient enrichment of organic materials and inorganic nutrients in rivers and lakes by waste waters, that promotes microbial and plankton growth which depletes oxygen in the water, resulting in the death of fish and other organisms. Nitrogen fertilizers on agricultural land and increased discharge of phosphates from sewage are main factors to cause eutrophication.

Excision   Enzymatic release or removal of a DNA segment from a chromosome or cloning vector, as a normal process within a cell or in-vitro protocol in molecular biology.

Exogenous   Derived from external or foreign source.

Exogenous DNA (heterologous or foreign DNA)   DNA (gene) derived from other organism for cloning into a vector for transferring into a new host cell.

Exon   A coding region of a gene, usually prevalent in eukaryotes.

Exon trapping   A molecular biology technique for isolating exons by using their flanking splice recognition sites.

Exonucleases   Enzymes that cleave nucleotides one by one from one end of a nucleic acid chain; they are specific for either the 5′ or 3′ end of DNA or RNA.

Expression   Transcription and then translation of a gene.

Expression library   A library of cloned DNA molecules in expression vectors which allow the transcription and translation of a foreign gene in a cell.

Ex-situ bioremediation   Contaminated soils or water is removed from the site of contamination for cleaning up at a separate location. (Cleaning up of pollutants at the contaminated site is known as in-situ bioremediation.

Extension (hanging end or sticky end of DNA)   A single stranded region hanging from the ends of a DNA molecule or its fragments, consisting of one or more nucleotides that help(s) in rejoining of DNA strands.

Extrachromosomal DNA   A replicable DNA element that is not part of an organism's chromosomal DNA; examples are plasmid and mitochondrial DNA.

Ex-vivo gene therapy   Gene therapy procedure (correction of defective gene) that involves removal of cells, such as blood stem cells, from an individual for introduction of normal or therapeutic gene and then placing back the cells into the individual.

Facultative anaerobe   An organism that can grow in the absence of oxygen, and also can utilize oxygen when available.

Fermentation A metabolic process of microbes (bacteria and yeast) that produces small amount of ATP from glucose (carbohydrate) in the absence of oxygen and ethyl alcohol (ethanol) or lactic acid (lactate) as by-product.

Fermenter (bioreactor) Sterile big container for growing microorganisms or mammalian cells under controlled conditions of temperature, pH, nutrient concentration, and cell density. This is utmost important for biotechnology industries.

Fertilization with nutrients Nitrogen and phosphorus added as nutrient fertilizer to polluted soil for growth of naturally occurring soil microorganisms for bioremediation.

5′ end The end of a strand of DNA or RNA which has a last nucleotide with free 5′ carbon of the pentose sugar, i.e., not attached to another nucleotide.

Flp recombinase (flaippase) Enzyme encoded by the 2 μm plasmid of yeast that catalyses recombination between inverted repeats (*frt* sites).

Fluorescein A fluorescent dye to label antibodies for visualizing binding of the antibodies with specific antigens in cells or tissues.

Fluorescence The process of absorption of light of one colour or wavelength by molecules and getting triggered to emit light of a longer wavelength (with lesser energy) and different colour for a short duration (for return of the excited molecules to their ground state of energy). When the absorbed photon is in the ultraviolet (UV) range, emitted light is in the visible range.

Fluorescent in-situ hybridization (FISH) Localization of gene (DNA sequences) on a chromosome by using single-stranded DNA or RNA probes labelled with fluorescent nucleotides that bind to the sites with high degree of sequence similarity, under fluorescent microscope.

Fluorophore Portion of a molecule with fluorescence property.

Footprinting A technique to identify the site on DNA to which some protein binds and protects it from degradation by nucleases.

Foreign DNA A DNA segment from an external source that is incorporated into either a cloning vector or a chromosomal site.

Forensic biotechnology Application of DNA and protein-sequence data as biologic evidence to solve crimes.

Fosmid A chimeric plasmid composed of origin of replication site from the F plasmid and a λ *cos* site; it is used for constructing stable libraries from complex genomes.

Frameshift mutation Mutation from insertion or deletion of nucleotides in chromosomal DNA causing a shift in the genetic code-reading frame of a gene.

Functional genomics Genetic sequencing of genes active during cellular functions.

Fungi Members of eukaryotic Kingdom Fungi that differs from the Plant Kingdom due to the ability of some of its member to synthesize chitin, a complex carbohydrate usually synthesized by some animal species.

Gamete A haploid (*n*) reproductive cell, sperm or ovum, arising from meiotic cell division, often called 'sex' cell. Male and female gametes fuse together during fertilization to form a diploid (2*n*) zygote.

Gel electrophoresis Separation of DNA, RNA, and proteins in a gel matrix on the basis of electrical charge and size of the molecules. Examples are agarose gel electrophoresis and polyacrylamide gel electrophoresis (PAGE).

Gel filtration (size exclusion or molecular sieve chromatography) Gel filtration chromatography is based on the penetration of solute molecules into small pores of column packing material of semipermeable porous resin beads (Sephadex, Sepharose, Acrylamide) on the basis of molecules' shape and size (technically hydrodynamic volume). It is usually used to separate protein molecules of disparate sizes.

GenBank An important public database of DNA sequences provided and accessed by researchers throughout the world for sharing and analysing DNA sequence information.

Gene A segment of nucleic acid that encodes a functional protein or RNA; an unit of inheritance.

Gene chip See DNA microarray.

Gene gun A microprojectile device to insert genes (DNA fragments) coated on metallic micro-particles (~1 μm diameter) into target cells.

Gene library (Genomic DNA library) A collection of clones representing the entire genomic DNA/cDNA of an organism.

Gene targeting A genetic technique to alter an endogenous gene through homologous recombination.

Genentech Name derived from genetic-engineering technology, for a California company founded in 1976; it is considered to be the world's first biotechnology company.

Gene therapy The insertion of a functional gene into an individual's cells and tissues to replace mutant allele causing a hereditary disease.

Generic drugs Copies of brand-name pharmaceutical products having same quality, effectiveness, and safety, but cheaper to the consumer than the branded ones.

Genetic engineering The process of altering gene(s) (DNA) of an organism by human manipulation through rDNA technology.

Genetic Information Non-discrimination Act This act is to protect the ethics that a person should not be discriminated on the basis of genetic information in any walk of life including health insurance and employment.

Genetically modified (GM) organisms Usually transgenic organisms fall under this category. GM crops that can resist pests, disease or extreme climates and with better products have already been introduced for human usages.

Gene specific primer Short-length nucleotide sequence complementary to the sequence at the beginning of a gene; it is helpful for duplication process of a specific gene.

Genetic code Triplet nucleotides in DNA (or RNA) specifying an amino acid for a protein.

Genetic complementation (mutant complementation) When two mutant genes or DNA molecules together produce a function that is not possible by the individual gene.

Genetic map (linkage map) The linear arrangement of genes on a chromosome, originally determined by recombinant frequencies between the genes; nowadays it is supported by DNA sequence studies.

Genome The total genes in an organism's DNA.

Genomics The study of DNA sequences of a genome, with the aid of computer.

Genomic library A set of cloned fragments together representing the entire genome of an organism.

Genotype The genetic constitution of an organism/individual.

Germ line cells Cells that produce gametes.

Germline genetic engineering Genetic alteration of germline cells (spems and eggs) by transferring a gene(s) to a fertilized egg or an early embryonic cell, so that the transferred gene(s) may be present in all the cells of the developing individuals including the reproductive (germ) cells.

Glycosidic bond A covalent bond formed between monosaccharides in a carbohydrate molecule.

Glycosylation A cellular process of adding short-chain carbohydrates (sugar) units to newly synthesized protein molecules in the lumen of the rough endoplasmic reticulum. This is usually done to most membrane proteins and secretory proteins.

Golden rice Genetically engineered variety of rice with high content of vitamin A.

Golgi complex (Golgi apparatus) A complex organelle of eukaryotic cells, consists of a stack of flattened membranous sacks, named after its Italian discoverer, Camillo Golgi, that functions in the processing and packaging of secretory vesicles.

Gram negative bacteria Bacteria that do not take the colour of crystal violet used as first stain in Gram staining, they get the pink colour of safranin, used for counterstain, when examined under a light microscope.

Gram positive bacteria The variety of bacteria that retains the colour of crystal violet of Gram stain as they have peptidoglycan in the cell wall; and do not take safranin counterstain.

Gratuitous inducer A substance capable of inducing transcription of gene(s) for enzyme(s) without being substrate of the enzyme(s).

Green fluorescence protein (GFP) A protein from bioluminescent jellyfish *Aequorea victoria*, whose green fluorescence with exposure to ultraviolet light allows use of the *GFP* gene as a reporter gene.

Growth hormone (GH) A peptide hormone synthesized in the pituitary gland, that stimulates the growth of bone, muscle, and other tissues.

GUS (b-glucuronidase) The bacterial enzyme β-glucuronidase; the gene encoding this enzyme is used as a reporter gene in the production of transgenic plant.

Half life The time required for the disappearance or decay of one-half of a given substance in a system. Also applicable for decay of radioactivity of an isotope.

Haploid A single set of chromosomes (*n*) of the organisms; in higher organisms one copy of each autosome and one sex chromosome (a set of 23 chromosomes in human).

Heat shock genes A set of genes at different loci that become active with increase in temperature and other stresses to the cells; that encode molecular chaperones to act on denatured/misfolded proteins.

Helicase An enzyme to separate the strands of the nucleic acid interwound in a helix and that utilizes energy from hydrolysis of ATP for its function.

Helper virus A virus that makes a mutated virus functional.

Hemimethylated site A palindromic sequence in DNA that is methylated on only one strand.

Herpes simplex virus DNA virus of *Herpetoviridae* family and replicates in the cell nucleus, that produces fever blisters and is related to herpes zoster (chickenpox or varicella). Virus particle has a diameter of 180 nm, and its natural host is man, but mice, guinea pigs, hamster, and rabbits are also infected.

Heterochromatin Highly condensed parts of genomic DNA, that are late replicating and do not transcribe; they are of two types – constitutive and facultative.

Heteroduplex DNA (hybrid DNA) DNA produced by hybridization of complementary single strands from two DNA molecules.

Heterokaryon A cell containing two nuclei in a common cytoplasm, usually produced by fusion of somatic cells, in course of formation of a hybridoma.

Heterologous probe A DNA probe from one organism to screen similar DNA sequence in a gene clone library obtained from another organism; this is often used for genomic studies.

Hetrozygote   An individual with different alleles at the same locus of two homologues.

Hfr strain   Bacterial strain that transfers chromosomal genes at high frequency for recombination due to integration of fertility (F+) or sex factor plasmid in the chromosome.

High performance liquid chromatography (HPLC)   An improved column chromatography technique to separate, identify, and quantitate chemical and biochemical substances; it uses stationary chromatographic packing material in the column of incompressible beads, a pump to generate pressure on mobile phase for separation of the substances in the column rapidly, and a computer-based detector to indicate the retention time of the different molecules.

Histidine tag (His tag)   Six histidine residues in tandem, tagged to proteins so that they can bind to nickel ions on a solid support for purification of the proteins.

Histone   DNA-binding proteins for compaction of double helix in a chromosome and play a role in gene activation.

Homogenization   The process of breaking apart cells to release cytoplasm and organelles; this is done by pestle and mortar (spreferably under liquid nitrogen), blender, and ultrasonicator.

Homologous chromosomes   Pairs of chromosomes derived from two parents; chromatids of a pair normally synapse and exchange chromatic material (recombination) during zygotene stage of meiotic cell division.

Homozygote   An individual having identical alleles at the same locus in its two homologous chromosomes.

Horizontal gene transfer (Lateral gene transfer)   The process of gene transfer in which genetic material is incorporated by an organism from another organism not being descendant of that organism; e.g., transformation in bacteria.

Hormone   The word was derived from the Greek verb *horman*, meaning 'to stir up or excite', and used by Ernest Henry Starling in 1905 to name a group of chemical substances synthesized in small amounts by endocrine tissue (organs) and carried by blood to act as messenger to regulate the wide range of functions of far-away target tissues or organs. Chemically, the endocrine hormones fall into four categories: amino acid derivatives, peptides, proteins, and lipid-like hormones (steroids). Hormones (e.g., indole acetic acid) also play roles in plants.

Hormone receptors   Protein molecules within or on the surface of target cells that bind to specific hormone molecules and initiate cellular responses.

Hotspot   A site in the genome at which mutations occur with high frequency.

Housekeeping genes (constitutive genes)   Theoretical assumption of genes that express in all cells for basic functions in all cell types.

Human cytomegalovirus   A virus of the herpes group, infecting humans.

Human genome project   An international research project with goals of finding the complete genomic DNA sequences of human (9.2 billion bp), identifying all human genes and determining their locations on respective chromosomes (mapping).

Human papilloma virus (HPV)   Virus (strain 16 and 18) that has been identified as causative agent for human cervical cancers.

Hybridization   The pairing of two polynucleotide strands by H bonding between complementary nucleotides; for instance, binding of a single stranded DNA probe or mRNA to the single strand of complementary DNA.

Hydrophilic   'Water-loving'; affinity of a molecule to water.

Hydrophobic   'Water-hating'; repelled by water.

Hydroxyapatite   Calcium phosphate, a structural component of bone and cartilage. Crystalline hydroxyapatite, precipitated from aqueous solution, is with large surface area in ratio to particle size, and used to separate single stranded nucleic acids from double stranded nucleic acids in hydroxyapatite column chromatography.

Icosahedral symmetry   A typical symmetry of viruses having polyhedron capsids.

Immunoaffinity chromatography   Antibody molecules bound on matrix in a column bind a specific protein and separate it from a complex mixture.

Immunoglobulin (antibody)   A class of globular glycoproteins produced by B lymphocytes in response to antigens.

Importin   A special type of *transport receptor* protein that binds to nuclear localization signal (NLS) sequences of cargo protein in cytoplasm and helps in recognition and transport of the complex by transporter at the centre of the nuclear pore from cytoplasm to the interior of the nucleus. [*Exportin* molecules do the similar job to export protein – RNA cargo from the nucleus to the cytoplasm].

In-situ (Latin for in original place) bioremediation   Cleaning up of pollutants at the site of pollution. *ex-situ* bioremediation is removing the bulk of polluted soil (or water) to another location.

In-situ hybridization (cytological hybridization)   Laboratory technique to identify specific gene sequences in a chromosome by allowing radiolabelled or fluorescent labelled single stranded RNA or DNA probes to hybridize with the denatured DNA of cells on a microscopic slide.

In-vitro (Latin, in glass)   'Outside of the living organism', or 'in test tubes', in culture.

In-vitro fertilization   Assisting reproduction technology in which sperms and egg cells are collected from individuals and cultured in a Petri dish under controlled condition to achieve fertilization.

In-vivo (Latin, in the living organism)   Occurring within a living organism.

In-vivo gene therapy   Introducing therapeutic gene(s) directly into an individual's tissue or organs (without removing them from the body) to correct a genetic disorder.

Indole 3-acetic acid (IAA)   An auxin, a plant hormone, derived from tryptophan.

Inducer   A small molecule that binds to a regulator protein for initiating gene transcription.

Inducible operon   A genetic operon consisting of structural genes that are expressed or transcribed in the presence of a specific inducer molecule.

Inducible promoter (regulatable promoter)   The promoter attached to a transgene that can be expressed only when some chemical agent (or environmental stimuli) is given transiently to induce the promoter.

Induction of prophage   The entry of a phage from integrated prophage (lysogenic dormant) condition in bacterial chromosome into the free and replicative lytic cycle, due to destruction of the lysogenic repressor.

Informed consent   Patients are informed about potential beneficial and harmful effects including risks of a particular treatment (e.g., new drug) for them to agree (consent) to participate.

Inoculums   The cells used to 'seed' (start) a new culture.

Insertion vector   A phage cloning vector that has non-essential genes removed to accommodate foreign DNA at a unique restriction enzyme site(s) as in plasmid.

Insulin   Protein hormones produced by β cells of islets of Langerhans in pancreas, to carry out glucose metabolism; deficiencies in insulin or insulin receptor production lead to different forms of diabetes. Interestingly, insulin is the first protein to have its complete amino acid sequence determined (in 1956 by Frederick Sanger), and human insulin was among the first proteins to be produced by genetic engineering.

Intron (intervening sequence)   A sequence of nucleotides intervening between functional sequences (or *exons*) in an eukaryotic gene, that is transcribed in the primary transcript (pre-mRNA) but excised before the transcript is translated.

Ion channel   Transmembrane protein molecules arranged in circular fashion with a central aqueous pore that selectively allows the passage of one type of ion through the membrane; generally regulated by either changes in membrane potential (voltage-gated channels) or binding of a specific ligand (ligand-gated channels).

Ion exchange chromatography   Technique for separation of molecules on the basis of ionic charge on them. The chromatographic column is a long tube filled with particles of a synthetic resin containing fixed charged groups; that with fixed anionic groups is called *cation-exchange* resin and that with fixed cationic groups, *anion-exchange* resin. The molecules bound by charge to the resin are eluted by buffer with different salt concentrations. This is used for separation of charged large proteins, small nucleotides, and amino acids.

Isoletric focussing   An electrophoretic method for separating macromolecules on the basis of their isoeletric point (pI).

Isoelectric point (pI) (or Isoelectric pH)   The pH at which a solute bears no net electric charge and does not move in an electric field.

Isoenzymes   See isozymes.

Isomers   Two molecules having same molecular formula differ in the arrangement of groups of atoms.

Isotopes   Atoms having the same number of protons and electrons but differ in the number of neutrons; used for radioactive labelling.

Isozymes (Isoenzymes)   Multiple forms of an enzyme that catalyse the same reaction but differ from each other in their amino acid sequence, substrate affinity, and reaction kinetics.

Ketose   A monosaccharide in which the carbonyl group is a ketone.

Ketosis   A state of abnormally high concentration of ketone bodies in the blood, tissues, and urine.

Kilo base (kb)   A unit of length of 1,000 nucleotide bases in a strand of nucleic acid, especially in RNA.

Kilobase pair (kbp)   A measure of length of 1,000 nucleotide base pairs in a DNA molecule.

Kinases   Enzymes that catalyse the phosphorylation of certain molecules by transfer of the phosphate group from ATP (to the hydroxyl group of a serine, threonine, or tyrosine residue in the protein).

Klenow fragment   The large fragment (68,000 Da) of DNA polymerase I from *E. coli* retaining both polymerase and 3′-proofreading exonuclease activity, but not 5′ exonuclease activity, that is obtained after mild protease treatment.

Knock-in (gene knock-in) technology   A genetic engineering method for insertion of a functional gene (cDNA sequence) at a particular locus following homologous recombination, to create a knock-in mouse strain.

Knock-out (gene knock-out) technology   Homologous recombination may also replace a normal functional gene with a mutant allele or a disrupted form of a gene to raise a transgenic knock-out mouse strain.

Krebs cycle (citric acid cycle or tricarboxylic acid (or TCA) cycle) In course of glucose metabolism, the cyclic series of reaction in the matrix of mitochondria to produce ATP, involves an important intermediate citrate which has three carboxylic acid groups. Each round of TCA cycle activity occurs with the entry of an acetyl CoA molecule from cytoplasm into mitochondria, and the release of two carbons as CO_2 and a molecule of ATP (or GTP) and the regeneration of oxaloacetate. In reference to citrate (tricarboxylic acid) the cyclical metabolic pathway was named citric acid or TCA cycle; it is also referred as the Krebs cycle in honour of Hans Krebs, whose laboratory played a central role in elucidating this cyclical sequence of biochemical reactions in the 1930s.

Lac operon A well-characterized bacterial operon that contains a series of functionally related three genes (*lac Z, Y, A*) for metabolism of the sugar lactose; the cluster of three genes is preceded by regulatory DNA sequences of inhibitory protein, promoter, and operator. This provides a beautiful system for studying the regulation of gene function.

Lagging strand During DNA replication, the strand of newly synthesized DNA that is copied by DNA polymerase 5' to 3' direction away from the replication fork in short discontinuous series of DNA pieces called Okazaki fragments, to be joined together later by DNA ligase to make a continuous new 3' to 5' DNA strand.

Leachate Soluble substance to wash with water or other liquids from the surface or near the surface to deeper layer of soil or downstream water source.

Leading strand During DNA replication, the new strand of DNA that is copied by DNA polymerase in a continuous fashion in 5' to 3' orientation towards the replication fork.

Leader sequence A short hydrophobic N-terminal sequence in newly synthesized polypeptide chain responsible for initiating passage of the chain through membrane, particularly reticulo-endothelium.

Lectins A group of plant proteins that can bind to specific oligosaccharides on the cell surface; they are usually obtained from seeds where they act as toxin towards certain pathogens.

Leukocytes White blood cells of five classes including B and T lymphocytes and macrophages, responsible for immune responses.

Ligand A molecule capable of binding to a receptor; binding to the receptors on the plasma membrane of a cell may activate some biochemical functions in the cytoplasm.

Limulus amoebocyte lysate (LAL) test An important test that involves amoebocyte type blood cells from the horseshoe crab, *Limulus polyphemus,* for detecting endotoxin and bacterial contamination in foods, medical instruments, etc.

Lipofection DNA, RNA, drugs or other compounds, encapsulated in artificial membrane (phospholipid) vesicle

liposome enter into eukaryotic cells after fusion of the vesicle with cell membrane.

Liposomes Small closed vesicles of about 0.1 μm diameter formed by lipid bilayer, devoid of membrane protein, that form on dispersal of cell membrane lipids in water. The polar groups and hydrophobic region of lipid bilayers remain in exactly same orientation as in a cell membrane; the hollow sphere of liposome with an aqueous cavity inside can hold different macromolecules and therapeutic agents for lipofection (see *lipofection*). Alec Bangham and his colleagues in 1961 first showed the formation of liposome.

Locus (pl. Loci) The position of a gene for a particular trait on a chromosome; two alleles of a gene occupy the same locus on two homologous chromosome.

Log phase The period during which bacteria and cells in culture divide at a constant doubling rate. This is preceded by lag phase and followed by stationary phase.

Long terminal repeat (LTR) sequences At both the ends of the linear RNA genomes of retroviruses are LTR sequences that are a few hundred nucleotides long. The sequences in LTRs are required for integration of retroviral DNA into the host cell DNA (chromosome) and the regulation of expression of viral genes. [Eukaryotic transposons (wandering genes) from yeast and *Drosophila* have been found to carry δ sequences at both the ends, which are structurally and functionally equivalent to LTRs, and are called retrotransposons.]

luc gene Gene that encodes luciferase in eukaryotes for bioluminescence, and is used as reporter gene.

Luciferase The enzyme that catalyses the oxidation of luciferin and emission of light in bioluminescent reactions, as in fireflies.

Luciferin The chemical substrate for luciferase enzyme in bioluminescent reactions.

lux gene Luciferase encoding gene from eubacteria, also used as reporter gene.

Lycopene A carotene photosynthetic pigment.

Lyophilization The process of freeze-drying, used for concentrating and preservation of proteins.

Lysis Refers to the breakage and destruction of cells by various means, such as viruses, antibody or cell mediated, chemical or physical treatment.

Lysogeny A relatively benign integration of viral (phage) genome into the bacterial chromosome where it is replicated passively with the host DNA for many generations. This is a way of survival of phage genome in a bacterium as a *prophage*.

Lysosome Membrane bound organelle in eukaryotic cells containing hydrolases capable of digesting all the major classes of biological macromolecules.

Lytic cycle   A process of viral infection and replication in a host cell and then rupturing (lysing) the host cell to release new viral particles.

Macrophage   (Literally means big eater) Phagocytic white blood cell derives from bone marrow monocyte. It engulfs and destroys dead cells and pathogens like bacteria and play accessory roles in immune reactions.

Major histocompatibility complex (MHC)   Multigenic multiallelic large complex of genes that encode the histocompatibility antigens expressed on cell surface of an individual and also encodes soluble complement factors and cytokines. The complex is located on chromosome 6 in man (HLA) and chromosome 17 in mouse (H-2). MHC encoded antigens between recipient and donor need to be 'matched' for successful organ transplantation.

Malaria   A parasitic disease accompanied by symptoms of recurrent chills, fever, and sweating, that is caused by the different species of protozoan parasite *Plasmodium*, of which *Plasmodium falciparum* is the most virulent and fatal. Usually mosquitoes transmit the parasites between individuals, as vector.

Mariculture   The cultivation or 'farming' of marine animals, particularly finfish and shellfish, and aquatic plants for commercial and recreational purposes.

Mass spectrometry (mass spec)   Technique for measuring the mass of molecular ions obtained from volatilized molecules.

Master plate   A culture plate with the original microbial colonies from which replica plates are made to carry on genetic studies.

Material balance   Material and energy balances help in quantifying the various flows in and out of a bioprocess engineering system, such as a bioreactor.

Meiosis   The type of cell division that occurs during formation of gametes in germinal tissue, to reduce the chromosome number from diploid ($2n$) to haploid (n). This is made possible in meiosis as one round of chromosomal replication is followed by two successive nuclear divisions.

Megabase pair (Mbp)   One million base pairs in DNA.

Messenger RNA (mRNA)   The category of RNA molecules that is synthesized from DNA by transcription and serves as a template (carrying the sequence of nucleotides as in gene) for synthesizing a protein chain, with the help of ribosomes.

Metabolic engineering   Modification of an energy-generating or energy-requiring chemical process, to improve energy-generation or to reduce energy utilization; this is done through biochemical or chemical process.

Metabolism   All the chemical transformations that occur in a cell or an organism, by a series of enzyme-catalysed reactions (a series of reactions constitute a metabolic pathway). Metabolism can be further characterized as *anabolic* (biosynthetic) and *catabolic* (degradative).

Metabolites   Hundreds of small organic molecules as intermediates in biosynthetic (anabolic) and degradative (catabolic) pathways of metabolism.

Metalloproteins (metallothioneins)   Proteins that contain a specific metal as prosthetic group; for example, ferritin containing iron, alcohol dehydrogenase containing zinc, calmodulin containing calcium, plastocyanin containing copper.

Methylation   Addition of a methyl group (CH_3) to a macromolecule, a protein or a nucleic acid, by methyltransferase (or methylase) enzyme.

Microbes (microorganisms)   Small organisms, usually imperceptible by naked eye, and viewed with the help of a microscope. Microbes are bacteria, viruses, fungi such as yeast and mold, algae, and single celled protozoa.

Microinjection   The technique of introduction of DNA or other compounds into an eukaryotic cell with a micro-needle operated under microscope. This is often used to produce transgenic animals.

MicroRNAs (miRNAs)   A rapidly growing family of small RNA molecules about 20–25 nucleotides long that play novel role in regulating gene expression in eukaryotic cells.

Microsatellite DNAs [short tandem repeats (STRs)]   Short repeating sequences of DNA (typically <10 bp), often one, two, or three-base sequences (for example, CACACA-CACA) in a genome; they are important markers for forensic DNA analysis.

Minimum inhibitory concentration (MIC)   The lowest concentration of an antimicrobial substance that prevents growth of a particular organism.

Ministellite DNA   Shorter regions of DNA, about 10^2 to 10^5 bp in length, composed of a tandem repeat unit of approximately 15 to 100 bp (typically >10 bp).

Mismatch   Lack of base pairing between one or more nucleotides of two hybridized nucleic acid strands.

Missense mutation   A mutation in DNA changing a codon to another codon that specifies a different amino acid.

Mitochondrial DNA   Small circular DNA in mitochondria of eukaryotic cells, codes for mitochondrial ribosomal RNA, tRNA for mitochondrial protein synthesis and mitochondrial proteins. This is a genome independent of nuclear genome. Mammalian mitochondrial DNA (mt DNA) typically have about 16,500 bp (16,569 bp and 5 μm in length in human), yeast mt DNA is roughly five times larger and plant mt DNA is even bigger. Mutations in mt DNA can be followed in lineage from mother to offspring because the ovum is the sole source of mitochondria.

Mitosis A process of equational division of nucleus in eukaryotic somatic cells and prokaryotes, providing equal amount of genomic DNA in daughter cells. (in eukaryotic cells, mitosis passes through four major stages—prophase, metaphase, anaphase, and telophase).

Molecular mass The sum of the masses of all the atoms in a molecule, expressed in the unit of Daltons or Kilo Daltons.

Molecular pharming The synthesis of pharmaceutical products in transgenic organisms for commercial interest.

Monoclonal antibody (mAb) Identical copies of antibody molecules with mono specificity produced by a clone of cells derived from a single cell. Nowadays, large quantities of monoclonal antibodies can be produced by the hybridoma technology in which a B lymphocyte capable of producing antibody to a specific antigen (epitope) is fused with a malignant myeloma cell.

Monocotyledonous plant A plant with a single cotyledon (embryonic seed leaf) as in maize.

Morula Latin of 'little mulberry', indicates the stage of embryonic development in animals represented by a solid ball of cells, produced by repeated cell division of the zygote.

Motifs Units of secondary structure of protein molecule, consists of small folded segments of an helix and or sheet connected by looped region of varying length The same motif may be present in different proteins, and usually functions similarly.

Multiple cloning site (MCS or polylinker) Biochemically synthesized DNA fragment containing recognition sequences for several restriction endonucleases, used in a vector for cloning of target DNA.

Mutagen A chemical or physical agent capable of inducing mutations.

Mutation An alteration in an existing DNA sequence; it may or may not have effect on the phenotype.

Myeloma Plasma cell (antibody secreting) tumour.

Narrow host range plasmid A plasmid capable of replication in one or only a few species of bacteria.

Nanotechnology Engineering structures, and technologies at the level of nanometre scale.

National Center for Bioinformatics Information (NCBI) National centre in USA that maintains public databases on DNA sequences of various organisms, protein sequences, mitochondrial DNA sequences, sequences of vector genomes, etc, which can be accessed by researchers from any corner of the world through computer programs.

Nick A gap in the sugar–phosphate backbone of one of the strands of DNA duplex due to cleavage of phosphodiester bond.

Nitrocellulose A membrane for transfer and immobilization of nucleic acids or proteins from gel by blotting technique.

Nonsense codon See Stop codon (termination codon).

Nonsense mutation Mutation that converts amino acid-coding codon into stop codon(s), which causes premature termination of translation.

Northern blotting A technique to detect mRNA as a measure of expression of an active gene by gel electrophoresis and transfer of mRNA from gel onto a matrix of nitrocellulose or nylon membrane by blotting and the presence of a specific type of RNA molecules is detected by hybridization with DNA probe on the membrane. The name of the technique is somewhat fanciful to keep parity with Southern blotting for detection of DNA.

Nuclear envelope The nucleus of an eukaryotic cell with chromosomes is separated from cytoplasm by a nuclear envelope which consists of two concentric membranes, called the *inner* and *outer nuclear membranes*. Uniquely the nuclear membranes have numerous *pores* for transport of many substances between the nucleus and cytoplasm.

Nuclear localization signal (NLS) A short amino acid signal sequences (8–30 amino acids) on the protein molecules synthesized in cytosol and destined for use in the nucleus of an eukaryotic cell. An NLS—containing 'cargo' protein binds to importin and the importin–cargo complex is then imported in the nucleus through the nuclear pore complex. (NLS is on the 'cargo' protein for transport through the nuclear pore complex).

Nuclear pore complex (NPC) A large, proteinaceous structure forming a channel across the inner and outer membranes of the nuclear envelope for bidirectional transport of molecules and macromolecules between the nucleus and the cytosol. The diameter of the entire pore complex is about 120 nm and has a mass of about 120 million Daltons and consists of dozens of different types of polypeptide subunits arranged in octagonal fashion.

Nucleic acids Molecules that encode genetic information and include DNA (deoxyribonucleic acid) and RNA (ribonucleic acid); these consist of a series of nitrogenous base nucleotides linked with phosphodiester bonds.

Nucleocapsid The genome of a virus with its surrounding protein coat.

Nucleoid A special region of prokaryotic cell, where the circular chromosome is located.

Nucleolus A prominent spherical structure in eukaryotic nucleus to produce rRNA for ribosomes. One or two nucleoli are usually present in a cell but multiples of them may also occur. Nucleolus is not bound with membrane, and consists of DNA fibrils containing multiple copies of rRNA genes arranged tandemly and granules; the granules are rRNA molecules being packaged with proteins (imported from the cytoplasm through nuclear pore complex) to form ribosomal subunits.

Nucleoside   A purine or pyrimidine base covalently linked to a five-carbon (pentose) sugar; sugar being ribose, the nucleoside becomes ribonucleoside, when the sugar is deoxyribose, the nucleoside is a deoxyribonucleoside. (Nucleoside plus phosphate group constitutes a nucleotide).

Nucleosome   Basic structural unit of eukaryotic chromosomes, consisting of about 200 base-pair length of DNA molecule wound around an octamer (8 subunits) of histones.

Nucleotide   The building block of nucleic acids, comprising a pentose sugar, a purine or pyrimidne nitrogenous base, and one or more phosphate groups (also called nucleoside monophosphate).

Nucleus   Large, spherical, double membrane-bound organelle that contains the chromosomal DNA of a eukaryotic cell. (The term eukaryote derives from *'eukaryon'* meaning 'true nucleus').

Nylon membrane   Used for transfer of nucleic acids or protein from gel by blotting, and it is stronger and more versatile than nitrocellulose.

Nutrigenomics   A new field of nutritional science for understanding interactions between diet and genes.

Okazaki fragments   Newly synthesized short fragments of lagging-strand DNA during DNA replication that are joined together in a strand by DNA ligase.

Oligonucleotide (Oligodeoxyribonucleotides or Oligomers)   A short, single-stranded synthetic DNA sequence, used in PCR (as primer) and as DNA probes.

Oligonucleotide array detector (DNA array, DNA chip)   Microchip used for simultaneous detection and identification of many short DNA fragments by DNA:DNA hybridization.

Oncogene   A mutated gene which can cause malignancy or cancer; when it arises by mutation from a normal cellular gene, the latter is called proto-oncogene.

Operon   Cluster of genes with related functions in bacteria that is under the control of a single operator and promoter which bind with regulatory molecules and thereby allow transcription of the genes in the cluster to be turned on or off together.

Opine   An arginine derivative, synthesized by plant cells infected with crown gall disease.

Origin of replication (ori or origin)   The nucleotide sequence or site in DNA where DNA replication is initiated.

Origin of transfer   Point on an F factor plasmid at which the transfer of the plasmid DNA from an F^+ donor bacterial cell to an F^- recipient cell begins in course of its replication during conjugation.

Organelles   Discrete intracellular structures, each of which performs a particular specialized function in an eukaryotic

cell. Examples of membrane bound organelles are the nucleus, mitochondria, endoplasmic reticulum, Golgi complex, and lysozomes. Ribosomes, microtubules, and microfilaments can be considered as organelles without membranes.

Organic molecules   Molecules that contain carbon and hydrogen.

Orthologs   Corresponding proteins in two species as defined by sequence homologies.

Osmolarity   Solute concentration on one side relative to that on the other side of the semi-permeable membrane that drives the osmotic movement of water across the membrane. (also See Osmosis)

Osmoprotectant   Small molecules acting as osmolytes that protect the cells to survive extreme osmotic stress, amino acids, betaines, and the sugar trehalose are examples.

Osmosis   The process of movement of water through a semipermeable membrane due to a difference in solute concentration on the two sides of the membrane.

Osteoporosis   Bone disorders that usually involve a progressive loss of bone mass.

Oxidation   A chemical reaction involving the removal of electron(s) from an atom or a molecule.

Oxidative phosphorylation   Electron flow from catabolic intermediates to molecular O_2 (during degradation of carbohydrates, fats, and amino acids in aerobic eukaryotic cells), yielding energy for the generation of ATP from ADP and Pi in mitochondria. Thus, oxidative phosphorylation is oxygen dependent electron transport, distinguishing it from *substrate level phosphorylation.*

p53 gene   Tumor suppressor gene in human that codes for the p53 protein, a transcription factor that prevents genetically (DNA) damaged cells proliferaling and leading them to apoptosis; this gene is mutated when most of the cancers occur in humans.

p arm   'Petit' or small/short arm of a chromosome.

Palindrome   A word or a phrase that reads same in forward or backward direction; e.g. 'a Toyota'. A DNA sequence with complementary sequences in two strands is a palindrome, reading same in both the directions; type II restriction endonucleases recognizes palindrome sequence to initiate cut.

Papain   Protein digesting enzyme, isolated from papaya.

Parthenogenesis   Creating an embryo from an ovum without fertilization by a sperm.

Patent   Legal right granted to an inventor or a researcher or an institute for a product for making, using or selling for certain years and that prohibits others to do so.

Pathogen   Disease-causing organisms, usually microorganisms.

P element   A type of transposon in *Drosophila melanogaster*, fruit fly, often used for recombinant DNA technology for inserting foreign gene(s).

P1-derived artificial chromosome (PAC)   PAC can be used as plasmid vector to insert a large DNA sequence in a host cell; it has the combination of origin of replication of P1 bacteriophage and bacterial artificial chromosome (BAC).

Paralogs   Proteins with higher degree of similarity, coded by the same genome.

Parasporal crystals   Tightly packed protoxin molecules produced by *Bacillus thuringenesis* during the formation of resting spores; these are highly toxic to the caterpillar of certain insects and act as biopesticides.

Pentose sugar   A five-carbon sugar; present in the sugar-phosphate backbone of nucleic acids.

Peptidoglycan (murein)   The major constituent of bacterial cell wall, comprising linear polymers of alternate units of specialized sugars N-acetylmuramic acid and N-acetylglucosamine, interconnected by short polypeptides.

P (peptidyl) site   Site of binding of peptidyl tRNA in ribosome during translation.

Peptidyl transferase   rRNA enzyme (ribozyme) as a part of large ribosomal subunit catalyses the formation of peptide bonds between amino acids brought in by tRNA during translation.

Perforin   A pore-forming protein produced by endotoxic T lymphocytes (CTLs) for killing the target cell. Perforin monomers undergo a Ca^{2+} induced conformational change leading to their insertion into the target cell membrane and polymerization to form transmembrane cylindrical pore.

Permease   Enzyme coded by the *lacY* gene of the *lac* (lactose) operon in bacteria; it participates in the transport of disaccharide lactose into bacterial cells.

Personalized genome   Sequencing a genome for an individual person.

Phage   see Bacteriophage.

Pharmacogenomics   A customized treatment for a particular health condition based on a person's genetic information.

Phenotype   The observable characteristics of an organism/individual.

Phosphodiester bond   Phosphate group (PO_4^-) that links covalently successive nucleotides in nucleic acid (DNA and RNA) strand.

Phosphodiesterase   An enzyme that hydrolyses phosphodiester bonds between nucleotides of DNA or RNA.

Photolithography (optical lithography)   A process that uses light to transfer a geometric pattern from a photomask (or mirror in maskless photolithography) to a light-sensitive chemical (photosensitive emulsion) on the substrate. This allows synthesis of a large number of different nucleotide arrays for making a DNA chip for detection of different sequences of DNA.

Phytohormones   Plant hormones that activates growth and other processes in plants; e.g. auxin, abscissic acid, cytokinin, gibberelin, and ethylene.

Phytoremediation   Bioremediation, cleaning pollutants in environment, by using plants.

Pilus (pl. pili)   A thin, flexible rod-like appendage, made up of pilin monomers (protein), a few on the surface of F^+ (male or Hfr) strain of bacterium, that serves as a conjugating tube for transfer of DNA from F^+ or Hfr strain of bacterium to a F^- (female) bacterium during the process of conjugation.

Placebo   A control group without treatment in a scientific investigation by the side of an experimental group subjected to a treatment, such as in a drug trial. The control group is given blank treatment with ineffective or sugar pill and injection of saline without medication.

Plaque   A clear, usually circular spot of dead bacteria appearing in an otherwise turbid background of bacterial culture plate, caused by infection and lysis of bacteria by bacteriophages. This is visible to the naked eyes after several hours of incubation of the bacterial culture plate.

Plasma (cell) membrane   A bilipid layer containing proteins and carbohydrates that surrounds a cell, and performs important functions like maintaining shape of the cells and being semipermeable, regulates transport of solute molecules, ions and water in and out of a cell.

Plasma cell   Antibody producing cell, large and with elaborate endoplasmic reticulum that develops from B lymphocytes after activation with specific (foreign) antigen.

Plasmid   Small, circular, self-replicating, extrachromosomal double-stranded DNA molecule found primarily in bacteria and in some yeasts, ofter harbours antibiotic resistant genes; specially engineered (rDNA technology) forms are used as vectors in gene cloning.

Plasmid curing   Removal of plasmids from bacteria.

Plating   Spreading of cells onto a solid (agar) nutrient medium in a Petridish.

Pluripotent   The term describes cells with the potential to develop into any type of cell, such as embryonic stem cells.

Plus strand of DNA   The non-template strand of the double helix DNA having the same sequences as that of the encoded RNA (same as coding strand).

Point mutation (single nucleotide mutation)   A single base (then a base pair) change in DNA sequence.

Polarity   Usually polarity refers to the 5′ and 3′ ends of DNA and RNA molecules.

Pollen Male gametes of plants in the form of microspores.

Polyacrylamide Electrically neutral polymer matrix used for separation of proteins or very small fragments of nucleic acid molecules in gel electrophoresis. [monomeric (unpolymerized form) acrylamide is a very potent neurotoxin].

Polyacrylamide gel electrophoresis (PAGE) Technique for separation of proteins and small fragments of DNA molecules by electrophoresis on a polyacrylamide gel. The resolving power of polyacrylamide gels is extremely high and more than that of agarose gels.

Polyadenylation Repeated addition of adenine (A) nucleotides to the 3′ end of an mRNA molecule in the form of a 'tail' by poly A polymerase enzyme. This occurs during RNA splicing in eukaryotic cells and poly (A) tail brings stability for mRNA in the cytoplasm.

Polycistronic mRNA An mRNA molecule that codes for more than one polypeptide.

Polyculture (integrated aquaculture) Raising more than one aquatic species (say fishes) in the same environment.

Polyethylene glycol (PEG) An inert, water-soluble flexible polymer that alters the plasma membrane of some types of cells to facilitate transformation, and also induce complete fusion of cells as performed in hybridoma technology for production of monoclonal antibodies.

Polymer A series of monomers covalently linked to form a polymer macromolecule.

Polymerase chain reaction (PCR) A technique for exponential amplification and cloning of a selected region (or gene) of a DNA molecule. This involves multiple cycle of heat denaturation for separation of two strands of DNA, primer hybridization to the single strands, and DNA polymerase synthesis of new double stranded DNA in a thermocycler (PCR machine).

Polynucleotide Twenty or more nucleotides linked in a linear series by phosphodiester bonds.

Polypeptide A chain of amino acids joined by peptide (covalent) bonds, and usually contains more than 50 amino acids.

Polyploidy A cell or an organism with extra complete set(s) of chromosomes to its normal 2n number of chromosomes.

Polyribosome (polysome) Cluster of two or more ribosomes simultaneously translating a single mRNA molecule.

Polysaccharide A carbohydrate polymer consisting of sugars (monosaccharides) and sugar derivatives linked together by glycosidic bonds.

Polytene chromosome Giant chromosome produced by repeated division of the same DNA molecule in the absence of cell division; that is, multiple copies of chromatin (DNA) thread (polytene) in a chromosome. The four (homologous chromosomes remain in a pair) giant chromosomes in the salivary gland cells of *Drosophila* larvae are examples of polytene chromosomes.

Post-translational modification The addition of phosphate groups, carbohydrates (glycosylation), or other molecules at specific sites to a protein after it is synthesized.

Pox (Vaccinia) viruses *Poxviridae* is a large family of *Vaccinia* and other pox viruses with a large linear ds DNA genome of ~3 kbp in size containing 150 to 200 genes. As a unique feature these viruses encode their own DNA replication, transcription and packaging machinery, and replicate in the cytoplasm of the infected host cells. (Most of the other DNA viruses replicate with the machinery of the host cell and nucleus as site). These viruses can be designed as vector by accommodating some foreign genes in addition to their regular genome or replacing some of their genes, for rDNA technology (genetic engineering).

Pre-implantation genetic testing PCR (polymerase chain reaction, to multiply small quantity of DNA for analysis), ASO (allele-specific oligonucleotide analysis, to detect single nucleotide change in any gene), and FISH (fluorescence in-situ hybridization, to detect defective chromosomes) are used for screening gene defects in a single cell from 8 to 32 cell-stage embryos created by in-vitro fertilization. Such tests allow prior selection of a healthy embryo without genetic defects for implantation.

Primary transcript (pre-mRNA) Newly synthesized mRNA molecule encoded by a gene in the nucleus of eukaryotic cells, having exons and introns; that undergoes processing for excision of introns to produce mature mRNA for entry into the cytoplasm.

Primase Enzyme that adds small RNA segments to a single strand of DNA as an early and necessary step for DNA replication on both the leading and lagging strands of DNA.

Primer A short synthesized oligonucleotide complementary to specific sequences of interest on template strand and provides a 3′OH end for the initiation of DNA synthesis; used in PCR technique to amplify particular DNA sequence.

Prions Infections protein particles responsible for neurological diseases such as scrapie in sheep and goat, bovine spongiform encephalitis or 'mad cow' disease, brain destroying human diseases – *kuru* and transformable spongiform encephalitis (TSE). Prion precursor proteins are found in normal mammalian cells (neurons) as a membrane glycoprotein; in diseased conditions conformational changes occur to this protein and altered prion protein involves and induces changes in normal proteins leading to accumulation of useless proteins that damage cells and cause diseases.

Probe Single-stranded labelled (radioactive or fluorescent nucleotides) DNA or RNA molecule used to bind to complementary sequence immobilized on a blot; this is to identify genes and their activity.

Progeny   The offsprings from a mating.

Prokaryotes   Single-cell organisms, mostly bacteria and *Archae*, lacking membrane bound nucleus and organelles.

Promoter   A specific base sequence, upstream and adjacent to a gene, where RNA polymerase binds to initiate transcription.

Pronuclear microinjection   Introduction of transgene DNA by microinjection (under microscope) to one of the pronucleus, usually contributed by sperm, before fusing with the other pronucleus during fertilization.

Proofreading   Mechanism to get rid off a wrong base or an amino acid assembled during synthesis of a nucleic acid or protein.

Prophage   A repressed or inactive state of a bacteriophage (virus) genome integrated in the bacterial chromosomal DNA.

Prosthetic group   A tightly bound non-peptide component of a protein, that may be lipids, carbohydrates, metal ions, or inorganic groups such as phosphates.

Prostrate-specific antigen (PSA)   A protein from the inflamed prostrate released to the bloodstream, elevated level of which can be a marker for prostrate inflammation or even imminent prostrate cancer.

Protease (proteinase or proteolytic enzyme)   Protein digesting enzyme that hydrolyses peptide bonds.

Proteasome   Multiprotein complex in cytoplasm that catalyses ATP-dependent degradation of proteins linked to ubiquitin.

Protein   Biological macromolecule consisting of amino acids joined by peptide bonds; major structural and functional molecules of cells.

Protein microarray (protein chip)   Microarray (arranged in rows and columns) of specific proteins immobilized at different spots on a glass slide or support for proteome analysis by fluorescent or radioactive labelling, similar to a DNA microarray.

Protein sequencing   Determination of amino acid sequence by cleaving one amino acid at a time.

Proteome   The entire set of proteins encoded by a genome or the total protein complement of an organism.

Proteomics   Study of proteins of different proteomes.

Protoplast   A naked bacterial, yeast, or plant cell with its cell wall removed either chemically or enzymatically.

Protoplast fusion   Fusion of protoplasts of two cells.

Protoxin   A latent inactive precursor of a toxin molecule.

Pseudomonas   A common, widely distributed Gram negative bacterial genus, often found in soil.

Purine   Combined pyrimidine and an imidazole ring, e.g., of bases—adenine and guanine.

Pyrimidine   A heterocyclic ring, e.g. of bases — thymine, cytosine, or uracil.

q arm   Long arm of a chromosome.

Quasi-automated x-ray crystallography   Fast x-ray crystallographic analysis of protein structure using algorithms and data analysis.

Quaternary structure   Structure of a protein molecule containing more than one polypeptide chain folded into a tertiary structure.

Quencher   Molecule that can prevent fluorescence by binding to the fluorophore and absorbing its activation energy.

Random primers   Chemically synthesized oligonucleotides of radom sequences, usually hexamers.

Read through   Transcription or translation beyond the normal termination point, due to absence or mutation of termination signal or codon.

Real time or quantitative PCR (qPCR)   Uses of primers with fluorescent dyes and specialized thermal cycler (PCR machine) allow to quantify amplification reactions as they occur and the amount of PCR product.

Rec A   A protein found in bacteria, that is needed for DNA repair and DNA recombination.

rec mutant   Mutation in *rec* gene in bacteria that does not allow general recombination.

Receptor   A transmembrane protein (or glycoprotein) on the plasma membrane that binds to a ligand by extracellular domain and causes changes in activity of the cytoplasmic domain. Thus, a cell can be activated or triggered by extracellular substances or signals.

Recessive gene   An allele at a locus that does not demonstrably contribute to the phenotype in a heterozygote.

Recognition site (restriction site/cognate DNA/host specificity site/target site)   Specific short sequences on ds DNA that are recognized by restriction enzymes for cut; different restriction enzymes recognizes different sequences.

Recombinant   An individual with two or more linked genes derived from one or more crossing overs.

Recombinant DNA molecule   DNA molecule having material from more than one source.

Recombinant protein   A protein with the amino acid sequence that is encoded by a cloned gene; e.g., recombinant insulin and growth hormone.

Recombination   New combinations of genes created by different processes.

Redox reaction   Combination of oxidation and reduction reactions.

Reduction   A chemical reaction in which addition of one or more electrons to a molecule takes place.

Regulator gene   Gene encoding a regulatory protein that controls the expression of other genes at transcriptional level.

Renaturation The reassociation of denatured complementary single strands of a DNA double helix (Reverse of denaturation of DNA molecule).

Rennin (chymosin) Protein degrading enzyme obtained from the stomach of milk-producing animals (cows and goats), used in the production of cheese; recombinant product is known as chymosin.

Repetitive DNA Similar sequences of DNA occurring repetitively in genome.

Replacement therapy The administration of metabolites, cofactors, or hormones to supplement their deficiency due to a genetic disease.

Replica plating Technique by which cells from bacterial colonies on one culture Petri plate are transferred to another plate to produce a replica plate with bacterial cells growing in the same location as on the original (master) Petri plate; this can also be done with yeast and virus cultures.

Replicon A replicating unit of genome with an origin for initiation of replication of DNA.

Replisome The multiprotein structure containing DNA polymerase and other enzymes and assembles at the bacterial replicating fork of DNA synthesis.

Reporter gene Gene that is used to track or monitor (for reporting) the expression of other genes; e.g., lux gene.

Repressor A protein capable of binding to the operator to prevent transcription by blocking the binding of RNA polymerase.

Restriction endonuclease (restriction enzyme) An enzyme that cleaves (cuts) the phosphodiester backbone of double stranded DNA at specific nucleotide sequence (restriction site); it is extensively used in genome mapping and rNDA technology.

Restriction fragment length polymorphism (RFLP) Due to difference in restriction sites (sites for cut by an endonuclease) between two related DNA molecules (may be from two individuals), an endonuclease produces restriction fragments of DNA of different lengths that are separated in gel electrophoresis and then detected with dyes or hybridization with DNA probes.

Restriction map A physical map of DNA showing the relative positions of the restriction sites for different restriction endonucleases.

Retroposon (retrotransposon) A transposon that acts through an initial RNA form; i.e., the DNA transposon element transcribes into RNA, which is then reverse-transcribed into DNA for insertion at a new location in the genome. (It differs from retroviruses by not being infective as the viral forms).

Retrovirus Member of eukaryotic RNA viruses in *Retroviridae* family that can form ss DNA copies of its RNA genome with the help of reverse transcriptase enzyme, the ss DNA is then replicated into ds DNA with DNA polymerase, which can integrate into chromosomes of the host cells.

Reverse transcriptase An RNA-dependent DNA polymerase that synthesizes complementary DNA strand on the template of an RNA molecule. (The process is known as *reverse transcription).*

Reverse transcription PCR (RT-PCR) Two-step technique that uses reverse transcriptase enzyme to synthesize cDNA from cellular RNA and then amplifies cDNA by the polymerase chain reaction (PCR); this is often used for studying gene expression.

Rho (r) factor A protein in *E. coli* that helps in termination of RNA synthesis at specific *rho*-dependent nucleotide sequence.

Ribonuclease (RNase) A nuclease that degrades RNA by catalysing the hydrolysis of certain internucleotide bonds.

Ribonucleic acid (RNA) Nucleic acid that differs from DNA having ribose sugar in place of deoxyribose and uracil in place of thymine. Basically it is single stranded and different types of RNA participate in protein synthesis and gene regulation.

Ribose sugar A pentose sugar present in RNA.

Ribosomal RNA (rRNA) A class of small molecular RNAs that are essential components of ribosomes.

Ribosome A supramolecular complex of rRNAs and proteins, found in both prokaryotes and eukaryotes (of different sizes subunits), that binds, to mRNA and tRNA molecules and serves as the site of protein synthesis.

Ribosome binding site (Shine-Dalgarno sequence) Pyrimidine nucleotide-rich sequence at the 3′ end of 16S rRNA of 30S ribosome subunit in bacteria for binding of purine-rich sequence at 5′ end of mRNA during initiation phase of protein synthesis.

Ribozymes Ribonucleic acid molecules having catalytic activities like enzymes.

RNA interference (RNAi) RNA based mechanisms of gene silencing. The short interfering RNAs (siRNAs) are bound by a protein-RNA complex called the RNA-induced silencing complex (RISC). RISC unwinds the double-stranded siRNAs, releasing single-stranded siRNAs which bind to complementary sequences in mRNA that interferes with ribosome binding and protein synthesis or leads to degradation of the mRNA.

RNA polymerases Enzymes that catalyse the formation of RNA from 5′ end of DNA template (DNA-dependent RNA polymerase); different forms of RNA polymerase synthesize different types of RNA.

RNA size markers RNAs of known molecular weight used in the gel for measuring the size of RNA of interest.

RNA splicing The process of removal (excision) of non-protein coding sequences or introns from pre-mRNA and joining of protein coding sequences or exons into a continuous final mRNA.

Robotic deposition Mechanized robots depositing pre-synthesized oligonucleotides on a glass slide for making DNA chips.

S1 nuclease Endonuclease obtained from *Aspergillus oryzae* that specifically degrades single stranded DNA.

Saccharomyces cerevisiae **(Yeast)** See Yeast.

Satellite DNA (Tandem repeats of simple sequence DNA) Tandem repeats of identical DNA sequences accounting 10–15 per cent of a typical mammalian genome, that are not usually transcribed, possibly impart special physical properties to certain regions of chromosome as in case of centromere and telomere. Depending on the length at any given site, satellite DNA can be of different types: *typical* – 10^5 – 10^7 bp long; *minisatellite* – about 10^2 to 10^5 bp, tandem repeat unit consisting of 15–100 bp; *microsatellite* – 10–100 bp per site with repeat unit of 1–4 bp. Mini – and microsatellite sequences are extremely useful for analysing *DNA fingerprints*.

Scaling up Conversion of laboratory version to industrial manufacturing modification, such as fermentation from a small scale to bioreactor set-up.

Scrapie Brain wasting disease in sheep, similar to 'mad cow disease' in cattle due to misfolding of normal protein molecules. (see Prion).

Second messenger Any of several small molecular substances, such as cyclic AMP, calcium ion, inositol triphosphate, and diacylglycerol, that is generated upon binding of ligands to the extracellular specific receptors for transmitting the activating signal to the cell interior (ultimately genes).

Sedimentation coefficient A measure to determine the relative size of an organelle or a macromolecule depending on how rapidly the particle sediments when subjected to centrifugation (in ultracentrifuge machine). It is expressed in Svedberg units(S), in honour of Theodor Svedberg, the Swedish chemist who developed the ultracentrifuge between 1920 and 1940.

Selectable marker A gene that allows cells containing it to be identified by the expression of a recognizable feature including survivality in a define medium.

Selectin Cell surface glycoprotein that mediates cell–cell adhesion by binding to specific carbohydrate groups (say *mucin* like molecules) present on the surface of a target cell.

Selective medium A medium having some particular components that favours the growth of a particular organism or cell type, often by suppressing the growth of other types; HAT is a selective medium for a cell type.

Self-replicating element Extrachromosomal DNA element that is with origin of replication for initiation of own DNA synthesis, such as plasmids in the prokaryotes.

Semiconservative replication Mode of DNA replication in which two parental (old) strands of DNA act as template and each newly formed DNA double helix molecule consists of one old strand and one newly synthesized strand.

Senescence Process of ageing.

Severe combined immunodeficiency (SCID) An autosomal recessive mutation causing the deficiency of a protein kinase enzyme, adenosine deaminase, which is necessary for double-stranded DNA repair, leads to *severe combined immunodeficiency disease* (SCID). The deficiency of the enzyme results in the abnormal joining of the immunoglobulin (Ig) and T cell receptor (TCR) gene segments in the course of recombination during development, and ultimately the failure of development of mature immunocompetent B and T cells. Affected individuals have no functional immune system and are prone to death even with minor infections. SCID mice accept foreign cells and grafts from other strains of mice or from other species.

Sex chromosomes A pair of chromosomes involved in determining the sex of an individual; in humans XX chromosome for females, XY for males.

Sex pilus (plural pili) See Pilus.

SH2 domain A region of a protein molecule that recognizes and binds to phosphorylated tyrosines in another protein molecule in different signal transduction pathways. The term SH2 derives from Src (sheep red cell) homology (domain) 2, having amino acid sequence strikingly similar to a portion of the Src protein.

Short interfering RNA (siRNA) Class of short double-stranded RNAs of about 21 to 23 nucleotides in length that silence gene expression, either by binding complementarily to mRNAs to promote their degradation, or similar binding to DNA (gene) to inhibit the transcription. (See also RNA interference)

Shot gun sequencing A genome is broken into random short fragments by endonuclease digestion for nucleotide sequencing; then the short fragments arranged in genomic sequence by computer program to find overlaps between individual short sequences. The technique was introduced by Craig Venter of Celera Genomics for faster sequencing of human genome.

Shuttle vector (bifunctional vector) Vector having two different origins of replication or a wide host-range origin of replication so that the vector can multiply in two or more different hosts.

Siderophore Low-molecular weight substance capable of binding tightly to iron that is synthesized by different soil microbes and plants for obtaining sufficient amounts of iron from the environment.

Sigma (s) factor Subunit of bacterial RNA polymerase that recognizes and binds to the promoter region of DNA for initiation of RNA synthesis or transcription.

Signal transduction The process of binding of a molecule or ligand to its specific receptors on cell membrane and transmission of the event through a cascade of biochemical reactions to the interior of the cell and/or genes.

Silent mutation Mutation causing base pair substitution with no effect on the structure and function of encoded protein.

Simian virus 40 (SV40) Small, spherical, double stranded DNA virus that infects monkeys and may cause malignancy in them by integrating its DNA into host genome.

Single cell protein (SCP) Dried cell mass of a protein-rich microorganism, used as feed for animals or as food of humans.

Single nucleotide polymorphisms (SNPs) A single nucleotide variation in DNA base sequence between individuals of the same species.

Single-stranded DNA binding protein (SSB) Protein that binds to the single strands of DNA at the replication fork to keep the DNA unwound for molecular replication machinery to work, and also protects ss DNA from attack by nucleases.

Site-specific mutagenesis Molecular technique to alter the DNA base sequence at a particular site of DNA, creating a specific mutation whose effects are then studied.

Size markers A set of macromolecules of known molecular masses run along with other macromolecules in electrophoresis to determine the molecular masses of other macromolecules.

Sludge A semisolid material produced in course of sewage treatment and consists mainly small particles of faeces, waste materials, and microorganisms.

Small nuclear RNA (snRNA) One small variety of RNA confined to the nucleus that binds to specific proteins to form small nuclear ribonucleoproteins (snRNPs), several of which assemble to form a spliceosome to splice off introns from pre-mRNA in the eukaryotes.

Smooth endoplasmic reticulum (smooth ER) Endoplasmic reticulum without attached ribosomes, that plays no direct role in protein synthesis, but is involved in the packaging of secretory proteins and synthesis of lipids.

SNARE (SNAP receptor) protein Two families of proteins that mediates fusion between vesicles and target membranes in eukaryotic cells; the v-SNARES are found on transport vesicles and the t-SNARES are on target membrane.

Sodium dodecyl sulphate (SDS) An anionic detergent that denatures proteins and imparts negative charge of sulphate group to the proteins. These proteins separate in polyacrylamide gel electrophoresis (SDS-PAGE) on the basis of molecular size.

Solid-phase bioremediation Ex-situ (taken out from the site) clean-up of polluted soil that involves composting, land-farming or biopiles; microbes degrade pollutants in the soil pile, usually in evaporating chemicals.

Solid-phase synthesis of oligonucleotides Synthesis of growing oligonucleotides after immobilizing or anchoring one end to a solid support.

Somatic cell A cell of a multicellular organism that does not produce gametes (sperm and egg cells).

Somatic cell gene therapy Correcting a gene defect by delivering a normal gene(s) to a tissue other than the reproductive cells of an organism.

Somatic cell nuclear transfer (SCNT) A technique for creating a clonal organism by transferring nucleus of a somatic cell in an enucleated ovum and allowing it to develop in a surrogate mother in case of mammals. The best example is creation of *Dolly* sheep.

Sonication (ultrasonication) Brief pulses of high-frequency sound waves generate hydrodynamic shearing force that disrupts cells, large genomic DNA, and it is also used for cleaning tips of micropipette which cannot be cleaned by brush normally.

Southern blotting Laboratory technique invented by Ed Southern that transfers denatured (two strands dissociated) DNA fragments separated by gel electrophoresis to nitrocellulose or nylon membrane under the capillary action generated by the stack of blotting papers on the membrane; on the membrane hybridization with radiolabelled DNA probe is carried out.

South-western blotting Technique that involves a DNA probe binding to a protein target, and is used to detect DNA-binding proteins.

Spectrophotometry A technique to find the concentration (or amount) of a substance in a solution on the basis of absorbance of light by the substance at a characteristic wavelength.

Sperm-mediated transfer Injection of foreign DNA directly to prominent sperm pronucleus before its fusion with egg or female nucleus during fertilization.

Spliceosome Protein–RNA complex that catalyses the excision of intron sequences from pre-mRNA.

Stem cell A precursor cell capable of multiples of division, the progenies differentiate into a variety of other cell types under the influence of microenvironments and types of cytokines. The embryonic stem cell derived from blastocoel of blastula can even differentiate into all the cell types forming a whole organism.

Sticky ends (cohesive ends or staggered end) See Cohesive ends.

Stone-age genomics Analysis of minuscule amounts of ancient DNA obtained from bone and other tissues and fossil samples that are tens of thousands of years old.

Stop codon (termination codon or nonsense codon) One of three triplet codons [UAG (amber), UAA (ochre), UGA (opal)] that stops the translation process of protein synthesis.

Strain Microorganisms or multicellular organisms that are a group of genetic variant of a standard parental stock.

Structural gene A sequence of DNA that encodes a protein.

Subcloning Transfer of a cloned fragment of DNA or gene from one vector to another for further investigation. This could be for detailed study of a short region of a large cloned fragment or for expression of the piece in another species of host.

Subculture Reinoculation of fresh culture medium with cells from an existing culture of microorganisms or eukaryotic cell.

Subspecies A population of organisms having certain characteristics that are not present in other populations of the same species.

Substrate A compound on which an enzyme acts to alter it. Sometimes the word is used to mean food source for growing cells or microorganisms.

Subtilisin A protease enzyme obtained from *Bacillus subtilis* and used in laundry detergents to remove protein stains from clothing.

Subunit vaccine Vaccine from purified components of a pathogen such as viral proteins or lipid molecules and that can also be produced from a cloned gene encoding pathogenic component.

Sucrose density gradient centrifugation A centrifugation procedure for fractionation of different sizes of mRNA or DNA fragments on differential concentration gradient of sucrose.

Superbug *Pseudomonas* bacterial strain developed by Ananda Mohan Chakraborty, combining into one organism hydrocarbon degrading genes on different plasmids from various strains, for degrading oil spill. This was the first genetically manipulated organism patented in the US after prolonged legal arguments.

Svedberg unit See Sedimentation coefficient.

SYBR Gold The most sensitive fluorescent stain for detecting ds or ss DNA or RNA in electrophoretic gels, with ultraviolet transilluminator.

SYBR Green I A ds DNA binding dye that fluoresceces only when bound.

Symbiosis A mutually beneficial association between organisms of two different species.

Synaptonemal complex An elaborate protein structure resembling a zipper holds tightly the two homologous chromosomes all along their lengths during meiotic prophase I (zygolene and pachytene stages). Formation of synaptonemal complexes is closely associated with the process of crossing over.

Syndrome A disease that is defined on the basis of number of symptoms, rather than a known etiology.

Tandem array Two or more identical nucleotide sequences adjacent to one another in series in a DNA molecule.

Taq DNA polymerase Thermostable DNA polymerase from *Thermus aquaticus*, a thermophilic Archae bacterium that lives in hot springs; the enzyme can withstand high temperatures (more than 90°C) without denaturation and is valuable for use in PCR.

T cell (T lymphocyte) A type of white blood cell that play an essential role in secreting cytokines and helping both B and other T lymphocytes for differentiation to recognize and respond to foreign antigens or pathogens; cytotoxic T lymphocytes play a critical role against virus infection and malignant cells.

TATA box (Hogness box) A conserved A-T-rich septamer sequence located about 25 base pairs (bp) 'upstream' in 5′ direction before the start point of eukaryotic RNA polymerase II transcription unit for a gene; this is involved in binding transcription factors and initiation of the polymerase function.

Telomerase The ribonucleoprotein enzyme that synthesizes repeating nucleotide units of one strand at the telomere of a chromosome, by adding bases to the DNA 3′ end, as directed by an RNA sequence in the RNA component of the enzyme.

Telomere The end of eukaryotic cheromosome consisting of specific tandem repeats of conserved DNA sequence. Human telomeres contain 250–1500 copies of the sequence TTAGGG.

Temperate phage A bacteriophage with a lysogenic stage when phage DNA gets intergrated with genome of host bacterium.

Template strand The strand of DNA double helix that serves as the template for RNA synthesis through complementary base pairing. [The non-template DNA strand is by convention called the *coding strand* (though it is not encoding mRNA) or *sense strand* because its sequence of nucleotides is similar to that in the single-stranded mRNA molecules that carry the coded message from template stand].

Terminator technology Genetic use restriction technology (GURT), generally called terminator technology refers to the technology for inclusion of gene restricting the propagation of transgenic plants by causing second-generation seeds to be sterile. This protects the market interest of firms producing transgenic seeds and plants.

Tertiary structure The unique three-dimensional conformation of a protein.

Thalidomide A compound initially used as a drug to combat morning sickness during pregnancy; certain alternate forms of thalidomide caused severe birth defects in human babies and the drug is withdrawn. Certain derivatives of thalidomide are used in research for treating cancer, HIV, and certain other diseases.

Thaumatin A plant protein with a sweet taste; some plants synthesize it in response to infection by pathogens.

Thermocycler Alternative name of polymerase chain reaction (PCR) machine, as because the machine achieves rapid shift of several temperature ranges in a pre-set cyclical order.

Thermophile A microorganism that normally grows at high temperatures; usually above $50^{\circ}C$; some thermophiles can grow at temperatures of $90^{\circ}c$ and above. *Thermus aquaticus* is such a bacterium, grows in hot springs and is the source of *Taq* polymerase enzyme (see *Taq* DNA polymerase).

Thin layer chromatography (TLC) Silica gel (silicic acid) bound as a thin layer on a glass or metal surface provides the medium for separation of compounds.

Ti plasmid (Tumor-inducing plasmid) A large DNA plasmid in *Agrobacterium tumerfaciens* soil bacterium that causes crown gall tumour formation in plants; this can be used as a cloning vector for introducing foreign genes into plant cells.

Tissue engineering Growing and designing tissues out of cells and scaffold (may be artificial) for using in medical applications.

Tm (Melting temperature) The temperature necessary to dissociate two strands of a nucleic acid hybrid, it is related to the G+C content and ionic strength of the buffer.

Tobacco mosaic virus (TMV) A filament-like ss RNA plant virus that is used as vector for gene transfer.

Topoisomerase An enzyme that creates swivel points by making single and double stranded breaks in DNA molecule and quickly reseal them to relieve supercoiling in unwinding DNA during its replication.

Totipotent Ability of a cell to divide and differentiate into all kinds of tissue cells and whole of the multicellular organism wherefrom it is originally derived. Embryonic stem cell is an ideal example of totipotent cell.

Tracking dye A low-molecular weight chemical with visible colour that moves along an ion front during gel electrophoresis.

Transcript An RNA molecule that is synthesized or transcribed from a specific DNA template.

Transcription The process of RNA synthesis by RNA polymerase from a DNA template strand.

Transcription factors Accessory protein molecules that are required for RNA polymerase to initiate transcription at specific promoter(s).

Transduction The process of bacteriophage-mediated transfer of genetic material between bacteria and allowing new recombinations.

Transfection Normally introduction of bacteriophage DNA into competent bacterial cell.

Also applicable for introduction of nucleic acids by non-viral methods in plant and animal cells.

Transformable spongiform encephalitis (TSE) See Prion.

Transformation The process of uptake of naked DNA from the environment and its integration into the bacterial genome. The term also defines change of a normal cell into a cancer cell.

Transgene A gene that is introduced into a cell from another organism.

Transgenic Organism that contains in it genome transegene(s) from other organism.

Transition A mutation when a purine nucleotide is replaced by another purine or a pyrimidine is replaced by a pyrimidine in DNA.

Translocation The transfer of a part of chromosome from one chromosome to another.

Transporter One type of receptor that binds molecules on its extracellular surface and transports the molecules across the plasma membrane into the cytoplasm.

Transposon (wandering gene) A DNA sequence that inserts itself at a new location in the genome and can modulate the function of neighbouring genes at the new site.

Transversion A mutation causing a purine replaced by a pyrimidine or a pyrimidine replaced by a purine in DNA molecule.

Tricarboxylic acid cycle (TCA cycle) Oxidation of acetyl CoA to carbon dioxide in the presence of oxygen in a cyclic metabolic pathway, generating ATP and the reduced coenzymes NADH and $FADH_2$. This is a component of aerobic respiration in a cell, and also known as the Krebs cycle after the pioneer researcher.

Triploid Eukaryotic organism with three sets of chromosomes ($3n$) in the cells.

Triton X-100 A non-ionic detergent.

Two-dimensional electrophoresis Separation with electrophoresis running in two electric fields at right angle to each other.

Type I, insulin-dependent diabetes mellitus Disease of elevated blood sugar levels (hyperglycaemia) due to lack of the pancreatic hormone insulin, which is needed for carbohydrate metabolism.

Ubiquitin Small protein molecule that links and marks the targeted protein for degradation by proteasome, cytoplasmic organelle.

UV transilluminator Transilluminator with UV light and UV filter for visualization and documentation of bands separated in gel electrophoresis.

Ultracentrifugation A centrifuge machine built to spin a rotor at very high speeds, capable of generating acceleration as high as 1,000,000 g (9,800 km/s^2), and with a cooling device.

Ultrafiltration An efficient type of membrane filtration in which hydrostatic pressure forces a liquid against a semipermeable membrane to filter very small particles or substances.

Upstream The region of nucleotides in template strand towards 5′ direction prior to the site of initiation of transcription; considering the first transcribing base as at +1 position the upstream nucleotides are designated by minus sign, such as −1, −10, −50.

Vaccine A preparation of dead or inactivated pathogens, their components or products that are used in small quantity to stimulate the protective immunity in higher organisms.

Vaccinia **virus** A large, complex, enveloped virus of poxvirus family, having a linear ds DNA of ~190 kbp (larger than usual viruses). It is used extensively for recombinant vaccines.

Variable number tandem repeats (VNTRs) Number of repeats of short nucleotide sequence in tandem, can be found on many chromosomes and show variations in length between individuals; each variant behaves as an inherited allele, and can be used for individual or parental identification. Their analysis is used in genetics, DNA fingerprinting, and forensics.

Vector (Cloning vector) A self-replicating DNA molecule capable of harbouring an inserted DNA (gene) sequence of foreign origin and multiplying along with it to produce a clone of the gene. Plasmid DNA and viruses are examples.

Virion An infective virus particle.

Virulence The degree of infectivity and pathogenicity of an organism.

Virus A submicroscopic, non-cellular infective agent, having RNA or DNA and protein, and depends on organelles and enzyme system of the host cell for its multiplication.

Void volume (Vo) The total volume of buffer that remains in between the beads in a gel, in other words the space surrounding the beads.

Western blotting Transfer of protein from a gel to a membrane by capillary action of buffer induced by a stack of blotting papers and detection of specific protein by reacting with antibody forming blot.

Wild type Commonly observed phenotype in nature or the normal variety, in contrast to a mutant condition.

Wobble hypothesis Flexibility in base-pairing between the third base of a codon and the corresponding base in its anticodon. This was proposed by Francis Crick.

Xenobiotic Manufactured chemical compound that is not produced by living organisms.

Xenotransplantation The transplantation of cells, tissues, or organs from one species to another species; such as a pig organ into a human.

X-ray diffraction (x-ray crystallography) Technique for determination of three-dimensional structure of macromolecules (DNA, protein) on the basis of the pattern produced by a beam of x-rays passing through a crystal of the macromolecule.

Yeast A unicellular eukaryotic fungus, used for fermenting beer and production of bread. Yeast cell (*Saccharomyces cerevisiae*) has ~12 Mbp linear ds DNA genome, larger than most bacteria, mechanisms of gene expression in yeast resemble those in human chromosome. Valuable model organism for studying eukaryotic chromosome structure, gene regulation, cell division, and cell cycle control. It is commonly known as Brewer's yeast/ Baker's yeast.

Yeast artificial chromosome (YAC) A YAC is constructed as a 'minimalist' eukaryotic chromosome that is a circular DNA molecule containing an origin of replication (ORI), a centromere, and two telomere sequences and two selectable marker genes. It has specific endonuclease site where a large piece of foreign DNA can be accommodated for cloning purpose. It is extensively used in human genome project.

Zygote Diploid cell formed by the union of two haploid gametes (egg and sperm cells) at the event of fertilization.

Bibliography

Adams, R. P. and J. E. Adams, (eds) 1992, *Conservation of Plant Genes, DNA Banking, and In Vitro Technology*, Academic Press, New York.

Alberts, B., A. Johnson, J. Lewis, M. Raff, K. Roberts, and P. Walter. 2002, *Molecular Biology. of the Cell*, Garland Science, New York and London.

Bains, W. 2001, *Biotechnology From A to Z*, 2nd ed. Oxford University Press, Oxford.

Balasubramanian, D., C.F.A. Bryce, K. Dharmalingam, J. Green, and K. Jayaraman. 2005, *Concepts in Biotechnology*, Universities Press (P) Ltd., Hyderabad.

Barritt, J. A., C. A. Brenner, H. E. Matter, et al. 2001, 'Mitochondria in Human Offspring Derived From Ooplasmic Transplantation'. *Forensic Science*, **46**, 513–516.

Becker, W. M., L. J. Kleinsmith, and J. Hardin. 2006, *The World of The Cell,* 6th ed. Pearson Education, Delhi.

Bhatia, S. C. 2005, *Textbook of Biotechnology*, Atlantic Publishers and Distributors. New Delhi.

Bhattacharya, B. and R. Banerjee. 2007, *Environmental Biotechnology*, Oxford University Press, New Delhi.

Brown, T. A. 2000, *Gene Cloning: An Introduction*, 4th ed. Chapman and Hall, London.

Bu'lock, J. D. and B. Kristiansen. 1987, *Basic Biotechnology*, Academic Press, London.

Burden, D. and D. Whitney. 1995, *Biotechnology: Proteins to PCR*, Birkhauser, Boston.

CAMLAB. 1991, *Operation of Municipal Waste Water Treatment Plants*, CAMLAB, Cambridge.

Campbell, A. M. and L. J. Heyer. 2007, *Discovering Genomics, Proteomics, and Bioinformatics*, 2nd ed. Benjamin Cummings, San Francisco.

Campbell, K. H., J. McWhir, W. A. Ritchie, and I. Wilmut. 1996, 'Sheep Cloned By Nuclear Transfer from a Cultured Cell Line'. *Nature*, **380,** 64–66.

Caruthers, M. H. et al. 1992, 'Chemical Synthesis of Deoxynucleotides and Deoxynucleotide Analogs'. *Methods Enzymology*, **211**, 3–20.

Casida, L. E. Jr. 1966, Industrial Microbiology. (Reprinted 1991). Wiley Eastern Ltd. New Delhi.

Chakravarty, A. K. 2006, *Immunology and Immunotechnology*, Oxford University Press, New Delhi.

Chan, A. W., K. Y. Chong, C. Martinovich, et al. 2001, 'Transgenic Monkeys Produced By Retroviral Gene Transfer into Mature Oocytes'. *Science*, **291**, 309–312.

Channarayappa. 2006, *Molecular Biotechnology: Principles and Practices*, Universities Press (India)(P) Ltd. Hyderabad.

Chaplin, M. F. and C. Bucke. 1990, *Enzyme Technology*, Cambridge University Press, Cambridge.

Chaudhuri, J. D. 2005, 'Genes Arrayed Out For You: The Amazing World of Microarrays'. *Medical Science Monitor*, **11**, RA52–62.

Chung, Shin-Ho, et al. 2007, *Biological Membrane Ion Channels: Dynamics, Structure and Applications*, Springer, New York.

Colavito, M. C. 2007, *Gene Therapy*, M.A. Palladino (ed). Benjamin Cummings, San Francisco.

Cowan, C. A., J. Atienza, D. A. Melton, et al. 2005, 'Nuclear Reprogramming of Somatic Cells After Fusion With Human Embryonic Stem Cells'. *Science*, **309**, 1369–1373.

Das, H. K. (ed.) 2005, *Textbook of Biotechnology*, 2nd ed. Wiley Dreamland India (P) Ltd., New Delhi.

Daugherty, E. 2012, *Biotechnology: Science for the New Millennium*, 1st ed., Revised. EMC Publishing, Saint Paul, Minnesota.

De Coppi, P., G. Bartsch, M. M. Siddiqui, et al. 2007, Isolation of Amniotic Stem Cell Lines With Potential for Therapy. *Nature Biotechnology*, **25**, 100–106.

De Grey, A. D. N. J. 2007, *Ending Aging: The Rejuvenation Breakthroughs That Could Reverse Human Aging in Our Lifetime*, St. Martin's, New York.

DeCosa, B., W. S. Moar, L. Fung-Bum, et al. 2001, 'Over-Expression of the *Bt Cry2aa2*, Operon in Chloroplasts Leads to Formation of Insecticidal Crystals'. *Nature Biotechnology*, **19**, 69–74.

Dykxhoorn, D. M. and J. Lieberman. 2006, 'Knocking Down Disease With Sirnas'. *Cell*, **126**, 231–235.

Edge, M. D. et al. 1981, 'Total Synthesis of a Human Leukocyte Interferon Gene'. *Nature*, **292**, 756–762.

Egorov, Alexei M. 2007, *New Aspects of Biotechnology and Medicine*, Nova Biomedical Books, New York.

Erlich, H. A. (ed.) 1989, *PCR Technology: Principles and Applications*, Stockton Press, New York.

Fletcher, G. L. and P. L. Davies. 1991, 'Transgenic Fish for Aquaculture', in J. K. Setlow (ed.) Genetic Engineering, Plant Biotechnology, Vol. I. Bleckie, Glasgow.

Fodor, S. 1997, 'DNA Sequencing: Massively Parallel Genomics'. *Science*, **277**, 393–395.

Freshney, R. I. 2000, *Culture of Animal Cells: A Manual of Basic Technique*, John Wiley and Sons, Inc., New York.

Friend, S. H. and R. B. Stoughton. 2002, 'The Magic of Microarrays'. *Scientific American*, **286**, 44–53.

Ganguli, P. 1998, 'Patenting Innovations: New Demands in Emerging Contexts'. *Current Science*, **75**, 433–439.

Ganguli, P. 1998, *Gearing up for Patents-The Indian Scenario*, Universities Press, Hyderabad.

Ghosh, P. K. 1997, Transgenic Plants and Biosafety Concerns in India. *Current Science*, **72**, 172–179.

Gilbert, D. M. 2004, 'The Future of Human Embryonic Stem Cell Research: Addressing Ethical Conflict with Responsible Scientific Research'. *Medical Science Monitor*, **10**, RA99–103.

Gilpin, A. 1994, *Environmental Impact Assessment*, Cambridge University Press, Cambridge.

Goding, J. W. 1993, *Monoclonal Antibodies: Principles and Practice*, 3rd ed. Academic Press, New York.

Greely, H. T. 2005, 'Banning Genetic Discrimination'. *New England Journal of Medicine*, **353**, 865–867.

Gresshoff, P. M. 1994, *Plant Genome Analysis*, CRC Press, Boca Raton, USA.

Guha, S. and S. C. Maheshwari. 1964, 'In Vitro Production of Embryos From Anthers of Datura'. *Nature*, **284**, 497.

Hanna, J., M. Wernig, S. Markoulaki, et al. 2007, Treatment of Sickle Cell Anemia Mouse Model with Ips Cells Generated From Autologous Skin. *Science*, **318**, 1920–1923.

Hannon, G. J. (ed.) 2003, *RNAi: A Guide to Gene Silencing*, Cold Spring Harbor Laboratory, New York.

Hercher, L. 2007, 'Diet Advice From DNA?' *Scientific American*, **294**, 47–54.

Hew, C. L. and G. L. Fletcher. 2001, The Role of Aquatic Biotechnology in Aquaculture. *Aquaculture*, **197**, 191–204.

Hogan, B., F. Constantini, and E. Lacy. 1994, *Manipulating the Mouse Embryo: A Laboratory Manual*, Cold Spring Harbor Laboratory, New York.

Huh, Won Ki, J. V. Falvo, L. C. Gerke, et al. 2003, 'Global Analysis of Protein Localization in Budding Yeast'. *Nature*, **425**, 686–691.

Human Genome Sequencing Consortium. 2001, 'Initial Sequencing and Analysis of the Human Genome'. *Nature*, **409**, 860–921.

Jensen, K. and F. Murray. 2005, 'Intellectual Property Landscape of the Human Genome'. *Science*, **310**, 239–240.

Jha, T. B. and B. Ghosh. 2005, *Plant Tissue Culture: Basic and Applied*, Universities Press (India)(P) Ltd., Hyderabad.

Karp, A., P. G. Isaac, and D. S. Ingram. 1998, *Molecular Tools for Screening Biodiversity*, Chapman and Hall, London.

Knopf. George K., Amarjeet, S. Bassi. 2007, *Smart Biosensor Technology*, CRC Press / Taylor & Francis, Boca Raton, Florida.

Koller, B. H. and O. Smithies. 1992, 'Altering Genes in Animals by Gene Targetting'. *Annual Review of Immunology*, **10**, 705–730.

Labhasetwar, Vinod, Diandra L. Leslie-Pelecky. 2007, *Biomedical Applications of Nanotechnology*, Wiley-Interscience, New Jersey.

Landecker, H. 2007, *Culturing Life: How Cells Became Technologies*, Harvard University, Cambridge, Massachusetts.

Langridge, W. H. H. 2000, 'Edible Vaccines'. *Scientific American*, September issue, pp. 26–31.

Lau, N. C. and D. P. Bartel. 2003, 'Censors of the Genome'. *Scientific American*, **289**, 34–41.

Leach, C. K. and M. C. E. Van Dam-Mieras (eds.) 1994, *Biotechnological Innovations in Energy and Environmental Management*, Butterworth-Heinemann Ltd., Oxford.

Lee, J. W. (ed.) 2013, *Advanced Biofuels and Bioproducts*, Springer, London.

Lesser, W. (ed.) 1989, *Animal Patents: The Legal, Economic and Social Issues*, Stockton Press, New York.

Levine, C. 2006, *Taking Sides: Clashing View on Controversial Bioethical Issues*, 11th ed. McGraw-Hill, Dubuque, USA.

Lewin, B. 2000, *Genes VII*, Oxford University Press, Oxford.

Lodish, A. Berk, S. L. Zipursky, P. Matsudaira, D. Baltimore, and J. F. Darnell. 2000, *Molecular Cell Biology*, 4th ed. W.H. Freeman, New York.

Lonberg, N. 2005, 'Human Antibodies From Transgenic Animals'. *Nature Biotechnology*, **23**, 1117–1125.

Losey, J. E., L. S. Rayor and M. E. Carter. 1999, 'Transgenic Paller Harms Monarch Larvae'. *Nature*, **399**, 214.

Lynd, L. R., J. H. Cushman, R. J. Nichols and C. E. Wyman. 1991, 'Fuel Ethanol From Cellulosic Biomass'. *Science*, **251**, 1318–1323.

MacQueen, B. D. 2002, 'The Moral and Ethical Quandary of Embryonic Stem Cell Research'. *Medical Science Monitor*, **8**, ED1–4.

Martin, D. K. 2007, *Nanobiotechnology of Biomimetic Membranes*, Springer, New York.

Mason, H. S., D. M. K. Lam, and C. J. Arntzen. 1992, 'Expression of Hepatitis B Surface Antigen in Transgenic Plants. *Proceedings National Academy Science*, USA, **89**, 11745–11749.

Mathews, C. K., K. E. van Holde, and K. G. Ahem. 1999, *Biochemistry*, 3rd ed. Pearson Benjamin Cummings, USA.

Mattick, J. S. 2004, 'The Hidden Genetic Program Of Complex Organisms'. *Scientific American*, **291**, 60–67.

Mattineau, B. 2001, *First Fruit: The Creation of the Flavr Savr^tm Tomato and the Birth of Genetically Engineered Foods*, McGraw-Hill, New York.

McCoy, H. 2001, *Zebrafish and Genomics*, *GEN*, **21**, 1.

Meissner, A., M. Wernig, and R. Jaenisch. 2007, 'Direct Reprogramming of Genetically Unmodified Fibroblasts into Pluripotent Stem Cells'. *Nature Biotechnology*, **25**, 1177–1181.

Meyers, R. A. 1995, *Molecular Biology and Biotechnology: A Comprehensive Desk Compendium*, VCH Publ. (UK) Ltd., Cambridge.

Micou, M. K. and D. Kilkenny. 2012, *A Laboratory Course in Tissue Engineering*, CRC Press / Taylor & Francis.

Morrison, A. R. 2003, 'Ethical Principles Guiding the Use of Animals in Research'. *The American Biology Teacher*, **65**, 105–108.

Mulligan, R. C. 1993, 'The Basic Science of Gene Therapy'. *Science*, **260**, 926–932.

Mullis, K. 1990, 'The Unusual Origin of the Polymerase Chain Reaction'. *Scientific American*, **262**, 56-61.

Natesh, S., V. L. Chopra, and S. Ramachandran. 1987, *Biotechnology in Agriculture*, Oxford & IBH, New Delhi.

Nelson, D. L., A. L. Lehninger, and M. M. Cox. 2000, *Principles of Biochemistry*, 3rd ed. Worth Publishers, Inc. USA.

NIH Guidelines for Recombinant DNA Technology Research. National Institute of Health, USA.

Palladino, M. A. 2006, *Understanding the Human Genome Project*, 2nd ed. Benjamin Cummings, San Francisco.

Parales, R. E., N. C. Bruce, A. Schmid, et al. 2002, 'Biodegradation, Biotransformation, and Biocatalysis (B3)'. *Applied and Environmental Microbiology*, **68**, 4699–4709.

Peterson, C. H., S. D. Rice, J. W. Short, et al. 2003, 'Long-Term Ecosystem Responses to the EXXON *Valdez* Oil Spill'. *Science*, **302**, 2082–2086.

Pinkert, C. A. 1994, *Transgenic Animal Technology: A Laboratory Handbook*, Academic Press, San Diego, California.

Pollard, J. W. and J. M. Walker (eds.) 1990, *Animal Cell Culture: Methods in Molecular Biology*, vol. 5. Humana Press, Clifton, New Jersey.

Ragauska, A. J., C. K. Williams, B. H. Davidson, et al. 2006, 'The Path Forward for Biofuels and Biomaterials'. *Science*, **311**, 484–489.

Ramani, K., Q. Hassan, et al. 1998, 'Site-Specific Gene Delivery In Vivo Through Engineered Sendai Viral Envelopes'. *Proceedings National Academy Science*, USA, **95**, 11886–11890.

Rashidi, H. H. and L. K. Buchler. 2000, *Bioinformatics Basics: Applications in Biological Science And Medicine*, CRC Press, London.

Rastogi, S. and N. Pathak. 2009, *Genetic Engineering*, Oxford University Press, New Delhi.

Romeis, J., M. Meissle, and F. Bigler. 2006, 'Transgenic Crops Expressing *Bacillus thuringiensis* Toxins and Biological Control'. *Nature, Biotechnology*, **24**, 63–71.

Sandel, M. J. 2007, *The Case Against Perfection: Ethics in the Age of Genetic Engineering*, Belknap Press, Cambridge, Massachusetts.

Seeman, N. D. 2004, 'Nanotechnology and the Double Helix'. *Scientific American*, **290**, 64–75.

Sensen, C. W. (ed.) 2002, *Essentials of Genomics and Bioinformatics*, Wiley-VCH Verlag GmbH, Weinheim.

Seshadri, R. L. Adrian, D. F. Fouts, et al. 2005, 'Genome Sequence of the PCE-Dechlorinating Bacterium *Dehalococcoides Ethenogenes*'. *Science*, **307**, 105–108.

Shuman, S. 1994, 'Novel Approach to Molecular Cloning and Polynucleotide Synthesis Using Vaccinia DNA Topoisomerase'. *Journal Biological Chemistry*, **269**, 32678–32684.

Singer, M. and P. Berg. 1991, *Genes and Genomes*, University Science Books, Mill Valley, California and Blackwell Science Publications, Oxford.

Singh, B. D. 2010, *Biotechnology Expanding Horizon*, 3rd ed. Kalyani Publishers, New Delhi.

Smith, H. O. and D. Nathans. 1973, 'A Suggested Nomenclature for Bacterial Host Modification and Restriction Systems and Their Enzymes'. *Journal Molecular Biology*, **81**, 419–423.

Smithies, O. 1993, 'Animal Models of Human Genetic Diseases'. *Trends Genetics*, **9**, 112–116.

Snustad, D. P. and M. J. Simmons. 2010, *Principles of Genetics*, 5th ed. John Wiley & Sons, Inc. Hoboken, New Jersey.

Staub, J. M. et al. 2000, 'High Yield Production of a Human Therapeutic Protein in Tobacco Chloroplasts'. *Nature Biotechnology*, **18**, 333–338.

Stix, G. A. 2005, 'Toxin Against Pain'. *Scientific American*, **292**, 88–93.

Stryer, L., J. M. Berg and J. L. Tymoczko. 2002, *Biochemistry*, 5th Revised ed. W. H. Freeman & Co. Ltd. New York.

Suzuki, Y. et al. 1997, 'Construction and Characterization of Full-Length Enriched and a 5′-End Enriched cDNA Library'. *Genetics*, **200**, 149–156.

TeBeest, D. O. 1996, 'Biological Control of Weeds With Plant Pathogens and Microbial Pesticides' *Advances in Agronomy*, **56**, 115.

The Indian Recombinant DNA Safety Guidelines and Regulations. Department of Biotechnology, New Delhi.

Thieman, W. J. and M. A. Palladino. 2011, *Introduction to Biotechnology*, 2nd ed. Pearson/Dorling Kindersley, New Delhi, India.

Tortora G. J., B. R. Funke, and C. L. Case. 2006, *Microbiology: An Introduction*, 9th ed. Benjamin Cummings, San Francisco.

Trevan, M. D. 1980, *Immobilized Enzymes: An Introduction and Applications in Biotechnology*, John Wiley & Sons, New York.

Trounson, A. and A. Conti. 1982, 'Research in Human In Vitro Fertilization and Embryo Transfer'. *British Medical Journal*, **285**, 244–248.

U.S. Department of Health and Human Services. 1993, *Biosafety in Microbiological and Biomedical Laboratories*, 3rd ed. Centre for Disease Control and Prevention.

Ulmer, J. B., U. Valley, and R. Rappuoli. 2006, 'Vaccine Manufacturing: Challenges and Solutions' *Nature Biotechnology*, **24**, 1377–1383.

Vasil, I. K. and T. A. Thorpe (eds.) 1994, *Plant Cell and Tissue Culture*, Kluwer Academic Publishers, London.

Vastag, B. 2006, 'New Clinical Trials Policy at FDA'. *Nature Biotechnology*, **24**, 1043.

Voet, D. and J. G. Voet. 2003, *Biochemistry*, 3rd ed. J. Wiley & Sons. USA.

Walsh, G. and D. Headon.1994, *Protein Biotechnology*, Wiley, New York.

Watson, J. D., T. A. Baker, S. P. Bell, A. Gann, M. Levine, and R. Losick. 2004, *Molecular Biology of the Gene*, 5th ed. Benjamin Cummings, San Francisco.

Westermeier, R. and T. Navan. 2000, *Proteomics in Practice: A Laboratory Manual of Proteome Analysis*, Wiley-VCH Verlag GmbH, Weinheim.

Willadsen, S. M. 1989, 'Cloning of Sheep and Cow Embryos'. *Genome*, **31**, 956–962.

Wilmut, I., K. Campbell and C. Tudge. 2000, *The Second Creation: Dolly and the Age of Biological Control*, Harvard University Press, Cambridge, Massachusetts.

Zhang, Z., J. Gildersleeve, Y. Y. Yang, et al. 2004, 'A New Strategy for the Synthesis of Glycoproteins'. *Science*, **303**, 371–373.

Index